CONCRETE MATHEMATICS

Dedicated to Leonhard Euler (1707–1783)

A Foundation for Computer Science

CONCRETE MATHEMATICS

Ronald L. Graham
AT&T Bell Laboratories

Donald E. Knuth
Stanford University

Oren Patashnik
Stanford University

ADDISON-WESLEY PUBLISHING COMPANY
Reading, Massachusetts   Menlo Park, California   New York
Don Mills, Ontario   Wokingham, England   Amsterdam   Bonn
Sydney   Singapore   Tokyo   Madrid   San Juan

Library of Congress Cataloging-in-Publication Data

Graham, Ronald Lewis, 1935-
Concrete mathematics : a foundation for computer science / Ronald L. Graham, Donald E. Knuth, Oren Patashnik.
xiii+625 p. 24 cm.
Bibliography: p. 578
Includes index.
ISBN 0-201-14236-8
1. Mathematics--1961- 2. Electronic data processing--Mathematics.
I. Knuth, Donald Ervin, 1938- . II. Patashnik, Oren. III. Title.
QA39.2.G733 1988
510--dc19 88-3779
CIP

Third printing, with corrections, May 1989

CDEFGHIJ–HA–89

Preface

"Audience, level, and treatment — a description of such matters is what prefaces are supposed to be about."
— P. R. Halmos [142]

THIS BOOK IS BASED on a course of the same name that has been taught annually at Stanford University since 1970. About fifty students have taken it each year — juniors and seniors, but mostly graduate students — and alumni of these classes have begun to spawn similar courses elsewhere. Thus the time seems ripe to present the material to a wider audience (including sophomores).

It was a dark and stormy decade when Concrete Mathematics was born. Long-held values were constantly being questioned during those turbulent years; college campuses were hotbeds of controversy. The college curriculum itself was challenged, and mathematics did not escape scrutiny. John Hammersley had just written a thought-provoking article "On the enfeeblement of mathematical skills by 'Modern Mathematics' and by similar soft intellectual trash in schools and universities" [145]; other worried mathematicians [272] even asked, "Can mathematics be saved?" One of the present authors had embarked on a series of books called *The Art of Computer Programming*, and in writing the first volume he (DEK) had found that there were mathematical tools missing from his repertoire; the mathematics he needed for a thorough, well-grounded understanding of computer programs was quite different from what he'd learned as a mathematics major in college. So he introduced a new course, teaching what he wished somebody had taught him.

"People do acquire a little brief authority by equipping themselves with jargon: they can pontificate and air a superficial expertise. But what we should ask of educated mathematicians is not what they can speechify about, nor even what they know about the existing corpus of mathematical knowledge, but rather what can they now do with their learning and whether they can actually solve mathematical problems arising in practice. In short, we look for deeds not words."
— J. Hammersley [145]

The course title "Concrete Mathematics" was originally intended as an antidote to "Abstract Mathematics," since concrete classical results were rapidly being swept out of the modern mathematical curriculum by a new wave of abstract ideas popularly called the "New Math." Abstract mathematics is a wonderful subject, and there's nothing wrong with it: It's beautiful, general, and useful. But its adherents had become deluded that the rest of mathematics was inferior and no longer worthy of attention. The goal of generalization had become so fashionable that a generation of mathematicians had become unable to relish beauty in the particular, to enjoy the challenge of solving quantitative problems, or to appreciate the value of technique. Abstract mathematics was becoming inbred and losing touch with reality; mathematical education needed a concrete counterweight in order to restore a healthy balance.

When DEK taught Concrete Mathematics at Stanford for the first time, he explained the somewhat strange title by saying that it was his attempt to teach a math course that was hard instead of soft. He announced that, contrary to the expectations of some of his colleagues, he was *not* going to teach the Theory of Aggregates, nor Stone's Embedding Theorem, nor even the Stone–Čech compactification. (Several students from the civil engineering department got up and quietly left the room.)

"The heart of mathematics consists of concrete examples and concrete problems."
—P. R. Halmos [141]

Although Concrete Mathematics began as a reaction against other trends, the main reasons for its existence were positive instead of negative. And as the course continued its popular place in the curriculum, its subject matter "solidified" and proved to be valuable in a variety of new applications. Meanwhile, independent confirmation for the appropriateness of the name came from another direction, when Z. A. Melzak published two volumes entitled *Companion to Concrete Mathematics* [214].

"It is downright sinful to teach the abstract before the concrete."
—Z. A. Melzak [214]

The material of concrete mathematics may seem at first to be a disparate bag of tricks, but practice makes it into a disciplined set of tools. Indeed, the techniques have an underlying unity and a strong appeal for many people. When another one of the authors (RLG) first taught the course in 1979, the students had such fun that they decided to hold a class reunion a year later.

But what exactly is Concrete Mathematics? It is a blend of CONtinuous and disCRETE mathematics. More concretely, it is the controlled manipulation of mathematical formulas, using a collection of techniques for solving problems. Once you, the reader, have learned the material in this book, all you will need is a cool head, a large sheet of paper, and fairly decent handwriting in order to evaluate horrendous-looking sums, to solve complex recurrence relations, and to discover subtle patterns in data. You will be so fluent in algebraic techniques that you will often find it easier to obtain exact results than to settle for approximate answers that are valid only in a limiting sense.

Concrete Mathematics is a bridge to abstract mathematics.

The major topics treated in this book include sums, recurrences, elementary number theory, binomial coefficients, generating functions, discrete probability, and asymptotic methods. The emphasis is on manipulative technique rather than on existence theorems or combinatorial reasoning; the goal is for each reader to become as familiar with discrete operations (like the greatest-integer function and finite summation) as a student of calculus is familiar with continuous operations (like the absolute-value function and infinite integration).

"The advanced reader who skips parts that appear too elementary may miss more than the less advanced reader who skips parts that appear too complex."
—G. Pólya [238]

Notice that this list of topics is quite different from what is usually taught nowadays in undergraduate courses entitled "Discrete Mathematics." Therefore the subject needs a distinctive name, and "Concrete Mathematics" has proved to be as suitable as any other.

(We're not bold enough to try Distinuous Mathematics.)

The original textbook for Stanford's course on concrete mathematics was the "Mathematical Preliminaries" section in *The Art of Computer Programming* [173]. But the presentation in those 110 pages is quite terse, so another author (OP) was inspired to draft a lengthy set of supplementary notes. The present book is an outgrowth of those notes; it is an expansion of, and a more leisurely introduction to, the material of Mathematical Preliminaries. Some of the more advanced parts have been omitted; on the other hand, several topics not found there have been included here so that the story will be complete.

". . . a concrete life preserver thrown to students sinking in a sea of abstraction."
— W. Gottschalk

The authors have enjoyed putting this book together because the subject began to jell and to take on a life of its own before our eyes; this book almost seemed to write itself. Moreover, the somewhat unconventional approaches we have adopted in several places have seemed to fit together so well, after these years of experience, that we can't help feeling that this book is a kind of manifesto about our favorite way to do mathematics. So we think the book has turned out to be a tale of mathematical beauty and surprise, and we hope that our readers will share at least ϵ of the pleasure we had while writing it.

Since this book was born in a university setting, we have tried to capture the spirit of a contemporary classroom by adopting an informal style. Some people think that mathematics is a serious business that must always be cold and dry; but we think mathematics is fun, and we aren't ashamed to admit the fact. Why should a strict boundary line be drawn between work and play? Concrete mathematics is full of appealing patterns; the manipulations are not always easy, but the answers can be astonishingly attractive. The joys and sorrows of mathematical work are reflected explicitly in this book because they are part of our lives.

Math graffiti:
Kilroy wasn't Haar.
Free the group.
Nuke the kernel.
Power to the n.
$N=1 \Rightarrow P=NP$.

Students always know better than their teachers, so we have asked the first students of this material to contribute their frank opinions, as "graffiti" in the margins. Some of these marginal markings are merely corny, some are profound; some of them warn about ambiguities or obscurities, others are typical comments made by wise guys in the back row; some are positive, some are negative, some are zero. But they all are real indications of feelings that should make the text material easier to assimilate. (The inspiration for such marginal notes comes from a student handbook entitled *Approaching Stanford*, where the official university line is counterbalanced by the remarks of outgoing students. For example, Stanford says, "There are a few things you cannot miss in this amorphous shape which is Stanford"; the margin says, "Amorphous ... what the h*** does that mean? Typical of the pseudo-intellectualism around here." Stanford: "There is no end to the potential of a group of students living together." Graffito: "Stanford dorms are like zoos without a keeper.")

I have only a marginal interest in this subject.

This was the most enjoyable course I've ever had. But it might be nice to summarize the material as you go along.

The margins also include direct quotations from famous mathematicians of past generations, giving the actual words in which they announced some of their fundamental discoveries. Somehow it seems appropriate to mix the words of Leibniz, Euler, Gauss, and others with those of the people who will be continuing the work. Mathematics is an ongoing endeavor for people everywhere; many strands are being woven into one rich fabric.

This book contains more than 500 exercises, divided into six categories:

I see: Concrete mathematics means drilling.

- ***Warmups*** are exercises that EVERY READER should try to do when first reading the material.
- ***Basics*** are exercises to develop facts that are best learned by trying one's own derivation rather than by reading somebody else's.
- ***Homework exercises*** are problems intended to deepen an understanding of material in the current chapter.
- ***Exam problems*** typically involve ideas from two or more chapters simultaneously; they are generally intended for use in take-home exams (not for in-class exams under time pressure).
- ***Bonus problems*** go beyond what an average student of concrete mathematics is expected to handle while taking a course based on this book; they extend the text in interesting ways.
- ***Research problems*** may or may not be humanly solvable, but the ones presented here seem to be worth a try (without time pressure).

The homework was tough but I learned a lot. It was worth every hour.

Take-home exams are vital — keep them.

Exams were harder than the homework led me to expect.

Answers to all the exercises appear in Appendix A, often with additional information about related results. (Of course, the "answers" to research problems are incomplete; but even in these cases, partial results or hints are given that might prove to be helpful.) Readers are encouraged to look at the answers, especially the answers to the warmup problems, but only AFTER making a serious attempt to solve the problem without peeking.

Cheaters may pass this course by just copying the answers, but they're only cheating themselves.

We have tried in Appendix C to give proper credit to the sources of each exercise, since a great deal of creativity and/or luck often goes into the design of an instructive problem. Mathematicians have unfortunately developed a tradition of borrowing exercises without any acknowledgment; we believe that the opposite tradition, practiced for example by books and magazines about chess (where names, dates, and locations of original chess problems are routinely specified) is far superior. However, we have not been able to pin down the sources of many problems that have become part of the folklore. If any reader knows the origin of an exercise for which our citation is missing or inaccurate, we would be glad to learn the details so that we can correct the omission in subsequent editions of this book.

Difficult exams don't take into account students who have other classes to prepare for.

The typeface used for mathematics throughout this book is a new design by Hermann Zapf [310], commissioned by the American Mathematical Society and developed with the help of a committee that included B. Beeton, R. P. Boas, L. K. Durst, D. E. Knuth, P. Murdock, R. S. Palais, P. Renz, E. Swanson, S. B. Whidden, and W. B. Woolf. The underlying philosophy of Zapf's design is to capture the flavor of mathematics as it might be written by a mathematician with excellent handwriting. A handwritten rather than mechanical style is appropriate because people generally create mathematics with pen, pencil, or chalk. (For example, one of the trademarks of the new design is the symbol for zero, '0', which is slightly pointed at the top because a handwritten zero rarely closes together smoothly when the curve returns to its starting point.) The letters are upright, not italic, so that subscripts, superscripts, and accents are more easily fitted with ordinary symbols. This new type family has been named *AMS Euler*, after the great Swiss mathematician Leonhard Euler (1707–1783) who discovered so much of mathematics as we know it today. The alphabets include Euler Text (Aa Bb Cc through Xx Yy Zz), Euler Fraktur ($\mathfrak{A}\mathfrak{a}\,\mathfrak{B}\mathfrak{b}\,\mathfrak{C}\mathfrak{c}$ through $\mathfrak{X}\mathfrak{x}\,\mathfrak{Y}\mathfrak{y}\,\mathfrak{Z}\mathfrak{z}$), and Euler Script Capitals ($\mathcal{A}\,\mathcal{B}\,\mathcal{C}$ through $\mathcal{X}\,\mathcal{Y}\,\mathcal{Z}$), as well as Euler Greek ($A\alpha\,B\beta\,\Gamma\gamma$ through $X\chi\,\Psi\psi\,\Omega\omega$) and special symbols such as $\wp$ and $\aleph$. We are especially pleased to be able to inaugurate the Euler family of typefaces in this book, because Leonhard Euler's spirit truly lives on every page: Concrete mathematics is Eulerian mathematics.

I'm unaccustomed to this face.

Dear prof: Thanks for (1) the puns, (2) the subject matter.

The authors are extremely grateful to Andrei Broder, Ernst Mayr, Andrew Yao, and Frances Yao, who contributed greatly to this book during the years that they taught Concrete Mathematics at Stanford. Furthermore we offer 1024 thanks to the teaching assistants who creatively transcribed what took place in class each year and who helped to design the examination questions; their names are listed in Appendix C. This book, which is essentially a compendium of sixteen years' worth of lecture notes, would have been impossible without their first-rate work.

I don't see how what I've learned will ever help me.

Many other people have helped to make this book a reality. For example, we wish to commend the students at Brown, Columbia, CUNY, Princeton, Rice, and Stanford who contributed the choice graffiti and helped to debug our first drafts. Our contacts at Addison-Wesley were especially efficient and helpful; in particular, we wish to thank our publisher (Peter Gordon), production supervisor (Bette Aaronson), designer (Roy Brown), and copy editor (Lyn Dupré). The National Science Foundation and the Office of Naval Research have given invaluable support. Cheryl Graham was tremendously helpful as we prepared the index. And above all, we wish to thank our wives (Fan, Jill, and Amy) for their patience, support, encouragement, and ideas.

I had a lot of trouble in this class, but I know it sharpened my math skills and my thinking skills.

We have tried to produce a perfect book, but we are imperfect authors. Therefore we solicit help in correcting any mistakes that we've made. A reward of \$2.56 will gratefully be paid to the first finder of any error, whether it is mathematical, historical, or typographical.

Murray Hill, New Jersey
and Stanford, California
May 1988

—RLG
DEK
OP

I would advise the casual student to stay away from this course.

A Note on Notation

SOME OF THE SYMBOLISM in this book has not (yet?) become standard. Here is a list of notations that might be unfamiliar to readers who have learned similar material from other books, together with the page numbers where these notations are explained:

Notation	*Name*	*Page*
$\ln x$	natural logarithm: $\log_e x$	262
$\lg x$	binary logarithm: $\log_2 x$	70
$\log x$	common logarithm: $\log_{10} x$	435
$\lfloor x \rfloor$	floor: $\max\{n \mid n \leqslant x,\ \text{integer } n\}$	67
$\lceil x \rceil$	ceiling: $\min\{n \mid n \geqslant x,\ \text{integer } n\}$	67
$x \bmod y$	remainder: $x - y\lfloor x/y \rfloor$	82
$\{x\}$	fractional part: $x \bmod 1$	70
$\sum f(x)\,\delta x$	indefinite summation	48
$\sum_a^b f(x)\,\delta x$	definite summation	49
$x^{\underline{n}}$	falling factorial power: $x!/(x-n)!$	47
$x^{\overline{n}}$	rising factorial power: $\Gamma(x+n)/\Gamma(x)$	48
$n¡$	subfactorial: $n!/0! - n!/1! + \cdots + (-1)^n n!/n!$	194
$\Re z$	real part: x, if $z = x + iy$	64
$\Im z$	imaginary part: y, if $z = x + iy$	64
H_n	harmonic number: $1/1 + \cdots + 1/n$	29
$H_n^{(x)}$	generalized harmonic number: $1/1^x + \cdots + 1/n^x$	263
$f^{(m)}(z)$	mth derivative of f at z	456

If you don't understand what the x denotes at the bottom of this page, try asking your Latin professor instead of your math professor.

$\left[{n \atop m}\right]$	Stirling cycle number (the "first kind")	245
$\left\{{n \atop m}\right\}$	Stirling subset number (the "second kind")	244
$\left\langle{n \atop m}\right\rangle$	Eulerian number	253
$\left\langle\!\!\left\langle{n \atop m}\right\rangle\!\!\right\rangle$	Second-order Eulerian number	256
$(a_m \ldots a_0)_b$	radix notation for $\sum_{k=0}^{m} a_k b^k$	11
$K(a_1, \ldots, a_n)$	continuant polynomial	288
$F\left({a, b \atop c} \middle| z\right)$	hypergeometric function	205
$\#A$	cardinality: number of elements in the set A	39
$[z^n]\, f(z)$	coefficient of z^n in $f(z)$	197
$[\alpha \,..\, \beta]$	closed interval: the set $\{x \mid \alpha \leqslant x \leqslant \beta\}$	73
$[m = n]$	1 if $m = n$, otherwise 0*	24
$[m \backslash n]$	1 if m divides n, otherwise 0*	102
$[m \backslash\backslash n]$	1 if m exactly divides n, otherwise 0*	146
$[m \perp n]$	1 if m is relatively prime to n, otherwise 0*	115

Prestressed concrete mathematics is concrete mathematics that's preceded by a bewildering list of notations.

*In general, if S is any statement that can be true or false, the bracketed notation [S] stands for 1 if S is true, 0 otherwise.

Throughout this text, we use single-quote marks ('...') to delimit text as it is *written*, double-quote marks ("...") for a phrase as it is *spoken*. Thus, the string of letters 'string' is sometimes called a "string."

Also 'nonstring' is a string.

An expression of the form 'a/bc' means the same as '$a/(bc)$'. Moreover, $\log x/\log y = (\log x)/(\log y)$ and $2n! = 2(n!)$.

Contents

1

Recurrent Problems

THIS CHAPTER EXPLORES three sample problems that give a feel for what's to come. They have two traits in common: They've all been investigated repeatedly by mathematicians; and their solutions all use the idea of *recurrence*, in which the solution to each problem depends on the solutions to smaller instances of the same problem.

1.1 THE TOWER OF HANOI

Let's look first at a neat little puzzle called the Tower of Hanoi, invented by the French mathematician Edouard Lucas in 1883. We are given a tower of eight disks, initially stacked in decreasing size on one of three pegs:

Raise your hand if you've never seen this. OK, the rest of you can cut to equation (1.1).

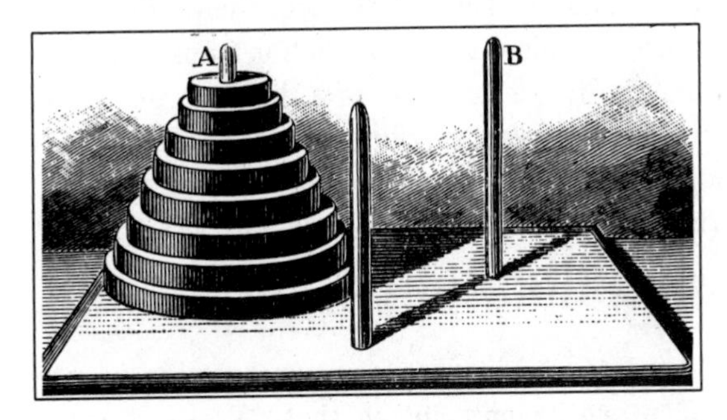

The objective is to transfer the entire tower to one of the other pegs, moving only one disk at a time and never moving a larger one onto a smaller.

Lucas [208] furnished his toy with a romantic legend about a much larger Tower of Brahma, which supposedly has 64 disks of pure gold resting on three diamond needles. At the beginning of time, he said, God placed these golden disks on the first needle and ordained that a group of priests should transfer them to the third, according to the rules above. The priests reportedly work day and night at their task. When they finish, the Tower will crumble and the world will end.

Gold — wow. Are our disks made of concrete?

It's not immediately obvious that the puzzle has a solution, but a little thought (or having seen the problem before) convinces us that it does. Now the question arises: What's the best we can do? That is, how many moves are necessary and sufficient to perform the task?

The best way to tackle a question like this is to generalize it a bit. The Tower of Brahma has 64 disks and the Tower of Hanoi has 8; let's consider what happens if there are n disks.

One advantage of this generalization is that we can scale the problem down even more. In fact, we'll see repeatedly in this book that it's advantageous to LOOK AT SMALL CASES first. It's easy to see how to transfer a tower that contains only one or two disks. And a small amount of experimentation shows how to transfer a tower of three.

The next step in solving the problem is to introduce appropriate notation: NAME AND CONQUER. Let's say that T_n is the minimum number of moves that will transfer n disks from one peg to another under Lucas's rules. Then T_1 is obviously 1, and $T_2 = 3$.

We can also get another piece of data for free, by considering the smallest case of all: Clearly $T_0 = 0$, because no moves at all are needed to transfer a tower of $n = 0$ disks! Smart mathematicians are not ashamed to think small, because general patterns are easier to perceive when the extreme cases are well understood (even when they are trivial).

But now let's change our perspective and try to think big; how can we transfer a large tower? Experiments with three disks show that the winning idea is to transfer the top two disks to the middle peg, then move the third, then bring the other two onto it. This gives us a clue for transferring n disks in general: We first transfer the $n-1$ smallest to a different peg (requiring T_{n-1} moves), then move the largest (requiring one move), and finally transfer the $n-1$ smallest back onto the largest (requiring another T_{n-1} moves). Thus we can transfer n disks (for $n > 0$) in at most $2T_{n-1} + 1$ moves:

$$T_n \leqslant 2T_{n-1} + 1, \qquad \text{for } n > 0.$$

This formula uses '$\leqslant$' instead of '$=$' because our construction proves only that $2T_{n-1} + 1$ moves suffice; we haven't shown that $2T_{n-1} + 1$ moves are necessary. A clever person might be able to think of a shortcut.

But is there a better way? Actually no. At some point we must move the largest disk. When we do, the $n-1$ smallest must be on a single peg, and it has taken at least T_{n-1} moves to put them there. We might move the largest disk more than once, if we're not too alert. But after moving the largest disk for the last time, we must transfer the $n-1$ smallest disks (which must again be on a single peg) back onto the largest; this too requires T_{n-1} moves. Hence

(Reference [7] in the bibliography, by Allardice and Fraser in 1884, contains the first published solution to this problem.)

$$T_n \geqslant 2T_{n-1} + 1, \qquad \text{for } n > 0.$$

These two inequalities, together with the trivial solution for $n = 0$, yield

$$\begin{aligned} T_0 &= 0\,; \\ T_n &= 2T_{n-1} + 1\,, \qquad \text{for } n > 0. \end{aligned} \tag{1.1}$$

(Notice that these formulas are consistent with the known values $T_1 = 1$ and $T_2 = 3$. Our experience with small cases has not only helped us to discover a general formula, it has also provided a convenient way to check that we haven't made a foolish error. Such checks will be especially valuable when we get into more complicated maneuvers in later chapters.)

Yeah, yeah... I seen that word before.

A set of equalities like (1.1) is called a *recurrence* (a.k.a. recurrence relation or recursion relation). It gives a boundary value and an equation for the general value in terms of earlier ones. Sometimes we refer to the general equation alone as a recurrence, although technically it needs a boundary value to be complete.

The recurrence allows us to compute T_n for any n we like. But nobody really likes to compute from a recurrence, when n is large; it takes too long. The recurrence only gives indirect, "local" information. A *solution to the recurrence* would make us much happier. That is, we'd like a nice, neat, "closed form" for T_n that lets us compute it quickly, even for large n. With a closed form, we can understand what T_n really is.

So how do we solve a recurrence? One way is to guess the correct solution, then to prove that our guess is correct. And our best hope for guessing the solution is to look (again) at small cases. So we compute, successively, $T_3 = 2\cdot3+1 = 7$; $T_4 = 2\cdot7+1 = 15$; $T_5 = 2\cdot15+1 = 31$; $T_6 = 2\cdot31+1 = 63$. Aha! It certainly looks as if

$$T_n = 2^n - 1\,, \qquad \text{for } n \geqslant 0. \tag{1.2}$$

At least this works for $n \leqslant 6$.

Mathematical induction is a general way to prove that some statement about the integer n is true for all $n \geqslant n_0$. First we prove the statement when n has its smallest value, n_0; this is called the *basis*. Then we prove the statement for $n > n_0$, assuming that it has already been proved for all values between n_0 and $n-1$, inclusive; this is called the *induction*. Such a proof gives infinitely many results with only a finite amount of work.

Mathematical induction proves that we can climb as high as we like on a ladder, by proving that we can climb onto the bottom rung (the basis) and that from each rung we can climb up to the next one (the induction).

Recurrences are ideally set up for mathematical induction. In our case, for example, (1.2) follows easily from (1.1): The basis is trivial, since $T_0 = 2^0 - 1 = 0$. And the induction follows for $n > 0$ if we assume that (1.2) holds when n is replaced by $n-1$:

$$T_n = 2T_{n-1} + 1 = 2(2^{n-1} - 1) + 1 = 2^n - 1\,.$$

Hence (1.2) holds for n as well. Good! Our quest for T_n has ended successfully.

Of course the priests' task hasn't ended; they're still dutifully moving disks, and will be for a while, because for $n = 64$ there are $2^{64} - 1$ moves (about 18 quintillion). Even at the impossible rate of one move per microsecond, they will need more than 5000 centuries to transfer the Tower of Brahma. Lucas's original puzzle is a bit more practical. It requires $2^8 - 1 = 255$ moves, which takes about four minutes for the quick of hand.

The Tower of Hanoi recurrence is typical of many that arise in applications of all kinds. In finding a closed-form expression for some quantity of interest like T_n we go through three stages:

1 Look at small cases. This gives us insight into the problem and helps us in stages 2 and 3.

2 Find and prove a mathematical expression for the quantity of interest. For the Tower of Hanoi, this is the recurrence (1.1) that allows us, given the inclination, to compute T_n for any n.

What is a proof? "One half of one percent pure alcohol."

3 Find and prove a closed form for our mathematical expression. For the Tower of Hanoi, this is the recurrence solution (1.2).

The third stage is the one we will concentrate on throughout this book. In fact, we'll frequently skip stages 1 and 2 entirely, because a mathematical expression will be given to us as a starting point. But even then, we'll be getting into subproblems whose solutions will take us through all three stages.

Our analysis of the Tower of Hanoi led to the correct answer, but it required an "inductive leap"; we relied on a lucky guess about the answer. One of the main objectives of this book is to explain how a person can solve recurrences *without* being clairvoyant. For example, we'll see that recurrence (1.1) can be simplified by adding 1 to both sides of the equations:

$$\begin{aligned} T_0 + 1 &= 1\,; \\ T_n + 1 &= 2T_{n-1} + 2\,, \qquad \text{for } n > 0. \end{aligned}$$

Now if we let $U_n = T_n + 1$, we have

Interesting: We get rid of the +1 in (1.1) by adding, not by subtracting.

$$\begin{aligned} U_0 &= 1\,; \\ U_n &= 2U_{n-1}\,, \qquad \text{for } n > 0. \end{aligned} \tag{1.3}$$

It doesn't take genius to discover that the solution to *this* recurrence is just $U_n = 2^n$; hence $T_n = 2^n - 1$. Even a computer could discover this.

1.2 LINES IN THE PLANE

Our second sample problem has a more geometric flavor: How many slices of pizza can a person obtain by making n straight cuts with a pizza knife? Or, more academically: What is the maximum number L_n of regions defined by n lines in the plane? This problem was first solved in 1826, by the Swiss mathematician Jacob Steiner [278].

(A pizza with Swiss cheese?)

Again we start by looking at small cases, remembering to begin with the smallest of all. The plane with no lines has one region; with one line it has two regions; and with two lines it has four regions:

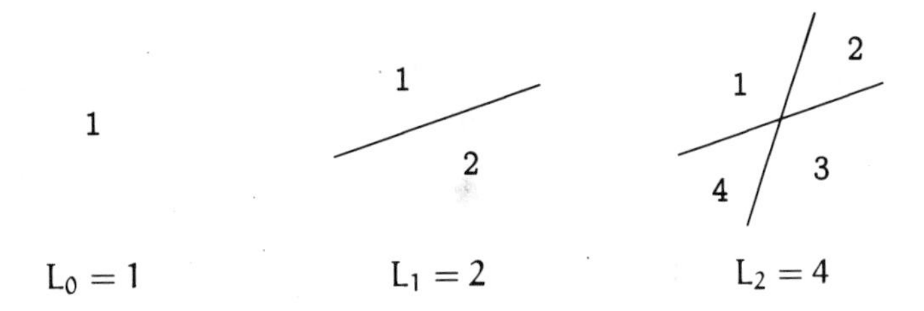

(Each line extends infinitely in both directions.)

Sure, we think, $L_n = 2^n$; of course! Adding a new line simply doubles the number of regions. Unfortunately this is wrong. We could achieve the doubling if the nth line would split each old region in two; certainly it can split an old region in at most two pieces, since each old region is convex. (A straight line can split a convex region into at most two new regions, which will also be convex.) But when we add the third line — the thick one in the diagram below — we soon find that it can split at most three of the old regions, no matter how we've placed the first two lines:

A region is convex if it includes all line segments between any two of its points. (That's not what my dictionary says, but it's what mathematicians believe.)

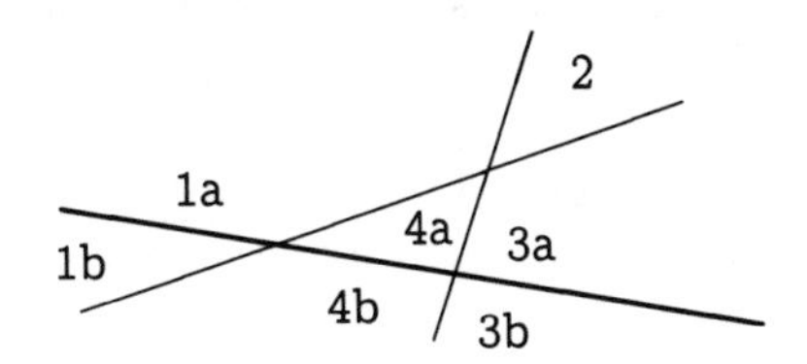

Thus $L_3 = 4 + 3 = 7$ is the best we can do.

And after some thought we realize the appropriate generalization. The nth line (for $n > 0$) increases the number of regions by k if and only if it splits k of the old regions, and it splits k old regions if and only if it hits the previous lines in $k-1$ different places. Two lines can intersect in at most one point. Therefore the new line can intersect the $n-1$ old lines in at most $n-1$ different points, and we must have $k \leqslant n$. We have established the upper bound

$$L_n \leqslant L_{n-1} + n, \qquad \text{for } n > 0.$$

Furthermore it's easy to show by induction that we can achieve equality in this formula. We simply place the nth line in such a way that it's not parallel to any of the others (hence it intersects them all), and such that it doesn't go through any of the existing intersection points (hence it intersects them all in different places). The recurrence is therefore

$$\begin{aligned} L_0 &= 1; \\ L_n &= L_{n-1} + n, \qquad \text{for } n > 0. \end{aligned} \tag{1.4}$$

The known values of L_1, L_2, and L_3 check perfectly here, so we'll buy this.

Now we need a closed-form solution. We could play the guessing game again, but 1, 2, 4, 7, 11, 16, ... doesn't look familiar; so let's try another tack. We can often understand a recurrence by "unfolding" or "unwinding" it all the way to the end, as follows:

$$\begin{aligned} L_n &= L_{n-1} + n \\ &= L_{n-2} + (n-1) + n \\ &= L_{n-3} + (n-2) + (n-1) + n \\ &\quad\vdots \\ &= L_0 + 1 + 2 + \cdots + (n-2) + (n-1) + n \\ &= 1 + S_n, \qquad \text{where } S_n = 1 + 2 + 3 + \cdots + (n-1) + n. \end{aligned}$$

Unfolding? I'd call this "plugging in."

In other words, L_n is one more than the sum S_n of the first n positive integers.

The quantity S_n pops up now and again, so it's worth making a table of small values. Then we might recognize such numbers more easily when we see them the next time:

n	1	2	3	4	5	6	7	8	9	10	11	12	13	14
S_n	1	3	6	10	15	21	28	36	45	55	66	78	91	105

These values are also called the *triangular numbers*, because S_n is the number of bowling pins in an n-row triangular array. For example, the usual four-row array ∵ has $S_4 = 10$ pins.

To evaluate S_n we can use a trick that Gauss reportedly came up with in 1786, when he was nine years old [73]:

It seems a lot of stuff is attributed to Gauss— either he was really smart or he had a great press agent.

Maybe he just had a magnetic personality.

$$\begin{array}{rcl} S_n &=& 1 + 2 + 3 + \cdots + (n-1) + n \\ +\,S_n &=& n + (n-1) + (n-2) + \cdots + 2 + 1 \\ \hline 2S_n &=& (n+1) + (n+1) + (n+1) + \cdots + (n+1) + (n+1) \end{array}$$

We merely add S_n to its reversal, so that each of the n columns on the right sums to $n+1$. Simplifying,

$$S_n = \frac{n(n+1)}{2}, \qquad \text{for } n \geqslant 0. \tag{1.5}$$

Actually Gauss is often called the greatest mathematician of all time. So it's nice to be able to understand at least one of his discoveries.

OK, we have our solution:

$$L_n = \frac{n(n+1)}{2} + 1, \qquad \text{for } n \geqslant 0. \tag{1.6}$$

As experts, we might be satisfied with this derivation and consider it a proof, even though we waved our hands a bit when doing the unfolding and reflecting. But students of mathematics should be able to meet stricter standards; so it's a good idea to construct a rigorous proof by induction. The key induction step is

$$L_n = L_{n-1} + n = \left(\tfrac{1}{2}(n-1)n + 1\right) + n = \tfrac{1}{2}n(n+1) + 1.$$

Now there can be no doubt about the closed form (1.6).

When in doubt, look at the words. Why is it "closed," as opposed to "open"? What image does it bring to mind? Answer: The equation is "closed," not defined in terms of itself—not leading to recurrence. The case is "closed"—it won't happen again. Metaphors are the key.

Incidentally we've been talking about "closed forms" without explicitly saying what we mean. Usually it's pretty clear. Recurrences like (1.1) and (1.4) are not in closed form—they express a quantity in terms of itself; but solutions like (1.2) and (1.6) are. Sums like $1 + 2 + \cdots + n$ are not in closed form—they cheat by using '$\cdots$'; but expressions like $n(n+1)/2$ are. We could give a rough definition like this: An expression for a quantity $f(n)$ is in closed form if we can compute it using at most a fixed number of "well known" standard operations, independent of n. For example, $2^n - 1$ and $n(n+1)/2$ are closed forms because they involve only addition, subtraction, multiplication, division, and exponentiation, in explicit ways.

The total number of simple closed forms is limited, and there are recurrences that don't have simple closed forms. When such recurrences turn out to be important, because they arise repeatedly, we add new operations to our repertoire; this can greatly extend the range of problems solvable in "simple" closed form. For example, the product of the first n integers, $n!$, has proved to be so important that we now consider it a basic operation. The formula '$n!$' is therefore in closed form, although its equivalent '$1 \cdot 2 \cdot \ldots \cdot n$' is not.

Is "zig" a technical term?

And now, briefly, a variation of the lines-in-the-plane problem: Suppose that instead of straight lines we use bent lines, each containing one "zig." What is the maximum number Z_n of regions determined by n such bent lines in the plane? We might expect Z_n to be about twice as big as L_n, or maybe three times as big. Let's see:

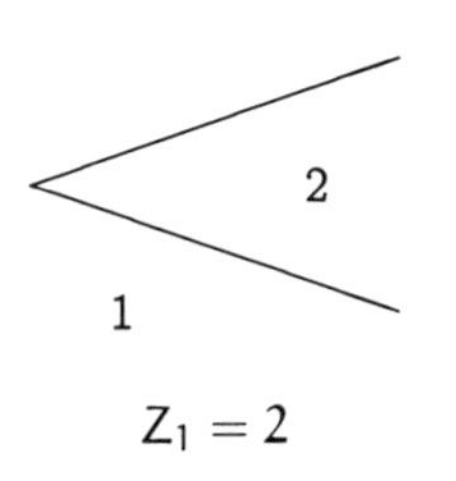

$Z_1 = 2$

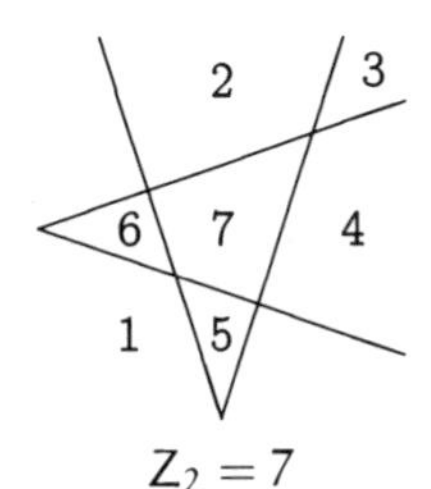

$Z_2 = 7$

From these small cases, and after a little thought, we realize that a bent line is like two straight lines except that regions merge when the "two" lines don't extend past their intersection point.

... and a little afterthought...

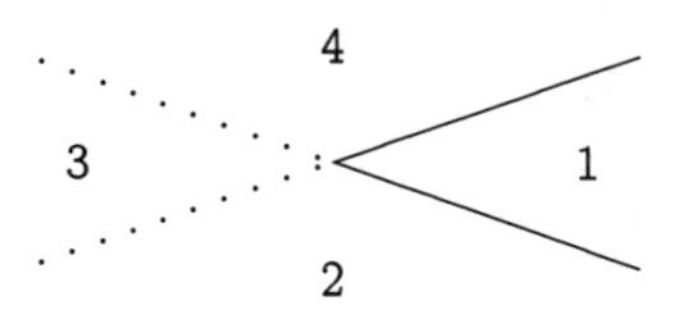

Regions 2, 3, and 4, which would be distinct with two lines, become a single region when there's a bent line; we lose two regions. However, if we arrange things properly — the zig point must lie "beyond" the intersections with the other lines — that's all we lose; that is, we lose only two regions per line. Thus

Exercise 18 has the details.

$$
\begin{aligned}
Z_n = L_{2n} - 2n &= 2n(2n+1)/2 + 1 - 2n \\
&= 2n^2 - n + 1\,, \qquad \text{for } n \geqslant 0.
\end{aligned}
\tag{1.7}
$$

Comparing the closed forms (1.6) and (1.7), we find that for large n,

$$
\begin{aligned}
L_n &\sim \tfrac{1}{2}n^2\,, \\
Z_n &\sim 2n^2\,;
\end{aligned}
$$

so we get about four times as many regions with bent lines as with straight lines. (In later chapters we'll be discussing how to analyze the approximate behavior of integer functions when n is large.)

1.3 THE JOSEPHUS PROBLEM

Our final introductory example is a variant of an ancient problem named for Flavius Josephus, a famous historian of the first century. Legend has it that Josephus wouldn't have lived to become famous without his mathematical talents. During the Jewish–Roman war, he was among a band of 41 Jewish rebels trapped in a cave by the Romans. Preferring suicide to capture, the rebels decided to form a circle and, proceeding around it, to kill every third remaining person until no one was left. But Josephus, along with an unindicted co-conspirator, wanted none of this suicide nonsense; so he quickly calculated where he and his friend should stand in the vicious circle.

(Ahrens [5, vol. 2] and Herstein and Kaplansky [156] discuss the interesting history of this problem. Josephus himself [166] is a bit vague.)

... thereby saving his tale for us to hear.

In our variation, we start with n people numbered 1 to n around a circle, and we eliminate every *second* remaining person until only one survives. For example, here's the starting configuration for $n = 10$:

Here's a case where $n = 0$ makes no sense.

The elimination order is 2, 4, 6, 8, 10, 3, 7, 1, 9, so 5 survives. The problem: Determine the survivor's number, $J(n)$.

We just saw that $J(10) = 5$. We might conjecture that $J(n) = n/2$ when n is even; and the case $n = 2$ supports the conjecture: $J(2) = 1$. But a few other small cases dissuade us — the conjecture fails for $n = 4$ and $n = 6$.

n	1	2	3	4	5	6
$J(n)$	1	1	3	1	3	5

Even so, a bad guess isn't a waste of time, because it gets us involved in the problem.

It's back to the drawing board; let's try to make a better guess. Hmmm ... $J(n)$ always seems to be odd. And in fact, there's a good reason for this: The first trip around the circle eliminates all the even numbers. Furthermore, if n itself is an even number, we arrive at a situation similar to what we began with, except that there are only half as many people, and their numbers have changed.

So let's suppose that we have $2n$ people originally. After the first go-round, we're left with

$2n-1$ 1 3
$2n-3$ 5
7
...

and 3 will be the next to go. This is just like starting out with n people, except that each person's number has been doubled and decreased by 1. That is,

This is the tricky part: We have $J(2n) = newnumber(J(n))$, where $newnumber(k) = 2k - 1$.

$$J(2n) = 2J(n) - 1\,, \qquad \text{for } n \geqslant 1.$$

We can now go quickly to large n. For example, we know that $J(10) = 5$, so

$$J(20) = 2J(10) - 1 = 2\cdot 5 - 1 = 9\,.$$

Similarly $J(40) = 17$, and we can deduce that $J(5\cdot 2^m) = 2^{m+1} + 1$.

But what about the odd case? With $2n+1$ people, it turns out that person number 1 is wiped out just after person number $2n$, and we're left with

Odd case? Hey, leave my brother out of it.

$2n+1$ 3 5

$2n-1$ 7

9

...

Again we almost have the original situation with n people, but this time their numbers are doubled and *increased* by 1. Thus

$$J(2n+1) \;=\; 2J(n)+1\,, \qquad \text{for } n \geqslant 1.$$

Combining these equations with $J(1)=1$ gives us a recurrence that defines J in all cases:

$$\begin{aligned} J(1) &= 1\,; \\ J(2n) &= 2J(n)-1\,, \qquad \text{for } n \geqslant 1; \\ J(2n+1) &= 2J(n)+1\,, \qquad \text{for } n \geqslant 1. \end{aligned} \tag{1.8}$$

Instead of getting $J(n)$ from $J(n-1)$, this recurrence is much more "efficient," because it reduces n by a factor of 2 or more each time it's applied. We could compute $J(1000000)$, say, with only 19 applications of (1.8). But still, we seek a closed form, because that will be even quicker and more informative. After all, this is a matter of life or death.

Our recurrence makes it possible to build a table of small values very quickly. Perhaps we'll be able to spot a pattern and guess the answer.

n	1	2	3	4	5	6	7	8	9	10	11	12	13	14	15	16
$J(n)$	1	1	3	1	3	5	7	1	3	5	7	9	11	13	15	1

Voilà! It seems we can group by powers of 2 (marked by vertical lines in the table); $J(n)$ is always 1 at the beginning of a group and it increases by 2 within a group. So if we write n in the form $n = 2^m + l$, where 2^m is the largest power of 2 not exceeding n and where l is what's left, the solution to our recurrence seems to be

$$J(2^m+l) \;=\; 2l+1\,, \qquad \text{for } m \geqslant 0 \text{ and } 0 \leqslant l < 2^m. \tag{1.9}$$

(Notice that if $2^m \leqslant n < 2^{m+1}$, the remainder $l = n - 2^m$ satisfies $0 \leqslant l < 2^{m+1} - 2^m = 2^m$.)

We must now prove (1.9). As in the past we use induction, but this time the induction is on m. When $m=0$ we must have $l=0$; thus the basis of

But there's a simpler way! The key fact is that $J(2^m) = 1$ *for all* m*, and this follows immediately from our first equation,*

$$J(2n) = 2J(n) - 1.$$

Hence we know that the first person will survive whenever n *is a power of 2. And in the general case, when* $n = 2^m + l$*, the number of people is reduced to a power of 2 after there have been* l *executions. The first remaining person at this point, the survivor, is number* $2l + 1$.

(1.9) reduces to $J(1) = 1$, which is true. The induction step has two parts, depending on whether l is even or odd. If $m > 0$ and $2^m + l = 2n$, then l is even and

$$J(2^m + l) = 2J(2^{m-1} + l/2) - 1 = 2(2l/2 + 1) - 1 = 2l + 1,$$

by (1.8) and the induction hypothesis; this is exactly what we want. A similar proof works in the odd case, when $2^m + l = 2n + 1$. We might also note that (1.8) implies the relation

$$J(2n+1) - J(2n) = 2.$$

Either way, the induction is complete and (1.9) is established.

To illustrate solution (1.9), let's compute $J(100)$. In this case we have $100 = 2^6 + 36$, so $J(100) = 2\cdot 36 + 1 = 73$.

Now that we've done the hard stuff (solved the problem) we seek the soft: Every solution to a problem can be generalized so that it applies to a wider class of problems. Once we've learned a technique, it's instructive to look at it closely and see how far we can go with it. Hence, for the rest of this section, we will examine the solution (1.9) and explore some generalizations of the recurrence (1.8). These explorations will uncover the structure that underlies all such problems.

Powers of 2 played an important role in our finding the solution, so it's natural to look at the radix 2 representations of n and $J(n)$. Suppose n's binary expansion is

$$n = (b_m\, b_{m-1} \ldots b_1\, b_0)_2;$$

that is,

$$n = b_m 2^m + b_{m-1} 2^{m-1} + \cdots + b_1 2 + b_0,$$

where each b_i is either 0 or 1 and where the leading bit b_m is 1. Recalling that $n = 2^m + l$, we have, successively,

$$\begin{aligned} n &= (1\, b_{m-1}\, b_{m-2} \ldots b_1\, b_0)_2, \\ l &= (0\, b_{m-1}\, b_{m-2} \ldots b_1\, b_0)_2, \\ 2l &= (b_{m-1}\, b_{m-2} \ldots b_1\, b_0\, 0)_2, \\ 2l+1 &= (b_{m-1}\, b_{m-2} \ldots b_1\, b_0\, 1)_2, \\ J(n) &= (b_{m-1}\, b_{m-2} \ldots b_1\, b_0\, b_m)_2. \end{aligned}$$

(The last step follows because $J(n) = 2l + 1$ and because $b_m = 1$.) We have proved that

$$J\big((b_m\, b_{m-1} \ldots b_1\, b_0)_2\big) = (b_{m-1} \ldots b_1\, b_0\, b_m)_2; \qquad (1.10)$$

that is, in the lingo of computer programming, we get $J(n)$ from n by doing a one-bit cyclic shift left! Magic. For example, if $n = 100 = (1100100)_2$ then $J(n) = J((1100100)_2) = (1001001)_2$, which is $64+8+1 = 73$. If we had been working all along in binary notation, we probably would have spotted this pattern immediately.

If we start with n and iterate the J function $m+1$ times, we're doing $m+1$ one-bit cyclic shifts; so, since n is an $(m+1)$-bit number, we might expect to end up with n again. But this doesn't quite work. For instance if $n = 13$ we have $J((1101)_2) = (1011)_2$, but then $J((1011)_2) = (111)_2$ and the process breaks down; the 0 disappears when it becomes the leading bit. In fact, $J(n)$ must always be $\leqslant n$ by definition, since $J(n)$ is the survivor's number; hence if $J(n) < n$ we can never get back up to n by continuing to iterate.

("Iteration" means applying a function to itself.)

Repeated application of J produces a sequence of decreasing values that eventually reach a "fixed point" where $J(n) = n$. The cyclic shift property makes it easy to see what that fixed point will be: Iterating the function m or more times will always produce a pattern of all 1's whose value is $2^{\nu(n)}-1$, where $\nu(n)$ is the number of 1 bits in the binary representation of n. Thus, since $\nu(13) = 3$, we have

$$\overbrace{J(J(\ldots J}^{\text{3 or more } J\text{'s}}(13)\ldots)) = 2^3 - 1 = 7;$$

similarly

$$\overbrace{J(J(\ldots J}^{\text{14 or more}}((1011011011011110)_2)\ldots)) = 2^{10} - 1 = 1023.$$

Curious, but true.

Curiously enough, if M is a compact C^∞ n-manifold ($n > 1$), there exists a differentiable immersion of M into $\mathbf{R}^{2n-\nu(n)}$ but not necessarily into $\mathbf{R}^{2n-\nu(n)-1}$. I wonder if Josephus was secretly a topologist?

Let's return briefly to our first guess, that $J(n) = n/2$ when n is even. This is obviously not true in general, but we can now determine exactly when it *is* true:

$$\begin{aligned} J(n) &= n/2, \\ 2l+1 &= (2^m + l)/2, \\ l &= \tfrac{1}{3}(2^m - 2). \end{aligned}$$

If this number $l = \frac{1}{3}(2^m - 2)$ is an integer, then $n = 2^m + l$ will be a solution, because l will be less than 2^m. It's not hard to verify that $2^m - 2$ is a multiple of 3 when m is odd, but not when m is even. (We will study such things in Chapter 4.) Therefore there are infinitely many solutions to the equation

$J(n) = n/2$, beginning as follows:

m	l	$n = 2^m + l$	$J(n) = 2l + 1 = n/2$	n (binary)
1	0	2	1	10
3	2	10	5	1010
5	10	42	21	101010
7	42	170	85	10101010

Notice the pattern in the rightmost column. These are the binary numbers for which cyclic-shifting one place left produces the same result as ordinary-shifting one place right (halving).

OK, we understand the J function pretty well; the next step is to generalize it. What would have happened if our problem had produced a recurrence that was something like (1.8), but with different constants? Then we might not have been lucky enough to guess the solution, because the solution might have been really weird. Let's investigate this by introducing constants α, β, and γ and trying to find a closed form for the more general recurrence

Looks like Greek to me.

$$\begin{aligned} f(1) &= \alpha; \\ f(2n) &= 2f(n) + \beta, && \text{for } n \geqslant 1; \\ f(2n+1) &= 2f(n) + \gamma, && \text{for } n \geqslant 1. \end{aligned} \tag{1.11}$$

(Our original recurrence had $\alpha = 1$, $\beta = -1$, and $\gamma = 1$.) Starting with $f(1) = \alpha$ and working our way up, we can construct the following general table for small values of n:

n	$f(n)$
1	α
2	$2\alpha + \beta$
3	$2\alpha + \gamma$
4	$4\alpha + 3\beta$
5	$4\alpha + 2\beta + \gamma$
6	$4\alpha + \beta + 2\gamma$
7	$4\alpha + 3\gamma$
8	$8\alpha + 7\beta$
9	$8\alpha + 6\beta + \gamma$

(1.12)

It seems that α's coefficient is n's largest power of 2. Furthermore, between powers of 2, β's coefficient decreases by 1 down to 0 and γ's increases by 1 up from 0. Therefore if we express $f(n)$ in the form

$$f(n) = A(n)\alpha + B(n)\beta + C(n)\gamma, \tag{1.13}$$

by separating out its dependence on α, β, and γ, it seems that

$$\begin{aligned} A(n) &= 2^m ; \\ B(n) &= 2^m - 1 - l ; \\ C(n) &= l . \end{aligned} \tag{1.14}$$

Here, as usual, $n = 2^m + l$ and $0 \leqslant l < 2^m$, for $n \geqslant 1$.

It's not terribly hard to prove (1.13) and (1.14) by induction, but the calculations are messy and uninformative. Fortunately there's a better way to proceed, by choosing particular values and then combining them. Let's illustrate this by considering the special case $\alpha = 1$, $\beta = \gamma = 0$, when $f(n)$ is supposed to be equal to $A(n)$: Recurrence (1.11) becomes

Hold onto your hats, this next part is new stuff.

$$\begin{aligned} A(1) &= 1 ; \\ A(2n) &= 2A(n) , \qquad \text{for } n \geqslant 1 ; \\ A(2n+1) &= 2A(n) , \qquad \text{for } n \geqslant 1 . \end{aligned}$$

Sure enough, it's true (by induction on m) that $A(2^m + l) = 2^m$.

Next, let's use recurrence (1.11) and solution (1.13) *in reverse*, by starting with a simple function $f(n)$ and seeing if there are any constants (α, β, γ) that will define it. Plugging in the constant function $f(n) = 1$ says that

A neat idea!

$$\begin{aligned} 1 &= \alpha ; \\ 1 &= 2 \cdot 1 + \beta ; \\ 1 &= 2 \cdot 1 + \gamma ; \end{aligned}$$

hence the values $(\alpha, \beta, \gamma) = (1, -1, -1)$ satisfying these equations will yield $A(n) - B(n) - C(n) = f(n) = 1$. Similarly, we can plug in $f(n) = n$:

$$\begin{aligned} 1 &= \alpha ; \\ 2n &= 2 \cdot n + \beta ; \\ 2n + 1 &= 2 \cdot n + \gamma ; \end{aligned}$$

These equations hold for all n when $\alpha = 1$, $\beta = 0$, and $\gamma = 1$, so we don't need to prove by induction that these parameters will yield $f(n) = n$. We already *know* that $f(n) = n$ will be the solution in such a case, because the recurrence (1.11) uniquely defines $f(n)$ for every value of n.

And now we're essentially done! We have shown that the functions $A(n)$, $B(n)$, and $C(n)$ of (1.13), which solve (1.11) in general, satisfy the equations

$$\begin{aligned} A(n) &= 2^m , \qquad \text{where } n = 2^m + l \text{ and } 0 \leqslant l < 2^m ; \\ A(n) - B(n) - C(n) &= 1 ; \\ A(n) + C(n) &= n . \end{aligned}$$

Our conjectures in (1.14) follow immediately, since we can solve these equations to get $C(n) = n - A(n) = l$ and $B(n) = A(n) - 1 - C(n) = 2^m - 1 - l$.

Beware: The authors are expecting us to figure out the idea of the repertoire method from seat-of-the-pants examples, instead of giving us a top-down presentation. The method works best with recurrences that are "linear," in the sense that their solutions can be expressed as a sum of arbitrary parameters multiplied by functions of n, as in (1.13). Equation (1.13) is the key.

This approach illustrates a surprisingly useful *repertoire method* for solving recurrences. First we find settings of general parameters for which we know the solution; this gives us a repertoire of special cases that we can solve. Then we obtain the general case by combining the special cases. We need as many independent special solutions as there are independent parameters (in this case three, for α, β, and γ). Exercises 16 and 20 provide further examples of the repertoire approach.

We know that the original J-recurrence has a magical solution, in binary:

$$J\big((b_m b_{m-1} \ldots b_1 b_0)_2\big) = (b_{m-1} \ldots b_1 b_0 b_m)_2, \qquad \text{where } b_m = 1.$$

Does the generalized Josephus recurrence admit of such magic?

Sure, why not? We can rewrite the generalized recurrence (1.11) as

$$\begin{aligned} f(1) &= \alpha; \\ f(2n+j) &= 2f(n) + \beta_j, \qquad \text{for } j = 0, 1 \quad \text{and} \quad n \geqslant 1, \end{aligned} \tag{1.15}$$

if we let $\beta_0 = \beta$ and $\beta_1 = \gamma$. And this recurrence unfolds, binary-wise:

$$\begin{aligned} f\big((b_m b_{m-1} \ldots b_1 b_0)_2\big) &= 2f\big((b_m b_{m-1} \ldots b_1)_2\big) + \beta_{b_0} \\ &= 4f\big((b_m b_{m-1} \ldots b_2)_2\big) + 2\beta_{b_1} + \beta_{b_0} \\ &\ \ \vdots \\ &= 2^m f\big((b_m)_2\big) + 2^{m-1}\beta_{b_{m-1}} + \cdots + 2\beta_{b_1} + \beta_{b_0} \\ &= 2^m\alpha + 2^{m-1}\beta_{b_{m-1}} + \cdots + 2\beta_{b_1} + \beta_{b_0}. \end{aligned}$$

('relax' = 'destroy')

Suppose we now relax the radix 2 notation to allow arbitrary digits instead of just 0 and 1. The derivation above tells us that

$$f\big((b_m b_{m-1} \ldots b_1 b_0)_2\big) = (\alpha\, \beta_{b_{m-1}}\, \beta_{b_{m-2}} \ldots \beta_{b_1}\, \beta_{b_0})_2. \tag{1.16}$$

Nice. We would have seen this pattern earlier if we had written (1.12) in another way:

n	$f(n)$
1	α
2	$2\alpha + \beta$
3	$2\alpha + \gamma$
4	$4\alpha + 2\beta + \beta$
5	$4\alpha + 2\beta + \gamma$
6	$4\alpha + 2\gamma + \beta$
7	$4\alpha + 2\gamma + \gamma$

I think I get it: The binary representations of $A(n)$, $B(n)$, and $C(n)$ have 1's in different positions.

For example, when $n = 100 = (1100100)_2$, our original Josephus values $\alpha = 1$, $\beta = -1$, and $\gamma = 1$ yield

$$\begin{array}{rcccccccccc} n = & (& 1 & 1 & 0 & 0 & 1 & 0 & 0\,)_2 & = & 100 \\ \hline f(n) = & (& 1 & 1 & -1 & -1 & 1 & -1 & -1\,)_2 & & \\ = & & +64 & +32 & -16 & -8 & +4 & -2 & -1 & = & 73 \end{array}$$

as before. The cyclic-shift property follows because each block of binary digits $(1\,0 \ldots 0\,0)_2$ in the representation of n is transformed into

$$(1\,{-1} \ldots {-1}\,{-1})_2 = (0\,0 \ldots 0\,1)_2 .$$

So our change of notation has given us the compact solution (1.16) to the general recurrence (1.15). If we're really uninhibited we can now generalize even more. The recurrence

There are two kinds of generalizations. One is cheap and the other is valuable. It is easy to generalize by diluting a little idea with a big terminology. It is much more difficult to prepare a refined and condensed extract from several good ingredients.
— G. Pólya [238]

$$\begin{aligned} f(j) &= \alpha_j , && \text{for } 1 \leqslant j < d; \\ f(dn + j) &= cf(n) + \beta_j , && \text{for } 0 \leqslant j < d \text{ and } n \geqslant 1, \end{aligned} \tag{1.17}$$

is the same as the previous one except that we start with numbers in radix d and produce values in radix c. That is, it has the radix-changing solution

$$f\big((b_m\, b_{m-1} \ldots b_1\, b_0)_d\big) = (\alpha_{b_m}\, \beta_{b_{m-1}}\, \beta_{b_{m-2}} \ldots \beta_{b_1}\, \beta_{b_0})_c . \tag{1.18}$$

For example, suppose that by some stroke of luck we're given the recurrence

$$\begin{aligned} f(1) &= 34 , \\ f(2) &= 5 , \\ f(3n) &= 10f(n) + 76 , && \text{for } n \geqslant 1, \\ f(3n+1) &= 10f(n) - 2 , && \text{for } n \geqslant 1, \\ f(3n+2) &= 10f(n) + 8 , && \text{for } n \geqslant 1, \end{aligned}$$

and suppose we want to compute $f(19)$. Here we have $d = 3$ and $c = 10$. Now $19 = (201)_3$, and the radix-changing solution tells us to perform a digit-by-digit replacement from radix 3 to radix 10. So the leading 2 becomes a 5, and the 0 and 1 become 76 and -2, giving

Perhaps this was a stroke of bad luck.

$$f(19) = f\big((201)_3\big) = (5\ 76\ {-2})_{10} = 1258 ,$$

which is our answer.

Thus Josephus and the Jewish–Roman war have led us to some interesting general recurrences.

But in general I'm against recurrences of war.

Exercises

Warmups

Please do all the warmups in all the chapters!
— The Mgm't

1 All horses are the same color; we can prove this by induction on the number of horses in a given set. Here's how: "If there's just one horse then it's the same color as itself, so the basis is trivial. For the induction step, assume that there are n horses numbered 1 to n. By the induction hypothesis, horses 1 through $n-1$ are the same color, and similarly horses 2 through n are the same color. But the middle horses, 2 through $n-1$, can't change color based on who they're grouped with; these are horses, not chameleons. So horses 1 and n must be the same color as well, by transitivity. Thus all n horses are the same color; QED." What, if anything, is wrong with this reasoning?

2 Find the shortest sequence of moves that transfers a tower of n disks from the left peg A to the right peg B, if direct moves between A and B are disallowed. (Each move must be to or from the middle peg. As usual, a larger disk must never appear above a smaller one.)

3 Show that, in the process of transferring a tower under the restrictions of the preceding exercise, we will actually encounter every properly stacked arrangement of n disks on three pegs.

4 Are there any starting and ending configurations of n disks on three pegs that are more than $2^n - 1$ moves apart, under Lucas's original rules?

5 A "Venn diagram" with three overlapping circles is often used to illustrate the eight possible subsets associated with three given sets:

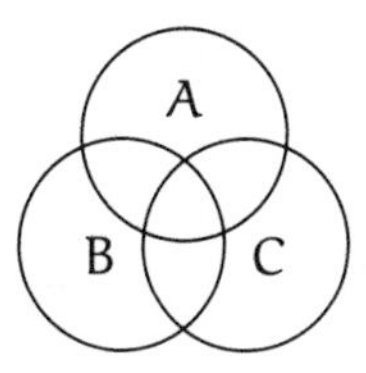

Can the sixteen possibilities that arise with four given sets be illustrated by four overlapping circles?

6 Some of the regions defined by n lines in the plane are infinite, while others are bounded. What's the maximum possible number of bounded regions?

7 Let $H(n) = J(n+1) - J(n)$. Equation (1.8) tells us that $H(2n) = 2$, and $H(2n+1) = J(2n+2) - J(2n+1) = \big(2J(n+1)-1\big) - \big(2J(n)+1\big) = 2H(n)-2$, for all $n \geqslant 1$. Therefore it seems possible to prove that $H(n) = 2$ for all n, by induction on n. What's wrong here?

Homework exercises

8 Solve the recurrence

$$Q_0 = \alpha; \qquad Q_1 = \beta;$$
$$Q_n = (1 + Q_{n-1})/Q_{n-2}, \qquad \text{for } n > 1.$$

Assume that $Q_n \neq 0$ for all $n \geqslant 0$. *Hint:* $Q_4 = (1 + \alpha)/\beta$.

9 Sometimes it's possible to use induction backwards, proving things from n to $n - 1$ instead of vice versa! For example, consider the statement

... now that's a horse of a different color.

$$P(n) : \quad x_1 \ldots x_n \leqslant \left(\frac{x_1 + \cdots + x_n}{n}\right)^n, \quad \text{if } x_1, \ldots, x_n \geqslant 0.$$

This is true when $n = 2$, since $(x_1 + x_2)^2 - 4x_1x_2 = (x_1 - x_2)^2 \geqslant 0$.

a By setting $x_n = (x_1 + \cdots + x_{n-1})/(n - 1)$, prove that $P(n)$ implies $P(n - 1)$ whenever $n > 1$.

b Show that $P(n)$ and $P(2)$ imply $P(2n)$.

c Explain why this implies the truth of $P(n)$ for all n.

10 Let Q_n be the minimum number of moves needed to transfer a tower of n disks from A to B if all moves must be *clockwise* — that is, from A to B, or from B to the other peg, or from the other peg to A. Also let R_n be the minimum number of moves needed to go from B back to A under this restriction. Prove that

$$Q_n = \begin{cases} 0, & \text{if } n = 0; \\ 2R_{n-1} + 1, & \text{if } n > 0; \end{cases} \qquad R_n = \begin{cases} 0, & \text{if } n = 0; \\ Q_n + Q_{n-1} + 1, & \text{if } n > 0. \end{cases}$$

(You need not solve these recurrences; we'll see how to do that in Chapter 7.)

11 A Double Tower of Hanoi contains $2n$ disks of n different sizes, two of each size. As usual, we're required to move only one disk at a time, without putting a larger one over a smaller one.

a How many moves does it take to transfer a double tower from one peg to another, if disks of equal size are indistinguishable from each other?

b What if we are required to reproduce the original top-to-bottom order of all the equal-size disks in the final arrangement?

12 Let's generalize exercise 11 even further, by assuming that there are m different sizes of disks and exactly n_k disks of size k. Determine $A(n_1, \ldots, n_m)$ and $B(n_1, \ldots, n_m)$, the minimum numbers of moves that are needed to transfer a tower when equal-size disks are considered to be (a) identical and (b) distinct.

13 What's the maximum number of regions definable by n zig-zag lines,

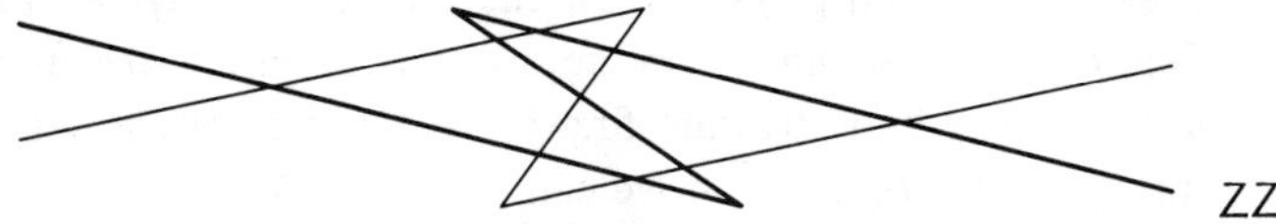

each of which consists of two parallel infinite half-lines joined by a straight segment?

14 How many pieces of cheese can you obtain from a single thick piece by making five straight slices? (The cheese must stay in its original position while you do all the cutting, and each slice must correspond to a plane in 3D.) Find a recurrence relation for P_n, the maximum number of three-dimensional regions that can be defined by n different planes.

Good luck keeping the cheese in position.

15 Josephus had a friend who was saved by getting into the next-to-last position. What is $I(n)$, the number of the penultimate survivor when every second person is executed?

16 Use the repertoire method to solve the general four-parameter recurrence

$$\begin{aligned} g(1) &= \alpha; \\ g(2n+j) &= 3g(n) + \gamma n + \beta_j\,, \qquad \text{for } j = 0,1 \quad \text{and} \quad n \geqslant 1. \end{aligned}$$

Hint: Try the function $g(n) = n$.

Exam problems

17 If W_n is the minimum number of moves needed to transfer a tower of n disks from one peg to another when there are four pegs instead of three, show that

$$W_{n(n+1)/2} \leqslant 2W_{n(n-1)/2} + T_n\,, \qquad \text{for } n > 0.$$

(Here $T_n = 2^n - 1$ is the ordinary three-peg number.) Use this to find a closed form $f(n)$ such that $W_{n(n+1)/2} \leqslant f(n)$ for all $n \geqslant 0$.

18 Show that the following set of n bent lines defines Z_n regions, where Z_n is defined in (1.7): The jth bent line, for $1 \leqslant j \leqslant n$, has its zig at $(n^{2j}, 0)$ and goes up through the points $(n^{2j} - n^j, 1)$ and $(n^{2j} - n^j - n^{-n}, 1)$.

19 Is it possible to obtain Z_n regions with n bent lines when the angle at each zig is $30°$?

20 Use the repertoire method to solve the general five-parameter recurrence

Is this like a five-star general recurrence?

$$\begin{aligned} h(1) &= \alpha; \\ h(2n+j) &= 4h(n) + \gamma_j n + \beta_j\,, \qquad \text{for } j = 0,1 \quad \text{and} \quad n \geqslant 1. \end{aligned}$$

Hint: Try the functions $h(n) = n$ and $h(n) = n^2$.

21 Suppose there are $2n$ people in a circle; the first n are "good guys" and the last n are "bad guys." Show that there is always an integer m (depending on n) such that, if we go around the circle executing every mth person, all the bad guys are first to go. (For example, when $n = 3$ we can take $m = 5$; when $n = 4$ we can take $m = 30$.)

Bonus problems

22 Show that it's possible to construct a Venn diagram for all 2^n possible subsets of n given sets, using n convex polygons that are congruent to each other and rotated about a common center.

23 Suppose that Josephus finds himself in a given position j, but he has a chance to name the elimination parameter q such that every qth person is executed. Can he always save himself?

Research problems

24 Find all recurrence relations of the form

$$X_n = \frac{a_0 + a_1 X_{n-1} + \cdots + a_k X_{n-k}}{b_1 X_{n-1} + \cdots + b_k X_{n-k}}$$

whose solution is periodic.

25 Solve the four-peg Tower of Hanoi problem by proving that equality holds in the relation of exercise 17.

26 Generalizing exercise 23, let's say that a *Josephus subset* of $\{1, 2, \ldots, n\}$ is a set of k numbers such that, for some q, the people with the other $n-k$ numbers will be eliminated first. (These are the k positions of the "good guys" Josephus wants to save.) It turns out that when $n = 9$, three of the 2^9 possible subsets are non-Josephus, namely $\{1, 2, 3, 6, 9\}$, $\{3, 4, 5, 6, 9\}$, and $\{3, 6, 7, 8, 9\}$. There are 13 non-Josephus sets when $n = 12$, none for any other values of $n \leqslant 12$. Are non-Josephus subsets rare for large n?

Yes, and well done if you find them.

2

Sums

SUMS ARE EVERYWHERE in mathematics, so we need basic tools to handle them. This chapter develops the notation and general techniques that make summation user-friendly.

2.1 NOTATION

In Chapter 1 we encountered the sum of the first n integers, which we wrote out as $1+2+3+\cdots+(n-1)+n$. The '$\cdots$' in such formulas tells us to complete the pattern established by the surrounding terms. Of course we have to watch out for sums like $1+7+\cdots+41.7$, which are meaningless without a mitigating context. On the other hand, the inclusion of terms like 3 and $(n-1)$ was a bit of overkill; the pattern would presumably have been clear if we had written simply $1+2+\cdots+n$. Sometimes we might even be so bold as to write just $1+\cdots+n$.

We'll be working with sums of the general form

$$a_1 + a_2 + \cdots + a_n \,, \tag{2.1}$$

where each a_k is a number that has been defined somehow. This notation has the advantage that we can "see" the whole sum, almost as if it were written out in full, if we have a good enough imagination.

A term is how long this course lasts.

Each element a_k of a sum is called a *term*. The terms are often specified implicitly as formulas that follow a readily perceived pattern, and in such cases we must sometimes write them in an expanded form so that the meaning is clear. For example, if

$$1+2+\cdots+2^{n-1}$$

is supposed to denote a sum of n terms, not of 2^{n-1}, we should write it more explicitly as

$$2^0+2^1+\cdots+2^{n-1}.$$

The three-dots notation has many uses, but it can be ambiguous and a bit long-winded. Other alternatives are available, notably the delimited form

$$\sum_{k=1}^{n} a_k , \qquad (2.2)$$

"Le signe $\sum_{i=1}^{i=\infty}$ indique que l'on doit donner au nombre entier i toutes ses valeurs 1, 2, 3, ..., et prendre la somme des termes."
—J. Fourier [102]

which is called Sigma-notation because it uses the Greek letter $\sum$ (upper-case sigma). This notation tells us to include in the sum precisely those terms a_k whose index k is an integer that lies between the lower and upper limits 1 and n, inclusive. In words, we "sum over k, from 1 to n." Joseph Fourier introduced this delimited $\sum$-notation in 1820, and it soon took the mathematical world by storm.

Incidentally, the quantity after $\sum$ (here a_k) is called the *summand*.

The index variable k is said to be *bound* to the $\sum$ sign in (2.2), because the k in a_k is unrelated to appearances of k outside the Sigma-notation. Any other letter could be substituted for k here without changing the meaning of (2.2). The letter i is often used (perhaps because it stands for "index"), but we'll generally sum on k since it's wise to keep i for $\sqrt{-1}$.

Well, I wouldn't want to use a or n as the index variable instead of k in (2.2); those letters are "free variables" that do have meaning outside the $\sum$ here.

It turns out that a generalized Sigma-notation is even more useful than the delimited form: We simply write one or more conditions under the $\sum$, to specify the set of indices over which summation should take place. For example, the sums in (2.1) and (2.2) can also be written as

$$\sum_{1\leqslant k\leqslant n} a_k . \qquad (2.3)$$

In this particular example there isn't much difference between the new form and (2.2), but the general form allows us to take sums over index sets that aren't restricted to consecutive integers. For example, we can express the sum of the squares of all odd positive integers below 100 as follows:

$$\sum_{\substack{1\leqslant k<100 \\ k \text{ odd}}} k^2 .$$

The delimited equivalent of this sum,

$$\sum_{k=0}^{49} (2k+1)^2 ,$$

is more cumbersome and less clear. Similarly, the sum of reciprocals of all prime numbers between 1 and N is

$$\sum_{\substack{p\leqslant N \\ p \text{ prime}}} \frac{1}{p} ;$$

the delimited form would require us to write

$$\sum_{k=1}^{\pi(N)} \frac{1}{p_k},$$

where p_k denotes the kth prime and $\pi(N)$ is the number of primes $\leqslant N$. (Incidentally, this sum gives the approximate average number of distinct prime factors of a random integer near N, since about $1/p$ of those integers are divisible by p. Its value for large N is approximately $\ln\ln N + 0.261972128$; $\ln x$ stands for the natural logarithm of x, and $\ln\ln x$ stands for $\ln(\ln x)$.)

The summation symbol looks like a distorted pacman.

The biggest advantage of general Sigma-notation is that we can manipulate it more easily than the delimited form. For example, suppose we want to change the index variable k to $k+1$. With the general form, we have

$$\sum_{1\leqslant k\leqslant n} a_k = \sum_{1\leqslant k+1\leqslant n} a_{k+1};$$

it's easy to see what's going on, and we can do the substitution almost without thinking. But with the delimited form, we have

$$\sum_{k=1}^{n} a_k = \sum_{k=0}^{n-1} a_{k+1};$$

it's harder to see what's happened, and we're more likely to make a mistake.

A tidy sum.

On the other hand, the delimited form isn't completely useless. It's nice and tidy, and we can write it quickly because (2.2) has seven symbols compared with (2.3)'s eight. Therefore we'll often use $\sum$ with upper and lower delimiters when we state a problem or present a result, but we'll prefer to work with relations-under-$\sum$ when we're manipulating a sum whose index variables need to be transformed.

That's nothing. You should see how many times Σ appears in The Iliad.

The $\sum$ sign occurs more than 1000 times in this book, so we should be sure that we know exactly what it means. Formally, we write

$$\sum_{P(k)} a_k \tag{2.4}$$

as an abbreviation for the sum of all terms a_k such that k is an integer satisfying a given property $P(k)$. (A "property $P(k)$" is any statement about k that can be either true or false.) For the time being, we'll assume that only finitely many integers k satisfying $P(k)$ have $a_k \neq 0$; otherwise infinitely many nonzero numbers are being added together, and things can get a bit tricky. At the other extreme, if $P(k)$ is false for all integers k, we have an "empty" sum; the value of an empty sum is defined to be zero.

A slightly modified form of (2.4) is used when a sum appears within the text of a paragraph rather than in a displayed equation: We write '$\sum_{P(k)} a_k$', attaching property $P(k)$ as a subscript of $\sum$, so that the formula won't stick out too much. Similarly, '$\sum_{k=1}^{n} a_k$' is a convenient alternative to (2.2) when we want to confine the notation to a single line.

People are often tempted to write

$$\sum_{k=2}^{n-1} k(k-1)(n-k) \qquad \text{instead of} \qquad \sum_{k=0}^{n} k(k-1)(n-k)$$

because the terms for $k = 0$, 1, and n in this sum are zero. Somehow it seems more efficient to add up $n-2$ terms instead of $n+1$ terms. But such temptations should be resisted; efficiency of computation is not the same as efficiency of understanding! We will find it advantageous to keep upper and lower bounds on an index of summation as simple as possible, because sums can be manipulated much more easily when the bounds are simple. Indeed, the form $\sum_{k=2}^{n-1}$ can even be dangerously ambiguous, because its meaning is not at all clear when $n = 0$ or $n = 1$ (see exercise 1). Zero-valued terms cause no harm, and they often save a lot of trouble.

So far the notations we've been discussing are quite standard, but now we are about to make a radical departure from tradition. Kenneth Iverson introduced a wonderful idea in his programming language APL [161, page 11], and we'll see that it greatly simplifies many of the things we want to do in this book. The idea is simply to enclose a true-or-false statement in brackets, and to say that the result is 1 if the statement is true, 0 if the statement is false. For example,

Hey: The "Kronecker delta" that I've seen in other books (I mean δ_{kn}, which is 1 if $k=n$, 0 otherwise) is just a special case of Iverson's convention: We can write $[k=n]$ instead.

$$[p \text{ prime}] = \begin{cases} 1, & \text{if } p \text{ is a prime number;} \\ 0, & \text{if } p \text{ is not a prime number.} \end{cases}$$

Iverson's convention allows us to express sums with no constraints whatever on the index of summation, because we can rewrite (2.4) in the form

$$\sum_{k} a_k \big[P(k)\big] . \tag{2.5}$$

If $P(k)$ is false, the term $a_k\big[P(k)\big]$ is zero, so we can safely include it among the terms being summed. This makes it easy to manipulate the index of summation, because we don't have to fuss with boundary conditions.

A slight technicality needs to be mentioned: Sometimes a_k isn't defined for all integers k. We get around this difficulty by assuming that $\big[P(k)\big]$ is "very strongly zero" when $P(k)$ is false; it's so much zero, it makes $a_k\big[P(k)\big]$ equal to zero even when a_k is undefined. For example, if we use Iverson's convention to write the sum of reciprocal primes $\leqslant N$ as

$$\sum_p [p \text{ prime}][p \leqslant N]/p\,,$$

there's no problem of division by zero when $p = 0$, because our convention tells us that $[0 \text{ prime}][0 \leqslant N]/0 = 0$.

Let's sum up what we've discussed so far about sums. There are two good ways to express a sum of terms: One way uses '$\cdots$', the other uses '$\sum$'. The three-dots form often suggests useful manipulations, particularly the combination of adjacent terms, since we might be able to spot a simplifying pattern if we let the whole sum hang out before our eyes. But too much detail can also be overwhelming. Sigma-notation is compact, impressive to family and friends, and often suggestive of manipulations that are not obvious in three-dots form. When we work with Sigma-notation, zero terms are not generally harmful; in fact, zeros often make $\sum$-manipulation easier.

... and it's less likely to lose points on an exam for "lack of rigor."

2.2 SUMS AND RECURRENCES

OK, we understand now how to express sums with fancy notation. But how does a person actually go about finding the value of a sum? One way is to observe that there's an intimate relation between sums and recurrences. The sum

$$S_n = \sum_{k=0}^{n} a_k$$

is equivalent to the recurrence

(Think of S_n as not just a single number, but as a sequence defined for all $n \geqslant 0$.)

$$\begin{aligned} S_0 &= a_0\,; \\ S_n &= S_{n-1} + a_n\,, \qquad \text{for } n > 0. \end{aligned} \tag{2.6}$$

Therefore we can evaluate sums in closed form by using the methods we learned in Chapter 1 to solve recurrences in closed form.

For example, if a_n is equal to a constant plus a multiple of n, the sum-recurrence (2.6) takes the following general form:

$$\begin{aligned} R_0 &= \alpha\,; \\ R_n &= R_{n-1} + \beta + \gamma n\,, \qquad \text{for } n > 0. \end{aligned} \tag{2.7}$$

Proceeding as in Chapter 1, we find $R_1 = \alpha + \beta + \gamma$, $R_2 = \alpha + 2\beta + 3\gamma$, and so on; in general the solution can be written in the form

$$R_n = A(n)\alpha + B(n)\beta + C(n)\gamma\,, \tag{2.8}$$

where $A(n)$, $B(n)$, and $C(n)$ are the coefficients of dependence on the general parameters α, β, and γ.

The repertoire method tells us to try plugging in simple functions of n for R_n, hoping to find constant parameters α, β, and γ where the solution is especially simple. Setting $R_n = 1$ implies $\alpha = 1$, $\beta = 0$, $\gamma = 0$; hence

$$A(n) = 1.$$

Setting $R_n = n$ implies $\alpha = 0$, $\beta = 1$, $\gamma = 0$; hence

$$B(n) = n.$$

Setting $R_n = n^2$ implies $\alpha = 0$, $\beta = -1$, $\gamma = 2$; hence

$$2C(n) - B(n) = n^2$$

and we have $C(n) = (n^2 + n)/2$. Easy as pie.

Actually easier; $\pi = \sum_{n\geq 0} \frac{8}{(4n+1)(4n+3)}$.

Therefore if we wish to evaluate

$$\sum_{k=0}^{n} (a + bk),$$

the sum-recurrence (2.6) boils down to (2.7) with $\alpha = \beta = a$, $\gamma = b$, and the answer is $aA(n) + aB(n) + bC(n) = a(n+1) + b(n+1)n/2$.

Conversely, many recurrences can be reduced to sums; therefore the special methods for evaluating sums that we'll be learning later in this chapter will help us solve recurrences that might otherwise be difficult. The Tower of Hanoi recurrence is a case in point:

$$\begin{aligned} T_0 &= 0; \\ T_n &= 2T_{n-1} + 1, \qquad \text{for } n > 0. \end{aligned}$$

It can be put into the special form (2.6) if we divide both sides by 2^n:

$$\begin{aligned} T_0/2^0 &= 0; \\ T_n/2^n &= T_{n-1}/2^{n-1} + 1/2^n, \qquad \text{for } n > 0. \end{aligned}$$

Now we can set $S_n = T_n/2^n$, and we have

$$\begin{aligned} S_0 &= 0; \\ S_n &= S_{n-1} + 2^{-n}, \qquad \text{for } n > 0. \end{aligned}$$

It follows that

$$S_n = \sum_{k=1}^{n} 2^{-k}.$$

(Notice that we've left the term for $k = 0$ out of this sum.) The sum of the geometric series $2^{-1} + 2^{-2} + \cdots + 2^{-n} = (\frac{1}{2})^1 + (\frac{1}{2})^2 + \cdots + (\frac{1}{2})^n$ will be derived later in this chapter; it turns out to be $1 - (\frac{1}{2})^n$. Hence $T_n = 2^n S_n = 2^n - 1$.

We have converted T_n to S_n in this derivation by noticing that the recurrence could be divided by 2^n. This trick is a special case of a general technique that can reduce virtually any recurrence of the form

$$a_n T_n = b_n T_{n-1} + c_n \tag{2.9}$$

to a sum. The idea is to multiply both sides by a *summation factor*, s_n:

$$s_n a_n T_n = s_n b_n T_{n-1} + s_n c_n .$$

This factor s_n is cleverly chosen to make

$$s_n b_n = s_{n-1} a_{n-1} .$$

Then if we write $S_n = s_n a_n T_n$ we have a sum-recurrence,

$$S_n = S_{n-1} + s_n c_n .$$

Hence

$$S_n = s_0 a_0 T_0 + \sum_{k=1}^{n} s_k c_k = s_1 b_1 T_0 + \sum_{k=1}^{n} s_k c_k ,$$

and the solution to the original recurrence (2.9) is

$$T_n = \frac{1}{s_n a_n}\left(s_1 b_1 T_0 + \sum_{k=1}^{n} s_k c_k\right). \tag{2.10}$$

(The value of s_1 cancels out, so it can be anything but zero.)

For example, when $n = 1$ we get $T_1 = (s_1 b_1 T_0 + s_1 c_1)/s_1 a_1 = (b_1 T_0 + c_1)/a_1$.

But how can we be clever enough to find the right s_n? No problem: The relation $s_n = s_{n-1} a_{n-1}/b_n$ can be unfolded to tell us that the fraction

$$s_n = \frac{a_{n-1} a_{n-2} \ldots a_1}{b_n b_{n-1} \ldots b_2} , \tag{2.11}$$

or any convenient constant multiple of this value, will be a suitable summation factor. For example, the Tower of Hanoi recurrence has $a_n = 1$ and $b_n = 2$; the general method we've just derived says that $s_n = 2^{-n}$ is a good thing to multiply by, if we want to reduce the recurrence to a sum. We don't need a brilliant flash of inspiration to discover this multiplier.

We must be careful, as always, not to divide by zero. The summation-factor method works whenever all the a's and all the b's are nonzero.

Let's apply these ideas to a recurrence that arises in the study of "quicksort," one of the most important methods for sorting data inside a computer. The average number of comparison steps made by quicksort when it is applied to n items in random order satisfies the recurrence

(Quicksort was invented by Hoare in 1962 [158].)

$$\begin{aligned} C_0 &= 0\,; \\ C_n &= n+1+\frac{2}{n}\sum_{k=0}^{n-1} C_k\,, \qquad \text{for } n>0. \end{aligned} \tag{2.12}$$

Hmmm. This looks much scarier than the recurrences we've seen before; it includes a sum over all previous values, and a division by n. Trying small cases gives us some data ($C_1 = 2$, $C_2 = 5$, $C_3 = \frac{26}{3}$) but doesn't do anything to quell our fears.

We can, however, reduce the complexity of (2.12) systematically, by first getting rid of the division and then getting rid of the $\sum$ sign. The idea is to multiply both sides by n, obtaining the relation

$$nC_n = n^2+n+2\sum_{k=0}^{n-1} C_k\,, \qquad \text{for } n>0;$$

hence, if we replace n by $n-1$,

$$(n-1)C_{n-1} = (n-1)^2+(n-1)+2\sum_{k=0}^{n-2} C_k\,, \qquad \text{for } n-1>0.$$

We can now subtract the second equation from the first, and the $\sum$ sign disappears:

$$nC_n-(n-1)C_{n-1} = 2n+2C_{n-1}\,, \qquad \text{for } n>1.$$

It turns out that this relation also holds when $n = 1$, because $C_1 = 2$. Therefore the original recurrence for C_n reduces to a much simpler one:

$$\begin{aligned} C_0 &= 0\,; \\ nC_n &= (n+1)C_{n-1}+2n\,, \qquad \text{for } n>0. \end{aligned}$$

Progress. We're now in a position to apply a summation factor, since this recurrence has the form of (2.9) with $a_n = n$, $b_n = n+1$, and $c_n = 2n$. The general method described on the preceding page tells us to multiply the recurrence through by some multiple of

$$s_n = \frac{a_{n-1}a_{n-2}\ldots a_1}{b_n b_{n-1}\ldots b_2} = \frac{(n-1)\cdot(n-2)\cdot\ldots\cdot 1}{(n+1)\cdot n\cdot\ldots\cdot 3} = \frac{2}{(n+1)n}\,.$$

We started with a $\sum$ in the recurrence, and worked hard to get rid of it. But then after applying a summation factor, we came up with another $\sum$. Are sums good, or bad, or what?

The solution, according to (2.10), is therefore

$$C_n = 2(n+1)\sum_{k=1}^{n}\frac{1}{k+1}.$$

The sum that remains is very similar to a quantity that arises frequently in applications. It arises so often, in fact, that we give it a special name and a special notation:

$$H_n = 1+\frac{1}{2}+\cdots+\frac{1}{n} = \sum_{k=1}^{n}\frac{1}{k}. \qquad (2.13)$$

The letter H stands for "harmonic"; H_n is a *harmonic number*, so called because the kth harmonic produced by a violin string is the fundamental tone produced by a string that is $1/k$ times as long.

We can complete our study of the quicksort recurrence (2.12) by putting C_n into closed form; this will be possible if we can express C_n in terms of H_n. The sum in our formula for C_n is

$$\sum_{k=1}^{n}\frac{1}{k+1} = \sum_{1\leqslant k\leqslant n}\frac{1}{k+1}.$$

We can relate this to H_n without much difficulty by changing k to $k-1$ and revising the boundary conditions:

$$\begin{aligned}\sum_{1\leqslant k\leqslant n}\frac{1}{k+1} &= \sum_{1\leqslant k-1\leqslant n}\frac{1}{k}\\ &= \sum_{2\leqslant k\leqslant n+1}\frac{1}{k}\\ &= \Bigl(\sum_{1\leqslant k\leqslant n}\frac{1}{k}\Bigr)-\frac{1}{1}+\frac{1}{n+1} = H_n-\frac{n}{n+1}.\end{aligned}$$

But your spelling is alwrong.

Alright! We have found the sum needed to complete the solution to (2.12): The average number of comparisons made by quicksort when it is applied to n randomly ordered items of data is

$$C_n = 2(n+1)H_n - 2n. \qquad (2.14)$$

As usual, we check that small cases are correct: $C_0 = 0$, $C_1 = 2$, $C_2 = 5$.

2.3 MANIPULATION OF SUMS

Not to be confused with finance.

The key to success with sums is an ability to change one $\sum$ into another that is simpler or closer to some goal. And it's easy to do this by learning a few basic rules of transformation and by practicing their use.

Let K be any finite set of integers. Sums over the elements of K can be transformed by using three simple rules:

$$\sum_{k\in K} ca_k = c\sum_{k\in K} a_k\,; \qquad \text{(distributive law)} \tag{2.15}$$

$$\sum_{k\in K}(a_k+b_k) = \sum_{k\in K} a_k + \sum_{k\in K} b_k\,; \qquad \text{(associative law)} \tag{2.16}$$

$$\sum_{k\in K} a_k = \sum_{p(k)\in K} a_{p(k)}\,. \qquad \text{(commutative law)} \tag{2.17}$$

The distributive law allows us to move constants in and out of a $\sum$. The associative law allows us to break a $\sum$ into two parts, or to combine two $\sum$'s into one. The commutative law says that we can reorder the terms in any way we please; here $p(k)$ is any permutation of the set of all integers. For example, if $K=\{-1,0,+1\}$ and if $p(k)=-k$, these three laws tell us respectively that

Why not call it permutative instead of commutative?

$$ca_{-1}+ca_0+ca_1 = c(a_{-1}+a_0+a_1)\,; \qquad \text{(distributive law)}$$

$$(a_{-1}+b_{-1})+(a_0+b_0)+(a_1+b_1) = (a_{-1}+a_0+a_1)+(b_{-1}+b_0+b_1)\,; \qquad \text{(associative law)}$$

$$a_{-1}+a_0+a_1 = a_1+a_0+a_{-1}\,. \qquad \text{(commutative law)}$$

Gauss's trick in Chapter 1 can be viewed as an application of these three basic laws. Suppose we want to compute the general sum of an *arithmetic progression*,

$$S = \sum_{0\le k\le n}(a+bk)\,.$$

By the commutative law we can replace k by $n-k$, obtaining

This is something like changing variables inside an integral, but easier.

$$S = \sum_{0\le n-k\le n}\bigl(a+b(n-k)\bigr) = \sum_{0\le k\le n}(a+bn-bk)\,.$$

These two equations can be added by using the associative law:

$$2S = \sum_{0\le k\le n}\bigl((a+bk)+(a+bn-bk)\bigr) = \sum_{0\le k\le n}(2a+bn)\,.$$

"What's one and one and one and one and one and one and one and one and one and one?" "I don't know," said Alice. "I lost count." "She can't do Addition." — *Lewis Carroll [44]*

And we can now apply the distributive law and evaluate a trivial sum:

$$2S = (2a+bn)\sum_{0\le k\le n} 1 = (2a+bn)(n+1).$$

Dividing by 2, we have proved that

$$\sum_{k=0}^{n}(a+bk) = (a+\tfrac{1}{2}bn)(n+1). \tag{2.18}$$

The right-hand side can be remembered as the average of the first and last terms, namely $\frac{1}{2}(a+(a+bn))$, times the number of terms, namely $(n+1)$.

It's important to bear in mind that the function $p(k)$ in the general commutative law (2.17) is supposed to be a permutation of all the integers. In other words, for every integer n there should be exactly one integer k such that $p(k)=n$. Otherwise the commutative law might fail; exercise 3 illustrates this with a vengeance. Transformations like $p(k)=k+c$ or $p(k)=c-k$, where c is an integer constant, are always permutations, so they always work.

On the other hand, we can relax the permutation restriction a little bit: We need to require only that there be exactly one integer k with $p(k)=n$ when n is an element of the index set K. If $n\notin K$ (that is, if n is not in K), it doesn't matter how often $p(k)=n$ occurs, because such k don't take part in the sum. Thus, for example, we can argue that

$$\sum_{\substack{k\in K\\ k\text{ even}}} a_k = \sum_{\substack{n\in K\\ n\text{ even}}} a_n = \sum_{\substack{2k\in K\\ 2k\text{ even}}} a_{2k} = \sum_{2k\in K} a_{2k}, \tag{2.19}$$

since there's exactly one k such that $2k=n$ when $n\in K$ and n is even.

Additional, eh?

Iverson's convention, which allows us to obtain the values 0 or 1 from logical statements in the middle of a formula, can be used together with the distributive, associative, and commutative laws to deduce additional properties of sums. For example, here is an important rule for combining different sets of indices: If K and K' are any sets of integers, then

$$\sum_{k\in K} a_k + \sum_{k\in K'} a_k = \sum_{k\in K\cap K'} a_k + \sum_{k\in K\cup K'} a_k. \tag{2.20}$$

This follows from the general formulas

$$\sum_{k\in K} a_k = \sum_k a_k\,[k\in K] \tag{2.21}$$

and

$$[k\in K]+[k\in K'] = [k\in K\cap K']+[k\in K\cup K']. \tag{2.22}$$

Typically we use rule (2.20) either to combine two almost-disjoint index sets, as in

$$\sum_{k=1}^{m} a_k + \sum_{k=m}^{n} a_k = a_m + \sum_{k=1}^{n} a_k, \qquad \text{for } 1 \leqslant m \leqslant n;$$

or to split off a single term from a sum, as in

$$\sum_{0\leqslant k\leqslant n} a_k = a_0 + \sum_{1\leqslant k\leqslant n} a_k, \qquad \text{for } n \geqslant 0. \tag{2.23}$$

(The two sides of (2.20) have been switched here.)

This operation of splitting off a term is the basis of a *perturbation method* that often allows us to evaluate a sum in closed form. The idea is to start with an unknown sum and call it S_n:

$$S_n = \sum_{0\leqslant k\leqslant n} a_k .$$

(Name and conquer.) Then we rewrite S_{n+1} in two ways, by splitting off both its last term and its first term:

$$\begin{aligned} S_n + a_{n+1} = \sum_{0\leqslant k\leqslant n+1} a_k &= a_0 + \sum_{1\leqslant k\leqslant n+1} a_k \\ &= a_0 + \sum_{1\leqslant k+1\leqslant n+1} a_{k+1} \\ &= a_0 + \sum_{0\leqslant k\leqslant n} a_{k+1} . \end{aligned} \tag{2.24}$$

Now we can work on this last sum and try to express it in terms of S_n. If we succeed, we obtain an equation whose solution is the sum we seek.

For example, let's use this approach to find the sum of a general *geometric progression*,

If it's geometric, there should be a geometric proof.

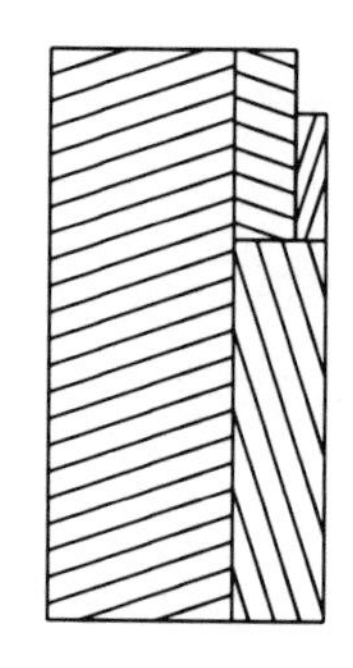

$$S_n = \sum_{0\leqslant k\leqslant n} ax^k .$$

The general perturbation scheme in (2.24) tells us that

$$S_n + ax^{n+1} = ax^0 + \sum_{0\leqslant k\leqslant n} ax^{k+1},$$

and the sum on the right is $x\sum_{0\leqslant k\leqslant n} ax^k = xS_n$ by the distributive law. Therefore $S_n + ax^{n+1} = a + xS_n$, and we can solve for S_n to obtain

$$\sum_{k=0}^{n} ax^k = \frac{a - ax^{n+1}}{1-x}, \qquad \text{for } x \neq 1. \tag{2.25}$$

(When $x = 1$, the sum is of course simply $a(n+1)$.) The right-hand side can be remembered as the first term included in the sum minus the first term excluded (the term after the last), divided by 1 minus the term ratio.

Ah yes, this formula was drilled into me in high school.

That was almost too easy. Let's try the perturbation technique on a slightly more difficult sum,

$$S_n = \sum_{0 \leqslant k \leqslant n} k 2^k .$$

In this case we have $S_0 = 0$, $S_1 = 2$, $S_2 = 10$, $S_3 = 34$, $S_4 = 98$; what is the general formula? According to (2.24) we have

$$S_n + (n+1)2^{n+1} = \sum_{0 \leqslant k \leqslant n} (k+1)2^{k+1} ;$$

so we want to express the right-hand sum in terms of S_n. Well, we can break it into two sums with the help of the associative law,

$$\sum_{0 \leqslant k \leqslant n} k 2^{k+1} + \sum_{0 \leqslant k \leqslant n} 2^{k+1} ,$$

and the first of the remaining sums is $2S_n$. The other sum is a geometric progression, which equals $(2 - 2^{n+2})/(1-2) = 2^{n+2} - 2$ by (2.25). Therefore we have $S_n + (n+1)2^{n+1} = 2S_n + 2^{n+2} - 2$, and algebra yields

$$\sum_{0 \leqslant k \leqslant n} k 2^k = (n-1)2^{n+1} + 2 .$$

Now we understand why $S_3 = 34$: It's $32 + 2$, not $2 \cdot 17$.

A similar derivation with x in place of 2 would have given us the equation $S_n + (n+1)x^{n+1} = xS_n + (x - x^{n+2})/(1-x)$; hence we can deduce that

$$\sum_{k=0}^{n} kx^k = \frac{x - (n+1)x^{n+1} + nx^{n+2}}{(1-x)^2} , \qquad \text{for } x \neq 1. \tag{2.26}$$

It's interesting to note that we could have derived this closed form in a completely different way, by using elementary techniques of differential calculus. If we start with the equation

$$\sum_{k=0}^{n} x^k = \frac{1 - x^{n+1}}{1 - x}$$

and take the derivative of both sides with respect to x, we get

$$\sum_{k=0}^{n} kx^{k-1} = \frac{(1-x)(-(n+1)x^n) + 1 - x^{n+1}}{(1-x)^2} = \frac{1 - (n+1)x^n + nx^{n+1}}{(1-x)^2} ,$$

because the derivative of a sum is the sum of the derivatives of its terms. We will see many more connections between calculus and discrete mathematics in later chapters.

2.4 MULTIPLE SUMS

The terms of a sum might be specified by two or more indices, not just by one. For example, here's a double sum of nine terms, governed by two indices j and k:

Oh no, a nine-term governor.

$$\sum_{1\leqslant j,k\leqslant 3} a_j b_k = a_1b_1 + a_1b_2 + a_1b_3 + a_2b_1 + a_2b_2 + a_2b_3 + a_3b_1 + a_3b_2 + a_3b_3\,.$$

Notice that this doesn't mean to sum over all $j \geqslant 1$ and all $k \leqslant 3$.

We use the same notations and methods for such sums as we do for sums with a single index. Thus, if $P(j,k)$ is a property of j and k, the sum of all terms $a_{j,k}$ such that $P(j,k)$ is true can be written in two ways, one of which uses Iverson's convention and sums over *all* pairs of integers j and k:

$$\sum_{P(j,k)} a_{j,k} = \sum_{j,k} a_{j,k}\,\bigl[P(j,k)\bigr]\,.$$

Only one $\sum$ sign is needed, although there is more than one index of summation; $\sum$ denotes a sum over all combinations of indices that apply.

We also have occasion to use two $\sum$'s, when we're talking about a sum of sums. For example,

$$\sum_j \sum_k a_{j,k}\,\bigl[P(j,k)\bigr]$$

is an abbreviation for

$$\sum_j \Bigl(\sum_k a_{j,k}\,\bigl[P(j,k)\bigr]\Bigr)\,,$$

which is the sum, over all integers j, of $\sum_k a_{j,k}\,\bigl[P(j,k)\bigr]$, the latter being the sum over all integers k of all terms $a_{j,k}$ for which $P(j,k)$ is true. In such cases we say that the double sum is "summed first on k." A sum that depends on more than one index can be summed first on any one of its indices.

Multiple Σ's are evaluated right to left (inside-out).

In this regard we have a basic law called *interchanging the order of summation*, which generalizes the associative law (2.16) we saw earlier:

$$\sum_j \sum_k a_{j,k}\bigl[P(j,k)\bigr] = \sum_{P(j,k)} a_{j,k} = \sum_k \sum_j a_{j,k}\bigl[P(j,k)\bigr]\,. \qquad (2.27)$$

The middle term of this law is a sum over two indices. On the left, $\sum_j \sum_k$ stands for summing first on k, then on j. On the right, $\sum_k \sum_j$ stands for summing first on j, then on k. In practice when we want to evaluate a double sum in closed form, it's usually easier to sum it first on one index rather than on the other; we get to choose whichever is more convenient.

Who's panicking? I think this rule is fairly obvious compared to some of the stuff in Chapter 1.

Sums of sums are no reason to panic, but they can appear confusing to a beginner, so let's do some more examples. The nine-term sum we began with provides a good illustration of the manipulation of double sums, because that sum can actually be simplified, and the simplification process is typical of what we can do with $\sum\sum$'s:

$$\begin{aligned}\sum_{1\leqslant j,k\leqslant 3} a_j b_k &= \sum_{j,k} a_j b_k [1\leqslant j,k\leqslant 3] = \sum_{j,k} a_j b_k [1\leqslant j\leqslant 3][1\leqslant k\leqslant 3] \\ &= \sum_j \sum_k a_j b_k [1\leqslant j\leqslant 3][1\leqslant k\leqslant 3] \\ &= \sum_j a_j [1\leqslant j\leqslant 3] \sum_k b_k [1\leqslant k\leqslant 3] \\ &= \sum_j a_j [1\leqslant j\leqslant 3] \Bigl(\sum_k b_k [1\leqslant k\leqslant 3]\Bigr) \\ &= \Bigl(\sum_j a_j [1\leqslant j\leqslant 3]\Bigr)\Bigl(\sum_k b_k [1\leqslant k\leqslant 3]\Bigr) \\ &= \Bigl(\sum_{j=1}^{3} a_j\Bigr)\Bigl(\sum_{k=1}^{3} b_k\Bigr).\end{aligned}$$

The first line here denotes a sum of nine terms in no particular order. The second line groups them in threes, $(a_1b_1 + a_1b_2 + a_1b_3) + (a_2b_1 + a_2b_2 + a_2b_3) + (a_3b_1 + a_3b_2 + a_3b_3)$. The third line uses the distributive law to factor out the a's, since a_j and $[1\leqslant j\leqslant 3]$ do not depend on k; this gives $a_1(b_1 + b_2 + b_3) + a_2(b_1 + b_2 + b_3) + a_3(b_1 + b_2 + b_3)$. The fourth line is the same as the third, but with a redundant pair of parentheses thrown in so that the fifth line won't look so mysterious. The fifth line factors out the $(b_1 + b_2 + b_3)$ that occurs for each value of j: $(a_1 + a_2 + a_3)(b_1 + b_2 + b_3)$. The last line is just another way to write the previous line. This method of derivation can be used to prove a *general distributive law*,

$$\sum_{\substack{j\in J\\ k\in K}} a_j b_k = \Bigl(\sum_{j\in J} a_j\Bigr)\Bigl(\sum_{k\in K} b_k\Bigr), \qquad (2.28)$$

valid for all sets of indices J and K.

The basic law (2.27) for interchanging the order of summation has many variations, which arise when we want to restrict the ranges of the indices instead of summing over all integers j and k. These variations come in two flavors, vanilla and rocky road. First, the vanilla version:

$$\sum_{j\in J}\sum_{k\in K} a_{j,k} = \sum_{\substack{j\in J\\ k\in K}} a_{j,k} = \sum_{k\in K}\sum_{j\in J} a_{j,k}. \tag{2.29}$$

This is just another way to write (2.27), since the Iversonian $[j\in J, k\in K]$ factors into $[j\in J][k\in K]$. The vanilla-flavored law applies whenever the ranges of j and k are independent of each other.

The rocky-road formula for interchange is a little trickier. It applies when the range of an inner sum depends on the index variable of the outer sum:

$$\sum_{j\in J}\sum_{k\in K(j)} a_{j,k} = \sum_{k\in K'}\sum_{j\in J'(k)} a_{j,k}. \tag{2.30}$$

Here the sets J, K(j), K′, and J′(k) must be related in such a way that

$$[j\in J]\bigl[k\in K(j)\bigr] = [k\in K']\bigl[j\in J'(k)\bigr].$$

A factorization like this is always possible in principle, because we can let $J = K'$ be the set of all integers and $K(j) = J'(k)$ be the basic property $P(j,k)$ that governs a double sum. But there are important special cases where the sets J, K(j), K′, and J′(k) have a simple form. These arise frequently in applications. For example, here's a particularly useful factorization:

$$[1\le j\le n][j\le k\le n] = [1\le j\le k\le n] = [1\le k\le n][1\le j\le k]. \tag{2.31}$$

This Iversonian equation allows us to write

$$\sum_{j=1}^{n}\sum_{k=j}^{n} a_{j,k} = \sum_{1\le j\le k\le n} a_{j,k} = \sum_{k=1}^{n}\sum_{j=1}^{k} a_{j,k}. \tag{2.32}$$

One of these two sums of sums is usually easier to evaluate than the other; we can use (2.32) to switch from the hard one to the easy one.

(Now is a good time to do warmup exercises 4 and 6.)

(Or to check out the Snickers bar languishing in the freezer.)

Let's apply these ideas to a useful example. Consider the array

$$\begin{bmatrix} a_1a_1 & a_1a_2 & a_1a_3 & \ldots & a_1a_n \\ a_2a_1 & a_2a_2 & a_2a_3 & \ldots & a_2a_n \\ a_3a_1 & a_3a_2 & a_3a_3 & \ldots & a_3a_n \\ \vdots & \vdots & \vdots & \ddots & \vdots \\ a_na_1 & a_na_2 & a_na_3 & \ldots & a_na_n \end{bmatrix}$$

of n^2 products a_ja_k. Our goal will be to find a simple formula for

$$S_{\triangledown} = \sum_{1\le j\le k\le n} a_ja_k,$$

the sum of all elements on or above the main diagonal of this array. Because $a_j a_k = a_k a_j$, the array is symmetrical about its main diagonal; therefore $S_\urcorner$ will be approximately half the sum of *all* the elements (except for a fudge factor that takes account of the main diagonal).

Does rocky road have fudge in it?

Such considerations motivate the following manipulations. We have

$$S_\urcorner = \sum_{1\leqslant j\leqslant k\leqslant n} a_j a_k = \sum_{1\leqslant k\leqslant j\leqslant n} a_k a_j = \sum_{1\leqslant k\leqslant j\leqslant n} a_j a_k = S_\llcorner\,,$$

because we can rename (j,k) as (k,j). Furthermore, since

$$[1\leqslant j\leqslant k\leqslant n]+[1\leqslant k\leqslant j\leqslant n] = [1\leqslant j,k\leqslant n]+[1\leqslant j=k\leqslant n]\,,$$

we have

$$2S_\urcorner = S_\urcorner + S_\llcorner = \sum_{1\leqslant j,k\leqslant n} a_j a_k + \sum_{1\leqslant j=k\leqslant n} a_j a_k\,.$$

The first sum is $\bigl(\sum_{j=1}^n a_j\bigr)\bigl(\sum_{k=1}^n a_k\bigr) = \bigl(\sum_{k=1}^n a_k\bigr)^2$, by the general distributive law (2.28). The second sum is $\sum_{k=1}^n a_k^2$. Therefore we have

$$S_\urcorner = \sum_{1\leqslant j\leqslant k\leqslant n} a_j a_k = \frac{1}{2}\Biggl(\biggl(\sum_{k=1}^{n} a_k\biggr)^2 + \sum_{k=1}^{n} a_k^2\Biggr)\,, \qquad (2.33)$$

an expression for the upper triangular sum in terms of simpler single sums.

Encouraged by such success, let's look at another double sum:

$$S = \sum_{1\leqslant j<k\leqslant n} (a_k - a_j)(b_k - b_j)\,.$$

Again we have symmetry when j and k are interchanged:

$$S = \sum_{1\leqslant k<j\leqslant n} (a_j - a_k)(b_j - b_k) = \sum_{1\leqslant k<j\leqslant n} (a_k - a_j)(b_k - b_j)\,.$$

So we can add S to itself, making use of the identity

$$[1\leqslant j<k\leqslant n]+[1\leqslant k<j\leqslant n] = [1\leqslant j,k\leqslant n]-[1\leqslant j=k\leqslant n]$$

to conclude that

$$2S = \sum_{1\leqslant j,k\leqslant n} (a_j - a_k)(b_j - b_k) - \sum_{1\leqslant j=k\leqslant n} (a_j - a_k)(b_j - b_k)\,.$$

The second sum here is zero; what about the first? It expands into four separate sums, each of which is vanilla flavored:

$$\sum_{1\le j,k\le n} a_j b_j \;-\; \sum_{1\le j,k\le n} a_j b_k \;-\; \sum_{1\le j,k\le n} a_k b_j \;+\; \sum_{1\le j,k\le n} a_k b_k$$
$$= 2\sum_{1\le j,k\le n} a_k b_k \;-\; 2\sum_{1\le j,k\le n} a_j b_k$$
$$= 2n\sum_{1\le k\le n} a_k b_k \;-\; 2\Bigl(\sum_{k=1}^{n} a_k\Bigr)\Bigl(\sum_{k=1}^{n} b_k\Bigr).$$

In the last step both sums have been simplified according to the general distributive law (2.28). If the manipulation of the first sum seems mysterious, here it is again in slow motion:

$$2\sum_{1\le j,k\le n} a_k b_k = 2\sum_{1\le k\le n}\;\sum_{1\le j\le n} a_k b_k$$
$$= 2\sum_{1\le k\le n} a_k b_k \sum_{1\le j\le n} 1$$
$$= 2\sum_{1\le k\le n} a_k b_k n \;=\; 2n\sum_{1\le k\le n} a_k b_k\,.$$

An index variable that doesn't appear in the summand (here j) can simply be eliminated if we multiply what's left by the size of that variable's index set (here n).

Returning to where we left off, we can now divide everything by 2 and rearrange things to obtain an interesting formula:

$$\Bigl(\sum_{k=1}^{n} a_k\Bigr)\Bigl(\sum_{k=1}^{n} b_k\Bigr) = n\sum_{k=1}^{n} a_k b_k - \sum_{1\le j<k\le n} (a_k - a_j)(b_k - b_j)\,. \qquad (2.34)$$

This identity yields *Chebyshev's summation inequalities* as a special case:

$$\Bigl(\sum_{k=1}^{n} a_k\Bigr)\Bigl(\sum_{k=1}^{n} b_k\Bigr) \le n\sum_{k=1}^{n} a_k b_k\,, \quad \text{if } a_1 \le \cdots \le a_n \text{ and } b_1 \le \cdots \le b_n;$$
$$\Bigl(\sum_{k=1}^{n} a_k\Bigr)\Bigl(\sum_{k=1}^{n} b_k\Bigr) \ge n\sum_{k=1}^{n} a_k b_k\,, \quad \text{if } a_1 \le \cdots \le a_n \text{ and } b_1 \ge \cdots \ge b_n.$$

(In general, if $a_1 \le \cdots \le a_n$ and if p is a permutation of $\{1,\ldots,n\}$, it's possible to prove that the largest value of $\sum_{k=1}^{n} a_k b_{p(k)}$ occurs when $b_{p(1)} \le \cdots \le b_{p(n)}$, and the smallest value occurs when $b_{p(1)} \ge \cdots \ge b_{p(n)}$.)

(Chebyshev actually proved the analogous result for integrals instead of sums: $(\int_a^b f(x)\,dx) \cdot (\int_a^b g(x)\,dx) \le (b-a) \cdot (\int_a^b f(x)g(x)\,dx)$, *if* $f(x)$ *and* $g(x)$ *are monotone nondecreasing functions.)*

Multiple summation has an interesting connection with the general operation of changing the index of summation in *single* sums. We know by the commutative law that

$$\sum_{k\in K} a_k = \sum_{p(k)\in K} a_{p(k)},$$

if $p(k)$ is any permutation of the integers. But what happens when we replace k by $f(j)$, where f is an arbitrary function

$$f:\ J \to K$$

that takes an integer $j \in J$ into an integer $f(j) \in K$? The general formula for index replacement is

$$\sum_{j\in J} a_{f(j)} = \sum_{k\in K} a_k\, \#f^-(k), \tag{2.35}$$

where $\#f^-(k)$ stands for the number of elements in the set

$$f^-(k) = \{\, j \mid f(j) = k \,\},$$

that is, the number of values of $j \in J$ such that $f(j)$ equals k.

It's easy to prove (2.35) by interchanging the order of summation,

$$\sum_{j\in J} a_{f(j)} = \sum_{\substack{j\in J\\ k\in K}} a_k\,[f(j)=k] = \sum_{k\in K} a_k \sum_{j\in J} [f(j)=k],$$

since $\sum_{j\in J}[f(j)=k] = \#f^-(k)$. In the special case that f is a one-to-one correspondence between J and K, we have $\#f^-(k) = 1$ for all k, and the general formula (2.35) reduces to

My other math teacher calls this a "bijection"; maybe I'll learn to love that word some day.

And then again . . .

$$\sum_{j\in J} a_{f(j)} = \sum_{f(j)\in K} a_{f(j)} = \sum_{k\in K} a_k .$$

This is the commutative law (2.17) we had before, slightly disguised.

Our examples of multiple sums so far have all involved general terms like a_k or b_k. But this book is supposed to be concrete, so let's take a look at a multiple sum that involves actual numbers:

$$S_n = \sum_{1\leqslant j<k\leqslant n} \frac{1}{k-j}.$$

For example, $S_1 = 0$; $S_2 = 1$; $S_3 = \frac{1}{2-1} + \frac{1}{3-1} + \frac{1}{3-2} = \frac{5}{2}$.

The normal way to evaluate a double sum is to sum first on j or first on k, so let's explore both options.

$$\begin{aligned}
S_n &= \sum_{1\le k\le n}\ \sum_{1\le j<k} \frac{1}{k-j} && \text{summing first on } j\\
&= \sum_{1\le k\le n}\ \sum_{1\le k-j<k} \frac{1}{j} && \text{replacing } j \text{ by } k-j\\
&= \sum_{1\le k\le n}\ \sum_{0<j\le k-1} \frac{1}{j} && \text{simplifying the bounds on } j\\
&= \sum_{1\le k\le n} H_{k-1} && \text{by (2.13), the definition of } H_{k-1}\\
&= \sum_{1\le k+1\le n} H_k && \text{replacing } k \text{ by } k+1\\
&= \sum_{0\le k<n} H_k\,. && \text{simplifying the bounds on } k
\end{aligned}$$

Alas! We don't know how to get a sum of harmonic numbers into closed form.

Get out the whip.

If we try summing first the other way, we get

$$\begin{aligned}
S_n &= \sum_{1\le j\le n}\ \sum_{j<k\le n} \frac{1}{k-j} && \text{summing first on } k\\
&= \sum_{1\le j\le n}\ \sum_{j<k+j\le n} \frac{1}{k} && \text{replacing } k \text{ by } k+j\\
&= \sum_{1\le j\le n}\ \sum_{0<k\le n-j} \frac{1}{k} && \text{simplifying the bounds on } k\\
&= \sum_{1\le j\le n} H_{n-j} && \text{by (2.13), the definition of } H_{n-j}\\
&= \sum_{1\le n-j\le n} H_j && \text{replacing } j \text{ by } n-j\\
&= \sum_{0\le j<n} H_j\,. && \text{simplifying the bounds on } j
\end{aligned}$$

We're back at the same impasse.

But there's *another* way to proceed, if we replace k by $k+j$ *before* deciding to reduce S_n to a sum of sums:

$$\begin{aligned}
S_n &= \sum_{1\le j<k\le n} \frac{1}{k-j} && \text{recopying the given sum}\\
&= \sum_{1\le j<k+j\le n} \frac{1}{k} && \text{replacing } k \text{ by } k+j
\end{aligned}$$

$$= \sum_{1 \le k \le n} \sum_{1 \le j \le n-k} \frac{1}{k} \qquad \text{summing first on } j$$

$$= \sum_{1 \le k \le n} \frac{n-k}{k} \qquad \text{the sum on } j \text{ is trivial}$$

$$= \sum_{1 \le k \le n} \frac{n}{k} - \sum_{1 \le k \le n} 1 \qquad \text{by the associative law}$$

$$= n\left(\sum_{1 \le k \le n} \frac{1}{k}\right) - n \qquad \text{by gosh}$$

$$= nH_n - n. \qquad \text{by (2.13), the definition of } H_n$$

It was smart to say $k \le n$ instead of $k \le n - 1$ in this derivation. Simple bounds save energy.

Aha! We've found S_n. Combining this with the false starts we made gives us a further identity as a bonus:

$$\sum_{0 \le k < n} H_k = nH_n - n. \qquad (2.36)$$

We can understand the trick that worked here in two ways, one algebraic and one geometric. (1) Algebraically, if we have a double sum whose terms involve $k+f(j)$, where f is an arbitrary function, this example indicates that it's a good idea to try replacing k by $k-f(j)$ and summing on j. (2) Geometrically, we can look at this particular sum S_n as follows, in the case $n = 4$:

	$k=1$	$k=2$	$k=3$	$k=4$
$j=1$		$\frac{1}{1}$ +	$\frac{1}{2}$ +	$\frac{1}{3}$
$j=2$			$\frac{1}{1}$ +	$\frac{1}{2}$
$j=3$				$\frac{1}{1}$
$j=4$				

Our first attempts, summing first on j (by columns) or on k (by rows), gave us $H_1 + H_2 + H_3 = H_3 + H_2 + H_1$. The winning idea was essentially to sum by diagonals, getting $\frac{3}{1} + \frac{2}{2} + \frac{1}{3}$.

2.5 GENERAL METHODS

Now let's consolidate what we've learned, by looking at a single example from several different angles. On the next few pages we're going to try to find a closed form for the sum of the first n squares, which we'll call $\Box_n$:

$$\Box_n = \sum_{0 \le k \le n} k^2, \qquad \text{for } n \ge 0. \qquad (2.37)$$

We'll see that there are at least seven different ways to solve this problem, and in the process we'll learn useful strategies for attacking sums in general.

First, as usual, we look at some small cases.

n	0	1	2	3	4	5	6	7	8	9	10	11	12
n^2	0	1	4	9	16	25	36	49	64	81	100	121	144
$\square_n$	0	1	5	14	30	55	91	140	204	285	385	506	650

No closed form for $\square_n$ is immediately evident; but when we do find one, we can use these values as a check.

Method 0: You could look it up.

A problem like the sum of the first n squares has probably been solved before, so we can most likely find the solution in a handy reference book. Sure enough, page 72 of the *CRC Standard Mathematical Tables* [24] has the answer:

$$\square_n = \frac{n(n+1)(2n+1)}{6}, \qquad \text{for } n \geqslant 0. \tag{2.38}$$

Just to make sure we haven't misread it, we check that this formula correctly gives $\square_5 = 5\cdot 6\cdot 11/6 = 55$. Incidentally, page 72 of the *CRC Tables* has further information about the sums of cubes, ..., tenth powers.

The definitive reference for mathematical formulas is the *Handbook of Mathematical Functions*, edited by Abramowitz and Stegun [2]. Pages 813–814 of that book list the values of $\square_n$ for $n \leqslant 100$; and pages 804 and 809 exhibit formulas equivalent to (2.38), together with the analogous formulas for sums of cubes, ..., fifteenth powers, with or without alternating signs.

(Harder sums can be found in Hansen's comprehensive table [147].)

But the best source for answers to questions about sequences is an amazing little book called the *Handbook of Integer Sequences*, by Sloane [270], which lists thousands of sequences by their numerical values. If you come up with a recurrence that you suspect has already been studied, all you have to do is compute enough terms to distinguish your recurrence from other famous ones; then chances are you'll find a pointer to the relevant literature in Sloane's *Handbook*. For example, 1, 5, 14, 30, ... turns out to be Sloane's sequence number 1574, and it's called the sequence of "square pyramidal numbers" (because there are $\square_n$ balls in a pyramid that has a square base of n^2 balls). Sloane gives three references, one of which is to the handbook of Abramowitz and Stegun that we've already mentioned.

Still another way to probe the world's store of accumulated mathematical wisdom is to use a computer program (such as MACSYMA) that provides tools for symbolic manipulation. Such programs are indispensable, especially for people who need to deal with large formulas.

It's good to be familiar with standard sources of information, because they can be extremely helpful. But Method 0 isn't really consistent with the spirit of this book, because we want to know how to figure out the answers

Or, at least to problems having the same answers as problems that other people have decided to consider.

by ourselves. The look-up method is limited to problems that other people have decided are worth considering; a new problem won't be there.

Method 1: Guess the answer, prove it by induction.

Perhaps a little bird has told us the answer to a problem, or we have arrived at a closed form by some other less-than-rigorous means. Then we merely have to prove that it is correct.

We might, for example, have noticed that the values of $\Box_n$ have rather small prime factors, so we may have come up with formula (2.38) as something that works for all small values of n. We might also have conjectured the equivalent formula

$$\Box_n = \frac{n(n+\frac{1}{2})(n+1)}{3}, \qquad \text{for } n \geqslant 0, \tag{2.39}$$

which is nicer because it's easier to remember. The preponderance of the evidence supports (2.39), but we must prove our conjectures beyond all reasonable doubt. Mathematical induction was invented for this purpose.

"Well, your honor, we know that $\Box_0 = 0 = 0(0+\frac{1}{2})(0+1)/3$, so the basis is easy. For the induction, suppose that $n > 0$, and assume that (2.39) holds when n is replaced by $n-1$. Since

$$\Box_n = \Box_{n-1} + n^2,$$

we have

$$\begin{aligned} 3\Box_n &= (n-1)(n-\tfrac{1}{2})(n) \;+\; 3n^2 \\ &= (n^3 - \tfrac{3}{2}n^2 + \tfrac{1}{2}n) \;+\; 3n^2 \\ &= (n^3 + \tfrac{3}{2}n^2 + \tfrac{1}{2}n) \\ &= n(n+\tfrac{1}{2})(n+1)\,. \end{aligned}$$

Therefore (2.39) indeed holds, beyond a reasonable doubt, for all $n \geqslant 0$." Judge Wapner, in his infinite wisdom, agrees.

Induction has its place, and it is somewhat more defensible than trying to look up the answer. But it's still not really what we're seeking. All of the other sums we have evaluated so far in this chapter have been conquered without induction; we should likewise be able to determine a sum like $\Box_n$ from scratch. Flashes of inspiration should not be necessary. We should be able to do sums even on our less creative days.

Method 2: Perturb the sum.

So let's go back to the perturbation method that worked so well for the geometric progression (2.25). We extract the first and last terms of $\Box_{n+1}$ in order to get an equation for $\Box_n$:

$$\begin{aligned}\Box_n + (n+1)^2 &= \sum_{0\le k\le n}(k+1)^2 = \sum_{0\le k\le n}(k^2+2k+1)\\ &= \sum_{0\le k\le n}k^2 + 2\sum_{0\le k\le n}k + \sum_{0\le k\le n}1\\ &= \Box_n \;+\; 2\sum_{0\le k\le n}k \;+\; (n+1)\,.\end{aligned}$$

Oops — the $\Box_n$'s cancel each other. Occasionally, despite our best efforts, the perturbation method produces something like $\Box_n = \Box_n$, so we lose.

Seems more like a draw.

On the other hand, this derivation is not a total loss; it does reveal a way to sum the first n integers in closed form,

$$2\sum_{0\le k\le n}k = (n+1)^2 - (n+1)\,,$$

even though we'd hoped to discover the sum of first integers squared. Could it be that if we start with the sum of the integers cubed, which we might call $⧈_n$, we will get an expression for the integers squared? Let's try it.

$$\begin{aligned}⧈_n + (n+1)^3 &= \sum_{0\le k\le n}(k+1)^3 = \sum_{0\le k\le n}(k^3+3k^2+3k+1)\\ &= ⧈_n + 3\Box_n + 3\frac{(n+1)n}{2} + (n+1)\,.\end{aligned}$$

Sure enough, the $⧈_n$'s cancel, and we have enough information to determine $\Box_n$ without relying on induction:

Method 2′: Perturb your TA.

$$\begin{aligned}3\Box_n &= (n+1)^3 - 3(n+1)n/2 - (n+1)\\ &= (n+1)(n^2+2n+1-\tfrac{3}{2}n-1) \;=\; (n+1)(n+\tfrac{1}{2})n\,.\end{aligned}$$

Method 3: Build a repertoire.

A slight generalization of the recurrence (2.7) will also suffice for summands involving n^2. The solution to

$$\begin{aligned}R_0 &= \alpha;\\ R_n &= R_{n-1} + \beta + \gamma n + \delta n^2\,, \qquad \text{for } n>0, \end{aligned} \tag{2.40}$$

will be of the general form

$$R_n = A(n)\alpha + B(n)\beta + C(n)\gamma + D(n)\delta\,; \tag{2.41}$$

and we have already determined $A(n)$, $B(n)$, and $C(n)$, because (2.41) is the same as (2.7) when $\delta = 0$. If we now plug in $R_n = n^3$, we find that n^3 is the solution when $\alpha = 0$, $\beta = 1$, $\gamma = -3$, $\delta = 3$. Hence

$$3D(n) - 3C(n) + B(n) \;=\; n^3\,;$$

this determines $D(n)$.

We're interested in the sum $\Box_n$, which equals $\Box_{n-1} + n^2$; thus we get $\Box_n = R_n$ if we set $\alpha = \beta = \gamma = 0$ and $\delta = 1$ in (2.41). Consequently $\Box_n = D(n)$. We needn't do the algebra to compute $D(n)$ from $B(n)$ and $C(n)$, since we already know what the answer will be; but doubters among us should be reassured to find that

$$3D(n) \;=\; n^3 + 3C(n) - B(n) \;=\; n^3 + 3\frac{(n+1)n}{2} - n \;=\; n(n+\tfrac{1}{2})(n+1)\,.$$

Method 4: Replace sums by integrals.

People who have been raised on calculus instead of discrete mathematics tend to be more familiar with $\int$ than with $\sum$, so they find it natural to try changing $\sum$ to $\int$. One of our goals in this book is to become so comfortable with $\sum$ that we'll think $\int$ is more difficult than $\sum$ (at least for exact results). But still, it's a good idea to explore the relation between $\sum$ and $\int$, since summation and integration are based on very similar ideas.

In calculus, an integral can be regarded as the area under a curve, and we can approximate this area by adding up the areas of long, skinny rectangles that touch the curve. We can also go the other way if a collection of long, skinny rectangles is given: Since $\Box_n$ is the sum of the areas of rectangles whose sizes are 1×1, 1×4, ..., $1 \times n^2$, it is approximately equal to the area under the curve $f(x) = x^2$ between 0 and n.

The horizontal scale here is ten times the vertical scale.

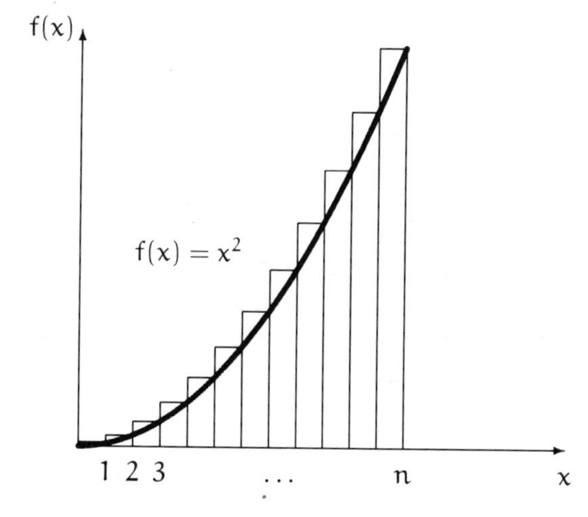

The area under this curve is $\int_0^n x^2\,dx = n^3/3$; therefore we know that $\Box_n$ is approximately $\frac{1}{3}n^3$.

One way to use this fact is to examine the error in the approximation, $E_n = \Box_n - \frac{1}{3}n^3$. Since $\Box_n$ satisfies the recurrence $\Box_n = \Box_{n-1} + n^2$, we find that E_n satisfies the simpler recurrence

$$\begin{aligned} E_n &= \Box_n - \tfrac{1}{3}n^3 = \Box_{n-1} + n^2 - \tfrac{1}{3}n^3 = E_{n-1} + \tfrac{1}{3}(n-1)^3 + n^2 - \tfrac{1}{3}n^3 \\ &= E_{n-1} + n - \tfrac{1}{3}\,. \end{aligned}$$

Another way to pursue the integral approach is to find a formula for E_n by summing the areas of the wedge-shaped error terms. We have

$$\begin{aligned} \Box_n - \int_0^n x^2\,dx &= \sum_{k=1}^{n}\left(k^2 - \int_{k-1}^{k} x^2\,dx\right) \\ &= \sum_{k=1}^{n}\left(k^2 - \frac{k^3-(k-1)^3}{3}\right) = \sum_{k=1}^{n}\left(k - \tfrac{1}{3}\right). \end{aligned}$$

This is for people addicted to calculus.

Either way, we could find E_n and then $\Box_n$.

Method 5: Expand and contract.

Yet another way to discover a closed form for $\Box_n$ is to replace the original sum by a seemingly more complicated double sum that can actually be simplified if we massage it properly:

$$\begin{aligned} \Box_n &= \sum_{1\leqslant k\leqslant n} k^2 = \sum_{1\leqslant j\leqslant k\leqslant n} k \\ &= \sum_{1\leqslant j\leqslant n}\ \sum_{j\leqslant k\leqslant n} k \\ &= \sum_{1\leqslant j\leqslant n}\left(\frac{j+n}{2}\right)(n-j+1) \\ &= \tfrac{1}{2}\sum_{1\leqslant j\leqslant n}\bigl(n(n+1)+j-j^2\bigr) \\ &= \tfrac{1}{2}n^2(n+1) + \tfrac{1}{4}n(n+1) - \tfrac{1}{2}\Box_n = \tfrac{1}{2}n(n+\tfrac{1}{2})(n+1) - \tfrac{1}{2}\Box_n\,. \end{aligned}$$

(The last step here is something like the last step of the perturbation method, because we get an equation with the unknown quantity on both sides.)

Going from a single sum to a double sum may appear at first to be a backward step, but it's actually progress, because it produces sums that are easier to work with. We can't expect to solve every problem by continually simplifying, simplifying, and simplifying: You can't scale the highest mountain peaks by climbing only uphill!

Method 6: Use finite calculus.

Method 7: Use generating functions.

Stay tuned for still more exciting calculations of $\Box_n = \sum_{k=0}^{n} k^2$, as we learn further techniques in the next section and in later chapters.

2.6 FINITE AND INFINITE CALCULUS

We've learned a variety of ways to deal with sums directly. Now it's time to acquire a broader perspective, by looking at the problem of summation from a higher level. Mathematicians have developed a "finite calculus," analogous to the more traditional infinite calculus, by which it's possible to approach summation in a nice, systematic fashion.

Infinite calculus is based on the properties of the *derivative* operator D, defined by

$$\mathrm{D}f(x) \;=\; \lim_{h\to 0}\frac{f(x+h)-f(x)}{h}\,.$$

Finite calculus is based on the properties of the *difference* operator Δ, defined by

$$\Delta f(x) \;=\; f(x+1)-f(x)\,. \qquad (2.42)$$

This is the finite analog of the derivative in which we restrict ourselves to positive integer values of h. Thus, $h = 1$ is the closest we can get to the "limit" as $h \to 0$, and $\Delta f(x)$ is the value of $\bigl(f(x+h)-f(x)\bigr)/h$ when $h = 1$.

The symbols D and Δ are called *operators* because they operate on functions to give new functions; they are functions of functions that produce functions. If f is a suitably smooth function of real numbers to real numbers, then $\mathrm{D}f$ is also a function from reals to reals. And if f is *any* real-to-real function, so is Δf. The values of the functions $\mathrm{D}f$ and Δf at a point x are given by the definitions above.

As opposed to a cassette function.

Early on in calculus we learn how D operates on the powers $f(x) = x^m$. In such cases $\mathrm{D}f(x) = mx^{m-1}$. We can write this informally with f omitted,

$$\mathrm{D}(x^m) \;=\; mx^{m-1}\,.$$

It would be nice if the Δ operator would produce an equally elegant result; unfortunately it doesn't. We have, for example,

$$\Delta(x^3) \;=\; (x+1)^3 - x^3 \;=\; 3x^2+3x+1\,.$$

Math power.

But there is a type of "mth power" that does transform nicely under Δ, and this is what makes finite calculus interesting. Such newfangled mth powers are defined by the rule

$$x^{\underline{m}} \;=\; \overbrace{x(x-1)\ldots(x-m+1)}^{m\text{ factors}}\,, \qquad \text{integer } m \geqslant 0. \qquad (2.43)$$

Notice the little straight line under the m; this implies that the m factors are supposed to go down and down, stepwise. There's also a corresponding definition where the factors go up and up:

$$x^{\overline{m}} = \overbrace{x(x+1)\ldots(x+m-1)}^{m \text{ factors}}, \qquad \text{integer } m \geqslant 0. \tag{2.44}$$

When $m = 0$, we have $x^{\underline{0}} = x^{\overline{0}} = 1$, because a product of no factors is conventionally taken to be 1 (just as a sum of no terms is conventionally 0).

The quantity $x^{\underline{m}}$ is called "x to the m falling," if we have to read it aloud; similarly, $x^{\overline{m}}$ is "x to the m rising." These functions are also called *falling factorial powers* and *rising factorial powers*, since they are closely related to the factorial function $n! = n(n-1)\ldots(1)$. In fact, $n! = n^{\underline{n}} = 1^{\overline{n}}$.

Several other notations for factorial powers appear in the mathematical literature, notably "Pochhammer's symbol" $(x)_m$ for $x^{\overline{m}}$ or $x^{\underline{m}}$; notations like $x^{(m)}$ or $x_{(m)}$ are also seen for $x^{\underline{m}}$. But the underline/overline convention is catching on, because it's easy to write, easy to remember, and free of redundant parentheses.

Mathematical terminology is sometimes crazy: Pochhammer [234] actually used the notation $(x)_m$ for the binomial coefficient $\binom{x}{m}$, not for factorial powers.

Falling powers $x^{\underline{m}}$ are especially nice with respect to Δ. We have

$$\begin{aligned} \Delta(x^{\underline{m}}) &= (x+1)^{\underline{m}} - x^{\underline{m}} \\ &= (x+1)x\ldots(x-m+2) \quad - \quad x\ldots(x-m+2)(x-m+1) \\ &= m\,x(x-1)\ldots(x-m+2), \end{aligned}$$

hence the finite calculus has a handy law to match $D(x^m) = mx^{m-1}$:

$$\Delta(x^{\underline{m}}) = mx^{\underline{m-1}}. \tag{2.45}$$

This is the basic factorial fact.

The operator D of infinite calculus has an inverse, the anti-derivative (or integration) operator $\int$. The Fundamental Theorem of Calculus relates D to $\int$:

$$g(x) = Df(x) \qquad \text{if and only if} \qquad \int g(x)\,dx = f(x) + C.$$

Here $\int g(x)\,dx$, the indefinite integral of $g(x)$, is the class of functions whose derivative is $g(x)$. Analogously, Δ has as an inverse, the anti-difference (or summation) operator $\sum$; and there's another Fundamental Theorem:

$$g(x) = \Delta f(x) \qquad \text{if and only if} \qquad \sum g(x)\,\delta x = f(x) + C. \tag{2.46}$$

"Quemadmodum ad differentiam denotandam usi sumus signo Δ, ita summam indicabimus signo Σ. ... ex quo æquatio $z = \Delta y$, si invertatur, dabit quoque $y = \Sigma z + C$."
—L. Euler [88]

Here $\sum g(x)\,\delta x$, the *indefinite sum* of $g(x)$, is the class of functions whose *difference* is $g(x)$. (Notice that the lowercase δ relates to uppercase Δ as d relates to D.) The "C" for indefinite integrals is an arbitrary constant; the "C" for indefinite sums is any function $p(x)$ such that $p(x+1) = p(x)$. For example, C might be the periodic function $a + b\sin 2\pi x$; such functions get washed out when we take differences, just as constants get washed out when we take derivatives. At integer values of x, the function C is constant.

Now we're almost ready for the punch line. Infinite calculus also has *definite* integrals: If $g(x) = \mathrm{D}f(x)$, then

$$\int_a^b g(x)\,dx = f(x)\Big|_a^b = f(b) - f(a)\,.$$

Therefore finite calculus — ever mimicking its more famous cousin — has definite *sums*: If $g(x) = \Delta f(x)$, then

$$\sum\nolimits_a^b g(x)\,\delta x = f(x)\Big|_a^b = f(b) - f(a)\,. \tag{2.47}$$

This formula gives a meaning to the notation $\sum_a^b g(x)\,\delta x$, just as the previous formula defines $\int_a^b g(x)\,dx$.

But what does $\sum_a^b g(x)\,\delta x$ really mean, intuitively? We've defined it by analogy, not by necessity. We want the analogy to hold, so that we can easily remember the rules of finite calculus; but the notation will be useless if we don't understand its significance. Let's try to deduce its meaning by looking first at some special cases, assuming that $g(x) = \Delta f(x) = f(x+1) - f(x)$. If $b = a$, we have

$$\sum\nolimits_a^a g(x)\,\delta x = f(a) - f(a) = 0\,.$$

Next, if $b = a + 1$, the result is

$$\sum\nolimits_a^{a+1} g(x)\,\delta x = f(a+1) - f(a) = g(a)\,.$$

More generally, if b increases by 1, we have

$$\begin{aligned}\sum\nolimits_a^{b+1} g(x)\,\delta x - \sum\nolimits_a^b g(x)\,\delta x &= \bigl(f(b+1) - f(a)\bigr) - \bigl(f(b) - f(a)\bigr)\\ &= f(b+1) - f(b) = g(b)\,.\end{aligned}$$

These observations, and mathematical induction, allow us to deduce exactly what $\sum_a^b g(x)\,\delta x$ means in general, when a and b are integers with $b \geqslant a$:

$$\sum\nolimits_a^b g(x)\,\delta x = \sum_{k=a}^{b-1} g(k) = \sum_{a\leqslant k<b} g(k)\,, \qquad \text{for integers } b \geqslant a. \tag{2.48}$$

You call this a punch line?

In other words, the definite sum is the same as an ordinary sum with limits, but excluding the value at the upper limit.

Let's try to recap this in a slightly different way. Suppose we've been given an unknown sum that's supposed to be evaluated in closed form, and suppose we can write it in the form $\sum_{a\leqslant k<b} g(k) = \sum_a^b g(x)\,\delta x$. The theory of finite calculus tells us that we can express the answer as $f(b) - f(a)$, if we can only find an indefinite sum or anti-difference function f such that $g(x) = f(x+1) - f(x)$. One way to understand this principle is to write $\sum_{a\leqslant k<b} g(k)$ out in full, using the three-dots notation:

$$\sum_{a\leqslant k<b} \bigl(f(k+1) - f(k)\bigr) = \bigl(f(a{+}1) - f(a)\bigr) + \bigl(f(a{+}2) - f(a{+}1)\bigr) + \cdots + \bigl(f(b{-}1) - f(b{-}2)\bigr) + \bigl(f(b) - f(b{-}1)\bigr).$$

Everything on the right-hand side cancels, except $f(b) - f(a)$; so $f(b) - f(a)$ is the value of the sum. (Sums of the form $\sum_{a\leqslant k<b}\bigl(f(k+1) - f(k)\bigr)$ are often called *telescoping*, by analogy with a collapsed telescope, because the thickness of a collapsed telescope is determined solely by the outer radius of the outermost tube and the inner radius of the innermost tube.)

And all this time I thought it was telescoping because it collapsed from a very long expression to a very short one.

But rule (2.48) applies only when $b \geqslant a$; what happens if $b < a$? Well, (2.47) says that we must have

$$\begin{aligned}\textstyle\sum_a^b g(x)\,\delta x &= f(b) - f(a)\\ &= -\bigl(f(a) - f(b)\bigr) = -\textstyle\sum_b^a g(x)\,\delta x.\end{aligned}$$

This is analogous to the corresponding equation for definite integration. A similar argument proves $\sum_a^b + \sum_b^c = \sum_a^c$, the summation analog of the identity $\int_a^b + \int_b^c = \int_a^c$. In full garb,

$$\sum\nolimits_a^b g(x)\,\delta x + \sum\nolimits_b^c g(x)\,\delta x = \sum\nolimits_a^c g(x)\,\delta x, \qquad (2.49)$$

for all integers a, b, and c.

At this point a few of us are probably starting to wonder what all these parallels and analogies buy us. Well for one, definite summation gives us a simple way to compute sums of falling powers: The basic laws (2.45), (2.47), and (2.48) imply the general law

Others have been wondering this for some time now.

$$\sum_{0\leqslant k<n} k^{\underline{m}} = \frac{k^{\underline{m+1}}}{m+1}\bigg|_0^n = \frac{n^{\underline{m+1}}}{m+1}, \qquad \text{for integers } m, n \geqslant 0. \qquad (2.50)$$

This formula is easy to remember because it's so much like the familiar $\int_0^n x^m\,dx = n^{m+1}/(m+1)$.

In particular, when $m = 1$ we have $k^{\underline{1}} = k$, so the principles of finite calculus give us an easy way to remember the fact that

$$\sum_{0 \leqslant k < n} k = \frac{n^{\underline{2}}}{2} = n(n-1)/2 .$$

The definite-sum method also gives us an inkling that sums over the range $0 \leqslant k < n$ often turn out to be simpler than sums over $1 \leqslant k \leqslant n$; the former are just $f(n) - f(0)$, while the latter must be evaluated as $f(n+1) - f(1)$.

Ordinary powers can also be summed in this new way, if we first express them in terms of falling powers. For example,

$$k^2 = k^{\underline{2}} + k^{\underline{1}},$$

hence

$$\sum_{0 \leqslant k < n} k^2 = \frac{n^{\underline{3}}}{3} + \frac{n^{\underline{2}}}{2} = \tfrac{1}{3}n(n-1)(n-2+\tfrac{3}{2}) = \tfrac{1}{3}n(n-\tfrac{1}{2})(n-1) .$$

Replacing n by $n+1$ gives us yet another way to compute the value of our old friend $\Box_n = \sum_{0 \leqslant k \leqslant n} k^2$ in closed form.

With friends like this...

Gee, that was pretty easy. In fact, it was easier than any of the umpteen other ways that beat this formula to death in the previous section. So let's try to go up a notch, from squares to *cubes*: A simple calculation shows that

$$k^3 = k^{\underline{3}} + 3k^{\underline{2}} + k^{\underline{1}} .$$

(It's always possible to convert between ordinary powers and factorial powers by using Stirling numbers, which we will study in Chapter 6.) Thus

$$\sum_{a \leqslant k < b} k^3 = \frac{k^{\underline{4}}}{4} + k^{\underline{3}} + \frac{k^{\underline{2}}}{2} \bigg|_a^b .$$

Falling powers are therefore very nice for sums. But do they have any other redeeming features? Must we convert our old friendly ordinary powers to falling powers before summing, but then convert back before we can do anything else? Well, no, it's often possible to work directly with factorial powers, because they have additional properties. For example, just as we have $(x+y)^2 = x^2 + 2xy + y^2$, it turns out that $(x+y)^{\underline{2}} = x^{\underline{2}} + 2x^{\underline{1}}y^{\underline{1}} + y^{\underline{2}}$, and the same analogy holds between $(x+y)^m$ and $(x+y)^{\underline{m}}$. (This "factorial binomial theorem" is proved in exercise 5.37.)

So far we've considered only falling powers that have nonnegative exponents. To extend the analogies with ordinary powers to negative exponents, we need an appropriate definition of $x^{\underline{m}}$ for $m < 0$. Looking at the sequence

$$\begin{aligned} x^{\underline{3}} &= x(x-1)(x-2)\,, \\ x^{\underline{2}} &= x(x-1)\,, \\ x^{\underline{1}} &= x\,, \\ x^{\underline{0}} &= 1\,, \end{aligned}$$

we notice that to get from $x^{\underline{3}}$ to $x^{\underline{2}}$ to $x^{\underline{1}}$ to $x^{\underline{0}}$ we divide by $x-2$, then by $x-1$, then by x. It seems reasonable (if not imperative) that we should divide by $x+1$ next, to get from $x^{\underline{0}}$ to $x^{\underline{-1}}$, thereby making $x^{\underline{-1}} = 1/(x+1)$. Continuing, the first few negative-exponent falling powers are

$$\begin{aligned} x^{\underline{-1}} &= \frac{1}{x+1}\,, \\ x^{\underline{-2}} &= \frac{1}{(x+1)(x+2)}\,, \\ x^{\underline{-3}} &= \frac{1}{(x+1)(x+2)(x+3)}\,, \end{aligned}$$

and our general definition for negative falling powers is

$$x^{\underline{-m}} = \frac{1}{(x+1)(x+2)\ldots(x+m)}\,, \qquad \text{for } m > 0. \tag{2.51}$$

(It's also possible to define falling powers for real or even complex m, but we will defer that until Chapter 5.)

How can a complex number be even?

With this definition, falling powers have additional nice properties. Perhaps the most important is a general law of exponents, analogous to the law

$$x^{m+n} = x^m x^n$$

for ordinary powers. The falling-power version is

$$x^{\underline{m+n}} = x^{\underline{m}}\,(x-m)^{\underline{n}}\,, \qquad \text{integers } m \text{ and } n. \tag{2.52}$$

For example, $x^{\underline{2+3}} = x^{\underline{2}}\,(x-2)^{\underline{3}}$; and with a negative n we have

$$x^{\underline{2-3}} = x^{\underline{2}}\,(x-2)^{\underline{-3}} = x(x-1)\frac{1}{(x-1)x(x+1)} = \frac{1}{x+1} = x^{\underline{-1}}\,.$$

If we had chosen to define $x^{\underline{-1}}$ as $1/x$ instead of as $1/(x+1)$, the law of exponents (2.52) would have failed in cases like $m = -1$ and $n = 1$. In fact, we could have used (2.52) to tell us exactly how falling powers ought to be defined in the case of negative exponents, by setting $m = -n$. When an existing notation is being extended to cover more cases, it's always best to formulate definitions in such a way that general laws continue to hold.

Laws have their exponents and their detractors.

Now let's make sure that the crucial difference property holds for our newly defined falling powers. Does $\Delta x^{\underline{m}} = mx^{\underline{m-1}}$ when $m < 0$? If $m = -2$, for example, the difference is

$$\begin{aligned}\Delta x^{\underline{-2}} &= \frac{1}{(x+2)(x+3)} - \frac{1}{(x+1)(x+2)} \\ &= \frac{(x+1)-(x+3)}{(x+1)(x+2)(x+3)} \\ &= -2x^{\underline{-3}}.\end{aligned}$$

Yes—it works! A similar argument applies for all $m < 0$.

Therefore the summation property (2.50) holds for negative falling powers as well as positive ones, as long as no division by zero occurs:

$$\sum\nolimits_a^b x^{\underline{m}}\,\delta x = \frac{x^{\underline{m+1}}}{m+1}\bigg|_a^b, \qquad \text{for } m \neq -1.$$

But what about when $m = -1$? Recall that for integration we use

$$\int_a^b x^{-1}\,dx = \ln x\bigg|_a^b$$

when $m = -1$. We'd like to have a finite analog of $\ln x$; in other words, we seek a function $f(x)$ such that

$$x^{\underline{-1}} = \frac{1}{x+1} = \Delta f(x) = f(x+1) - f(x).$$

It's not too hard to see that

$$f(x) = \frac{1}{1} + \frac{1}{2} + \cdots + \frac{1}{x}$$

is such a function, when x is an integer, and this quantity is just the harmonic number H_x of (2.13). Thus H_x is the discrete analog of the continuous $\ln x$. (We will define H_x for noninteger x in Chapter 6, but integer values are good enough for present purposes. We'll also see in Chapter 9 that, for large x, the value of $H_x - \ln x$ is approximately $0.577 + 1/(2x)$. Hence H_x and $\ln x$ are not only analogous, their values usually differ by less than 1.)

0.577 exactly? Maybe they mean $1/\sqrt{3}$. Then again, maybe not.

We can now give a complete description of the sums of falling powers:

$$\sum\nolimits_a^b x^{\underline{m}}\,\delta x = \begin{cases} \dfrac{x^{\underline{m+1}}}{m+1}\bigg|_a^b, & \text{if } m \neq -1; \\ H_x\bigg|_a^b, & \text{if } m = -1. \end{cases} \qquad (2.53)$$

This formula indicates why harmonic numbers tend to pop up in the solutions to discrete problems like the analysis of quicksort, just as so-called natural logarithms arise naturally in the solutions to continuous problems.

Now that we've found an analog for $\ln x$, let's see if there's one for e^x. What function $f(x)$ has the property that $\Delta f(x) = f(x)$, corresponding to the identity $De^x = e^x$? Easy:

$$f(x+1) - f(x) = f(x) \qquad \Longleftrightarrow \qquad f(x+1) = 2f(x);$$

so we're dealing with a simple recurrence, and we can take $f(x) = 2^x$ as the discrete exponential function.

The difference of c^x is also quite simple, for arbitrary c, namely

$$\Delta(c^x) = c^{x+1} - c^x = (c-1)c^x .$$

Hence the anti-difference of c^x is $c^x/(c-1)$, if $c \neq 1$. This fact, together with the fundamental laws (2.47) and (2.48), gives us a tidy way to understand the general formula for the sum of a geometric progression:

$$\sum_{a \leqslant k < b} c^k = \sum\nolimits_a^b c^x \,\delta x = \frac{c^x}{c-1}\bigg|_a^b = \frac{c^b - c^a}{c-1}, \qquad \text{for } c \neq 1.$$

Every time we encounter a function f that might be useful as a closed form, we can compute its difference $\Delta f = g$; then we have a function g whose indefinite sum $\sum g(x)\,\delta x$ is known. Table 55 is the beginning of a table of difference/anti-difference pairs useful for summation.

'Table 55' is on page 55. Get it?

Despite all the parallels between continuous and discrete math, some continuous notions have no discrete analog. For example, the chain rule of infinite calculus is a handy rule for the derivative of a function of a function; but there's no corresponding chain rule of finite calculus, because there's no nice form for $\Delta f\big(g(x)\big)$. Discrete change-of-variables is hard, except in certain cases like the replacement of x by $c \pm x$.

However, $\Delta\big(f(x)\,g(x)\big)$ *does* have a fairly nice form, and it provides us with a rule for *summation by parts*, the finite analog of what infinite calculus calls integration by parts. Let's recall that the formula

$$D(uv) = u\,Dv + v\,Du$$

of infinite calculus leads to the rule for integration by parts,

$$\int u\,Dv = uv - \int v\,Du,$$

Table 55 What's the difference?

$f = \Sigma g$	$\Delta f = g$	$f = \Sigma g$	$\Delta f = g$
$x^{\underline{0}} = 1$	0	2^x	2^x
$x^{\underline{1}} = x$	1	c^x	$(c-1)c^x$
$x^{\underline{2}} = x(x-1)$	$2x$	$c^x/(c-1)$	c^x
$x^{\underline{m}}$	$mx^{\underline{m-1}}$	cf	$c\Delta f$
$x^{\underline{m+1}}/(m+1)$	$x^{\underline{m}}$	$f+g$	$\Delta f + \Delta g$
H_x	$x^{\underline{-1}} = 1/(x+1)$	fg	$f\Delta g + \mathrm{E}g\Delta f$

after integration and rearranging terms; we can do a similar thing in finite calculus.

We start by applying the difference operator to the product of two functions $u(x)$ and $v(x)$:

$$\begin{aligned}\Delta\big(u(x)\,v(x)\big) &= u(x{+}1)\,v(x{+}1) - u(x)\,v(x)\\ &= u(x{+}1)\,v(x{+}1) - u(x)\,v(x{+}1)\\ &\qquad\qquad + u(x)\,v(x{+}1) - u(x)\,v(x)\\ &= u(x)\,\Delta v(x) + v(x{+}1)\,\Delta u(x)\,. \end{aligned} \tag{2.54}$$

This formula can be put into a convenient form using the *shift operator* E, defined by

$$\mathrm{E}f(x) = f(x+1)\,.$$

Substituting this for $v(x{+}1)$ yields a compact rule for the difference of a product:

$$\Delta(uv) = u\,\Delta v + \mathrm{E}v\,\Delta u\,. \tag{2.55}$$

Infinite calculus avoids E here by letting $1 \to 0$.

(The E is a bit of a nuisance, but it makes the equation correct.) Taking the indefinite sum on both sides of this equation, and rearranging its terms, yields the advertised rule for summation by parts:

$$\sum u\,\Delta v = uv - \sum \mathrm{E}v\,\Delta u \tag{2.56}$$

As with infinite calculus, limits can be placed on all three terms, making the indefinite sums definite.

I guess $e^x = 2^x$, for small values of 1.

This rule is useful when the sum on the left is harder to evaluate than the one on the right. Let's look at an example. The function $\int xe^x\,dx$ is typically integrated by parts; its discrete analog is $\sum x2^x\,\delta x$, which we encountered earlier this chapter in the form $\sum_{k=0}^{n} k\,2^k$. To sum this by parts, we let

$u(x) = x$ and $\Delta v(x) = 2^x$; hence $\Delta u(x) = 1$, $v(x) = 2^x$, and $Ev(x) = 2^{x+1}$. Plugging into (2.56) gives

$$\sum x2^x\,\delta x \;=\; x2^x - \sum 2^{x+1}\,\delta x \;=\; x2^x - 2^{x+1} + C\,.$$

And we can use this to evaluate the sum we did before, by attaching limits:

$$\begin{aligned}\sum_{k=0}^{n} k2^k &= \sum\nolimits_0^{n+1} x2^x\,\delta x\\ &= x2^x - 2^{x+1}\,\Big|_0^{n+1}\\ &= \big((n+1)2^{n+1} - 2^{n+2}\big) - (0\cdot 2^0 - 2^1) \;=\; (n-1)2^{n+1} + 2\,.\end{aligned}$$

It's easier to find the sum this way than to use the perturbation method, because we don't have to think.

The ultimate goal of mathematics is to eliminate all need for intelligent thought.

We stumbled across a formula for $\sum_{0\le k<n} H_k$ earlier in this chapter, and counted ourselves lucky. But we could have found our formula (2.36) systematically, if we had known about summation by parts. Let's demonstrate this assertion by tackling a sum that looks even harder, $\sum_{0\le k<n} kH_k$. The solution is not difficult if we are guided by analogy with $\int x\ln x\,dx$: We take $u(x) = H_x$ and $\Delta v(x) = x = x^{\underline{1}}$, hence $\Delta u(x) = x^{\underline{-1}}$, $v(x) = x^{\underline{2}}/2$, $Ev(x) = (x+1)^{\underline{2}}/2$, and we have

$$\begin{aligned}\sum xH_x\,\delta x &= \frac{x^{\underline{2}}}{2}H_x - \sum \frac{(x+1)^{\underline{2}}}{2}\,x^{\underline{-1}}\,\delta x\\ &= \frac{x^{\underline{2}}}{2}H_x - \frac{1}{2}\sum x^{\underline{1}}\,\delta x\\ &= \frac{x^{\underline{2}}}{2}H_x - \frac{x^{\underline{2}}}{4} + C\,.\end{aligned}$$

(In going from the first line to the second, we've combined two falling powers $(x+1)^{\underline{2}}\,x^{\underline{-1}}$ by using the law of exponents (2.52) with $m=-1$ and $n=2$.) Now we can attach limits and conclude that

$$\sum_{0\le k<n} kH_k \;=\; \sum\nolimits_0^n xH_x\,\delta x \;=\; \frac{n^{\underline{2}}}{2}\Big(H_n - \tfrac{1}{2}\Big)\,. \tag{2.57}$$

2.7 INFINITE SUMS

When we defined $\sum$-notation at the beginning of this chapter, we finessed the question of infinite sums by saying, in essence, "Wait until later. For now, we can assume that all the sums we meet have only finitely many nonzero terms." But the time of reckoning has finally arrived; we must face

This is finesse?

the fact that sums can be infinite. And the truth is that infinite sums are bearers of both good news and bad news.

First, the bad news: It turns out that the methods we've used for manipulating $\sum$'s are *not* always valid when infinite sums are involved. But next, the good news: There is a large, easily understood class of infinite sums for which all the operations we've been performing are perfectly legitimate. The reasons underlying both these news items will be clear after we have looked more closely at the underlying meaning of summation.

Everybody knows what a finite sum is: We add up a bunch of terms, one by one, until they've all been added. But an infinite sum needs to be defined more carefully, lest we get into paradoxical situations.

For example, it seems natural to define things so that the infinite sum

$$S = 1 + \tfrac{1}{2} + \tfrac{1}{4} + \tfrac{1}{8} + \tfrac{1}{16} + \tfrac{1}{32} + \cdots$$

is equal to 2, because if we double it we get

$$2S = 2 + 1 + \tfrac{1}{2} + \tfrac{1}{4} + \tfrac{1}{8} + \tfrac{1}{16} + \cdots = 2 + S\,.$$

On the other hand, this same reasoning suggests that we ought to define

$$T = 1 + 2 + 4 + 8 + 16 + 32 + \cdots$$

to be -1, for if we double it we get

Sure: $1 + 2 + 4 + 8 + \cdots$ is the "infinite precision" representation of the number -1, in a binary computer with infinite word size.

$$2T = 2 + 4 + 8 + 16 + 32 + 64 + \cdots = T - 1\,.$$

Something funny is going on; how can we get a negative number by summing positive quantities? It seems better to leave T undefined; or perhaps we should say that $T = \infty$, since the terms being added in T become larger than any fixed, finite number. (Notice that ∞ is another "solution" to the equation $2T = T - 1$; it also "solves" the equation $2S = 2 + S$.)

Let's try to formulate a good definition for the value of a general sum $\sum_{k \in K} a_k$, where K might be infinite. For starters, let's assume that all the terms a_k are *nonnegative*. Then a suitable definition is not hard to find: If there's a bounding constant A such that

$$\sum_{k \in F} a_k \leqslant A$$

for all *finite* subsets $F \subset K$, then we define $\sum_{k \in K} a_k$ to be the *least* such A. (It follows from well-known properties of the real numbers that the set of all such A always contains a smallest element.) But if there's no bounding constant A, we say that $\sum_{k \in K} a_k = \infty$; this means that if A is any real number, there's a set of finitely many terms a_k whose sum exceeds A.

The definition in the previous paragraph has been formulated carefully so that it doesn't depend on any order that might exist in the index set K. Therefore the arguments we are about to make will apply to multiple sums with many indices $k_1, k_2, \ldots$, not just to sums over the set of integers.

In the special case that K is the set of nonnegative integers, our definition for nonnegative terms a_k implies that

$$\sum_{k\geqslant 0} a_k \;=\; \lim_{n\to\infty} \sum_{k=0}^{n} a_k \,.$$

Here's why: Any nondecreasing sequence of real numbers has a limit (possibly ∞). If the limit is A, and if F is any finite set of nonnegative integers whose elements are all $\leqslant n$, we have $\sum_{k\in F} a_k \leqslant \sum_{k=0}^{n} a_k \leqslant A$; hence $A=\infty$ or A is a bounding constant. And if A' is any number less than the stated limit A, then there's an n such that $\sum_{k=0}^{n} a_k > A'$; hence the finite set $F=\{0,1,\ldots,n\}$ witnesses to the fact that A' is not a bounding constant.

The set K might even be uncountable. But only a countable number of terms can be nonzero, if a bounding constant A exists, because at most nA terms are $\geqslant 1/n$.

We can now easily compute the value of certain infinite sums, according to the definition just given. For example, if $a_k = x^k$, we have

$$\sum_{k\geqslant 0} x^k \;=\; \lim_{n\to\infty} \frac{1-x^{n+1}}{1-x} \;=\; \begin{cases} 1/(1-x), & \text{if } 0\leqslant x<1; \\ \infty, & \text{if } x\geqslant 1. \end{cases}$$

In particular, the infinite sums S and T considered a minute ago have the respective values 2 and ∞, just as we suspected. Another interesting example is

$$\begin{aligned} \sum_{k\geqslant 0} \frac{1}{(k+1)(k+2)} \;&=\; \sum_{k\geqslant 0} k^{\underline{-2}} \\ &=\; \lim_{n\to\infty} \sum_{k=0}^{n} k^{\underline{-2}} \;=\; \lim_{n\to\infty} \frac{k^{\underline{-1}}}{-1}\bigg|_0^n \;=\; 1\,. \end{aligned}$$

Now let's consider the case that the sum might have negative terms as well as nonnegative ones. What, for example, should be the value of

$$\sum_{k\geqslant 0} (-1)^k \;=\; 1-1+1-1+1-1+\cdots\,?$$

If we group the terms in pairs, we get

$$(1-1)+(1-1)+(1-1)+\cdots \;=\; 0+0+0+\cdots\,,$$

so the sum comes out zero; but if we start the pairing one step later, we get

$$1-(1-1)-(1-1)-(1-1)-\cdots \;=\; 1-0-0-0-\cdots\,;$$

the sum is 1.

"Aggregatum quantitatum $a-a+a-a+a-a$ etc. nunc est $=a$, nunc $=0$, adeoque continuata in infinitum serie ponendus $=a/2$, fateor acumen et veritatem animadversionis tuæ."
—G. Grandi [133]

We might also try setting $x = -1$ in the formula $\sum_{k\geq 0} x^k = 1/(1-x)$, since this formula is known to be correct when $0 \leq x < 1$; but then we are forced to conclude that the infinite sum is $\frac{1}{2}$, although it's a sum of integers!

Another interesting example is the doubly infinite $\sum_k a_k$ where $a_k = 1/(k+1)$ for $k \geq 0$ and $a_k = 1/(k-1)$ for $k < 0$. We can write this as

$$\cdots + (-\tfrac{1}{4}) + (-\tfrac{1}{3}) + (-\tfrac{1}{2}) + 1 + \tfrac{1}{2} + \tfrac{1}{3} + \tfrac{1}{4} + \cdots . \qquad (2.58)$$

If we evaluate this sum by starting at the "center" element and working outward,

$$\cdots + \Bigl(-\tfrac{1}{4} + (-\tfrac{1}{3} + (-\tfrac{1}{2} + (1) + \tfrac{1}{2}) + \tfrac{1}{3}) + \tfrac{1}{4}\Bigr) + \cdots ,$$

we get the value 1; and we obtain the same value 1 if we shift all the parentheses one step to the left,

$$\cdots + \Bigl(-\tfrac{1}{5} + (-\tfrac{1}{4} + (-\tfrac{1}{3} + (-\tfrac{1}{2}) + 1) + \tfrac{1}{2}) + \tfrac{1}{3}\Bigr) + \cdots ,$$

because the sum of all numbers inside the innermost n parentheses is

$$-\frac{1}{n+1} - \frac{1}{n} - \cdots - \frac{1}{2} + 1 + \frac{1}{2} + \cdots + \frac{1}{n-1} = 1 - \frac{1}{n} - \frac{1}{n+1} .$$

A similar argument shows that the value is 1 if these parentheses are shifted any fixed amount to the left or right; this encourages us to believe that the sum is indeed 1. On the other hand, if we group terms in the following way,

$$\cdots + \Bigl(-\tfrac{1}{4} + (-\tfrac{1}{3} + (-\tfrac{1}{2} + 1 + \tfrac{1}{2}) + \tfrac{1}{3} + \tfrac{1}{4}) + \tfrac{1}{5} + \tfrac{1}{6}\Bigr) + \cdots ,$$

the nth pair of parentheses from inside out contains the numbers

$$-\frac{1}{n+1} - \frac{1}{n} - \cdots - \frac{1}{2} + 1 + \frac{1}{2} + \cdots + \frac{1}{2n-1} + \frac{1}{2n} = 1 + H_{2n} - H_{n+1} .$$

We'll prove in Chapter 9 that $\lim_{n\to\infty}(H_{2n} - H_{n+1}) = \ln 2$; hence this grouping suggests that the doubly infinite sum should really be equal to $1 + \ln 2$.

There's something flaky about a sum that gives different values when its terms are added up in different ways. Advanced texts on analysis have a variety of definitions by which meaningful values can be assigned to such pathological sums; but if we adopt those definitions, we cannot operate with $\sum$-notation as freely as we have been doing. We don't need the delicate refinements of "conditional convergence" for the purposes of this book; therefore we'll stick to a definition of infinite sums that preserves the validity of all the operations we've been doing in this chapter.

Is this the first page with no graffiti?

In fact, our definition of infinite sums is quite simple. Let K be any set, and let a_k be a real-valued term defined for each $k \in K$. (Here 'k' might actually stand for several indices $k_1, k_2, \ldots$, and K might therefore be multidimensional.) Any real number x can be written as the difference of its positive and negative parts,

$$x = x^+ - x^-, \qquad \text{where } x^+ = x\cdot[x>0] \text{ and } x^- = -x\cdot[x<0].$$

(Either $x^+ = 0$ or $x^- = 0$.) We've already explained how to define values for the infinite sums $\sum_{k\in K} a_k^+$ and $\sum_{k\in K} a_k^-$, because a_k^+ and a_k^- are nonnegative. Therefore our general definition is

$$\sum_{k\in K} a_k = \sum_{k\in K} a_k^+ - \sum_{k\in K} a_k^-, \qquad (2.59)$$

unless the right-hand sums are both equal to ∞. In the latter case, we leave $\sum_{k\in K} a_k$ undefined.

Let $A^+ = \sum_{k\in K} a_k^+$ and $A^- = \sum_{k\in K} a_k^-$. If A^+ and A^- are both finite, the sum $\sum_{k\in K} a_k$ is said to *converge absolutely* to the value $A = A^+ - A^-$. If $A^+ = \infty$ but A^- is finite, the sum $\sum_{k\in K} a_k$ is said to *diverge* to $+\infty$. Similarly, if $A^- = \infty$ but A^+ is finite, $\sum_{k\in K} a_k$ is said to diverge to $-\infty$. If $A^+ = A^- = \infty$, all bets are off.

In other words, absolute convergence means that the sum of absolute values converges.

We started with a definition that worked for nonnegative terms, then we extended it to real-valued terms. If the terms a_k are complex numbers, we can extend the definition once again, in the obvious way: The sum $\sum_{k\in K} a_k$ is defined to be $\sum_{k\in K} \Re a_k + i\sum_{k\in K} \Im a_k$, where $\Re a_k$ and $\Im a_k$ are the real and imaginary parts of a_k —provided that both of those sums are defined. Otherwise $\sum_{k\in K} a_k$ is undefined. (See exercise 18.)

The bad news, as stated earlier, is that some infinite sums must be left undefined, because the manipulations we've been doing can produce inconsistencies in all such cases. (See exercise 34.) The good news is that all of the manipulations of this chapter are perfectly valid whenever we're dealing with sums that converge absolutely, as just defined.

We can verify the good news by showing that each of our transformation rules preserves the value of all absolutely convergent sums. This means, more explicitly, that we must prove the distributive, associative, and commutative laws, plus the rule for summing first on one index variable; everything else we've done has been derived from those four basic operations on sums.

The distributive law (2.15) can be formulated more precisely as follows: If $\sum_{k\in K} a_k$ converges absolutely to A and if c is any complex number, then $\sum_{k\in K} ca_k$ converges absolutely to cA. We can prove this by breaking the sum into real and imaginary, positive and negative parts as above, and by proving the special case in which $c > 0$ and each term a_k is nonnegative. The proof in this special case works because $\sum_{k\in F} ca_k = c\sum_{k\in F} a_k$ for all finite sets F; the latter fact follows by induction on the size of F.

The associative law (2.16) can be stated as follows: If $\sum_{k\in K} a_k$ and $\sum_{k\in K} b_k$ converge absolutely to A and B, respectively, then $\sum_{k\in K}(a_k + b_k)$ converges absolutely to $A + B$. This turns out to be a special case of a more general theorem that we will prove shortly.

The commutative law (2.17) doesn't really need to be proved, because we have shown in the discussion following (2.35) how to derive it as a special case of a general rule for interchanging the order of summation.

The main result we need to prove is the fundamental principle of multiple sums: *Absolutely convergent sums over two or more indices can always be summed first with respect to any one of those indices.* Formally, we shall prove that if J and the elements of $\{K_j \mid j \in J\}$ are any sets of indices such that

Best to skim this page the first time you get here.
— Your friendly TA

$$\sum_{\substack{j\in J\\ k\in K_j}} a_{j,k} \quad \text{converges absolutely to } A\,,$$

then there exist complex numbers A_j for each $j \in J$ such that

$$\sum_{k\in K_j} a_{j,k} \quad \text{converges absolutely to } A_j, \quad \text{and}$$

$$\sum_{j\in J} A_j \quad \text{converges absolutely to } A\,.$$

It suffices to prove this assertion when all terms are nonnegative, because we can prove the general case by breaking everything into real and imaginary, positive and negative parts as before. Let's assume therefore that $a_{j,k} \geqslant 0$ for all pairs $(j,k) \in M$, where M is the master index set $\{(j,k) \mid j \in J, k \in K_j\}$.

We are given that $\sum_{(j,k)\in M} a_{j,k}$ is finite, namely that

$$\sum_{(j,k)\in F} a_{j,k} \leqslant A$$

for all finite subsets $F \subseteq M$, and that A is the least such upper bound. If j is any element of J, each sum of the form $\sum_{k\in F_j} a_{j,k}$ where F_j is a finite subset of K_j is bounded above by A. Hence these finite sums have a least upper bound $A_j \geqslant 0$, and $\sum_{k\in K_j} a_{j,k} = A_j$ by definition.

We still need to prove that A is the least upper bound of $\sum_{j\in G} A_j$, for all finite subsets $G \subseteq J$. Suppose that G is a finite subset of J with $\sum_{j\in G} a_{j,k} = A' > A$. We can find finite subsets $F_j \subseteq K_j$ such that $\sum_{k\in F_j} a_{j,k} > (A/A')A_j$ for each $j \in G$ with $A_j > 0$. There is at least one such j. But then $\sum_{j\in G, k\in F_j} a_{j,k} > (A/A')\sum_{j\in G} A_j = A$, contradicting the fact that we have

$\sum_{(j,k)\in F} a_{j,k} \leqslant A$ for all finite subsets $F \subseteq M$. Hence $\sum_{j\in G} A_j \leqslant A$, for all finite subsets $G \subseteq J$.

Finally, let A' be any real number less than A. Our proof will be complete if we can find a finite set $G \subseteq J$ such that $\sum_{j\in G} A_j > A'$. We know that there's a finite set $F \subseteq M$ such that $\sum_{(j,k)\in F} a_{j,k} > A'$; let G be the set of j's in this F, and let $F_j = \{k \mid (j,k) \in F\}$. Then $\sum_{j\in G} A_j \geqslant \sum_{j\in G} \sum_{k\in F_j} a_{j,k} = \sum_{(j,k)\in F} a_{j,k} > A'$; QED.

OK, we're now legitimate! Everything we've been doing with infinite sums is justified, as long as there's a finite bound on all finite sums of the absolute values of the terms. Since the doubly infinite sum (2.58) gave us two different answers when we evaluated it in two different ways, its positive terms $1 + \frac{1}{2} + \frac{1}{3} + \cdots$ must diverge to ∞; otherwise we would have gotten the same answer no matter how we grouped the terms.

So why have I been hearing a lot lately about "harmonic convergence"?

Exercises

Warmups

1 What does the notation

$$\sum_{k=4}^{0} q_k$$

mean?

2 Simplify the expression $x \cdot \big([x>0] - [x<0]\big)$.

3 Demonstrate your understanding of $\sum$-notation by writing out the sums

$$\sum_{0\leqslant k\leqslant 5} a_k \qquad \text{and} \qquad \sum_{0\leqslant k^2\leqslant 5} a_{k^2}$$

in full. (Watch out — the second sum is a bit tricky.)

4 Express the triple sum

$$\sum_{1\leqslant i<j<k\leqslant 4} a_{ijk}$$

as a three-fold summation (with three $\sum$'s),

a summing first on k, then j, then i;

b summing first on i, then j, then k.

Also write your triple sums out in full without the $\sum$-notation, using parentheses to show what is being added together first.

5 What's wrong with the following derivation?

$$\Bigl(\sum_{j=1}^{n} a_j\Bigr)\Bigl(\sum_{k=1}^{n} \frac{1}{a_k}\Bigr) = \sum_{j=1}^{n}\sum_{k=1}^{n} \frac{a_j}{a_k} = \sum_{k=1}^{n}\sum_{k=1}^{n} \frac{a_k}{a_k} = \sum_{k=1}^{n} n = n^2 .$$

6 What is the value of $\sum_k [1 \leqslant j \leqslant k \leqslant n]$, as a function of j and n?

Yield to the rising power.

7 Let $\nabla f(x) = f(x) - f(x-1)$. What is $\nabla(x^{\overline{m}})$?

8 What is the value of $0^{\underline{m}}$, when m is a given integer?

9 What is the law of exponents for rising factorial powers, analogous to (2.52)? Use this to define $x^{\overline{-n}}$.

10 The text derives the following formula for the difference of a product:

$$\Delta(uv) = u\,\Delta v + Ev\,\Delta u .$$

How can this formula be correct, when the left-hand side is symmetric with respect to u and v but the right-hand side is not?

Basics

11 The general rule (2.56) for summation by parts is equivalent to

$$\sum_{0\leqslant k<n} (a_{k+1} - a_k)b_k = a_n b_n - a_0 b_0 - \sum_{0\leqslant k<n} a_{k+1}(b_{k+1} - b_k), \quad \text{for } n \geqslant 0.$$

Prove this formula directly by using the distributive, associative, and commutative laws.

12 Show that the function $p(k) = k + (-1)^k c$ is a permutation of the set of all integers, whenever c is an integer.

13 Use the repertoire method to find a closed form for $\sum_{k=0}^{n} (-1)^k k^2$.

14 Evaluate $\sum_{k=1}^{n} k2^k$ by rewriting it as the multiple sum $\sum_{1\leqslant j\leqslant k\leqslant n} 2^k$.

15 Evaluate $⧉_n = \sum_{k=1}^{n} k^3$ by the text's Method 5 as follows: First write $⧉_n + \Box_n = 2\sum_{1\leqslant j\leqslant k\leqslant n} jk$; then apply (2.33).

16 Prove that $x^{\underline{m}}/(x-n)^{\underline{m}} = x^{\underline{n}}/(x-m)^{\underline{n}}$, unless one of the denominators is zero.

17 Show that the following formulas can be used to convert between rising and falling factorial powers, for all integers m:

$$\begin{aligned} x^{\overline{m}} &= (-1)^m(-x)^{\underline{m}} = (x+m-1)^{\underline{m}} = 1/(x-1)^{\underline{-m}};\\ x^{\underline{m}} &= (-1)^m(-x)^{\overline{m}} = (x-m+1)^{\overline{m}} = 1/(x+1)^{\overline{-m}}. \end{aligned}$$

(The answer to exercise 9 defines $x^{\overline{-m}}$.)

18 Let $\Re z$ and $\Im z$ be the real and imaginary parts of the complex number z. The absolute value $|z|$ is $\sqrt{(\Re z)^2 + (\Im z)^2}$. A sum $\sum_{k \in K} a_k$ of complex terms a_k is said to converge absolutely when the real-valued sums $\sum_{k \in K} \Re a_k$ and $\sum_{k \in K} \Im a_k$ both converge absolutely. Prove that $\sum_{k \in K} a_k$ converges absolutely if and only if there is a bounding constant B such that $\sum_{k \in F} |a_k| \leqslant B$ for all finite subsets $F \subseteq K$.

Homework exercises

19 Use a summation factor to solve the recurrence

$$\begin{aligned} T_0 &= 5; \\ 2T_n &= nT_{n-1} + 3 \cdot n!, \qquad \text{for } n > 0. \end{aligned}$$

20 Try to evaluate $\sum_{k=0}^{n} kH_k$ by the perturbation method, but deduce the value of $\sum_{k=0}^{n} H_k$ instead.

21 Evaluate the sums $S_n = \sum_{k=0}^{n}(-1)^{n-k}$, $T_n = \sum_{k=0}^{n}(-1)^{n-k}k$, and $U_n = \sum_{k=0}^{n}(-1)^{n-k}k^2$ by the perturbation method, assuming that $n \geqslant 0$.

22 Prove *Lagrange's identity* (without using induction):

It's hard to prove the identity of somebody who's been dead for 175 years.

$$\sum_{1 \leqslant j < k \leqslant n} (a_j b_k - a_k b_j)^2 = \Bigl(\sum_{k=1}^{n} a_k^2\Bigr)\Bigl(\sum_{k=1}^{n} b_k^2\Bigr) - \Bigl(\sum_{k=1}^{n} a_k b_k\Bigr)^2 .$$

This, incidentally, implies *Cauchy's inequality*,

$$\Bigl(\sum_{k=1}^{n} a_k b_k\Bigr)^2 \leqslant \Bigl(\sum_{k=1}^{n} a_k^2\Bigr)\Bigl(\sum_{k=1}^{n} b_k^2\Bigr).$$

23 Evaluate the sum $\sum_{k=1}^{n}(2k+1)/\bigl(k(k+1)\bigr)$ in two ways:

a Replace $1/k(k+1)$ by the "partial fractions" $1/k - 1/(k+1)$.

b Sum by parts.

24 What is $\sum_{0 \leqslant k < n} H_k/(k+1)(k+2)$? *Hint:* Generalize the derivation of (2.57).

25 The notation $\prod_{k \in K} a_k$ means the product of the numbers a_k for all $k \in K$. Assume for simplicity that $a_k \neq 1$ for only finitely many k; hence infinite products need not be defined. What laws does this $\prod$-notation satisfy, analogous to the distributive, associative, and commutative laws that hold for $\sum$?

This notation was introduced by Jacobi in 1829 [162].

26 Express the double product $\prod_{1 \leqslant j \leqslant k \leqslant n} a_j a_k$ in terms of the single product $\prod_{k=1}^{n} a_k$ by manipulating $\prod$-notation. (This exercise gives us a product analog of the upper-triangle identity (2.33).)

27 Compute $\Delta(c^{\underline{x}})$, and use it to deduce the value of $\sum_{k=1}^{n}(-2)^{\underline{k}}/k$.

28 At what point does the following derivation go astray?

$$\begin{aligned}
1 = \sum_{k\geqslant 1}\frac{1}{k(k+1)} &= \sum_{k\geqslant 1}\left(\frac{k}{k+1}-\frac{k-1}{k}\right)\\
&= \sum_{k\geqslant 1}\sum_{j\geqslant 1}\left(\frac{k}{j}[j=k+1]-\frac{j}{k}[j=k-1]\right)\\
&= \sum_{j\geqslant 1}\sum_{k\geqslant 1}\left(\frac{k}{j}[j=k+1]-\frac{j}{k}[j=k-1]\right)\\
&= \sum_{j\geqslant 1}\sum_{k\geqslant 1}\left(\frac{k}{j}[k=j-1]-\frac{j}{k}[k=j+1]\right)\\
&= \sum_{j\geqslant 1}\left(\frac{j-1}{j}-\frac{j}{j+1}\right) = \sum_{j\geqslant 1}\frac{-1}{j(j+1)} = -1\,.
\end{aligned}$$

Exam problems

29 Evaluate the sum $\sum_{k\geqslant 1}(-1)^k k/(4k^2-1)$.

30 Cribbage players have long been aware that $15 = 7+8 = 4+5+6 = 1+2+3+4+5$. Find the number of ways to represent 1050 as a sum of consecutive positive integers. (The trivial representation '1050' by itself counts as one way; thus there are four, not three, ways to represent 15 as a sum of consecutive positive integers. Incidentally, a knowledge of cribbage rules is of no use in this problem.)

31 Riemann's zeta function $\zeta(k)$ is defined to be the infinite sum

$$1+\frac{1}{2^k}+\frac{1}{3^k}+\cdots = \sum_{j\geqslant 1}\frac{1}{j^k}\,.$$

Prove that $\sum_{k\geqslant 2}(\zeta(k)-1) = 1$. What is the value of $\sum_{k\geqslant 1}(\zeta(2k)-1)$?

32 Let $a \dot{-} b = \max(0, a-b)$. Prove that

$$\sum_{k\geqslant 0}\min(k, x\dot{-}k) = \sum_{k\geqslant 0}(x\dot{-}(2k+1))$$

for all real $x \geqslant 0$.

Bonus problems

33 Let $\bigwedge_{k\in K} a_k$ denote the minimum of the numbers a_k (or their greatest lower bound, if K is infinite), assuming that each a_k is either real or $\pm\infty$. What laws are valid for $\bigwedge$-notation, analogous to those that work for $\sum$ and $\prod$? (See exercise 25.)

The laws of the jungle.

34 Prove that if the sum $\sum_{k \in K} a_k$ is undefined according to (2.59), then it is extremely flaky in the following sense: If A^- and A^+ are any given real numbers, it's possible to find a sequence of finite subsets $F_1 \subset F_2 \subset F_3 \subset \cdots$ of K such that

$$\sum_{k \in F_n} a_k \leqslant A^-, \text{ when } n \text{ is odd}; \qquad \sum_{k \in F_n} a_k \geqslant A^+, \text{ when } n \text{ is even}.$$

35 Prove Goldbach's theorem

$$1 = \frac{1}{3} + \frac{1}{7} + \frac{1}{8} + \frac{1}{15} + \frac{1}{24} + \frac{1}{26} + \frac{1}{31} + \frac{1}{35} + \cdots = \sum_{k \in P} \frac{1}{k-1},$$

where P is the set of "perfect powers" defined recursively as follows:

$$P = \{ m^n \mid m \geqslant 2, n \geqslant 2, m \notin P \}.$$

Perfect power corrupts perfectly.

36 Solomon Golomb's "self-describing sequence" $\langle f(1), f(2), f(3), \ldots \rangle$ is the only nondecreasing sequence of positive integers with the property that it contains exactly $f(k)$ occurrences of k for each k. A few moments' thought reveals that the sequence must begin as follows:

n	1	2	3	4	5	6	7	8	9	10	11	12
$f(n)$	1	2	2	3	3	4	4	4	5	5	5	6

Let $g(n)$ be the largest integer m such that $f(m) = n$. Show that

a $g(n) = \sum_{k=1}^{n} f(k)$.

b $g(g(n)) = \sum_{k=1}^{n} kf(k)$.

c $g(g(g(n))) = \frac{1}{2}ng(n)\bigl(g(n)+1\bigr) - \frac{1}{2}\sum_{k=1}^{n-1} g(k)\bigl(g(k)+1\bigr)$.

Research problem

37 Will all the $1/k$ by $1/(k+1)$ rectangles, for $k \geqslant 1$, fit together inside a 1 by 1 square? (Recall that their areas sum to 1.)

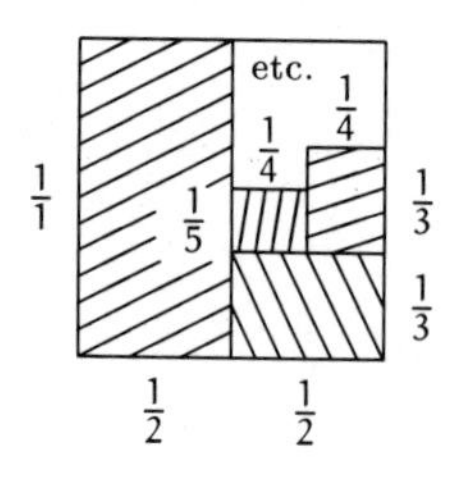

3

Integer Functions

WHOLE NUMBERS constitute the backbone of discrete mathematics, and we often need to convert from fractions or arbitrary real numbers to integers. Our goal in this chapter is to gain familiarity and fluency with such conversions and to learn some of their remarkable properties.

3.1 FLOORS AND CEILINGS

We start by covering the floor (greatest integer) and ceiling (least integer) functions, which are defined for all real x as follows:

$$\begin{aligned} \lfloor x \rfloor &= \text{the greatest integer less than or equal to } x\,; \\ \lceil x \rceil &= \text{the least integer greater than or equal to } x\,. \end{aligned} \tag{3.1}$$

Kenneth E. Iverson introduced this notation, as well as the names "floor" and "ceiling," early in the 1960s [161, page 12]. He found that typesetters could handle the symbols by shaving the tops and bottoms off of '[' and ']'. His notation has become sufficiently popular that floor and ceiling brackets can now be used in a technical paper without an explanation of what they mean. Until recently, people had most often been writing '$[x]$' for the greatest integer $\leqslant x$, without a good equivalent for the least integer function. Some authors had even tried to use '$]x[$'—with a predictable lack of success.

)Ouch.(

Besides variations in notation, there are variations in the functions themselves. For example, some pocket calculators have an INT function, defined as $\lfloor x \rfloor$ when x is positive and $\lceil x \rceil$ when x is negative. The designers of these calculators probably wanted their INT function to satisfy the identity $\text{INT}(-x) = -\text{INT}(x)$. But we'll stick to our floor and ceiling functions, because they have even nicer properties than this.

One good way to become familiar with the floor and ceiling functions is to understand their graphs, which form staircase-like patterns above and below the line $f(x) = x$:

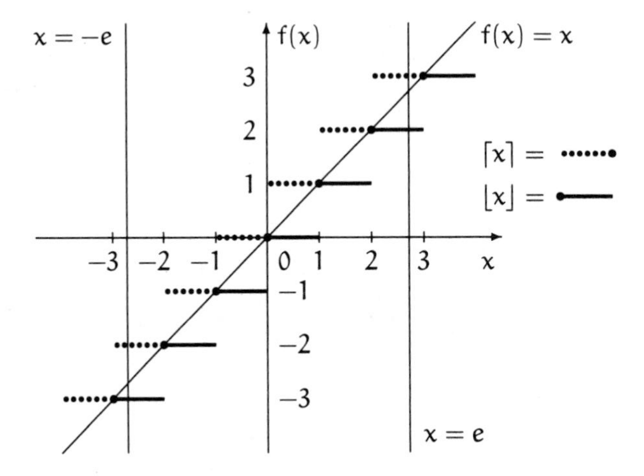

We see from the graph that, for example,

$$\lfloor e \rfloor = 2\,, \qquad \lfloor -e \rfloor = -3\,,$$
$$\lceil e \rceil = 3\,, \qquad \lceil -e \rceil = -2\,,$$

since $e = 2.71828\ldots$.

By staring at this illustration we can observe several facts about floors and ceilings. First, since the floor function lies on or below the diagonal line $f(x) = x$, we have $\lfloor x \rfloor \leqslant x$; similarly $\lceil x \rceil \geqslant x$. (This, of course, is quite obvious from the definition.) The two functions are equal precisely at the integer points:

$$\lfloor x \rfloor = x \quad\iff\quad x \text{ is an integer} \quad\iff\quad \lceil x \rceil = x\,.$$

(We use the notation '$\iff$' to mean "if and only if.") Furthermore, when they differ the ceiling is exactly 1 higher than the floor:

$$\lceil x \rceil - \lfloor x \rfloor = [x \text{ is not an integer}]\,. \tag{3.2}$$

Cute.
By Iverson's parenthesis convention, this is a complete equation.

If we shift the diagonal line down one unit, it lies completely below the floor function, so $x - 1 < \lfloor x \rfloor$; similarly $x + 1 > \lceil x \rceil$. Combining these observations gives us

$$x - 1 < \lfloor x \rfloor \leqslant x \leqslant \lceil x \rceil < x + 1\,. \tag{3.3}$$

Finally, the functions are reflections of each other about both axes:

$$\lfloor -x \rfloor = -\lceil x \rceil\,; \qquad \lceil -x \rceil = -\lfloor x \rfloor\,. \tag{3.4}$$

Thus each is easily expressible in terms of the other. This fact helps to explain why the ceiling function once had no notation of its own. But we see ceilings often enough to warrant giving them special symbols, just as we have adopted special notations for rising powers as well as falling powers. Mathematicians have long had both sine and cosine, tangent and cotangent, secant and cosecant, max and min; now we also have both floor and ceiling.

Next week we're getting walls.

To actually prove properties about the floor and ceiling functions, rather than just to observe such facts graphically, the following four rules are especially useful:

$$\begin{aligned} \lfloor x \rfloor = n \quad &\iff \quad n \leqslant x < n+1\,, &&\text{(a)}\\ \lfloor x \rfloor = n \quad &\iff \quad x-1 < n \leqslant x\,, &&\text{(b)}\\ \lceil x \rceil = n \quad &\iff \quad n-1 < x \leqslant n\,, &&\text{(c)}\\ \lceil x \rceil = n \quad &\iff \quad x \leqslant n < x+1\,. &&\text{(d)} \end{aligned} \qquad (3.5)$$

(We assume in all four cases that n is an integer and that x is real.) Rules (a) and (c) are immediate consequences of definition (3.1); rules (b) and (d) are the same but with the inequalities rearranged so that n is in the middle.

It's possible to move an integer term in or out of a floor (or ceiling):

$$\lfloor x+n \rfloor \;=\; \lfloor x \rfloor + n\,, \qquad \text{integer } n. \qquad (3.6)$$

(Because rule (3.5(a)) says that this assertion is equivalent to the inequalities $\lfloor x \rfloor + n \leqslant x+n < \lfloor x \rfloor + n + 1$.) But similar operations, like moving out a constant factor, cannot be done in general. For example, we have $\lfloor nx \rfloor \neq n\lfloor x \rfloor$ when $n = 2$ and $x = 1/2$. This means that floor and ceiling brackets are comparatively inflexible. We are usually happy if we can get rid of them or if we can prove anything at all when they are present.

It turns out that there are many situations in which floor and ceiling brackets are redundant, so that we can insert or delete them at will. For example, any inequality between a real and an integer is equivalent to a floor or ceiling inequality between integers:

$$\begin{aligned} x < n \quad &\iff \quad \lfloor x \rfloor < n\,, &&\text{(a)}\\ n < x \quad &\iff \quad n < \lceil x \rceil\,, &&\text{(b)}\\ x \leqslant n \quad &\iff \quad \lceil x \rceil \leqslant n\,, &&\text{(c)}\\ n \leqslant x \quad &\iff \quad n \leqslant \lfloor x \rfloor\,. &&\text{(d)} \end{aligned} \qquad (3.7)$$

These rules are easily proved. For example, if $x < n$ then surely $\lfloor x \rfloor < n$, since $\lfloor x \rfloor \leqslant x$. Conversely, if $\lfloor x \rfloor < n$ then we must have $x < n$, since $x < \lfloor x \rfloor + 1$ and $\lfloor x \rfloor + 1 \leqslant n$.

It would be nice if the four rules in (3.7) were as easy to remember as they are to prove. Each inequality without floor or ceiling corresponds to the same inequality with floor or with ceiling; but we need to think twice before deciding which of the two is appropriate.

The difference between x and $\lfloor x \rfloor$ is called the *fractional part* of x, and it arises often enough in applications to deserve its own notation:

$$\{x\} \;=\; x - \lfloor x \rfloor \,. \tag{3.8}$$

Hmmm. We'd better not write $\{x\}$ for the fractional part when it could be confused with the set containing x as its only element.

We sometimes call $\lfloor x \rfloor$ the *integer part* of x, since $x = \lfloor x \rfloor + \{x\}$. If a real number x can be written in the form $x = n + \theta$, where n is an integer and $0 \leqslant \theta < 1$, we can conclude by (3.5(a)) that $n = \lfloor x \rfloor$ and $\theta = \{x\}$.

Identity (3.6) doesn't hold if n is an arbitrary real. But we can deduce that there are only two possibilities for $\lfloor x + y \rfloor$ in general: If we write $x = \lfloor x \rfloor + \{x\}$ and $y = \lfloor y \rfloor + \{y\}$, then we have $\lfloor x + y \rfloor = \lfloor x \rfloor + \lfloor y \rfloor + \lfloor \{x\} + \{y\} \rfloor$. And since $0 \leqslant \{x\} + \{y\} < 2$, we find that sometimes $\lfloor x + y \rfloor$ is $\lfloor x \rfloor + \lfloor y \rfloor$, otherwise it's $\lfloor x \rfloor + \lfloor y \rfloor + 1$.

The second case occurs if and only if there's a "carry" at the position of the decimal point, when the fractional parts $\{x\}$ and $\{y\}$ are added together.

3.2 FLOOR/CEILING APPLICATIONS

We've now seen the basic tools for handling floors and ceilings. Let's put them to use, starting with an easy problem: What's $\lceil \lg 35 \rceil$? (We use 'lg' to denote the base-2 logarithm.) Well, since $2^5 < 35 \leqslant 2^6$, we can take logs to get $5 < \lg 35 \leqslant 6$; so (3.5(c)) tells us that $\lceil \lg 35 \rceil = 6$.

Note that the number 35 is six bits long when written in radix 2 notation: $35 = (100011)_2$. Is it always true that $\lceil \lg n \rceil$ is the length of n written in binary? Not quite. We also need six bits to write $32 = (100000)_2$. So $\lceil \lg n \rceil$ is the wrong answer to the problem. (It fails only when n is a power of 2, but that's infinitely many failures.) We can find a correct answer by realizing that it takes m bits to write each number n such that $2^{m-1} \leqslant n < 2^m$; thus (3.5(a)) tells us that $m - 1 = \lfloor \lg n \rfloor$, so $m = \lfloor \lg n \rfloor + 1$. That is, we need $\lfloor \lg n \rfloor + 1$ bits to express n in binary, for all $n > 0$. Alternatively, a similar derivation yields the answer $\lceil \lg(n+1) \rceil$; this formula holds for $n = 0$ as well, if we're willing to say that it takes zero bits to write $n = 0$ in binary.

Let's look next at expressions with several floors or ceilings. What is $\lceil \lfloor x \rfloor \rceil$? Easy — since $\lfloor x \rfloor$ is an integer, $\lceil \lfloor x \rfloor \rceil$ is just $\lfloor x \rfloor$. So is any other expression with an innermost $\lfloor x \rfloor$ surrounded by any number of floors or ceilings.

Here's a tougher problem: Prove or disprove the assertion

$$\bigl\lfloor \sqrt{\lfloor x \rfloor} \bigr\rfloor \;=\; \lfloor \sqrt{x} \rfloor \,, \qquad \text{real } x \geqslant 0. \tag{3.9}$$

Equality obviously holds when x is an integer, because $x = \lfloor x \rfloor$. And there's equality in the special cases $\pi = 3.14159\ldots$, $e = 2.71828\ldots$, and $\phi = (1+\sqrt{5})/2 = 1.61803\ldots$, because we get $1 = 1$. Our failure to find a counterexample suggests that equality holds in general, so let's try to prove it.

(Of course π, e, and ϕ are the obvious first real numbers to try, aren't they?)

Incidentally, when we're faced with a "prove or disprove," we're usually better off trying first to disprove with a counterexample, for two reasons: A disproof is potentially easier (we need just one counterexample); and nit-picking arouses our creative juices. Even if the given assertion is true, our search for a counterexample often leads us to a proof, as soon as we see why a counterexample is impossible. Besides, it's healthy to be skeptical.

Skepticism is healthy only to a limited extent. Being skeptical about proofs and programs (particularly your own) will probably keep your grades healthy and your job fairly secure. But applying that much skepticism will probably also keep you shut away working all the time, instead of letting you get out for exercise and relaxation. Too much skepticism is an open invitation to the state of rigor mortis, where you become so worried about being correct and rigorous that you never get anything finished.
—A skeptic

If we try to prove that $\lfloor\sqrt{\lfloor x\rfloor}\rfloor = \lfloor\sqrt{x}\rfloor$ with the help of calculus, we might start by decomposing x into its integer and fractional parts $\lfloor x\rfloor + \{x\} = n + \theta$ and then expanding the square root using the binomial theorem: $(n+\theta)^{1/2} = n^{1/2} + n^{-1/2}\theta/2 - n^{-3/2}\theta^2/8 + \cdots$. But this approach gets pretty messy.

It's much easier to use the tools we've developed. Here's a possible strategy: Somehow strip off the outer floor and square root of $\lfloor\sqrt{\lfloor x\rfloor}\rfloor$, then remove the inner floor, then add back the outer stuff to get $\lfloor\sqrt{x}\rfloor$. OK. We let $m = \lfloor\sqrt{\lfloor x\rfloor}\rfloor$ and invoke (3.5(a)), giving $m \leqslant \sqrt{\lfloor x\rfloor} < m+1$. That removes the outer floor bracket without losing any information. Squaring, since all three expressions are nonnegative, we have $m^2 \leqslant \lfloor x\rfloor < (m+1)^2$. That gets rid of the square root. Next we remove the floor, using (3.7(d)) for the left inequality and (3.7(a)) for the right: $m^2 \leqslant x < (m+1)^2$. It's now a simple matter to retrace our steps, taking square roots to get $m \leqslant \sqrt{x} < m+1$ and invoking (3.5(a)) to get $m = \lfloor\sqrt{x}\rfloor$. Thus $\lfloor\sqrt{\lfloor x\rfloor}\rfloor = m = \lfloor\sqrt{x}\rfloor$; the assertion is true. Similarly, we can prove that

$$\lceil\sqrt{\lceil x\rceil}\rceil = \lceil\sqrt{x}\rceil, \qquad \text{real } x \geqslant 0.$$

The proof we just found doesn't rely heavily on the properties of square roots. A closer look shows that we can generalize the ideas and prove much more: Let $f(x)$ be any continuous, monotonically increasing function with the property that

$$f(x) = \text{integer} \qquad \Longrightarrow \qquad x = \text{integer}.$$

(The symbol '$\Longrightarrow$' means "implies.") Then we have

(This observation was made by Robert J. McEliece when he was an undergrad.)

$$\lfloor f(x)\rfloor = \lfloor f(\lfloor x\rfloor)\rfloor \qquad \text{and} \qquad \lceil f(x)\rceil = \lceil f(\lceil x\rceil)\rceil, \tag{3.10}$$

whenever $f(x)$, $f(\lfloor x\rfloor)$, and $f(\lceil x\rceil)$ are defined. Let's prove this general property for ceilings, since we did floors earlier and since the proof for floors is almost the same. If $x = \lceil x\rceil$, there's nothing to prove. Otherwise $x < \lceil x\rceil$, and $f(x) < f(\lceil x\rceil)$ since f is increasing. Hence $\lceil f(x)\rceil \leqslant \lceil f(\lceil x\rceil)\rceil$, since $\lceil\,\rceil$ is nondecreasing. If $\lceil f(x)\rceil < \lceil f(\lceil x\rceil)\rceil$, there must be a number y such that $x \leqslant y < \lceil x\rceil$ and $f(y) = \lceil f(x)\rceil$, since f is continuous. This y is an integer, because of f's special property. But there cannot be an integer strictly between x and $\lceil x\rceil$. This contradiction implies that we must have $\lceil f(x)\rceil = \lceil f(\lceil x\rceil)\rceil$.

An important special case of this theorem is worth noting explicitly:

$$\left\lfloor \frac{x+m}{n} \right\rfloor = \left\lfloor \frac{\lfloor x \rfloor + m}{n} \right\rfloor \quad \text{and} \quad \left\lceil \frac{x+m}{n} \right\rceil = \left\lceil \frac{\lceil x \rceil + m}{n} \right\rceil , \qquad (3.11)$$

if m and n are integers and the denominator n is positive. For example, let $m = 0$; we have $\lfloor \lfloor \lfloor x/10 \rfloor /10 \rfloor /10 \rfloor = \lfloor x/1000 \rfloor$. Dividing thrice by 10 and throwing off digits is the same as dividing by 1000 and tossing the remainder.

Let's try now to prove or disprove another statement:

$$\left\lceil \sqrt{\lfloor x \rfloor} \right\rceil \stackrel{?}{=} \lceil \sqrt{x} \rceil , \qquad \text{real } x \geqslant 0.$$

This works when $x = \pi$ and $x = e$, but it fails when $x = \phi$; so we know that it isn't true in general.

Before going any further, let's digress a minute to discuss different "levels" of questions that can be asked in books about mathematics:

Level 1. Given an explicit object x and an explicit property $P(x)$, *prove* that $P(x)$ is true. For example, "Prove that $\lfloor \pi \rfloor = 3$." Here the problem involves finding a proof of some purported fact.

Level 2. Given an explicit set X and an explicit property $P(x)$, prove that $P(x)$ is true *for all* $x \in X$. For example, "Prove that $\lfloor x \rfloor \leqslant x$ for all real x." Again the problem involves finding a proof, but the proof this time must be general. We're doing algebra, not just arithmetic.

Level 3. Given an explicit set X and an explicit property $P(x)$, prove *or disprove* that $P(x)$ is true for all $x \in X$. For example, "Prove or disprove that $\left\lceil \sqrt{\lfloor x \rfloor} \right\rceil = \lceil \sqrt{x} \rceil$ for all real $x \geqslant 0$." Here there's an additional level of uncertainty; the outcome might go either way. This is closer to the real situation a mathematician constantly faces: Assertions that get into books tend to be true, but new things have to be looked at with a jaundiced eye. If the statement is false, our job is to find a counterexample. If the statement is true, we must find a proof as in level 2.

In my other texts "prove or disprove" seems to mean the same as "prove," about 99.44% of the time; but not in this book.

Level 4. Given an explicit set X and an explicit property $P(x)$, find a *necessary and sufficient condition* $Q(x)$ that $P(x)$ is true. For example, "Find a necessary and sufficient condition that $\lfloor x \rfloor \geqslant \lceil x \rceil$." The problem is to find Q such that $P(x) \iff Q(x)$. Of course, there's always a trivial answer; we can take $Q(x) = P(x)$. But the implied requirement is to find a condition that's as simple as possible. Creativity is required to discover a simple condition that will work. (For example, in this case, "$\lfloor x \rfloor \geqslant \lceil x \rceil \iff x$ is an integer.") The extra element of discovery needed to find $Q(x)$ makes this sort of problem more difficult, but it's more typical of what mathematicians must do in the "real world." Finally, of course, a proof must be given that $P(x)$ is true if and only if $Q(x)$ is true.

But no simpler.
—A. Einstein

Level 5. Given an explicit set X, find *an interesting property* $P(x)$ of its elements. Now we're in the scary domain of pure research, where students might think that total chaos reigns. This is real mathematics. Authors of textbooks rarely dare to ask level 5 questions.

End of digression. But let's convert our last question from level 3 to level 4: What is a necessary and sufficient condition that $\lceil\sqrt{\lfloor x\rfloor}\rceil = \lceil\sqrt{x}\rceil$? We have observed that equality holds when $x = 3.142$ but not when $x = 1.618$; further experimentation shows that it fails also when x is between 9 and 10. Oho. Yes. We see that bad cases occur whenever $m^2 < x < m^2+1$, since this gives m on the left and $m+1$ on the right. In all other cases where $\sqrt{x}$ is defined, namely when $x = 0$ or $m^2+1 \leqslant x \leqslant (m+1)^2$, we get equality. The following statement is therefore necessary and sufficient for equality: Either x is an integer or $\sqrt{\lfloor x\rfloor}$ isn't.

Home of the Toledo Mudhens.

For our next problem let's consider a handy new notation, suggested by Lyle Ramshaw, for intervals of the real line: $[\alpha\,..\,\beta]$ denotes the set of real numbers x such that $\alpha \leqslant x \leqslant \beta$. This set is called a *closed interval* because it contains both endpoints α and β. The interval containing neither endpoint, denoted by $(\alpha\,..\,\beta)$, consists of all x such that $\alpha < x < \beta$; this is called an *open interval.* And the intervals $[\alpha\,..\,\beta)$ and $(\alpha\,..\,\beta]$, which contain just one endpoint, are defined similarly and called *half-open.*

(Or, by pessimists, half-closed.)

How many integers are contained in such intervals? The half-open intervals are easier, so we start with them. In fact half-open intervals are almost always nicer than open or closed intervals. For example, they're additive — we can combine the half-open intervals $[\alpha\,..\,\beta)$ and $[\beta\,..\,\gamma)$ to form the half-open interval $[\alpha\,..\,\gamma)$. This wouldn't work with open intervals because the point β would be excluded, and it could cause problems with closed intervals because β would be included twice.

Back to our problem. The answer is easy if α and β are integers: Then $[\alpha\,..\,\beta)$ contains the $\beta-\alpha$ integers $\alpha, \alpha+1, \ldots, \beta-1$, assuming that $\alpha \leqslant \beta$. Similarly $(\alpha\,..\,\beta]$ contains $\beta-\alpha$ integers in such a case. But our problem is harder, because α and β are arbitrary reals. We can convert it to the easier problem, though, since

$$\begin{aligned} \alpha \leqslant n < \beta \quad &\Longleftrightarrow \quad \lceil\alpha\rceil \leqslant n < \lceil\beta\rceil\,, \\ \alpha < n \leqslant \beta \quad &\Longleftrightarrow \quad \lfloor\alpha\rfloor < n \leqslant \lfloor\beta\rfloor\,, \end{aligned}$$

when n is an integer, according to (3.7). The intervals on the right have integer endpoints and contain the same number of integers as those on the left, which have real endpoints. So the interval $[\alpha\,..\,\beta)$ contains exactly $\lceil\beta\rceil - \lceil\alpha\rceil$ integers, and $(\alpha\,..\,\beta]$ contains $\lfloor\beta\rfloor - \lfloor\alpha\rfloor$. This is a case where we actually want to introduce floor or ceiling brackets, instead of getting rid of them.

By the way, there's a mnemonic for remembering which case uses floors and which uses ceilings: Half-open intervals that include the left endpoint but not the right (such as $0 \leqslant \theta < 1$) are slightly more common than those that include the right endpoint but not the left; and floors are slightly more common than ceilings. So by Murphy's Law, the correct rule is the opposite of what we'd expect — ceilings for $[\alpha\,..\,\beta)$ and floors for $(\alpha\,..\,\beta]$.

Just like we can remember the date of Columbus's departure by singing, "In fourteen hundred and ninety-three/ Columbus sailed the deep blue sea."

Similar analyses show that the closed interval $[\alpha\,..\,\beta]$ contains exactly $\lfloor\beta\rfloor-\lceil\alpha\rceil+1$ integers and that the open interval $(\alpha\,..\,\beta)$ contains $\lceil\beta\rceil-\lfloor\alpha\rfloor-1$; but we place the additional restriction $\alpha \neq \beta$ on the latter so that the formula won't ever embarrass us by claiming that an empty interval $(\alpha\,..\,\alpha)$ contains a total of -1 integers. To summarize, we've deduced the following facts:

$$\begin{array}{lcl} \text{interval} & \text{integers contained} & \text{restrictions} \\ [\alpha\,..\,\beta] & \lfloor\beta\rfloor - \lceil\alpha\rceil + 1 & \alpha \leqslant \beta\,, \\ [\alpha\,..\,\beta) & \lceil\beta\rceil - \lceil\alpha\rceil & \alpha \leqslant \beta\,, \\ (\alpha\,..\,\beta] & \lfloor\beta\rfloor - \lfloor\alpha\rfloor & \alpha \leqslant \beta\,, \\ (\alpha\,..\,\beta) & \lceil\beta\rceil - \lfloor\alpha\rfloor - 1 & \alpha < \beta\,. \end{array} \tag{3.12}$$

Now here's a problem we can't refuse. The Concrete Math Club has a casino (open only to purchasers of this book) in which there's a roulette wheel with one thousand slots, numbered 1 to 1000. If the number n that comes up on a spin is divisible by the floor of its cube root, that is, if

$$\lfloor\sqrt[3]{n}\rfloor \backslash n\,,$$

then it's a winner and the house pays us \$5; otherwise it's a loser and we must pay \$1. (The notation $a\backslash b$, read "a divides b," means that b is an exact multiple of a; Chapter 4 investigates this relation carefully.) Can we expect to make money if we play this game?

(A poll of the class at this point showed that 28 students thought it was a bad idea to play, 13 wanted to gamble, and the rest were too confused to answer.)

(So we hit them with the Concrete Math Club.)

We can compute the average winnings — that is, the amount we'll win (or lose) per play — by first counting the number W of winners and the number $L = 1000 - W$ of losers. If each number comes up once during 1000 plays, we win $5W$ dollars and lose L dollars, so the average winnings will be

$$\frac{5W - L}{1000} = \frac{5W - (1000 - W)}{1000} = \frac{6W - 1000}{1000}\,.$$

If there are 167 or more winners, we have the advantage; otherwise the advantage is with the house.

How can we count the number of winners among 1 through 1000? It's not hard to spot a pattern. The numbers from 1 through $2^3 - 1 = 7$ are all winners because $\lfloor\sqrt[3]{n}\rfloor = 1$ for each. Among the numbers $2^3 = 8$ through $3^3 - 1 = 26$, only the even numbers are winners. And among $3^3 = 27$ through $4^3 - 1 = 63$, only those divisible by 3 are. And so on.

The whole setup can be analyzed systematically if we use the summation techniques of Chapter 2, taking advantage of Iverson's convention about logical statements evaluating to 0 or 1:

$$\begin{aligned}
W &= \sum_{n=1}^{1000} [n \text{ is a winner}] \\
&= \sum_{1\leqslant n\leqslant 1000} \big[\lfloor\sqrt[3]{n}\rfloor \backslash n\big] = \sum_{k,n} \big[k=\lfloor\sqrt[3]{n}\rfloor\big][k\backslash n][1\leqslant n\leqslant 1000] \\
&= \sum_{k,m,n} \big[k^3 \leqslant n < (k+1)^3\big][n=km][1\leqslant n\leqslant 1000] \\
&= 1 + \sum_{k,m} \big[k^3 \leqslant km < (k+1)^3\big][1\leqslant k<10] \\
&= 1 + \sum_{k,m} \big[m \in \big[k^2 \,..\, (k+1)^3/k\big)\big][1\leqslant k<10] \\
&= 1 + \sum_{1\leqslant k<10} \big(\lceil k^2+3k+3+1/k\rceil - \lceil k^2\rceil\big) \\
&= 1 + \sum_{1\leqslant k<10} (3k+4) = 1 + \frac{7+31}{2}\cdot 9 = 172\,.
\end{aligned}$$

This derivation merits careful study. Notice that line 6 uses our formula (3.12) for the number of integers in a half-open interval. The only "difficult" maneuver is the decision made between lines 3 and 4 to treat $n = 1000$ as a special case. (The inequality $k^3 \leqslant n < (k+1)^3$ does not combine easily with $1 \leqslant n \leqslant 1000$ when $k = 10$.) In general, boundary conditions tend to be the most critical part of $\sum$-manipulations.

True.

The bottom line says that $W = 172$; hence our formula for average winnings per play reduces to $(6\cdot 172 - 1000)/1000$ dollars, which is 3.2 cents. We can expect to be about \$3.20 richer after making 100 bets of \$1 each. (Of course, the house may have made some numbers more equal than others.)

Where did you say this casino is?

The casino problem we just solved is a dressed-up version of the more mundane question, "How many integers n, where $1 \leqslant n \leqslant 1000$, satisfy the relation $\lfloor\sqrt[3]{n}\rfloor \backslash n$?" Mathematically the two questions are the same. But sometimes it's a good idea to dress up a problem. We get to use more vocabulary (like "winners" and "losers"), which helps us to understand what's going on.

Let's get general. Suppose we change 1000 to 1000000, or to an even larger number, N. (We assume that the casino has connections and can get a bigger wheel.) Now how many winners are there?

The same argument applies, but we need to deal more carefully with the largest value of k, which we can call K for convenience:

$$K = \lfloor\sqrt[3]{N}\rfloor\,.$$

(Previously K was 10.) The total number of winners for general N comes to

$$\begin{aligned} W &= \sum_{1\leqslant k<K} (3k+4) + \sum_m [K^3 \leqslant Km \leqslant N] \\ &= \tfrac{1}{2}(7+3K+1)(K-1) + \sum_m [m \in [K^2\,..\,N/K]] \\ &= \tfrac{3}{2}K^2 + \tfrac{5}{2}K - 4 + \sum_m [m \in [K^2\,..\,N/K]]\,. \end{aligned}$$

We know that the remaining sum is $\lfloor N/K \rfloor - \lceil K^2 \rceil + 1 = \lfloor N/K \rfloor - K^2 + 1$; hence the formula

$$W = \lfloor N/K \rfloor + \tfrac{1}{2}K^2 + \tfrac{5}{2}K - 3\,, \qquad K = \lfloor \sqrt[3]{N} \rfloor \tag{3.13}$$

gives the general answer for a wheel of size N.

The first two terms of this formula are approximately $N^{2/3} + \frac{1}{2}N^{2/3} = \frac{3}{2}N^{2/3}$, and the other terms are much smaller in comparison, when N is large. In Chapter 9 we'll learn how to derive expressions like

$$W = \tfrac{3}{2}N^{2/3} + O(N^{1/3})\,,$$

where $O(N^{1/3})$ stands for a quantity that is no more than a constant times $N^{1/3}$. Whatever the constant is, we know that it's independent of N; so for large N the contribution of the O-term to W will be quite small compared with $\frac{3}{2}N^{2/3}$. For example, the following table shows how close $\frac{3}{2}N^{2/3}$ is to W:

N	$\frac{3}{2}N^{2/3}$	W	% error
1,000	150.0	172	12.791
10,000	696.2	746	6.670
100,000	3231.7	3343	3.331
1,000,000	15000.0	15247	1.620
10,000,000	69623.8	70158	0.761
100,000,000	323165.2	324322	0.357
1,000,000,000	1500000.0	1502496	0.166

It's a pretty good approximation.

Approximate formulas are useful because they're simpler than formulas with floors and ceilings. However, the exact truth is often important, too, especially for the smaller values of N that tend to occur in practice. For example, the casino owner may have falsely assumed that there are only $\frac{3}{2}N^{2/3} = 150$ winners when $N = 1000$ (in which case there would be a 10¢ advantage for the house).

Our last application in this section looks at so-called spectra. We define the *spectrum* of a real number α to be an infinite multiset of integers,

$$\mathrm{Spec}(\alpha) = \{\lfloor\alpha\rfloor, \lfloor 2\alpha\rfloor, \lfloor 3\alpha\rfloor, \ldots\}.$$

(A multiset is like a set but it can have repeated elements.) For example, the spectrum of $1/2$ starts out $\{0, 1, 1, 2, 2, 3, 3, \ldots\}$.

It's easy to prove that no two spectra are equal—that $\alpha \neq \beta$ implies $\mathrm{Spec}(\alpha) \neq \mathrm{Spec}(\beta)$. For, assuming without loss of generality that $\alpha < \beta$, there's a positive integer m such that $m(\beta - \alpha) \geqslant 1$. (In fact, any $m \geqslant \lceil 1/(\beta - \alpha)\rceil$ will do; but we needn't show off our knowledge of floors and ceilings all the time.) Hence $m\beta - m\alpha \geqslant 1$, and $\lfloor m\beta\rfloor > \lfloor m\alpha\rfloor$. Thus $\mathrm{Spec}(\beta)$ has fewer than m elements $\leqslant \lfloor m\alpha\rfloor$, while $\mathrm{Spec}(\alpha)$ has at least m.

... without lots of generality...

Spectra have many beautiful properties. For example, consider the two multisets

$$\begin{aligned}\mathrm{Spec}(\sqrt{2}\,) &= \{1, 2, 4, 5, 7, 8, 9, 11, 12, 14, 15, 16, 18, 19, 21, 22, 24, \ldots\},\\ \mathrm{Spec}(2 + \sqrt{2}\,) &= \{3, 6, 10, 13, 17, 20, 23, 27, 30, 34, 37, 40, 44, 47, 51, \ldots\}.\end{aligned}$$

It's easy to calculate $\mathrm{Spec}(\sqrt{2}\,)$ with a pocket calculator, and the nth element of $\mathrm{Spec}(2+\sqrt{2}\,)$ is just $2n$ more than the nth element of $\mathrm{Spec}(\sqrt{2}\,)$, by (3.6). A closer look shows that these two spectra are also related in a much more surprising way: It seems that any number missing from one is in the other, but that no number is in both! And it's true: The positive integers are the disjoint union of $\mathrm{Spec}(\sqrt{2}\,)$ and $\mathrm{Spec}(2+\sqrt{2}\,)$. We say that these spectra form a *partition* of the positive integers.

"If x be an incommensurable number less than unity, one of the series of quantities m/x, $m/(1-x)$, where m is a whole number, can be found which shall lie between any given consecutive integers, and but one such quantity can be found."
—Rayleigh [245]

To prove this assertion, we will count how many of the elements of $\mathrm{Spec}(\sqrt{2}\,)$ are $\leqslant n$, and how many of the elements of $\mathrm{Spec}(2+\sqrt{2}\,)$ are $\leqslant n$. If the total is n, for each n, these two spectra do indeed partition the integers.

Right, because exactly one of the counts must increase when n increases by 1.

Let α be positive. The number of elements in $\mathrm{Spec}(\alpha)$ that are $\leqslant n$ is

$$\begin{aligned} N(\alpha, n) &= \sum_{k>0} \bigl[\lfloor k\alpha\rfloor \leqslant n\bigr]\\ &= \sum_{k>0} \bigl[\lfloor k\alpha\rfloor < n + 1\bigr]\\ &= \sum_{k>0} [k\alpha < n + 1]\\ &= \sum_{k} [0 < k < (n + 1)/\alpha)\\ &= \lceil (n + 1)/\alpha\rceil - 1\,. \end{aligned} \tag{3.14}$$

This derivation has two special points of interest. First, it uses the law

$$m \leqslant n \quad\iff\quad m < n+1\,, \qquad \text{integers } m \text{ and } n \tag{3.15}$$

to change '$\leqslant$' to '$<$', so that the floor brackets can be removed by (3.7). Also — and this is more subtle — it sums over the range $k > 0$ instead of $k \geqslant 1$, because $(n+1)/\alpha$ might be less than 1 for certain n and α. If we had tried to apply (3.12) to determine the number of integers in $[1\,..\,(n+1)/\alpha)$, rather than the number of integers in $(0\,..\,(n+1)/\alpha)$, we would have gotten the right answer; but our derivation would have been faulty because the conditions of applicability wouldn't have been met.

Good, we have a formula for $N(\alpha, n)$. Now we can test whether or not $\mathrm{Spec}(\sqrt{2}\,)$ and $\mathrm{Spec}(2+\sqrt{2}\,)$ partition the positive integers, by testing whether or not $N(\sqrt{2}, n) + N(2+\sqrt{2}, n) = n$ for all integers $n > 0$, using (3.14):

$$\begin{aligned}
&\left\lceil \frac{n+1}{\sqrt{2}} \right\rceil - 1 + \left\lceil \frac{n+1}{2+\sqrt{2}} \right\rceil - 1 = n \\
&\iff \left\lfloor \frac{n+1}{\sqrt{2}} \right\rfloor + \left\lfloor \frac{n+1}{2+\sqrt{2}} \right\rfloor = n\,, && \text{by (3.2);} \\
&\iff \frac{n+1}{\sqrt{2}} - \left\{ \frac{n+1}{\sqrt{2}} \right\} + \frac{n+1}{2+\sqrt{2}} - \left\{ \frac{n+1}{2+\sqrt{2}} \right\} = n\,, && \text{by (3.8).}
\end{aligned}$$

Everything simplifies now because of the neat identity

$$\frac{1}{\sqrt{2}} + \frac{1}{2+\sqrt{2}} = 1\,;$$

our condition reduces to testing whether or not

$$\left\{ \frac{n+1}{\sqrt{2}} \right\} + \left\{ \frac{n+1}{2+\sqrt{2}} \right\} = 1\,,$$

for all $n > 0$. And we win, because these are the fractional parts of two noninteger numbers that add up to the integer $n+1$. A partition it is.

3.3 FLOOR/CEILING RECURRENCES

Floors and ceilings add an interesting new dimension to the study of recurrence relations. Let's look first at the recurrence

$$\begin{aligned}
K_0 &= 1\,; \\
K_{n+1} &= 1 + \min(2K_{\lfloor n/2 \rfloor}, 3K_{\lfloor n/3 \rfloor})\,, \qquad \text{for } n \geqslant 0.
\end{aligned} \tag{3.16}$$

Thus, for example, K_1 is $1 + \min(2K_0, 3K_0) = 3$; the sequence begins 1, 3, 3, 4, 7, 7, 7, 9, 9, 10, 13, ... One of the authors of this book has modestly decided to call these the Knuth numbers.

Exercise 25 asks for a proof or disproof that $K_n \geqslant n$, for all $n \geqslant 0$. The first few K's just listed do satisfy the inequality, so there's a good chance that it's true in general. Let's try an induction proof: The basis $n = 0$ comes directly from the defining recurrence. For the induction step, we assume that the inequality holds for all values up through some fixed nonnegative n, and we try to show that $K_{n+1} \geqslant n+1$. From the recurrence we know that $K_{n+1} = 1 + \min(2K_{\lfloor n/2 \rfloor}, 3K_{\lfloor n/3 \rfloor})$. The induction hypothesis tells us that $2K_{\lfloor n/2 \rfloor} \geqslant 2\lfloor n/2 \rfloor$ and $3K_{\lfloor n/3 \rfloor} \geqslant 3\lfloor n/3 \rfloor$. However, $2\lfloor n/2 \rfloor$ can be as small as $n-1$, and $3\lfloor n/3 \rfloor$ can be as small as $n-2$. The most we can conclude from our induction hypothesis is that $K_{n+1} \geqslant 1 + (n-2)$; this falls far short of $K_{n+1} \geqslant n+1$.

We now have reason to worry about the truth of $K_n \geqslant n$, so let's try to disprove it. If we can find an n such that either $2K_{\lfloor n/2 \rfloor} < n$ or $3K_{\lfloor n/3 \rfloor} < n$, or in other words such that

$$K_{\lfloor n/2 \rfloor} < n/2 \qquad \text{or} \qquad K_{\lfloor n/3 \rfloor} < n/3\,,$$

we will have $K_{n+1} < n+1$. Can this be possible? We'd better not give the answer away here, because that will spoil exercise 25.

Recurrence relations involving floors and/or ceilings arise often in computer science, because algorithms based on the important technique of "divide and conquer" often reduce a problem of size n to the solution of similar problems of integer sizes that are fractions of n. For example, one way to sort n records, if $n > 1$, is to divide them into two approximately equal parts, one of size $\lceil n/2 \rceil$ and the other of size $\lfloor n/2 \rfloor$. (Notice, incidentally, that

$$n = \lceil n/2 \rceil + \lfloor n/2 \rfloor\,; \tag{3.17}$$

this formula comes in handy rather often.) After each part has been sorted separately (by the same method, applied recursively), we can merge the records into their final order by doing at most $n-1$ further comparisons. Therefore the total number of comparisons performed is at most $f(n)$, where

$$\begin{aligned} f(1) &= 0\,; \\ f(n) &= f(\lceil n/2 \rceil) + f(\lfloor n/2 \rfloor) + n - 1\,, \qquad \text{for } n > 1. \end{aligned} \tag{3.18}$$

A solution to this recurrence appears in exercise 34.

The Josephus problem of Chapter 1 has a similar recurrence, which can be cast in the form

$$\begin{aligned} J(1) &= 1\,; \\ J(n) &= 2J(\lfloor n/2 \rfloor) - (-1)^n\,, \qquad \text{for } n > 1. \end{aligned}$$

We've got more tools to work with than we had in Chapter 1, so let's consider the more authentic Josephus problem in which every third person is eliminated, instead of every second. If we apply the methods that worked in Chapter 1 to this more difficult problem, we wind up with a recurrence like

$$J_3(n) = \left\lceil \tfrac{3}{2} J_3\left(\lfloor \tfrac{2}{3} n \rfloor\right) + a_n \right\rceil \bmod n + 1\,,$$

where 'mod' is a function that we will be studying shortly, and where we have $a_n = -2$, $+1$, or $-\frac{1}{2}$ according as $n \bmod 3 = 0$, 1, or 2. But this recurrence is too horrible to pursue.

There's another approach to the Josephus problem that gives a much better setup. Whenever a person is passed over, we can assign a new number. Thus, 1 and 2 become $n+1$ and $n+2$, then 3 is executed; 4 and 5 become $n+3$ and $n+4$, then 6 is executed; ...; $3k+1$ and $3k+2$ become $n+2k+1$ and $n+2k+2$, then $3k+3$ is executed; ... then $3n$ is executed (or left to survive). For example, when $n = 10$ the numbers are

1	2	3	4	5	6	7	8	9	10
11	12		13	14		15	16		17
18			19	20			21		22
			23	24					25
			26						27
			28						
			29						
			30						

The kth person eliminated ends up with number $3k$. So we can figure out who the survivor is if we can figure out the original number of person number $3n$.

If $N > n$, person number N must have had a previous number, and we can find it as follows: We have $N = n + 2k + 1$ or $N = n + 2k + 2$, hence $k = \lfloor (N - n - 1)/2 \rfloor$; the previous number was $3k+1$ or $3k+2$, respectively. That is, it was $3k + (N - n - 2k) = k + N - n$. Hence we can calculate the survivor's number $J_3(n)$ as follows:

$$N := 3n;$$

$$\textbf{while } N > n \textbf{ do } N := \left\lfloor \frac{N - n - 1}{2} \right\rfloor + N - n;$$

$$J_3(n) := N.$$

This is not a closed form for $J_3(n)$; it's not even a recurrence. But at least it tells us how to calculate the answer reasonably fast, if n is large.

"Not too slow, not too fast."
—L. Armstrong

Fortunately there's a way to simplify this algorithm if we use the variable $D = 3n + 1 - N$ in place of N. (This change in notation corresponds to assigning numbers from $3n$ down to 1, instead of from 1 up to $3n$; it's sort of like a countdown.) Then the complicated assignment to N becomes

$$\begin{aligned} D &:= 3n+1-\left(\left\lfloor\frac{(3n+1-D)-n-1}{2}\right\rfloor + (3n+1-D)-n\right) \\ &= n+D-\left\lfloor\frac{2n-D}{2}\right\rfloor = D-\left\lfloor\frac{-D}{2}\right\rfloor = D+\left\lceil\frac{D}{2}\right\rceil = \left\lceil\tfrac{3}{2}D\right\rceil, \end{aligned}$$

and we can rewrite the algorithm as follows:

$$\begin{aligned} &D := 1\,; \\ &\textbf{while } D \leqslant 2n \textbf{ do } D := \left\lceil\tfrac{3}{2}D\right\rceil; \\ &J_3(n) := 3n+1-D\,. \end{aligned}$$

Aha! This looks much nicer, because n enters the calculation in a very simple way. In fact, we can show by the same reasoning that the survivor $J_q(n)$ when every qth person is eliminated can be calculated as follows:

$$\begin{aligned} &D := 1\,; \\ &\textbf{while } D \leqslant (q-1)n \textbf{ do } D := \left\lceil\tfrac{q}{q-1}D\right\rceil; \\ &J_q(n) := qn+1-D\,. \end{aligned} \tag{3.19}$$

In the case $q = 2$ that we know so well, this makes D grow to 2^{m+1} when $n = 2^m + l$; hence $J_2(n) = 2(2^m + l) + 1 - 2^{m+1} = 2l + 1$. Good.

The recipe in (3.19) computes a sequence of integers that can be defined by the following recurrence:

$$\begin{aligned} D_0^{(q)} &= 1\,; \\ D_n^{(q)} &= \left\lceil\frac{q}{q-1}D_{n-1}^{(q)}\right\rceil \qquad \text{for } n > 0. \end{aligned} \tag{3.20}$$

These numbers don't seem to relate to any familiar functions in a simple way, except when $q = 2$; hence they probably don't have a nice closed form. But if we're willing to accept the sequence $D_n^{(q)}$ as "known," then it's easy to describe the solution to the generalized Josephus problem: The survivor $J_q(n)$ is $qn + 1 - D_k^{(q)}$, where k is as small as possible such that $D_k^{(q)} > (q-1)n$.

"Known" like, say, harmonic numbers.

3.4 'MOD': THE BINARY OPERATION

The quotient of n divided by m is $\lfloor n/m \rfloor$, when m and n are positive integers. It's handy to have a simple notation also for the remainder of this division, and we call it '$n \bmod m$'. The basic formula

$$n = \underbrace{m\lfloor n/m \rfloor}_{\text{quotient}} + \underbrace{n \bmod m}_{\text{remainder}}$$

tells us that we can express $n \bmod m$ as $n - m\lfloor n/m \rfloor$. We can generalize this to negative integers, and in fact to arbitrary real numbers:

$$x \bmod y = x - y\lfloor x/y \rfloor, \qquad \text{for } y \neq 0. \tag{3.21}$$

This defines 'mod' as a binary operation, just as addition and subtraction are binary operations. Mathematicians have used mod this way informally for a long time, taking various quantities mod 10, mod 2π, and so on, but only in the last twenty years has it caught on formally. Old notion, new notation.

Why do they call it 'mod': The Binary Operation? Stay tuned to find out in the next, exciting, chapter!

We can easily grasp the intuitive meaning of $x \bmod y$, when x and y are positive real numbers, if we imagine a circle of circumference y whose points have been assigned real numbers in the interval $[0\,..\,y)$. If we travel a distance x around the circle, starting at 0, we end up at $x \bmod y$. (And the number of times we encounter 0 as we go is $\lfloor x/y \rfloor$.)

When x or y is negative, we need to look at the definition carefully in order to see exactly what it means. Here are some integer-valued examples:

Beware of computer languages that use another definition.

$$\begin{aligned} 5 \bmod 3 &= 5 - 3\lfloor 5/3 \rfloor &&= 2\,; \\ 5 \bmod -3 &= 5 - (-3)\lfloor 5/(-3) \rfloor &&= -1\,; \\ -5 \bmod 3 &= -5 - 3\lfloor -5/3 \rfloor &&= 1\,; \\ -5 \bmod -3 &= -5 - (-3)\lfloor -5/(-3) \rfloor &&= -2\,. \end{aligned}$$

The number after 'mod' is called the *modulus*; nobody has yet decided what to call the number before 'mod'. In applications, the modulus is usually positive, but the definition makes perfect sense when the modulus is negative. In both cases the value of $x \bmod y$ is between 0 and the modulus:

How about calling the other number the modumor?

$$\begin{aligned} 0 &\leqslant x \bmod y < y\,, && \text{for } y > 0; \\ 0 &\geqslant x \bmod y > y\,, && \text{for } y < 0. \end{aligned}$$

What about $y = 0$? Definition (3.21) leaves this case undefined, in order to avoid division by zero, but to be complete we can define

$$x \bmod 0 = x\,. \tag{3.22}$$

This convention preserves the property that $x \bmod y$ always differs from x by a multiple of y. (It might seem more natural to make the function continuous at 0, by defining $x \bmod 0 = \lim_{y \to 0} x \bmod y = 0$. But we'll see in Chapter 4 that this would be much less useful. Continuity is not an important aspect of the mod operation.)

We've already seen one special case of mod in disguise, when we wrote x in terms of its integer and fractional parts, $x = \lfloor x \rfloor + \{x\}$. The fractional part can also be written $x \bmod 1$, because we have

$$x = \lfloor x \rfloor + x \bmod 1 .$$

Notice that parentheses aren't needed in this formula; we take mod to bind more tightly than addition or subtraction.

The floor function has been used to define mod, and the ceiling function hasn't gotten equal time. We could perhaps use the ceiling to define a mod analog like

$$x \operatorname{mumble} y = y\lceil x/y \rceil - x ;$$

There was a time in the 70s when 'mod' was the fashion. Maybe the new mumble function should be called 'punk'?

No—I like 'mumble'.

in our circle analogy this represents the distance the traveler needs to continue, after going a distance x, to get back to the starting point 0. But of course we'd need a better name than 'mumble'. If sufficient applications come along, an appropriate name will probably suggest itself.

The distributive law is mod's most important algebraic property: We have

$$c(x \bmod y) = (cx) \bmod (cy) \tag{3.23}$$

for all real c, x, and y. (Those who like mod to bind less tightly than multiplication may remove the parentheses from the right side here, too.) It's easy to prove this law from definition (3.21), since

$$c(x \bmod y) = c(x - y\lfloor x/y \rfloor) = cx - cy\lfloor cx/cy \rfloor = cx \bmod cy ,$$

if $cy \neq 0$; and the zero-modulus cases are trivially true. Our four examples using ± 5 and ± 3 illustrate this law twice, with $c = -1$. An identity like (3.23) is reassuring, because it gives us reason to believe that 'mod' has not been defined improperly.

The remainder, eh?

In the remainder of this section, we'll consider an application in which 'mod' turns out to be helpful although it doesn't play a central role. The problem arises frequently in a variety of situations: We want to partition n things into m groups as equally as possible.

Suppose, for example, that we have n short lines of text that we'd like to arrange in m columns. For æsthetic reasons, we want the columns to be arranged in decreasing order of length (actually nonincreasing order); and the lengths should be approximately the same—no two columns should differ by more than one line's worth of text. If 37 lines of text are being divided into five columns, we would therefore prefer the arrangement on the right:

8	8	8	8	5
line 1	line 9	line 17	line 25	line 33
line 2	line 10	line 18	line 26	line 34
line 3	line 11	line 19	line 27	line 35
line 4	line 12	line 20	line 28	line 36
line 5	line 13	line 21	line 29	line 37
line 6	line 14	line 22	line 30	
line 7	line 15	line 23	line 31	
line 8	line 16	line 24	line 32	

8	8	7	7	7
line 1	line 9	line 17	line 24	line 31
line 2	line 10	line 18	line 25	line 32
line 3	line 11	line 19	line 26	line 33
line 4	line 12	line 20	line 27	line 34
line 5	line 13	line 21	line 28	line 35
line 6	line 14	line 22	line 29	line 36
line 7	line 15	line 23	line 30	line 37
line 8	line 16			

Furthermore we want to distribute the lines of text columnwise — first deciding how many lines go into the first column and then moving on to the second, the third, and so on — because that's the way people read. Distributing row by row would give us the correct number of lines in each column, but the ordering would be wrong. (We would get something like the arrangement on the right, but column 1 would contain lines 1, 6, 11, ..., 36, instead of lines 1, 2, 3, ..., 8 as desired.)

A row-by-row distribution strategy can't be used, but it does tell us how many lines to put in each column. If n is not a multiple of m, the row-by-row procedure makes it clear that the long columns should each contain $\lceil n/m \rceil$ lines, and the short columns should each contain $\lfloor n/m \rfloor$. There will be exactly $n \bmod m$ long columns (and, as it turns out, there will be exactly n mumble m short ones).

Let's generalize the terminology and talk about 'things' and 'groups' instead of 'lines' and 'columns'. We have just decided that the first group should contain $\lceil n/m \rceil$ things; therefore the following sequential distribution scheme ought to work: To distribute n things into m groups, when $m > 0$, put $\lceil n/m \rceil$ things into one group, then use the same procedure recursively to put the remaining $n' = n - \lceil n/m \rceil$ things into $m' = m - 1$ additional groups.

For example, if $n = 314$ and $m = 6$, the distribution goes like this:

remaining things	remaining groups	$\lceil$things/groups$\rceil$
314	6	53
261	5	53
208	4	52
156	3	52
104	2	52
52	1	52

It works. We get groups of approximately the same size, even though the divisor keeps changing.

Why does it work? In general we can suppose that $n = qm + r$, where $q = \lfloor n/m \rfloor$ and $r = n \bmod m$. The process is simple if $r = 0$: We put $\lceil n/m \rceil = q$ things into the first group and replace n by $n' = n - q$, leaving

$n' = qm'$ things to put into the remaining $m' = m - 1$ groups. And if $r > 0$, we put $\lceil n/m \rceil = q + 1$ things into the first group and replace n by $n' = n - q - 1$, leaving $n' = qm' + r - 1$ things for subsequent groups. The new remainder is $r' = r - 1$, but q stays the same. It follows that there will be r groups with $q + 1$ things, followed by $m - r$ groups with q things.

How many things are in the kth group? We'd like a formula that gives $\lceil n/m \rceil$ when $k \leqslant n \bmod m$, and $\lfloor n/m \rfloor$ otherwise. It's not hard to verify that

$$\left\lceil \frac{n-k+1}{m} \right\rceil$$

has the desired properties, because this reduces to $q + \lceil (r-k+1)/m \rceil$ if we write $n = qm + r$ as in the preceding paragraph; here $q = \lfloor n/m \rfloor$. We have $\lceil (r-k+1)/m \rceil = [k \leqslant r]$, if $1 \leqslant k \leqslant m$ and $0 \leqslant r < m$. Therefore we can write an identity that expresses the partition of n into m as-equal-as-possible parts in nonincreasing order:

$$n = \left\lceil \frac{n}{m} \right\rceil + \left\lceil \frac{n-1}{m} \right\rceil + \cdots + \left\lceil \frac{n-m+1}{m} \right\rceil . \tag{3.24}$$

This identity is valid for all positive integers m, and for all integers n (whether positive, negative, or zero). We have already encountered the case $m = 2$ in (3.17), although we wrote it in a slightly different form, $n = \lceil n/2 \rceil + \lfloor n/2 \rfloor$.

If we had wanted the parts to be in nondecreasing order, with the small groups coming before the larger ones, we could have proceeded in the same way but with $\lfloor n/m \rfloor$ things in the first group. Then we would have derived the corresponding identity

$$n = \left\lfloor \frac{n}{m} \right\rfloor + \left\lfloor \frac{n+1}{m} \right\rfloor + \cdots + \left\lfloor \frac{n+m-1}{m} \right\rfloor . \tag{3.25}$$

It's possible to convert between (3.25) and (3.24) by using either (3.4) or the identity of exercise 12.

Some claim that it's too dangerous to replace anything by an mx.

Now if we replace n in (3.25) by $\lfloor mx \rfloor$, and apply rule (3.11) to remove floors inside of floors, we get an identity that holds for all real x:

$$\lfloor mx \rfloor = \lfloor x \rfloor + \left\lfloor x + \frac{1}{m} \right\rfloor + \cdots + \left\lfloor x + \frac{m-1}{m} \right\rfloor . \tag{3.26}$$

This is rather amazing, because the floor function is an integer approximation of a real value, but the single approximation on the left equals the sum of a bunch of them on the right. If we assume that $\lfloor x \rfloor$ is roughly $x - \frac{1}{2}$ on the average, the left-hand side is roughly $mx - \frac{1}{2}$, while the right-hand side comes to roughly $(x - \frac{1}{2}) + (x - \frac{1}{2} + \frac{1}{m}) + \cdots + (x - \frac{1}{2} + \frac{m-1}{m}) = mx - \frac{1}{2}$; the sum of all these rough approximations turns out to be exact!

3.5 FLOOR/CEILING SUMS

Equation (3.26) demonstrates that it's possible to get a closed form for at least one kind of sum that involves $\lfloor\ \rfloor$. Are there others? Yes. The trick that usually works in such cases is to get rid of the floor or ceiling by introducing a new variable.

For example, let's see if it's possible to do the sum

$$\sum_{0\le k<n} \lfloor\sqrt{k}\rfloor$$

in closed form. One idea is to introduce the variable $m = \lfloor\sqrt{k}\rfloor$; we can do this "mechanically" by proceeding as we did in the roulette problem:

$$\begin{aligned}\sum_{0\le k<n} \lfloor\sqrt{k}\rfloor &= \sum_{k,m\ge 0} m[k<n]\bigl[m=\lfloor\sqrt{k}\rfloor\bigr]\\ &= \sum_{k,m\ge 0} m[k<n]\bigl[m\le\sqrt{k}<m+1\bigr]\\ &= \sum_{k,m\ge 0} m[k<n]\bigl[m^2\le k<(m+1)^2\bigr]\\ &= \sum_{k,m\ge 0} m\bigl[m^2\le k<(m+1)^2\le n\bigr]\\ &\qquad + \sum_{k,m\ge 0} m\bigl[m^2\le k<n<(m+1)^2\bigr].\end{aligned}$$

Once again the boundary conditions are a bit delicate. Let's assume first that $n = a^2$ is a perfect square. Then the second sum is zero, and the first can be evaluated by our usual routine:

$$\begin{aligned}&\sum_{k,m\ge 0} m\bigl[m^2\le k<(m+1)^2\le a^2\bigr]\\ &\qquad = \sum_{m\ge 0} m\bigl((m+1)^2-m^2\bigr)[m+1\le a]\\ &\qquad = \sum_{m\ge 0} m(2m+1)[m<a]\\ &\qquad = \sum_{m\ge 0} (2m^{\underline{2}}+3m^{\underline{1}})[m<a]\\ &\qquad = \sum\nolimits_0^a (2m^{\underline{2}}+3m^{\underline{1}})\,\delta m\\ &\qquad = \tfrac{2}{3}a(a-1)(a-2)+\tfrac{3}{2}a(a-1) \;=\; \tfrac{1}{6}(4a+1)a(a-1).\end{aligned}$$

Falling powers make the sum come tumbling down.

In the general case we can let $a = \lfloor\sqrt{n}\rfloor$; then we merely need to add the terms for $a^2 \leqslant k < n$, which are all equal to a, so they sum to $(n - a^2)a$. This gives the desired closed form,

$$\sum_{0\leqslant k<n} \lfloor\sqrt{k}\rfloor = na - \tfrac{1}{3}a^3 - \tfrac{1}{2}a^2 - \tfrac{1}{6}a\,, \qquad a = \lfloor\sqrt{n}\rfloor. \tag{3.27}$$

Another approach to such sums is to replace an expression of the form $\lfloor x\rfloor$ by $\sum_j [1 \leqslant j \leqslant x]$; this is legal whenever $x \geqslant 0$. Here's how that method works in the sum of $\lfloor$square roots$\rfloor$, if we assume for convenience that $n = a^2$:

$$\begin{aligned}\sum_{0\leqslant k<n} \lfloor\sqrt{k}\rfloor &= \sum_{j,k} [1 \leqslant j \leqslant \sqrt{k}\,][0 \leqslant k < a^2] \\ &= \sum_{1\leqslant j<a} \sum_{k} [j^2 \leqslant k < a^2] \\ &= \sum_{1\leqslant j<a} (a^2 - j^2) = a^3 - \tfrac{1}{3}a(a + \tfrac{1}{2})(a + 1)\,.\end{aligned}$$

Now here's another example where a change of variable leads to a transformed sum. A remarkable theorem was discovered independently by three mathematicians — Bohl [28], Sierpiński [265], and Weyl [300] — at about the same time in 1909: If α is irrational then the fractional parts $\{n\alpha\}$ are very uniformly distributed between 0 and 1, as $n \to \infty$. One way to state this is that

$$\lim_{n\to\infty} \frac{1}{n} \sum_{0\leqslant k<n} f\bigl(\{k\alpha\}\bigr) = \int_0^1 f(x)\,dx \tag{3.28}$$

for all irrational α and all functions f that are continuous almost everywhere. For example, the *average* value of $\{n\alpha\}$ can be found by setting $f(x) = x$; we get $\frac{1}{2}$. (That's exactly what we might expect; but it's nice to know that it is really, provably true, no matter how irrational α is.)

Warning: This stuff is fairly advanced. Better skim the next two pages on first reading; they aren't crucial.
— Friendly TA

The theorem of Bohl, Sierpiński, and Weyl is proved by approximating $f(x)$ above and below by "step functions," which are linear combinations of the simple functions

$$f_v(x) = [0 \leqslant x < v]$$

when $0 \leqslant v \leqslant 1$. Our purpose here is not to prove the theorem; that's a job for calculus books. But let's try to figure out the basic reason why it holds, by seeing how well it works in the special case $f(x) = f_v(x)$. In other words, let's try to see how close the sum

Start Skimming

$$\sum_{0\leqslant k<n} \bigl[\{k\alpha\} < v\bigr]$$

gets to the "ideal" value nv, when n is large and α is irrational.

For this purpose we define the *discrepancy* $D(\alpha, n)$ to be the maximum absolute value, over all $0 \leqslant v \leqslant 1$, of the sum

$$s(\alpha, n, v) = \sum_{0 \leqslant k < n} \Bigl(\bigl[\{k\alpha\} < v\bigr] - v\Bigr). \tag{3.29}$$

Our goal is to show that $D(\alpha, n)$ is "not too large" when compared with n, by showing that $|s(\alpha, n, v)|$ is always reasonably small.

First we can rewrite $s(\alpha, n, v)$ in simpler form, then introduce a new index variable j:

$$\begin{aligned}\sum_{0 \leqslant k < n} \Bigl(\bigl[\{k\alpha\} < v\bigr] - v\Bigr) &= \sum_{0 \leqslant k < n} \bigl(\lfloor k\alpha \rfloor - \lfloor k\alpha - v \rfloor - v\bigr) \\ &= -nv + \sum_{0 \leqslant k < n} \sum_{j} [k\alpha - v < j \leqslant k\alpha] \\ &= -nv + \sum_{0 \leqslant j < \lceil n\alpha \rceil} \sum_{k < n} \bigl[j\alpha^{-1} \leqslant k < (j+v)\alpha^{-1}\bigr].\end{aligned}$$

If we're lucky, we can do the sum on k. But we ought to introduce some new variables, so that the formula won't be such a mess. Without loss of generality, we can assume that $0 < \alpha < 1$; let us write

Right, name and conquer.
The change of variable from k to j is the main point.
—Friendly TA

$$\begin{aligned} a &= \lfloor \alpha^{-1} \rfloor, & \alpha^{-1} &= a + \alpha'; \\ b &= \lceil v\alpha^{-1} \rceil, & v\alpha^{-1} &= b - v'. \end{aligned}$$

Thus $\alpha' = \{\alpha^{-1}\}$ is the fractional part of α^{-1}, and v' is the mumble-fractional part of $v\alpha^{-1}$.

Once again the boundary conditions are our only source of grief. For now, let's forget the restriction '$k < n$' and evaluate the sum on k without it:

$$\begin{aligned}\sum_{k} \Bigl[k \in \bigl[j\alpha^{-1} \mathinner{\ldotp\ldotp} (j+v)\alpha^{-1}\bigr)\Bigr] &= \bigl\lceil (j+v)(a+\alpha') \bigr\rceil - \bigl\lceil j(a+\alpha') \bigr\rceil \\ &= b + \lceil j\alpha' - v' \rceil - \lceil j\alpha' \rceil.\end{aligned}$$

OK, that's pretty simple; we plug it in and plug away:

$$s(\alpha, n, v) = -nv + \lceil n\alpha \rceil b + \sum_{0 \leqslant j < \lceil n\alpha \rceil} \bigl(\lceil j\alpha' - v' \rceil - \lceil j\alpha' \rceil\bigr) - S, \tag{3.30}$$

where S is a correction for the cases with $k \geqslant n$ that we have failed to exclude. The quantity $j\alpha'$ will never be an integer, since α (hence α') is irrational; and $j\alpha' - v'$ will be an integer for at most one value of j. So we can change the ceiling terms to floors:

$$s(\alpha,n,v) \;=\; -nv + \lceil n\alpha\rceil b - \sum_{0\le j<\lceil n\alpha\rceil} \bigl(\lfloor j\alpha'\rfloor - \lfloor j\alpha' - v'\rfloor\bigr) - S + [0 \text{ or } 1]\,.$$

(The formula [0 or 1] stands for something that's either 0 or 1; we needn't commit ourselves, because the details don't really matter.)

Interesting. Instead of a closed form, we're getting a sum that looks rather like $s(\alpha,n,v)$ but with different parameters: α' instead of α, $\lceil n\alpha\rceil$ instead of n, and v' instead of v. So we'll have a recurrence for $s(\alpha,n,v)$, which (hopefully) will lead to a recurrence for the discrepancy $D(\alpha,n)$. This means we want to get

$$s(\alpha',\lceil n\alpha\rceil,v') \;=\; \sum_{0\le j<\lceil n\alpha\rceil} \bigl(\lfloor j\alpha'\rfloor - \lfloor j\alpha' - v'\rfloor - v'\bigr)$$

into the act:

$$s(\alpha,n,v) \;=\; -nv + \lceil n\alpha\rceil b - \lceil n\alpha\rceil v' - s(\alpha',\lceil n\alpha\rceil,v') - S + [0 \text{ or } 1]\,.$$

Recalling that $b - v' = v\alpha^{-1}$, we see that everything will simplify beautifully if we replace $\lceil n\alpha\rceil(b-v')$ by $n\alpha(b-v') = nv$:

$$s(\alpha,n,v) \;=\; -s(\alpha',\lceil n\alpha\rceil,v') - S + \epsilon + [0 \text{ or } 1]\,.$$

Here ϵ is a positive error of at most $v\alpha^{-1}$. Exercise 18 proves that S is, likewise, between 0 and α^{-1}. We can also remove the term for $j = \lceil n\alpha\rceil - 1 = \lfloor n\alpha\rfloor$ from the sum, since it contributes either v' or $v' - 1$. Hence, if we take the maximum of absolute values over all v, we get

$$D(\alpha,n) \;\le\; D(\alpha',\lfloor \alpha n\rfloor) + \alpha^{-1} + 2\,. \qquad (3.31)$$

The methods we'll learn in succeeding chapters will allow us to conclude from this recurrence that $D(\alpha,n)$ is always much smaller than n, when n is sufficiently large. Hence the theorem (3.28) is not only true, it can also be strengthened: Convergence to the limit is very fast.

Stop Skimming

Whew; that was quite an exercise in manipulation of sums, floors, and ceilings. Readers who are not accustomed to "proving that errors are small" might find it hard to believe that anybody would have the courage to keep going, when faced with such weird-looking sums. But actually, a second look shows that there's a simple motivating thread running through the whole calculation. The main idea is that a certain sum $s(\alpha,n,v)$ of n terms can be reduced to a similar sum of at most αn terms. Everything else cancels out except for a small residual left over from terms near the boundaries.

Let's take a deep breath now and do one more sum, which is not trivial but has the great advantage (compared with what we've just been doing) that it comes out in closed form so that we can easily check the answer. Our goal now will be to generalize the sum in (3.26) by finding an expression for

Is this a harder sum of floors, or a sum of harder floors?

$$\sum_{0\leqslant k<m} \left\lfloor \frac{nk+x}{m} \right\rfloor, \qquad \text{integer } m>0, \quad \text{integer } n.$$

Finding a closed form for this sum is tougher than what we've done so far (except perhaps for the discrepancy problem we just looked at). But it's instructive, so we'll hack away at it for the rest of this chapter.

Be forewarned: This is the beginning of a pattern, in that the last part of the chapter consists of the solution of some long, difficult problem, with little more motivation than curiosity.
—Students

As usual, especially with tough problems, we start by looking at small cases. The special case $n = 1$ is (3.26), with x replaced by x/m:

$$\left\lfloor \frac{x}{m} \right\rfloor + \left\lfloor \frac{1+x}{m} \right\rfloor + \cdots + \left\lfloor \frac{m-1+x}{m} \right\rfloor = \lfloor x \rfloor .$$

And as in Chapter 1, we find it useful to get more data by generalizing downwards to the case $n = 0$:

$$\left\lfloor \frac{x}{m} \right\rfloor + \left\lfloor \frac{x}{m} \right\rfloor + \cdots + \left\lfloor \frac{x}{m} \right\rfloor = m \left\lfloor \frac{x}{m} \right\rfloor .$$

Touché. But c'mon, gang, do you always need to be told about applications before you can get interested in something? This sum arises, for example, in the study of random number generation and testing. But mathematicians looked at it long before computers came along, because they found it natural to ask if there's a way to sum arithmetic progressions that have been "floored."
—Your instructor

Our problem has two parameters, m and n; let's look at some small cases for m. When $m = 1$ there's just a single term in the sum and its value is $\lfloor x \rfloor$. When $m = 2$ the sum is $\lfloor x/2 \rfloor + \lfloor (x+n)/2 \rfloor$. We can remove the interaction between x and n by removing n from inside the floor function, but to do that we must consider even and odd n separately. If n is even, $n/2$ is an integer, so we can remove it from the floor:

$$\left\lfloor \frac{x}{2} \right\rfloor + \left(\left\lfloor \frac{x}{2} \right\rfloor + \frac{n}{2} \right) = 2\left\lfloor \frac{x}{2} \right\rfloor + \frac{n}{2} .$$

If n is odd, $(n-1)/2$ is an integer so we get

$$\left\lfloor \frac{x}{2} \right\rfloor + \left(\left\lfloor \frac{x+1}{2} \right\rfloor + \frac{n-1}{2} \right) = \lfloor x \rfloor + \frac{n-1}{2} .$$

The last step follows from (3.26) with $m = 2$.

These formulas for even and odd n slightly resemble those for $n = 0$ and 1, but no clear pattern has emerged yet; so we had better continue exploring some more small cases. For $m = 3$ the sum is

$$\left\lfloor \frac{x}{3} \right\rfloor + \left\lfloor \frac{x+n}{3} \right\rfloor + \left\lfloor \frac{x+2n}{3} \right\rfloor ,$$

and we consider three cases for n: Either it's a multiple of 3, or it's 1 more than a multiple, or it's 2 more. That is, $n \bmod 3 = 0$, 1, or 2. If $n \bmod 3 = 0$

then $n/3$ and $2n/3$ are integers, so the sum is

$$\left\lfloor\frac{x}{3}\right\rfloor + \left(\left\lfloor\frac{x}{3}\right\rfloor + \frac{n}{3}\right) + \left(\left\lfloor\frac{x}{3}\right\rfloor + \frac{2n}{3}\right) = 3\left\lfloor\frac{x}{3}\right\rfloor + n .$$

If $n \bmod 3 = 1$ then $(n-1)/3$ and $(2n-2)/3$ are integers, so we have

$$\left\lfloor\frac{x}{3}\right\rfloor + \left(\left\lfloor\frac{x+1}{3}\right\rfloor + \frac{n-1}{3}\right) + \left(\left\lfloor\frac{x+2}{3}\right\rfloor + \frac{2n-2}{3}\right) = \lfloor x\rfloor + n - 1 .$$

Again this last step follows from (3.26), this time with $m = 3$. And finally, if $n \bmod 3 = 2$ then

$$\left\lfloor\frac{x}{3}\right\rfloor + \left(\left\lfloor\frac{x+2}{3}\right\rfloor + \frac{n-2}{3}\right) + \left(\left\lfloor\frac{x+1}{3}\right\rfloor + \frac{2n-1}{3}\right) = \lfloor x\rfloor + n - 1 .$$

"Inventive genius requires pleasurable mental activity as a condition for its vigorous exercise. 'Necessity is the mother of invention' is a silly proverb. 'Necessity is the mother of futile dodges' is much nearer to the truth. The basis of the growth of modern invention is science, and science is almost wholly the outgrowth of pleasurable intellectual curiosity."
—A. N. Whitehead [303]

The left hemispheres of our brains have finished the case $m = 3$, but the right hemispheres still can't recognize the pattern, so we proceed to $m = 4$:

$$\left\lfloor\frac{x}{4}\right\rfloor + \left\lfloor\frac{x+n}{4}\right\rfloor + \left\lfloor\frac{x+2n}{4}\right\rfloor + \left\lfloor\frac{x+3n}{4}\right\rfloor .$$

At least we know enough by now to consider cases based on $n \bmod m$. If $n \bmod 4 = 0$ then

$$\left\lfloor\frac{x}{4}\right\rfloor + \left(\left\lfloor\frac{x}{4}\right\rfloor + \frac{n}{4}\right) + \left(\left\lfloor\frac{x}{4}\right\rfloor + \frac{2n}{4}\right) + \left(\left\lfloor\frac{x}{4}\right\rfloor + \frac{3n}{4}\right) = 4\left\lfloor\frac{x}{4}\right\rfloor + \frac{3n}{2} .$$

And if $n \bmod 4 = 1$,

$$\left\lfloor\frac{x}{4}\right\rfloor + \left(\left\lfloor\frac{x+1}{4}\right\rfloor + \frac{n-1}{4}\right) + \left(\left\lfloor\frac{x+2}{4}\right\rfloor + \frac{2n-2}{4}\right) + \left(\left\lfloor\frac{x+3}{4}\right\rfloor + \frac{3n-3}{4}\right)$$
$$= \lfloor x\rfloor + \frac{3n}{2} - \frac{3}{2} .$$

The case $n \bmod 4 = 3$ turns out to give the same answer. Finally, in the case $n \bmod 4 = 2$ we get something a bit different, and this turns out to be an important clue to the behavior in general:

$$\left\lfloor\frac{x}{4}\right\rfloor + \left(\left\lfloor\frac{x+2}{4}\right\rfloor + \frac{n-2}{4}\right) + \left(\left\lfloor\frac{x}{4}\right\rfloor + \frac{2n}{4}\right) + \left(\left\lfloor\frac{x+2}{4}\right\rfloor + \frac{3n-2}{4}\right)$$
$$= 2\left(\left\lfloor\frac{x}{4}\right\rfloor + \left\lfloor\frac{x+2}{4}\right\rfloor\right) + \frac{3n}{2} - 1 = 2\left\lfloor\frac{x}{2}\right\rfloor + \frac{3n}{2} - 1 .$$

This last step simplifies something of the form $\lfloor y/2\rfloor + \lfloor (y+1)/2\rfloor$, which again is a special case of (3.26).

To summarize, here's the value of our sum for small m:

m	$n \bmod m = 0$	$n \bmod m = 1$	$n \bmod m = 2$	$n \bmod m = 3$
1	$\lfloor x \rfloor$			
2	$2\left\lfloor \frac{x}{2} \right\rfloor + \frac{n}{2}$	$\lfloor x \rfloor + \frac{n}{2} - \frac{1}{2}$		
3	$3\left\lfloor \frac{x}{3} \right\rfloor + n$	$\lfloor x \rfloor + n - 1$	$\lfloor x \rfloor + n - 1$	
4	$4\left\lfloor \frac{x}{4} \right\rfloor + \frac{3n}{2}$	$\lfloor x \rfloor + \frac{3n}{2} - \frac{3}{2}$	$2\left\lfloor \frac{x}{2} \right\rfloor + \frac{3n}{2} - 1$	$\lfloor x \rfloor + \frac{3n}{2} - \frac{3}{2}$

It looks as if we're getting something of the form

$$a\left\lfloor \frac{x}{a} \right\rfloor + bn + c\,,$$

where a, b, and c somehow depend on m and n. Even the myopic among us can see that b is probably $(m-1)/2$. It's harder to discern an expression for a; but the case $n \bmod 4 = 2$ gives us a hint that a is probably $\gcd(m,n)$, the greatest common divisor of m and n. This makes sense because $\gcd(m,n)$ is the factor we remove from m and n when reducing the fraction n/m to lowest terms, and our sum involves the fraction n/m. (We'll look carefully at gcd operations in Chapter 4.) The value of c seems more mysterious, but perhaps it will drop out of our proofs for a and b.

In computing the sum for small m, we've effectively rewritten each term of the sum as

$$\left\lfloor \frac{x+kn}{m} \right\rfloor = \left\lfloor \frac{x + kn \bmod m}{m} \right\rfloor + \frac{kn}{m} - \frac{kn \bmod m}{m}\,,$$

because $(kn - kn \bmod m)/m$ is an integer that can be removed from inside the floor brackets. Thus the original sum can be expanded into the following tableau:

$$\begin{array}{rcccccc}
 & \left\lfloor \dfrac{x}{m} \right\rfloor & + & \dfrac{0}{m} & - & \dfrac{0 \bmod m}{m} \\
+ & \left\lfloor \dfrac{x + n \bmod m}{m} \right\rfloor & + & \dfrac{n}{m} & - & \dfrac{n \bmod m}{m} \\
+ & \left\lfloor \dfrac{x + 2n \bmod m}{m} \right\rfloor & + & \dfrac{2n}{m} & - & \dfrac{2n \bmod m}{m} \\
 & \vdots & & \vdots & & \vdots \\
+ & \left\lfloor \dfrac{x + (m-1)n \bmod m}{m} \right\rfloor & + & \dfrac{(m-1)n}{m} & - & \dfrac{(m-1)n \bmod m}{m}\,.
\end{array}$$

When we experimented with small values of m, these three columns led respectively to $a\lfloor x/a \rfloor$, bn, and c.

In particular, we can see how b arises. The second column is an arithmetic progression, whose sum we know — it's the average of the first and last terms, times the number of terms:

$$\frac{1}{2}\left(0 + \frac{(m-1)n}{m}\right) \cdot m \;=\; \frac{(m-1)n}{2}\,.$$

So our guess that $b = (m-1)/2$ has been verified.

The first and third columns seem tougher; to determine a and c we must take a closer look at the sequence of numbers

$$0 \bmod m,\quad n \bmod m,\quad 2n \bmod m,\quad \ldots,\quad (m-1)n \bmod m.$$

Suppose, for example, that $m = 12$ and $n = 5$. If we think of the sequence as times on a clock, the numbers are 0 o'clock (we take 12 o'clock to be 0 o'clock), then 5 o'clock, 10 o'clock, 3 o'clock (= 15 o'clock), 8 o'clock, and so on. It turns out that we hit every hour exactly once.

Now suppose $m = 12$ and $n = 8$. The numbers are 0 o'clock, 8 o'clock, 4 o'clock (= 16 o'clock), but then 0, 8, and 4 repeat. Since both 8 and 12 are multiples of 4, and since the numbers start at 0 (also a multiple of 4), there's no way to break out of this pattern — they must all be multiples of 4.

Lemma now, dilemma later.

In these two cases we have $\gcd(12,5) = 1$ and $\gcd(12,8) = 4$. The general rule, which we will prove next chapter, states that if $d = \gcd(m,n)$ then we get the numbers $0, d, 2d, \ldots, m-d$ in some order, followed by $d-1$ more copies of the same sequence. For example, with $m = 12$ and $n = 8$ the pattern 0, 8, 4 occurs four times.

The first column of our sum now makes complete sense. It contains d copies of the terms $\lfloor x/m \rfloor$, $\lfloor (x+d)/m \rfloor$, $\ldots$, $\lfloor (x+m-d)/m \rfloor$, in some order, so its sum is

$$\begin{aligned} d&\left(\left\lfloor\frac{x}{m}\right\rfloor + \left\lfloor\frac{x+d}{m}\right\rfloor + \cdots + \left\lfloor\frac{x+m-d}{m}\right\rfloor\right) \\ &= d\left(\left\lfloor\frac{x/d}{m/d}\right\rfloor + \left\lfloor\frac{x/d+1}{m/d}\right\rfloor + \cdots + \left\lfloor\frac{x/d+m/d-1}{m/d}\right\rfloor\right) \\ &= d\left\lfloor\frac{x}{d}\right\rfloor\,. \end{aligned}$$

This last step is yet another application of (3.26). Our guess for a has been verified:

$$a \;=\; d \;=\; \gcd(m,n)\,.$$

Also, as we guessed, we can now compute c, because the third column has become easy to fathom. It contains d copies of the arithmetic progression $0/m$, d/m, $2d/m$, $\ldots$, $(m-d)/m$, so its sum is

$$d\left(\frac{1}{2}\left(0+\frac{m-d}{m}\right)\cdot\frac{m}{d}\right) = \frac{m-d}{2};$$

the third column is actually subtracted, not added, so we have

$$c = \frac{d-m}{2}.$$

End of mystery, end of quest. The desired closed form is

$$\sum_{0\le k<m}\left\lfloor\frac{nk+x}{m}\right\rfloor = d\left\lfloor\frac{x}{d}\right\rfloor + \frac{m-1}{2}n + \frac{d-m}{2},$$

where $d=\gcd(m,n)$. As a check, we can make sure this works in the special cases $n=0$ and $n=1$ that we knew before: When $n=0$ we get $d=\gcd(m,0)=m$; the last two terms of the formula are zero so the formula properly gives $m\lfloor x/m\rfloor$. And for $n=1$ we get $d=\gcd(m,1)=1$; the last two terms cancel nicely, and the sum is just $\lfloor x\rfloor$.

By manipulating the closed form a bit, we can actually make it symmetric in m and n:

$$\begin{aligned}\sum_{0\le k<m}\left\lfloor\frac{nk+x}{m}\right\rfloor &= d\left\lfloor\frac{x}{d}\right\rfloor + \frac{m-1}{2}n + \frac{d-m}{2}\\ &= d\left\lfloor\frac{x}{d}\right\rfloor + \frac{(m-1)(n-1)}{2} + \frac{m-1}{2} + \frac{d-m}{2}\\ &= d\left\lfloor\frac{x}{d}\right\rfloor + \frac{(m-1)(n-1)}{2} + \frac{d-1}{2}.\end{aligned} \tag{3.32}$$

This is astonishing, because there's no reason to suspect that such a sum should be symmetrical. We have proved a "reciprocity law,"

Yup, I'm floored.

$$\sum_{0\le k<m}\left\lfloor\frac{nk+x}{m}\right\rfloor = \sum_{0\le k<n}\left\lfloor\frac{mk+x}{n}\right\rfloor, \qquad \text{integers } m,n>0.$$

For example, if $m=41$ and $n=127$, the left sum has 41 terms and the right has 127; but they still come out equal, for all real x.

Exercises

Warmups

1 When we analyzed the Josephus problem in Chapter 1, we represented an arbitrary positive integer n in the form $n = 2^m + l$, where $0 \leqslant l < 2^m$. Give explicit formulas for l and m as functions of n, using floor and/or ceiling brackets.

2 What is a formula for the nearest integer to a given real number x? In case of ties, when x is exactly halfway between two integers, give an expression that rounds (a) up — that is, to $\lceil x \rceil$; (b) down — that is, to $\lfloor x \rfloor$.

3 Evaluate $\lfloor \lfloor m\alpha \rfloor n/\alpha \rfloor$, when m and n are positive integers and α is an irrational number greater than n.

4 The text describes problems at levels 1 through 5. What is a level 0 problem? (This, by the way, is *not* a level 0 problem.)

5 Find a necessary and sufficient condition that $\lfloor nx \rfloor = n\lfloor x \rfloor$, when n is a positive integer. (Your condition should involve $\{x\}$.)

6 Can something interesting be said about $\lfloor f(x) \rfloor$ when $f(x)$ is a continuous, monotonically *decreasing* function that takes integer values only when x is an integer?

7 Solve the recurrence

$$\begin{aligned} X_n &= n\,, && \text{for } 0 \leqslant n < m; \\ X_n &= X_{n-m} + 1\,, && \text{for } n \geqslant m. \end{aligned}$$

You know you're in college when the book doesn't tell you how to pronounce 'Dirichlet'.

8 Prove the *Dirichlet box principle*: If n objects are put into m boxes, some box must contain $\geqslant \lceil n/m \rceil$ objects, and some box must contain $\leqslant \lfloor n/m \rfloor$.

9 Egyptian mathematicians in 1800 B.C. represented rational numbers between 0 and 1 as sums of unit fractions $1/x_1 + \cdots + 1/x_k$, where the x's were distinct positive integers. For example, they wrote $\frac{1}{3} + \frac{1}{15}$ instead of $\frac{2}{5}$. Prove that it is always possible to do this in a systematic way: If $0 < m/n < 1$, then

$$\frac{m}{n} = \frac{1}{q} + \left\{\text{representation of } \frac{m}{n} - \frac{1}{q}\right\}, \qquad q = \left\lceil \frac{n}{m} \right\rceil.$$

(This is *Fibonacci's algorithm*, due to Leonardo Fibonacci, A.D. 1202.)

Basics

10 Show that the expression

$$\left\lceil \frac{2x+1}{2} \right\rceil - \left\lceil \frac{2x+1}{4} \right\rceil + \left\lfloor \frac{2x+1}{4} \right\rfloor$$

is always either $\lfloor x \rfloor$ or $\lceil x \rceil$. In what circumstances does each case arise?

11 Give details of the proof alluded to in the text, that the open interval $(\alpha \mathinner{..} \beta)$ contains exactly $\lceil \beta \rceil - \lfloor \alpha \rfloor - 1$ integers when $\alpha < \beta$. Why does the case $\alpha = \beta$ have to be excluded in order to make the proof correct?

12 Prove that

$$\left\lceil \frac{n}{m} \right\rceil = \left\lfloor \frac{n+m-1}{m} \right\rfloor,$$

for all integers n and all positive integers m. [This identity gives us another way to convert ceilings to floors and vice versa, instead of using the reflective law (3.4).]

13 Let α and β be positive real numbers. Prove that $\mathrm{Spec}(\alpha)$ and $\mathrm{Spec}(\beta)$ partition the positive integers if and only if α and β are irrational and $1/\alpha + 1/\beta = 1$.

14 Prove or disprove:

$$(x \bmod ny) \bmod y = x \bmod y, \qquad \text{integer } n.$$

15 Is there an identity analogous to (3.26) that uses ceilings instead of floors?

16 Prove that $n \bmod 2 = \bigl(1-(-1)^n\bigr)/2$. Find and prove a similar expression for $n \bmod 3$ in the form $a+b\omega^n+c\omega^{2n}$, where ω is the complex number $(-1+i\sqrt{3})/2$. *Hint:* $\omega^3 = 1$ and $1+\omega+\omega^2 = 0$.

17 Evaluate the sum $\sum_{0 \leqslant k < m} \lfloor x + k/m \rfloor$ in the case $x \geqslant 0$ by substituting $\sum_j [1 \leqslant j \leqslant x + k/m]$ for $\lfloor x + k/m \rfloor$ and summing first on k. Does your answer agree with (3.26)?

18 Prove that the boundary-value error term S in (3.18) is at most $\alpha^{-1}v$. *Hint:* Show that small values of j are not involved.

Homework exercises

19 Find a necessary and sufficient condition on the real number $b > 1$ such that

$$\lfloor \log_b x \rfloor = \lfloor \log_b \lfloor x \rfloor \rfloor$$

for all real $x \geqslant 1$.

20 Find the sum of all multiples of x in the closed interval $[\alpha\,..\,\beta]$, when $x > 0$.

21 How many of the numbers 2^m, for $0 \leqslant m \leqslant M$, have leading digit 1 in decimal notation?

22 Evaluate the sums $S_n = \sum_{k\geqslant 1}\lfloor n/2^k + \frac{1}{2}\rfloor$ and $T_n = \sum_{k\geqslant 1} 2^k\lfloor n/2^k + \frac{1}{2}\rfloor^2$.

23 Show that the nth element of the sequence

$$1, 2, 2, 3, 3, 3, 4, 4, 4, 4, 5, 5, 5, 5, 5, \ldots$$

is $\lfloor\sqrt{2n} + \frac{1}{2}\rfloor$. (The sequence contains exactly m occurrences of m.)

24 Exercise 13 establishes an interesting relation between the two multisets $\mathrm{Spec}(\alpha)$ and $\mathrm{Spec}\big(\alpha/(\alpha-1)\big)$, when α is any irrational number > 1, because $1/\alpha + (\alpha-1)/\alpha = 1$. Find (and prove) an interesting relation between the two multisets $\mathrm{Spec}(\alpha)$ and $\mathrm{Spec}\big(\alpha/(\alpha+1)\big)$, when α is any positive real number.

25 Prove or disprove that the Knuth numbers, defined by (3.16), satisfy $K_n \geqslant n$ for all nonnegative n.

26 Show that the auxiliary Josephus numbers (3.20) satisfy

$$\left(\frac{q}{q-1}\right)^n \leqslant D_n^{(q)} \leqslant q\left(\frac{q}{q-1}\right)^n, \qquad \text{for } n \geqslant 0.$$

27 Prove that infinitely many of the numbers $D_n^{(3)}$ defined by (3.20) are even, and that infinitely many are odd.

28 Solve the recurrence

$$\begin{aligned} a_0 &= 1\,; \\ a_n &= a_{n-1} + \lfloor\sqrt{a_{n-1}}\rfloor, \qquad \text{for } n > 0. \end{aligned}$$

29 Show that, in addition to (3.31), we have

$$D(\alpha, n) \geqslant D\big(\alpha', \lfloor\alpha n\rfloor\big) - \alpha^{-1} - 2\,.$$

30 Show that the recurrence

$$\begin{aligned} X_0 &= m\,, \\ X_n &= X_{n-1}^2 - 2\,, \qquad \text{for } n > 0, \end{aligned}$$

has the solution $X_n = \lceil\alpha^{2^n}\rceil$, if m is an integer greater than 2, where $\alpha + \alpha^{-1} = m$ and $\alpha > 1$. For example, if $m = 3$ the solution is

$$X_n = \lceil\phi^{2^{n+1}}\rceil, \qquad \phi = \frac{1+\sqrt{5}}{2}, \qquad \alpha = \phi^2\,.$$

31 Prove or disprove: $\lfloor x\rfloor + \lfloor y\rfloor + \lfloor x+y\rfloor \leqslant \lfloor 2x\rfloor + \lfloor 2y\rfloor$.

32 Let $\|x\| = \min(x - \lfloor x\rfloor, \lceil x\rceil - x)$ denote the distance from x to the nearest integer. What is the value of

$$\sum_k 2^k \|x/2^k\|^2\,?$$

(Note that this sum can be doubly infinite. For example, when $x = 1/3$ the terms are nonzero as $k \to -\infty$ and also as $k \to +\infty$.)

Exam problems

33 A circle, $2n-1$ units in diameter, has been drawn symmetrically on a $2n \times 2n$ chessboard, illustrated here for $n = 3$:

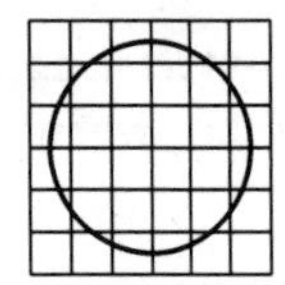

- **a** How many cells of the board contain a segment of the circle?
- **b** Find a function $f(k)$ such that exactly $\sum_{k=1}^{n-1} f(k)$ cells of the board lie entirely within the circle.

34 Let $f(n) = \sum_{k=1}^{n} \lceil \lg k\rceil$.
- **a** Find a closed form for $f(n)$, when $n \geqslant 1$.
- **b** Prove that $f(n) = n - 1 + f(\lceil n/2\rceil) + f(\lfloor n/2\rfloor)$ for all $n \geqslant 1$.

35 Simplify the formula $\lfloor (n+1)^2 n!\, e\rfloor \bmod n$.

Simplify it, but don't change the value.

36 Assuming that n is a nonnegative integer, find a closed form for the sum

$$\sum_{1<k<2^{2^n}} \frac{1}{2^{\lfloor \lg k\rfloor} 4^{\lfloor \lg\lg k\rfloor}}\,.$$

37 Prove the identity

$$\sum_{0\leqslant k<m} \left(\left\lfloor \frac{m+k}{n}\right\rfloor - \left\lfloor \frac{k}{n}\right\rfloor\right) = \left\lfloor \frac{m^2}{n}\right\rfloor - \left\lfloor \frac{\min(m \bmod n, (-m) \bmod n)^2}{n}\right\rfloor$$

for all positive integers m and n.

38 Let $x_1, \ldots, x_n$ be real numbers such that the identity

$$\sum_{k=1}^{n} \lfloor m x_k\rfloor = \left\lfloor m \sum_{1\leqslant k\leqslant n} x_k \right\rfloor$$

holds for all positive integers m. Prove something interesting about $x_1, \ldots, x_n$.

39 Prove that the double sum $\sum_{0\leqslant k\leqslant \log_b x} \sum_{0<j<b} \lceil (x + jb^k)/b^{k+1} \rceil$ equals $(b-1)(\lfloor \log_b x \rfloor + 1) + \lceil x \rceil - 1$, for every real number $x \geqslant 1$ and every integer $b > 1$.

40 The spiral function $\sigma(n)$, indicated in the diagram below, maps a non-negative integer n onto an ordered pair of integers $(x(n), y(n))$. For example, it maps $n = 9$ onto the ordered pair $(1, 2)$.

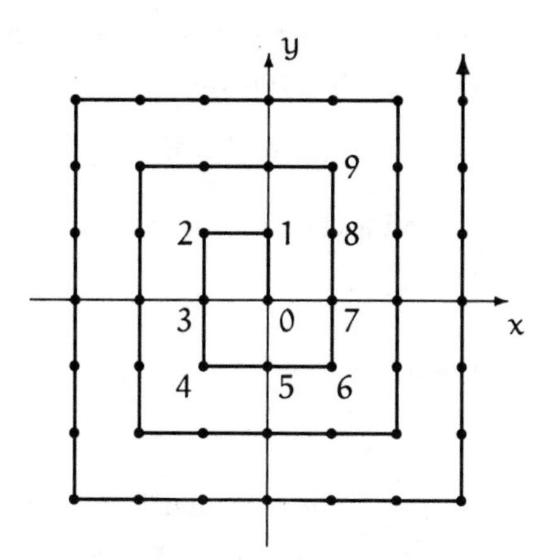

a Prove that if $m = \lfloor \sqrt{n} \rfloor$,

$$x(n) = (-1)^m \Big(\big(n - m(m+1)\big) \cdot \big[\lfloor 2\sqrt{n} \rfloor \text{ is even}\big] + \lceil \tfrac{1}{2} m \rceil \Big),$$

and find a similar formula for $y(n)$. *Hint:* Classify the spiral into segments W_k, S_k, E_k, N_k according as $\lfloor 2\sqrt{n} \rfloor = 4k-2$, $4k-1$, $4k$, $4k+1$.

b Prove that, conversely, we can determine n from $\sigma(n)$ by a formula of the form

$$n = (2k)^2 \pm \big(2k + x(n) + y(n)\big), \quad k = \max\big(|x(n)|, |y(n)|\big).$$

Give a rule for when the sign is $+$ and when the sign is $-$.

Bonus problems

41 Let f and g be increasing functions such that the sets $\{f(1), f(2), \ldots\}$ and $\{g(1), g(2), \ldots\}$ partition the positive integers. Suppose that f and g are related by the condition $g(n) = f\big(f(n)\big) + 1$ for all $n > 0$. Prove that $f(n) = \lfloor n\phi \rfloor$ and $g(n) = \lfloor n\phi^2 \rfloor$, where $\phi = (1 + \sqrt{5})/2$.

42 Do there exist real numbers α, β, and γ such that $\mathrm{Spec}(\alpha)$, $\mathrm{Spec}(\beta)$, and $\mathrm{Spec}(\gamma)$ together partition the set of positive integers?

43 Find an interesting interpretation of the Knuth numbers, by unfolding the recurrence (3.16).

44 Show that there are integers $a_n^{(q)}$ and $d_n^{(q)}$ such that

$$a_n^{(q)} = \frac{D_{n-1}^{(q)} + d_n^{(q)}}{q-1} = \frac{D_n^{(q)} + d_n^{(q)}}{q}, \qquad \text{for } n > 0,$$

when $D_n^{(q)}$ is the solution to (3.20). Use this fact to obtain another form of the solution to the generalized Josephus problem:

$$J_q(n) = 1 + d_k^{(q)} + q(n - a_k^{(q)}), \qquad \text{for } a_k^{(q)} \leqslant n < a_{k+1}^{(q)}.$$

45 Extend the trick of exercise 30 to find a closed-form solution to

$$Y_0 = m,$$
$$Y_n = 2Y_{n-1}^2 - 1, \qquad \text{for } n > 0,$$

if m is a positive integer.

46 Prove that if $n = \lfloor(\sqrt{2}^{\,l} + \sqrt{2}^{\,l-1})m\rfloor$, where m and l are nonnegative integers, then $\lfloor\sqrt{2n(n+1)}\rfloor = \lfloor(\sqrt{2}^{\,l+1} + \sqrt{2}^{\,l})m\rfloor$. Use this remarkable property to find a closed form solution to the recurrence

$$L_0 = a, \qquad \text{integer } a > 0;$$
$$L_n = \left\lfloor \sqrt{2L_{n-1}(L_{n-1}+1)} \right\rfloor, \qquad \text{for } n > 0.$$

Hint: $\lfloor\sqrt{2n(n+1)}\rfloor = \lfloor\sqrt{2}(n+\frac{1}{2})\rfloor$.

47 The function $f(x)$ is said to be *replicative* if it satisfies

$$f(mx) = f(x) + f\left(x + \frac{1}{m}\right) + \cdots + f\left(x + \frac{m-1}{m}\right)$$

for every positive integer m. Find necessary and sufficient conditions on the real number c for the following functions to be replicative:

a $f(x) = x + c$.

b $f(x) = [x + c \text{ is an integer}]$.

c $f(x) = \max(\lfloor x\rfloor, c)$.

d $f(x) = x + c\lfloor x\rfloor - \frac{1}{2}[x \text{ is not an integer}]$.

48 Find a necessary and sufficient condition on the real numbers $0 \leqslant \alpha < 1$ and $\beta \geqslant 0$ such that we can determine α and β from the infinite multiset of values

$$\{\, \lfloor n\alpha\rfloor + \lfloor n\beta\rfloor \mid n > 0 \,\}.$$

Research problems

49 Find a necessary and sufficient condition on the nonnegative real numbers α and β such that we can determine α and β from the infinite multiset of values

$$\{\, \lfloor \lfloor n\alpha \rfloor \beta \rfloor \mid n > 0 \,\}.$$

50 Let x be a real number $\geqslant \phi = \frac{1}{2}(1+\sqrt{5})$. The solution to the recurrence

$$\begin{aligned} Z_0(x) &= x\,, \\ Z_n(x) &= Z_{n-1}(x)^2 - 1\,, \qquad \text{for } n > 0, \end{aligned}$$

can be written $Z_n(x) = \lceil f(x)^{2^n} \rceil$, if x is an integer, where

$$f(x) = \lim_{n \to \infty} Z_n(x)^{1/2^n}\,,$$

because $Z_n(x) - 1 < f(x)^{2^n} < Z_n(x)$. What interesting properties does this function $f(x)$ have?

51 Given nonnegative real numbers α and β, let

$$\mathrm{Spec}(\alpha;\beta) = \{\lfloor \alpha+\beta \rfloor, \lfloor 2\alpha+\beta \rfloor, \lfloor 3\alpha+\beta \rfloor, \ldots\}$$

be a multiset that generalizes $\mathrm{Spec}(\alpha) = \mathrm{Spec}(\alpha;0)$. Prove or disprove: If the $m \geqslant 3$ multisets $\mathrm{Spec}(\alpha_1;\beta_1)$, $\mathrm{Spec}(\alpha_2;\beta_2)$, $\ldots$, $\mathrm{Spec}(\alpha_m;\beta_m)$ partition the positive integers, and if the parameters $\alpha_1 < \alpha_2 < \cdots < \alpha_m$ are rational, then

$$\alpha_k = \frac{2^m - 1}{2^{k-1}}\,, \qquad \text{for } 1 \leqslant k \leqslant m.$$

52 Fibonacci's algorithm (exercise 9) is "greedy" in the sense that it chooses the least conceivable q at every step. A more complicated algorithm is known by which every fraction m/n with n odd can be represented as a sum of distinct unit fractions $1/q_1 + \cdots + 1/q_k$ with *odd* denominators. Does the greedy algorithm for such a representation always terminate?

4

Number Theory

INTEGERS ARE CENTRAL to the discrete mathematics we are emphasizing in this book. Therefore we want to explore the *theory of numbers*, an important branch of mathematics concerned with the properties of integers.

We tested the number theory waters in the previous chapter, by introducing binary operations called 'mod' and 'gcd'. Now let's plunge in and really immerse ourselves in the subject.

In other words, be prepared to drown.

4.1 DIVISIBILITY

We say that m divides n (or n is divisible by m) if $m > 0$ and the ratio n/m is an integer. This property underlies all of number theory, so it's convenient to have a special notation for it. We therefore write

$$m \backslash n \quad \iff \quad m > 0 \text{ and } n = mk \text{ for some integer } k. \tag{4.1}$$

(The notation '$m|n$' is actually much more common than '$m\backslash n$' in current mathematics literature. But vertical lines are overused — for absolute values, set delimiters, conditional probabilities, etc. — and backward slashes are underused. Moreover, '$m\backslash n$' gives an impression that m is the denominator of an implied ratio. So we shall boldly let our divisibility symbol lean leftward.)

If m does not divide n we write '$m \nmid n$'.

There's a similar relation, "n is a multiple of m," which means almost the same thing except that m doesn't have to be positive. In this case we simply mean that $n = mk$ for some integer k. Thus, for example, there's only one multiple of 0 (namely 0), but nothing is divisible by 0. Every integer is a multiple of -1, but no integer is divisible by -1 (strictly speaking). These definitions apply when m and n are any real numbers; for example, 2π is divisible by π. But we'll almost always be using them when m and n are integers. After all, this is number theory.

"... no integer is divisible by -1 (strictly speaking)." — Graham, Knuth, and Patashnik [131]

In Britain we call this 'hcf' (highest common factor).

The *greatest common divisor* of two integers m and n is the largest integer that divides them both:

$$\gcd(m,n) \;=\; \max\{\,k \mid k\backslash m \quad\text{and}\quad k\backslash n\,\}. \tag{4.2}$$

For example, $\gcd(12,18) = 6$. This is a familiar notion, because it's the common factor that fourth graders learn to take out of a fraction m/n when reducing it to lowest terms: $12/18 = (12/6)/(18/6) = 2/3$. Notice that if $n > 0$ we have $\gcd(0,n) = n$, because any positive number divides 0, and because n is the largest divisor of itself. The value of $\gcd(0,0)$ is undefined.

Not to be confused with the greatest common multiple.

Another familiar notion is the *least common multiple*,

$$\operatorname{lcm}(m,n) \;=\; \min\{\,k \mid k>0,\quad m\backslash k \quad\text{and}\quad n\backslash k\,\}; \tag{4.3}$$

this is undefined if $m \leqslant 0$ or $n \leqslant 0$. Students of arithmetic recognize this as the least common denominator, which is used when adding fractions with denominators m and n. For example, $\operatorname{lcm}(12,18) = 36$, and fourth graders know that $\frac{7}{12} + \frac{1}{18} = \frac{21}{36} + \frac{2}{36} = \frac{23}{36}$. The lcm is somewhat analogous to the gcd, but we don't give it equal time because the gcd has nicer properties.

One of the nicest properties of the gcd is that it is easy to compute, using a 2300-year-old method called *Euclid's algorithm*. To calculate $\gcd(m,n)$, for given values $0 \leqslant m < n$, Euclid's algorithm uses the recurrence

$$\begin{aligned} \gcd(0,n) \;&=\; n\,; \\ \gcd(m,n) \;&=\; \gcd(n \bmod m,\ m)\,, \qquad \text{for } m > 0. \end{aligned} \tag{4.4}$$

Thus, for example, $\gcd(12,18) = \gcd(6,12) = \gcd(0,6) = 6$. The stated recurrence is valid, because any common divisor of m and n must also be a common divisor of both m and the number $n \bmod m$, which is $n - \lfloor n/m \rfloor m$. There doesn't seem to be any recurrence for $\operatorname{lcm}(m,n)$ that's anywhere near as simple as this. (See exercise 2.)

Euclid's algorithm also gives us more: We can extend it so that it will compute integers m' and n' satisfying

$$m'm + n'n \;=\; \gcd(m,n)\,. \tag{4.5}$$

(Remember that m' or n' can be negative.)

Here's how. If $m = 0$, we simply take $m' = 0$ and $n' = 1$. Otherwise we let $r = n \bmod m$ and apply the method recursively with r and m in place of m and n, computing $\bar{r}$ and $\overline{m}$ such that

$$\bar{r}\,r + \overline{m}\,m \;=\; \gcd(r,m)\,.$$

Since $r = n - \lfloor n/m \rfloor m$ and $\gcd(r,m) = \gcd(m,n)$, this equation tells us that

$$\bar{r}\,\bigl(n - \lfloor n/m \rfloor m\bigr) + \overline{m}\,m \;=\; \gcd(m,n)\,.$$

The left side can be rewritten to show its dependency on m and n:

$$(\overline{m} - \lfloor n/m \rfloor \overline{r})\, m + \overline{r}\, n \;=\; \gcd(m,n)\,;$$

hence $m' = \overline{m} - \lfloor n/m \rfloor \overline{r}$ and $n' = \overline{r}$ are the integers we need in (4.5). For example, in our favorite case $m = 12$, $n = 18$, this method gives $6 = 0{\cdot}0+1{\cdot}6 = 1\cdot 6 + 0\cdot 12 = (-1)\cdot 12 + 1\cdot 18$.

But why is (4.5) such a neat result? The main reason is that there's a sense in which the numbers m' and n' actually *prove* that Euclid's algorithm has produced the correct answer in any particular case. Let's suppose that our computer has told us after a lengthy calculation that $\gcd(m,n) = d$ and that $m'm + n'n = d$; but we're skeptical and think that there's really a greater common divisor, which the machine has somehow overlooked. This cannot be, however, because any common divisor of m and n has to divide $m'm + n'n$; so it has to divide d; so it has to be $\leqslant d$. Furthermore we can easily check that d does divide both m and n. (Algorithms that output their own proofs of correctness are called *self-certifying*.)

We'll be using (4.5) a lot in the rest of this chapter. One of its important consequences is the following mini-theorem:

$$k\backslash m \quad\text{and}\quad k\backslash n \qquad\Longleftrightarrow\qquad k\backslash \gcd(m,n)\,. \tag{4.6}$$

(Proof: If k divides both m and n, it divides $m'm + n'n$, so it divides $\gcd(m,n)$. Conversely, if k divides $\gcd(m,n)$, it divides a divisor of m and a divisor of n, so it divides both m and n.) We always knew that any common divisor of m and n must be *less than or equal to* their gcd; that's the definition of greatest common divisor. But now we know that any common divisor is, in fact, *a divisor of* their gcd.

Sometimes we need to do sums over all divisors of n. In this case it's often useful to use the handy rule

$$\sum_{m\backslash n} a_m \;=\; \sum_{m\backslash n} a_{n/m}\,, \qquad \text{integer } n > 0, \tag{4.7}$$

which holds since n/m runs through all divisors of n when m does. For example, when $n = 12$ this says that $a_1 + a_2 + a_3 + a_4 + a_6 + a_{12} = a_{12} + a_6 + a_4 + a_3 + a_2 + a_1$.

There's also a slightly more general identity,

$$\sum_{m\backslash n} a_m \;=\; \sum_k \sum_{m>0} a_m [n = mk]\,, \tag{4.8}$$

which is an immediate consequence of the definition (4.1). If n is positive, the right-hand side of (4.8) is $\sum_{k\backslash n} a_{n/k}$; hence (4.8) implies (4.7). And equation

(4.8) works also when n is negative. (In such cases, the nonzero terms on the right occur when k is the negative of a divisor of n.)

Moreover, a double sum over divisors can be "interchanged" by the law

$$\sum_{m\backslash n}\sum_{k\backslash m} a_{k,m} = \sum_{k\backslash n}\sum_{l\backslash(n/k)} a_{k,kl}\,. \tag{4.9}$$

For example, this law takes the following form when $n = 12$:

$$\begin{aligned}
&a_{1,1} + (a_{1,2}+a_{2,2}) + (a_{1,3}+a_{3,3})\\
&\qquad + (a_{1,4}+a_{2,4}+a_{4,4}) + (a_{1,6}+a_{2,6}+a_{3,6}+a_{6,6})\\
&\qquad + (a_{1,12}+a_{2,12}+a_{3,12}+a_{4,12}+a_{6,12}+a_{12,12})\\
&\quad = (a_{1,1}+a_{1,2}+a_{1,3}+a_{1,4}+a_{1,6}+a_{1,12})\\
&\qquad + (a_{2,2}+a_{2,4}+a_{2,6}+a_{2,12}) + (a_{3,3}+a_{3,6}+a_{3,12})\\
&\qquad + (a_{4,4}+a_{4,12}) + (a_{6,6}+a_{6,12}) + a_{12,12}\,.
\end{aligned}$$

We can prove (4.9) with Iversonian manipulation. The left-hand side is

$$\sum_{j,l}\sum_{k,m>0} a_{k,m}[n=jm][m=kl] = \sum_{j}\sum_{k,l>0} a_{k,kl}[n=jkl]\,;$$

the right-hand side is

$$\sum_{j,m}\sum_{k,l>0} a_{k,kl}[n=jk][n/k=ml] = \sum_{m}\sum_{k,l>0} a_{k,kl}[n=mlk]\,,$$

which is the same except for renaming the indices. This example indicates that the techniques we've learned in Chapter 2 will come in handy as we study number theory.

4.2 PRIMES

A positive integer p is called *prime* if it has just two divisors, namely 1 and p. *Throughout the rest of this chapter, the letter* p *will always stand for a prime number, even when we don't say so explicitly.* By convention, 1 isn't prime, so the sequence of primes starts out like this:

How about the p in 'explicitly'?

$$2, 3, 5, 7, 11, 13, 17, 19, 23, 29, 31, 37, 41, \ldots\,.$$

Some numbers look prime but aren't, like 91 $(= 7\cdot 13)$ and 161 $(= 7\cdot 23)$. These numbers and others that have three or more divisors are called *composite*. Every integer greater than 1 is either prime or composite, but not both.

Primes are of great importance, because they're the fundamental building blocks of all the positive integers. Any positive integer n can be written as a product of primes,

$$n = p_1 \ldots p_m = \prod_{k=1}^{m} p_k, \qquad p_1 \leqslant \cdots \leqslant p_m. \tag{4.10}$$

For example, $12 = 2\cdot2\cdot3$; $11011 = 7\cdot11\cdot11\cdot13$; $11111 = 41\cdot271$. (Products denoted by $\prod$ are analogous to sums denoted by $\sum$, as explained in exercise 2.25. If $m = 0$, we consider this to be an empty product, whose value is 1 by definition; that's the way $n = 1$ gets represented by (4.10).) Such a factorization is always possible because if $n > 1$ is not prime it has a divisor n_1 such that $1 < n_1 < n$; thus we can write $n = n_1 \cdot n_2$, and (by induction) we know that n_1 and n_2 can be written as products of primes.

Moreover, the expansion in (4.10) is *unique*: There's only one way to write n as a product of primes in nondecreasing order. This statement is called the Fundamental Theorem of Arithmetic, and it seems so obvious that we might wonder why it needs to be proved. How could there be two different sets of primes with the same product? Well, there can't, but the reason *isn't* simply "by definition of prime numbers." For example, if we consider the set of all real numbers of the form $m + n\sqrt{10}$ when m and n are integers, the product of any two such numbers is again of the same form, and we can call such a number "prime" if it can't be factored in a nontrivial way. The number 6 has two representations, $2\cdot3 = (4+\sqrt{10}\,)(4-\sqrt{10}\,)$; yet exercise 36 shows that 2, 3, $4+\sqrt{10}$, and $4-\sqrt{10}$ are all "prime" in this system.

Therefore we should prove rigorously that (4.10) is unique. There is certainly only one possibility when $n = 1$, since the product must be empty in that case; so let's suppose that $n > 1$ and that all smaller numbers factor uniquely. Suppose we have two factorizations

$$n = p_1 \ldots p_m = q_1 \ldots q_k, \qquad p_1 \leqslant \cdots \leqslant p_m \quad \text{and} \quad q_1 \leqslant \cdots \leqslant q_k,$$

where the p's and q's are all prime. We will prove that $p_1 = q_1$. If not, we can assume that $p_1 < q_1$, making p_1 smaller than all the q's. Since p_1 and q_1 are prime, their gcd must be 1; hence Euclid's self-certifying algorithm gives us integers a and b such that $ap_1 + bq_1 = 1$. Therefore

$$ap_1q_2 \ldots q_k + bq_1q_2 \ldots q_k = q_2 \ldots q_k.$$

Now p_1 divides both terms on the left, since $q_1q_2 \ldots q_k = n$; hence p_1 divides the right-hand side, $q_2 \ldots q_k$. But $q_2 \ldots q_k < n$, so it has a unique factorization (by induction); it cannot be divisible by p_1. This contradiction shows that p_1 must be equal to q_1 after all. Therefore we can divide both of n's factorizations by p_1, obtaining $p_2 \ldots p_m = q_2 \ldots q_k < n$. The other factors must likewise be equal (by induction), so our proof of uniqueness is complete.

It's the factorization, not the theorem, that's unique.

Sometimes it's more useful to state the Fundamental Theorem in another way: *Every positive integer can be written uniquely in the form*

$$n \;=\; \prod_p p^{n_p}\,, \qquad \text{where each } n_p \geqslant 0. \tag{4.11}$$

The right-hand side is a product over infinitely many primes; but for any particular n all but a few exponents are zero, so the corresponding factors are 1. Therefore it's really a finite product, just as many "infinite" sums are really finite because their terms are mostly zero.

Formula (4.11) represents n uniquely, so we can think of the sequence $\langle n_2, n_3, n_5, \ldots \rangle$ as a *number system* for positive integers. For example, the prime-exponent representation of 12 is $\langle 2, 1, 0, 0, \ldots \rangle$ and the prime-exponent representation of 18 is $\langle 1, 2, 0, 0, \ldots \rangle$. To multiply two numbers, we simply add their representations. In other words,

$$k = mn \qquad \iff \qquad k_p = m_p + n_p \quad \text{for all } p. \tag{4.12}$$

This implies that

$$m \backslash n \qquad \iff \qquad m_p \leqslant n_p \quad \text{for all } p, \tag{4.13}$$

and it follows immediately that

$$k = \gcd(m, n) \qquad \iff \qquad k_p = \min(m_p, n_p) \quad \text{for all } p; \tag{4.14}$$

$$k = \operatorname{lcm}(m, n) \qquad \iff \qquad k_p = \max(m_p, n_p) \quad \text{for all } p. \tag{4.15}$$

For example, since $12 = 2^2 \cdot 3^1$ and $18 = 2^1 \cdot 3^2$, we can get their gcd and lcm by taking the min and max of common exponents:

$$\begin{aligned} \gcd(12, 18) &= 2^{\min(2,1)} \cdot 3^{\min(1,2)} = 2^1 \cdot 3^1 = 6; \\ \operatorname{lcm}(12, 18) &= 2^{\max(2,1)} \cdot 3^{\max(1,2)} = 2^2 \cdot 3^2 = 36. \end{aligned}$$

If the prime p divides a product mn then it divides either m or n, perhaps both, because of the unique factorization theorem. But composite numbers do not have this property. For example, the nonprime 4 divides $60 = 6 \cdot 10$, but it divides neither 6 nor 10. The reason is simple: In the factorization $60 = 6 \cdot 10 = (2 \cdot 3)(2 \cdot 5)$, the two prime factors of $4 = 2 \cdot 2$ have been split into two parts, hence 4 divides neither part. But a prime is unsplittable, so it must divide one of the original factors.

4.3 PRIME EXAMPLES

How many primes are there? A lot. In fact, infinitely many. Euclid proved this long ago in his Theorem 9:20, as follows. Suppose there were only finitely many primes, say k of them—2, 3, 5, ..., P_k. Then, said Euclid, we should consider the number

$$M = 2 \cdot 3 \cdot 5 \cdot \ldots \cdot P_k + 1 .$$

None of the k primes can divide M, because each divides $M - 1$. Thus there must be some other prime that divides M; perhaps M itself is prime. This contradicts our assumption that 2, 3, ..., P_k are the only primes, so there must indeed be infinitely many.

"Οἱ πρῶτοι ἀριθμοὶ πλείους εἰσὶ παντὸς τοῦ προτεθέντος πλήθους πρώτων ἀριθμῶν."
—Euclid [80]

[Translation: "There are more primes than in any given set of primes."]

Euclid's proof suggests that we define *Euclid numbers* by the recurrence

$$\begin{aligned} e_1 &= 2; \\ e_n &= e_1 e_2 \ldots e_{n-1} + 1 . \end{aligned} \tag{4.16}$$

The sequence starts out

$$\begin{aligned} e_2 &= 2+1 = 3; \\ e_3 &= 2 \cdot 3 + 1 = 7; \\ e_4 &= 2 \cdot 3 \cdot 7 + 1 = 43; \end{aligned}$$

these are all prime. But the next case, e_5, is $1807 = 13 \cdot 139$. It turns out that $e_6 = 3263443$ is prime, while

$$\begin{aligned} e_7 &= 547 \cdot 607 \cdot 1033 \cdot 31051; \\ e_8 &= 29881 \cdot 67003 \cdot 9119521 \cdot 6212157481 . \end{aligned}$$

It is known that $e_9, \ldots, e_{17}$ are composite, and the remaining e_n are probably composite as well. However, the Euclid numbers are all *relatively prime* to each other; that is,

$$\gcd(e_m, e_n) = 1, \qquad \text{when } m \neq n.$$

Euclid's algorithm (what else?) tells us this in three short steps, because $e_n \bmod e_m = 1$ when $n > m$:

$$\gcd(e_m, e_n) = \gcd(1, e_m) = \gcd(0, 1) = 1 .$$

Therefore, if we let q_j be the smallest factor of e_j for all $j \geqslant 1$, the primes q_1, q_2, q_3, ... are all different. This is a sequence of infinitely many primes.

Let's pause to consider the Euclid numbers from the standpoint of Chapter 1. Can we express e_n in closed form? Recurrence (4.16) can be simplified by removing the three dots:

$$e_n = e_1 \ldots e_{n-2} e_{n-1} + 1 = (e_{n-1} - 1) e_{n-1} + 1 = e_{n-1}^2 - e_{n-1} + 1 .$$

Thus e_n has about twice as many decimal digits as e_{n-1}. Exercise 37 proves that there's a constant $E \approx 1.264$ such that

$$e_n = \lfloor E^{2^n} + \tfrac{1}{2} \rfloor . \tag{4.17}$$

And exercise 60 provides a similar formula that gives nothing but primes:

$$p_n = \lfloor P^{3^n} \rfloor , \tag{4.18}$$

for some constant P. But equations like (4.17) and (4.18) cannot really be considered to be in closed form, because the constants E and P are computed from the numbers e_n and p_n in a sort of sneaky way. No independent relation is known (or likely) that would connect them with other constants of mathematical interest.

Indeed, nobody knows *any* useful formula that gives arbitrarily large primes but only primes. Computer scientists at Chevron Geosciences did, however, strike mathematical oil in 1984. Using a program developed by David Slowinski, they discovered the largest known prime,

$$2^{216091} - 1 ,$$

while testing a new Cray X-MP supercomputer. It's easy to compute this number in a few milliseconds on a personal computer, because modern computers work in binary notation and this number is simply $(11\ldots1)_2$. All 216,091 of its bits are '1'. But it's much harder to prove that this number is prime. In fact, just about any computation with it takes a lot of time, because it's so large. For example, even a sophisticated algorithm requires several minutes just to convert $2^{216091} - 1$ to radix 10 on a PC. When printed out, its 65,050 decimal digits require 65 cents U.S. postage to mail first class.

Or probably more, by the time you read this.

Incidentally, $2^{216091} - 1$ is the number of moves necessary to solve the Tower of Hanoi problem when there are 216,091 disks. Numbers of the form

$$2^p - 1$$

(where p is prime, as always in this chapter) are called *Mersenne numbers*, after Father Marin Mersenne who investigated some of their properties in the seventeenth century. The Mersenne primes known to date occur for $p = 2$, 3, 5, 7, 13, 17, 19, 31, 61, 89, 107, 127, 521, 607, 1279, 2203, 2281, 3217, 4253, 4423, 9689, 9941, 11213, 19937, 21701, 23209, 44497, 86243, 110503, 132049, and 216091.

The number $2^n - 1$ can't possibly be prime if n is composite, because $2^{mn} - 1$ has $2^m - 1$ as a factor:

$$2^{mn} - 1 = (2^m - 1)(2^{m(n-1)} + 2^{m(n-2)} + \cdots + 1) .$$

But $2^p - 1$ isn't always prime when p is prime; $2^{11} - 1 = 2047 = 23 \cdot 89$ is the smallest such nonprime. (Mersenne knew this.)

Factoring and primality testing of large numbers are hot topics nowadays. A summary of what was known up to 1981 appears in Section 4.5.4 of [174], and many new results continue to be discovered. Pages 391–394 of that book explain a special way to test Mersenne numbers for primality.

For most of the last two hundred years, the largest known prime has been a Mersenne prime, although only 31 Mersenne primes are known. Many people are trying to find larger ones, but it's getting tough. So those really interested in fame (if not fortune) and a spot in *The Guinness Book of World Records* might instead try numbers of the form $2^n k + 1$, for small values of k like 3 or 5. These numbers can be tested for primality almost as quickly as Mersenne numbers can; exercise 4.5.4–27 of [174] gives the details.

We haven't fully answered our original question about how many primes there are. There are infinitely many, but some infinite sets are "denser" than others. For instance, among the positive integers there are infinitely many even numbers and infinitely many perfect squares, yet in several important senses there are more even numbers than perfect squares. One such sense looks at the size of the nth value. The nth even integer is $2n$ and the nth perfect square is n^2; since $2n$ is much less than n^2 for large n, the nth even integer occurs much sooner than the nth perfect square, so we can say there are many more even integers than perfect squares. A similar sense looks at the number of values not exceeding x. There are $\lfloor x/2 \rfloor$ such even integers and $\lfloor \sqrt{x} \rfloor$ perfect squares; since $x/2$ is much larger than $\sqrt{x}$ for large x, again we can say there are many more even integers.

Weird. I thought there were the same number of even integers as perfect squares, since there's a one-to-one correspondence between them.

What can we say about the primes in these two senses? It turns out that the nth prime, P_n, is about n times the natural log of n:

$$P_n \sim n \ln n .$$

(The symbol '$\sim$' can be read "is asymptotic to"; it means that the limit of the ratio $P_n/n \ln n$ is 1 as n goes to infinity.) Similarly, for the number of primes $\pi(x)$ not exceeding x we have what's known as the prime number theorem:

$$\pi(x) \sim \frac{x}{\ln x} .$$

Proving these two facts is beyond the scope of this book, although we can show easily that each of them implies the other. In Chapter 9 we will discuss the rates at which functions approach infinity, and we'll see that the function $n \ln n$, our approximation to P_n, lies between $2n$ and n^2 asymptotically. Hence there are fewer primes than even integers, but there are more primes than perfect squares.

These formulas, which hold only in the limit as n or $x \to \infty$, can be replaced by more exact estimates. For example, Rosser and Schoenfeld [253] have established the handy bounds

$$\ln x - \tfrac{3}{2} < \tfrac{x}{\pi(x)} < \ln x - \tfrac{1}{2}, \qquad \text{for } x \geqslant 67; \quad (4.19)$$

$$n(\ln n + \ln\ln n - \tfrac{3}{2}) < P_n < n(\ln n + \ln\ln n - \tfrac{1}{2}), \quad \text{for } n \geqslant 20. \quad (4.20)$$

If we look at a "random" integer n, the chances of its being prime are about one in $\ln n$. For example, if we look at numbers near 10^{16}, we'll have to examine about $16 \ln 10 \approx 36.8$ of them before finding a prime. (It turns out that there are exactly 10 primes between $10^{16} - 370$ and $10^{16} - 1$.) Yet the distribution of primes has many irregularities. For example, all the numbers between $P_1 P_2 \ldots P_n + 2$ and $P_1 P_2 \ldots P_n + P_{n+1} - 1$ inclusive are composite. Many examples of "twin primes" p and $p + 2$ are known (5 and 7, 11 and 13, 17 and 19, 29 and 31, ..., 9999999999999641 and 9999999999999643, ...), yet nobody knows whether or not there are infinitely many pairs of twin primes. (See Hardy and Wright [150, §1.4 and §2.8].)

One simple way to calculate all $\pi(x)$ primes $\leqslant x$ is to form the so-called sieve of Eratosthenes: First write down all integers from 2 through x. Next circle 2, marking it prime, and cross out all other multiples of 2. Then repeatedly circle the smallest uncircled, uncrossed number and cross out its other multiples. When everything has been circled or crossed out, the circled numbers are the primes. For example when $x = 10$ we write down 2 through 10, circle 2, then cross out its multiples 4, 6, 8, and 10. Next 3 is the smallest uncircled, uncrossed number, so we circle it and cross out 6 and 9. Now 5 is smallest, so we circle it and cross out 10. Finally we circle 7. The circled numbers are 2, 3, 5, and 7; so these are the $\pi(10) = 4$ primes not exceeding 10.

"Je me sers de la notation très simple $n!$ pour désigner le produit de nombres décroissans depuis n jusqu'à l'unité, savoir $n(n-1)(n-2)\ldots .3.2.1$. L'emploi continuel de l'analyse combinatoire que je fais dans la plupart de mes démonstrations, a rendu cette notation indispensable."
—Ch. Kramp [186]

4.4 FACTORIAL FACTORS

Now let's take a look at the factorization of some interesting highly composite numbers, the factorials:

$$n! = 1 \cdot 2 \cdot \ldots \cdot n = \prod_{k=1}^{n} k, \qquad \text{integer } n \geqslant 0. \quad (4.21)$$

According to our convention for an empty product, this defines $0!$ to be 1. Thus $n! = (n-1)!\,n$ for every positive integer n. This is the number of permutations of n distinct objects. That is, it's the number of ways to arrange n things in a row: There are n choices for the first thing; for each choice of first thing, there are $n-1$ choices for the second; for each of these $n(n-1)$ choices, there are $n-2$ for the third; and so on, giving $n(n-1)(n-2)\ldots(1)$

arrangements in all. Here are the first few values of the factorial function.

n	0	1	2	3	4	5	6	7	8	9	10
$n!$	1	1	2	6	24	120	720	5040	40320	362880	3628800

It's useful to know a few factorial facts, like the first six or so values, and the fact that 10! is about $3\frac{1}{2}$ million plus change; another interesting fact is that the number of digits in $n!$ exceeds n when $n \geqslant 25$.

We can prove that $n!$ is plenty big by using something like Gauss's trick of Chapter 1:

$$n!^2 = (1 \cdot 2 \cdot \ldots \cdot n)(n \cdot \ldots \cdot 2 \cdot 1) = \prod_{k=1}^{n} k(n+1-k).$$

We have $n \leqslant k(n+1-k) \leqslant \frac{1}{4}(n+1)^2$, since the quadratic polynomial $k(n+1-k) = \frac{1}{4}(n+1)^2 - \bigl(k - \frac{1}{2}(n+1)\bigr)^2$ has its smallest value at $k = 1$ and its largest value at $k = \frac{1}{2}(n+1)$. Therefore

$$\prod_{k=1}^{n} n \leqslant n!^2 \leqslant \prod_{k=1}^{n} \frac{(n+1)^2}{4};$$

that is,

$$n^{n/2} \leqslant n! \leqslant \frac{(n+1)^n}{2^n}. \qquad (4.22)$$

by Robbins 1955, Feller 1968 $\left(\sqrt{2\pi}\, n^{n+1/2} e^{-n+1/(12n+1)} < n! < \sqrt{2\pi}\, n^{n+1/2} e^{-n+\frac{1}{12n}}\right)$

This relation tells us that the factorial function grows exponentially!!

To approximate $n!$ more accurately for large n we can use Stirling's formula, which we will derive in Chapter 9:

$$n! \sim \sqrt{2\pi n}\left(\frac{n}{e}\right)^n. \qquad (4.23)$$

And a still more precise approximation tells us the asymptotic relative error: Stirling's formula undershoots $n!$ by a factor of about $1/(12n)$. Even for fairly small n this more precise estimate is pretty good. For example, Stirling's approximation (4.23) gives a value near 3598696 when $n = 10$, and this is about $0.83\% \approx 1/120$ too small. Good stuff, asymptotics.

But let's get back to primes. We'd like to determine, for any given prime p, the largest power of p that divides $n!$; that is, we want the exponent of p in $n!$'s unique factorization. We denote this number by $\epsilon_p(n!)$, and we start our investigations with the small case $p = 2$ and $n = 10$. Since 10! is the product of ten numbers, $\epsilon_2(10!)$ can be found by summing the powers-of-2 contributions of those ten numbers; this calculation corresponds to summing the columns of the following array:

	1	2	3	4	5	6	7	8	9	10	powers of 2
divisible by 2		x		x		x		x		x	$5 = \lfloor 10/2 \rfloor$
divisible by 4				x				x			$2 = \lfloor 10/4 \rfloor$
divisible by 8								x			$1 = \lfloor 10/8 \rfloor$
powers of 2	0	1	0	2	0	1	0	3	0	1	8

(The column sums form what's sometimes called the *ruler function* $\rho(k)$, because of their similarity to '┌┬┬┬┬┬┬┬┬┬┬┬┬┬┬┐', the lengths of lines marking fractions of an inch.) The sum of these ten sums is 8; hence 2^8 divides $10!$ but 2^9 doesn't.

There's also another way: We can sum the contributions of the rows. The first row marks the numbers that contribute a power of 2 (and thus are divisible by 2); there are $\lfloor 10/2 \rfloor = 5$ of them. The second row marks those that contribute an additional power of 2; there are $\lfloor 10/4 \rfloor = 2$ of them. And the third row marks those that contribute yet another; there are $\lfloor 10/8 \rfloor = 1$ of them. These account for all contributions, so we have $\epsilon_2(10!) = 5+2+1 = 8$.

For general n this method gives

$$\epsilon_2(n!) = \left\lfloor \frac{n}{2} \right\rfloor + \left\lfloor \frac{n}{4} \right\rfloor + \left\lfloor \frac{n}{8} \right\rfloor + \cdots = \sum_{k \geqslant 1} \left\lfloor \frac{n}{2^k} \right\rfloor .$$

This sum is actually finite, since the summand is zero when $2^k > n$. Therefore it has only $\lfloor \lg n \rfloor$ nonzero terms, and it's computationally quite easy. For instance, when $n = 100$ we have

$$\epsilon_2(100!) = 50 + 25 + 12 + 6 + 3 + 1 = 97 .$$

Each term is just the floor of half the previous term. This is true for all n, because as a special case of (3.11) we have $\lfloor n/2^{k+1} \rfloor = \lfloor \lfloor n/2^k \rfloor / 2 \rfloor$. It's especially easy to see what's going on here when we write the numbers in binary:

$$\begin{aligned}
100 &= (1100100)_2 = 100 \\
\lfloor 100/2 \rfloor &= (110010)_2 = 50 \\
\lfloor 100/4 \rfloor &= (11001)_2 = 25 \\
\lfloor 100/8 \rfloor &= (1100)_2 = 12 \\
\lfloor 100/16 \rfloor &= (110)_2 = 6 \\
\lfloor 100/32 \rfloor &= (11)_2 = 3 \\
\lfloor 100/64 \rfloor &= (1)_2 = 1
\end{aligned}$$

We merely drop the least significant bit from one term to get the next.

The binary representation also shows us how to derive another formula,

$$\epsilon_2(n!) = n - \nu_2(n), \tag{4.24}$$

where $\nu_2(n)$ is the number of 1's in the binary representation of n. This simplification works because each 1 that contributes 2^m to the value of n contributes $2^{m-1} + 2^{m-2} + \cdots + 2^0 = 2^m - 1$ to the value of $\epsilon_2(n!)$.

Generalizing our findings to an arbitrary prime p, we have

$$\epsilon_p(n!) = \left\lfloor \frac{n}{p} \right\rfloor + \left\lfloor \frac{n}{p^2} \right\rfloor + \left\lfloor \frac{n}{p^3} \right\rfloor + \cdots = \sum_{k \geqslant 1} \left\lfloor \frac{n}{p^k} \right\rfloor \tag{4.25}$$

by the same reasoning as before.

About how large is $\epsilon_p(n!)$? We get an easy (but good) upper bound by simply removing the floor from the summand and then summing an infinite geometric progression:

$$\begin{aligned}
\epsilon_p(n!) &< \frac{n}{p} + \frac{n}{p^2} + \frac{n}{p^3} + \cdots \\
&= \frac{n}{p}\left(1 + \frac{1}{p} + \frac{1}{p^2} + \cdots\right) \\
&= \frac{n}{p}\left(\frac{p}{p-1}\right) \\
&= \frac{n}{p-1}.
\end{aligned}$$

For $p = 2$ and $n = 100$ this inequality says that $97 < 100$. Thus the upper bound 100 is not only correct, it's also close to the true value 97. In fact, the true value $n - \nu_2(n)$ is $\sim n$ in general, because $\nu_2(n) \leqslant \lceil \lg n \rceil$ is asymptotically much smaller than n.

When $p = 2$ and 3 our formulas give $\epsilon_2(n!) \sim n$ and $\epsilon_3(n!) \sim n/2$, so it seems reasonable that every once in awhile $\epsilon_3(n!)$ should be exactly half as big as $\epsilon_2(n!)$. For example, this happens when $n = 6$ and $n = 7$, because $6! = 2^4 \cdot 3^2 \cdot 5 = 7!/7$. But nobody has yet proved that such coincidences happen infinitely often.

The bound on $\epsilon_p(n!)$ in turn gives us a bound on $p^{\epsilon_p(n!)}$, which is p's contribution to $n!$:

$$p^{\epsilon_p(n!)} < p^{n/(p-1)}.$$

And we can simplify this formula (at the risk of greatly loosening the upper bound) by noting that $p \leqslant 2^{p-1}$; hence $p^{n/(p-1)} \leqslant (2^{p-1})^{n/(p-1)} = 2^n$. In other words, the contribution that any prime makes to $n!$ is less than 2^n.

We can use this observation to get another proof that there are infinitely many primes. For if there were only the k primes 2, 3, ..., P_k, then we'd have $n! < (2^n)^k = 2^{nk}$ for all $n > 1$, since each prime can contribute at most a factor of $2^n - 1$. But we can easily contradict the inequality $n! < 2^{nk}$ by choosing n large enough, say $n = 2^{2k}$. Then

$$n! \;<\; 2^{nk} \;=\; 2^{2^{2k}k} \;=\; n^{n/2},$$

contradicting the inequality $n! \geqslant n^{n/2}$ that we derived in (4.22). There are infinitely many primes, still.

We can even beef up this argument to get a crude bound on $\pi(n)$, the number of primes not exceeding n. Every such prime contributes a factor of less than 2^n to $n!$; so, as before,

$$n! \;<\; 2^{n\pi(n)}.$$

If we replace $n!$ here by Stirling's approximation (4.23), which is a lower bound, and take logarithms, we get

$$n\pi(n) \;>\; n\lg(n/e) + \tfrac{1}{2}\lg(2\pi n);$$

hence

$$\pi(n) \;>\; \lg(n/e).$$

This lower bound is quite weak, compared with the actual value $\pi(n) \sim n/\ln n$, because $\log n$ is much smaller than $n/\log n$ when n is large. But we didn't have to work very hard to get it, and a bound is a bound.

4.5 RELATIVE PRIMALITY

When $\gcd(m, n) = 1$, the integers m and n have no prime factors in common and we say that they're *relatively prime.*

This concept is so important in practice, we ought to have a special notation for it; but alas, number theorists haven't come up with a very good one yet. Therefore we cry: HEAR US, O MATHEMATICIANS OF THE WORLD! LET US NOT WAIT ANY LONGER! WE CAN MAKE MANY FORMULAS CLEARER BY DEFINING A NEW NOTATION NOW! LET US AGREE TO WRITE '$m \perp n$', AND TO SAY "m IS PRIME TO n," IF m AND n ARE RELATIVELY PRIME. In other words, let us declare that

Like perpendicular lines don't have a common direction, perpendicular numbers don't have common factors.

$$m \perp n \quad\iff\quad m, n \text{ are integers and } \gcd(m, n) = 1. \qquad (4.26)$$

A fraction m/n is in lowest terms if and only if $m \perp n$. Since we reduce fractions to lowest terms by casting out the largest common factor of numerator and denominator, we suspect that, in general,

$$m/\gcd(m,n) \quad \perp \quad n/\gcd(m,n)\,; \tag{4.27}$$

and indeed this is true. It follows from a more general law, $\gcd(km, kn) = k\gcd(m,n)$, proved in exercise 14.

The $\perp$ relation has a simple formulation when we work with the prime-exponent representations of numbers, because of the gcd rule (4.14):

$$m \perp n \qquad \iff \qquad \min(m_p, n_p) = 0 \quad \text{for all } p. \tag{4.28}$$

Furthermore, since m_p and n_p are nonnegative, we can rewrite this as

The dot product is zero, like orthogonal vectors.

$$m \perp n \qquad \iff \qquad m_p n_p = 0 \quad \text{for all } p. \tag{4.29}$$

And now we can prove an important law by which we can split and combine two $\perp$ relations with the same left-hand side:

$$k \perp m \quad \text{and} \quad k \perp n \qquad \iff \qquad k \perp mn\,. \tag{4.30}$$

In view of (4.29), this law is another way of saying that $k_p m_p = 0$ and $k_p n_p = 0$ if and only if $k_p(m_p + n_p) = 0$, when m_p and n_p are nonnegative.

There's a beautiful way to construct the set of all nonnegative fractions m/n with $m \perp n$, called the *Stern–Brocot tree* because it was discovered independently by Moriz Stern [279], a German mathematician, and Achille Brocot [35], a French clockmaker. The idea is to start with the two fractions $(\frac{0}{1}, \frac{1}{0})$ and then to repeat the following operation as many times as desired:

Interesting how mathematicians will say "discovered" when absolutely anyone else would have said "invented."

$$\text{Insert } \frac{m+m'}{n+n'} \text{ between two adjacent fractions } \frac{m}{n} \text{ and } \frac{m'}{n'}\,.$$

The new fraction $(m+m')/(n+n')$ is called the *mediant* of m/n and m'/n'. For example, the first step gives us one new entry between $\frac{0}{1}$ and $\frac{1}{0}$,

$$\tfrac{0}{1}, \tfrac{1}{1}, \tfrac{1}{0};$$

and the next gives two more:

$$\tfrac{0}{1}, \tfrac{1}{2}, \tfrac{1}{1}, \tfrac{2}{1}, \tfrac{1}{0}.$$

The next gives four more,

$$\tfrac{0}{1}, \tfrac{1}{3}, \tfrac{1}{2}, \tfrac{2}{3}, \tfrac{1}{1}, \tfrac{3}{2}, \tfrac{2}{1}, \tfrac{3}{1}, \tfrac{1}{0};$$

and then we'll get 8, 16, and so on. The entire array can be regarded as an infinite binary tree structure whose top levels look like this:

I guess $1/0$ is infinity, "in lowest terms."

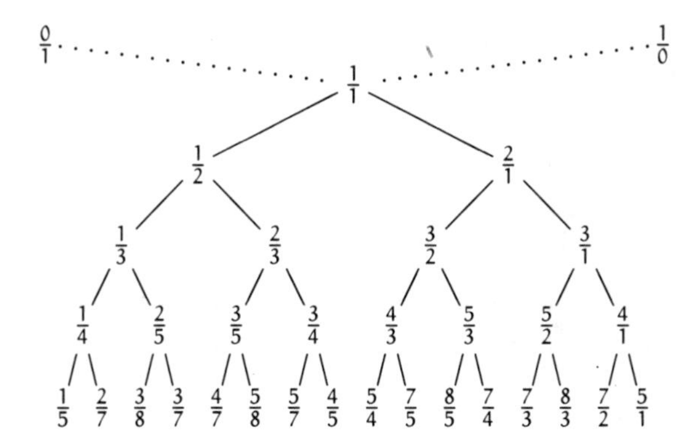

Each fraction is $\frac{m+m'}{n+n'}$, where $\frac{m}{n}$ is the nearest ancestor above and to the left, and $\frac{m'}{n'}$ is the nearest ancestor above and to the right. (An "ancestor" is a fraction that's reachable by following the branches upward.) Many patterns can be observed in this tree.

Why does this construction work? Why, for example, does each mediant fraction $(m+m')/(n+n')$ turn out to be in lowest terms when it appears in this tree? (If m, m', n, and n' were all odd, we'd get even/even; somehow the construction guarantees that fractions with odd numerators and denominators never appear next to each other.) And why do all possible fractions m/n occur exactly once? Why can't a particular fraction occur twice, or not at all?

Conserve parody.

All of these questions have amazingly simple answers, based on the following fundamental fact: *If m/n and m'/n' are consecutive fractions at any stage of the construction, we have*

$$m'n - mn' = 1. \tag{4.31}$$

This relation is true initially ($1\cdot 1 - 0\cdot 0 = 1$); and when we insert a new mediant $(m+m')/(n+n')$, the new cases that need to be checked are

$$(m+m')n - m(n+n') = 1;$$
$$m'(n+n') - (m+m')n' = 1.$$

Both of these equations are equivalent to the original condition (4.31) that they replace. Therefore (4.31) is invariant at all stages of the construction.

Furthermore, if $m/n < m'/n'$ and if all values are nonnegative, it's easy to verify that

$$m/n < (m+m')/(n+n') < m'/n'.$$

A mediant fraction isn't halfway between its progenitors, but it does lie somewhere in between. Therefore the construction preserves order, and we couldn't possibly get the same fraction in two different places.

True, but if you get a compound fracture you'd better go see a doctor.

One question still remains. Can any positive fraction a/b with $a \perp b$ possibly be omitted? The answer is no, because we can confine the construction to the immediate neighborhood of a/b, and in this region the behavior is easy to analyze: Initially we have

$$\tfrac{m}{n} = \tfrac{0}{1} < \left(\tfrac{a}{b}\right) < \tfrac{1}{0} = \tfrac{m'}{n'},$$

where we put parentheses around $\frac{a}{b}$ to indicate that it's not really present yet. Then if at some stage we have

$$\tfrac{m}{n} < \left(\tfrac{a}{b}\right) < \tfrac{m'}{n'},$$

the construction forms $(m+m')/(n+n')$ and there are three cases. Either $(m+m')/(n+n') = a/b$ and we win; or $(m+m')/(n+n') < a/b$ and we can set $m \leftarrow m+m'$, $n \leftarrow n+n'$; or $(m+m')/(n+n') > a/b$ and we can set $m' \leftarrow m+m'$, $n' \leftarrow n+n'$. This process cannot go on indefinitely, because the conditions

$$\tfrac{a}{b} - \tfrac{m}{n} > 0 \qquad \text{and} \qquad \tfrac{m'}{n'} - \tfrac{a}{b} > 0$$

imply that

$$an - bm \geqslant 1 \qquad \text{and} \qquad bm' - an' \geqslant 1;$$

hence

$$(m'+n')(an-bm) + (m+n)(bm'-an') \geqslant m'+n'+m+n;$$

and this is the same as $a+b \geqslant m'+n'+m+n$ by (4.31). Either m or n or m' or n' increases at each step, so we must win after at most $a+b$ steps.

The *Farey series* of order N, denoted by $\mathcal{F}_N$, is the set of all reduced fractions between 0 and 1 whose denominators are N or less, arranged in increasing order. For example, if $N = 6$ we have

$$\mathcal{F}_6 = \tfrac{0}{1}, \tfrac{1}{6}, \tfrac{1}{5}, \tfrac{1}{4}, \tfrac{1}{3}, \tfrac{2}{5}, \tfrac{1}{2}, \tfrac{3}{5}, \tfrac{2}{3}, \tfrac{3}{4}, \tfrac{4}{5}, \tfrac{5}{6}, \tfrac{1}{1}.$$

We can obtain $\mathcal{F}_N$ in general by starting with $\mathcal{F}_1 = \frac{0}{1}, \frac{1}{1}$ and then inserting mediants whenever it's possible to do so without getting a denominator that is too large. We don't miss any fractions in this way, because we know that the Stern–Brocot construction doesn't miss any, and because a mediant with denominator $\leqslant N$ is never formed from a fraction whose denominator is $> N$. (In other words, $\mathcal{F}_N$ defines a *subtree* of the Stern–Brocot tree, obtained by pruning off unwanted branches.) It follows that $m'n - mn' = 1$ whenever m/n and m'/n' are consecutive elements of a Farey series.

This method of construction reveals that $\mathcal{F}_N$ can be obtained in a simple way from $\mathcal{F}_{N-1}$: We simply insert the fraction $(m + m')/N$ between consecutive fractions m/n, m'/n' of $\mathcal{F}_{N-1}$ whose denominators sum to N. For example, it's easy to obtain $\mathcal{F}_7$ from the elements of $\mathcal{F}_6$, by inserting $\frac{1}{7}$, $\frac{2}{7}$, ..., $\frac{6}{7}$ according to the stated rule:

$$\mathcal{F}_7 = \tfrac{0}{1}, \tfrac{1}{7}, \tfrac{1}{6}, \tfrac{1}{5}, \tfrac{1}{4}, \tfrac{2}{7}, \tfrac{1}{3}, \tfrac{2}{5}, \tfrac{3}{7}, \tfrac{1}{2}, \tfrac{4}{7}, \tfrac{3}{5}, \tfrac{2}{3}, \tfrac{5}{7}, \tfrac{3}{4}, \tfrac{4}{5}, \tfrac{5}{6}, \tfrac{6}{7}, \tfrac{1}{1}.$$

When N is prime, $N - 1$ new fractions will appear; but otherwise we'll have fewer than $N - 1$, because this process generates only numerators that are relatively prime to N.

Long ago in (4.5) we proved—in different words—that whenever $m \perp n$ and $0 < m \leqslant n$ we can find integers a and b such that

$$ma - nb = 1. \qquad (4.32)$$

(Actually we said $m'm + n'n = \gcd(m, n)$, but we can write 1 for $\gcd(m, n)$, a for m', and b for $-n'$.) The Farey series gives us another proof of (4.32), because we can let b/a be the fraction that precedes m/n in $\mathcal{F}_n$. Thus (4.5) is just (4.31) again. For example, one solution to $3a - 7b = 1$ is $a = 5$, $b = 2$, since $\frac{2}{5}$ precedes $\frac{3}{7}$ in $\mathcal{F}_7$. This construction implies that we can always find a solution to (4.32) with $0 \leqslant b < a < n$, if $0 < m \leqslant n$. Similarly, if $0 \leqslant n < m$ and $m \perp n$, we can solve (4.32) with $0 < a \leqslant b \leqslant m$ by letting a/b be the fraction that *follows* n/m in $\mathcal{F}_m$.

Sequences of three consecutive terms in a Farey series have an amazing property that is proved in exercise 61. But we had better not discuss the Farey series any further, because the entire Stern–Brocot tree turns out to be even more interesting.

Farey 'nough.

We can, in fact, regard the Stern–Brocot tree as a *number system* for representing rational numbers, because each positive, reduced fraction occurs exactly once. Let's use the letters L and R to stand for going down to the left or right branch as we proceed from the root of the tree to a particular fraction; then a string of L's and R's uniquely identifies a place in the tree. For example, LRRL means that we go left from $\frac{1}{1}$ down to $\frac{1}{2}$, then right to $\frac{2}{3}$, then right to $\frac{3}{4}$, then left to $\frac{5}{7}$. We can consider LRRL to be a representation of $\frac{5}{7}$. Every positive fraction gets represented in this way as a unique string of L's and R's.

Well, actually there's a slight problem: The fraction $\frac{1}{1}$ corresponds to the *empty* string, and we need a notation for that. Let's agree to call it I, because that looks something like 1 and it stands for "identity."

This representation raises two natural questions: (1) Given positive integers m and n with $m \perp n$, what is the string of L's and R's that corresponds to m/n? (2) Given a string of L's and R's, what fraction corresponds to it? Question 2 seems easier, so let's work on it first. We define

$$f(S) = \text{fraction corresponding to } S$$

when S is a string of L's and R's. For example, $f(LRRL) = \frac{5}{7}$.

According to the construction, $f(S) = (m+m')/(n+n')$ if m/n and m'/n' are the closest fractions preceding and following S in the upper levels of the tree. Initially $m/n = 0/1$ and $m'/n' = 1/0$; then we successively replace either m/n or m'/n' by the mediant $(m+m')/(n+n')$ as we move right or left in the tree, respectively.

How can we capture this behavior in mathematical formulas that are easy to deal with? A bit of experimentation suggests that the best way is to maintain a 2×2 matrix

$$M(S) = \begin{pmatrix} n & n' \\ m & m' \end{pmatrix}$$

that holds the four quantities involved in the ancestral fractions m/n and m'/n' enclosing S. We could put the m's on top and the n's on the bottom, fractionwise; but this upside-down arrangement works out more nicely because we have $M(I) = \binom{1\,0}{0\,1}$ when the process starts, and $\binom{1\,0}{0\,1}$ is traditionally called the identity matrix I.

A step to the left replaces n' by $n+n'$ and m' by $m+m'$; hence

$$M(SL) = \begin{pmatrix} n & n+n' \\ m & m+m' \end{pmatrix} = \begin{pmatrix} n & n' \\ m & m' \end{pmatrix} \begin{pmatrix} 1 & 1 \\ 0 & 1 \end{pmatrix} = M(S) \begin{pmatrix} 1 & 1 \\ 0 & 1 \end{pmatrix}.$$

(This is a special case of the general rule

$$\begin{pmatrix} a & b \\ c & d \end{pmatrix} \begin{pmatrix} w & x \\ y & z \end{pmatrix} = \begin{pmatrix} aw+by & ax+bz \\ cw+dy & cx+dz \end{pmatrix}$$

for multiplying 2×2 matrices.) Similarly it turns out that

If you're clueless about matrices, don't panic; this book uses them only here.

$$M(SR) = \begin{pmatrix} n+n' & n' \\ m+m' & m' \end{pmatrix} = M(S) \begin{pmatrix} 1 & 0 \\ 1 & 1 \end{pmatrix}.$$

Therefore if we define L and R as 2×2 matrices,

$$L = \begin{pmatrix} 1 & 1 \\ 0 & 1 \end{pmatrix}, \quad R = \begin{pmatrix} 1 & 0 \\ 1 & 1 \end{pmatrix}, \tag{4.33}$$

we get the simple formula $M(S) = S$, by induction on the length of S. Isn't that nice? (The letters L and R serve dual roles, as matrices and as letters in the string representation.) For example,

$$M(LRRL) = LRRL = \binom{1\,1}{0\,1}\binom{1\,0}{1\,1}\binom{1\,0}{1\,1}\binom{1\,1}{0\,1} = \binom{2\,1}{1\,1}\binom{1\,1}{1\,2} = \binom{3\,4}{2\,3};$$

the ancestral fractions that enclose $LRRL = \frac{5}{7}$ are $\frac{2}{3}$ and $\frac{3}{4}$. And this construction gives us the answer to Question 2:

$$f(S) = f\left(\begin{pmatrix} n & n' \\ m & m' \end{pmatrix}\right) = \frac{m+m'}{n+n'}. \tag{4.34}$$

How about Question 1? That's easy, now that we understand the fundamental connection between tree nodes and 2×2 matrices. Given a pair of positive integers m and n, with $m \perp n$, we can find the position of m/n in the Stern–Brocot tree by "binary search" as follows:

$S := I;$
while $m/n \neq f(S)$ **do**
  if $m/n < f(S)$ **then** (output(L); $S := SL$)
          else (output(R); $S := SR$).

This outputs the desired string of L's and R's.

There's also another way to do the same job, by changing m and n instead of maintaining the state S. If S is any 2×2 matrix, we have

$$f(RS) = f(S) + 1$$

because RS is like S but with the top row added to the bottom row. (Let's look at it in slow motion:

$$S = \begin{pmatrix} n & n' \\ m & m' \end{pmatrix}; \qquad RS = \begin{pmatrix} n & n' \\ m+n & m'+n' \end{pmatrix};$$

hence $f(S) = (m+m')/(n+n')$ and $f(RS) = \bigl((m+n)+(m'+n')\bigr)/(n+n')$.) If we carry out the binary search algorithm on a fraction m/n with $m > n$, the first output will be R; hence the subsequent behavior of the algorithm will have $f(S)$ exactly 1 greater than if we had begun with $(m-n)/n$ instead of m/n. A similar property holds for L, and we have

$$\frac{m}{n} = f(RS) \iff \frac{m-n}{n} = f(S), \quad \text{when } m > n;$$

$$\frac{m}{n} = f(LS) \iff \frac{m}{n-m} = f(S), \quad \text{when } m < n.$$

This means that we can transform the binary search algorithm to the following matrix-free procedure:

$$\begin{aligned}&\textbf{while}\ m \ne n\ \textbf{do}\\ &\qquad \textbf{if}\ m < n\ \textbf{then}\ \big(\mathrm{output}(L);\ n := n - m\big)\\ &\qquad\qquad\qquad \textbf{else}\ \big(\mathrm{output}(R);\ m := m - n\big)\,.\end{aligned}$$

For example, given $m/n = 5/7$, we have successively

$$\begin{array}{rccccc} m = & 5 & 5 & 3 & 1 & 1\\ n = & 7 & 2 & 2 & 2 & 1\\ \text{output} & L & R & R & L & \end{array}$$

in the simplified algorithm.

Irrational numbers don't appear in the Stern–Brocot tree, but all the rational numbers that are "close" to them do. For example, if we try the binary search algorithm with the number $e = 2.71828\ldots$, instead of with a fraction m/n, we'll get an infinite string of L's and R's that begins

$$\mathrm{RRLRRLRLLLLRLRRRRRRLRLLLLLLLLRLR}\ldots\,.$$

We can consider this infinite string to be the representation of e in the Stern–Brocot number system, just as we can represent e as an infinite decimal $2.718281828459\ldots$ or as an infinite binary fraction $(10.101101111110\ldots)_2$. Incidentally, it turns out that e's representation has a regular pattern in the Stern–Brocot system:

$$e = \mathrm{RL^0RLR^2LRL^4RLR^6LRL^8RLR^{10}LRL^{12}RL}\ldots\,;$$

this is equivalent to a special case of something that Euler [84] discovered when he was 24 years old.

From this representation we can deduce that the fractions

$$\begin{array}{ccccccccccccccccccc} R & R & L & R & R & L & R & L & L & L & L & R & L & R & R & R & R & R & R\\ \frac{1}{1}, & \frac{2}{1}, & \frac{3}{1}, & \frac{5}{2}, & \frac{8}{3}, & \frac{11}{4}, & \frac{19}{7}, & \frac{30}{11}, & \frac{49}{18}, & \frac{68}{25}, & \frac{87}{32}, & \frac{106}{39}, & \frac{193}{71}, & \frac{299}{110}, & \frac{492}{181}, & \frac{685}{252}, & \frac{878}{323}, & \frac{1071}{394}, & \frac{1264}{465}, \ldots \end{array}$$

are the simplest rational upper and lower approximations to e. For if m/n does not appear in this list, then some fraction in this list whose numerator is $\le m$ and whose denominator is $\le n$ lies between m/n and e. For example, $\frac{27}{10}$ is not as simple an approximation as $\frac{19}{7} = 2.714\ldots$, which appears in the list and is closer to e. We can see this because the Stern–Brocot tree not only includes all rationals, it includes them in order, and because all fractions with small numerator and denominator appear above all less simple ones. Thus, $\frac{27}{10} = \mathrm{RRLRRLL}$ is less than $\frac{19}{7} = \mathrm{RRLRRL}$, which is less than

$e = \mathrm{RRLRRLR}\ldots$. Excellent approximations can be found in this way. For example, $\frac{1264}{465} \approx 2.718280$ agrees with e to six decimal places; we obtained this fraction from the first 19 letters of e's Stern–Brocot representation, and the accuracy is about what we would get with 19 bits of e's binary representation.

We can find the infinite representation of an irrational number α by a simple modification of the matrix-free binary search procedure:

$$\begin{aligned} &\textbf{if } \alpha < 1 \textbf{ then } \bigl(\text{output}(\mathrm{L});\ \ \alpha := \alpha/(1-\alpha)\bigr) \\ &\qquad\qquad \textbf{else } \ \bigl(\text{output}(\mathrm{R});\ \ \alpha := \alpha - 1\bigr)\,. \end{aligned}$$

(These steps are to be repeated infinitely many times, or until we get tired.) If α is rational, the infinite representation obtained in this way is the same as before but with RL^∞ appended at the right of α's (finite) representation. For example, if $\alpha = 1$, we get $\mathrm{RLLL}\ldots$, corresponding to the infinite sequence of fractions $\frac{1}{1}, \frac{2}{1}, \frac{3}{2}, \frac{4}{3}, \frac{5}{4}, \ldots$, which approach 1 in the limit. This situation is exactly analogous to ordinary binary notation, if we think of L as 0 and R as 1: Just as every real number x in $[0,1)$ has an infinite binary representation $(.b_1b_2b_3\ldots)_2$ not ending with all 1's, every real number α in $[0,\infty)$ has an infinite Stern–Brocot representation $B_1B_2B_3\ldots$ not ending with all R's. Thus we have a one-to-one order-preserving correspondence between $[0,1)$ and $[0,\infty)$ if we let $0 \leftrightarrow \mathrm{L}$ and $1 \leftrightarrow \mathrm{R}$.

There's an intimate relationship between Euclid's algorithm and the Stern–Brocot representations of rationals. Given $\alpha = m/n$, we get $\lfloor m/n \rfloor$ R's, then $\lfloor n/(m \bmod n) \rfloor$ L's, then $\lfloor (m \bmod n)/\bigl(n \bmod (m \bmod n)\bigr) \rfloor$ R's, and so on. These numbers $m \bmod n$, $n \bmod (m \bmod n)$, ... are just the values examined in Euclid's algorithm. (A little fudging is needed at the end to make sure that there aren't infinitely many R's.) We will explore this relationship further in Chapter 6.

4.6 'MOD': THE CONGRUENCE RELATION

Modular arithmetic is one of the main tools provided by number theory. We got a glimpse of it in Chapter 3 when we used the binary operation 'mod', usually as one operation amidst others in an expression. In this chapter we will use 'mod' also with entire equations, for which a slightly different notation is more convenient:

"Numerorum congruentiam hoc signo, $\equiv$, in posterum denotabimus, modulum udi opus erit in clausulis adiungentes, $-16 \equiv 9$ (mod. 5), $-7 \equiv 15$ (modo 11)."
—C. F. Gauss [115]

$$a \equiv b \pmod m \qquad \iff \qquad a \bmod m = b \bmod m\,. \tag{4.35}$$

For example, $9 \equiv -16 \pmod 5$, because $9 \bmod 5 = 4 = (-16) \bmod 5$. The formula '$a \equiv b \pmod m$' can be read "$a$ is congruent to b modulo m." The definition makes sense when a, b, and m are arbitrary real numbers, but we almost always use it with integers only.

Since $x \bmod m$ differs from x by a multiple of m, we can understand congruences in another way:

$$a \equiv b \pmod{m} \qquad \iff \qquad a - b \text{ is a multiple of } m. \tag{4.36}$$

For if $a \bmod m = b \bmod m$, then the definition of 'mod' in (3.21) tells us that $a - b = a \bmod m + km - (b \bmod m + lm) = (k - l)m$ for some integers k and l. Conversely if $a - b = km$, then $a = b$ if $m = 0$; otherwise

$$\begin{aligned} a \bmod m \;=\; a - \lfloor a/m \rfloor m \;&=\; b + km - \lfloor (b + km)/m \rfloor m \\ &=\; b - \lfloor b/m \rfloor m \;=\; b \bmod m\,. \end{aligned}$$

The characterization of $\equiv$ in (4.36) is often easier to apply than (4.35). For example, we have $8 \equiv 23 \pmod 5$ because $8 - 23 = -15$ is a multiple of 5; we don't have to compute both $8 \bmod 5$ and $23 \bmod 5$.

The congruence sign '$\equiv$' looks conveniently like '$=$', because congruences are almost like equations. For example, congruence is an *equivalence relation*; that is, it satisfies the reflexive law '$a \equiv a$', the symmetric law '$a \equiv b \;\Rightarrow\; b \equiv a$', and the transitive law '$a \equiv b \equiv c \;\Rightarrow\; a \equiv c$'. All these properties are easy to prove, because any relation '$\equiv$' that satisfies '$a \equiv b \iff f(a) = f(b)$' for some function f is an equivalence relation. (In our case, $f(x) = x \bmod m$.) Moreover, we can add and subtract congruent elements without losing congruence:

"I feel fine today modulo a slight headache."
— *The Hacker's Dictionary [277]*

$$\begin{aligned} a \equiv b \quad\text{and}\quad c \equiv d \qquad &\Longrightarrow \qquad a + c \equiv b + d \pmod{m}\,; \\ a \equiv b \quad\text{and}\quad c \equiv d \qquad &\Longrightarrow \qquad a - c \equiv b - d \pmod{m}\,. \end{aligned}$$

For if $a - b$ and $c - d$ are both multiples of m, so are $(a + c) - (b + d) = (a - b) + (c - d)$ and $(a - c) - (b - d) = (a - b) - (c - d)$. Incidentally, it isn't necessary to write '(mod m)' once for every appearance of '$\equiv$'; if the modulus is constant, we need to name it only once in order to establish the context. This is one of the great conveniences of congruence notation.

Multiplication works too, provided that we are dealing with integers:

$$a \equiv b \quad\text{and}\quad c \equiv d \qquad \Longrightarrow \qquad ac \equiv bd \pmod{m}\,, \quad \text{integers } b, c.$$

Proof: $ac - bd = (a - b)c + b(c - d)$. Repeated application of this multiplication property now allows us to take powers:

$$a \equiv b \qquad \Longrightarrow \qquad a^n \equiv b^n \pmod{m}\,, \qquad \text{integers } a, b;\ \text{integer } n \geqslant 0.$$

For example, since $2 \equiv -1 \pmod 3$, we have $2^n \equiv (-1)^n \pmod 3$; this means that $2^n - 1$ is a multiple of 3 if and only if n is even.

Thus, most of the algebraic operations that we customarily do with equations can also be done with congruences. Most, but not all. The operation of division is conspicuously absent. If $ad \equiv bd \pmod m$, we can't always conclude that $a \equiv b$. For example, $3 \cdot 2 \equiv 5 \cdot 2 \pmod 4$, but $3 \not\equiv 5$.

We can salvage the cancellation property for congruences, however, in the common case that d and m are relatively prime:

$$ad \equiv bd \iff a \equiv b \pmod m, \quad \text{integers } a, b, d, m \text{ and } d \perp m. \tag{4.37}$$

For example, it's legit to conclude from $15 \equiv 35 \pmod m$ that $3 \equiv 7 \pmod m$, unless the modulus m is a multiple of 5.

To prove this property, we use the extended gcd law (4.5) again, finding d' and m' such that $d'd + m'm = 1$. Then if $ad \equiv bd$ we can multiply both sides of the congruence by d', obtaining $ad'd \equiv bd'd$. Since $d'd \equiv 1$, we have $ad'd \equiv a$ and $bd'd \equiv b$; hence $a \equiv b$. This proof shows that the number d' acts almost like $1/d$ when congruences are considered $\pmod m$; therefore we call it the "inverse of d modulo m."

Another way to apply division to congruences is to divide the modulus as well as the other numbers:

$$ad \equiv bd \pmod{md} \iff a \equiv b \pmod m, \quad \text{for } d \neq 0. \tag{4.38}$$

This law holds for all real a, b, d, and m, because it depends only on the distributive law $(a \bmod m)d = ad \bmod md$: We have $a \bmod m = b \bmod m \iff (a \bmod m)d = (b \bmod m)d \iff ad \bmod md = bd \bmod md$. Thus, for example, from $3 \cdot 2 \equiv 5 \cdot 2 \pmod 4$ we conclude that $3 \equiv 5 \pmod 2$.

We can combine (4.37) and (4.38) to get a general law that changes the modulus as little as possible:

$$\begin{aligned} &ad \equiv bd \pmod m \\ &\qquad \iff a \equiv b \left(\bmod \frac{m}{\gcd(d, m)}\right), \quad \text{integer } a, b, d, m. \end{aligned} \tag{4.39}$$

For we can multiply $ad \equiv bd$ by d', where $d'd + m'm = \gcd(d, m)$; this gives the congruence $a \cdot \gcd(d, m) \equiv b \cdot \gcd(d, m) \pmod m$, which can be divided by $\gcd(d, m)$.

Let's look a bit further into this idea of changing the modulus. If we know that $a \equiv b \pmod{100}$, then we also must have $a \equiv b \pmod{10}$, or modulo any divisor of 100. It's stronger to say that $a - b$ is a multiple of 100 than to say that it's a multiple of 10. In general,

$$a \equiv b \pmod{md} \quad\Longrightarrow\quad a \equiv b \pmod{m}, \quad \text{integer } d, \tag{4.40}$$

because any multiple of md is a multiple of m.

Conversely, if we know that $a \equiv b$ with respect to two small moduli, can we conclude that $a \equiv b$ with respect to a larger one? Yes; the rule is

Modulitos?

$$\begin{aligned} &a \equiv b \pmod{m} \quad\text{and}\quad a \equiv b \pmod{n} \\ &\qquad\Longleftrightarrow\quad a \equiv b \pmod{\operatorname{lcm}(m,n)}, \qquad \text{integer } m,n > 0. \end{aligned} \tag{4.41}$$

For example, if we know that $a \equiv b$ modulo 12 and 18, we can safely conclude that $a \equiv b \pmod{36}$. The reason is that if $a - b$ is a common multiple of m and n, it is a multiple of $\operatorname{lcm}(m,n)$. This follows from the principle of unique factorization.

The special case $m \perp n$ of this law is extremely important, because $\operatorname{lcm}(m,n) = mn$ when m and n are relatively prime. Therefore we will state it explicitly:

$$\begin{aligned} &a \equiv b \pmod{mn} \\ &\quad\Longleftrightarrow\quad a \equiv b \pmod{m} \quad\text{and}\quad a \equiv b \pmod{n}, \quad \text{if } m \perp n. \end{aligned} \tag{4.42}$$

For example, $a \equiv b \pmod{100}$ if and only if $a \equiv b \pmod{25}$ and $a \equiv b \pmod{4}$. Saying this another way, if we know $x \bmod 25$ and $x \bmod 4$, then we have enough facts to determine $x \bmod 100$. This is a special case of the *Chinese Remainder Theorem* (see exercise 30), so called because it was discovered by Sun Tsŭ in China, about A.D. 350.

The moduli m and n in (4.42) can be further decomposed into relatively prime factors until every distinct prime has been isolated. Therefore

$$a \equiv b \pmod{m} \qquad\Longleftrightarrow\qquad a \equiv b \pmod{p^{m_p}} \;\text{ for all } p,$$

if the prime factorization (4.11) of m is $\prod_p p^{m_p}$. Congruences modulo powers of primes are the building blocks for all congruences modulo integers.

4.7 INDEPENDENT RESIDUES

One of the important applications of congruences is a *residue number system*, in which an integer x is represented as a sequence of residues (or remainders) with respect to moduli that are prime to each other:

$$\operatorname{Res}(x) = (x \bmod m_1, \ldots, x \bmod m_r), \qquad \text{if } m_j \perp m_k \text{ for } 1 \leqslant j < k \leqslant r.$$

Knowing $x \bmod m_1$, ..., $x \bmod m_r$ doesn't tell us everything about x. But it does allow us to determine $x \bmod m$, where m is the product $m_1 \ldots m_r$.

In practical applications we'll often know that x lies in a certain range; then we'll know everything about x if we know $x \bmod m$ and if m is large enough.

For example, let's look at a small case of a residue number system that has only two moduli, 3 and 5:

$x \bmod 15$	$x \bmod 3$	$x \bmod 5$
0	0	0
1	1	1
2	2	2
3	0	3
4	1	4
5	2	0
6	0	1
7	1	2
8	2	3
9	0	4
10	1	0
11	2	1
12	0	2
13	1	3
14	2	4

Each ordered pair $(x \bmod 3, x \bmod 5)$ is different, because $x \bmod 3 = y \bmod 3$ and $x \bmod 5 = y \bmod 5$ if and only if $x \bmod 15 = y \bmod 15$.

We can perform addition, subtraction, and multiplication on the two components *independently*, because of the rules of congruences. For example, if we want to multiply $7 = (1,2)$ by $13 = (1,3)$ modulo 15, we calculate $1{\cdot}1 \bmod 3 = 1$ and $2{\cdot}3 \bmod 5 = 1$. The answer is $(1,1) = 1$; hence $7{\cdot}13 \bmod 15$ must equal 1. Sure enough, it does.

This independence principle is useful in computer applications, because different components can be worked on separately (for example, by different computers). If each modulus m_k is a distinct prime p_k, chosen to be slightly less than 2^{31}, then a computer whose basic arithmetic operations handle integers in the range $[-2^{31}, 2^{31})$ can easily compute sums, differences, and products modulo p_k. A set of r such primes makes it possible to add, subtract, and multiply "multiple-precision numbers" of up to almost $31r$ bits, and the residue system makes it possible to do this faster than if such large numbers were added, subtracted, or multiplied in other ways.

For example, the Mersenne prime
$2^{31} - 1$
works well.

We can even do division, in appropriate circumstances. For example, suppose we want to compute the exact value of a large determinant of integers. The result will be an integer D, and bounds on $|D|$ can be given based on the size of its entries. But the only fast ways known for calculating determinants require division, and this leads to fractions (and loss of accuracy, if we resort to binary approximations). The remedy is to evaluate $D \bmod p_k = D_k$, for various large primes p_k. We can safely divide modulo p_k unless the divisor happens to be a multiple of p_k. That's very unlikely, but if it does happen we can choose another prime. Finally, knowing D_k for sufficiently many primes, we'll have enough information to determine D.

But we haven't explained how to get from a given sequence of residues $(x \bmod m_1, \ldots, x \bmod m_r)$ back to $x \bmod m$. We've shown that this conversion can be done in principle, but the calculations might be so formidable that they might rule out the idea in practice. Fortunately, there is a reasonably simple way to do the job, and we can illustrate it in the situation $(x \bmod 3, x \bmod 5)$ shown in our little table. The key idea is to solve the problem in the two cases $(1,0)$ and $(0,1)$; for if $(1,0) = a$ and $(0,1) = b$, then $(x,y) = (ax + by) \bmod 15$, since congruences can be multiplied and added.

In our case $a = 10$ and $b = 6$, by inspection of the table; but how could we find a and b when the moduli are huge? In other words, if $m \perp n$, what is a good way to find numbers a and b such that the equations

$$a \bmod m = 1, \quad a \bmod n = 0, \quad b \bmod m = 0, \quad b \bmod n = 1$$

all hold? Once again, (4.5) comes to the rescue: With Euclid's algorithm, we can find m' and n' such that

$$m'm + n'n = 1 .$$

Therefore we can take $a = n'n$ and $b = m'm$, reducing them both mod mn if desired.

Further tricks are needed in order to minimize the calculations when the moduli are large; the details are beyond the scope of this book, but they can be found in [174, page 274]. Conversion from residues to the corresponding original numbers is feasible, but it is sufficiently slow that we save total time only if a sequence of operations can all be done in the residue number system before converting back.

Let's firm up these congruence ideas by trying to solve a little problem: How many solutions are there to the congruence

$$x^2 \equiv 1 \pmod{m}, \qquad (4.43)$$

if we consider two solutions x and x' to be the same when $x \equiv x'$?

According to the general principles explained earlier, we should consider first the case that m is a prime power, p^k, where $k > 0$. Then the congruence $x^2 \equiv 1$ can be written

$$(x-1)(x+1) \equiv 0 \pmod{p^k},$$

so p must divide either $x-1$ or $x+1$, or both. But p can't divide both $x-1$ and $x+1$ unless $p=2$; we'll leave that case for later. If $p>2$, then $p^k \backslash (x-1)(x+1) \iff p^k \backslash (x-1)$ or $p^k \backslash (x+1)$; so there are exactly two solutions, $x \equiv +1$ and $x \equiv -1$.

The case $p=2$ is a little different. If $2^k \backslash (x-1)(x+1)$ then either $x-1$ or $x+1$ is divisible by 2 but not by 4, so the other one must be divisible by 2^{k-1}. This means that we have four solutions when $k \geqslant 3$, namely $x \equiv \pm 1$ and $x \equiv 2^{k-1} \pm 1$. (For example, when $p^k = 8$ the four solutions are $x \equiv 1, 3, 5, 7 \pmod 8$; it's often useful to know that *the square of any odd integer has the form* $8n+1$.)

Now $x^2 \equiv 1 \pmod m$ if and only if $x^2 \equiv 1 \pmod{p^{m_p}}$ for all primes p with $m_p > 0$ in the complete factorization of m. Each prime is independent of the others, and there are exactly two possibilities for $x \bmod p^{m_p}$ except when $p=2$. Therefore if m has exactly r different prime divisors, the total number of solutions to $x^2 \equiv 1$ is 2^r, except for a correction when m is even. The exact number in general is

All primes are odd except 2, which is the oddest of all.

$$2^{r+[8\backslash m]+[4\backslash m]-[2\backslash m]}. \tag{4.44}$$

For example, there are four "square roots of unity modulo 12," namely 1, 5, 7, and 11. When $m=15$ the four are those whose residues mod 3 and mod 5 are ± 1, namely $(1,1)$, $(1,4)$, $(2,1)$, and $(2,4)$ in the residue number system. These solutions are 1, 4, 11, and 14 in the ordinary (decimal) number system.

4.8 ADDITIONAL APPLICATIONS

There's some unfinished business left over from Chapter 3: We wish to prove that the m numbers

$$0 \bmod m,\quad n \bmod m,\quad 2n \bmod m,\quad \ldots,\quad (m-1)n \bmod m \tag{4.45}$$

consist of precisely d copies of the m/d numbers

$$0,\quad d,\quad 2d,\quad \ldots,\quad m-d$$

in some order, where $d = \gcd(m,n)$. For example, when $m=12$ and $n=8$ we have $d=4$, and the numbers are 0, 8, 4, 0, 8, 4, 0, 8, 4, 0, 8, 4.

The first part of the proof—to show that we get d copies of the first m/d values—is now trivial. We have

Mathematicians love to say that things are trivial.

$$jn \equiv kn \pmod m \qquad \iff \qquad j(n/d) \equiv k(n/d) \pmod{m/d}$$

by (4.38); hence we get d copies of the values that occur when $0 \leqslant k < m/d$.

Now we must show that those m/d numbers are $\{0, d, 2d, \ldots, m-d\}$ in some order. Let's write $m = m'd$ and $n = n'd$. Then $kn \bmod m = d(kn' \bmod m')$, by the distributive law (3.23); so the values that occur when $0 \leqslant k < m'$ are d times the numbers

$$0 \bmod m', \quad n' \bmod m', \quad 2n' \bmod m', \quad \ldots, \quad (m'-1)n' \bmod m'.$$

But we know that $m' \perp n'$ by (4.27); we've divided out their gcd. Therefore we need only consider the case $d = 1$, namely the case that m and n are relatively prime.

So let's assume that $m \perp n$. In this case it's easy to see that the numbers (4.45) are just $\{0, 1, \ldots, m-1\}$ in some order, by using the "pigeonhole principle." This principle states that if m pigeons are put into m pigeonholes, there is an empty hole if and only if there's a hole with more than one pigeon. (Dirichlet's box principle, proved in exercise 3.8, is similar.) We know that the numbers (4.45) are distinct, because

$$jn \equiv kn \pmod m \qquad \iff \qquad j \equiv k \pmod m$$

when $m \perp n$; this is (4.37). Therefore the m different numbers must fill all the pigeonholes $0, 1, \ldots, m-1$. Therefore the unfinished business of Chapter 3 is finished.

The proof is complete, but we can prove even more if we use a direct method instead of relying on the indirect pigeonhole argument. If $m \perp n$ and if a value $j \in [0, m)$ is given, we can explicitly compute $k \in [0, m)$ such that $kn \bmod m = j$ by solving the congruence

$$kn \equiv j \pmod m$$

for k. We simply multiply both sides by n', where $m'm + n'n = 1$, to get

$$k \equiv jn' \pmod m;$$

hence $k = jn' \bmod m$.

We can use the facts just proved to establish an important result discovered by Pierre de Fermat in 1640. Fermat was a great mathematician who contributed to the discovery of calculus and many other parts of mathematics. He left notebooks containing dozens of theorems stated without proof, and each of those theorems has subsequently been verified—except one. The one that remains, now called "Fermat's Last Theorem," states that

$$a^n + b^n \neq c^n \tag{4.46}$$

NEWS FLASH

Euler [93] conjectured that $a^4 + b^4 + c^4 \neq d^4$, *but Noam Elkies found infinitely many solutions in August, 1987.*

Now Roger Frye has done an exhaustive computer search, proving (after about 110 hours on a Connection Machine) that the smallest solution is: $95800^4 + 217519^4 + 414560^4 = 422481^4$.

for all positive integers a, b, c, and n, when $n > 2$. (Of course there are lots of solutions to the equations $a + b = c$ and $a^2 + b^2 = c^2$.) This conjecture has been verified for all $n \leqslant 150000$ by Tanner and Wagstaff [285].

Fermat's theorem of 1640 is one of the many that turned out to be provable. It's now called Fermat's Little Theorem (or just Fermat's theorem, for short), and it states that

$$n^{p-1} \equiv 1 \pmod p, \qquad \text{if } n \perp p. \tag{4.47}$$

Proof: As usual, we assume that p denotes a prime. We know that the $p-1$ numbers $n \bmod p$, $2n \bmod p$, ..., $(p-1)n \bmod p$ are the numbers 1, 2, ..., $p-1$ in some order. Therefore if we multiply them together we get

$$\begin{aligned} n \cdot (2n) \cdot \ldots \cdot \big((p-1)n\big) &\equiv (n \bmod p) \cdot (2n \bmod p) \cdot \ldots \cdot \big((p-1)n \bmod p\big) \\ &\equiv (p-1)!\,, \end{aligned}$$

where the congruence is modulo p. This means that

$$(p-1)!\, n^{p-1} \equiv (p-1)! \pmod p,$$

and we can cancel the $(p-1)!$ since it's not divisible by p. QED.

An alternative form of Fermat's theorem is sometimes more convenient:

$$n^p \equiv n \pmod p, \qquad \text{integer } n. \tag{4.48}$$

This congruence holds for all integers n. The proof is easy: If $n \perp p$ we simply multiply (4.47) by n. If not, $p \backslash n$, so $n^p \equiv 0 \equiv n$.

In the same year that he discovered (4.47), Fermat wrote a letter to Mersenne, saying he suspected that the number

$$f_n = 2^{2^n} + 1$$

"... laquelle proposition, si elle est vraie, est de très grand usage."
—P. de Fermat [97]

would turn out to be prime for all $n \geqslant 0$. He knew that the first five cases gave primes:

$$2^1+1 = 3;\; 2^2+1 = 5;\; 2^4+1 = 17;\; 2^8+1 = 257;\; 2^{16}+1 = 65537;$$

but he couldn't see how to prove that the next case, $2^{32} + 1 = 4294967297$, would be prime.

It's interesting to note that Fermat could have proved that $2^{32}+1$ is *not* prime, using his own recently discovered theorem, if he had taken time to perform a few dozen multiplications: We can set $n = 3$ in (4.47), deducing that

$$3^{2^{32}} \equiv 1 \pmod{2^{32}+1}, \qquad \text{if } 2^{32}+1 \text{ is prime.}$$

And it's possible to test this relation by hand, beginning with 3 and squaring 32 times, keeping only the remainders mod $2^{32}+1$. First we have $3^2 = 9$, then $3^{2^2} = 81$, then $3^{2^3} = 6561$, and so on until we reach

If this is Fermat's Little Theorem, the other one was last but not least.

$$3^{2^{32}} \equiv 3029026160 \pmod{2^{32}+1}.$$

The result isn't 1, so $2^{32}+1$ isn't prime. This method of disproof gives us no clue about what the factors might be, but it does prove that factors exist. (They are 641 and 6700417.)

If $3^{2^{32}}$ had turned out to be 1, modulo $2^{32}+1$, the calculation wouldn't have proved that $2^{32}+1$ is prime; it just wouldn't have disproved it. But exercise 47 discusses a converse to Fermat's theorem by which we *can* prove that large prime numbers are prime, without doing an enormous amount of laborious arithmetic.

We proved Fermat's theorem by cancelling $(p-1)!$ from both sides of a congruence. It turns out that $(p-1)!$ is always congruent to -1, modulo p; this is part of a classical result known as Wilson's theorem:

$$(n-1)! \equiv -1 \pmod n \qquad \Longleftrightarrow \qquad n \text{ is prime}, \qquad \text{if } n > 1. \tag{4.49}$$

One half of this theorem is trivial: If $n > 1$ is not prime, it has a prime divisor p that appears as a factor of $(n-1)!$, so $(n-1)!$ cannot be congruent to -1. (If $(n-1)!$ were congruent to -1 modulo n, it would also be congruent to -1 modulo p, but it isn't.)

The other half of Wilson's theorem states that $(p-1)! \equiv -1 \pmod p$. We can prove this half by pairing up numbers with their inverses mod p. If $n \perp p$, we know that there exists n' such that

$$n'n \equiv 1 \pmod p;$$

here n' is the inverse of n, and n is also the inverse of n'. Any two inverses of n must be congruent to each other, since $nn' \equiv nn''$ implies $n' \equiv n''$.

If p is prime, is p' prime prime?

Now suppose we pair up each number between 1 and $p-1$ with its inverse. Since the product of a number and its inverse is congruent to 1, the product of all the numbers in all pairs of inverses is also congruent to 1; so it seems that $(p-1)!$ is congruent to 1. Let's check, say for $p = 5$. We get $4! = 24$; but this is congruent to 4, not 1, modulo 5. Oops — what went wrong? Let's take a closer look at the inverses:

$$1' = 1, \qquad 2' = 3, \qquad 3' = 2, \qquad 4' = 4.$$

Ah so; 2 and 3 pair up but 1 and 4 don't — they're their own inverses.

To resurrect our analysis we must determine which numbers are their own inverses. If x is its own inverse, then $x^2 \equiv 1 \pmod p$; and we have already proved that this congruence has exactly two roots when $p > 2$. (If $p = 2$ it's obvious that $(p-1)! \equiv -1$, so we needn't worry about that case.) The roots are 1 and $p-1$, and the other numbers (between 1 and $p-1$) pair up; hence

$$(p-1)! \;\equiv\; 1 \cdot (p-1) \;\equiv\; -1\,,$$

as desired.

Unfortunately, we can't compute factorials efficiently, so Wilson's theorem is of no use as a practical test for primality. It's just a theorem.

4.9 PHI AND MU

How many of the integers $\{0, 1, \ldots, m-1\}$ are relatively prime to m? This is an important quantity called $\varphi(m)$, the "totient" of m (so named by J. J. Sylvester [284], a British mathematician who liked to invent new words). We have $\varphi(1) = 1$, $\varphi(p) = p-1$, and $\varphi(m) < m-1$ for all composite numbers m.

The φ function is called *Euler's totient function*, because Euler was the first person to study it. Euler discovered, for example, that Fermat's theorem (4.47) can be generalized to nonprime moduli in the following way:

"Si fuerit N ad x numerus primes et n numerus partium ad N primarum, tum potestas x^n unitate minuta semper per numerum N erit divisibilis."
—L. Euler [89]

$$n^{\varphi(m)} \equiv 1 \pmod m\,, \qquad \text{if } n \perp m. \tag{4.50}$$

(Exercise 32 asks for a proof of Euler's theorem.)

If m is a prime power p^k, it's easy to compute $\varphi(m)$, because $n \perp p^k \iff p\nmid n$. The multiples of p in $\{0, 1, \ldots, p^k-1\}$ are $\{0, p, 2p, \ldots, p^k-p\}$; hence there are p^{k-1} of them, and $\varphi(p^k)$ counts what is left:

$$\varphi(p^k) \;=\; p^k - p^{k-1}\,.$$

Notice that this formula properly gives $\varphi(p) = p-1$ when $k = 1$.

If $m > 1$ is not a prime power, we can write $m = m_1 m_2$ where $m_1 \perp m_2$. Then the numbers $0 \leqslant n < m$ can be represented in a residue number system as $(n \bmod m_1, n \bmod m_2)$. We have

$$n \perp m \qquad\iff\qquad n \bmod m_1 \perp m_1 \quad\text{and}\quad n \bmod m_2 \perp m_2$$

by (4.30) and (4.4). Hence, $n \bmod m$ is "good" if and only if $n \bmod m_1$ and $n \bmod m_2$ are both "good," if we consider relative primality to be a virtue. The total number of good values modulo m can now be computed, recursively: It is $\varphi(m_1)\varphi(m_2)$, because there are $\varphi(m_1)$ good ways to choose the first component $n \bmod m_1$ and $\varphi(m_2)$ good ways to choose the second component $n \bmod m_2$ in the residue representation.

For example, $\varphi(12) = \varphi(4)\varphi(3) = 2\cdot 2 = 4$, because n is prime to 12 if and only if $n \bmod 4 = (1 \text{ or } 3)$ and $n \bmod 3 = (1 \text{ or } 2)$. The four values prime to 12 are $(1,1)$, $(1,2)$, $(3,1)$, $(3,2)$ in the residue number system; they are 1, 5, 7, 11 in ordinary decimal notation. Euler's theorem states that $n^4 \equiv 1 \pmod{12}$ whenever $n \perp 12$.

"Si sint A et B numeri inter se primi et numerus partium ad A primarum sit = a, numerus vero partium ad B primarum sit = b, tum numerus partium ad productum AB primarum erit = ab."
—L. Euler [89]

A function $f(m)$ of positive integers is called *multiplicative* if $f(1) = 1$ and

$$f(m_1 m_2) = f(m_1)f(m_2) \qquad \text{whenever } m_1 \perp m_2. \tag{4.51}$$

We have just proved that $\varphi(m)$ is multiplicative. We've also seen another instance of a multiplicative function earlier in this chapter: The number of incongruent solutions to $x^2 \equiv 1 \pmod m$ is multiplicative. Still another example is $f(m) = m^\alpha$ for any power α.

A multiplicative function is defined completely by its values at prime powers, because we can decompose any positive integer m into its prime-power factors, which are relatively prime to each other. The general formula

$$f(m) = \prod_p f(p^{m_p}), \qquad \text{if } m = \prod_p p^{m_p} \tag{4.52}$$

holds if and only if f is multiplicative.

In particular, this formula gives us the value of Euler's totient function for general m:

$$\varphi(m) = \prod_{p\backslash m}(p^{m_p} - p^{m_p-1}) = m\prod_{p\backslash m}\Bigl(1 - \frac{1}{p}\Bigr). \tag{4.53}$$

For example, $\varphi(12) = (4-2)(3-1) = 12(1-\frac{1}{2})(1-\frac{1}{3})$.

Now let's look at an application of the φ function to the study of rational numbers mod 1. We say that the fraction m/n is *basic* if $0 \leqslant m < n$. Therefore $\varphi(n)$ is the number of reduced basic fractions with denominator n; and the Farey series $\mathcal{F}_n$ contains all the reduced basic fractions with denominator n or less, as well as the non-basic fraction $\frac{1}{1}$.

The set of *all* basic fractions with denominator 12, before reduction to lowest terms, is

$$\tfrac{0}{12}, \tfrac{1}{12}, \tfrac{2}{12}, \tfrac{3}{12}, \tfrac{4}{12}, \tfrac{5}{12}, \tfrac{6}{12}, \tfrac{7}{12}, \tfrac{8}{12}, \tfrac{9}{12}, \tfrac{10}{12}, \tfrac{11}{12}.$$

Reduction yields

$$\tfrac{0}{1}, \tfrac{1}{12}, \tfrac{1}{6}, \tfrac{1}{4}, \tfrac{1}{3}, \tfrac{5}{12}, \tfrac{1}{2}, \tfrac{7}{12}, \tfrac{2}{3}, \tfrac{3}{4}, \tfrac{5}{6}, \tfrac{11}{12},$$

and we can group these fractions by their denominators:

$$\tfrac{0}{1};\quad \tfrac{1}{2};\quad \tfrac{1}{3},\tfrac{2}{3};\quad \tfrac{1}{4},\tfrac{3}{4};\quad \tfrac{1}{6},\tfrac{5}{6};\quad \tfrac{1}{12},\tfrac{5}{12},\tfrac{7}{12},\tfrac{11}{12}.$$

What can we make of this? Well, every divisor d of 12 occurs as a denominator, together with all $\varphi(d)$ of its numerators. The only denominators that occur are divisors of 12. Thus

$$\varphi(1)+\varphi(2)+\varphi(3)+\varphi(4)+\varphi(6)+\varphi(12) \;=\; 12.$$

A similar thing will obviously happen if we begin with the unreduced fractions $\frac{0}{m}, \frac{1}{m}, \ldots, \frac{m-1}{m}$ for any m, hence

$$\sum_{d\backslash m} \varphi(d) \;=\; m. \tag{4.54}$$

We said near the beginning of this chapter that problems in number theory often require sums over the divisors of a number. Well, (4.54) is one such sum, so our claim is vindicated. (We will see other examples.)

Now here's a curious fact: If f is any function such that the sum

$$g(m) \;=\; \sum_{d\backslash m} f(d)$$

is multiplicative, then f itself is multiplicative. (This result, together with (4.54) and the fact that $g(m)=m$ is obviously multiplicative, gives another reason why $\varphi(m)$ is multiplicative.) We can prove this curious fact by induction on m: The basis is easy because $f(1)=g(1)=1$. Let $m>1$, and assume that $f(m_1m_2)=f(m_1)f(m_2)$ whenever $m_1 \perp m_2$ and $m_1m_2<m$. If $m=m_1m_2$ and $m_1 \perp m_2$, we have

$$g(m_1m_2) \;=\; \sum_{d\backslash m_1m_2} f(d) \;=\; \sum_{d_1\backslash m_1}\,\sum_{d_2\backslash m_2} f(d_1d_2)\,,$$

and $d_1 \perp d_2$ since all divisors of m_1 are relatively prime to all divisors of m_2. By the induction hypothesis, $f(d_1d_2)=f(d_1)f(d_2)$ except possibly when $d_1=m_1$ and $d_2=m_2$; hence we obtain

$$\Bigl(\sum_{d_1\backslash m_1} f(d_1) \sum_{d_2\backslash m_2} f(d_2)\Bigr) - f(m_1)f(m_2) + f(m_1m_2)$$
$$= \; g(m_1)g(m_2) - f(m_1)f(m_2) + f(m_1m_2)\,.$$

But this equals $g(m_1m_2)=g(m_1)g(m_2)$, so $f(m_1m_2)=f(m_1)f(m_2)$.

Conversely, if $f(m)$ is multiplicative, the corresponding sum-over-divisors function $g(m) = \sum_{d\backslash m} f(d)$ is always multiplicative. In fact, exercise 33 shows that even more is true. Hence the curious fact is a fact.

The *Möbius function* $\mu(m)$, named after the nineteenth-century mathematician August Möbius who also had a famous band, is defined for all $m \geqslant 1$ by the equation

$$\sum_{d\backslash m} \mu(d) = [m=1]. \tag{4.55}$$

This equation is actually a recurrence, since the left-hand side is a sum consisting of $\mu(m)$ and certain values of $\mu(d)$ with $d < m$. For example, if we plug in $m = 1, 2, \ldots, 12$ successively we can compute the first twelve values:

n	1	2	3	4	5	6	7	8	9	10	11	12
$\mu(n)$	1	−1	−1	0	−1	1	−1	0	0	1	−1	0

Möbius came up with the recurrence formula (4.55) because he noticed that it corresponds to the following important "inversion principle":

$$g(m) = \sum_{d\backslash m} f(d) \iff f(m) = \sum_{d\backslash m} \mu(d)\, g\Bigl(\frac{m}{d}\Bigr). \tag{4.56}$$

According to this principle, the μ function gives us a new way to understand any function $f(m)$ for which we know $\sum_{d\backslash m} f(d)$.

Now is a good time to try warmup exercise 11.

The proof of (4.56) uses two tricks (4.7) and (4.9) that we described near the beginning of this chapter: If $g(m) = \sum_{d\backslash m} f(d)$ then

$$\begin{aligned}
\sum_{d\backslash m} \mu(d)\, g\Bigl(\frac{m}{d}\Bigr) &= \sum_{d\backslash m} \mu\Bigl(\frac{m}{d}\Bigr) g(d) \\
&= \sum_{d\backslash m} \mu\Bigl(\frac{m}{d}\Bigr) \sum_{k\backslash d} f(k) \\
&= \sum_{k\backslash m} \sum_{d\backslash (m/k)} \mu\Bigl(\frac{m}{kd}\Bigr) f(k) \\
&= \sum_{k\backslash m} \sum_{d\backslash (m/k)} \mu(d) f(k) \\
&= \sum_{k\backslash m} [m/k=1] f(k) = f(m).
\end{aligned}$$

The other half of (4.56) is proved similarly (see exercise 12).

Relation (4.56) gives us a useful property of the Möbius function, and we have tabulated the first twelve values; but what is the value of $\mu(m)$ when

m is large? How can we solve the recurrence (4.55)? Well, the function $g(m) = [m=1]$ is obviously multiplicative—after all, it's zero except when $m = 1$. So the Möbius function defined by (4.55) must be multiplicative, by what we proved a minute or two ago. Therefore we can figure out what $\mu(m)$ is if we compute $\mu(p^k)$.

Depending on how fast you read.

When $m = p^k$, (4.55) says that

$$\mu(1) + \mu(p) + \mu(p^2) + \cdots + \mu(p^k) = 0$$

for all $k \geqslant 1$, since the divisors of p^k are $1, \ldots, p^k$. It follows that

$$\mu(p) = -1; \qquad \mu(p^k) = 0 \quad \text{for } k > 1.$$

Therefore by (4.52), we have the general formula

$$\mu(m) = \prod_{p \backslash m} \mu(p^{m_p}) = \begin{cases} (-1)^r, & \text{if } m = p_1 p_2 \ldots p_r; \\ 0, & \text{if } m \text{ is divisible by some } p^2. \end{cases} \tag{4.57}$$

That's μ.

If we regard (4.54) as a recurrence for the function $\varphi(m)$, we can solve that recurrence by applying Möbius's rule (4.56). The resulting solution is

$$\varphi(m) = \sum_{d \backslash m} \mu(d) \frac{m}{d}. \tag{4.58}$$

For example,

$$\begin{aligned} \varphi(12) &= \mu(1)\cdot 12 + \mu(2)\cdot 6 + \mu(3)\cdot 4 + \mu(4)\cdot 3 + \mu(6)\cdot 2 + \mu(12)\cdot 1 \\ &= 12 - 6 - 4 + 0 + 2 + 0 = 4. \end{aligned}$$

If m is divisible by r different primes, say $\{p_1, \ldots, p_r\}$, the sum (4.58) has only 2^r nonzero terms, because the μ function is often zero. Thus we can see that (4.58) checks with formula (4.53), which reads

$$\varphi(m) = m\Bigl(1 - \frac{1}{p_1}\Bigr) \ldots \Bigl(1 - \frac{1}{p_r}\Bigr);$$

if we multiply out the r factors $(1 - 1/p_j)$, we get precisely the 2^r nonzero terms of (4.58). The advantage of the Möbius function is that it applies in many situations besides this one.

For example, let's try to figure out how many fractions are in the Farey series $\mathcal{F}_n$. This is the number of reduced fractions in $[0, 1]$ whose denominators do not exceed n, so it is 1 greater than $\Phi(n)$ where we define

$$\Phi(x) = \sum_{1 \leqslant k \leqslant x} \varphi(k). \tag{4.59}$$

(We must add 1 to $\Phi(n)$ because of the final fraction $\frac{1}{1}$.) The sum in (4.59) looks difficult, but we can determine $\Phi(x)$ indirectly by observing that

$$\sum_{d\geqslant 1}\Phi\left(\frac{x}{d}\right) = \frac{1}{2}\lfloor x\rfloor\lfloor 1+x\rfloor \tag{4.60}$$

for all real $x \geqslant 0$. Why does this identity hold? Well, it's a bit awesome yet not really beyond our ken. There are $\frac{1}{2}\lfloor x\rfloor\lfloor 1+x\rfloor$ basic fractions m/n with $0 \leqslant m < n \leqslant x$, counting both reduced and unreduced fractions; that gives us the right-hand side. The number of such fractions with $\gcd(m,n) = d$ is $\Phi(x/d)$, because such fractions are m'/n' with $0 \leqslant m' < n' \leqslant x/d$ after replacing m by $m'd$ and n by $n'd$. So the left-hand side counts the same fractions in a different way, and the identity must be true.

Let's look more closely at the situation, so that equations (4.59) and (4.60) become clearer. The definition of $\Phi(x)$ implies that $\Phi(x) = \Phi(\lfloor x\rfloor)$; but it turns out to be convenient to define $\Phi(x)$ for arbitrary real values, not just for integers. At integer values we have the table

(This extension to real values is a useful trick for many recurrences that arise in the analysis of algorithms.)

n	0	1	2	3	4	5	6	7	8	9	10	11	12
$\varphi(n)$	–	1	1	2	2	4	2	6	4	6	4	10	4
$\Phi(n)$	0	1	2	4	6	10	12	18	22	28	32	42	46

and we can check (4.60) when $x = 12$:

$$\begin{aligned}&\Phi(12)+\Phi(6)+\Phi(4)+\Phi(3)+\Phi(2)+\Phi(2)+6\cdot\Phi(1)\\ &\qquad = 46+12+6+4+2+2+6 = 78 = \tfrac{1}{2}\cdot 12\cdot 13\,.\end{aligned}$$

Amazing.

Identity (4.60) can be regarded as an implicit recurrence for $\Phi(x)$; for example, we've just seen that we could have used it to calculate $\Phi(12)$ from certain values of $\Phi(m)$ with $m < 12$. And we can solve such recurrences by using another beautiful property of the Möbius function:

$$g(x) = \sum_{d\geqslant 1} f(x/d) \qquad\iff\qquad f(x) = \sum_{d\geqslant 1}\mu(d)g\left(\frac{x}{d}\right). \tag{4.61}$$

This inversion law holds for all functions f such that $\sum_{k,d\geqslant 1}\bigl|f(x/kd)\bigr| < \infty$; we can prove it as follows. Suppose $g(x) = \sum_{d\geqslant 1} f(x/d)$. Then

$$\begin{aligned}\sum_{d\geqslant 1}\mu(d)g(x/d) &= \sum_{d\geqslant 1}\mu(d)\sum_{k\geqslant 1} f(x/kd)\\ &= \sum_{m\geqslant 1} f(x/m)\sum_{d,k\geqslant 1}\mu(d)[m=kd]\end{aligned}$$

$$= \sum_{m\geqslant 1} f(x/m) \sum_{d\backslash m} \mu(d) \;=\; \sum_{m\geqslant 1} f(x/m)[m=1] \;=\; f(x)\,.$$

The proof in the other direction is essentially the same.

So now we can solve the recurrence (4.60) for $\Phi(x)$:

$$\Phi(x) \;=\; \frac{1}{2}\sum_{d\geqslant 1} \mu(d)\lfloor x/d\rfloor\lfloor 1+x/d\rfloor\,. \tag{4.62}$$

This is always a finite sum. For example,

$$\begin{aligned}\Phi(12) \;&=\; \tfrac{1}{2}(12\cdot 13 - 6\cdot 7 - 4\cdot 5 + 0 - 2\cdot 3 + 2\cdot 3 \\ &\qquad\qquad - 1\cdot 2 + 0 + 0 + 1\cdot 2 - 1\cdot 2 + 0) \\ &=\; 78 - 21 - 10 - 3 + 3 - 1 + 1 - 1 \;=\; 46\,.\end{aligned}$$

In Chapter 9 we'll see how to use (4.62) to get a good approximation to $\Phi(x)$; in fact, we'll prove that

$$\Phi(x) \;=\; \frac{3}{\pi^2}x^2 + O(x\log x)\,.$$

Therefore the function $\Phi(x)$ grows "smoothly"; it averages out the erratic behavior of $\varphi(k)$.

In keeping with the tradition established last chapter, let's conclude this chapter with a problem that illustrates much of what we've just seen and that also points ahead to the next chapter. Suppose we have beads of n different colors; our goal is to count how many different ways there are to string them into circular necklaces of length m. We can try to "name and conquer" this problem by calling the number of possible necklaces $N(m,n)$.

For example, with two colors of beads R and B, we can make necklaces of length 4 in $N(4,2) = 6$ different ways:

$$\begin{array}{ccc} & R & \\ R & & R \\ & R & \end{array}\qquad \begin{array}{ccc} & R & \\ R & & R \\ & B & \end{array}\qquad \begin{array}{ccc} & R & \\ R & & B \\ & B & \end{array}\qquad \begin{array}{ccc} & R & \\ B & & B \\ & R & \end{array}\qquad \begin{array}{ccc} & R & \\ B & & B \\ & B & \end{array}\qquad \begin{array}{ccc} & B & \\ B & & B \\ & B & \end{array}$$

All other ways are equivalent to one of these, because rotations of a necklace do not change it. However, reflections are considered to be different; in the case $m = 6$, for example,

$$\begin{array}{ccc} & B & \\ R & & R \\ | & & | \\ R & & B \\ & B & \end{array}\qquad \text{is different from} \qquad \begin{array}{ccc} & B & \\ R & & R \\ | & & | \\ B & & R \\ & B & \end{array}\,.$$

The problem of counting these configurations was first solved by P. A. MacMahon in 1892 [212].

There's no obvious recurrence for $N(m,n)$, but we can count the necklaces by breaking them each into linear strings in m ways and considering the resulting fragments. For example, when $m = 4$ and $n = 2$ we get

RRRR	RRRR	RRRR	RRRR
RRBR	RRRB	BRRR	RBRR
RBBR	RRBB	BRRB	BBRR
RBRB	BRBR	RBRB	BRBR
RBBB	BRBB	BBRB	BBBR
BBBB	BBBB	BBBB	BBBB

Each of the n^m possible patterns appears at least once in this array of $mN(m,n)$ strings, and some patterns appear more than once. How many times does a pattern $a_0 \ldots a_{m-1}$ appear? That's easy: It's the number of cyclic shifts $a_k \ldots a_{m-1} a_0 \ldots a_{k-1}$ that produce the same pattern as the original $a_0 \ldots a_{m-1}$. For example, BRBR occurs twice, because the four ways to cut the necklace formed from BRBR produce four cyclic shifts (BRBR, RBRB, BRBR, RBRB); two of these coincide with BRBR itself. This argument shows that

$$\begin{aligned} mN(m,n) &= \sum_{a_0,\ldots,a_{m-1}\in S_n} \sum_{0\leqslant k<m} [a_0 \ldots a_{m-1} = a_k \ldots a_{m-1} a_0 \ldots a_{k-1}] \\ &= \sum_{0\leqslant k<m} \sum_{a_0,\ldots,a_{m-1}\in S_n} [a_0 \ldots a_{m-1} = a_k \ldots a_{m-1} a_0 \ldots a_{k-1}] \,. \end{aligned}$$

Here S_n is a set of n different colors.

Let's see how many patterns satisfy $a_0 \ldots a_{m-1} = a_k \ldots a_{m-1} a_0 \ldots a_{k-1}$, when k is given. For example, if $m = 12$ and $k = 8$, we want to count the number of solutions to

$$a_0 a_1 a_2 a_3 a_4 a_5 a_6 a_7 a_8 a_9 a_{10} a_{11} \;=\; a_8 a_9 a_{10} a_{11} a_0 a_1 a_2 a_3 a_4 a_5 a_6 a_7 \,.$$

This means $a_0 = a_8 = a_4$; $a_1 = a_9 = a_5$; $a_2 = a_{10} = a_6$; and $a_3 = a_{11} = a_7$. So the values of a_0, a_1, a_2, and a_3 can be chosen in n^4 ways, and the remaining a's depend on them. Does this look familiar? In general, the solution to

$$a_j \;=\; a_{(j+k) \bmod m} \,, \qquad \text{for } 0 \leqslant j < m$$

makes us equate a_j with $a_{(j+kl) \bmod m}$ for $l = 1, 2, \ldots$; and we know that the multiples of k modulo m are $\{0, d, 2d, \ldots, m-d\}$, where $d = \gcd(k,m)$. Therefore the general solution is to choose $a_0, \ldots, a_{d-1}$ independently and then to set $a_j = a_{j-d}$ for $d \leqslant j < m$. There are n^d solutions.

We have just proved that

$$mN(m,n) \;=\; \sum_{0\leqslant k<m} n^{\gcd(k,m)}\,.$$

This sum can be simplified, since it includes only terms n^d where $d\backslash m$. Substituting $d=\gcd(k,m)$ yields

$$\begin{aligned} N(m,n) \;&=\; \frac{1}{m}\sum_{d\backslash m} n^d \sum_{0\leqslant k<m} \bigl[d=\gcd(k,m)\bigr] \\ &=\; \frac{1}{m}\sum_{d\backslash m} n^d \sum_{0\leqslant k<m} \bigl[k/d\perp m/d\bigr] \\ &=\; \frac{1}{m}\sum_{d\backslash m} n^d \sum_{0\leqslant k<m/d} \bigl[k\perp m/d\bigr]\,. \end{aligned}$$

(We are allowed to replace k/d by k because k must be a multiple of d.) Finally, we have $\sum_{0\leqslant k<m/d}[k\perp m/d] = \varphi(m/d)$ by definition, so we obtain MacMahon's formula:

$$N(m,n) \;=\; \frac{1}{m}\sum_{d\backslash m} n^d\,\varphi\Bigl(\frac{m}{d}\Bigr) \;=\; \frac{1}{m}\sum_{d\backslash m} \varphi(d)\,n^{m/d}\,. \tag{4.63}$$

When $m=4$ and $n=2$, for example, the number of necklaces is $\frac{1}{4}(1\cdot 2^4 + 1\cdot 2^2 + 2\cdot 2^1) = 6$, just as we suspected.

It's not immediately obvious that the value $N(m,n)$ defined by MacMahon's sum is an integer! Let's try to prove directly that

$$\sum_{d\backslash m} \varphi(d)\,n^{m/d} \;\equiv\; 0 \qquad (\text{mod } m)\,, \tag{4.64}$$

without using the clue that this is related to necklaces. In the special case that m is prime, this congruence reduces to $n^p + (p-1)n \equiv 0 \pmod p$; that is, it reduces to $n^p \equiv n$. We've seen in (4.48) that this congruence is an alternative form of Fermat's theorem. Therefore (4.64) holds when $m = p$; we can regard it as a generalization of Fermat's theorem to the case when the modulus is not prime. (Euler's generalization (4.50) is different.)

We've proved (4.64) for all prime moduli, so let's look at the smallest case left, $m = 4$. We must prove that

$$n^4 + n^2 + 2n \;\equiv\; 0 \qquad (\text{mod } 4)\,.$$

The proof is easy if we consider even and odd cases separately. If n is even, all three terms on the left are congruent to 0 modulo 4, so their sum is too. If

n is odd, n^4 and n^2 are each congruent to 1, and $2n$ is congruent to 2; hence the left side is congruent to $1+1+2$ and thus to 0 modulo 4, and we're done.

Next, let's be a bit daring and try $m = 12$. This value of m ought to be interesting because it has lots of factors, including the square of a prime, yet it is fairly small. (Also there's a good chance we'll be able to generalize a proof for 12 to a proof for general m.) The congruence we must prove is

$$n^{12} + n^6 + 2n^4 + 2n^3 + 2n^2 + 4n \equiv 0 \pmod{12}.$$

Now what? By (4.42) this congruence holds if and only if it also holds modulo 3 and modulo 4. So let's prove that it holds modulo 3. Our congruence (4.64) holds for primes, so we have $n^3 + 2n \equiv 0 \pmod 3$. Careful scrutiny reveals that we can use this fact to group terms of the larger sum:

$$\begin{aligned} n^{12} &+ n^6 + 2n^4 + 2n^3 + 2n^2 + 4n \\ &= (n^{12} + 2n^4) + (n^6 + 2n^2) + 2(n^3 + 2n) \\ &\equiv 0 + 0 + 2\cdot 0 \equiv 0 \pmod 3. \end{aligned}$$

So it works modulo 3.

We're half done. To prove congruence modulo 4 we use the same trick. We've proved that $n^4 + n^2 + 2n \equiv 0 \pmod 4$, so we use this pattern to group:

$$\begin{aligned} n^{12} &+ n^6 + 2n^4 + 2n^3 + 2n^2 + 4n \\ &= (n^{12} + n^6 + 2n^3) + 2(n^4 + n^2 + 2n) \\ &\equiv 0 + 2\cdot 0 \equiv 0 \pmod 4. \end{aligned}$$

QED for the case $m = 12$.

QED: Quite Easily Done.

So far we've proved our congruence for prime m, for $m = 4$, and for $m = 12$. Now let's try to prove it for prime powers. For concreteness we may suppose that $m = p^3$ for some prime p. Then the left side of (4.64) is

$$\begin{aligned} n^{p^3} &+ \varphi(p)n^{p^2} + \varphi(p^2)n^p + \varphi(p^3)n \\ &= n^{p^3} + (p-1)n^{p^2} + (p^2-p)n^p + (p^3-p^2)n \\ &= (n^{p^3} - n^{p^2}) + p(n^{p^2} - n^p) + p^2(n^p - n) + p^3 n. \end{aligned}$$

We can show that this is congruent to 0 modulo p^3 if we can prove that $n^{p^3} - n^{p^2}$ is divisible by p^3, that $n^{p^2} - n^p$ is divisible by p^2, and that $n^p - n$ is divisible by p, because the whole thing will then be divisible by p^3. By the alternative form of Fermat's theorem we have $n^p \equiv n \pmod p$, so p divides $n^p - n$; hence there is an integer q such that

$$n^p = n + pq.$$

Now we raise both sides to the pth power, expand the right side according to the binomial theorem (which we'll meet in Chapter 5), and regroup, giving

$$\begin{aligned} n^{p^2} &= (n+pq)^p = n^p + (pq)^1 n^{p-1}\binom{p}{1} + (pq)^2 n^{p-2}\binom{p}{2} + \cdots \\ &= n^p + p^2 Q \end{aligned}$$

for some other integer Q. We're able to pull out a factor of p^2 here because $\binom{p}{1} = p$ in the second term, and because a factor of $(pq)^2$ appears in all the terms that follow. So we find that p^2 divides $n^{p^2} - n^p$.

Again we raise both sides to the pth power, expand, and regroup, to get

$$\begin{aligned} n^{p^3} &= (n^p + p^2 Q)^p \\ &= n^{p^2} + (p^2 Q)^1 n^{p(p-1)}\binom{p}{1} + (p^2 Q)^2 n^{p(p-2)}\binom{p}{2} + \cdots \\ &= n^{p^2} + p^3 \mathbf{Q} \end{aligned}$$

for yet another integer $\mathbf{Q}$. So p^3 divides $n^{p^3} - n^{p^2}$. This finishes the proof for $m = p^3$, because we've shown that p^3 divides the left-hand side of (4.64).

Moreover we can prove by induction that

$$n^{p^k} = n^{p^{k-1}} + p^k \mathfrak{Q}$$

for some final integer $\mathfrak{Q}$ (final because we're running out of fonts); hence

$$n^{p^k} \equiv n^{p^{k-1}} \pmod{p^k}, \qquad \text{for } k > 0. \tag{4.65}$$

Thus the left side of (4.64), which is

$$(n^{p^k} - n^{p^{k-1}}) + p(n^{p^{k-1}} - n^{p^{k-2}}) + \cdots + p^{k-1}(n^p - n) + p^k n,$$

is divisible by p^k and so is congruent to 0 modulo p^k.

We're almost there. Now that we've proved (4.64) for prime powers, all that remains is to prove it when $m = m_1 m_2$, where $m_1 \perp m_2$, assuming that the congruence is true for m_1 and m_2. Our examination of the case $m = 12$, which factored into instances of $m = 3$ and $m = 4$, encourages us to think that this approach will work.

We know that the φ function is multiplicative, so we can write

$$\begin{aligned} \sum_{d \backslash m} \varphi(d)\, n^{m/d} &= \sum_{d_1 \backslash m_1,\, d_2 \backslash m_2} \varphi(d_1 d_2)\, n^{m_1 m_2 / d_1 d_2} \\ &= \sum_{d_1 \backslash m_1} \varphi(d_1) \Bigl(\sum_{d_2 \backslash m_2} \varphi(d_2) (n^{m_1/d_1})^{m_2/d_2} \Bigr). \end{aligned}$$

But the inner sum is congruent to 0 modulo m_2, because we've assumed that (4.64) holds for m_2; so the entire sum is congruent to 0 modulo m_2. By a symmetric argument, we find that the entire sum is congruent to 0 modulo m_1 as well. Thus by (4.42) it's congruent to 0 modulo m. QED.

Exercises

Warmups

1 What is the smallest positive integer that has exactly k divisors, for $1 \leqslant k \leqslant 6$?

2 Prove that $\gcd(m, n) \cdot \operatorname{lcm}(m, n) = m \cdot n$, and use this identity to express $\operatorname{lcm}(m, n)$ in terms of $\operatorname{lcm}(n \bmod m, m)$, when $n \bmod m \neq 0$. *Hint:* Use (4.12), (4.14), and (4.15).

3 Let $\pi(x)$ be the number of primes not exceeding x. Prove or disprove:

$$\pi(x) - \pi(x-1) \;=\; [x \text{ is prime}] .$$

4 What would happen if the Stern–Brocot construction started with the five fractions $\left(\frac{0}{1}, \frac{1}{0}, \frac{0}{-1}, \frac{-1}{0}, \frac{0}{1}\right)$ instead of with $\left(\frac{0}{1}, \frac{1}{0}\right)$?

5 Find simple formulas for L^k and R^k, when L and R are the 2×2 matrices of (4.33).

6 What does '$a \equiv b \pmod 0$' mean?

7 Ten people numbered 1 to 10 are lined up in a circle as in the Josephus problem, and every mth person is executed. (The value of m may be much larger than 10.) Prove that the first three people to go cannot be 10, k, and $k+1$ (in this order), for any k.

8 The residue number system $(x \bmod 3, x \bmod 5)$ considered in the text has the curious property that 13 corresponds to $(1, 3)$, which looks almost the same. Explain how to find all instances of such a coincidence, without calculating all fifteen pairs of residues. In other words, find all solutions to the congruences

$$10x + y \;\equiv\; x \pmod 3 , \qquad 10x + y \;\equiv\; y \pmod 5 .$$

Hint: Use the facts that $10u + 6v \equiv u \pmod 3$ and $10u + 6v \equiv v \pmod 5$.

9 Show that $(3^{77} - 1)/2$ is odd and composite. *Hint:* What is $3^{77} \bmod 4$?

10 Compute $\varphi(999)$.

11 Find a function $\sigma(n)$ with the property that

$$g(n) = \sum_{0\leqslant k\leqslant n} f(k) \qquad \Longleftrightarrow \qquad f(n) = \sum_{0\leqslant k\leqslant n} \sigma(k)\, g(n-k).$$

(This is analogous to the Möbius function; see (4.56).)

12 Simplify the formula $\sum_{d\backslash m}\sum_{k\backslash d}\mu(k)\,g(d/k)$.

13 A positive integer n is called *squarefree* if it is not divisible by m^2 for any $m > 1$. Find a necessary and sufficient condition that n is squarefree,

a in terms of the prime-exponent representation (4.11) of n;

b in terms of $\mu(n)$.

Basics

14 Prove or disprove:

a $\gcd(km, kn) = k\gcd(m, n)$;

b $\operatorname{lcm}(km, kn) = k\operatorname{lcm}(m, n)$.

15 Does every prime occur as a factor of some Euclid number e_n?

16 What is the sum of the reciprocals of the first n Euclid numbers?

17 Let f_n be the "Fermat number" $2^{2^n} + 1$. Prove that $f_m \perp f_n$ if $m < n$.

18 Show that if $2^n + 1$ is prime then n is a power of 2.

19 For every positive integer n there's a prime p such that $n < p \leqslant 2n$. (This is essentially "Bertrand's postulate," which Joseph Bertrand verified for $n < 3000000$ in 1845 and Chebyshev proved for all n in 1850.) Use Bertrand's postulate to prove that there's a constant $b \approx 1.25$ such that the numbers

$$\lfloor 2^b \rfloor,\ \lfloor 2^{2^b} \rfloor,\ \lfloor 2^{2^{2^b}} \rfloor,\ \ldots$$

are all prime.

20 Let P_n be the nth prime number. Find a constant K such that

$$\lfloor (10^{n^2} K) \bmod 10^n \rfloor = P_n.$$

21 Prove the following identities when n is a positive integer:

$$\sum_{1\leqslant k<n} \left\lfloor \frac{\varphi(k+1)}{k} \right\rfloor = \sum_{1<m\leqslant n} \left\lfloor \Bigl(\sum_{1\leqslant k<m} \bigl\lfloor (m/k)/\lceil m/k \rceil \bigr\rfloor \Bigr)^{-1} \right\rfloor$$

$$= n - 1 - \sum_{k=1}^{n} \left\lceil \left\{ \frac{(k-1)!+1}{k} \right\} \right\rceil.$$

Hint: This is a trick question and the answer is pretty easy.

22 The number 1111111111111111111 is prime. Prove that, in any radix b, $(11\ldots1)_b$ can be prime only if the number of 1's is prime.

Is this a test for strabismus?

23 State a recurrence for $\rho(k)$, the ruler function in the text's discussion of $\epsilon_2(n!)$. Show that there's a connection between $\rho(k)$ and the disk that's moved at step k when an n-disk Tower of Hanoi is being transferred in 2^n-1 moves, for $1 \leqslant k \leqslant 2^n-1$.

24 Express $\epsilon_p(n!)$ in terms of $\nu_p(n)$, the sum of the digits in the radix p representation of n, thereby generalizing (4.24).

Look, ma, sideways addition.

25 We say that m *exactly divides* n, written $m\backslash\backslash n$, if $m\backslash n$ and $m \perp n/m$. For example, in the text's discussion of factorial factors, $p^{\epsilon_p(n!)}\backslash\backslash n!$. Prove or disprove the following:

a $k\backslash\backslash n$ and $m\backslash\backslash n \iff km\backslash\backslash n$, if $k \perp m$.

b For all $m, n > 0$, either $\gcd(m,n)\backslash\backslash m$ or $\gcd(m,n)\backslash\backslash n$.

26 Consider the sequence $\mathcal{G}_N$ of all nonnegative reduced fractions m/n such that $mn \leqslant N$. For example,

$$\mathcal{G}_{10} = \tfrac{0}{1}, \tfrac{1}{10}, \tfrac{1}{9}, \tfrac{1}{8}, \tfrac{1}{7}, \tfrac{1}{6}, \tfrac{1}{5}, \tfrac{1}{4}, \tfrac{1}{3}, \tfrac{2}{5}, \tfrac{1}{2}, \tfrac{2}{3}, \tfrac{1}{1}, \tfrac{3}{2}, \tfrac{2}{1}, \tfrac{5}{2}, \tfrac{3}{1}, \tfrac{4}{1}, \tfrac{5}{1}, \tfrac{6}{1}, \tfrac{7}{1}, \tfrac{8}{1}, \tfrac{9}{1}, \tfrac{10}{1}.$$

Is it true that $m'n - mn' = 1$ whenever m/n immediately precedes m'/n' in $\mathcal{G}_N$?

27 Give a simple rule for comparing rational numbers based on their representations as L's and R's in the Stern–Brocot number system.

28 The Stern–Brocot representation of π is

$$\pi = R^3L^7R^{15}LR^{292}LRLR^2LR^3LR^{14}L^2R\ldots;$$

use it to find all the simplest rational approximations to π whose denominators are less than 50. Is $\frac{22}{7}$ one of them?

29 The text describes a correspondence between binary real numbers $x = (.b_1b_2b_3\ldots)_2$ in $[0,1)$ and Stern–Brocot real numbers $\alpha = B_1B_2B_3\ldots$ in $[0,\infty)$. If x corresponds to α and $x \neq 0$, what number corresponds to $1-x$?

30 Prove the following statement (the Chinese Remainder Theorem): Let $m_1, \ldots, m_r$ be integers with $m_j \perp m_k$ for $1 \leqslant j < k \leqslant r$; let $m = m_1\ldots m_r$; and let $a_1, \ldots, a_r$, A be integers. Then there is exactly one integer a such that

$$a \equiv a_k \pmod{m_k} \text{ for } 1 \leqslant k \leqslant r \quad\text{and}\quad A \leqslant a < A+m.$$

31 A number in decimal notation is divisible by 3 if and only if the sum of its digits is divisible by 3. Prove this well-known rule, and generalize it.

Why is "Euler" pronounced "Oiler" when "Euclid" is "Yooklid"?

32 Prove Euler's theorem (4.50) by generalizing the proof of (4.47).

33 Show that if $f(m)$ and $g(m)$ are multiplicative functions, then so is $h(m) = \sum_{d\backslash m} f(d)\, g(m/d)$.

34 Prove that (4.56) is a special case of (4.61).

Homework exercises

35 Let $I(m, n)$ be a function that satisfies the relation

$$I(m,n)m + I(n,m)n \;=\; \gcd(m,n)\,,$$

when m and n are nonnegative integers with $m \neq n$. Thus, $I(m,n) = m'$ and $I(n,m) = n'$ in (4.5); the value of $I(m,n)$ is an *inverse* of m with respect to n. Find a recurrence that defines $I(m,n)$.

36 Consider the set $Z(\sqrt{10}\,) = \{\, m + n\sqrt{10} \mid \text{integer } m, n\,\}$. The number $m + n\sqrt{10}$ is called a *unit* if $m^2 - 10n^2 = \pm 1$, since it has an inverse (that is, since $(m+n\sqrt{10}\,)\cdot \pm(m-n\sqrt{10}\,) = 1$). For example, $3 + \sqrt{10}$ is a unit, and so is $19 - 6\sqrt{10}$. Pairs of cancelling units can be inserted into any factorization, so we ignore them. Nonunit numbers of $Z(\sqrt{10}\,)$ are called prime if they cannot be written as a product of two nonunits. Show that 2, 3, and $4 \pm \sqrt{10}$ are primes of $Z(\sqrt{10}\,)$. *Hint:* If $2 = (k + l\sqrt{10}\,) \times (m + n\sqrt{10}\,)$ then $4 = (k^2 - 10l^2)(m^2 - 10n^2)$. Furthermore, the square of any integer mod 10 is 0, 1, 4, 5, 6, or 9.

37 Prove (4.17). *Hint:* Show that $e_n - \frac{1}{2} = (e_{n-1} - \frac{1}{2})^2 + \frac{1}{4}$, and consider $2^{-n}\log(e_n - \frac{1}{2})$.

38 Prove that if $a \perp b$ and $a > b$ then

$$\gcd(a^m - b^m,\, a^n - b^n) \;=\; a^{\gcd(m,n)} - b^{\gcd(m,n)}\,, \qquad 0 \leqslant m < n.$$

(All variables are integers.) *Hint:* Use Euclid's algorithm.

39 Let $S(m)$ be the smallest positive integer n for which there exists an increasing sequence of integers

$$m = a_1 < a_2 < \cdots < a_t = n$$

such that $a_1 a_2 \ldots a_t$ is a perfect square. (If m is a perfect square, we can let $t = 1$ and $n = m$.) For example, $S(2) = 6$ because the best such sequence is $2\cdot 3\cdot 6$. We have

n	1	2	3	4	5	6	7	8	9	10	11	12
$S(n)$	1	6	8	4	10	12	14	15	9	18	22	20

Prove that $S(m) \neq S(m')$ whenever $0 < m < m'$.

40 If the radix p representation of n is $(a_m \ldots a_1 a_0)_p$, prove that

$$n!/p^{\epsilon_p(n!)} \equiv (-1)^{\epsilon_p(n!)} a_m! \ldots a_1!\, a_0! \qquad (\text{mod } p).$$

(The left side is simply $n!$ with all p factors removed. When $n = p$ this reduces to Wilson's theorem.)

Wilson's theorem: "Martha, that boy is a menace."

41 **a** Show that if $p \bmod 4 = 3$, there is no integer n such that p divides $n^2 + 1$. *Hint:* Use Fermat's theorem.

b But show that if $p \bmod 4 = 1$, there is such an integer. *Hint:* Write $(p-1)!$ as $\bigl(\prod_{k=1}^{(p-1)/2} k(p-k)\bigr)$ and think about Wilson's theorem.

42 Consider two fractions m/n and m'/n' in lowest terms. Prove that when the sum $m/n+m'/n'$ is reduced to lowest terms, the denominator will be nn' if and only if $n \perp n'$. (In other words, $(mn'+m'n)/nn'$ will already be in lowest terms if and only if n and n' have no common factor.)

43 There are 2^k nodes at level k of the Stern–Brocot tree, corresponding to the matrices L^k, $L^{k-1}R$, ..., R^k. Show that this sequence can be obtained by starting with L^k and then multiplying successively by

$$\begin{pmatrix} 0 & -1 \\ 1 & 2\rho(n)+1 \end{pmatrix}$$

for $1 \leqslant n < 2^k$, where $\rho(n)$ is the ruler function.

44 Prove that a baseball player whose batting average is .316 must have batted at least 19 times. (If he has m hits in n times at bat, then $m/n \in [.3155, .3165)$.)

Radio announcer: "... pitcher Mark LeChiffre hits a two-run single! Mark was batting only .080, so he gets his second hit of the year." Anything wrong?

45 The number 9376 has the peculiar self-reproducing property that

$$9376^2 = 87909376.$$

How many 4-digit numbers x satisfy the equation $x^2 \bmod 10000 = x$? How many n-digit numbers x satisfy the equation $x^2 \bmod 10^n = x$?

46 **a** Prove that if $n^j \equiv 1$ and $n^k \equiv 1$ (mod m), then $n^{\gcd(j,k)} \equiv 1$.

b Show that $2^n \not\equiv 1$ (mod n), if $n > 1$. *Hint:* Consider the least prime factor of n.

47 Show that if $n^{m-1} \equiv 1$ (mod m) and if $n^{(m-1)/p} \not\equiv 1$ (mod m) for all primes such that $p\backslash(m-1)$, then m is prime. *Hint:* Show that if this condition holds, the numbers $n^k \bmod m$ are distinct, for $1 \leqslant k < m$.

The proof that large numbers are prime is very easy: Let x be a large prime number; then x is prime, QED.

48 Generalize Wilson's theorem (4.49) by ascertaining the value of the expression $\bigl(\prod_{1\leqslant n<m,\, n\perp m} n\bigr) \bmod m$, when $m > 1$.

49 Let $R(N)$ be the number of pairs of integers (m, n) such that $0 \leqslant m < N$, $0 \leqslant n < N$, and $m \perp n$.

a Express $R(N)$ in terms of the Φ function.

b Prove that $R(N) = \sum_{d \geqslant N} \lfloor N/d \rfloor^2 \mu(d)$.

50 Let m be a positive integer and let

$$\omega = e^{2\pi i/m} = \cos(2\pi/m) + i\sin(2\pi/m).$$

What are the roots of disunity?

We say that ω is an mth *root of unity*, since $\omega^m = e^{2\pi i} = 1$. In fact, each of the m complex numbers $\omega^0, \omega^1, \ldots, \omega^{m-1}$ is an mth root of unity, because $(\omega^k)^m = e^{2\pi k i} = 1$; therefore $z - \omega^k$ is a factor of the polynomial $z^m - 1$, for $0 \leqslant k < m$. Since these factors are distinct, the complete factorization of $z^m - 1$ over the complex numbers must be

$$z^m - 1 = \prod_{0 \leqslant k < m} (z - \omega^k).$$

a Let $\Psi_m(z) = \prod_{0 \leqslant k < m,\, k \perp m} (z - \omega^k)$. (This polynomial of degree $\varphi(m)$ is called the *cyclotomic polynomial of order* m.) Prove that

$$z^m - 1 = \prod_{d \backslash m} \Psi_d(z).$$

b Prove that $\Psi_m(z) = \prod_{d \backslash m} (z^d - 1)^{\mu(m/d)}$.

Exam problems

51 Prove Fermat's theorem (4.48) by expanding $(1 + 1 + \cdots + 1)^p$ via the multinomial theorem.

52 Let n and x be positive integers such that x has no divisors $\leqslant n$ (except 1), and let p be a prime number. Prove that at least $\lfloor n/p \rfloor$ of the numbers $\{x - 1, x^2 - 1, \ldots, x^{n-1} - 1\}$ are multiples of p.

53 Find all positive integers n such that $n \backslash \lfloor (n-1)!/(n+1) \rfloor$.

54 Determine the value of $1000! \bmod 10^{250}$ by hand calculation.

55 Let P_n be the product of the first n factorials, $\prod_{k=1}^{n} k!$. Prove that P_{2n}/P_n^4 is an integer, for all positive integers n.

56 Show that

$$\left(\prod_{k=1}^{2n-1} k^{\min(k,\, 2n-k)}\right) \Big/ \left(\prod_{k=1}^{n-1} (2k+1)^{2n-2k-1}\right)$$

is a power of 2.

57 Let $S(m,n)$ be the set of all integers k such that

$$m \bmod k + n \bmod k \geqslant k.$$

For example, $S(7,9) = \{2,4,5,8,10,11,12,13,14,15,16\}$. Prove that

$$\sum_{k \in S(m,n)} \varphi(k) = mn.$$

Hint: Prove first that $\sum_{1 \leqslant m \leqslant n} \sum_{d \backslash m} \varphi(d) = \sum_{d \geqslant 1} \varphi(d) \lfloor n/d \rfloor$. Then consider $\lfloor (m+n)/d \rfloor - \lfloor m/d \rfloor - \lfloor n/d \rfloor$.

58 Let $f(m) = \sum_{d \backslash m} d$. Find a necessary and sufficient condition that $f(m)$ is a power of 2.

Bonus problems

59 Prove that if $x_1, \ldots, x_n$ are positive integers with $1/x_1 + \cdots + 1/x_n = 1$, then $\max(x_1, \ldots, x_n) < e_n$. *Hint:* Prove the following stronger result by induction: "If $1/x_1 + \cdots + 1/x_n + 1/\alpha = 1$, where $x_1, \ldots, x_n$ are positive integers and α is a rational number $\geqslant \max(x_1, \ldots, x_n)$, then $\alpha + 1 \leqslant e_{n+1}$ and $x_1 \ldots x_n(\alpha + 1) \leqslant e_1 \ldots e_n e_{n+1}$." (The proof is nontrivial.)

60 Prove that there's a constant P such that (4.18) gives only primes. You may use the following (highly nontrivial) fact: There is a prime between p and $p + cp^\theta$, for some constant c and all sufficiently large p, where $\theta = \frac{1051}{1920}$.

61 Prove that if m/n, m'/n', and m''/n'' are consecutive elements of $\mathcal{F}_N$, then

$$m'' = \lfloor (n+N)/n' \rfloor m' - m,$$
$$n'' = \lfloor (n+N)/n' \rfloor n' - n.$$

(This recurrence allows us to compute the elements of $\mathcal{F}_N$ in order, starting with $\frac{0}{1}$ and $\frac{1}{N}$.)

62 What binary number corresponds to e, in the binary $\leftrightarrow$ Stern–Brocot correspondence? (Express your answer as an infinite sum; you need not evaluate it in closed form.)

63 Show that if Fermat's Last Theorem (4.46) is false, the least n for which it fails is prime. (You may assume that the result holds when $n = 4$.) Furthermore, if $a^p + b^p = c^p$ and $a \perp b$, show that there exists an integer m such that

$$a + b = \begin{cases} m^p, & \text{if } p \nmid c; \\ p^{p-1} m^p, & \text{if } p \backslash c. \end{cases}$$

Thus c must be really huge. *Hint:* Let $x = a + b$, and note that $\gcd\bigl(x, (a^p + (x-a)^p)/x\bigr) = \gcd(x, pa^{p-1})$.

64 The *Peirce sequence* $\mathcal{P}_N$ *of order* N is an infinite string of fractions separated by '<' or '=' signs, containing all the nonnegative fractions m/n with $m \geqslant 0$ and $n \leqslant N$ (including fractions that are not reduced). It is defined recursively by starting with

$$\mathcal{P}_1 = \tfrac{0}{1}<\tfrac{1}{1}<\tfrac{2}{1}<\tfrac{3}{1}<\tfrac{4}{1}<\tfrac{5}{1}<\tfrac{6}{1}<\tfrac{7}{1}<\tfrac{8}{1}<\tfrac{9}{1}<\tfrac{10}{1}<\cdots.$$

For $N \geqslant 1$, we form $\mathcal{P}_{N+1}$ by inserting two symbols just before the kNth symbol of $\mathcal{P}_N$, for all $k > 0$. The two inserted symbols are

$$\begin{array}{lll} \dfrac{k-1}{N+1} & = , & \text{if } kN \text{ is odd;} \\ \mathcal{P}_{N,kN} & \dfrac{k-1}{N+1}, & \text{if } kN \text{ is even.} \end{array}$$

Here $\mathcal{P}_{N,j}$ denotes the jth symbol of $\mathcal{P}_N$, which will be either '<' or '=' when j is even; it will be a fraction when j is odd. For example,

$$\begin{aligned}
\mathcal{P}_2 &= \tfrac{0}{2}=\tfrac{0}{1}<\tfrac{1}{2}<\tfrac{2}{2}=\tfrac{1}{1}<\tfrac{3}{2}<\tfrac{4}{2}=\tfrac{2}{1}<\tfrac{5}{2}<\tfrac{6}{2}=\tfrac{3}{1}<\tfrac{7}{2}<\tfrac{8}{2}=\tfrac{4}{1}<\tfrac{9}{2}<\tfrac{10}{2}=\tfrac{5}{1}<\cdots; \\
\mathcal{P}_3 &= \tfrac{0}{2}=\tfrac{0}{3}=\tfrac{0}{1}<\tfrac{1}{3}<\tfrac{1}{2}<\tfrac{2}{3}<\tfrac{2}{2}=\tfrac{3}{3}=\tfrac{1}{1}<\tfrac{4}{3}<\tfrac{3}{2}<\tfrac{5}{3}<\tfrac{4}{2}=\tfrac{6}{3}=\tfrac{2}{1}<\tfrac{7}{3}<\tfrac{5}{2}<\cdots; \\
\mathcal{P}_4 &= \tfrac{0}{2}=\tfrac{0}{4}=\tfrac{0}{3}=\tfrac{0}{1}<\tfrac{1}{4}<\tfrac{1}{3}<\tfrac{2}{4}=\tfrac{1}{2}<\tfrac{2}{3}<\tfrac{3}{4}<\tfrac{2}{2}=\tfrac{4}{4}=\tfrac{3}{3}=\tfrac{1}{1}<\tfrac{5}{4}<\tfrac{4}{3}<\tfrac{6}{4}=\cdots; \\
\mathcal{P}_5 &= \tfrac{0}{2}=\tfrac{0}{4}=\tfrac{0}{5}=\tfrac{0}{3}=\tfrac{0}{1}<\tfrac{1}{5}<\tfrac{1}{4}<\tfrac{1}{3}<\tfrac{2}{5}<\tfrac{2}{4}=\tfrac{1}{2}<\tfrac{2}{5}<\tfrac{2}{3}<\tfrac{3}{4}<\tfrac{4}{5}<\tfrac{2}{2}=\tfrac{4}{4}=\cdots; \\
\mathcal{P}_6 &= \tfrac{0}{2}=\tfrac{0}{4}=\tfrac{0}{6}=\tfrac{0}{5}=\tfrac{0}{3}=\tfrac{0}{1}<\tfrac{1}{6}<\tfrac{1}{5}<\tfrac{1}{4}<\tfrac{2}{6}=\tfrac{1}{3}<\tfrac{2}{5}<\tfrac{2}{4}=\tfrac{3}{6}=\tfrac{1}{2}<\tfrac{3}{5}<\tfrac{4}{6}=\cdots.
\end{aligned}$$

(Equal elements occur in a slightly peculiar order.) Prove that the '<' and '=' signs defined by the rules above correctly describe the relations between adjacent fractions in the Peirce sequence.

Research problems

65 Are the Euclid numbers e_n all squarefree?

66 Are the Mersenne numbers $2^p - 1$ all squarefree?

67 Prove or disprove that $\max_{1\leqslant j<k\leqslant n} a_k/\gcd(a_j, a_k) \geqslant n$, for all sequences of integers $0 < a_1 < \cdots < a_n$.

68 Is there a constant Q such that $\lfloor Q^{2^n} \rfloor$ is prime for all $n \geqslant 0$?

69 Let P_n denote the nth prime. Prove or disprove that $P_{n+1} - P_n = O(\log P_n)^2$.

70 Does $\epsilon_3(n!) = \epsilon_2(n!)/2$ for infinitely many n?

71 Prove or disprove: If $k \neq 1$ there exists $n > 1$ such that $2^n \equiv k \pmod{n}$. Are there infinitely many such n?

72 Prove or disprove: For all integers a, there exist infinitely many n such that $\varphi(n)\backslash(n + a)$.

73 If the $\Phi(n)+1$ terms of the Farey series

$$\mathcal{F}_n = \langle \mathcal{F}_n(0), \mathcal{F}_n(1), \ldots, \mathcal{F}_n(\Phi(n)) \rangle$$

were fairly evenly distributed, we would expect $\mathcal{F}_n(k) \approx k/\Phi(n)$. Therefore the sum $D(n) = \sum_{k=0}^{\Phi(n)} \left|\mathcal{F}_n(k) - k/\Phi(n)\right|$ measures the "deviation of $\mathcal{F}_n$ from uniformity." Is it true that $D(n) = O(n^{1/2+\epsilon})$ for all $\epsilon > 0$?

74 Approximately how many distinct values are there in the set $\{0! \bmod p, 1! \bmod p, \ldots, (p-1)! \bmod p\}$, as $p \to \infty$?

5

Binomial Coefficients

Lucky us!

LET'S TAKE A BREATHER. The previous chapters have seen some heavy going, with sums involving floor, ceiling, mod, phi, and mu functions. Now we're going to study binomial coefficients, which turn out to be (a) more important in applications, and (b) easier to manipulate, than all those other quantities.

5.1 BASIC IDENTITIES

Otherwise known as combinations of n things, k at a time.

The symbol $\binom{n}{k}$ is a binomial coefficient, so called because of an important property we look at later this section, the binomial theorem. But we read the symbol "n choose k." This incantation arises from its combinatorial interpretation—it is the number of ways to choose a k-element subset from an n-element set. For example, from the set $\{1,2,3,4\}$ we can choose two elements in six ways,

$$\{1,2\}, \quad \{1,3\}, \quad \{1,4\}, \quad \{2,3\}, \quad \{2,4\}, \quad \{3,4\};$$

so $\binom{4}{2} = 6$.

To express the number $\binom{n}{k}$ in more familiar terms it's easiest to first determine the number of k-element *sequences*, rather than subsets, chosen from an n-element set; for sequences, the order of the elements counts. We use the same argument we used in Chapter 4 to show that $n!$ is the number of permutations of n objects. There are n choices for the first element of the sequence; for each, there are $n-1$ choices for the second; and so on, until there are $n-k+1$ choices for the kth. This gives $n(n-1)\ldots(n-k+1) = n^{\underline{k}}$ choices in all. And since each k-element subset has exactly $k!$ different orderings, this number of *sequences* counts each *subset* exactly $k!$ times. To get our answer, we simply divide by $k!$:

$$\binom{n}{k} = \frac{n(n-1)\ldots(n-k+1)}{k(k-1)\ldots(1)}.$$

For example,

$$\binom{4}{2} = \frac{4\cdot 3}{2\cdot 1} = 6;$$

this agrees with our previous enumeration.

We call n the *upper index* and k the *lower index*. The indices are restricted to be nonnegative integers by the combinatorial interpretation, because sets don't have negative or fractional numbers of elements. But the binomial coefficient has many uses besides its combinatorial interpretation, so we will remove some of the restrictions. It's most useful, it turns out, to allow an arbitrary real (or even complex) number to appear in the upper index, and to allow an arbitrary integer in the lower. Our formal definition therefore takes the following form:

$$\binom{r}{k} = \begin{cases} \dfrac{r(r-1)\ldots(r-k+1)}{k(k-1)\ldots(1)} = \dfrac{r^{\underline{k}}}{k!}, & \text{integer } k \geqslant 0; \\ 0, & \text{integer } k < 0. \end{cases} \tag{5.1}$$

This definition has several noteworthy features. First, the upper index is called r, not n; the letter r emphasizes the fact that binomial coefficients make sense when any real number appears in this position. For instance, we have $\binom{-1}{3} = (-1)(-2)(-3)/(3\cdot 2\cdot 1) = -1$. There's no combinatorial interpretation here, but $r = -1$ turns out to be an important special case. A noninteger index like $r = -1/2$ also turns out to be useful.

Second, we can view $\binom{r}{k}$ as a kth-degree polynomial in r. We'll see that this viewpoint is often helpful.

Third, we haven't defined binomial coefficients for noninteger lower indices. A reasonable definition can be given, but actual applications are rare, so we will defer this generalization to later in the chapter.

Final note: We've listed the restrictions 'integer $k \geqslant 0$' and 'integer $k < 0$' at the right of the definition. Such restrictions will be listed in all the identities we will study, so that the range of applicability will be clear. In general the fewer restrictions the better, because an unrestricted identity is most useful; still, any restrictions that apply are an important part of the identity. When we manipulate binomial coefficients, it's easier to ignore difficult-to-remember restrictions temporarily and to check later that nothing has been violated. But the check needs to be made.

For example, almost every time we encounter $\binom{n}{n}$ it equals 1, so we can get lulled into thinking that it's always 1. But a careful look at definition (5.1) tells us that $\binom{n}{n}$ is 1 only when $n \geqslant 0$ (assuming that n is an integer); when $n < 0$ we have $\binom{n}{n} = 0$. Traps like this can (and will) make life adventuresome.

Before getting to the identities that we will use to tame binomial coefficients, let's take a peek at some small values. The numbers in Table 155 form the beginning of *Pascal's triangle*, named after Blaise Pascal (1623–1662)

Table 155 Pascal's triangle.

n	$\binom{n}{0}$	$\binom{n}{1}$	$\binom{n}{2}$	$\binom{n}{3}$	$\binom{n}{4}$	$\binom{n}{5}$	$\binom{n}{6}$	$\binom{n}{7}$	$\binom{n}{8}$	$\binom{n}{9}$	$\binom{n}{10}$
0	1										
1	1	1									
2	1	2	1								
3	1	3	3	1							
4	1	4	6	4	1						
5	1	5	10	10	5	1					
6	1	6	15	20	15	6	1				
7	1	7	21	35	35	21	7	1			
8	1	8	28	56	70	56	28	8	1		
9	1	9	36	84	126	126	84	36	9	1	
10	1	10	45	120	210	252	210	120	45	10	1

Binomial coefficients were well known in Asia, many centuries before Pascal was born [74], but he had no way to know that.

because he wrote an influential treatise about them [227]. The empty entries in this table are actually 0's, because of a zero in the numerator of (5.1); for example, $\binom{1}{2} = (1\cdot 0)/(2\cdot 1) = 0$. These entries have been left blank simply to help emphasize the rest of the table.

It's worthwhile to memorize formulas for the first three columns,

$$\binom{r}{0} = 1, \qquad \binom{r}{1} = r, \qquad \binom{r}{2} = \frac{r(r-1)}{2}; \tag{5.2}$$

these hold for arbitrary reals. (Recall that $\binom{n+1}{2} = \frac{1}{2}n(n+1)$ is the formula we derived for triangular numbers in Chapter 1; triangular numbers are conspicuously present in the $\binom{n}{2}$ column of Table 155.) It's also a good idea to memorize the first five rows or so of Pascal's triangle, so that when the pattern 1, 4, 6, 4, 1 appears in some problem we will have a clue that binomial coefficients probably lurk nearby.

In Italy it's called Tartaglia's triangle.

The numbers in Pascal's triangle satisfy, practically speaking, infinitely many identities, so it's not too surprising that we can find some surprising relationships by looking closely. For example, there's a curious "hexagon property," illustrated by the six numbers 56, 28, 36, 120, 210, 126 that surround 84 in the lower right portion of Table 155. Both ways of multiplying alternate numbers from this hexagon give the same product: $56\cdot 36\cdot 210 = 28\cdot 120\cdot 126 = 423360$. The same thing holds if we extract such a hexagon from any other part of Pascal's triangle.

"C'est une chose estrange combien il est fertile en proprietez."
—B. Pascal [227]

And now the identities. Our goal in this section will be to learn a few simple rules by which we can solve the vast majority of practical problems involving binomial coefficients.

Definition (5.1) can be recast in terms of factorials in the common case that the upper index r is an integer, n, that's greater than or equal to the lower index k:

$$\binom{n}{k} = \frac{n!}{k!\,(n-k)!}, \qquad \text{integers } n \geqslant k \geqslant 0. \tag{5.3}$$

To get this formula, we just multiply the numerator and denominator of (5.1) by $(n-k)!$. It's occasionally useful to expand a binomial coefficient into this factorial form (for example, when proving the hexagon property). And we often want to go the other way, changing factorials into binomials.

The factorial representation hints at a symmetry in Pascal's triangle: Each row reads the same left-to-right as right-to-left. The identity reflecting this—called the *symmetry* identity—is obtained by changing k to $n-k$:

$$\binom{n}{k} = \binom{n}{n-k}, \qquad \begin{array}{c}\text{integer } n \geqslant 0,\\ \text{integer } k.\end{array} \tag{5.4}$$

This formula makes combinatorial sense, because by specifying the k chosen things out of n we're in effect specifying the $n-k$ unchosen things.

The restriction that n and k be integers in identity (5.4) is obvious, since each lower index must be an integer. But why can't n be negative? Suppose, for example, that $n=-1$. Is

$$\binom{-1}{k} \stackrel{?}{=} \binom{-1}{-1-k}$$

a valid equation? No. For instance, when $k=0$ we get 1 on the left and 0 on the right. In fact, for any integer $k \geqslant 0$ the left side is

$$\binom{-1}{k} = \frac{(-1)(-2)\ldots(-k)}{k!} = (-1)^k,$$

which is either 1 or -1; but the right side is 0, because the lower index is negative. And for negative k the left side is 0 but the right side is

$$\binom{-1}{-1-k} = (-1)^{-1-k},$$

which is either 1 or -1. So the equation '$\binom{-1}{k} = \binom{-1}{-1-k}$' is always false!

The symmetry identity fails for all other negative integers n, too. But unfortunately it's all too easy to forget this restriction, since the expression in the upper index is sometimes negative only for obscure (but legal) values

I just hope I don't fall into this trap during the midterm.

of its variables. Everyone who's manipulated binomial coefficients much has fallen into this trap at least three times.

But the symmetry identity does have a big redeeming feature: It works for all values of k, even when $k < 0$ or $k > n$. (Because both sides are zero in such cases.) Otherwise $0 \leqslant k \leqslant n$, and symmetry follows immediately from (5.3):

$$\binom{n}{k} = \frac{n!}{k!\,(n-k)!} = \frac{n!}{\bigl(n-(n-k)\bigr)!\,\,(n-k)!} = \binom{n}{n-k}.$$

Our next important identity lets us move things in and out of binomial coefficients:

$$\binom{r}{k} = \frac{r}{k}\binom{r-1}{k-1}, \qquad \text{integer } k \neq 0. \tag{5.5}$$

The restriction on k prevents us from dividing by 0 here. We call (5.5) an *absorption* identity, because we often use it to absorb a variable into a binomial coefficient when that variable is a nuisance outside. The equation follows from definition (5.1), because $r^{\underline{k}} = r(r-1)^{\underline{k-1}}$ and $k! = k(k-1)!$ when $k > 0$; both sides are zero when $k < 0$.

If we multiply both sides of (5.5) by k, we get an absorption identity that works even when $k = 0$:

$$k\binom{r}{k} = r\binom{r-1}{k-1}, \qquad \text{integer } k. \tag{5.6}$$

This one also has a companion that keeps the lower index intact:

$$(r-k)\binom{r}{k} = r\binom{r-1}{k}, \qquad \text{integer } k. \tag{5.7}$$

We can derive (5.7) by sandwiching an application of (5.6) between two applications of symmetry:

$$\begin{aligned}
(r-k)\binom{r}{k} &= (r-k)\binom{r}{r-k} && \text{(by symmetry)} \\
&= r\binom{r-1}{r-k-1} && \text{(by (5.6))} \\
&= r\binom{r-1}{k}. && \text{(by symmetry)}
\end{aligned}$$

But wait a minute. We've claimed that the identity holds for *all* real r, yet the derivation we just gave holds only when r is a positive integer. (The upper index $r-1$ must be a nonnegative integer if we're to use the symmetry property (5.4) with impunity.) Have we been cheating? No. It's true that the derivation is valid only for positive integers r; but we can claim that the identity holds for all values of r, because both sides of (5.7) are polynomials in r of degree $k+1$. A nonzero polynomial of degree d or less can have at most d distinct zeros; therefore the difference of two such polynomials, which also has degree d or less, cannot be zero at more than d points unless it is identically zero. In other words, if two polynomials of degree d or less agree at more than d points, they must agree everywhere. We have shown that $(r-k)\binom{r}{k} = r\binom{r-1}{k}$ whenever r is a positive integer; so these two polynomials agree at infinitely many points, and they must be identically equal.

(Well, not here anyway.)

The proof technique in the previous paragraph, which we will call the *polynomial argument*, is useful for extending many identities from integers to reals; we'll see it again and again. Some equations, like the symmetry identity (5.4), are not identities between polynomials, so we can't always use this method. But many identities do have the necessary form.

For example, here's another polynomial identity, perhaps the most important binomial identity of all, known as the *addition formula*:

$$\binom{r}{k} = \binom{r-1}{k} + \binom{r-1}{k-1}, \qquad \text{integer } k. \tag{5.8}$$

When r is a positive integer, the addition formula tells us that every number in Pascal's triangle is the sum of two numbers in the previous row, one directly above it and the other just to the left. And the formula applies also when r is negative, real, or complex; the only restriction is that k be an integer, so that the binomial coefficients are defined.

One way to prove the addition formula is to assume that r is a positive integer and to use the combinatorial interpretation. Recall that $\binom{r}{k}$ is the number of possible k-element subsets chosen from an r-element set. If we have a set of r eggs that includes exactly one bad egg, there are $\binom{r}{k}$ ways to select k of the eggs. Exactly $\binom{r-1}{k}$ of these selections involve nothing but good eggs; and $\binom{r-1}{k-1}$ of them contain the bad egg, because such selections have $k-1$ of the $r-1$ good eggs. Adding these two numbers together gives (5.8). This derivation assumes that r is a positive integer, and that $k \geqslant 0$. But both sides of the identity are zero when $k < 0$, and the polynomial argument establishes (5.8) in all remaining cases.

We can also derive (5.8) by adding together the two absorption identities (5.7) and (5.6):

$$(r-k)\binom{r}{k} + k\binom{r}{k} = r\binom{r-1}{k} + r\binom{r-1}{k-1};$$

the left side is $r\binom{r}{k}$, and we can divide through by r. This derivation is valid for everything but $r=0$, and it's easy to check that remaining case.

Those of us who tend not to discover such slick proofs, or who are otherwise into tedium, might prefer to derive (5.8) by a straightforward manipulation of the definition. If $k > 0$,

$$\binom{r-1}{k} + \binom{r-1}{k-1} = \frac{(r-1)^{\underline{k}}}{k!} + \frac{(r-1)^{\underline{k-1}}}{(k-1)!}$$
$$= \frac{(r-1)^{\underline{k-1}}(r-k)}{k!} + \frac{(r-1)^{\underline{k-1}}k}{k!}$$
$$= \frac{(r-1)^{\underline{k-1}}r}{k!} = \frac{r^{\underline{k}}}{k!} = \binom{r}{k}.$$

Again, the cases for $k \leqslant 0$ are easy to handle.

We've just seen three rather different proofs of the addition formula. This is not surprising; binomial coefficients have many useful properties, several of which are bound to lead to proofs of an identity at hand.

The addition formula is essentially a recurrence for the numbers of Pascal's triangle, so we'll see that it is especially useful for proving other identities by induction. We can also get a new identity immediately by unfolding the recurrence. For example,

$$\binom{5}{3} = \binom{4}{3} + \binom{4}{2}$$
$$= \binom{4}{3} + \binom{3}{2} + \binom{3}{1}$$
$$= \binom{4}{3} + \binom{3}{2} + \binom{2}{1} + \binom{2}{0}$$
$$= \binom{4}{3} + \binom{3}{2} + \binom{2}{1} + \binom{1}{0} + \binom{1}{-1}.$$

Since $\binom{1}{-1} = 0$, that term disappears and we can stop. This method yields the general formula

$$\sum_{k \leqslant n} \binom{r+k}{k} = \binom{r}{0} + \binom{r+1}{1} + \cdots + \binom{r+n}{n}$$
$$= \binom{r+n+1}{n}, \qquad \text{integer } n. \qquad (5.9)$$

Notice that we don't need the lower limit $k \geqslant 0$ on the index of summation, because the terms with $k < 0$ are zero.

This formula expresses one binomial coefficient as the sum of others whose upper and lower indices stay the same distance apart. We found it by repeatedly expanding the binomial coefficient with the smallest lower index: first

$\binom{5}{3}$, then $\binom{4}{2}$, then $\binom{3}{1}$, then $\binom{2}{0}$. What happens if we unfold the other way, repeatedly expanding the one with largest lower index? We get

$$\begin{aligned}\binom{5}{3} &= \binom{4}{3}+\binom{4}{2}\\ &= \binom{3}{3}+\binom{3}{2}+\binom{4}{2}\\ &= \binom{2}{3}+\binom{2}{2}+\binom{3}{2}+\binom{4}{2}\\ &= \binom{1}{3}+\binom{1}{2}+\binom{2}{2}+\binom{3}{2}+\binom{4}{2}\\ &= \binom{0}{3}+\binom{0}{2}+\binom{1}{2}+\binom{2}{2}+\binom{3}{2}+\binom{4}{2}.\end{aligned}$$

Now $\binom{0}{3}$ is zero (so are $\binom{0}{2}$ and $\binom{1}{2}$, but these make the identity nicer), and we can spot the general pattern:

$$\begin{aligned}\sum_{0\leqslant k\leqslant n}\binom{k}{m} &= \binom{0}{m}+\binom{1}{m}+\cdots+\binom{n}{m}\\ &= \binom{n+1}{m+1}, \qquad \text{integers } m,n\geqslant 0. \end{aligned}\tag{5.10}$$

This identity, which we call *summation on the upper index*, expresses a binomial coefficient as the sum of others whose lower indices are constant. In this case the sum needs the lower limit $k \geqslant 0$, because the terms with $k < 0$ *aren't* zero. Also, m and n can't in general be negative.

Identity (5.10) has an interesting combinatorial interpretation. If we want to choose $m+1$ tickets from a set of $n+1$ tickets numbered 0 through n, there are $\binom{k}{m}$ ways to do this when the largest ticket selected is number k.

We can prove both (5.9) and (5.10) by induction using the addition formula, but we can also prove them from each other. For example, let's prove (5.9) from (5.10); our proof will illustrate some common binomial coefficient manipulations. Our general plan will be to massage the left side $\sum\binom{r+k}{k}$ of (5.9) so that it looks like the left side $\sum\binom{k}{m}$ of (5.10); then we'll invoke that identity, replacing the sum by a single binomial coefficient; finally we'll transform that coefficient into the right side of (5.9).

We can assume for convenience that r and n are nonnegative integers; the general case of (5.9) follows from this special case, by the polynomial argument. Let's write m instead of r, so that this variable looks more like a nonnegative integer. The plan can now be carried out systematically as follows:

$$\begin{aligned}\sum_{k\leqslant n}\binom{m+k}{k} &= \sum_{-m\leqslant k\leqslant n}\binom{m+k}{k}\\ &= \sum_{-m\leqslant k\leqslant n}\binom{m+k}{m}\\ &= \sum_{0\leqslant k\leqslant m+n}\binom{k}{m}\\ &= \binom{m+n+1}{m+1} \;=\; \binom{m+n+1}{n}.\end{aligned}$$

Let's look at this derivation blow by blow. The key step is in the second line, where we apply the symmetry law (5.4) to replace $\binom{m+k}{k}$ by $\binom{m+k}{m}$. We're allowed to do this only when $m + k \geqslant 0$, so our first step restricts the range of k by discarding the terms with $k < -m$. (This is legal because those terms are zero.) Now we're almost ready to apply (5.10); the third line sets this up, replacing k by $k - m$ and tidying up the range of summation. This step, like the first, merely plays around with $\sum$-notation. Now k appears by itself in the upper index and the limits of summation are in the proper form, so the fourth line applies (5.10). One more use of symmetry finishes the job.

Certain sums that we did in Chapters 1 and 2 were actually special cases of (5.10), or disguised versions of this identity. For example, the case $m = 1$ gives the sum of the nonnegative integers up through n:

$$\binom{0}{1}+\binom{1}{1}+\cdots+\binom{n}{1} = 0+1+\cdots+n = \frac{(n+1)n}{2} = \binom{n+1}{2}.$$

And the general case is equivalent to Chapter 2's rule

$$\sum_{0\leqslant k\leqslant n} k^{\underline{m}} = \frac{(n+1)^{\underline{m+1}}}{m+1}, \qquad \text{integers } m, n \geqslant 0,$$

if we divide both sides of this formula by $m!$. In fact, the addition formula (5.8) tells us that

$$\Delta\left(\binom{x}{m}\right) = \binom{x+1}{m} - \binom{x}{m} = \binom{x}{m-1},$$

if we replace r and k respectively by $x + 1$ and m. Hence the methods of Chapter 2 give us the handy indefinite summation formula

$$\sum \binom{x}{m}\,\delta x = \binom{x}{m+1} + C. \tag{5.11}$$

Binomial coefficients get their name from the *binomial theorem*, which deals with powers of the binomial expression $x+y$. Let's look at the smallest cases of this theorem:

"At the age of twenty-one he [Moriarty] wrote a treatise upon the Binomial Theorem, which has had a European vogue. On the strength of it, he won the Mathematical Chair at one of our smaller Universities."
—S. Holmes [71]

$$\begin{aligned}
(x+y)^0 &= 1x^0y^0 \\
(x+y)^1 &= 1x^1y^0 + 1x^0y^1 \\
(x+y)^2 &= 1x^2y^0 + 2x^1y^1 + 1x^0y^2 \\
(x+y)^3 &= 1x^3y^0 + 3x^2y^1 + 3x^1y^2 + 1x^0y^3 \\
(x+y)^4 &= 1x^4y^0 + 4x^3y^1 + 6x^2y^2 + 4x^1y^3 + 1x^0y^4 .
\end{aligned}$$

It's not hard to see why these coefficients are the same as the numbers in Pascal's triangle: When we expand the product

$$(x+y)^n = \overbrace{(x+y)(x+y)\ldots(x+y)}^{n \text{ factors}},$$

every term is itself the product of n factors, each either an x or y. The number of such terms with k factors of x and $n-k$ factors of y is the coefficient of x^ky^{n-k} after we combine like terms. And this is exactly the number of ways to choose k of the n binomials from which an x will be contributed; that is, it's $\binom{n}{k}$.

Some textbooks leave the quantity 0^0 undefined, because the functions x^0 and 0^x have different limiting values when x decreases to 0. But this is a mistake. We must define

$$x^0 = 1, \qquad \text{for all } x,$$

if the binomial theorem is to be valid when $x=0$, $y=0$, and/or $x=-y$. The theorem is too important to be arbitrarily restricted! By contrast, the function 0^x is quite unimportant.

But what exactly is the binomial theorem? In its full glory it is the following identity:

$$(x+y)^r = \sum_k \binom{r}{k} x^k y^{r-k}, \qquad \begin{array}{l}\text{integer } r \geqslant 0 \\ \text{or } |x/y| < 1.\end{array} \tag{5.12}$$

The sum is over all integers k; but it is really a finite sum when r is a nonnegative integer, because all terms are zero except those with $0 \leqslant k \leqslant r$. On the other hand, the theorem is also valid when r is negative, or even when r is an arbitrary real or complex number. In such cases the sum really is infinite, and we must have $|x/y| < 1$ to guarantee the sum's absolute convergence.

Two special cases of the binomial theorem are worth special attention, even though they are extremely simple. If $x = y = 1$ and $r = n$ is nonnegative, we get

$$2^n = \binom{n}{0} + \binom{n}{1} + \cdots + \binom{n}{n}, \qquad \text{integer } n \geqslant 0.$$

This equation tells us that row n of Pascal's triangle sums to 2^n. And when x is -1 instead of $+1$, we get

$$0^n = \binom{n}{0} - \binom{n}{1} + \cdots + (-1)^n\binom{n}{n}, \qquad \text{integer } n \geqslant 0.$$

For example, $1 - 4 + 6 - 4 + 1 = 0$; the elements of row n sum to zero if we give them alternating signs, except in the top row (when $n = 0$ and $0^0 = 1$).

When r is not a nonnegative integer, we most often use the binomial theorem in the special case $y = 1$. Let's state this special case explicitly, writing z instead of x to emphasize the fact that an arbitrary complex number can be involved here:

$$(1+z)^r = \sum_k \binom{r}{k} z^k, \qquad |z| < 1. \tag{5.13}$$

The general formula in (5.12) follows from this one if we set $z = x/y$ and multiply both sides by y^r.

We have proved the binomial theorem only when r is a nonnegative integer, by using a combinatorial interpretation. We can't deduce the general case from the nonnegative-integer case by using the polynomial argument, because the sum is infinite in the general case. But when r is arbitrary, we can use Taylor's theorem from infinite calculus:

$$\begin{aligned} f(z) &= \frac{f(0)}{0!}z^0 + \frac{f'(0)}{1!}z^1 + \frac{f''(0)}{2!}z^2 + \cdots \\ &= \sum_{k \geqslant 0} \frac{f^{(k)}(0)}{k!} z^k . \end{aligned}$$

The derivatives of the function $f(z) = (1+z)^r$ are easily evaluated; in fact, $f^{(k)}(z) = r^{\underline{k}}(1+z)^{r-k}$. Setting $z = 0$ gives (5.13).

(Chapter 9 tells the meaning of O.)

We also need to prove that the infinite sum converges, when $|z| < 1$. It does, because $\binom{r}{k} = O(k^{-1-r})$ by equation (5.93) below.

Now let's look more closely at the values of $\binom{n}{k}$ when n is a negative integer. One way to approach these values is to use the addition law (5.8) to fill in the entries that lie above the numbers in Table 155, thereby obtaining Table 164. For example, we must have $\binom{-1}{0} = 1$, since $\binom{0}{0} = \binom{-1}{0} + \binom{-1}{-1}$ and $\binom{-1}{-1} = 0$; then we must have $\binom{-1}{1} = -1$, since $\binom{0}{1} = \binom{-1}{1} + \binom{-1}{0}$; and so on.

Table 164 Pascal's triangle, extended upward.

n	$\binom{n}{0}$	$\binom{n}{1}$	$\binom{n}{2}$	$\binom{n}{3}$	$\binom{n}{4}$	$\binom{n}{5}$	$\binom{n}{6}$	$\binom{n}{7}$	$\binom{n}{8}$	$\binom{n}{9}$	$\binom{n}{10}$
−4	1	−4	10	−20	35	−56	84	−120	165	−220	286
−3	1	−3	6	−10	15	−21	28	−36	45	−55	66
−2	1	−2	3	−4	5	−6	7	−8	9	−10	11
−1	1	−1	1	−1	1	−1	1	−1	1	−1	1
0	1	0	0	0	0	0	0	0	0	0	0

All these numbers are familiar. Indeed, the rows and columns of Table 164 appear as columns in Table 155 (but minus the minus signs). So there must be a connection between the values of $\binom{n}{k}$ for negative n and the values for positive n. The general rule is

$$\binom{r}{k} = (-1)^k \binom{k-r-1}{k}, \qquad \text{integer } k; \tag{5.14}$$

it is easily proved, since

$$\begin{aligned} r^{\underline{k}} &= r(r-1)\ldots(r-k+1) \\ &= (-1)^k(-r)(1-r)\ldots(k-1-r) = (-1)^k(k-r-1)^{\underline{k}} \end{aligned}$$

when $k \geqslant 0$, and both sides are zero when $k < 0$.

Identity (5.14) is particularly valuable because it holds without any restriction. (Of course, the lower index must be an integer so that the binomial coefficients are defined.) The transformation in (5.14) is called *negating the upper index*, or "upper negation."

But how can we remember this important formula? The other identities we've seen — symmetry, absorption, addition, etc. — are pretty simple, but this one looks rather messy. Still, there's a mnemonic that's not too bad: To negate the upper index, we begin by writing down $(-1)^k$, where k is the lower index. (The lower index doesn't change.) Then we immediately write k again, twice, in both lower and upper index positions. Then we negate the original upper index by *subtracting* it from the new upper index. And we complete the job by *subtracting* 1 more (always subtracting, not adding, because this is a negation process).

You call this a mnemonic? I'd call it pneumatic — full of air. It does help me remember, though.

Let's negate the upper index twice in succession, for practice. We get

(Now is a good time to do warmup exercise 4.)

$$\begin{aligned} \binom{r}{k} &= (-1)^k \binom{k-r-1}{k} \\ &= (-1)^{2k} \binom{k-(k-r-1)-1}{k} = \binom{r}{k}, \end{aligned}$$

so we're right back where we started. This is probably not what the framers of the identity intended; but it's reassuring to know that we haven't gone astray.

It's also frustrating, if we're trying to get somewhere else.

Some applications of (5.14) are, of course, more useful than this. We can use upper negation, for example, to move quantities between upper and lower index positions. The identity has a symmetric formulation,

$$(-1)^m\binom{-n-1}{m} = (-1)^n\binom{-m-1}{n}, \qquad \text{integers } m,n \geqslant 0, \qquad (5.15)$$

which holds because both sides are equal to $\binom{m+n}{n}$.

Upper negation can also be used to derive the following interesting sum:

$$\sum_{k\leqslant m}\binom{r}{k}(-1)^k = \binom{r}{0}-\binom{r}{1}+\cdots+(-1)^m\binom{r}{m}$$
$$= (-1)^m\binom{r-1}{m}, \qquad \text{integer } m. \qquad (5.16)$$

The idea is to negate the upper index, then apply (5.9), and negate again:

(Here double negation helps, because we've sandwiched another operation in between.)

$$\sum_{k\leqslant m}\binom{r}{k}(-1)^k = \sum_{k\leqslant m}\binom{k-r-1}{k}$$
$$= \binom{-r+m}{m}$$
$$= (-1)^m\binom{r-1}{m}.$$

This formula gives us a partial sum of the rth row of Pascal's triangle, provided that the entries of the row have been given alternating signs. For instance, if $r=5$ and $m=2$ the formula gives $1-5+10=6=(-1)^2\binom{4}{2}$.

Notice that if $m \geqslant r$, (5.16) gives the alternating sum of the entire row, and this sum is zero when r is a positive integer. We proved this before, when we expanded $(1-1)^r$ by the binomial theorem; it's interesting to know that the partial sums of this expression can also be evaluated in closed form.

How about the simpler partial sum,

$$\sum_{k\leqslant m}\binom{n}{k} = \binom{n}{0}+\binom{n}{1}+\cdots+\binom{n}{m}; \qquad (5.17)$$

surely if we can evaluate the corresponding sum with alternating signs, we ought to be able to do this one? But no; there is no closed form for the partial sum of a row of Pascal's triangle. We can do columns — that's (5.10) — but not rows. Curiously, however, there is a way to partially sum the row elements if they have been multiplied by their distance from the center:

$$\sum_{k\leqslant m}\binom{r}{k}\Bigl(\frac{r}{2}-k\Bigr) = \frac{m+1}{2}\binom{r}{m+1}, \qquad \text{integer } m. \tag{5.18}$$

(This formula is easily verified by induction on m.) The relation between these partial sums with and without the factor of $(r/2-k)$ in the summand is analogous to the relation between the integrals

$$\int_{-\infty}^{\alpha} xe^{-x^2}\,dx = -\tfrac{1}{2}e^{-\alpha^2} \qquad\text{and}\qquad \int_{-\infty}^{\alpha} e^{-x^2}\,dx\,.$$

The apparently more complicated integral on the left, with the factor of x, has a closed form, while the simpler-looking integral on the right, without the factor, has none. Appearances can be deceiving.

(Well, it actually equals $\frac{1}{2}\sqrt{\pi}\,\mathrm{erf}\,\alpha$, a multiple of the "error function" of α, if we're willing to accept that as a closed form.)

At the end of this chapter, we'll study a method by which it's possible to determine whether or not there is a closed form for the partial sums of a given series involving binomial coefficients, in a fairly general setting. This method is capable of discovering identities (5.16) and (5.18), and it also will tell us that (5.17) is a dead end.

Partial sums of the binomial series lead to a curious relationship of another kind:

$$\sum_{k\leqslant m}\binom{m+r}{k}x^k y^{m-k} = \sum_{k\leqslant m}\binom{-r}{k}(-x)^k(x+y)^{m-k}, \quad \text{integer } m. \tag{5.19}$$

This identity isn't hard to prove by induction: Both sides are zero when $m<0$ and 1 when $m=0$. If we let S_m stand for the sum on the left, we can apply the addition formula (5.8) and show easily that

$$S_m = \sum_{k\leqslant m}\binom{m-1+r}{k}x^k y^{m-k} + \sum_{k\leqslant m}\binom{m-1+r}{k-1}x^k y^{m-k};$$

and

$$\sum_{k\leqslant m}\binom{m-1+r}{k}x^k y^{m-k} = yS_{m-1} + \binom{m-1+r}{m}x^m,$$

$$\sum_{k\leqslant m}\binom{m-1+r}{k-1}x^k y^{m-k} = xS_{m-1},$$

when $m>0$. Hence

$$S_m = (x+y)S_{m-1} + \binom{-r}{m}(-x)^m,$$

and this recurrence is satisfied also by the right-hand side of (5.19). By induction, both sides must be equal; QED.

But there's a neater proof. When r is an integer in the range $0 \geqslant r \geqslant -m$, the binomial theorem tells us that both sides of (5.19) are $(x+y)^{m+r}y^{-r}$. And since both sides are polynomials in r of degree m or less, agreement at $m+1$ different values is enough (but just barely!) to prove equality in general.

It may seem foolish to have an identity where one sum equals another. Neither side is in closed form. But sometimes one side turns out to be easier to evaluate than the other. For example, if we set $x = -1$ and $y = 1$, we get

$$\sum_{k \leqslant m} \binom{m+r}{k} (-1)^k = \binom{-r}{m}, \qquad \text{integer } m \geqslant 0,$$

an alternative form of identity (5.16). And if we set $x = y = 1$ and $r = m+1$, we get

$$\sum_{k \leqslant m} \binom{2m+1}{k} = \sum_{k \leqslant m} \binom{m+k}{k} 2^{m-k}.$$

The left-hand side sums just half of the binomial coefficients with upper index $2m+1$, and these are equal to their counterparts in the other half because Pascal's triangle has left-right symmetry. Hence the left-hand side is just $\frac{1}{2}2^{2m+1} = 2^{2m}$. This yields a formula that is quite unexpected,

$$\sum_{k \leqslant m} \binom{m+k}{k} 2^{-k} = 2^m, \qquad \text{integer } m \geqslant 0. \tag{5.20}$$

Let's check it when $m = 2$: $\binom{2}{0} + \frac{1}{2}\binom{3}{1} + \frac{1}{4}\binom{4}{2} = 1 + \frac{3}{2} + \frac{6}{4} = 4$. Astounding.

So far we've been looking either at binomial coefficients by themselves or at sums of terms in which there's only one binomial coefficient per term. But many of the challenging problems we face involve products of two or more binomial coefficients, so we'll spend the rest of this section considering how to deal with such cases.

Here's a handy rule that often helps to simplify the product of two binomial coefficients:

$$\binom{r}{m}\binom{m}{k} = \binom{r}{k}\binom{r-k}{m-k}, \qquad \text{integers } m, k. \tag{5.21}$$

We've already seen the special case $k = 1$; it's the absorption identity (5.6). Although both sides of (5.21) are products of binomial coefficients, one side often is easier to sum because of interactions with the rest of a formula. For example, the left side uses m twice, the right side uses it only once. Therefore we usually want to replace $\binom{r}{m}\binom{m}{k}$ by $\binom{r}{k}\binom{r-k}{m-k}$ when summing on m.

Equation (5.21) holds primarily because of cancellation between $m!$'s in the factorial representations of $\binom{r}{m}$ and $\binom{m}{k}$. If all variables are integers and $r \geqslant m \geqslant k \geqslant 0$, we have

$$\begin{aligned}\binom{r}{m}\binom{m}{k} &= \frac{r!}{m!\,(r-m)!}\,\frac{m!}{k!\,(m-k)!}\\ &= \frac{r!}{k!\,(m-k)!\,(r-m)!}\\ &= \frac{r!}{k!\,(r-k)!}\,\frac{(r-k)!}{(m-k)!\,(r-m)!} = \binom{r}{k}\binom{r-k}{m-k}.\end{aligned}$$

That was easy. Furthermore, if $m < k$ or $k < 0$, both sides of (5.21) are zero; so the identity holds for all integers m and k. Finally, the polynomial argument extends its validity to all real r.

Yeah, right.

A binomial coefficient $\binom{r}{k} = r!/(r-k)!\,k!$ can be written in the form $(a+b)!/a!\,b!$ after a suitable renaming of variables. Similarly, the quantity in the middle of the derivation above, $r!/k!\,(m-k)!\,(r-m)!$, can be written in the form $(a+b+c)!/a!\,b!\,c!$. This is a "trinomial coefficient," which arises in the "trinomial theorem":

$$\begin{aligned}(x+y+z)^n &= \sum_{\substack{0\leqslant a,b,c\leqslant n\\ a+b+c=n}} \frac{(a+b+c)!}{a!\,b!\,c!}x^a y^b z^c\\ &= \sum_{\substack{0\leqslant a,b,c\leqslant n\\ a+b+c=n}} \binom{a+b+c}{b+c}\binom{b+c}{c} x^a y^b z^c.\end{aligned}$$

"Excogitavi autem olim mirabilem regulam pro numeris coefficientibus potestatum, non tantum a binomio $x+y$, sed et a trinomio $x+y+z$, imo a polynomio quocunque, ut data potentia gradus cujuscunque v. gr. decimi, et potentia in ejus valore comprehensa, ut $x^5y^3z^2$, possim statim assignare numerum coefficientem, quem habere debet, sine ulla Tabula jam calculata."
— G. W. Leibniz [200]

So $\binom{r}{m}\binom{m}{k}$ is really a trinomial coefficient in disguise. Trinomial coefficients pop up occasionally in applications, and we can conveniently write them as

$$\binom{a+b+c}{a,\,b,\,c} = \frac{(a+b+c)!}{a!\,b!\,c!}$$

in order to emphasize the symmetry present.

Binomial and trinomial coefficients generalize to *multinomial coefficients*, which are always expressible as products of binomial coefficients:

$$\begin{aligned}\binom{a_1+a_2+\cdots+a_m}{a_1,a_2,\ldots,a_m} &= \frac{(a_1+a_2+\cdots+a_m)!}{a_1!\,a_2!\,\ldots\,a_m!}\\ &= \binom{a_1+a_2+\cdots+a_m}{a_2+\cdots+a_m}\ldots\binom{a_{m-1}+a_m}{a_m}.\end{aligned}$$

Therefore, when we run across such a beastie, our standard techniques apply.

Table 169 Sums of products of binomial coefficients.

$$\sum_k \binom{r}{m+k}\binom{s}{n-k} = \binom{r+s}{m+n}, \qquad \text{integers } m,n. \tag{5.22}$$

$$\sum_k \binom{l}{m+k}\binom{s}{n+k} = \binom{l+s}{l-m+n}, \qquad \text{integer } l \geqslant 0, \text{ integers } m,n. \tag{5.23}$$

$$\sum_k \binom{l}{m+k}\binom{s+k}{n}(-1)^k = (-1)^{l+m}\binom{s-m}{n-l}, \qquad \text{integer } l \geqslant 0, \text{ integers } m,n. \tag{5.24}$$

$$\sum_{k\leqslant l} \binom{l-k}{m}\binom{s}{k-n}(-1)^k = (-1)^{l+m}\binom{s-m-1}{l-m-n}, \qquad \text{integers } l,m,n \geqslant 0. \tag{5.25}$$

$$\sum_{0\leqslant k\leqslant l} \binom{l-k}{m}\binom{q+k}{n} = \binom{l+q+1}{m+n+1}, \qquad \text{integers } l,m \geqslant 0, \text{ integers } n \geqslant q \geqslant 0. \tag{5.26}$$

Now we come to Table 169, which lists identities that are among the most important of our standard techniques. These are the ones we rely on when struggling with a sum involving a product of two binomial coefficients. Each of these identities is a sum over k, with one appearance of k in each binomial coefficient; there also are four nearly independent parameters, called m, n, r, etc., one in each index position. Different cases arise depending on whether k appears in the upper or lower index, and on whether it appears with a plus or minus sign. Sometimes there's an additional factor of $(-1)^k$, which is needed to make the terms summable in closed form.

Fold down the corner on this page, so you can find the table quickly later. You'll need it!

Table 169 is far too complicated to memorize in full; it is intended only for reference. But the first identity in this table is by far the most memorable, and it should be remembered. It states that the sum (over all integers k) of the product of two binomial coefficients, in which the upper indices are constant and the lower indices have a constant sum for all k, is the binomial coefficient obtained by summing both lower and upper indices. This identity is known as *Vandermonde's convolution*, because Alexandre Vandermonde wrote a significant paper about it in the late 1700s [293]; it was, however, known to Chu Shih-Chieh in China as early as 1303. All of the other identities in Table 169 can be obtained from Vandermonde's convolution by doing things like negating upper indices or applying the symmetry law, etc., with care; therefore Vandermonde's convolution is the most basic of all.

We can prove Vandermonde's convolution by giving it a nice combinatorial interpretation. If we replace k by $k-m$ and n by $n-m$, we can assume that $m = 0$; hence the identity to be proved is

$$\sum_k \binom{r}{k}\binom{s}{n-k} = \binom{r+s}{n}, \qquad \text{integer } n. \tag{5.27}$$

Let r and s be nonnegative integers; the general case then follows by the polynomial argument. On the right side, $\binom{r+s}{n}$ is the number of ways to choose n people from among r men and s women. On the left, each term of the sum is the number of ways to choose k of the men and $n-k$ of the women. Summing over all k counts each possibility exactly once.

Sexist! You mentioned men first.

Much more often than not we use these identities left to right, since that's the direction of simplification. But every once in a while it pays to go the other direction, temporarily making an expression more complicated. When this works, we've usually created a double sum for which we can interchange the order of summation and then simplify.

Before moving on let's look at proofs for two more of the identities in Table 169. It's easy to prove (5.23); all we need to do is replace the first binomial coefficient by $\binom{l}{l-m-k}$, then Vandermonde's (5.22) applies.

The next one, (5.24), is a bit more difficult. We can reduce it to Vandermonde's convolution by a sequence of transformations, but we can just as easily prove it by resorting to the old reliable technique of mathematical induction. Induction is often the first thing to try when nothing else obvious jumps out at us, and induction on l works just fine here.

For the basis $l = 0$, all terms are zero except when $k = -m$; so both sides of the equation are $(-1)^m\binom{s-m}{n}$. Now suppose that the identity holds for all values less than some fixed l, where $l > 0$. We can use the addition formula to replace $\binom{l}{m+k}$ by $\binom{l-1}{m+k} + \binom{l-1}{m+k-1}$; the original sum now breaks into two sums, each of which can be evaluated by the induction hypothesis:

$$\sum_k \binom{l-1}{m+k}\binom{s+k}{n}(-1)^k + \sum_k \binom{l-1}{m+k-1}\binom{s+k}{n}(-1)^k$$
$$= (-1)^{l-1+m}\binom{s-m}{n-l+1} + (-1)^{l+m}\binom{s-m+1}{n-l+1}.$$

And this simplifies to the right-hand side of (5.24), if we apply the addition formula once again.

Two things about this derivation are worthy of note. First, we see again the great convenience of summing over all integers k, not just over a certain range, because there's no need to fuss over boundary conditions. Second, the addition formula works nicely with mathematical induction, because it's a recurrence for binomial coefficients. A binomial coefficient whose upper index is l is expressed in terms of two whose upper indices are $l-1$, and that's exactly what we need to apply the induction hypothesis.

So much for Table 169. What about sums with three or more binomial coefficients? If the index of summation is spread over all the coefficients, our chances of finding a closed form aren't great: Only a few closed forms are known for sums of this kind, hence the sum we need might not match the given specs. One of these rarities, proved in exercise 43, is

$$\sum_k \binom{m-r+s}{k}\binom{n+r-s}{n-k}\binom{r+k}{m+n} = \binom{r}{m}\binom{s}{n}, \qquad \text{integers } m,n \geqslant 0. \tag{5.28}$$

Here's another, more symmetric example:

$$\sum_k \binom{a+b}{a+k}\binom{b+c}{b+k}\binom{c+a}{c+k}(-1)^k = \frac{(a+b+c)!}{a!\,b!\,c!}, \qquad \text{integers } a,b,c \geqslant 0. \tag{5.29}$$

This one has a two-coefficient counterpart,

$$\sum_k \binom{a+b}{a+k}\binom{b+a}{b+k}(-1)^k = \frac{(a+b)!}{a!\,b!}, \qquad \text{integers } a,b \geqslant 0, \tag{5.30}$$

which incidentally doesn't appear in Table 169. The analogous four-coefficient sum doesn't have a closed form, but a similar sum does:

$$\sum_k (-1)^k \binom{a+b}{a+k}\binom{b+c}{b+k}\binom{c+d}{c+k}\binom{d+a}{d+k}\bigg/\binom{2a+2b+2c+2d}{a+b+c+d+k}$$
$$= \frac{(a+b+c+d)!\,(a+b+c)!\,(a+b+d)!\,(a+c+d)!\,(b+c+d)!}{(2a+2b+2c+2d)!\,(a+c)!\,(b+d)!\,a!\,b!\,c!\,d!}, \qquad \text{integers } a,b,c,d \geqslant 0.$$

This was discovered by John Dougall [69] early in the twentieth century.

Is Dougall's identity the hairiest sum of binomial coefficients known? No! The champion so far is

$$\sum_{k_{ij}} (-1)^{\Sigma_{1\leqslant i<j<n} k_{ij}} \Biggl(\prod_{1\leqslant i<j<n} \binom{a_i+a_j}{a_j+k_{ij}}\Biggr)\Biggl(\prod_{1\leqslant j<n} \binom{a_j+a_n}{a_n+\Sigma_{i<j}k_{ij}-\Sigma_{i>j}k_{ji}}\Biggr)$$
$$= \binom{a_1+\cdots+a_n}{a_1,a_2,\ldots,a_n}, \qquad \text{integers } a_1,a_2,\ldots,a_n \geqslant 0. \tag{5.31}$$

Here the sum is over $\binom{n-1}{2}$ index variables k_{ij} for $1 \leqslant i < j < n$. Equation (5.29) is the special case $n = 3$; the case $n = 4$ can be written out as follows, if we use (a, b, c, d) for (a_1, a_2, a_3, a_4) and (i, j, k) for (k_{12}, k_{13}, k_{23}):

$$\sum_{i,j,k}(-1)^{i+j+k}\binom{a+b}{b+i}\binom{a+c}{c+j}\binom{b+c}{c+k}\binom{a+d}{d-i-j}\binom{b+d}{d+i-k}\binom{c+d}{d+j+k}$$
$$= \frac{(a+b+c+d)!}{a!\,b!\,c!\,d!}, \qquad \text{integers } a, b, c, d \geqslant 0.$$

The left side of (5.31) is the coefficient of $z_1^0 z_2^0 \ldots z_n^0$ after the product of $n(n-1)$ fractions

$$\prod_{\substack{1 \leqslant i,j \leqslant n \\ i \neq j}} \left(1 - \frac{z_i}{z_j}\right)^{a_i}$$

has been fully expanded into positive and negative powers of the z's. The right side of (5.31) was conjectured by Freeman Dyson in 1962 and proved by several people shortly thereafter. Exercise 86 gives a "simple" proof of (5.31).

Another noteworthy identity involving lots of binomial coefficients is

$$\sum_{j,k}(-1)^{j+k}\binom{j+k}{j}\binom{r}{j}\binom{n}{k}\binom{m+n-j-k}{m-j}$$
$$= \binom{n+r}{n}\binom{m-r}{m-n}, \qquad \text{integers } m, n \geqslant 0. \tag{5.32}$$

This one, proved in exercise 83, even has a chance of arising in practical applications. But we're getting far afield from our theme of "basic identities," so we had better stop and take stock of what we've learned.

We've seen that binomial coefficients satisfy an almost bewildering variety of identities. Some of these, fortunately, are easily remembered, and we can use the memorable ones to derive most of the others in a few steps. Table 174 collects ten of the most useful formulas, all in one place; these are the best identities to know.

5.2 BASIC PRACTICE

In the previous section we derived a bunch of identities by manipulating sums and plugging in other identities. It wasn't too tough to find those derivations — we knew what we were trying to prove, so we could formulate a general plan and fill in the details without much trouble. Usually, however, out in the real world, we're not faced with an identity to prove; we're faced with a sum to simplify. And we don't know what a simplified form might look like (or even if one exists). By tackling many such sums in this section and the next, we will hone our binomial coefficient tools.

To start, let's try our hand at a few sums involving a single binomial coefficient.

Problem 1: A sum of ratios.

Algorithm *self-teach:*
1 read problem
2 attempt solution
3 skim book solution
4 if attempt failed goto 1 else goto next problem

We'd like to have a closed form for

$$\sum_{k=0}^{m} \binom{m}{k} \Big/ \binom{n}{k}, \qquad \text{integers } n \geqslant m \geqslant 0.$$

At first glance this sum evokes panic, because we haven't seen any identities that deal with a quotient of binomial coefficients. (Furthermore the sum involves two binomial coefficients, which seems to contradict the sentence preceding this problem.) However, just as we can use the factorial representations to reexpress a product of binomial coefficients as another product — that's how we got identity (5.21) — we can do likewise with a quotient. In fact we can avoid the grubby factorial representations by letting $r = n$ and dividing both sides of equation (5.21) by $\binom{n}{k}\binom{n}{m}$; this yields

Unfortunately, that algorithm can put you in an infinite loop.
Suggested patches:
0 set $c \leftarrow 0$
3a set $c \leftarrow c+1$
3b if $c = N$ *goto your TA*

$$\binom{m}{k} \Big/ \binom{n}{k} = \binom{n-k}{m-k} \Big/ \binom{n}{m}.$$

So we replace the quotient on the left, which appears in our sum, by the one on the right; the sum becomes

$$\sum_{k=0}^{m} \binom{n-k}{m-k} \Big/ \binom{n}{m}.$$

We still have a quotient, but the binomial coefficient in the denominator doesn't involve the index of summation k, so we can remove it from the sum. We'll restore it later.

— E. W. Dijkstra

We can also simplify the boundary conditions by summing over all $k \geqslant 0$; the terms for $k > m$ are zero. The sum that's left isn't so intimidating:

$$\sum_{k \geqslant 0} \binom{n-k}{m-k}.$$

It's similar to the one in identity (5.9), because the index k appears twice with the same sign. But here it's $-k$ and in (5.9) it's not. The next step should therefore be obvious; there's only one reasonable thing to do:

. . . But this subchapter is called BASIC practice.

$$\sum_{k \geqslant 0} \binom{n-k}{m-k} = \sum_{m-k \geqslant 0} \binom{n-(m-k)}{m-(m-k)} = \sum_{k \leqslant m} \binom{n-m+k}{k}.$$

Table 174 The top ten binomial coefficient identities.

$\binom{n}{k} = \dfrac{n!}{k!\,(n-k)!}$,	integers $n \geqslant k \geqslant 0$.	*factorial expansion*
$\binom{n}{k} = \binom{n}{n-k}$,	integer $n \geqslant 0$, integer k.	*symmetry*
$\binom{r}{k} = \dfrac{r}{k}\binom{r-1}{k-1}$,	integer $k \neq 0$.	*absorption/extraction*
$\binom{r}{k} = \binom{r-1}{k} + \binom{r-1}{k-1}$,	integer k.	*addition/induction*
$\binom{r}{k} = (-1)^k\binom{k-r-1}{k}$,	integer k.	*upper negation*
$\binom{r}{m}\binom{m}{k} = \binom{r}{k}\binom{r-k}{m-k}$,	integers m, k.	*trinomial revision*
$\sum_k \binom{r}{k} x^k y^{r-k} = (x+y)^r$,	integer $r \geqslant 0$, or $\lvert x/y \rvert < 1$.	*binomial theorem*
$\sum_{k \leqslant n} \binom{r+k}{k} = \binom{r+n+1}{n}$,	integer n.	*parallel summation*
$\sum_{0 \leqslant k \leqslant n} \binom{k}{m} = \binom{n+1}{m+1}$,	integers $m, n \geqslant 0$.	*upper summation*
$\sum_k \binom{r}{k}\binom{s}{n-k} = \binom{r+s}{n}$,	integer n.	*Vandermonde convolution*

And now we can apply the parallel summation identity, (5.9):

$$\sum_{k \leqslant m} \binom{n-m+k}{k} = \binom{(n-m)+m+1}{m} = \binom{n+1}{m}.$$

Finally we reinstate the $\binom{n}{m}$ in the denominator that we removed from the sum earlier, and then apply (5.7) to get the desired closed form:

$$\binom{n+1}{m} \Big/ \binom{n}{m} = \frac{n+1}{n+1-m}.$$

This derivation actually works for any real value of n, as long as no division by zero occurs; that is, as long as n isn't one of the integers $0, 1, \ldots, m-1$.

The more complicated the derivation, the more important it is to check the answer. This one wasn't too complicated but we'll check anyway. In the small case $m = 2$ and $n = 4$ we have

$$\binom{2}{0}\Big/\binom{4}{0} + \binom{2}{1}\Big/\binom{4}{1} + \binom{2}{2}\Big/\binom{4}{2} = 1 + \frac{1}{2} + \frac{1}{6} = \frac{5}{3};$$

yes, this agrees perfectly with our closed form $(4+1)/(4+1-2)$.

Problem 2: From the literature of sorting.

Our next sum appeared way back in ancient times (the early 1970s) before people were fluent with binomial coefficients. A paper that introduced an improved merging technique [165] concludes with the following remarks: "It can be shown that the expected number of saved transfers ... is given by the expression

$$T = \sum_{r=0}^{n} r \frac{{}_{m-r-1}C_{m-n-1}}{{}_{m}C_{n}}$$

Here m and n are as defined above, and ${}_mC_n$ is the symbol for the number of combinations of m objects taken n at a time. ... The author is grateful to the referee for reducing a more complex equation for expected transfers saved to the form given here."

We'll see that this is definitely not a final answer to the author's problem. It's not even a midterm answer.

Please, don't remind me of the midterm.

First we should translate the sum into something we can work with; the ghastly notation ${}_{m-r-1}C_{m-n-1}$ is enough to stop anybody, save the enthusiastic referee (please). In our language we'd write

$$T = \sum_{k=0}^{n} k\binom{m-k-1}{m-n-1}\Big/\binom{m}{n}, \qquad \text{integers } m > n \geqslant 0.$$

The binomial coefficient in the denominator doesn't involve the index of summation, so we can remove it and work with the new sum

$$S = \sum_{k=0}^{n} k\binom{m-k-1}{m-n-1}.$$

What next? The index of summation appears in the upper index of the binomial coefficient but not in the lower index. So if the other k weren't there, we could massage the sum and apply summation on the upper index (5.10). With the extra k, though, we can't. If we could somehow absorb that k into the binomial coefficient, using one of our absorption identities, we could then sum on the upper index. Unfortunately those identities don't work here. But if the k were instead $m-k$, we could use absorption identity (5.6):

$$(m-k)\binom{m-k-1}{m-n-1} = (m-n)\binom{m-k}{m-n}.$$

So here's the key: We'll rewrite k as $m-(m-k)$ and split the sum S into two sums:

$$\begin{aligned}\sum_{k=0}^{n} k\binom{m-k-1}{m-n-1} &= \sum_{k=0}^{n}(m-(m-k))\binom{m-k-1}{m-n-1}\\ &= \sum_{k=0}^{n} m\binom{m-k-1}{m-n-1} - \sum_{k=0}^{n}(m-k)\binom{m-k-1}{m-n-1}\\ &= m\sum_{k=0}^{n}\binom{m-k-1}{m-n-1} - \sum_{k=0}^{n}(m-n)\binom{m-k}{m-n}\\ &= mA-(m-n)B\,,\end{aligned}$$

where

$$A = \sum_{k=0}^{n}\binom{m-k-1}{m-n-1}, \qquad B = \sum_{k=0}^{n}\binom{m-k}{m-n}.$$

The sums A and B that remain are none other than our old friends in which the upper index varies while the lower index stays fixed. Let's do B first, because it looks simpler. A little bit of massaging is enough to make the summand match the left side of (5.10):

$$\begin{aligned}\sum_{0\le k\le n}\binom{m-k}{m-n} &= \sum_{0\le m-k\le n}\binom{m-(m-k)}{m-n}\\ &= \sum_{m-n\le k\le m}\binom{k}{m-n}\\ &= \sum_{0\le k\le m}\binom{k}{m-n}.\end{aligned}$$

In the last step we've included the terms with $0\le k<m-n$ in the sum; they're all zero, because the upper index is less than the lower. Now we sum on the upper index, using (5.10), and get

$$B = \sum_{0\le k\le m}\binom{k}{m-n} = \binom{m+1}{m-n+1}.$$

The other sum A is the same, but with m replaced by $m-1$. Hence we have a closed form for the given sum S, which can be further simplified:

$$\begin{aligned} S = mA - (m-n)B &= m\binom{m}{m-n} - (m-n)\binom{m+1}{m-n+1} \\ &= \left(m - (m-n)\frac{m+1}{m-n+1}\right)\binom{m}{m-n} \\ &= \left(\frac{n}{m-n+1}\right)\binom{m}{m-n}. \end{aligned}$$

And this gives us a closed form for the original sum:

$$\begin{aligned} T &= S \bigg/ \binom{m}{n} \\ &= \frac{n}{m-n+1}\binom{m}{m-n} \bigg/ \binom{m}{n} \\ &= \frac{n}{m-n+1}. \end{aligned}$$

Even the referee can't simplify this.

Again we use a small case to check the answer. When $m=4$ and $n=2$, we have

$$T = 0\cdot\tbinom{3}{1}/\tbinom{4}{2} + 1\cdot\tbinom{2}{1}/\tbinom{4}{2} + 2\cdot\tbinom{1}{1}/\tbinom{4}{2} = 0 + \tfrac{2}{6} + \tfrac{2}{6} = \tfrac{2}{3},$$

which agrees with our formula $2/(4-2+1)$.

Problem 3: From an old exam.

Do old exams ever die?

Let's do one more sum that involves a single binomial coefficient. This one, unlike the last, originated in the halls of academia; it was a problem on a take-home test. We want the value of $Q_{1000000}$, when

$$Q_n = \sum_{k \leqslant 2^n} \binom{2^n - k}{k}(-1)^k, \qquad \text{integer } n \geqslant 0.$$

This one's harder than the others; we can't apply *any* of the identities we've seen so far. And we're faced with a sum of $2^{1000000}$ terms, so we can't just add them up. The index of summation k appears in both indices, upper and lower, but with opposite signs. Negating the upper index doesn't help, either; it removes the factor of $(-1)^k$, but it introduces a $2k$ in the upper index.

When nothing obvious works, we know that it's best to look at small cases. If we can't spot a pattern and prove it by induction, at least we'll have some data for checking our results. Here are the nonzero terms and their sums for the first four values of n.

n		Q_n
0	$\binom{1}{0} = 1$	$= 1$
1	$\binom{2}{0} - \binom{1}{1} = 1 - 1$	$= 0$
2	$\binom{4}{0} - \binom{3}{1} + \binom{2}{2} = 1 - 3 + 1$	$= -1$
3	$\binom{8}{0} - \binom{7}{1} + \binom{6}{2} - \binom{5}{3} + \binom{4}{4} = 1 - 7 + 15 - 10 + 1$	$= 0$

We'd better not try the next case, $n = 4$; the chances of making an arithmetic error are too high. (Computing terms like $\binom{12}{4}$ and $\binom{11}{5}$ by hand, let alone combining them with the others, is worthwhile only if we're desperate.)

So the pattern starts out 1, 0, -1, 0. Even if we knew the next term or two, the closed form wouldn't be obvious. But if we could find and prove a recurrence for Q_n we'd probably be able to guess and prove its closed form. To find a recurrence, we need to relate Q_n to Q_{n-1} (or to $Q_{\text{smaller values}}$); but to do this we need to relate a term like $\binom{128-13}{13}$, which arises when $n = 7$ and $k = 13$, to terms like $\binom{64-13}{13}$. This doesn't look promising; we don't know any neat relations between entries in Pascal's triangle that are 64 rows apart. The addition formula, our main tool for induction proofs, only relates entries that are one row apart.

But this leads us to a key observation: There's no need to deal with entries that are 2^{n-1} rows apart. The variable n never appears by itself, it's always in the context 2^n. So the 2^n is a red herring! If we replace 2^n by m, all we need to do is find a closed form for the more general (but easier) sum

Oh, the sneakiness of the instructor who set that exam.

$$R_m = \sum_{k \leqslant m} \binom{m-k}{k} (-1)^k, \qquad \text{integer } m \geqslant 0;$$

then we'll also have a closed form for $Q_n = R_{2^n}$. And there's a good chance that the addition formula will give us a recurrence for the sequence R_m.

Values of R_m for small m can be read from Table 155, if we alternately add and subtract values that appear in a southwest-to-northeast diagonal. The results are:

m	0	1	2	3	4	5	6	7	8	9	10
R_m	1	1	0	-1	-1	0	1	1	0	-1	-1

There seems to be a lot of cancellation going on.

Let's look now at the formula for R_m and see if it defines a recurrence. Our strategy is to apply the addition formula (5.8) and to find sums that have the form R_k in the resulting expression, somewhat as we did in the perturbation method of Chapter 2:

$$\begin{aligned}
R_m &= \sum_{k\leqslant m}\binom{m-k}{k}(-1)^k \\
&= \sum_{k\leqslant m}\binom{m-1-k}{k}(-1)^k + \sum_{k\leqslant m}\binom{m-1-k}{k-1}(-1)^k \\
&= \sum_{k\leqslant m}\binom{m-1-k}{k}(-1)^k + \sum_{k+1\leqslant m}\binom{m-2-k}{k}(-1)^{k+1} \\
&= \sum_{k\leqslant m-1}\binom{m-1-k}{k}(-1)^k + \binom{-1}{m}(-1)^m \\
&\qquad - \sum_{k\leqslant m-2}\binom{m-2-k}{k}(-1)^k - \binom{-1}{m-1}(-1)^{m-1} \\
&= R_{m-1} + (-1)^{2m} - R_{m-2} - (-1)^{2(m-1)} = R_{m-1} - R_{m-2}\,.
\end{aligned}$$

(In the next-to-last step we've used the formula $\binom{-1}{m} = (-1)^m$, which we know is true when $m \geqslant 0$.) This derivation is valid for $m \geqslant 2$.

Anyway those of us who've done warmup exercise 4 know it.

From this recurrence we can generate values of R_m quickly, and we soon perceive that the sequence is periodic. Indeed,

$$R_m = \begin{cases} 1 \\ 1 \\ 0 \\ -1 \\ -1 \\ 0 \end{cases} \qquad \text{if } m \bmod 6 = \begin{cases} 0 \\ 1 \\ 2 \\ 3 \\ 4 \\ 5 \end{cases}.$$

The proof by induction is by inspection. Or, if we must give a more academic proof, we can unfold the recurrence one step to obtain

$$R_m = (R_{m-2} - R_{m-3}) - R_{m-2} = -R_{m-3}\,,$$

whenever $m \geqslant 3$. Hence $R_m = R_{m-6}$ whenever $m \geqslant 6$.

Finally, since $Q_n = R_{2^n}$, we can determine Q_n by determining $2^n \bmod 6$ and using the closed form for R_m. When $n = 0$ we have $2^0 \bmod 6 = 1$; after that we keep multiplying by 2 (mod 6), so the pattern 2, 4 repeats. Thus

$$Q_n = R_{2^n} = \begin{cases} R_1 = 1, & \text{if } n = 0; \\ R_2 = 0, & \text{if } n \text{ is odd}; \\ R_4 = -1, & \text{if } n > 0 \text{ is even}. \end{cases}$$

This closed form for Q_n agrees with the first four values we calculated when we started on the problem. We conclude that $Q_{1000000} = R_4 = -1$.

Problem 4: A sum involving two binomial coefficients.

Our next task is to find a closed form for

$$\sum_{k=0}^{n} k\binom{m-k-1}{m-n-1}, \qquad \text{integers } m > n \geqslant 0.$$

Wait a minute. Where's the second binomial coefficient promised in the title of this problem? And why should we try to simplify a sum we've already simplified? (This is the sum S from Problem 2.)

Well, this is a sum that's easier to simplify if we view the summand as a product of two binomial coefficients, and then use one of the general identities found in Table 169. The second binomial coefficient materializes when we rewrite k as $\binom{k}{1}$:

$$\sum_{k=0}^{n} k\binom{m-k-1}{m-n-1} = \sum_{0\leqslant k\leqslant n} \binom{k}{1}\binom{m-k-1}{m-n-1}.$$

And identity (5.26) is the one to apply, since its index of summation appears in both upper indices and with opposite signs.

But our sum isn't quite in the correct form yet. The upper limit of summation should be $m-1$, if we're to have a perfect match with (5.26). No problem; the terms for $n < k \leqslant m-1$ are zero. So we can plug in, with $(l, m, n, q) \leftarrow (m-1, m-n-1, 1, 0)$; the answer is

$$S = \binom{m}{m-n+1}.$$

This is cleaner than the formula we got before. We can convert it to the previous formula by using (5.7):

$$\binom{m}{m-n+1} = \frac{n}{m-n+1}\binom{m}{m-n}.$$

Similarly, we can get interesting results by plugging special values into the other general identities we've seen. Suppose, for example, that we set $m = n = 1$ and $q = 0$ in (5.26). Then the identity reads

$$\sum_{0\leqslant k\leqslant l} (l-k)k = \binom{l+1}{3}.$$

The left side is $l\big((l+1)l/2\big) - (1^2+2^2+\cdots+l^2)$, so this gives us a brand new way to solve the sum-of-squares problem that we beat to death in Chapter 2.

The moral of this story is: Special cases of very general sums are sometimes best handled in the general form. When learning general forms, it's wise to learn their simple specializations.

Problem 5: A sum with three factors.

Here's another sum that isn't too bad. We wish to simplify

$$\sum_k \binom{n}{k}\binom{s}{k}k, \qquad \text{integer } n \geqslant 0.$$

The index of summation k appears in both lower indices and with the same sign; therefore identity (5.23) in Table 169 looks close to what we need. With a bit of manipulation, we should be able to use it.

The biggest difference between (5.23) and what we have is the extra k in our sum. But we can absorb k into one of the binomial coefficients by using one of the absorption identities:

$$\sum_k \binom{n}{k}\binom{s}{k}k = \sum_k \binom{n}{k}\binom{s-1}{k-1}s$$
$$= s\sum_k \binom{n}{k}\binom{s-1}{k-1}.$$

We don't care that the s appears when the k disappears, because it's constant. And now we're ready to apply the identity and get the closed form,

$$s\sum_k \binom{n}{k}\binom{s-1}{k-1} = s\binom{n+s-1}{n-1}.$$

If we had chosen in the first step to absorb k into $\binom{n}{k}$, not $\binom{s}{k}$, we wouldn't have been allowed to apply (5.23) directly, because $n-1$ might be negative; the identity requires a nonnegative value in at least one of the upper indices.

Problem 6: A sum with menacing characteristics.

The next sum is more challenging. We seek a closed form for

$$\sum_{k\geqslant 0} \binom{n+k}{2k}\binom{2k}{k}\frac{(-1)^k}{k+1}, \qquad \text{integer } n \geqslant 0.$$

So we should deep six this sum, right?

One useful measure of a sum's difficulty is the number of times the index of summation appears. By this measure we're in deep trouble — k appears six times. Furthermore, the key step that worked in the previous problem — to absorb something outside the binomial coefficients into one of them — won't work here. If we absorb the $k+1$ we just get another occurrence of k in its place. And not only that: Our index k is twice shackled with the coefficient 2 inside a binomial coefficient. Multiplicative constants are usually harder to remove than additive constants.

We're lucky this time, though. The 2k's are right where we need them for identity (5.21) to apply, so we get

$$\sum_{k\geqslant 0}\binom{n+k}{2k}\binom{2k}{k}\frac{(-1)^k}{k+1} = \sum_{k\geqslant 0}\binom{n+k}{k}\binom{n}{k}\frac{(-1)^k}{k+1}.$$

The two 2's disappear, and so does one occurrence of k. So that's one down and five to go.

The $k+1$ in the denominator is the most troublesome characteristic left, and now we can absorb it into $\binom{n}{k}$ using identity (5.6):

$$\begin{aligned}\sum_{k\geqslant 0}\binom{n+k}{k}\binom{n}{k}\frac{(-1)^k}{k+1} &= \sum_{k}\binom{n+k}{k}\binom{n+1}{k+1}\frac{(-1)^k}{n+1}\\ &= \frac{1}{n+1}\sum_{k}\binom{n+k}{k}\binom{n+1}{k+1}(-1)^k.\end{aligned}$$

(Recall that $n \geqslant 0$.) Two down, four to go.

To eliminate another k we have two promising options. We could use symmetry on $\binom{n+k}{k}$; or we could negate the upper index $n+k$, thereby eliminating that k as well as the factor $(-1)^k$. Let's explore both possibilities, starting with the symmetry option:

$$\frac{1}{n+1}\sum_{k}\binom{n+k}{k}\binom{n+1}{k+1}(-1)^k = \frac{1}{n+1}\sum_{k}\binom{n+k}{n}\binom{n+1}{k+1}(-1)^k.$$

Third down, three to go, and we're in position to make a big gain by plugging into (5.24): Replacing (l,m,n,s) by $(n+1,1,n,n)$, we get

For a minute I thought we'd have to punt.

$$\frac{1}{n+1}\sum_{k}\binom{n+k}{n}\binom{n+1}{k+1}(-1)^k = \frac{1}{n+1}(-1)^n\binom{n-1}{-1} = 0.$$

Zero, eh? After all that work? Let's check it when $n=2$: $\binom{2}{0}\binom{0}{0}\frac{1}{1}-\binom{3}{2}\binom{2}{1}\frac{1}{2}+\binom{4}{4}\binom{4}{2}\frac{1}{3} = 1-\frac{6}{2}+\frac{6}{3}=0$. It checks.

Just for the heck of it, let's explore our other option, negating the upper index of $\binom{n+k}{k}$:

$$\frac{1}{n+1}\sum_{k}\binom{n+k}{k}\binom{n+1}{k+1}(-1)^k = \frac{1}{n+1}\sum_{k}\binom{-n-1}{k}\binom{n+1}{k+1}.$$

Now (5.23) applies, with $(l,m,n,s) \leftarrow (n+1,1,0,-n-1)$, and

$$\frac{1}{n+1}\sum_{k}\binom{-n-1}{k}\binom{n+1}{k+1} = \frac{1}{n+1}\binom{0}{n}.$$

Hey wait. This is zero when $n > 0$, but it's 1 when $n = 0$. Our other path to the solution told us that the sum was zero in all cases! What gives? The sum actually does turn out to be 1 when $n = 0$, so the correct answer is '$[n=0]$'. We must have made a mistake in the previous derivation.

Try binary search: Replay the middle formula first, to see if the mistake was early or late.

Let's do an instant replay on that derivation when $n = 0$, in order to see where the discrepancy first arises. Ah yes; we fell into the old trap mentioned earlier: We tried to apply symmetry when the upper index could be negative! We were not justified in replacing $\binom{n+k}{k}$ by $\binom{n+k}{n}$ when k ranges over all integers, because this converts zero into a nonzero value when $k < -n$. (Sorry about that.)

The other factor in the sum, $\binom{n+1}{k+1}$, turns out to be zero when $k < -n$, except when $n = 0$ and $k = -1$. Hence our error didn't show up when we checked the case $n = 2$. Exercise 6 explains what we should have done.

Problem 7: A new obstacle.

This one's even tougher; we want a closed form for

$$\sum_{k\geqslant 0}\binom{n+k}{m+2k}\binom{2k}{k}\frac{(-1)^k}{k+1}, \qquad \text{integers } m, n > 0.$$

If m were 0 we'd have the sum from the problem we just finished. But it's not, and we're left with a real mess—nothing we used in Problem 6 works here. (Especially not the crucial first step.)

However, if we could somehow get rid of the m, we could use the result just derived. So our strategy is: Replace $\binom{n+k}{m+2k}$ by a sum of terms like $\binom{l+k}{2k}$ for some nonnegative integer l; the summand will then look like the summand in Problem 6, and we can interchange the order of summation.

What should we substitute for $\binom{n+k}{m+2k}$? A painstaking examination of the identities derived earlier in this chapter turns up only one suitable candidate, namely equation (5.26) in Table 169. And one way to use it is to replace the parameters (l, m, n, q, k) by $(n+k-1, 2k, m-1, 0, j)$, respectively:

$$\begin{aligned}
&\sum_{k\geqslant 0}\binom{n+k}{m+2k}\binom{2k}{k}\frac{(-1)^k}{k+1}\\
&\quad= \sum_{k\geqslant 0}\ \sum_{0\leqslant j\leqslant n+k-1}\binom{n+k-1-j}{2k}\binom{j}{m-1}\binom{2k}{k}\frac{(-1)^k}{k+1}\\
&\quad= \sum_{j\geqslant 0}\binom{j}{m-1}\sum_{\substack{k\geqslant j-n+1\\ k\geqslant 0}}\binom{n+k-1-j}{2k}\binom{2k}{k}\frac{(-1)^k}{k+1}.
\end{aligned}$$

In the last step we've changed the order of summation, manipulating the conditions below the $\sum$'s according to the rules of Chapter 2.

We can't quite replace the inner sum using the result of Problem 6, because it has the extra condition $k \geqslant j-n+1$. But this extra condition is superfluous unless $j-n+1>0$; that is, unless $j \geqslant n$. And when $j \geqslant n$, the first binomial coefficient of the inner sum is zero, because its upper index is between 0 and $k-1$, thus strictly less than the lower index $2k$. We may therefore place the additional restriction $j<n$ on the outer sum, without affecting which nonzero terms are included. This makes the restriction $k \geqslant j-n+1$ superfluous, and we can use the result of Problem 6. The double sum now comes tumbling down:

$$\sum_{j\geqslant 0}\binom{j}{m-1}\sum_{\substack{k\geqslant j-n+1\\ k\geqslant 0}}\binom{n+k-1-j}{2k}\binom{2k}{k}\frac{(-1)^k}{k+1}$$
$$=\sum_{0\leqslant j<n}\binom{j}{m-1}\sum_{k\geqslant 0}\binom{n+k-1-j}{2k}\binom{2k}{k}\frac{(-1)^k}{k+1}$$
$$=\sum_{0\leqslant j<n}\binom{j}{m-1}[n-1-j=0] \;=\; \binom{n-1}{m-1}.$$

The inner sums vanish except when $j=n-1$, so we get a simple closed form as our answer.

Problem 8: A different obstacle.

Let's branch out from Problem 6 in another way by considering the sum

$$S_m = \sum_{k\geqslant 0}\binom{n+k}{2k}\binom{2k}{k}\frac{(-1)^k}{k+1+m}, \qquad \text{integers } m,n\geqslant 0.$$

Again, when $m=0$ we have the sum we did before; but now the m occurs in a different place. This problem is a bit harder yet than Problem 7, but (fortunately) we're getting better at finding solutions. We can begin as in Problem 6,

$$S_m = \sum_{k\geqslant 0}\binom{n+k}{k}\binom{n}{k}\frac{(-1)^k}{k+1+m}.$$

Now (as in Problem 7) we try to expand the part that depends on m into terms that we know how to deal with. When m was zero, we absorbed $k+1$ into $\binom{n}{k}$; if $m>0$, we can do the same thing if we expand $1/(k+1+m)$ into absorbable terms. And our luck still holds: We proved a suitable identity

$$\sum_{j=0}^{m}\binom{m}{j}\binom{r}{j}^{-1} = \frac{r+1}{r+1-m}, \qquad \begin{array}{c}\text{integer } m\geqslant 0,\\ r\notin\{0,1,\ldots,m-1\}.\end{array} \tag{5.33}$$

in Problem 1. Replacing r by $-k-2$ gives the desired expansion,

$$S_m = \sum_{k\geqslant 0}\binom{n+k}{k}\binom{n}{k}\frac{(-1)^k}{k+1}\sum_{j\geqslant 0}\binom{m}{j}\binom{-k-2}{j}^{-1}.$$

Now the $(k+1)^{-1}$ can be absorbed into $\binom{n}{k}$, as planned. In fact, it could also be absorbed into $\binom{-k-2}{j}^{-1}$. Double absorption suggests that even more cancellation might be possible behind the scenes. Yes — expanding everything in our new summand into factorials and going back to binomial coefficients gives a formula that we can sum on k:

They expect us to check this on a sheet of scratch paper.

$$\begin{aligned} S_m &= \frac{m!\,n!}{(m+n+1)!}\sum_{j\geqslant 0}(-1)^j\binom{m+n+1}{n+1+j}\sum_k\binom{n+1+j}{k+j+1}\binom{-n-1}{k} \\ &= \frac{m!\,n!}{(m+n+1)!}\sum_{j\geqslant 0}(-1)^j\binom{m+n+1}{n+1+j}\binom{j}{n}. \end{aligned}$$

The sum over all integers j is zero, by (5.24). Hence $-S_m$ is the sum for $j<0$.

To evaluate $-S_m$ for $j<0$, let's replace j by $-k-1$ and sum for $k\geqslant 0$:

$$\begin{aligned} S_m &= \frac{m!\,n!}{(m+n+1)!}\sum_{k\geqslant 0}(-1)^k\binom{m+n+1}{n-k}\binom{-k-1}{n} \\ &= \frac{m!\,n!}{(m+n+1)!}\sum_{k\leqslant n}(-1)^{n-k}\binom{m+n+1}{k}\binom{k-n-1}{n} \\ &= \frac{m!\,n!}{(m+n+1)!}\sum_{k\leqslant n}(-1)^k\binom{m+n+1}{k}\binom{2n-k}{n} \\ &= \frac{m!\,n!}{(m+n+1)!}\sum_{k\leqslant 2n}(-1)^k\binom{m+n+1}{k}\binom{2n-k}{n}. \end{aligned}$$

Finally (5.25) applies, and we have our answer:

$$S_m = (-1)^n\frac{m!\,n!}{(m+n+1)!}\binom{m}{n} = (-1)^n m^{\underline{n}} m^{\underline{-n-1}}.$$

Whew; we'd better check it. When $n=2$ we find

$$S_m = \frac{1}{m+1}-\frac{6}{m+2}+\frac{6}{m+3} = \frac{m(m-1)}{(m+1)(m+2)(m+3)}.$$

Our derivation requires m to be an integer, but the result holds for all real m, because $(m+1)^{\overline{n}}S_m$ is a polynomial in m of degree $<n$.

5.3 TRICKS OF THE TRADE

Let's look next at three techniques that significantly amplify the methods we have already learned.

Trick 1: Going halves.

This should really be called Trick 1/2.

Many of our identities involve an arbitrary real number r. When r has the special form "integer minus one half," the binomial coefficient $\binom{r}{k}$ can be written as a quite different-looking product of binomial coefficients. This leads to a new family of identities that can be manipulated with surprising ease.

One way to see how this works is to begin with the *duplication formula*

$$r^{\underline{k}}(r-\tfrac{1}{2})^{\underline{k}} = (2r)^{\underline{2k}}/2^{2k}, \qquad \text{integer } k \geqslant 0. \tag{5.34}$$

This identity is obvious if we expand the falling powers and interleave the factors on the left side:

$$r(r-\tfrac{1}{2})(r-1)(r-\tfrac{3}{2})\ldots(r-k+1)(r-k+\tfrac{1}{2}) \\ = \frac{(2r)(2r-1)\ldots(2r-2k+1)}{2\cdot 2\cdot\ldots\cdot 2}.$$

Now we can divide both sides by $k!^2$, and we get

$$\binom{r}{k}\binom{r-1/2}{k} = \binom{2r}{2k}\binom{2k}{k}\Big/2^{2k}, \qquad \text{integer } k. \tag{5.35}$$

If we set $k = r = n$, where n is an integer, this yields

$$\binom{n-1/2}{n} = \binom{2n}{n}\Big/2^{2n}, \qquad \text{integer } n. \tag{5.36}$$

And negating the upper index gives yet another useful formula,

$$\binom{-1/2}{n} = \left(\frac{-1}{4}\right)^n\binom{2n}{n}, \qquad \text{integer } n. \tag{5.37}$$

For example, when $n = 4$ we have

. . . we halve . . .

$$\begin{aligned}\binom{-1/2}{4} &= \frac{(-1/2)(-3/2)(-5/2)(-7/2)}{4!} \\ &= \left(\frac{-1}{2}\right)^4\frac{1\cdot3\cdot5\cdot7}{1\cdot2\cdot3\cdot4} \\ &= \left(\frac{-1}{4}\right)^4\frac{1\cdot3\cdot5\cdot7\cdot2\cdot4\cdot6\cdot8}{1\cdot2\cdot3\cdot4\cdot1\cdot2\cdot3\cdot4} = \left(\frac{-1}{4}\right)^4\binom{8}{4}.\end{aligned}$$

Notice how we've changed a product of odd numbers into a factorial.

Identity (5.35) has an amusing corollary. Let $r = \frac{1}{2}n$, and take the sum over all integers k. The result is

$$\sum_k \binom{n}{2k}\binom{2k}{k}2^{-2k} = \sum_k \binom{n/2}{k}\binom{(n-1)/2}{k}$$
$$= \binom{n-1/2}{\lfloor n/2 \rfloor}, \qquad \text{integer } n \geqslant 0 \qquad (5.38)$$

by (5.23), because either $n/2$ or $(n-1)/2$ is $\lfloor n/2 \rfloor$, a nonnegative integer!

We can also use Vandermonde's convolution (5.27) to deduce that

$$\sum_k \binom{-1/2}{k}\binom{-1/2}{n-k} = \binom{-1}{n} = (-1)^n, \qquad \text{integer } n \geqslant 0.$$

Plugging in the values from (5.37) gives

$$\binom{-1/2}{k}\binom{-1/2}{n-k} = \left(\frac{-1}{4}\right)^k\binom{2k}{k}\left(\frac{-1}{4}\right)^{n-k}\binom{2(n-k)}{n-k}$$
$$= \frac{(-1)^n}{4^n}\binom{2k}{k}\binom{2n-2k}{n-k};$$

this is what sums to $(-1)^n$. Hence we have a remarkable property of the "middle" elements of Pascal's triangle:

$$\sum_k \binom{2k}{k}\binom{2n-2k}{n-k} = 4^n, \qquad \text{integer } n \geqslant 0. \qquad (5.39)$$

For example, $\binom{0}{0}\binom{6}{3}+\binom{2}{1}\binom{4}{2}+\binom{4}{2}\binom{2}{1}+\binom{6}{3}\binom{0}{0} = 1\cdot 20+2\cdot 6+6\cdot 2+20\cdot 1 = 64 = 4^3$.

These illustrations of our first trick indicate that it's wise to try changing binomial coefficients of the form $\binom{2k}{k}$ into binomial coefficients of the form $\binom{n-1/2}{k}$, where n is some appropriate integer (usually 0, 1, or k); the resulting formula might be much simpler.

Trick 2: High-order differences.

We saw earlier that it's possible to evaluate partial sums of the series $\binom{n}{k}(-1)^k$, but not of the series $\binom{n}{k}$. It turns out that there are many important applications of binomial coefficients with alternating signs, $\binom{n}{k}(-1)^k$. One of the reasons for this is that such coefficients are intimately associated with the difference operator Δ defined in Section 2.6.

The difference Δf of a function f at the point x is

$$\Delta f(x) = f(x+1) - f(x);$$

if we apply Δ again, we get the second difference

$$\begin{aligned}\Delta^2 f(x) &= \Delta f(x+1) - \Delta f(x) = \bigl(f(x+2) - f(x+1)\bigr) - \bigl(f(x+1) - f(x)\bigr)\\ &= f(x+2) - 2f(x+1) + f(x),\end{aligned}$$

which is analogous to the second derivative. Similarly, we have

$$\begin{aligned}\Delta^3 f(x) &= f(x+3) - 3f(x+2) + 3f(x+1) - f(x);\\ \Delta^4 f(x) &= f(x+4) - 4f(x+3) + 6f(x+2) - 4f(x+1) + f(x);\end{aligned}$$

and so on. Binomial coefficients enter these formulas with alternating signs.

In general, the nth difference is

$$\Delta^n f(x) = \sum_k \binom{n}{k}(-1)^{n-k} f(x+k), \qquad \text{integer } n \geqslant 0. \tag{5.40}$$

This formula is easily proved by induction, but there's also a nice way to prove it directly using the elementary theory of operators. Recall that Section 2.6 defines the shift operator E by the rule

$$Ef(x) = f(x+1);$$

hence the operator Δ is $E - 1$, where 1 is the identity operator defined by the rule $1f(x) = f(x)$. By the binomial theorem,

$$\Delta^n = (E-1)^n = \sum_k \binom{n}{k} E^k (-1)^{n-k}.$$

This is an equation whose elements are operators; it is equivalent to (5.40), since E^k is the operator that takes $f(x)$ into $f(x+k)$.

An interesting and important case arises when we consider negative falling powers. Let $f(x) = (x-1)^{\underline{-1}} = 1/x$. Then, by rule (2.45), we have $\Delta f(x) = (-1)(x-1)^{\underline{-2}}$, $\Delta^2 f(x) = (-1)(-2)(x-1)^{\underline{-3}}$, and in general

$$\Delta^n\bigl((x-1)^{\underline{-1}}\bigr) = (-1)^{\underline{n}}(x-1)^{\underline{-n-1}} = (-1)^n \frac{n!}{x(x+1)\ldots(x+n)}.$$

Equation (5.40) now tells us that

$$\begin{aligned}\sum_k \binom{n}{k}\frac{(-1)^k}{x+k} &= \frac{n!}{x(x+1)\ldots(x+n)}\\ &= x^{-1}\binom{x+n}{n}^{-1}, \qquad x \notin \{0, -1, \ldots, -n\}.\end{aligned} \tag{5.41}$$

For example,

$$\frac{1}{x} - \frac{4}{x+1} + \frac{6}{x+2} - \frac{4}{x+3} + \frac{1}{x+4}$$
$$= \frac{4!}{x(x+1)(x+2)(x+3)(x+4)} = 1\Big/x\binom{x+4}{4}.$$

The sum in (5.41) is the partial fraction expansion of $n!/\bigl(x(x+1)\ldots(x+n)\bigr)$.

Significant results can be obtained from positive falling powers too. If $f(x)$ is a polynomial of degree d, the difference $\Delta f(x)$ is a polynomial of degree $d-1$; therefore $\Delta^d f(x)$ is a constant, and $\Delta^n f(x) = 0$ if $n > d$. This extremely important fact simplifies many formulas.

A closer look gives further information: Let

$$f(x) = a_d x^d + a_{d-1}x^{d-1} + \cdots + a_1 x^1 + a_0 x^0$$

be any polynomial of degree d. We will see in Chapter 6 that we can express ordinary powers as sums of falling powers (for example, $x^2 = x^{\underline{2}} + x^{\underline{1}}$); hence there are coefficients $b_d, b_{d-1}, \ldots, b_1, b_0$ such that

$$f(x) = b_d x^{\underline{d}} + b_{d-1}x^{\underline{d-1}} + \cdots + b_1 x^{\underline{1}} + b_0 x^{\underline{0}}.$$

(It turns out that $b_d = a_d$ and $b_0 = a_0$, but the intervening coefficients are related in a more complicated way.) Let $c_k = k!\, b_k$ for $0 \leqslant k \leqslant d$. Then

$$f(x) = c_d\binom{x}{d} + c_{d-1}\binom{x}{d-1} + \cdots + c_1\binom{x}{1} + c_0\binom{x}{0};$$

thus, any polynomial can be represented as a sum of multiples of binomial coefficients. Such an expansion is called the *Newton series* of $f(x)$, because Isaac Newton used it extensively.

We observed earlier in this chapter that the addition formula implies

$$\Delta\left(\binom{x}{k}\right) = \binom{x}{k-1}.$$

Therefore, by induction, the nth difference of a Newton series is very simple:

$$\Delta^n f(x) = c_d\binom{x}{d-n} + c_{d-1}\binom{x}{d-1-n} + \cdots + c_1\binom{x}{1-n} + c_0\binom{x}{-n}.$$

If we now set $x = 0$, all terms $c_k\binom{x}{k-n}$ on the right side are zero, except the term with $k - n = 0$; hence

$$\Delta^n f(0) = \begin{cases} c_n, & \text{if } n \leqslant d; \\ 0, & \text{if } n > d. \end{cases}$$

The Newton series for $f(x)$ is therefore

$$f(x) = \Delta^d f(0)\binom{x}{d} + \Delta^{d-1} f(0)\binom{x}{d-1} + \cdots + \Delta f(0)\binom{x}{1} + f(0)\binom{x}{0}.$$

For example, suppose $f(x) = x^3$. It's easy to calculate

$$\begin{array}{cccc} f(0) = 0, & f(1) = 1, & f(2) = 8, & f(3) = 27; \\ \Delta f(0) = 1, & \Delta f(1) = 7, & \Delta f(2) = 19; & \\ \Delta^2 f(0) = 6, & \Delta^2 f(1) = 12; & & \\ \Delta^3 f(0) = 6. & & & \end{array}$$

So the Newton series is $x^3 = 6\binom{x}{3} + 6\binom{x}{2} + 1\binom{x}{1} + 0\binom{x}{0}$.

Our formula $\Delta^n f(0) = c_n$ can also be stated in the following way, using (5.40) with $x = 0$:

$$\sum_k \binom{n}{k}(-1)^k\left(c_0\binom{k}{0} + c_1\binom{k}{1} + c_2\binom{k}{2} + \cdots\right) = (-1)^n c_n\,, \qquad \text{integer } n \geqslant 0.$$

Here $\langle c_0, c_1, c_2, \ldots\rangle$ is an arbitrary sequence of coefficients; the infinite sum $c_0\binom{k}{0} + c_1\binom{k}{1} + c_2\binom{k}{2} + \cdots$ is actually finite for all $k \geqslant 0$, so convergence is not an issue. In particular, we can prove the important identity

$$\sum_k \binom{n}{k}(-1)^k(a_0 + a_1 k + \cdots + a_n k^n) = (-1)^n n!\, a_n\,, \qquad \text{integer } n \geqslant 0, \tag{5.42}$$

because the polynomial $a_0 + a_1 k + \cdots + a_n k^n$ can always be written as a Newton series $c_0\binom{k}{0} + c_1\binom{k}{1} + \cdots + c_n\binom{k}{n}$ with $c_n = n!\, a_n$.

Many sums that appear to be hopeless at first glance can actually be summed almost trivially by using the idea of nth differences. For example, let's consider the identity

$$\sum_k \binom{n}{k}\binom{r - sk}{n}(-1)^k = s^n\,, \qquad \text{integer } n \geqslant 0. \tag{5.43}$$

This looks very impressive, because it's quite different from anything we've seen so far. But it really is easy to understand, once we notice the telltale factor $\binom{n}{k}(-1)^k$ in the summand, because the function

$$f(k) = \binom{r - sk}{n} = \frac{1}{n!}(-1)^n s^n k^n + \cdots = (-1)^n s^n \binom{k}{n} + \cdots$$

is a polynomial in k of degree n, with leading coefficient $(-1)^n s^n/n!$. Therefore (5.43) is nothing more than an application of (5.42).

We have discussed Newton series under the assumption that $f(x)$ is a polynomial. But we've also seen that infinite Newton series

$$f(x) = c_0\binom{x}{0} + c_1\binom{x}{1} + c_2\binom{x}{2} + \cdots$$

make sense too, because such sums are always finite when x is a nonnegative integer. Our derivation of the formula $\Delta^n f(0) = c_n$ works in the infinite case, just as in the polynomial case; so we have the general identity

$$f(x) = f(0)\binom{x}{0} + \Delta f(0)\binom{x}{1} + \Delta^2 f(0)\binom{x}{2} + \Delta^3 f(0)\binom{x}{3} + \cdots, \qquad \text{integer } x \geqslant 0. \tag{5.44}$$

This formula is valid for any function $f(x)$ that is defined for nonnegative integers x. Moreover, if the right-hand side converges for other values of x, it defines a function that "interpolates" $f(x)$ in a natural way. (There are infinitely many ways to interpolate function values, so we cannot assert that (5.44) is true for all x that make the infinite series converge. For example, if we let $f(x) = \sin(\pi x)$, we have $f(x) = 0$ at all integer points, so the right-hand side of (5.44) is identically zero; but the left-hand side is nonzero at all noninteger x.)

A Newton series is finite calculus's answer to infinite calculus's Taylor series. Just as a Taylor series can be written

$$g(a+x) = \frac{g(a)}{0!}x^0 + \frac{g'(a)}{1!}x^1 + \frac{g''(a)}{2!}x^2 + \frac{g'''(a)}{3!}x^3 + \cdots,$$

(Since $E = 1 + \Delta$, $E^x = \sum_k \binom{x}{k}\Delta^k$; and $E^x g(a) = g(a+x)$.)

the Newton series for $f(x) = g(a+x)$ can be written

$$g(a+x) = \frac{g(a)}{0!}x^{\underline{0}} + \frac{\Delta g(a)}{1!}x^{\underline{1}} + \frac{\Delta^2 g(a)}{2!}x^{\underline{2}} + \frac{\Delta^3 g(a)}{3!}x^{\underline{3}} + \cdots. \tag{5.45}$$

(This is the same as (5.44), because $\Delta^n f(0) = \Delta^n g(a)$ for all $n \geqslant 0$ when $f(x) = g(a+x)$.) Both the Taylor and Newton series are finite when g is a polynomial, or when $x = 0$; in addition, the Newton series is finite when x is a positive integer. Otherwise the sums may or may not converge for particular values of x. If the Newton series converges when x is not a nonnegative integer, it might actually converge to a value that's *different* from $g(a+x)$, because the Newton series (5.45) depends only on the spaced-out function values $g(a)$, $g(a+1)$, $g(a+2)$, ...

One example of a convergent Newton series is provided by the binomial theorem. Let $g(x) = (1+z)^x$, where z is a fixed complex number such that $|z| < 1$. Then $\Delta g(x) = (1+z)^{x+1} - (1+z)^x = z(1+z)^x$, hence $\Delta^n g(x) = z^n(1+z)^x$. In this case the infinite Newton series

$$g(a+x) = \sum_n \Delta^n g(a)\binom{x}{n} = (1+z)^a \sum_n \binom{x}{n} z^n$$

converges to the "correct" value $(1+z)^{a+x}$, for all x.

James Stirling tried to use Newton series to generalize the factorial function to noninteger values. First he found coefficients S_n such that

$$x! = \sum_n S_n \binom{x}{n} = S_0\binom{x}{0} + S_1\binom{x}{1} + S_2\binom{x}{2} + \cdots \qquad (5.46)$$

is an identity for $x = 0$, $x = 1$, $x = 2$, etc. But he discovered that the resulting series doesn't converge except when x is a nonnegative integer. So he tried again, this time writing

"Forasmuch as these terms increase very fast, their differences will make a diverging progression, which hinders the ordinate of the parabola from approaching to the truth; therefore in this and the like cases, I interpolate the logarithms of the terms, whose differences constitute a series swiftly converging."
—J. Stirling [281]

$$\ln x! = \sum_n s_n \binom{x}{n} = s_0\binom{x}{0} + s_1\binom{x}{1} + s_2\binom{x}{2} + \cdots. \qquad (5.47)$$

Now $\Delta(\ln x!) = \ln(x+1)! - \ln x! = \ln(x+1)$, hence

$$\begin{aligned} s_n &= \Delta^n(\ln x!)\big|_{x=0} \\ &= \Delta^{n-1}(\ln(x+1))\big|_{x=0} \\ &= \sum_k \binom{n-1}{k}(-1)^{n-1-k}\ln(k+1) \end{aligned}$$

by (5.40). The coefficients are therefore $s_0 = s_1 = 0$; $s_2 = \ln 2$; $s_3 = \ln 3 - 2\ln 2 = \ln\frac{3}{4}$; $s_4 = \ln 4 - 3\ln 3 + 3\ln 2 = \ln\frac{32}{27}$; etc. In this way Stirling obtained a series that does converge (although he didn't prove it); in fact, his series converges for all $x > -1$. He was thereby able to evaluate $\frac{1}{2}!$ satisfactorily. Exercise 88 tells the rest of the story.

(Proofs of convergence were not invented until the nineteenth century.)

Trick 3: Inversion.

A special case of the rule (5.45) we've just derived for Newton's series can be rewritten in the following way:

$$g(n) = \sum_k \binom{n}{k}(-1)^k f(k) \iff f(n) = \sum_k \binom{n}{k}(-1)^k g(k). \qquad (5.48)$$

This dual relationship between f and g is called an *inversion formula*; it's rather like the Möbius inversion formulas (4.56) and (4.61) that we encountered in Chapter 4. Inversion formulas tell us how to solve "implicit recurrences," where an unknown sequence is embedded in a sum.

Invert this: 'zınb ppo'.

For example, $g(n)$ might be a known function, and $f(n)$ might be unknown; and we might have found a way to prove that $g(n) = \sum_k \binom{n}{k}(-1)^k f(k)$. Then (5.48) lets us express $f(n)$ as a sum of known values.

We can prove (5.48) directly by using the basic methods at the beginning of this chapter. If $g(n) = \sum_k \binom{n}{k}(-1)^k f(k)$ for all $n \geqslant 0$, then

$$\begin{aligned}\sum_k \binom{n}{k}(-1)^k g(k) &= \sum_k \binom{n}{k}(-1)^k \sum_j \binom{k}{j}(-1)^j f(j)\\ &= \sum_j f(j) \sum_k \binom{n}{k}(-1)^{k+j}\binom{k}{j}\\ &= \sum_j f(j) \sum_k \binom{n}{j}(-1)^{k+j}\binom{n-j}{k-j}\\ &= \sum_j f(j)\binom{n}{j} \sum_k (-1)^k \binom{n-j}{k}\\ &= \sum_j f(j)\binom{n}{j}[n-j=0] \quad = \quad f(n).\end{aligned}$$

The proof in the other direction is, of course, the same, because the relation between f and g is symmetric.

Let's illustrate (5.48) by applying it to the "football victory problem": A group of n fans of the winning football team throw their hats high into the air. The hats come back randomly, one hat to each of the n fans. How many ways $h(n,k)$ are there for exactly k fans to get their own hats back?

For example, if $n = 4$ and if the hats and fans are named A, B, C, D, the $4! = 24$ possible ways for hats to land generate the following numbers of rightful owners:

ABCD	4	BACD	2	CABD	1	DABC	0
ABDC	2	BADC	0	CADB	0	DACB	1
ACBD	2	BCAD	1	CBAD	2	DBAC	1
ACDB	1	BCDA	0	CBDA	1	DBCA	2
ADBC	1	BDAC	0	CDAB	0	DCAB	0
ADCB	2	BDCA	1	CDBA	0	DCBA	0

Therefore $h(4,4) = 1$; $h(4,3) = 0$; $h(4,2) = 6$; $h(4,1) = 8$; $h(4,0) = 9$.

We can determine $h(n,k)$ by noticing that it is the number of ways to choose k lucky hat owners, namely $\binom{n}{k}$, times the number of ways to arrange the remaining $n-k$ hats so that none of them goes to the right owner, namely $h(n-k,0)$. A permutation is called a *derangement* if it moves every item, and the number of derangements of n objects is sometimes denoted by the symbol '$n¡$', read "n subfactorial." Therefore $h(n-k,0) = (n-k)¡$, and we have the general formula

$$h(n,k) \;=\; \binom{n}{k}(n-k)¡\,.$$

(Subfactorial notation isn't standard, and it's not clearly a great idea; but let's try it awhile to see if we grow to like it. We can always resort to 'D_n' or something, if '$n¡$' doesn't work out.)

Our problem would be solved if we had a closed form for $n¡$, so let's see what we can find. There's an easy way to get a recurrence, because the sum of $h(n,k)$ for all k is the total number of permutations of n hats:

$$\begin{aligned} n! \;=\; \sum_k h(n,k) \;&=\; \sum_k \binom{n}{k}(n-k)¡ \\ &=\; \sum_k \binom{n}{k} k¡\,, \qquad \text{integer } n \geqslant 0. \end{aligned} \tag{5.49}$$

(We've changed k to $n-k$ and $\binom{n}{n-k}$ to $\binom{n}{k}$ in the last step.) With this implicit recurrence we can compute all the $h(n,k)$'s we like:

n	$h(n,0)$	$h(n,1)$	$h(n,2)$	$h(n,3)$	$h(n,4)$	$h(n,5)$	$h(n,6)$
0	1						
1	0	1					
2	1	0	1				
3	2	3	0	1			
4	9	8	6	0	1		
5	44	45	20	10	0	1	
6	265	264	135	40	15	0	1

For example, here's how the row for $n=4$ can be computed: The two rightmost entries are obvious—there's just one way for all hats to land correctly, and there's no way for just three fans to get their own. (Whose hat would the fourth fan get?) When $k=2$ and $k=1$, we can use our equation for $h(n,k)$, giving $h(4,2) = \binom{4}{2}h(2,0) = 6\cdot 1 = 6$, and $h(4,1) = \binom{4}{1}h(3,0) = 4\cdot 2 = 8$. We can't use this equation for $h(4,0)$; rather, we can, but it gives us $h(4,0) = \binom{4}{0}h(4,0)$, which is true but useless. Taking another tack, we can use the relation $h(4,0)+8+6+0+1 = 4!$ to deduce that $h(4,0) = 9$; this is the value of $4¡$. Similarly $n¡$ depends on the values of $k¡$ for $k < n$.

The art of mathematics, as of life, is knowing which truths are useless.

How can we solve a recurrence like (5.49)? Easy; it has the form of (5.48), with $g(n) = n!$ and $f(k) = (-1)^k k¡$. Hence its solution is

$$n¡ = (-1)^n \sum_k \binom{n}{k} (-1)^k k!\,.$$

Well, this isn't really a solution; it's a sum that should be put into closed form if possible. But it's better than a recurrence. The sum can be simplified, since $k!$ cancels with a hidden $k!$ in $\binom{n}{k}$, so let's try that: We get

$$n¡ = \sum_{0 \le k \le n} \frac{n!}{(n-k)!} (-1)^{n+k} = n! \sum_{0 \le k \le n} \frac{(-1)^k}{k!}\,. \qquad (5.50)$$

The remaining sum converges rapidly to the number $\sum_{k \ge 0} (-1)^k/k! = e^{-1}$. In fact, the terms that are excluded from the sum are

$$\begin{aligned} n! \sum_{k > n} \frac{(-1)^k}{k!} &= \frac{(-1)^{n+1}}{n+1} \sum_{k \ge 0} (-1)^k \frac{(n+1)!}{(k+n+1)!} \\ &= \frac{(-1)^{n+1}}{n+1} \left(1 - \frac{1}{n+2} + \frac{1}{(n+2)(n+3)} - \cdots \right), \end{aligned}$$

and the parenthesized quantity lies between 1 and $1 - \frac{1}{n+2} = \frac{n+1}{n+2}$. Therefore the difference between $n¡$ and $n!/e$ is roughly $1/n$ in absolute value; more precisely, it lies between $1/(n+1)$ and $1/(n+2)$. But $n¡$ is an integer. Therefore it must be what we get when we round $n!/e$ to the nearest integer, if $n > 0$. So we have the closed form we seek:

$$n¡ = \left\lfloor \frac{n!}{e} + \frac{1}{2} \right\rfloor + [n = 0]\,. \qquad (5.51)$$

This is the number of ways that no fan gets the right hat back. When n is large, it's more meaningful to know the *probability* that this happens. If we assume that each of the $n!$ arrangements is equally likely—because the hats were thrown extremely high—this probability is

Baseball fans: .367 is also Ty Cobb's lifetime batting average, the all-time record. Can this be a coincidence?

$$\frac{n¡}{n!} = \frac{n!/e + O(1)}{n!} \sim \frac{1}{e} = .367\ldots\,.$$

So when n gets large the probability that all hats are misplaced is almost 37%.

(Hey wait, you're fudging. Cobb's average was 4191/11429 ≈ .366699, while 1/e ≈ .367879. But maybe if Wade Boggs has a few really good seasons. . .)

Incidentally, recurrence (5.49) for subfactorials is exactly the same as (5.46), the first recurrence considered by Stirling when he was trying to generalize the factorial function. Hence $S_k = k¡$. These coefficients are so large, it's no wonder the infinite series (5.46) diverges for noninteger x.

Before leaving this problem, let's look briefly at two interesting patterns that leap out at us in the table of small $h(n,k)$. First, it seems that the numbers 1, 3, 6, 10, 15, ... below the all-0 diagonal are the triangular numbers.

This observation is easy to prove, since those table entries are the $h(n, n-2)$'s, and we have

$$h(n,n-2) \;=\; \binom{n}{n-2} 2¡ \;=\; \binom{n}{2}.$$

It also seems that the numbers in the first two columns differ by ± 1. Is this always true? Yes,

$$\begin{aligned} h(n,0) - h(n,1) &= n¡ - n(n-1)¡ \\ &= \Bigl(n! \sum_{0 \le k \le n} \frac{(-1)^k}{k!}\Bigr) - \Bigl(n(n-1)! \sum_{0 \le k \le n-1} \frac{(-1)^k}{k!}\Bigr) \\ &= n! \frac{(-1)^n}{n!} \;=\; (-1)^n. \end{aligned}$$

In other words, $n¡ = n(n-1)¡ + (-1)^n$. This is a much simpler recurrence for the derangement numbers than we had before.

Now let's invert something else. If we apply inversion to the formula

But inversion is the source of smog.

$$\sum_k \binom{n}{k} \frac{(-1)^k}{x+k} \;=\; \frac{1}{x} \binom{x+n}{n}^{-1}$$

that we derived in (5.41), we find

$$\frac{x}{x+n} \;=\; \sum_{k \ge 0} \binom{n}{k} (-1)^k \binom{x+k}{k}^{-1}.$$

This is interesting, but not really new. If we negate the upper index in $\binom{x+k}{k}$, we have merely discovered identity (5.33) again.

5.4 GENERATING FUNCTIONS

We come now to the most important idea in this whole book, the notion of a *generating function*. An infinite sequence $\langle a_0, a_1, a_2, \ldots \rangle$ that we wish to deal with in some way can conveniently be represented as a *power series* in an auxiliary variable z,

$$A(z) \;=\; a_0 + a_1 z + a_2 z^2 + \cdots \;=\; \sum_{k \ge 0} a_k z^k. \tag{5.52}$$

It's appropriate to use the letter z as the name of the auxiliary variable, because we'll often be thinking of z as a complex number. The theory of complex variables conventionally uses 'z' in its formulas; power series (a.k.a. analytic functions or holomorphic functions) are central to that theory.

We will be seeing lots of generating functions in subsequent chapters. Indeed, Chapter 7 is entirely devoted to them. Our present goal is simply to introduce the basic concepts, and to demonstrate the relevance of generating functions to the study of binomial coefficients.

A generating function is useful because it's a single quantity that represents an entire infinite sequence. We can often solve problems by first setting up one or more generating functions, then by fooling around with those functions until we know a lot about them, and finally by looking again at the coefficients. With a little bit of luck, we'll know enough about the function to understand what we need to know about its coefficients.

If $A(z)$ is any power series $\sum_{k\geqslant 0} a_k z^k$, we will find it convenient to write

$$[z^n]\,A(z) \;=\; a_n\,; \tag{5.53}$$

in other words, $[z^n]\,A(z)$ denotes the coefficient of z^n in $A(z)$.

Let $A(z)$ be the generating function for $\langle a_0, a_1, a_2, \ldots\rangle$ as in (5.52), and let $B(z)$ be the generating function for another sequence $\langle b_0, b_1, b_2, \ldots\rangle$. Then the product $A(z)B(z)$ is the power series

$$\begin{aligned}(a_0 + a_1 z + a_2 z^2 + \cdots)&(b_0 + b_1 z + b_2 z^2 + \cdots)\\ &= a_0 b_0 \;+\; (a_0 b_1 + a_1 b_0)z \;+\; (a_0 b_2 + a_1 b_1 + a_2 b_0)z^2 \;+\; \cdots\,;\end{aligned}$$

the coefficient of z^n in this product is

$$a_0 b_n + a_1 b_{n-1} + \cdots + a_n b_0 \;=\; \sum_{k=0}^{n} a_k b_{n-k}\,.$$

Therefore if we wish to evaluate any sum that has the general form

$$c_n \;=\; \sum_{k=0}^{n} a_k b_{n-k}\,, \tag{5.54}$$

and if we know the generating functions $A(z)$ and $B(z)$, we have

$$c_n \;=\; [z^n]\,A(z)B(z)\,.$$

The sequence $\langle c_n\rangle$ defined by (5.54) is called the *convolution* of the sequences $\langle a_n\rangle$ and $\langle b_n\rangle$; two sequences are "convolved" by forming the sums of all products whose subscripts add up to a given amount. The gist of the previous paragraph is that convolution of sequences corresponds to multiplication of their generating functions.

Generating functions give us powerful ways to discover and/or prove identities. For example, the binomial theorem tells us that $(1+z)^r$ is the generating function for the sequence $\langle\binom{r}{0},\binom{r}{1},\binom{r}{2},\ldots\rangle$:

$$(1+z)^r = \sum_{k\geq 0}\binom{r}{k}z^k .$$

Similarly,

$$(1+z)^s = \sum_{k\geq 0}\binom{s}{k}z^k .$$

If we multiply these together, we get another generating function:

$$(1+z)^r(1+z)^s = (1+z)^{r+s} .$$

And now comes the punch line: Equating coefficients of z^n on both sides of this equation gives us

$$\sum_{k=0}^{n}\binom{r}{k}\binom{s}{n-k} = \binom{r+s}{n} .$$

We've discovered Vandermonde's convolution, (5.27)!

(5.27)! = (5.27)(4.27) (3.27)(2.27) (1.27)(0.27)!.

That was nice and easy; let's try another. This time we use $(1-z)^r$, which is the generating function for the sequence $\langle(-1)^n\binom{r}{n}\rangle = \langle\binom{r}{0},-\binom{r}{1},\binom{r}{2},\ldots\rangle$. Multiplying by $(1+z)^r$ gives another generating function whose coefficients we know:

$$(1-z)^r(1+z)^r = (1-z^2)^r .$$

Equating coefficients of z^n now gives the equation

$$\sum_{k=0}^{n}\binom{r}{k}\binom{r}{n-k}(-1)^k = (-1)^{n/2}\binom{r}{n/2}[n \text{ even}] . \qquad (5.55)$$

We should check this on a small case or two. When $n=3$, for example, the result is

$$\binom{r}{0}\binom{r}{3}-\binom{r}{1}\binom{r}{2}+\binom{r}{2}\binom{r}{1}-\binom{r}{3}\binom{r}{0} = 0 .$$

Each positive term is cancelled by a corresponding negative term. And the same thing happens whenever n is odd, in which case the sum isn't very interesting. But when n is even, say $n = 2$, we get a nontrivial sum that's different from Vandermonde's convolution:

$$\binom{r}{0}\binom{r}{2} - \binom{r}{1}\binom{r}{1} + \binom{r}{2}\binom{r}{0} = 2\binom{r}{2} - r^2 = -r.$$

So (5.55) checks out fine when $n = 2$. It turns out that (5.30) is a special case of our new identity (5.55).

Binomial coefficients also show up in some other generating functions, most notably the following important identities in which the lower index stays fixed and the upper index varies:

If you have a highlighter pen, these two equations have got to be marked.

$$\frac{1}{(1-z)^{n+1}} = \sum_{k\geqslant 0}\binom{n+k}{n} z^k, \qquad \text{integer } n \geqslant 0 \tag{5.56}$$

$$\frac{z^n}{(1-z)^{n+1}} = \sum_{k\geqslant 0}\binom{k}{n} z^k, \qquad \text{integer } n \geqslant 0. \tag{5.57}$$

The second identity here is just the first one multiplied by z^n, that is, "shifted right" by n places. The first identity is just a special case of the binomial theorem in slight disguise: If we expand $(1-z)^{-n-1}$ by (5.13), the coefficient of z^k is $\binom{-n-1}{k}(-1)^k$, which can be rewritten as $\binom{k+n}{k}$ or $\binom{n+k}{n}$ by negating the upper index. These special cases are worth noting explicitly, because they arise so frequently in applications.

When $n = 0$ we get a special case of a special case, the geometric series:

$$\frac{1}{1-z} = 1 + z + z^2 + z^3 + \cdots = \sum_{k\geqslant 0} z^k.$$

This is the generating function for the sequence $\langle 1, 1, 1, \ldots \rangle$, and it is especially useful because the convolution of any other sequence with this one is the sequence of sums: When $b_k = 1$ for all k, (5.54) reduces to

$$c_n = \sum_{k=0}^{n} a_k.$$

Therefore if $A(z)$ is the generating function for the summands $\langle a_0, a_1, a_2, \ldots \rangle$, then $A(z)/(1-z)$ is the generating function for the sums $\langle c_0, c_1, c_2, \ldots \rangle$.

The problem of derangements, which we solved by inversion in connection with hats and football fans, can be resolved with generating functions in an interesting way. The basic recurrence

$$n! = \sum_k \binom{n}{k}(n-k)¡$$

can be put into the form of a convolution if we expand $\binom{n}{k}$ in factorials and divide both sides by $n!$:

$$1 = \sum_{k=0}^{n} \frac{1}{k!}\,\frac{(n-k){¡}}{(n-k)!}\,.$$

The generating function for the sequence $\langle \frac{1}{0!}, \frac{1}{1!}, \frac{1}{2!}, \ldots \rangle$ is e^z; hence if we let

$$D(z) = \sum_{k\geqslant 0} \frac{k{¡}}{k!} z^k\,,$$

the convolution/recurrence tells us that

$$\frac{1}{1-z} = e^z\,D(z)\,.$$

Solving for $D(z)$ gives

$$D(z) = \frac{1}{1-z}\,e^{-z} = \frac{1}{1-z}\left(\frac{1}{0!}z^0 - \frac{1}{1!}z^1 + \frac{1}{2!}z^2 + \cdots\right).$$

Equating coefficients of z^n now tells us that

$$\frac{n{¡}}{n!} = \sum_{k=0}^{n} \frac{(-1)^k}{k!}\,;$$

this is the formula we derived earlier by inversion.

So far our explorations with generating functions have given us slick proofs of things that we already knew how to derive by more cumbersome methods. But we haven't used generating functions to obtain any new results, except for (5.55). Now we're ready for something new and more surprising. There are two families of power series that generate an especially rich class of binomial coefficient identities: Let us define the *generalized binomial series* $\mathcal{B}_t(z)$ and the *generalized exponential series* $\mathcal{E}_t(z)$ as follows:

$$\mathcal{B}_t(z) = \sum_{k\geqslant 0} (tk)^{\underline{k-1}}\frac{z^k}{k!}\,; \qquad \mathcal{E}_t(z) = \sum_{k\geqslant 0} (tk+1)^{k-1}\frac{z^k}{k!}\,. \tag{5.58}$$

It can be shown that these functions satisfy the identities

$$\mathcal{B}_t(z)^{1-t} - \mathcal{B}_t(z)^{-t} = z\,; \qquad \mathcal{E}_t(z)^{-t}\ln\mathcal{E}_t(z) = z\,. \tag{5.59}$$

In the special case $t = 0$, we have

$$\mathcal{B}_0(z) = 1+z\,; \qquad \mathcal{E}_0(z) = e^z\,;$$

this explains why the series with parameter t are called "generalized" binomials and exponentials.

The following pairs of identities are valid for all real r:

$$\mathcal{B}_t(z)^r = \sum_{k\geq 0}\binom{tk+r}{k}\frac{r}{tk+r}z^k;$$

$$\mathcal{E}_t(z)^r = \sum_{k\geq 0} r\frac{(tk+r)^{k-1}}{k!}z^k; \qquad (5.60)$$

$$\frac{\mathcal{B}_t(z)^r}{1-t+t\mathcal{B}_t(z)^{-1}} = \sum_{k\geq 0}\binom{tk+r}{k}z^k;$$

$$\frac{\mathcal{E}_t(z)^r}{1-zt\mathcal{E}_t(z)^t} = \sum_{k\geq 0}\frac{(tk+r)^k}{k!}z^k. \qquad (5.61)$$

(When $tk+r=0$, we have to be a little careful about how the coefficient of z^k is interpreted; each coefficient is a polynomial in r. For example, the constant term of $\mathcal{E}_t(z)^r$ is $r(0+r)^{-1}$, and this is equal to 1 even when $r=0$.)

Since equations (5.60) and (5.61) hold for all r, we get very general identities when we multiply together the series that correspond to different powers r and s. For example,

$$\begin{aligned}\mathcal{B}_t(z)^r\frac{\mathcal{B}_t(z)^s}{1-t+t\mathcal{B}_t(z)^{-1}} &= \sum_{k\geq 0}\binom{tk+r}{k}\frac{r}{tk+r}z^k\sum_{j\geq 0}\binom{tj+s}{j}z^j\\ &= \sum_{n\geq 0}z^n\sum_{k\geq 0}\binom{tk+r}{k}\frac{r}{tk+r}\binom{t(n-k)+s}{n-k}.\end{aligned}$$

This power series must equal

$$\frac{\mathcal{B}_t(z)^{r+s}}{1-t+t\mathcal{B}_t(z)^{-1}} = \sum_{n\geq 0}\binom{tn+r+s}{n}z^n;$$

hence we can equate coefficients of z^n and get the identity

$$\sum_k\binom{tk+r}{k}\binom{t(n-k)+s}{n-k}\frac{r}{tk+r} = \binom{tn+r+s}{n}, \qquad \text{integer } n,$$

valid for all real r, s, and t. When $t=0$ this identity reduces to Vandermonde's convolution. (If by chance $tk+r$ happens to equal zero in this formula, the denominator factor $tk+r$ should be considered to cancel with the $tk+r$ in the numerator of the binomial coefficient. Both sides of the identity are polynomials in r, s, and t.) Similar identities hold when we multiply $\mathcal{B}_t(z)^r$ by $\mathcal{B}_t(z)^s$, etc.; Table 202 presents the results.

Table 202 General convolution identities, valid for integer $n \geqslant 0$.

$$\sum_k \binom{tk+r}{k}\binom{tn-tk+s}{n-k}\frac{r}{tk+r} = \binom{tn+r+s}{n}. \tag{5.62}$$

$$\sum_k \binom{tk+r}{k}\binom{tn-tk+s}{n-k}\frac{r}{tk+r}\cdot\frac{s}{tn-tk+s}$$
$$= \binom{tn+r+s}{n}\frac{r+s}{tn+r+s}. \tag{5.63}$$

$$\sum_k \binom{n}{k}(tk+r)^k(tn-tk+s)^{n-k}\frac{r}{tk+r} = (tn+r+s)^n. \tag{5.64}$$

$$\sum_k \binom{n}{k}(tk+r)^k(tn-tk+s)^{n-k}\frac{r}{tk+r}\cdot\frac{s}{tn-tk+s}$$
$$= (tn+r+s)^n\frac{r+s}{tn+r+s}. \tag{5.65}$$

We have learned that it's generally a good idea to look at special cases of general results. What happens, for example, if we set $t = 1$? The generalized binomial $\mathcal{B}_1(z)$ is very simple—it's just

$$\mathcal{B}_1(z) = \sum_{k\geqslant 0} z^k = \frac{1}{1-z};$$

therefore $\mathcal{B}_1(z)$ doesn't give us anything we didn't already know from Vandermonde's convolution. But $\mathcal{E}_1(z)$ is an important function,

$$\mathcal{E}(z) = \sum_{k\geqslant 0}(k+1)^{k-1}\frac{z^k}{k!} = 1+z+\frac{3}{2}z^2+\frac{8}{3}z^3+\frac{125}{24}z^4+\cdots \tag{5.66}$$

that we haven't seen before; it satisfies the basic identity

$$\mathcal{E}(z) = e^{z\mathcal{E}(z)}. \tag{5.67}$$

Aha! This is the iterated power function $\mathcal{E}(\ln z) = z^{z^{z^{\cdot^{\cdot^{\cdot}}}}}$ that I've often wondered about.

This function, first studied by Eisenstein [75], arises in many applications.

The special cases $t = 2$ and $t = -1$ of the generalized binomial are of particular interest, because their coefficients occur again and again in problems that have a recursive structure. Therefore it's useful to display these

Zzzzzz...

series explicitly for future reference:

$$\mathcal{B}_2(z) = \sum_k \binom{2k}{k} \frac{z^k}{1+k}$$

$$= \sum_k \binom{2k+1}{k} \frac{z^k}{1+2k} = \frac{1-\sqrt{1-4z}}{2z}. \tag{5.68}$$

$$\mathcal{B}_{-1}(z) = \sum_k \binom{1-k}{k} \frac{z^k}{1-k}$$

$$= \sum_k \binom{2k-1}{k} \frac{(-z)^k}{1-2k} = \frac{1+\sqrt{1+4z}}{2}. \tag{5.69}$$

$$\mathcal{B}_2(z)^r = \sum_k \binom{2k+r}{k} \frac{r}{2k+r} z^k. \tag{5.70}$$

$$\mathcal{B}_{-1}(z)^r = \sum_k \binom{r-k}{k} \frac{r}{r-k} z^k. \tag{5.71}$$

$$\frac{\mathcal{B}_2(z)^r}{\sqrt{1-4z}} = \sum_k \binom{2k+r}{k} z^k. \tag{5.72}$$

$$\frac{\mathcal{B}_{-1}(z)^{r+1}}{\sqrt{1+4z}} = \sum_k \binom{r-k}{k} z^k. \tag{5.73}$$

The coefficients $\binom{2n}{n}\frac{1}{n+1}$ of $\mathcal{B}_2(z)$ are called the *Catalan numbers* C_n, because Eugène Catalan wrote an influential paper about them in the 1830s [46]. The sequence begins as follows:

n	0	1	2	3	4	5	6	7	8	9	10
C_n	1	1	2	5	14	42	132	429	1430	4862	16796

The coefficients of $\mathcal{B}_{-1}(z)$ are essentially the same, but there's an extra 1 at the beginning and the other numbers alternate in sign: $\langle 1, 1, -1, 2, -5, 14, \ldots \rangle$. Thus $\mathcal{B}_{-1}(z) = 1 + z\mathcal{B}_2(-z)$. We also have $\mathcal{B}_{-1}(z) = \mathcal{B}_2(-z)^{-1}$.

Let's close this section by deriving an important consequence of (5.72) and (5.73), a relation that shows further connections between the functions $\mathcal{B}_{-1}(z)$ and $\mathcal{B}_2(-z)$:

$$\frac{\mathcal{B}_{-1}(z)^{n+1} - (-z)^{n+1}\mathcal{B}_2(-z)^{n+1}}{\sqrt{1+4z}} = \sum_{k \le n} \binom{n-k}{k} z^k.$$

This holds because the coefficient of z^k in $(-z)^{n+1}\mathcal{B}_2(-z)^{n+1}/\sqrt{1+4z}$ is

$$\begin{aligned}
[z^k]\,\frac{(-z)^{n+1}\mathcal{B}_2(-z)^{n+1}}{\sqrt{1+4z}} &= (-1)^{n+1}[z^{k-n-1}]\,\frac{\mathcal{B}_2(-z)^{n+1}}{\sqrt{1+4z}} \\
&= (-1)^{n+1}(-1)^{k-n-1}[z^{k-n-1}]\,\frac{\mathcal{B}_2(z)^{n+1}}{\sqrt{1-4z}} \\
&= (-1)^k\binom{2(k-n-1)+n+1}{k-n-1} \\
&= (-1)^k\binom{2k-n-1}{k-n-1} = (-1)^k\binom{2k-n-1}{k} \\
&= \binom{n-k}{k} = [z^k]\,\frac{\mathcal{B}_{-1}(z)^{n+1}}{\sqrt{1+4z}}
\end{aligned}$$

when $k > n$. The terms nicely cancel each other out. We can now use (5.68) and (5.69) to obtain the closed form

$$\sum_{k\leqslant n}\binom{n-k}{k}z^k = \frac{1}{\sqrt{1+4z}}\left(\left(\frac{1+\sqrt{1+4z}}{2}\right)^{n+1} - \left(\frac{1-\sqrt{1+4z}}{2}\right)^{n+1}\right),$$

$$\text{integer } n \geqslant 0. \qquad (5.74)$$

(The special case $z = -1$ came up in Problem 3 of Section 5.2. Since the numbers $\frac{1}{2}(1 \pm \sqrt{-3}\,)$ are sixth roots of unity, the sums $\sum_{k\leqslant n}\binom{n-k}{k}(-1)^k$ have the periodic behavior we observed in that problem.) Similarly we can combine (5.70) with (5.71) to cancel the large coefficients and get

$$\sum_{k<n}\binom{n-k}{k}\frac{n}{n-k}z^k = \left(\frac{1+\sqrt{1+4z}}{2}\right)^n + \left(\frac{1-\sqrt{1+4z}}{2}\right)^n,$$

$$\text{integer } n > 0. \qquad (5.75)$$

5.5 HYPERGEOMETRIC FUNCTIONS

The methods we've been applying to binomial coefficients are very effective, when they work, but we must admit that they often appear to be ad hoc — more like tricks than techniques. When we're working on a problem, we often have many directions to pursue, and we might find ourselves going around in circles. Binomial coefficients are like chameleons, changing their appearance easily. Therefore it's natural to ask if there isn't some unifying principle that will systematically handle a great variety of binomial coefficient summations all at once. Fortunately, the answer is yes. The unifying principle is based on the theory of certain infinite sums called *hypergeometric series*.

They're even more versatile than chameleons; we can dissect them and put them back together in different ways.

The study of hypergeometric series was launched many years ago by Euler, Gauss, and Riemann; such series, in fact, are still the subject of considerable research. But hypergeometrics have a somewhat formidable notation, which takes a little time to get used to.

Anything that has survived for centuries with such awesome notation must be really useful.

The general hypergeometric series is a power series in z with $m+n$ parameters, and it is defined as follows in terms of rising factorial powers:

$$F\left({a_1, \ldots, a_m \atop b_1, \ldots, b_n} \middle| z\right) = \sum_{k \geqslant 0} \frac{a_1^{\overline{k}} \ldots a_m^{\overline{k}}}{b_1^{\overline{k}} \ldots b_n^{\overline{k}}} \frac{z^k}{k!} . \tag{5.76}$$

To avoid division by zero, none of the b's may be zero or a negative integer. Other than that, the a's and b's may be anything we like. The notation '$F(a_1, \ldots, a_m; b_1, \ldots, b_n; z)$' is also used as an alternative to the two-line form (5.76), since a one-line form sometimes works better typographically. The a's are said to be *upper parameters*; they occur in the numerator of the terms of F. The b's are *lower parameters*, and they occur in the denominator. The final quantity z is called the *argument*.

Standard reference books often use '${}_mF_n$' instead of 'F' as the name of a hypergeometric with m upper parameters and n lower parameters. But the extra subscripts tend to clutter up the formulas and waste our time, if we're compelled to write them over and over. We can count how many parameters there are, so we usually don't need extra additional unnecessary redundancy.

Many important functions occur as special cases of the general hypergeometric; indeed, that's why hypergeometrics are so powerful. For example, the simplest case occurs when $m = n = 0$: There are no parameters at all, and we get the familiar series

$$F\left({} \middle| z\right) = \sum_{k \geqslant 0} \frac{z^k}{k!} = e^z .$$

Actually the notation looks a bit unsettling when m or n is zero. We can add an extra '1' above and below in order to avoid this:

$$F\left({1 \atop 1} \middle| z\right) = e^z .$$

In general we don't change the function if we cancel a parameter that occurs in both numerator and denominator, or if we insert two identical parameters.

The next simplest case has $m = 1$, $a_1 = 1$, and $n = 0$; we change the parameters to $m = 2$, $a_1 = a_2 = 1$, $n = 1$, and $b_1 = 1$, so that $n > 0$. This series also turns out to be familiar, because $1^{\overline{k}} = k!$:

$$F\left({1,1 \atop 1} \middle| z\right) = \sum_{k \geqslant 0} z^k = \frac{1}{1-z} .$$

It's our old friend, the geometric series; $F(a_1, \ldots, a_m; b_1, \ldots, b_n; z)$ is called hypergeometric because it includes the geometric series $F(1, 1; 1; z)$ as a very special case.

The general case $m = 1$ and $n = 0$ is, in fact, easy to sum in closed form,

$$F\left({a, 1 \atop 1} \,\middle|\, z\right) = \sum_{k \geqslant 0} a^{\overline{k}} \frac{z^k}{k!} = \sum_k \binom{a+k-1}{k} z^k = \frac{1}{(1-z)^a}, \qquad (5.77)$$

using (5.56). If we replace a by $-a$ and z by $-z$, we get the binomial theorem,

$$F\left({-a, 1 \atop 1} \,\middle|\, -z\right) = (1+z)^a.$$

A negative integer as upper parameter causes the infinite series to become finite, since $(-a)^{\overline{k}} = 0$ whenever $k > a \geqslant 0$ and a is an integer.

The general case $m = 0$, $n = 1$ is another famous series, but it's not as well known in the literature of discrete mathematics:

$$F\left({1 \atop b, 1} \,\middle|\, z\right) = \sum_{k \geqslant 0} \frac{(b-1)!}{(b-1+k)!} \frac{z^k}{k!} = I_{b-1}(2\sqrt{z}) \frac{(b-1)!}{z^{(b-1)/2}}. \qquad (5.78)$$

This function I_{b-1} is called a "modified Bessel function" of order $b-1$. The special case $b = 1$ gives us $F\left({1 \atop 1,1} \,\middle|\, z\right) = I_0(2\sqrt{z})$, which is the interesting series $\sum_{k \geqslant 0} z^k/k!^2$.

The special case $m = n = 1$ is called a "confluent hypergeometric series" and often denoted by the letter M:

$$F\left({a \atop b} \,\middle|\, z\right) = \sum_{k \geqslant 0} \frac{a^{\overline{k}}}{b^{\overline{k}}} \frac{z^k}{k!} = M(a, b, z). \qquad (5.79)$$

This function, which has important applications to engineering, was introduced by Ernst Kummer.

By now a few of us are wondering why we haven't discussed convergence of the infinite series (5.76). The answer is that we can ignore convergence if we are using z simply as a formal symbol. It is not difficult to verify that formal infinite sums of the form $\sum_{k \geqslant n} \alpha_k z^k$ form a field, if the coefficients α_k lie in a field. We can add, subtract, multiply, divide, differentiate, and do functional composition on such formal sums without worrying about convergence; any identities we derive will still be formally true. For example, the hypergeometric $F\left({1,1,1 \atop 1} \,\middle|\, z\right) = \sum_{k \geqslant 0} k!\, z^k$ doesn't converge for any nonzero z; yet we'll see in Chapter 7 that we can still use it to solve problems. On the other hand, whenever we replace z by a particular numerical value, we do have to be sure that the infinite sum is well defined.

The next step up in complication is actually the most famous hypergeometric of all. In fact, it was *the* hypergeometric series until about 1870, when everything was generalized to arbitrary m and n. This one has two upper parameters and one lower parameter:

$$F\left({a, b \atop c} \middle| z \right) = \sum_{k \geqslant 0} \frac{a^{\overline{k}} b^{\overline{k}} z^k}{c^{\overline{k}} k!}. \tag{5.80}$$

It is often called the Gaussian hypergeometric, because many of its subtle properties were first proved by Gauss in his doctoral dissertation of 1812 [116], although Euler [95] and Pfaff [233] had already discovered some remarkable things about it. One of its important special cases is

"There must be many universities to-day where 95 per cent, if not 100 per cent, of the functions studied by physics, engineering, and even mathematics students, are covered by this single symbol $F(a, b; c; x)$."
— W. W. Sawyer [257]

$$\begin{aligned} \ln(1+z) &= zF\left({1,1 \atop 2} \middle| -z \right) = z \sum_{k \geqslant 0} \frac{k!\,k!}{(k+1)!} \frac{(-z)^k}{k!} \\ &= z - \frac{z^2}{2} + \frac{z^3}{3} - \frac{z^4}{4} + \cdots . \end{aligned}$$

Notice that $z^{-1}\ln(1+z)$ is a hypergeometric function, but $\ln(1+z)$ itself cannot be hypergeometric, since a hypergeometric series always has the value 1 when $z = 0$.

So far hypergeometrics haven't actually done anything for us except provide an excuse for name-dropping. But we've seen that several very different functions can all be regarded as hypergeometric; this will be the main point of interest in what follows. We'll see that a large class of sums can be written as hypergeometric series in a "canonical" way, hence we will have a good filing system for facts about binomial coefficients.

What series are hypergeometric? It's easy to answer this question if we look at the ratio between consecutive terms:

$$F\left({a_1, \ldots, a_m \atop b_1, \ldots, b_n} \middle| z \right) = \sum_{k \geqslant 0} t_k, \qquad t_k = \frac{a_1^{\overline{k}} \ldots a_m^{\overline{k}} z^k}{b_1^{\overline{k}} \ldots b_n^{\overline{k}} k!}.$$

The first term is $t_0 = 1$, and the other terms have ratios given by

$$\begin{aligned} \frac{t_{k+1}}{t_k} &= \frac{a_1^{\overline{k+1}} \ldots a_m^{\overline{k+1}}}{a_1^{\overline{k}} \ldots a_m^{\overline{k}}} \frac{b_1^{\overline{k}} \ldots b_n^{\overline{k}}}{b_1^{\overline{k+1}} \ldots b_n^{\overline{k+1}}} \frac{k!}{(k+1)!} \frac{z^{k+1}}{z^k} \\ &= \frac{(k+a_1) \ldots (k+a_m)\, z}{(k+b_1) \ldots (k+b_n)(k+1)}. \end{aligned} \tag{5.81}$$

This is a *rational function* of k, that is, a quotient of polynomials in k. Any rational function of k can be factored over the complex numbers and put into this form. The a's are the negatives of the roots of the polynomial in the numerator, and the b's are the negatives of the roots of the polynomial in the denominator. If the denominator doesn't already contain the special factor $(k+1)$, we can include $(k+1)$ in both numerator and denominator. A constant factor remains, and we can call it z. Therefore hypergeometric series are precisely those series whose first term is 1 and whose term ratio t_{k+1}/t_k is a rational function of k.

Suppose, for example, that we're given an infinite series with term ratio

$$\frac{t_{k+1}}{t_k} = \frac{k^2+7k+10}{4k^2+1},$$

a rational function of k. The numerator polynomial splits nicely into two factors, $(k+2)(k+5)$, and the denominator is $4(k+i/2)(k-i/2)$. Since the denominator is missing the required factor $(k+1)$, we write the term ratio as

$$\frac{t_{k+1}}{t_k} = \frac{(k+2)(k+5)(k+1)(1/4)}{(k+i/2)(k-i/2)(k+1)},$$

and we can read off the results: The given series is

$$\sum_{k\geqslant 0} t_k = t_0\, F\left({2,\,5,\,1 \atop i/2,\,-i/2}\middle|\,1/4\right).$$

Thus, we have a general method for finding the hypergeometric representation of a given quantity S, when such a representation is possible: First we write S as an infinite series whose first term is nonzero. We choose a notation so that the series is $\sum_{k\geqslant 0} t_k$ with $t_0 \neq 0$. Then we calculate t_{k+1}/t_k. If the term ratio is not a rational function of k, we're out of luck. Otherwise we express it in the form (5.81); this gives parameters $a_1, \ldots, a_m, b_1, \ldots, b_n$, and an argument z, such that $S = t_0\, F(a_1, \ldots, a_m; b_1, \ldots, b_n; z)$.

(Now is a good time to do warmup exercise 11.)

Gauss's hypergeometric series can be written in the recursively factored form

$$F\left({a,b \atop c}\middle|\,z\right) = 1 + \frac{a}{1}\frac{b}{c}z\left(1 + \frac{a+1}{2}\frac{b+1}{c+1}z\left(1 + \frac{a+2}{3}\frac{b+2}{c+2}z(1+\cdots)\right)\right)$$

if we wish to emphasize the importance of term ratios.

Let's try now to reformulate the binomial coefficient identities derived earlier in this chapter, expressing them as hypergeometrics. For example, let's figure out what the parallel summation law,

$$\sum_{k\leqslant n}\binom{r+k}{k} = \binom{r+n+1}{n}, \qquad \text{integer } n,$$

looks like in hypergeometric notation. We need to write the sum as an infinite series that starts at $k = 0$, so we replace k by $n - k$:

$$\sum_{k \geqslant 0} \binom{r+n-k}{n-k} = \sum_{k \geqslant 0} \frac{(r+n-k)!}{r!\,(n-k)!} = \sum_{k \geqslant 0} t_k \,.$$

This series is formally infinite but actually finite, because the $(n-k)!$ in the denominator will make $t_k = 0$ when $k > n$. (We'll see later that $1/x!$ is defined for all x, and that $1/x! = 0$ when x is a negative integer. But for now, let's blithely disregard such technicalities until we gain more hypergeometric experience.) The term ratio is

$$\frac{t_{k+1}}{t_k} = \frac{(r+n-k-1)!\,r!\,(n-k)!}{r!\,(n-k-1)!\,(r+n-k)!} = \frac{n-k}{r+n-k} = \frac{(k+1)(k-n)(1)}{(k-n-r)(k+1)} \,.$$

Furthermore $t_0 = \binom{r+n}{n}$. Hence the parallel summation law is equivalent to the hypergeometric identity

$$\binom{r+n}{n} F\left({1, -n \atop -n-r} \middle| \, 1 \right) = \binom{r+n+1}{n} .$$

Dividing through by $\binom{r+n}{n}$ gives a slightly simpler version,

$$F\left({1, -n \atop -n-r} \middle| \, 1 \right) = \frac{r+n+1}{r+1} , \qquad \text{if } \binom{r+n}{n} \neq 0. \tag{5.82}$$

Let's do another one. The term ratio of identity (5.16),

$$\sum_{k \leqslant m} \binom{r}{k} (-1)^k = (-1)^m \binom{r-1}{m} , \qquad \text{integer } m,$$

is $(k-m)/(r-m+k+1) = (k+1)(k-m)(1)/(k-m+r+1)(k+1)$, after we replace k by $m-k$; hence (5.16) gives a closed form for

$$F\left({1, -m \atop -m+r+1} \middle| \, 1 \right) .$$

This is essentially the same as the hypergeometric function on the left of (5.82), but with m in place of n and $r+1$ in place of $-r$. Therefore identity (5.16) could have been derived from (5.82), the hypergeometric version of (5.9). (No wonder we found it easy to prove (5.16) by using (5.9).)

First derangements, now degenerates.

Before we go further, we should think about degenerate cases, because hypergeometrics are not defined when a lower parameter is zero or a negative integer. We usually apply the parallel summation identity when r and n are positive integers; but then $-n-r$ is a negative integer and the hypergeometric (5.76) is undefined. How then can we consider (5.82) to be legitimate? The answer is that we can take the limit of $F\left({1,-n \atop -n-r+\epsilon}\middle|1\right)$ as $\epsilon \to 0$.

We will look at such things more closely later in this chapter, but for now let's just be aware that some denominators can be dynamite. It is interesting, however, that the very first sum we've tried to express hypergeometrically has turned out to be degenerate.

(We proved the identities originally for integer r, and used the polynomial argument to show that they hold in general. Now we're proving them first for irrational r, and using a limiting argument to show that they hold for integers!)

Another possibly sore point in our derivation of (5.82) is that we expanded $\binom{r+n-k}{n-k}$ as $(r+n-k)!/r!\,(n-k)!$. This expansion fails when r is a negative integer, because $(-m)!$ has to be ∞ if the law

$$0! \;=\; 0\cdot(-1)\cdot(-2)\cdot\ldots\cdot(-m+1)\cdot(-m)!$$

is going to hold. Again, we need to approach integer results by considering a limit of $r+\epsilon$ as $\epsilon \to 0$.

But we defined the factorial representation $\binom{r}{k} = r!/k!\,(r-k)!$ only when r is an integer! If we want to work effectively with hypergeometrics, we need a factorial function that is defined for all complex numbers. Fortunately there is such a function, and it can be defined in many ways. Here's one of the most useful definitions of $z!$, actually a definition of $1/z!$:

$$\frac{1}{z!} \;=\; \lim_{n\to\infty}\binom{n+z}{n}n^{-z}. \qquad (5.83)$$

(Euler [81] discovered this when he was 22 years old.) The limit can be shown to exist for all complex z, and it is zero only when z is a negative integer. Another significant definition is

$$z! \;=\; \int_0^\infty t^z e^{-t}\,dt\,, \qquad \text{if } \Re z > -1. \qquad (5.84)$$

This integral exists only when the real part of z exceeds -1, but we can use the formula

$$z! \;=\; z\,(z-1)! \qquad (5.85)$$

to extend (5.84) to all complex z (except negative integers). Still another definition comes from Stirling's interpolation of $\ln z!$ in (5.47). All of these approaches lead to the same generalized factorial function.

There's a very similar function called the *Gamma function*, which relates to ordinary factorials somewhat as rising powers relate to falling powers. Standard reference books often use factorials and Gamma functions simultaneously, and it's convenient to convert between them if necessary using the following formulas:

$$\Gamma(z+1) = z!\,; \tag{5.86}$$

$$(-z)!\,\Gamma(z) = \frac{\pi}{\sin \pi z}\,. \tag{5.87}$$

How do you write z to the $\overline{w}$ power, when $\overline{w}$ is the complex conjugate of w?

$z^{(\overline{w})}$.

We can use these generalized factorials to define generalized factorial powers, when z and w are arbitrary complex numbers:

$$z^{\underline{w}} = \frac{z!}{(z-w)!}\,; \tag{5.88}$$

$$z^{\overline{w}} = \frac{\Gamma(z+w)}{\Gamma(z)}\,. \tag{5.89}$$

The only proviso is that we must use appropriate limiting values when these formulas give ∞/∞. (The formulas never give $0/0$, because factorials and Gamma-function values are never zero.) A binomial coefficient can be written

$$\binom{z}{w} = \lim_{\zeta\to z}\lim_{\omega\to w}\frac{\zeta!}{\omega!\,(\zeta-\omega)!} \tag{5.90}$$

when z and w are any complex numbers whatever.

I see, the lower index arrives at its limit first. That's why $\binom{z}{w}$ is zero when w is a negative integer.

Armed with generalized factorial tools, we can return to our goal of reducing the identities derived earlier to their hypergeometric essences. The binomial theorem (5.13) turns out to be neither more nor less than (5.77), as we might expect. So the next most interesting identity to try is Vandermonde's convolution (5.27):

$$\sum_k \binom{r}{k}\binom{s}{n-k} = \binom{r+s}{n}, \qquad \text{integer } n.$$

The kth term here is

$$t_k = \frac{r!}{(r-k)!\,k!}\,\frac{s!}{(s-n+k)!\,(n-k)!}\,,$$

and we are no longer too shy to use generalized factorials in these expressions. Whenever t_k contains a factor like $(\alpha+k)!$, with a plus sign before the k, we get $(\alpha+k+1)!/(\alpha+k)! = k+\alpha+1$ in the term ratio t_{k+1}/t_k, by (5.85); this contributes the parameter '$\alpha+1$' to the corresponding hypergeometric — as an upper parameter if $(\alpha+k)!$ was in the numerator of t_k, but as a lower parameter otherwise. Similarly, a factor like $(\alpha-k)!$ leads to $(\alpha-k-1)!/(\alpha-k)! = (-1)/(k-\alpha)$; this contributes '$-\alpha$' to the opposite set of parameters (reversing the roles of upper and lower), and negates the hypergeometric argument. Factors like $r!$, which are independent of k, go into t_0 but disappear from the term ratio. Using such tricks we can predict without further calculation that the term ratio of (5.27) is

$$\frac{t_{k+1}}{t_k} = \frac{k-r}{k+1}\,\frac{k-n}{k+s-n+1}$$

times $(-1)^2 = 1$, and Vandermonde's convolution becomes

$$\binom{s}{n} F\left({-r,\,-n \atop s-n+1} \,\middle|\, 1\right) = \binom{r+s}{n}. \tag{5.91}$$

We can use this equation to determine $F(a,b;c;z)$ in general, when $z = 1$ and when b is a negative integer.

Let's rewrite (5.91) in a form so that table lookup is easy when a new sum needs to be evaluated. The result turns out to be

$$F\left({a,b \atop c} \,\middle|\, 1\right) = \frac{\Gamma(c-a-b)\,\Gamma(c)}{\Gamma(c-a)\,\Gamma(c-b)}; \qquad \begin{matrix}\text{integer } b \leqslant 0\\ \text{or } \Re c > \Re a + \Re b.\end{matrix} \tag{5.92}$$

Vandermonde's convolution (5.27) covers only the case that one of the upper parameters, say b, is a nonpositive integer; but Gauss proved that (5.92) is valid also when a, b, c are complex numbers whose real parts satisfy $\Re c > \Re a + \Re b$. In other cases, the infinite series $F\left({a,b \atop c}\middle| 1\right)$ doesn't converge. When $b = -n$, the identity can be written more conveniently with factorial powers instead of Gamma functions:

$$F\left({a,\,-n \atop c} \,\middle|\, 1\right) = \frac{(c-a)^{\overline{n}}}{c^{\overline{n}}} = \frac{(a-c)^{\underline{n}}}{(-c)^{\underline{n}}}, \qquad \text{integer } n \geqslant 0. \tag{5.93}$$

It turns out that all five of the identities in Table 169 are special cases of Vandermonde's convolution; formula (5.93) covers them all, when proper attention is paid to degenerate situations.

Notice that (5.82) is just the special case $a = 1$ of (5.93). Therefore we don't really need to remember (5.82); and we don't really need the identity (5.9) that led us to (5.82), even though Table 174 said that it was memorable. A computer program for formula manipulation, faced with the problem of evaluating $\sum_{k\leqslant n}\binom{r+k}{k}$, could convert the sum to a hypergeometric and plug into the general identity for Vandermonde's convolution.

Problem 1 in Section 5.2 asked for the value of

$$\sum_{k\geqslant 0}\binom{m}{k}\Big/\binom{n}{k}.$$

This problem is a natural for hypergeometrics, and after a bit of practice any hypergeometer can read off the parameters immediately as $F(1,-m;-n;1)$. Hmmm; that problem was yet another special takeoff on Vandermonde!

The sum in Problem 2 and Problem 4 likewise yields $F(2, 1-n; 2-m; 1)$. (We need to replace k by $k+1$ first.) And the "menacing" sum in Problem 6 turns out to be just $F(n+1, -n; 2; 1)$. Is there nothing more to sum, besides disguised versions of Vandermonde's powerful convolution?

Well, yes, Problem 3 is a bit different. It deals with a special case of the general sum $\sum_k \binom{n-k}{k} z^k$ considered in (5.74), and this leads to a closed-form expression for

$$F\left({1+2\lceil n/2\rceil,\ -n \atop 1/2}\,\middle|\, -z/4\right).$$

We also proved something new in (5.55), when we looked at the coefficients of $(1-z)^r(1+z)^r$:

$$F\left({1-c-2n,\ -2n \atop c}\,\middle|\, -1\right) = (-1)^n \frac{(2n)!}{n!}\frac{(c-1)!}{(c+n-1)!}\,, \qquad \text{integer } n \geqslant 0.$$

Kummer was a summer.

This is called *Kummer's formula* when it's generalized to complex numbers:

$$F\left({a,\ b \atop 1+b-a}\,\middle|\, -1\right) = \frac{(b/2)!}{b!}(b-a)^{\underline{b/2}}. \tag{5.94}$$

The summer of '36.

(Ernst Kummer [187] proved this in 1836.)

It's interesting to compare these two formulas. Replacing c by $1-2n-a$, we find that the results are consistent if and only if

$$(-1)^n \frac{(2n)!}{n!} = \lim_{b\to -2n} \frac{(b/2)!}{b!} = \lim_{x\to -n} \frac{x!}{(2x)!} \tag{5.95}$$

when n is a positive integer. Suppose, for example, that $n = 3$; then we should have $-6!/3! = \lim_{x\to -3} x!/(2x)!$. We know that $(-3)!$ and $(-6)!$ are both infinite; but we might choose to ignore that difficulty and to imagine that $(-3)! = (-3)(-4)(-5)(-6)!$, so that the two occurrences of $(-6)!$ will cancel. Such temptations must, however, be resisted, because they lead to the wrong answer! The limit of $x!/(2x)!$ as $x \to -3$ is not $(-3)(-4)(-5)$ but rather $-6!/3! = (-4)(-5)(-6)$, according to (5.95).

The right way to evaluate the limit in (5.95) is to use equation (5.87), which relates negative-argument factorials to positive-argument Gamma functions. If we replace x by $-n+\epsilon$ and let $\epsilon \to 0$, two applications of (5.87) give

$$\frac{(-n-\epsilon)!}{(-2n-2\epsilon)!}\frac{\Gamma(n+\epsilon)}{\Gamma(2n+2\epsilon)} = \frac{\sin(2n+2\epsilon)\pi}{\sin(n+\epsilon)\pi}\,.$$

Now $\sin(x+y) = \sin x \cos y + \cos x \sin y$; so this ratio of sines is

$$\frac{\cos 2n\pi \sin 2\epsilon\pi}{\cos n\pi \sin \epsilon\pi} = (-1)^n\bigl(2+O(\epsilon)\bigr),$$

by the methods of Chapter 9. Therefore, by (5.86), we have

$$\lim_{\epsilon\to 0}\frac{(-n-\epsilon)!}{(-2n-2\epsilon)!} = 2(-1)^n\frac{\Gamma(2n)}{\Gamma(n)} = 2(-1)^n\frac{(2n-1)!}{(n-1)!} = (-1)^n\frac{(2n)!}{n!},$$

as desired.

Let's complete our survey by restating the other identities we've seen so far in this chapter, clothing them in hypergeometric garb. The triple-binomial sum in (5.29) can be written

$$F\left({1-a-2n,\ 1-b-2n,\ -2n \atop a,\ b}\middle|\,1\right) = (-1)^n\frac{(2n)!}{n!}\frac{(a+b+2n-2)^{\overline{n}}}{a^{\overline{n}}\,b^{\overline{n}}},\qquad \text{integer } n \geqslant 0.$$

When this one is generalized to complex numbers, it is called *Dixon's formula*:

$$F\left({a,\ b,\ c \atop 1+c-a,\ 1+c-b}\middle|\,1\right) = \frac{(c/2)!}{c!}\frac{(c-a)^{\underline{c/2}}(c-b)^{\underline{c/2}}}{(c-a-b)^{\underline{c/2}}},\qquad \Re a + \Re b < 1 + \Re c/2. \tag{5.96}$$

One of the most general formulas we've encountered is the triple-binomial sum (5.28), which yields *Saalschütz's identity*:

$$F\left({a,\ b,\ -n \atop c,\ a+b-c-n+1}\middle|\,1\right) = \frac{(c-a)^{\overline{n}}(c-b)^{\overline{n}}}{c^{\overline{n}}(c-a-b)^{\overline{n}}} = \frac{(a-c)^{\underline{n}}(b-c)^{\underline{n}}}{(-c)^{\underline{n}}(a+b-c)^{\underline{n}}},\qquad \text{integer } n \geqslant 0. \tag{5.97}$$

This formula gives the value at $z = 1$ of the general hypergeometric series with three upper parameters and two lower parameters, provided that one of the upper parameters is a nonpositive integer and that $b_1 + b_2 = a_1 + a_2 + a_3 + 1$. (If the sum of the lower parameters exceeds the sum of the upper parameters by 2 instead of by 1, the formula of exercise 25 can be used to express $F(a_1, a_2, a_3; b_1, b_2; 1)$ in terms of two hypergeometrics that satisfy Saalschütz's identity.)

Our hard-won identity in Problem 8 of Section 5.2 reduces to

$$\frac{1}{1+x}F\left({x+1,\ n+1,\ -n \atop 1,\ x+2}\middle|\,1\right) = (-1)^n x^{\underline{n}} x^{\underline{-n-1}}.$$

Sigh. This is just the special case $c = 1$ of Saalschütz's identity (5.97), so we could have saved a lot of work by going to hypergeometrics directly!

What about Problem 7? That extra-menacing sum gives us the formula

$$F\left({n+1,\ m-n,\ 1,\ \frac{1}{2} \atop \frac{1}{2}m+1,\ \frac{1}{2}m+\frac{1}{2},\ 2}\,\middle|\,1\right) = \frac{m}{n},$$

which is the first case we've seen with three lower parameters. So it looks new. But it really isn't; the left-hand side can be replaced by

$$F\left({n,\ m-n-1,\ -\frac{1}{2} \atop \frac{1}{2}m,\ \frac{1}{2}m-\frac{1}{2}}\,\middle|\,1\right) - 1,$$

using exercise 26, and Saalschütz's identity wins again.

(Historical note: The great relevance of hypergeometric series to binomial coefficient identities was first pointed out by George Andrews in 1974 [9, section 5].)

Well, that's another deflating experience, but it's also another reason to appreciate the power of hypergeometric methods.

The convolution identities in Table 202 do not have hypergeometric equivalents, because their term ratios are rational functions of k only when t is an integer. Equations (5.64) and (5.65) aren't hypergeometric even when $t = 1$. But we can take note of what (5.62) tells us when t has small integer values:

$$F\left({\frac{1}{2}r,\ \frac{1}{2}r+\frac{1}{2},\ -n,\ -n-s \atop r+1,\ -n-\frac{1}{2}s,\ -n-\frac{1}{2}s+\frac{1}{2}}\,\middle|\,1\right) = \binom{r+s+2n}{n}\Big/\binom{s+2n}{n};$$

$$F\left({\frac{1}{3}r,\ \frac{1}{3}r+\frac{1}{3},\ \frac{1}{3}r+\frac{2}{3},\ -n,\ -n-\frac{1}{2}s,\ -n-\frac{1}{2}s-\frac{1}{2} \atop \frac{1}{2}r+\frac{1}{2},\ \frac{1}{2}r+1,\ -n-\frac{1}{3}s,\ -n-\frac{1}{3}s+\frac{1}{3},\ -n-\frac{1}{3}s+\frac{2}{3}}\,\middle|\,1\right)$$
$$= \binom{r+s+3n}{n}\Big/\binom{s+3n}{n}.$$

The first of these formulas gives the result of Problem 7 again, when the quantities (r, s, n) are replaced respectively by $(1, 2n+1-m, -1-n)$.

Finally, the "unexpected" sum (5.20) gives us an unexpected hypergeometric identity that turns out to be quite instructive. Let's look at it in slow motion. First we convert to an infinite sum,

$$\sum_{k\leqslant m}\binom{m+k}{k}2^{-k} = 2^m \quad\Longleftrightarrow\quad \sum_{k\geqslant 0}\binom{2m-k}{m-k}2^k = 2^{2m}.$$

The term ratio from $(2m-k)!\,2^k/m!\,(m-k)!$ is $2(k-m)/(k-2m)$, so we have a hypergeometric identity with $z = 2$:

$$\binom{2m}{m}F\left({1,\ -m \atop -2m}\,\middle|\,2\right) = 2^{2m}, \qquad \text{integer } m \geqslant 0. \tag{5.98}$$

But look at the lower parameter '$-2m$'. Negative integers are verboten, so this identity is undefined!

It's high time to look at such limiting cases carefully, as promised earlier, because degenerate hypergeometrics can often be evaluated by approaching them from nearby nondegenerate points. We must be careful when we do this, because different results can be obtained if we take limits in different ways. For example, here are two limits that turn out to be quite different when one of the upper parameters is increased by ϵ:

$$\begin{aligned}
\lim_{\epsilon\to 0} F\left({-1+\epsilon,\,-3 \atop -2+\epsilon}\,\middle|\,1\right) &= \lim_{\epsilon\to 0}\Bigl(1+\tfrac{(-1+\epsilon)(-3)}{(-2+\epsilon)\,1!}+\tfrac{(-1+\epsilon)(\epsilon)(-3)(-2)}{(-2+\epsilon)(-1+\epsilon)\,2!}\\
&\qquad\qquad +\tfrac{(-1+\epsilon)(\epsilon)(1+\epsilon)(-3)(-2)(-1)}{(-2+\epsilon)(-1+\epsilon)(\epsilon)\,3!}\Bigr)\\
&= 1-\tfrac{3}{2}+0+\tfrac{1}{2} = 0\,;\\
\lim_{\epsilon\to 0} F\left({-1,-3 \atop -2+\epsilon}\,\middle|\,1\right) &= \lim_{\epsilon\to 0}\Bigl(1+\tfrac{(-1)(-3)}{(-2+\epsilon)\,1!}+0+0\Bigr)\\
&= 1-\tfrac{3}{2}+0+0 = -\tfrac{1}{2}\,.
\end{aligned}$$

Similarly, we have defined $\binom{-1}{-1} = 0 = \lim_{\epsilon\to 0}\binom{-1+\epsilon}{-1}$; this is not the same as $\lim_{\epsilon\to 0}\binom{-1+\epsilon}{-1+\epsilon} = 1$. The proper way to treat (5.98) as a limit is to realize that the upper parameter $-m$ is being used to make all terms of the series $\sum_{k\geqslant 0}\binom{2m-k}{m-k}2^k$ zero for $k > m$; this means that we want to make the following more precise statement:

$$\binom{2m}{m}\lim_{\epsilon\to 0} F\left({1,\,-m \atop -2m+\epsilon}\,\middle|\,2\right) = 2^{2m}\,, \qquad \text{integer } m \geqslant 0. \tag{5.99}$$

Each term of this limit is well defined, because the denominator factor $(-2m)^{\overline{k}}$ does not become zero until $k > 2m$. Therefore this limit gives us exactly the sum (5.20) we began with.

5.6 HYPERGEOMETRIC TRANSFORMATIONS

It should be clear by now that a database of known hypergeometric closed forms is a useful tool for doing sums of binomial coefficients. We simply convert any given sum into its canonical hypergeometric form, then look it up in the table. If it's there, fine, we've got the answer. If not, we can add it to the database if the sum turns out to be expressible in closed form. We might also include entries in the table that say, "This sum does not have a simple closed form in general." For example, the sum $\sum_{k\leqslant m}\binom{n}{k}$ corresponds to the hypergeometric

$$F\left({1,\,-m \atop n-m+1}\,\middle|\,-1\right), \qquad \text{integers } n \geqslant m \geqslant 0; \tag{5.100}$$

this has a simple closed form only if m is near 0, $\frac{1}{2}n$, or n.

But there's more to the story, since hypergeometric functions also obey identities of their own. This means that every closed form for hypergeometrics leads to additional closed forms and to additional entries in the database. For example, the identities in exercises 25 and 26 tell us how to transform one hypergeometric into two others with similar but different parameters. These can in turn be transformed again.

The hypergeometric database should really be a "knowledge base."

In 1793, J. F. Pfaff discovered a surprising *reflection law*,

$$\frac{1}{(1-z)^a}\,F\left({a,b \atop c}\,\middle|\,\frac{-z}{1-z}\right) = F\left({a,\,c-b \atop c}\,\middle|\,z\right), \tag{5.101}$$

which is a transformation of another type. This is a formal identity in power series, if the quantity $(-z)^k/(1-z)^{k+a}$ is replaced by the infinite series $(-z)^k\bigl(1+\binom{k+a}{1}z+\binom{k+a+1}{2}z^2+\cdots\bigr)$ when the left-hand side is expanded (see exercise 50). We can use this law to derive new formulas from the identities we already know, when $z \neq 1$.

For example, Kummer's formula (5.94) can be combined with the reflection law (5.101) if we choose the parameters so that both identities apply:

$$\begin{aligned} 2^{-a}\,F\left({a,\,1-a \atop 1+b-a}\,\middle|\,\frac{1}{2}\right) &= F\left({a,\,b \atop 1+b-a}\,\middle|\,-1\right) \\ &= \frac{(b/2)!}{b!}(b-a)^{\underline{b/2}}. \end{aligned} \tag{5.102}$$

We can now set $a=-n$ and go back from this equation to a new identity in binomial coefficients that we might need some day:

$$\begin{aligned} \sum_{k\geqslant 0}\frac{(-n)^{\overline{k}}\,(1+n)^{\overline{k}}}{(1+b+n)^{\overline{k}}}\,\frac{2^{-k}}{k!} &= \sum_k \binom{n}{k}\left(\frac{-1}{2}\right)^k \binom{n+k}{k}\Big/\binom{n+b+k}{k} \\ &= 2^{-n}\,\frac{(b/2)!\,(b+n)!}{b!\,(b/2+n)!}, \qquad \text{integer } n \geqslant 0. \end{aligned} \tag{5.103}$$

For example, when $n=3$ this identity says that

$$\begin{aligned} 1 - 3\,\frac{4}{2(4+b)} + 3\,\frac{4\cdot 5}{4(4+b)(5+b)} &- \frac{4\cdot 5\cdot 6}{8(4+b)(5+b)(6+b)} \\ &= \frac{(b+3)(b+2)(b+1)}{(b+6)(b+4)(b+2)}. \end{aligned}$$

It's almost unbelievable, but true, for all b. (Except when a factor in the denominator vanishes.)

This is fun; let's try again. Maybe we'll find a formula that will really astonish our friends. What does Pfaff's reflection law tell us if we apply it to the strange form (5.99), where $z = 2$? In this case we set $a = -m$, $b = 1$, and $c = -2m + \epsilon$, obtaining

$$
\begin{aligned}
(-1)^m \lim_{\epsilon \to 0} F\left({-m, 1 \atop -2m+\epsilon} \middle| 2\right) &= \lim_{\epsilon \to 0} F\left({-m, -2m-1+\epsilon \atop -2m+\epsilon} \middle| 2\right) \\
&= \lim_{\epsilon \to 0} \sum_{k \geqslant 0} \frac{(-m)^{\overline{k}}\,(-2m-1+\epsilon)^{\overline{k}}}{(-2m+\epsilon)^{\overline{k}}} \frac{2^k}{k!} \\
&= \sum_{k \leqslant m} \binom{m}{k} \frac{(2m+1)^{\underline{k}}}{(2m)^{\underline{k}}} (-2)^k ,
\end{aligned}
$$

because none of the limiting terms is close to zero. This leads to another miraculous formula,

$$
\begin{aligned}
\sum_{k \leqslant m} \binom{m}{k} \frac{2m+1}{2m+1-k} (-2)^k &= (-1)^m 2^{2m} \Big/ \binom{2m}{m} \\
&= 1 \Big/ \binom{-1/2}{m}, \qquad \text{integer } m \geqslant 0. \qquad (5.104)
\end{aligned}
$$

When $m = 3$, for example, the sum is

$$
1 - 7 + \frac{84}{5} - 14 = -\frac{16}{5},
$$

and $\binom{-1/2}{3}$ is indeed equal to $-\frac{5}{16}$.

When we looked at our binomial coefficient identities and converted them to hypergeometric form, we overlooked (5.19) because it was a relation between two sums instead of a closed form. But now we can regard (5.19) as an identity between hypergeometric series. If we differentiate it n times with respect to y and then replace k by $m - n - k$, we get

$$
\begin{aligned}
&\sum_{k \geqslant 0} \binom{m+r}{m-n-k} \binom{n+k}{n} x^{m-n-k} y^k \\
&\qquad = \sum_{k \geqslant 0} \binom{-r}{m-n-k} \binom{n+k}{n} (-x)^{m-n-k} (x+y)^k .
\end{aligned}
$$

This yields the following hypergeometric transformation:

$$
F\left({a, -n \atop c} \middle| z\right) = \frac{(a-c)^{\underline{n}}}{(-c)^{\underline{n}}} F\left({a, -n \atop 1-n+a-c} \middle| 1-z\right), \qquad \text{integer } n \geqslant 0. \qquad (5.105)
$$

Notice that when $z = 1$ this reduces to Vandermonde's convolution, (5.93).

Differentiation seems to be useful, if this example is any indication; we also found it helpful in Chapter 2, when summing $x + 2x^2 + \cdots + nx^n$. Let's see what happens when a general hypergeometric series is differentiated with respect to z:

$$\begin{aligned}\frac{d}{dz} F\left({a_1, \ldots, a_m \atop b_1, \ldots, b_n} \middle| z\right) &= \sum_{k \geq 1} \frac{a_1^{\overline{k}} \ldots a_m^{\overline{k}} \, z^{k-1}}{b_1^{\overline{k}} \ldots b_n^{\overline{k}} \, (k-1)!} \\ &= \sum_{k+1 \geq 1} \frac{a_1^{\overline{k+1}} \ldots a_m^{\overline{k+1}} \, z^k}{b_1^{\overline{k+1}} \ldots b_n^{\overline{k+1}} \, k!} \\ &= \sum_{k \geq 0} \frac{a_1 (a_1{+}1)^{\overline{k}} \ldots a_m (a_m{+}1)^{\overline{k}} \, z^k}{b_1 (b_1{+}1)^{\overline{k}} \ldots b_n (b_n{+}1)^{\overline{k}} \, k!} \\ &= \frac{a_1 \ldots a_m}{b_1 \ldots b_n} F\left({a_1{+}1, \ldots, a_m{+}1 \atop b_1{+}1, \ldots, b_n{+}1} \middle| z\right). \end{aligned} \tag{5.106}$$

The parameters move out and shift up.

It's also possible to use differentiation to tweak just one of the parameters while holding the rest of them fixed. For this we use the operator

How do you pronounce ϑ?

(Dunno, but TeX calls it 'vartheta'.)

$$\vartheta = z \frac{d}{dz},$$

which acts on a function by differentiating it and then multiplying by z. This operator gives

$$\vartheta F\left({a_1, \ldots, a_m \atop b_1, \ldots, b_n} \middle| z\right) = z \sum_{k \geq 1} \frac{a_1^{\overline{k}} \ldots a_m^{\overline{k}} \, z^{k-1}}{b_1^{\overline{k}} \ldots b_n^{\overline{k}} \, (k-1)!} = \sum_{k \geq 0} \frac{k \, a_1^{\overline{k}} \ldots a_m^{\overline{k}} \, z^k}{b_1^{\overline{k}} \ldots b_n^{\overline{k}} \, k!},$$

which by itself isn't too useful. But if we multiply F by one of its upper parameters, say a_1, and add ϑF, we get

$$\begin{aligned}(\vartheta + a_1) F\left({a_1, \ldots, a_m \atop b_1, \ldots, b_n} \middle| z\right) &= \sum_{k \geq 0} \frac{(k + a_1) a_1^{\overline{k}} \ldots a_m^{\overline{k}} \, z^k}{b_1^{\overline{k}} \ldots b_n^{\overline{k}} \, k!}, \\ &= \sum_{k \geq 0} \frac{a_1 (a_1{+}1)^{\overline{k}} \, a_2^{\overline{k}} \ldots a_m^{\overline{k}} \, z^k}{b_1^{\overline{k}} \ldots b_n^{\overline{k}} \, k!} \\ &= a_1 F\left({a_1{+}1, \, a_2, \, \ldots, \, a_m \atop b_1, \, \ldots, \, b_n} \middle| z\right). \end{aligned}$$

Only one parameter has been shifted.

A similar trick works with lower parameters, but in this case things shift down instead of up:

$$(\vartheta + b_1 - 1)\,F\left({a_1,\ldots,a_m \atop b_1,\ldots,b_n}\,\middle|\,z\right) = \sum_{k\geq 0} \frac{(k+b_1-1)a_1^{\overline{k}}\ldots a_m^{\overline{k}}\,z^k}{b_1^{\overline{k}}\ldots b_n^{\overline{k}}\,k!}\,,$$

$$= \sum_{k\geq 0} \frac{(b_1-1)\,a_1^{\overline{k}}\ldots a_m^{\overline{k}}\,z^k}{(b_1-1)^{\overline{k}}\,b_2^{\overline{k}}\ldots b_n^{\overline{k}}\,k!}$$

$$= (b_1-1)\,F\left({a_1,\;\ldots,\;a_m \atop b_1-1,\;b_2,\;\ldots,\;b_n}\,\middle|\,z\right).$$

We can now combine all these operations and make a mathematical "pun" by expressing the same quantity in two different ways. Namely, we have

Ever hear the one about the brothers who named their cattle ranch Focus, because it's where the sons raise meat?

$$(\vartheta + a_1)\ldots(\vartheta + a_m)F = a_1\ldots a_m\;F\left({a_1+1,\;\ldots,\;a_m+1 \atop b_1,\;\ldots,\;b_n}\,\middle|\,z\right),$$

and

$$(\vartheta + b_1 - 1)\ldots(\vartheta + b_n - 1)F$$
$$= (b_1-1)\ldots(b_n-1)\,F\left({a_1,\;\ldots,\;a_m \atop b_1-1,\;\ldots,\;b_n-1}\,\middle|\,z\right),$$

where $F = F(a_1,\ldots,a_m;b_1,\ldots,b_n;z)$. And (5.106) tells us that the top line is the derivative of the bottom line. Therefore the general hypergeometric function F satisfies the differential equation

$$D(\vartheta + b_1 - 1)\ldots(\vartheta + b_n - 1)F = (\vartheta + a_1)\ldots(\vartheta + a_m)F\,, \tag{5.107}$$

where D is the operator $\frac{d}{dz}$.

This cries out for an example. Let's find the differential equation satisfied by the standard 2-over-1 hypergeometric series $F(z) = F(a,b;c;z)$. According to (5.107), we have

$$D(\vartheta + c - 1)F = (\vartheta + a)(\vartheta + b)F\,.$$

What does this mean in ordinary notation? Well, $(\vartheta + c - 1)F$ is $zF'(z) + (c-1)F(z)$, and the derivative of this gives the left-hand side,

$$F'(z) + zF''(z) + (c-1)F'(z)\,.$$

On the right-hand side we have

$$\begin{aligned}(\vartheta+a)\bigl(zF'(z)+bF(z)\bigr) &= z\frac{d}{dz}\bigl(zF'(z)+bF(z)\bigr) + a\bigl(zF'(z)+bF(z)\bigr)\\ &= zF'(z)+z^2F''(z)+bzF'(z)+azF'(z)+abF(z)\,.\end{aligned}$$

Equating the two sides tells us that

$$z(1-z)F''(z) + \bigl(c - z(a+b+1)\bigr)F'(z) - abF(z) \;=\; 0\,. \tag{5.108}$$

This equation is equivalent to the factored form (5.107).

Conversely, we can go back from the differential equation to the power series. Let's assume that $F(z) = \sum_{k\geqslant 0} t_k z^k$ is a power series satisfying (5.107). A straightforward calculation shows that we must have

$$\frac{t_{k+1}}{t_k} \;=\; \frac{(k+a_1)\ldots(k+a_m)}{(k+b_1)\ldots(k+b_n)(k+1)}\,;$$

hence $F(z)$ must be $t_0\,F(a_1,\ldots,a_m;b_1,\ldots,b_n;z)$. We've proved that the hypergeometric series (5.76) is the only formal power series that satisfies the differential equation (5.107) and has the constant term 1.

It would be nice if hypergeometrics solved all the world's differential equations, but they don't quite. The right-hand side of (5.107) always expands into a sum of terms of the form $\alpha_k z^k F^{(k)}(z)$, where $F^{(k)}(z)$ is the kth derivative $D^kF(k)$; the left-hand side always expands into a sum of terms of the form $\beta_k z^{k-1}F^{(k)}(z)$ with $k>0$. So the differential equation (5.107) always takes the special form

$$z^{n-1}(\beta_n - z\alpha_n)F^{(n)}(z) + \cdots + (\beta_1 - z\alpha_1)F'(z) - \alpha_0F(z) \;=\; 0\,.$$

Equation (5.108) illustrates this in the case $n=2$. Conversely, we will prove in exercise 6.13 that any differential equation of this form can be factored in terms of the ϑ operator, to give an equation like (5.107). So these are the differential equations whose solutions are power series with rational term ratios.

Multiplying both sides of (5.107) by z dispenses with the D operator and gives us an instructive all-ϑ form,

$$\vartheta(\vartheta+b_1-1)\ldots(\vartheta+b_n-1)F \;=\; z(\vartheta+a_1)\ldots(\vartheta+a_m)F\,. \tag{5.109}$$

The first factor $\vartheta = (\vartheta+1-1)$ on the left corresponds to the $(k+1)$ in the term ratio (5.81), which corresponds to the $k!$ in the denominator of the kth term in a general hypergeometric series. The other factors $(\vartheta+b_j-1)$ correspond to the denominator factor $(k+b_j)$, which corresponds to $b_j^{\overline{k}}$ in (5.76). On the right, the z corresponds to z^k, and $(\vartheta+a_j)$ corresponds to $a_j^{\overline{k}}$.

One use of this differential theory is to find and prove new transformations. For example, we can readily verify that both of the hypergeometrics

$$F\left({2a, 2b \atop a+b+\frac{1}{2}} \,\Big|\, z\right) \quad\text{and}\quad F\left({a, b \atop a+b+\frac{1}{2}} \,\Big|\, 4z(1-z)\right)$$

satisfy the differential equation

$$z(1-z)F''(z) + (a+b+\tfrac{1}{2})(1-2z)F'(z) - 4abF(z) = 0;$$

hence *Gauss's identity* [116, equation 102]

$$F\left({2a, 2b \atop a+b+\frac{1}{2}} \,\Big|\, z\right) = F\left({a, b \atop a+b+\frac{1}{2}} \,\Big|\, 4z(1-z)\right) \tag{5.110}$$

must be true. In particular,

$$F\left({2a, 2b \atop a+b+\frac{1}{2}} \,\Big|\, \frac{1}{2}\right) = F\left({a, b \atop a+b+\frac{1}{2}} \,\Big|\, 1\right), \tag{5.111}$$

whenever both infinite sums converge.

(Caution: We can't use (5.110) safely when $|z| > 1/2$, unless both sides are polynomials; see exercise 53.)

Every new identity for hypergeometrics has consequences for binomial coefficients, and this one is no exception. Let's consider the sum

$$\sum_{k \leqslant m} \binom{m-k}{n}\binom{m+n+1}{k}\left(\frac{-1}{2}\right)^k, \qquad \text{integers } m \geqslant n \geqslant 0.$$

The terms are nonzero for $0 \leqslant k \leqslant m-n$, and with a little delicate limit-taking as before we can express this sum as the hypergeometric

$$\lim_{\epsilon \to 0} \binom{m}{n} F\left({n-m,\ -n-m-1+\alpha\epsilon \atop -m+\epsilon} \,\Big|\, \frac{1}{2}\right).$$

The value of α doesn't affect the limit, since the nonpositive upper parameter $n-m$ cuts the sum off early. We can set $\alpha = 2$, so that (5.111) applies. The limit can now be evaluated because the right-hand side is a special case of (5.92). The result can be expressed in simplified form,

$$\sum_{k \leqslant m} \binom{m-k}{n}\binom{m+n+1}{k}\left(\frac{-1}{2}\right)^k = \binom{(m+n)/2}{n} 2^{n-m} [m+n \text{ is even}], \qquad \begin{array}{c}\text{integers}\\ m \geqslant n \geqslant 0,\end{array} \tag{5.112}$$

as shown in exercise 54. For example, when $m = 5$ and $n = 2$ we get $\binom{5}{2}\binom{8}{0} - \binom{4}{2}\binom{8}{1}/2 + \binom{3}{2}\binom{8}{2}/4 - \binom{2}{2}\binom{8}{3}/8 = 10 - 24 + 21 - 7 = 0$; when $m = 4$ and $n = 2$, both sides give $\frac{3}{4}$.

We can also find cases where (5.110) gives binomial sums when $z = -1$, but these are really weird. If we set $a = \frac{1}{6} - \frac{n}{3}$ and $b = -n$, we get the monstrous formula

$$F\left({\frac{1}{3}-\frac{2}{3}n,\ -2n \atop \frac{2}{3}-\frac{4}{3}n}\Big|\,-1\right) = F\left({\frac{1}{6}-\frac{1}{3}n,\ -n \atop \frac{2}{3}-\frac{4}{3}n}\Big|\,-8\right).$$

These hypergeometrics are nondegenerate polynomials when $n \not\equiv 2 \pmod 3$; and the parameters have been cleverly chosen so that the left-hand side can be evaluated by (5.94). We are therefore led to a truly mind-boggling result,

$$\sum_k \binom{n}{k}\binom{\frac{1}{3}n-\frac{1}{6}}{k}8^k \Big/ \binom{\frac{4}{3}n-\frac{2}{3}}{k} = \binom{2n}{n}\Big/\binom{\frac{4}{3}n-\frac{2}{3}}{n}, \quad \text{integer } n \geqslant 0,\ n \not\equiv 2 \pmod 3. \qquad (5.113)$$

This is the most startling identity in binomial coefficients that we've seen. Small cases of the identity aren't even easy to check by hand. (It turns out that both sides do give $\frac{81}{7}$ when $n = 3$.) But the identity is completely useless, of course; surely it will never arise in a practical problem.

The only use of (5.113) is to demonstrate the existence of incredibly useless identities.

So that's our hype for hypergeometrics. We've seen that hypergeometric series provide a high-level way to understand what's going on in binomial coefficient sums. A great deal of additional information can be found in the classic book by Wilfred N. Bailey [15] and its sequel by Lucy Joan Slater [269].

5.7 PARTIAL HYPERGEOMETRIC SUMS

Most of the sums we've evaluated in this chapter range over all indices $k \geqslant 0$, but sometimes we've been able to find a closed form that works over a general range $0 \leqslant k < m$. For example, we know from (5.16) that

$$\sum_{k<m} \binom{n}{k}(-1)^k = (-1)^{m-1}\binom{n-1}{m-1}, \qquad \text{integer } m. \qquad (5.114)$$

The theory in Chapter 2 gives us a nice way to understand formulas like this: If $f(k) = \Delta g(k) = g(k+1) - g(k)$, then we've agreed to write $\sum f(k)\,\delta k = g(k) + C$, and

$$\sum\nolimits_a^b f(k)\,\delta k = g(k)\Big|_a^b = g(b) - g(a).$$

Furthermore, when a and b are integers with $a \leqslant b$, we have

$$\sum\nolimits_a^b f(k)\,\delta k = \sum_{a\leqslant k<b} f(k) = g(b) - g(a).$$

Therefore identity (5.114) corresponds to the indefinite summation formula

$$\sum \binom{n}{k}(-1)^k \,\delta k = (-1)^{k-1}\binom{n-1}{k-1} + C,$$

and to the difference formula

$$\Delta\left((-1)^k\binom{n}{k}\right) = (-1)^{k+1}\binom{n+1}{k+1}.$$

It's easy to start with a function $g(k)$ and to compute $\Delta g(k) = f(k)$, a function whose sum will be $g(k)+C$. But it's much harder to start with $f(k)$ and to figure out its indefinite sum $\sum f(k)\,\delta k = g(k) + C$; this function g might not have a simple form. For example, there is apparently no simple form for $\sum \binom{n}{k}\,\delta k$; otherwise we could evaluate sums like $\sum_{k\leqslant n/3}\binom{n}{k}$, about which we're clueless.

In 1977, R. W. Gosper [124] discovered a beautiful way to decide whether a given function is indefinitely summable with respect to a general class of functions called *hypergeometric terms.* Let us write

$$F\left({a_1, \ldots, a_m \atop b_1, \ldots, b_n} \middle| z\right)_k = \frac{a_1^{\overline{k}} \ldots a_m^{\overline{k}}}{b_1^{\overline{k}} \ldots b_n^{\overline{k}}} \frac{z^k}{k!} \tag{5.115}$$

for the kth term of the hypergeometric series $F(a_1, \ldots, a_m; b_1, \ldots, b_n; z)$. We will regard $F(a_1, \ldots, a_m; b_1, \ldots, b_n; z)_k$ as a function of k, not of z. Gosper's decision procedure allows us to decide if there exist parameters c, A_1, ..., A_M, B_1, ..., B_N, and Z such that

$$\sum F\left({a_1, \ldots, a_m \atop b_1, \ldots, b_n} \middle| z\right)_k \delta k = c\,F\left({A_1, \ldots, A_M \atop B_1, \ldots, B_N} \middle| Z\right)_k + C, \tag{5.116}$$

given a_1, ..., a_m, b_1, ..., b_n, and z. We will say that a given function $F(a_1, \ldots, a_m; b_1, \ldots, b_n; z)_k$ is *summable in hypergeometric terms* if such constants c, A_1, ..., A_M, B_1, ..., B_N, Z exist.

Let's write $t(k)$ and $T(k)$ as abbreviations for $F(a_1, \ldots, a_m; b_1, \ldots, b_n; z)_k$ and $F(A_1, \ldots, A_M; B_1, \ldots, B_N; Z)_k$, respectively. The first step in Gosper's decision procedure is to express the term ratio

$$\frac{t(k+1)}{t(k)} = \frac{(k+a_1)\ldots(k+a_m)\,z}{(k+b_1)\ldots(k+b_n)(k+1)}$$

in the special form

$$\frac{t(k+1)}{t(k)} = \frac{p(k+1)}{p(k)}\,\frac{q(k)}{r(k+1)}, \tag{5.117}$$

where p, q, and r are polynomials subject to the following condition:

(Divisibility of polynomials is analogous to divisibility of integers. For example, $(k+\alpha)\backslash q(k)$ means that the quotient $q(k)/(k+\alpha)$ is a polynomial. It's well known that $(k+\alpha)\backslash q(k)$ if and only if $q(-\alpha)=0$.)

$$(k+\alpha)\backslash q(k) \quad \text{and} \quad (k+\beta)\backslash r(k) \implies \alpha-\beta \text{ is not a positive integer.} \tag{5.118}$$

This condition is easy to achieve: We start by provisionally setting $p(k)=1$, $q(k)=(k+a_1)\ldots(k+a_m)z$, and $r(k)=(k+b_1-1)\ldots(k+b_n-1)k$; then we check if (5.118) is violated. If q and r have factors $(k+\alpha)$ and $(k+\beta)$ where $\alpha-\beta=N>0$, we divide them out of q and r and replace $p(k)$ by

$$p(k)(k+\alpha-1)^{\underline{N-1}} = p(k)(k+\alpha-1)(k+\alpha-2)\ldots(k+\beta+1).$$

The new p, q, and r still satisfy (5.117), and we can repeat this process until (5.118) holds.

Our goal is to find a hypergeometric term $T(k)$ such that

$$t(k) = cT(k+1)-cT(k) \tag{5.119}$$

for some constant c. Let's write

$$cT(k) = \frac{r(k)\,s(k)\,t(k)}{p(k)}, \tag{5.120}$$

where $s(k)$ is a secret function that must be discovered somehow. Plugging (5.120) into (5.117) and (5.119) gives us the equation that $s(k)$ must satisfy:

(Exercise 55 explains why we might want to make this magic substitution.)

$$p(k) = q(k)s(k+1)-r(k)s(k). \tag{5.121}$$

If we can find $s(k)$ satisfying this recurrence, we've found $\sum t(k)\,\delta k$.

We're assuming that $T(k+1)/T(k)$ is a rational function of k. Therefore, by (5.120) and (5.119), $r(k)s(k)/p(k)=T(k)/\bigl(T(k+1)-T(k)\bigr)$ is a rational function of k, and $s(k)$ itself must be a quotient of polynomials:

$$s(k) = f(k)/g(k). \tag{5.122}$$

But in fact we can prove that $s(k)$ is itself a polynomial. For if $g(k)\neq 1$, and if $f(k)$ and $g(k)$ have no common factors, let N be the largest integer such that $(k+\beta)$ and $(k+\beta+N-1)$ both occur as factors of $g(k)$ for some complex number β. The value of N is positive, since $N=1$ always satisfies this condition. Equation (5.121) can be rewritten

$$p(k)g(k{+}1)g(k) = q(k)f(k{+}1)g(k)-r(k)g(k{+}1)f(k),$$

and if we set $k=-\beta$ and $k=-\beta-N$ we get

$$r(-\beta)g(1{-}\beta)f(-\beta) = 0 = q(-\beta{-}N)f(1{-}\beta{-}N)g(-\beta{-}N).$$

Now $f(-\beta) \neq 0$ and $f(1-\beta-N) \neq 0$, because f and g have no common roots. Also $g(1-\beta) \neq 0$ and $g(-\beta-N) \neq 0$, because $g(k)$ would otherwise contain the factor $(k+\beta-1)$ or $(k+\beta+N)$, contrary to the maximality of N. Therefore

$$r(-\beta) \;=\; q(-\beta-N) \;=\; 0\,.$$

But this contradicts condition (5.118). Hence $s(k)$ must be a polynomial.

The remaining task is to decide whether there exists a polynomial $s(k)$ satisfying (5.121), when $p(k)$, $q(k)$, and $r(k)$ are given polynomials. It's easy to decide this for polynomials of any particular degree d, since we can write

$$s(k) \;=\; \alpha_d k^d + \alpha_{d-1}k^{d-1} + \cdots + \alpha_0\,, \qquad \alpha_d \neq 0$$

for unknown coefficients $(\alpha_d, \ldots, \alpha_0)$ and plug this expression into the defining equation. The polynomial $s(k)$ will satisfy the recurrence if and only if the α's satisfy certain linear equations, because each power of k must have the same coefficient on both sides of (5.121).

But how can we determine the degree of s? It turns out that there actually are at most two possibilities. We can rewrite (5.121) in the form

$$\begin{aligned} 2p(k) \;&=\; Q(k)\big(s(k+1)+s(k)\big) + R(k)\big(s(k+1)-s(k)\big)\,, \\ \text{where}\quad Q(k) \;&=\; q(k)-r(k) \quad\text{and}\quad R(k) \;=\; q(k)+r(k)\,. \end{aligned} \tag{5.123}$$

If $s(k)$ has degree d, then the sum $s(k+1)+s(k) = 2\alpha_d k^d + \cdots$ also has degree d, while the difference $s(k+1)-s(k) = \Delta s(k) = d\alpha_d k^{d-1} + \cdots$ has degree $d-1$. (The zero polynomial can be assumed to have degree -1.) Let's write $\deg(p)$ for the degree of a polynomial p. If $\deg(Q) \geqslant \deg(R)$, then the degree of the right-hand side of (5.123) is $\deg(Q)+d$, so we must have $d = \deg(p) - \deg(Q)$. On the other hand if $\deg(Q) < \deg(R) = d'$, we can write $Q(k) = \beta k^{d'-1} + \cdots$ and $R(k) = \gamma k^{d'} + \cdots$ where $\gamma \neq 0$; the right-hand side of (5.123) has the form

$$(2\beta\alpha_d + \gamma d\,\alpha_d)k^{d+d'-1} + \cdots\,.$$

Ergo, two possibilities: Either $2\beta+\gamma d \neq 0$, and $d = \deg(p) - \deg(R) + 1$; or $2\beta + \gamma d = 0$, and $d > \deg(p) - \deg(R) + 1$. The second case needs to be examined only if $-2\beta/\gamma$ is an integer d greater than $\deg(p) - \deg(R) + 1$.

Thus we have enough facts to decide if a suitable polynomial $s(k)$ exists. If so, we can plug it into (5.120) and we have our T. If not, we've proved that $\sum t(k)\,\delta k$ is not a hypergeometric term.

Time for an example. Let's try the partial sum (5.114); Gosper's method should be able to deduce the value of

$$\sum \binom{n}{k}(-1)^k\,\delta k$$

for any fixed n. Ignoring factors that don't involve k, we want the sum of

$$t(k) = F\left({1,-n \atop 1}\,\Big|\,1\right)_k .$$

The first step is to put the term ratio into the required form (5.117); we have

$$\frac{t(k+1)}{t(k)} = \frac{(k-n)}{(k+1)} = \frac{p(k+1)\,q(k)}{p(k)\,r(k+1)}$$

Why isn't it $r(k) = k+1$? Oh, I see.

so we simply take $p(k) = 1$, $q(k) = k-n$, and $r(k) = k$. This choice of p, q, and r satisfies (5.118), unless n is a negative integer; let's suppose it isn't. According to (5.123), we should consider the polynomials $Q(k) = -n$ and $R(k) = 2k-n$. Since R has larger degree than Q, we need to look at two cases. Either $d = \deg(p) - \deg(R) + 1$, which is 0; or $d = -2\beta/\gamma$ where $\beta = -n$ and $\gamma = 2$, hence $d = n$. The first case is nicer, so let's try it first: Equation (5.121) is

$$1 = (k-n)\alpha_0 - k\alpha_0$$

and so we choose $\alpha_0 = -1/n$. This satisfies the required conditions and gives

$$\begin{aligned} cT(k) &= \frac{r(k)\,s(k)\,t(k)}{p(k)} \\ &= k\cdot\frac{(-1)}{n}\cdot\binom{n}{k}(-1)^k \\ &= \binom{n-1}{k-1}(-1)^{k-1}, \end{aligned}$$

which is the answer we were hoping to confirm.

If we apply the same method to find the indefinite sum $\sum\binom{n}{k}\,\delta k$, without the $(-1)^k$, everything will be almost the same except that $q(k)$ will be $n-k$; hence $Q(k) = n-2k$ will have greater degree than $R(k) = n$, and we will conclude that d has the impossible value $\deg(p) - \deg(Q) = -1$. Therefore the function $\binom{n}{k}$ is not summable in hypergeometric terms.

However, once we have eliminated the impossible, whatever remains — however improbable — must be the truth (according to S. Holmes [70]). When we defined p, q, and r we decided to ignore the possibility that n might be a negative integer. What if it is? Let's set $n = -N$, where N is positive. Then the term ratio for $\sum \binom{n}{k} \delta k$ is

$$\frac{t(k+1)}{t(k)} = \frac{-(k+N)}{(k+1)} = \frac{p(k+1)}{p(k)} \frac{q(k)}{r(k+1)}$$

and it should be represented by $p(k) = (k+1)^{\overline{N-1}}$, $q(k) = -1$, $r(k) = 1$. Gosper's method now tells us to look for a polynomial $s(k)$ of degree $d = N-1$; maybe there's hope after all. For example, when $N = 2$ we want to solve

$$k+1 = -\bigl((k+1)\alpha_1 + \alpha_0\bigr) - (k\alpha_1 + \alpha_0)\,.$$

Equating coefficients of k and 1 tells us that

$$1 = -\alpha_1 - \alpha_1; \qquad 1 = -\alpha_1 - \alpha_0 - \alpha_0;$$

hence $s(k) = -\frac{1}{2}k - \frac{1}{4}$ is a solution, and

$$cT(k) = \frac{1 \cdot \left(-\frac{1}{2}k - \frac{1}{4}\right) \cdot \binom{-2}{k}}{k+1} = (-1)^{k-1}\frac{2k+1}{4}\,.$$

Can this be the desired sum? Yes, it checks out:

"Excellent, Holmes!"
"Elementary, my dear Watson."

$$(-1)^k\frac{2k+3}{4} - (-1)^{k-1}\frac{2k+1}{4} = (-1)^k(k+1) = \binom{-2}{k}\,.$$

We can write the summation formula in another form,

$$\begin{aligned}\sum_{k<m}\binom{-2}{k} &= (-1)^{k-1}\frac{2k+1}{4}\bigg|_0^m \\ &= \frac{(-1)^{m-1}}{2}\left(m + \frac{1-(-1)^m}{2}\right) \\ &= (-1)^{m-1}\left\lceil\frac{m}{2}\right\rceil\,.\end{aligned}$$

This representation conceals the fact that $\binom{-2}{k}$ is summable in hypergeometric terms, because $\lceil m/2 \rceil$ is not a hypergeometric term.

A catalog of summable hypergeometric terms makes a useful addition to the database of hypergeometric sums mentioned earlier in this chapter. Let's try to compile a list of the sums-in-hypergeometric-terms that we know. The geometric series $\sum z^k \delta k$ is a very special case, which can be written $\sum z^k \delta k = (z-1)^{-1}z^k + C$ or

$$\sum F\left({1,1 \atop 1}\bigg|\, z\right)_k \delta k = \frac{1}{z-1} F\left({1,1 \atop 1}\bigg|\, z\right)_k + C\,. \tag{5.124}$$

We also computed $\sum kz^k\,\delta k$ in Chapter 2. This summand is zero when $k = 0$, so we get a more suitable hypergeometric term by considering the sum $\sum(k+1)z^k\,\delta k$ instead. The appropriate formula turns out to be

$$\sum F\left(\begin{matrix}2,1\\1\end{matrix}\middle|\,z\right)_k \delta k = \frac{-1}{(1-z)^2}\,F\left(\begin{matrix}1+1/(1-z),\,1\\1/(1-z)\end{matrix}\middle|\,z\right)_k \tag{5.125}$$

in hypergeometric notation.

There's also the formula $\sum\binom{k}{n}\delta k = \binom{k}{n+1}$, equation (5.10); we write it $\sum\binom{k+n+1}{n}\delta k = \binom{k+n+1}{n+1}$, to avoid division by zero, and get

$$\sum F\left(\begin{matrix}n+2,1\\2\end{matrix}\middle|\,1\right)_k \delta k = \frac{1}{n+1}\,F\left(\begin{matrix}n+2,1\\1\end{matrix}\middle|\,1\right)_k, \qquad n \ne -1. \tag{5.126}$$

Identity (5.9) turns out to be equivalent to this, when we express it hypergeometrically.

In general if we have a summation formula of the form

$$\sum F\left(\begin{matrix}a_1,\ldots,a_m,1\\b_1,\ldots,b_n\end{matrix}\middle|\,z\right)_k \delta k = c\,F\left(\begin{matrix}A_1,\ldots,A_M,1\\B_1,\ldots,B_N\end{matrix}\middle|\,z\right)_k, \tag{5.127}$$

then we also have

$$\sum F\left(\begin{matrix}a_1,\ldots,a_m,1\\b_1,\ldots,b_n\end{matrix}\middle|\,z\right)_{k+l} \delta k = c\,F\left(\begin{matrix}A_1,\ldots,A_M,1\\B_1,\ldots,B_N\end{matrix}\middle|\,z\right)_{k+l},$$

for any integer l. There's a general formula for shifting the index by l:

$$F\left(\begin{matrix}a_1,\ldots,a_m\\b_1,\ldots,b_n\end{matrix}\middle|\,z\right)_{k+l} = \frac{a_1^{\overline{l}}\ldots a_m^{\overline{l}}\,z^l}{b_1^{\overline{l}}\ldots b_n^{\overline{l}}\,l!}\,F\left(\begin{matrix}a_1+l,\,\ldots,\,a_m+l,\,1\\b_1+l,\,\ldots,\,b_n+l,\,l+1\end{matrix}\middle|\,z\right)_k.$$

Hence any given identity (5.127) has an infinite number of shifted forms:

$$\begin{aligned}&\sum F\left(\begin{matrix}a_1+l,\,\ldots,\,a_m+l,\,1\\b_1+l,\,\ldots,\,b_n+l\end{matrix}\middle|\,z\right)_k \delta k\\ &\qquad = c\,\frac{b_1^{\overline{l}}\ldots b_n^{\overline{l}}}{a_1^{\overline{l}}\ldots a_m^{\overline{l}}}\,\frac{A_1^{\overline{l}}\ldots A_M^{\overline{l}}}{B_1^{\overline{l}}\ldots B_N^{\overline{l}}}\,F\left(\begin{matrix}A_1+l,\,\ldots,\,A_M+l,\,1\\B_1+l,\,\ldots,\,B_N+l\end{matrix}\middle|\,z\right)_k.\end{aligned} \tag{5.128}$$

There's usually a fair amount of cancellation among the a's, A's, b's, and B's here. For example, if we apply this shift formula to (5.126), we get the general identity

$$\sum F\left(\begin{matrix}n+l+2,\,1\\l+2\end{matrix}\middle|\,1\right)_k \delta k = \frac{l+1}{n+1}\,F\left(\begin{matrix}n+l+2,1\\l+1\end{matrix}\middle|\,1\right)_k, \tag{5.129}$$

valid for all $n \neq 1$. The shifted version of (5.125) is

$$\sum F\left({l+2,\,1 \atop l+1}\,\Big|\,z\right)_k \delta k = \frac{-1}{(1-z)^2}\,\frac{l+1/(1-z)}{l+1}\,F\left({l+1+1/(1-z),\,1 \atop l+1/(1-z)}\,\Big|\,z\right)_k. \qquad (5.130)$$

With a bit of patience, we can compute a few more indefinite summation identities that are potentially useful:

$$\sum F\left({a,\,2+(1-a)z/(1-z),\,1 \atop 1+(1-a)z/(1-z),\,2}\,\Big|\,z\right)_k \delta k = \frac{1}{az-1}\,F\left({a,\,1 \atop 1}\,\Big|\,z\right)_k; \qquad (5.131)$$

$$\sum F\left({a,\,b,\,1+(c-ab)/(c-a-b+1),\,1 \atop c+1,\,(c-ab)/(c-a-b+1),\,2}\,\Big|\,1\right)_k \delta k = \frac{c}{ab-c}\,F\left({a,\,b \atop c}\,\Big|\,1\right)_k; \qquad (5.132)$$

$$\sum F\left({a,\,b,\,1 \atop c+1,\,a+b-c+1}\,\Big|\,1\right)_k \delta k = \frac{(c)(c-b-a)}{(c-a)(c-b)}\,F\left({a,\,b,\,1 \atop c,\,a+b-c}\,\Big|\,1\right)_k. \qquad (5.133)$$

Exercises

Warmups

1 What is 11^4? Why is this number easy to compute, for a person who knows binomial coefficients?

2 For which value(s) of k is $\binom{n}{k}$ a maximum, when n is a given positive integer? Prove your answer.

3 Prove the hexagon property, $\binom{n-1}{k-1}\binom{n}{k+1}\binom{n+1}{k} = \binom{n-1}{k}\binom{n+1}{k+1}\binom{n}{k-1}$.

4 Evaluate $\binom{-1}{k}$ by negating (actually un-negating) its upper index.

5 Let p be prime. Show that $\binom{p}{k} \bmod p = 0$ for $0 < k < p$. What does this imply about the binomial coefficients $\binom{p-1}{k}$?

6 Fix up the text's derivation in Problem 6, Section 5.2, by correctly applying symmetry.

A case of mistaken identity.

7 Is (5.34) true also when $k < 0$?

8 Evaluate $\sum_k \binom{n}{k}(-1)^k(1-k/n)^n$. What is the approximate value of this sum, when n is very large? *Hint:* This sum is $\Delta^n f(0)$ for some function f.

9 Show that the generalized exponentials of (5.58) obey the law

$$\mathcal{E}_t(z) = \mathcal{E}(tz)^{1/t}, \qquad \text{if } t \neq 0,$$

where $\mathcal{E}(z)$ is an abbreviation for $\mathcal{E}_1(z)$.

10 Show that $-2\bigl(\ln(1-z)+z\bigr)/z^2$ is a hypergeometric function.

11 Express the two functions

$$\sin z = z - \frac{z^3}{3!} + \frac{z^5}{5!} - \frac{z^7}{7!} + \cdots$$
$$\arcsin z = z + \frac{1\cdot z^3}{2\cdot 3} + \frac{1\cdot 3\cdot z^5}{2\cdot 4\cdot 5} + \frac{1\cdot 3\cdot 5\cdot z^7}{2\cdot 4\cdot 6\cdot 7} + \cdots$$

in terms of hypergeometric series.

12 Which of the following functions of k is a "hypergeometric term," in the sense of (5.115)? Explain why or why not.

a n^k.

b k^n.

c $\bigl(k! + (k+1)!\bigr)/2$.

d H_k, that is, $\frac{1}{1} + \frac{1}{2} + \cdots + \frac{1}{k}$.

(Here t and T aren't necessarily related as in (5.119).)

e $t(k)T(n-k)/T(n)$, when t and T are hypergeometric terms.

f $\bigl(t(k)+T(k)\bigr)/2$, when t and T are hypergeometric terms.

g $\bigl(a\,t(k) + b\,t(k{+}1) + c\,t(k{+}2)\bigr)/\bigl(a + b\,t(1) + c\,t(2)\bigr)$, when t is a hypergeometric term.

Basics

13 Find relations between the superfactorial function $P_n = \prod_{k=1}^n k!$ of exercise 4.55, the hyperfactorial function $Q_n = \prod_{k=1}^n k^k$, and the product $R_n = \prod_{k=0}^n \binom{n}{k}$.

14 Prove identity (5.25) by negating the upper index in Vandermonde's convolution (5.22). Then show that another negation yields (5.26).

15 What is $\sum_k \binom{n}{k}^3(-1)^k$? *Hint:* See (5.29).

16 Evaluate the sum

$$\sum_k \binom{2a}{a+k}\binom{2b}{b+k}\binom{2c}{c+k}(-1)^k$$

when a, b, c are nonnegative integers.

17 Find a simple relation between $\binom{2n-1/2}{n}$ and $\binom{2n-1/2}{2n}$.

18 Find an alternative form analogous to (5.35) for the product

$$\binom{r}{k}\binom{r-1/3}{k}\binom{r-2/3}{k}.$$

19 Show that the generalized binomials of (5.58) obey the law

$$\mathcal{B}_t(z) = \mathcal{B}_{1-t}(-z)^{-1}.$$

20 Define a "generalized bloopergeometric series" by the formula

$$G\left({a_1, \ldots, a_m \atop b_1, \ldots, b_n}\middle|\, z\right) = \sum_{k\geq 0} \frac{a_1^{\underline{k}} \ldots a_m^{\underline{k}}}{b_1^{\underline{k}} \ldots b_n^{\underline{k}}} \frac{z^k}{k!},$$

using falling powers instead of the rising ones in (5.76). Explain how G is related to F.

21 Show that Euler's definition of factorials is consistent with the ordinary definition, by showing that the limit in (5.83) is $1/((m-1)\ldots(1))$ when $z = m$ is a positive integer.

22 Use (5.83) to prove the *factorial duplication formula*:

$$x!\,(x-\tfrac{1}{2})! = (2x)!\,(-\tfrac{1}{2})!/2^{2x}.$$

By the way, $(-\frac{1}{2})! = \sqrt{\pi}$.

23 What is the value of $F(-n, 1; ; 1)$?

24 Find $\sum_k \binom{n}{m+k}\binom{m+k}{2k}4^k$ by using hypergeometric series.

25 Show that

$$(a_1 - b_1)\, F\left({a_1,\, a_2,\, \ldots,\, a_m \atop b_1{+}1,\, b_2,\, \ldots,\, b_n}\middle|\, z\right)$$
$$= a_1\, F\left({a_1{+}1,\, a_2,\, \ldots,\, a_m \atop b_1{+}1,\, b_2,\, \ldots,\, b_n}\middle|\, z\right) - b_1\, F\left({a_1,\, a_2,\, \ldots,\, a_m \atop b_1,\, b_2,\, \ldots,\, b_n}\middle|\, z\right).$$

Find a similar relation between the hypergeometrics $F(a_1, a_2, a_3 \ldots, a_m; b_1, \ldots, b_n; z)$, $F(a_1+1, a_2, a_3 \ldots, a_m; b_1, \ldots, b_n; z)$, and $F(a_1, a_2+1, a_3 \ldots, a_m; b_1, \ldots, b_n; z)$.

26 Express the function $G(z)$ in the formula

$$F\left({a_1, \ldots, a_m \atop b_1, \ldots, b_n}\middle|\, z\right) = 1 + G(z)$$

as a multiple of a hypergeometric series.

27 Prove that

$$F\left({a_1,\, a_1+\frac12,\, \ldots,\, a_m,\, a_m+\frac12 \atop b_1,\, b_1+\frac12,\, \ldots,\, b_n,\, b_n+\frac12,\, \frac12} \,\Big|\, (2^{m-n-1}z)^2\right)$$
$$= \frac12\left(F\left({2a_1,\ldots,2a_m \atop 2b_1,\ldots,2b_n}\,\Big|\,z\right) + F\left({2a_1,\ldots,2a_m \atop 2b_1,\ldots,2b_n}\,\Big|\,-z\right)\right).$$

28 Prove *Euler's identity*

$$F\left({a,b \atop c}\,\Big|\,z\right) = (1-z)^{c-a-b}\,F\left({c-a,\, c-b \atop c}\,\Big|\,z\right)$$

by applying Pfaff's reflection law (5.101) twice.

29 Show that confluent hypergeometrics satisfy

$$e^z F\left({a \atop b}\,\Big|\,-z\right) = F\left({b-a \atop b}\,\Big|\,z\right).$$

30 What hypergeometric series F satisfies $zF'(z) + F(z) = 1/(1-z)$?

31 Show that if $f(k)$ is any function summable in hypergeometric terms, then f itself is a multiple of a hypergeometric term. In other words, if $\sum f(k)\,\delta k = cF(A_1,\ldots,A_M; B_1,\ldots,B_N; Z)_k + C$, then there exist constants $a_1, \ldots, a_m, b_1, \ldots, b_n$, and z such that $f(k)$ is a constant times $F(a_1,\ldots,a_m; b_1,\ldots,b_n; z)_k$.

32 Find $\sum k^2\,\delta k$ by Gosper's method.

33 Use Gosper's method to find $\sum \delta k/(k^2-1)$.

34 Show that a partial hypergeometric sum can always be represented as a limit of ordinary hypergeometrics:

$$\sum_{k\leqslant c} F\left({a_1,\, \ldots,\, a_m \atop b_1,\, \ldots,\, b_n}\,\Big|\,z\right)_k = \lim_{\epsilon\to 0} F\left({-c,\, a_1,\, \ldots,\, a_m \atop \epsilon-c,\, b_1,\, \ldots,\, b_n}\,\Big|\,z\right),$$

when c is a nonnegative integer. Use this idea to evaluate $\sum_{k\leqslant m}\binom{n}{k}(-1)^k$.

Homework exercises

35 The notation $\sum_{k\leqslant n}\binom{n}{k}2^{k-n}$ is ambiguous without context. Evaluate it

a as a sum on k;

b as a sum on n.

36 Let p^k be the largest power of the prime p that divides $\binom{m+n}{m}$, when m and n are nonnegative integers. Prove that k is the number of carries that occur when m is added to n in the radix p number system. *Hint:* Exercise 4.24 helps here.

37 Show that an analog of the binomial theorem holds for factorial powers. That is, prove the identities

$$(x+y)^{\underline{n}} = \sum_k \binom{n}{k} x^{\underline{k}}\, y^{\underline{n-k}},$$

$$(x+y)^{\overline{n}} = \sum_k \binom{n}{k} x^{\overline{k}}\, y^{\overline{n-k}},$$

for all nonnegative integers n.

38 Show that all nonnegative integers n can be represented uniquely in the form $n = \binom{a}{1}+\binom{b}{2}+\binom{c}{3}$ where a, b, and c are integers with $0 \leqslant a < b < c$. (This is called the *binomial number system*.)

39 Show that if $xy = ax + by$ then $x^n y^n = \sum_{k=1}^{n} \binom{2n-1-k}{n-1}(a^n b^{n-k} x^k + a^{n-k} b^n y^k)$ for all $n > 0$. Find a similar formula for the more general product $x^m y^n$.

40 Find a closed form for

$$\sum_{j=1}^{m} (-1)^{j+1} \binom{r}{j} \sum_{k=1}^{n} \binom{-j+rk+s}{m-j}, \qquad \text{integers } m, n \geqslant 0.$$

41 Evaluate $\sum_k \binom{n}{k} k!/(n+1+k)!$ when n is a nonnegative integer.

42 Find the indefinite sum $\sum \bigl((-1)^x / \binom{n}{x}\bigr)\,\delta x$, and use it to compute the sum $\sum_{k=0}^{n} (-1)^k / \binom{n}{k}$ in closed form.

43 Prove the triple-binomial identity (5.28). *Hint:* First replace $\binom{r+k}{m+n}$ by $\sum_j \binom{r}{m+n-j}\binom{k}{j}$.

44 Use identity (5.32) to find closed forms for the double sums

$$\sum_{j,k} (-1)^{j+k} \binom{j+k}{j} \binom{a}{j} \binom{b}{k} \binom{m+n-j-k}{m-j} \quad \text{and}$$

$$\sum_{j,k \geqslant 0} (-1)^{j+k} \binom{a}{j} \binom{m}{j} \binom{b}{k} \binom{n}{k} \Big/ \binom{m+n}{j+k},$$

given integers $m \geqslant a \geqslant 0$ and $n \geqslant b \geqslant 0$.

45 Find a closed form for $\sum_{k \leqslant n} \binom{2k}{k} 4^{-k}$.

46 Evaluate the following sum in closed form, when n is a positive integer:

$$\sum_k \binom{2k-1}{k} \binom{4n-2k-1}{2n-k} \frac{(-1)^{k-1}}{(2k-1)(4n-2k-1)}.$$

Hint: Generating functions win again.

47 The sum $\sum_k \binom{rk+s}{k}\binom{rn-rk-s}{n-k}$ is a polynomial in r and s. Show that it doesn't depend on s.

48 The identity $\sum_{k\leqslant n}\binom{n+k}{n}2^{-k} = 2^n$ can be combined with $\sum_{k\geqslant 0}\binom{n+k}{n}z^k = 1/(1-z)^{n+1}$ to yield $\sum_{k>n}\binom{n+k}{n}2^{-k} = 2^n$. What is the hypergeometric form of the latter identity?

49 Use the hypergeometric method to evaluate

$$\sum_k (-1)^k \binom{x}{k}\binom{x+n-k}{n-k}\frac{y}{y+n-k}.$$

50 Prove Pfaff's reflection law (5.101) by comparing the coefficients of z^n on both sides of the equation.

51 The derivation of (5.104) shows that

$$\lim_{\epsilon\to 0} F(-m, -2m-1+\epsilon; -2m+\epsilon; 2) \;=\; 1\big/\tbinom{-1/2}{m}.$$

In this exercise we will see that slightly different limiting processes lead to distinctly different answers for the degenerate hypergeometric series $F(-m, -2m-1; -2m; 2)$.

a Show that $\lim_{\epsilon\to 0} F(-m+\epsilon, -2m-1; -2m+2\epsilon; 2) = 0$, by using Pfaff's reflection law to prove the identity $F(a, -2m-1; 2a; 2) = 0$ for all integers $m \geqslant 0$.

b What is $\lim_{\epsilon\to 0} F(-m+\epsilon, -2m-1; -2m+\epsilon; 2)$?

52 Prove that if N is a nonnegative integer,

$$\begin{aligned} b_1^{\overline{N}}\ldots b_n^{\overline{N}}\, F\!\left({a_1,\ldots,a_m,-N \atop b_1,\ldots,b_n}\,\middle|\, z\right) \\ = a_1^{\overline{N}}\ldots a_m^{\overline{N}}(-z)^N\, F\!\left({1-b_1-N,\ldots,1-b_n-N,-N \atop 1-a_1-N,\ldots,1-a_m-N}\,\middle|\,\frac{(-1)^{m+n}}{z}\right). \end{aligned}$$

53 If we put $b = -\frac{1}{2}$ and $z = 1$ in Gauss's identity (5.110), the left side reduces to -1 while the right side is $+1$. Why doesn't this prove that $-1 = +1$?

54 Explain how the right-hand side of (5.112) was obtained.

55 If the hypergeometric terms $t(k) = F(a_1,\ldots,a_m;\, b_1,\ldots,b_n;\, z)_k$ and $T(k) = F(A_1,\ldots,A_M; B_1,\ldots,B_N; Z)_k$ satisfy $t(k) = c\big(T(k+1) - T(k)\big)$ for all $k \geqslant 0$, show that $z = Z$ and $m - n = M - N$.

56 Find a general formula for $\sum \binom{-3}{k}\,\delta k$ using Gosper's method. Show that $(-1)^{k-1}\lfloor\frac{k+1}{2}\rfloor\lfloor\frac{k+2}{2}\rfloor$ is also a solution.

57 Use Gosper's method to find a constant θ such that

$$\sum \binom{n}{k} z^k (k+\theta)\,\delta k$$

is summable in hypergeometric terms.

58 If m and n are integers with $0 \leqslant m \leqslant n$, let

$$T_{m,n} = \sum_{0 \leqslant k < n} \binom{k}{m} \frac{1}{n-k}.$$

Find a relation between $T_{m,n}$ and $T_{m-1,n-1}$, then solve your recurrence by applying a summation factor.

Exam problems

59 Find a closed form for

$$\sum_{k \geqslant 1} \binom{n}{\lfloor \log_m k \rfloor}$$

when m and n are positive integers.

60 Use Stirling's approximation (4.23) to estimate $\binom{m+n}{n}$ when m and n are both large. What does your formula reduce to when $m = n$?

61 Prove that when p is prime, we have

$$\binom{n}{m} \equiv \binom{\lfloor n/p \rfloor}{\lfloor m/p \rfloor} \binom{n \bmod p}{m \bmod p} \pmod{p},$$

for all nonnegative integers m and n.

62 Assuming that p is prime and that m and n are positive integers, determine the value of $\binom{np}{mp} \bmod p^2$. *Hint:* You may wish to use the following generalization of Vandermonde's convolution:

$$\sum_{k_1+k_2+\cdots+k_m=n} \binom{r_1}{k_1}\binom{r_2}{k_2}\ldots\binom{r_m}{k_m} = \binom{r_1+r_2+\cdots+r_m}{n}.$$

63 Find a closed form for

$$\sum_{k=0}^{n} (-4)^k \binom{n+k}{2k},$$

given an integer $n \geqslant 0$.

64 Evaluate $\sum_{k=0}^{n} \binom{n}{k} \Big/ \left\lceil \frac{k+1}{2} \right\rceil$, given an integer $n \geqslant 0$.

65 Prove that

$$\sum_k \binom{n-1}{k} n^{-k}(k+1)! \;=\; n\,.$$

66 Evaluate "Harry's double sum,"

$$\sum_{0 \leqslant j \leqslant k} \binom{-1}{j - \lfloor\sqrt{k-j}\rfloor} \binom{j}{m} \frac{1}{2^j}\,, \qquad \text{integer } m \geqslant 0,$$

as a function of m. (The sum is over both j and k.)

67 Find a closed form for

$$\sum_{k=0}^{n} \binom{\binom{k}{2}}{2} \binom{2n-k}{n}\,, \qquad \text{integer } n \geqslant 0.$$

68 Find a closed form for

$$\sum_k \binom{n}{k} \min(k, n-k)\,, \qquad \text{integer } n \geqslant 0.$$

69 Find a closed form for

$$\min_{\substack{k_1,\ldots,k_m \geqslant 0 \\ k_1+\cdots+k_m = n}} \sum_{j=1}^{m} \binom{k_j}{2}$$

as a function of m and n.

70 Find a closed form for

$$\sum_k \binom{n}{k} \binom{2k}{k} \left(\frac{-1}{2}\right)^k, \qquad \text{integer } n \geqslant 0.$$

71 Let

$$S_n \;=\; \sum_{k \geqslant 0} \binom{n+k}{m+2k} a_k\,,$$

where m and n are nonnegative integers, and let $A(z) = \sum_{k\geqslant 0} a_k z^k$ be the generating function for the sequence $\langle a_0, a_1, a_2, \ldots \rangle$.

a Express the generating function $S(z) = \sum_{n\geqslant 0} S_n z^n$ in terms of $A(z)$.

b Use this technique to solve Problem 7 in Section 5.2.

72 Prove that, if m, n, and k are integers and $n > 0$,

$$\binom{m/n}{k} n^{2k-\nu(k)} \quad \text{is an integer,}$$

where $\nu(k)$ is the number of 1's in the binary representation of k.

73 Use the repertoire method to solve the recurrence

$$X_0 = \alpha; \qquad X_1 = \beta;$$
$$X_n = (n-1)(X_{n-1} + X_{n-2}), \qquad \text{for } n > 1.$$

Hint: Both $n!$ and $n¡$ satisfy this recurrence.

74 This problem concerns a deviant version of Pascal's triangle in which the sides consist of the numbers 1, 2, 3, 4, ... instead of all 1's, although the interior numbers still satisfy the addition formula:

```
        1
      2   2
    3   4   3
  4   7   7   4
5  11  14  11   5
.  .   .   .  .  .
```

If $\left(\!\binom{n}{k}\!\right)$ denotes the kth number in row n, for $1 \leqslant k \leqslant n$, we have $\left(\!\binom{n}{1}\!\right) = \left(\!\binom{n}{n}\!\right) = n$, and $\left(\!\binom{n}{k}\!\right) = \left(\!\binom{n-1}{k}\!\right) + \left(\!\binom{n-1}{k-1}\!\right)$ for $1 < k < n$. Express the quantity $\left(\!\binom{n}{k}\!\right)$ in closed form.

75 Find a relation between the functions

$$S_0(n) = \sum_k \binom{n}{3k},$$

$$S_1(n) = \sum_k \binom{n}{3k+1},$$

$$S_2(n) = \sum_k \binom{n}{3k+2}$$

and the quantities $\lfloor 2^n/3 \rfloor$ and $\lceil 2^n/3 \rceil$.

76 Solve the following recurrence for $n, k \geqslant 0$:

$$Q_{n,0} = 1; \qquad Q_{0,k} = [k=0];$$

$$Q_{n,k} = Q_{n-1,k} + Q_{n-1,k-1} + \binom{n}{k}, \qquad \text{for } n, k > 0.$$

77 What is the value of

$$\sum_{0\leqslant k_1,\ldots,k_m\leqslant n}\ \prod_{1\leqslant j<m}\binom{k_{j+1}}{k_j}, \qquad \text{if } m>1?$$

78 Find a closed form for

$$\sum_{k=0}^{2m^2}\binom{k \bmod m}{(2k+1)\bmod(2m+1)},$$

assuming that m is a positive integer.

79 What is the greatest common divisor of $\binom{2n}{1}, \binom{2n}{3}, \ldots, \binom{2n}{2n-1}$? *Hint:* Consider the sum of these n numbers.

80 Show that the least common multiple of $\binom{n}{0}, \binom{n}{1}, \ldots, \binom{n}{n}$ is equal to $L(n+1)/(n+1)$, where $L(n)=\operatorname{lcm}(1,2,\ldots,n)$.

81 If $0<\theta<1$ and $0\leqslant x\leqslant 1$, and if l, m, n are nonnegative integers with $m<n$, prove the inequality

$$(-1)^{n-m-1}\sum_k\binom{l}{k}\binom{m+\theta}{n+k}x^k > 0.$$

Hint: Consider taking the derivative with respect to x.

Bonus problems

82 Prove that Pascal's triangle has an even more surprising hexagon property than the one cited in the text:

$$\gcd\left(\binom{n-1}{k-1},\binom{n}{k+1},\binom{n+1}{k}\right) = \gcd\left(\binom{n-1}{k},\binom{n+1}{k+1},\binom{n}{k-1}\right),$$

if $0<k<n$. For example, $\gcd(56,36,210)=\gcd(28,120,126)=2$.

83 Prove the amazing identity (5.32) by first showing that it's true whenever the right-hand side is zero.

84 Show that the second pair of convolution formulas, (5.61), follows from the first pair, (5.60). *Hint:* Differentiate with respect to z.

85 Prove that

$$\sum_{m=1}^{n}(-1)^m\sum_{1\leqslant k_1<k_2<\cdots<k_m\leqslant n}\binom{k_1^3+k_2^3+\cdots+k_m^3+2^n}{n} = (-1)^n n!^3-\binom{2^n}{n}.$$

(The left side is a sum of 2^n-1 terms.) *Hint:* Much more is true.

86 Let $a_1, \ldots, a_n$ be nonnegative integers, and let $C(a_1, \ldots, a_n)$ be the coefficient of the constant term $z_1^0 \ldots z_n^0$ when the $n(n-1)$ factors

$$\prod_{\substack{1\leqslant i,j\leqslant n \\ i\neq j}} \left(1 - \frac{z_i}{z_j}\right)^{a_i}$$

are fully expanded into positive and negative powers of the complex variables $z_1, \ldots, z_n$.

a Prove that $C(a_1, \ldots, a_n)$ equals the left-hand side of (5.31).

b Prove that if $z_1, \ldots, z_n$ are distinct complex numbers, then the polynomial

$$f(z) = \sum_{k=1}^{n} \prod_{\substack{1\leqslant j\leqslant n \\ j\neq k}} \frac{z - z_j}{z_k - z_j}$$

is identically equal to 1.

c Multiply the original product of $n(n-1)$ factors by $f(0)$ and deduce that $C(a_1, a_2, \ldots, a_n)$ is equal to

$$C(a_1 - 1, a_2, \ldots, a_n) + C(a_1, a_2 - 1, \ldots, a_n) \\ + \cdots + C(a_1, a_2, \ldots, a_n - 1).$$

(This recurrence defines multinomial coefficients, so $C(a_1, \ldots, a_n)$ must equal the right-hand side of (5.31).)

87 Let m be a positive integer and let $\zeta = e^{\pi i/m}$. Show that

$$\sum_{k\leqslant n/m} \binom{n - mk}{k} z^{mk} \\ = \frac{\mathcal{B}_{-m}(z^m)^{n+1}}{(1+m)\mathcal{B}_{-m}(z^m) - m} \\ - \sum_{0\leqslant j<m} \frac{\left(\zeta^{2j+1} z\, \mathcal{B}_{1+1/m}(\zeta^{2j+1} z)^{1/m}\right)^{n+1}}{(m+1)\mathcal{B}_{1+1/m}(\zeta^{2j+1} z)^{-1} - 1}.$$

(This reduces to (5.74) in the special case $m = 1$.)

88 Prove that the coefficients s_k in (5.47) are equal to

$$(-1)^k \int_0^\infty e^{-t}(1 - e^{-t})^{k-1} \frac{dt}{t},$$

for all $k > 1$; hence $|s_k| < 1/(k-1)$.

89 Prove that (5.19) has an infinite counterpart,

$$\sum_{k>m}\binom{m+r}{k}x^k y^{m-k} = \sum_{k>m}\binom{-r}{k}(-x)^k(x+y)^{m-k}, \qquad \text{integer } m,$$

if $|x| < |y|$ and $|x| < |x+y|$. Differentiate this identity n times with respect to y and express it in terms of hypergeometrics; what relation do you get?

90 Problem 1 in Section 5.2 considers $\sum_{k\geqslant 0}\binom{r}{k}/\binom{s}{k}$ when r and s are integers with $s \geqslant r \geqslant 0$. What is the value of this sum if r and s aren't integers?

91 Prove *Whipple's identity,*

$$F\left({\tfrac{1}{2}a,\ \tfrac{1}{2}a+\tfrac{1}{2},\ 1-a-b-c \atop 1+a-b,\ 1+a-c}\,\middle|\,\frac{-4z}{(1-z)^2}\right)$$
$$= (1-z)^a\, F\left({a,\ b,\ c \atop 1+a-b,\ 1+a-c}\,\middle|\,z\right),$$

by showing that both sides satisfy the same differential equation.

92 Prove *Clausen's product identities*

$$F\left({a,\ b \atop a+b+\tfrac{1}{2}}\,\middle|\,z\right)^2 = F\left({2a,\ a+b,\ 2b \atop 2a+2b,\ a+b+\tfrac{1}{2}}\,\middle|\,z\right);$$

$$F\left({\tfrac{1}{4}+a,\ \tfrac{1}{4}+b \atop 1+a+b}\,\middle|\,z\right) F\left({\tfrac{1}{4}-a,\ \tfrac{1}{4}-b \atop 1-a-b}\,\middle|\,z\right)$$
$$= F\left({\tfrac{1}{2},\ \tfrac{1}{2}+a-b,\ \tfrac{1}{2}-a+b \atop 1+a+b,\ 1-a-b}\,\middle|\,z\right).$$

What identities result when the coefficients of z^n on both sides of these formulas are equated?

93 Show that the indefinite sum

$$\sum\left(\prod_{j=1}^{k-1}(f(j)+\alpha)\Big/\prod_{j=1}^{k}f(j)\right)\delta k$$

has a (fairly) simple form, given any function f and any constant α.

94 Show that if $\omega = e^{2\pi i/3}$ we have

$$\sum_{k+l+m=3n}\binom{3n}{k,l,m}^2\omega^{l+2m} = \binom{4n}{n,n,2n}, \qquad \text{integer } n \geqslant 0.$$

Research problems

95 Let $q(n)$ be the smallest odd prime factor of the middle binomial coefficient $\binom{2n}{n}$. According to exercise 36, the odd primes that do *not* divide $\binom{2n}{n}$ are those for which all digits in n's radix p representation are $(p-1)/2$ or less. Computer experiments have shown that $q(n) \leqslant 11$ for all $n < 10^{10000}$, except that $q(3160) = 13$.

a Is $q(n) \leqslant 11$ for all $n > 3160$?

b Is $q(n) = 11$ for infinitely many n?

A reward of $\$\binom{5}{1}\binom{5}{2}\binom{5}{3}$ is offered for a solution to either (a) or (b).

96 Is $\binom{2n}{n}$ divisible by the square of a prime, for all $n > 4$?

97 For what values of n is $\binom{2n}{n} \equiv (-1)^n \pmod{2n+1}$?

6

Special Numbers

SOME SEQUENCES of numbers arise so often in mathematics that we recognize them instantly and give them special names. For example, everybody who learns arithmetic knows the sequence of square numbers $\langle 1, 4, 9, 16, \ldots \rangle$. In Chapter 1 we encountered the triangular numbers $\langle 1, 3, 6, 10, \ldots \rangle$; in Chapter 4 we studied the prime numbers $\langle 2, 3, 5, 7, \ldots \rangle$; in Chapter 5 we looked briefly at the Catalan numbers $\langle 1, 2, 5, 14, \ldots \rangle$.

In the present chapter we'll get to know a few other important sequences. First on our agenda will be the Stirling numbers ${n \brace k}$ and ${n \brack k}$, and the Eulerian numbers $\left\langle {n \atop k} \right\rangle$; these form triangular patterns of coefficients analogous to the binomial coefficients $\binom{n}{k}$ in Pascal's triangle. Then we'll take a good look at the harmonic numbers H_n, and the Bernoulli numbers B_n; these differ from the other sequences we've been studying because they're fractions, not integers. Finally, we'll examine the fascinating Fibonacci numbers F_n and some of their important generalizations.

6.1 STIRLING NUMBERS

We begin with some close relatives of the binomial coefficients, the Stirling numbers, named after James Stirling (1692–1770). These numbers come in two flavors, traditionally called by the no-frills names "Stirling numbers of the first and second kind." Although they have a venerable history and numerous applications, they still lack a standard notation. We will write ${n \brace k}$ for Stirling numbers of the second kind and ${n \brack k}$ for Stirling numbers of the first kind, because these symbols turn out to be more user-friendly than the many other notations that people have tried.

Tables 244 and 245 show what ${n \brace k}$ and ${n \brack k}$ look like when n and k are small. A problem that involves the numbers "1, 7, 6, 1" is likely to be related to ${n \brace k}$, and a problem that involves "6, 11, 6, 1" is likely to be related to ${n \brack k}$, just as we assume that a problem involving "1, 4, 6, 4, 1" is likely to be related to $\binom{n}{k}$; these are the trademark sequences that appear when $n = 4$.

Table 244 Stirling's triangle for subsets.

n	$\left\{{n \atop 0}\right\}$	$\left\{{n \atop 1}\right\}$	$\left\{{n \atop 2}\right\}$	$\left\{{n \atop 3}\right\}$	$\left\{{n \atop 4}\right\}$	$\left\{{n \atop 5}\right\}$	$\left\{{n \atop 6}\right\}$	$\left\{{n \atop 7}\right\}$	$\left\{{n \atop 8}\right\}$	$\left\{{n \atop 9}\right\}$
0	1									
1	0	1								
2	0	1	1							
3	0	1	3	1						
4	0	1	7	6	1					
5	0	1	15	25	10	1				
6	0	1	31	90	65	15	1			
7	0	1	63	301	350	140	21	1		
8	0	1	127	966	1701	1050	266	28	1	
9	0	1	255	3025	7770	6951	2646	462	36	1

Stirling numbers of the second kind show up more often than those of the other variety, so let's consider last things first. The symbol $\left\{{n \atop k}\right\}$ stands for the number of ways to partition a set of n things into k nonempty subsets. For example, there are seven ways to split a four-element set into two parts:

(Stirling himself considered this kind first in his book [281].)

$$\begin{array}{llll}\{1,2,3\}\cup\{4\}, & \{1,2,4\}\cup\{3\}, & \{1,3,4\}\cup\{2\}, & \{2,3,4\}\cup\{1\}, \\ \{1,2\}\cup\{3,4\}, & \{1,3\}\cup\{2,4\}, & \{1,4\}\cup\{2,3\}; & \end{array} \tag{6.1}$$

thus $\left\{{4 \atop 2}\right\} = 7$. Notice that curly braces are used to denote sets as well as the numbers $\left\{{n \atop k}\right\}$. This notational kinship helps us remember the meaning of $\left\{{n \atop k}\right\}$, which can be read "$n$ subset k."

Let's look at small k. There's just one way to put n elements into a single nonempty set; hence $\left\{{n \atop 1}\right\} = 1$, for all $n > 0$. On the other hand $\left\{{0 \atop 1}\right\} = 0$, because a 0-element set is empty.

The case $k = 0$ is a bit tricky. Things work out best if we agree that there's just one way to partition an empty set into zero nonempty parts; hence $\left\{{0 \atop 0}\right\} = 1$. But a nonempty set needs at least one part, so $\left\{{n \atop 0}\right\} = 0$ for $n > 0$.

What happens when $k = 2$? Certainly $\left\{{0 \atop 2}\right\} = 0$. If a set of $n > 0$ objects is divided into two nonempty parts, one of those parts contains the last object and some subset of the first $n-1$ objects. There are 2^{n-1} ways to choose the latter subset, since each of the first $n-1$ objects is either in it or out of it; but we mustn't put all of those objects in it, because we want to end up with two nonempty parts. Therefore we subtract 1:

$$\left\{{n \atop 2}\right\} = 2^{n-1} - 1, \qquad \text{integer } n > 0. \tag{6.2}$$

(This tallies with our enumeration of $\left\{{4 \atop 2}\right\} = 7 = 2^3 - 1$ ways above.)

Table 245 Stirling's triangle for cycles.

n	$\left[{n \atop 0}\right]$	$\left[{n \atop 1}\right]$	$\left[{n \atop 2}\right]$	$\left[{n \atop 3}\right]$	$\left[{n \atop 4}\right]$	$\left[{n \atop 5}\right]$	$\left[{n \atop 6}\right]$	$\left[{n \atop 7}\right]$	$\left[{n \atop 8}\right]$	$\left[{n \atop 9}\right]$
0	1									
1	0	1								
2	0	1	1							
3	0	2	3	1						
4	0	6	11	6	1					
5	0	24	50	35	10	1				
6	0	120	274	225	85	15	1			
7	0	720	1764	1624	735	175	21	1		
8	0	5040	13068	13132	6769	1960	322	28	1	
9	0	40320	109584	118124	67284	22449	4536	546	36	1

A modification of this argument leads to a recurrence by which we can compute $\left\{{n \atop k}\right\}$ for all k: Given a set of $n > 0$ objects to be partitioned into k nonempty parts, we either put the last object into a class by itself (in $\left\{{n-1 \atop k-1}\right\}$ ways), or we put it together with some nonempty subset of the first $n-1$ objects. There are $k\left\{{n-1 \atop k}\right\}$ possibilities in the latter case, because each of the $\left\{{n-1 \atop k}\right\}$ ways to distribute the first $n-1$ objects into k nonempty parts gives k subsets that the nth object can join. Hence

$$\left\{{n \atop k}\right\} = k\left\{{n-1 \atop k}\right\} + \left\{{n-1 \atop k-1}\right\}, \qquad \text{integer } n > 0. \tag{6.3}$$

This is the law that generates Table 244; without the factor of k it would reduce to the addition formula (5.8) that generates Pascal's triangle.

And now, Stirling numbers of the first kind. These are somewhat like the others, but $\left[{n \atop k}\right]$ counts the number of ways to arrange n objects into k *cycles* instead of subsets. We verbalize '$\left[{n \atop k}\right]$' by saying "n cycle k."

Cycles are cyclic arrangements, like the necklaces we considered in Chapter 4. The cycle

```
 ┌A┐
D   B
 └C┘
```

can be written more compactly as '$[A, B, C, D]$', with the understanding that

$$[A, B, C, D] = [B, C, D, A] = [C, D, A, B] = [D, A, B, C];$$

a cycle "wraps around" because its end is joined to its beginning. On the other hand, the cycle $[A, B, C, D]$ is not the same as $[A, B, D, C]$ or $[D, C, B, A]$.

There are eleven different ways to make two cycles from four elements:

$$\begin{array}{llll} [1,2,3]\,[4]\,, & [1,2,4]\,[3]\,, & [1,3,4]\,[2]\,, & [2,3,4]\,[1]\,, \\ [1,3,2]\,[4]\,, & [1,4,2]\,[3]\,, & [1,4,3]\,[2]\,, & [2,4,3]\,[1]\,, \\ [1,2]\,[3,4]\,, & [1,3]\,[2,4]\,, & [1,4]\,[2,3]\,; & \end{array} \tag{6.4}$$

"There are nine and sixty ways of constructing tribal lays, And-every-single-one-of-them-is-right."
— Rudyard Kipling

hence $\left[{4 \atop 2}\right] = 11$.

A singleton cycle (that is, a cycle with only one element) is essentially the same as a singleton set (a set with only one element). Similarly, a 2-cycle is like a 2-set, because we have $[A,B] = [B,A]$ just as $\{A,B\} = \{B,A\}$. But there are two *different* 3-cycles, $[A,B,C]$ and $[A,C,B]$. Notice, for example, that the eleven cycle pairs in (6.4) can be obtained from the seven set pairs in (6.1) by making two cycles from each of the 3-element sets.

In general, $n!/n = (n-1)!$ cycles can be made from any n-element set, whenever $n > 0$. (There are $n!$ permutations, and each cycle corresponds to n of them because any one of its elements can be listed first.) Therefore we have

$$\left[{n \atop 1}\right] = (n-1)!\,, \qquad \text{integer } n > 0. \tag{6.5}$$

This is much larger than the value $\left\{{n \atop 1}\right\} = 1$ we had for Stirling subset numbers. In fact, it is easy to see that the cycle numbers must be at least as large as the subset numbers,

$$\left[{n \atop k}\right] \geqslant \left\{{n \atop k}\right\}, \qquad \text{integers } n,k \geqslant 0, \tag{6.6}$$

because every partition into nonempty subsets leads to at least one arrangement of cycles.

Equality holds in (6.6) when all the cycles are necessarily singletons or doubletons, because cycles are equivalent to subsets in such cases. This happens when $k = n$ and when $k = n-1$; hence

$$\left[{n \atop n}\right] = \left\{{n \atop n}\right\}; \qquad \left[{n \atop n-1}\right] = \left\{{n \atop n-1}\right\}.$$

In fact, it is easy to see that

$$\left[{n \atop n}\right] = \left\{{n \atop n}\right\} = 1\,; \qquad \left[{n \atop n-1}\right] = \left\{{n \atop n-1}\right\} = \binom{n}{2}. \tag{6.7}$$

(The number of ways to arrange n objects into $n-1$ cycles or subsets is the number of ways to choose the two objects that will be in the same cycle or subset.) The triangular numbers $\binom{n}{2} = 1, 3, 6, 10, \ldots$ are conspicuously present in both Table 244 and Table 245.

We can derive a recurrence for $\left[{n \atop k}\right]$ by modifying the argument we used for $\left\{{n \atop k}\right\}$. Every arrangement of n objects in k cycles either puts the last object into a cycle by itself (in $\left[{n-1 \atop k-1}\right]$ ways) or inserts that object into one of the $\left[{n-1 \atop k}\right]$ cycle arrangements of the first $n-1$ objects. In the latter case, there are $n-1$ different ways to do the insertion. (This takes some thought, but it's not hard to verify that there are j ways to put a new element into a j-cycle in order to make a $(j+1)$-cycle. When $j = 3$, for example, the cycle $[A, B, C]$ leads to

$$[A, B, C, D]\,, \qquad [A, B, D, C]\,, \qquad \text{or} \qquad [A, D, B, C]$$

when we insert a new element D, and there are no other possibilities. Summing over all j gives a total of $n-1$ ways to insert an nth object into a cycle decomposition of $n-1$ objects.) The desired recurrence is therefore

$$\left[{n \atop k}\right] = (n-1)\left[{n-1 \atop k}\right] + \left[{n-1 \atop k-1}\right], \qquad \text{integer } n > 0. \tag{6.8}$$

This is the addition-formula analog that generates Table 245.

Comparison of (6.8) and (6.3) shows that the first term on the right side is multiplied by its upper index $(n-1)$ in the case of Stirling cycle numbers, but by its lower index k in the case of Stirling subset numbers. We can therefore perform "absorption" in terms like $n\left[{n \atop k}\right]$ and $k\left\{{n \atop k}\right\}$, when we do proofs by mathematical induction.

Every permutation is equivalent to a set of cycles. For example, consider the permutation that takes 123456789 into 384729156. We can conveniently represent it in two rows,

$$\begin{matrix} 1\,2\,3\,4\,5\,6\,7\,8\,9 \\ 3\,8\,4\,7\,2\,9\,1\,5\,6\,, \end{matrix}$$

showing that 1 goes to 3 and 2 goes to 8, etc. The cycle structure comes about because 1 goes to 3, which goes to 4, which goes to 7, which goes back to 1; that's the cycle $[1, 3, 4, 7]$. Another cycle in this permutation is $[2, 8, 5]$; still another is $[6, 9]$. Therefore the permutation 384729156 is equivalent to the cycle arrangement

$$[1, 3, 4, 7]\,[2, 8, 5]\,[6, 9]\,.$$

If we have any permutation $\pi_1 \pi_2 \ldots \pi_n$ of $\{1, 2, \ldots, n\}$, every element is in a unique cycle. For if we start with $m_0 = m$ and look at $m_1 = \pi_{m_0}$, $m_2 = \pi_{m_1}$, etc., we must eventually come back to $m_k = m_0$. (The numbers must repeat sooner or later, and the first number to reappear must be m_0 because we know the unique predecessors of the other numbers m_1, m_2, $\ldots$, m_{k-1}.) Therefore every permutation defines a cycle arrangement. Conversely, every cycle arrangement obviously defines a permutation if we reverse the construction, and this one-to-one correspondence shows that permutations and cycle arrangements are essentially the same thing.

Therefore $\left[{n \atop k}\right]$ is the number of permutations of n objects that contain exactly k cycles. If we sum $\left[{n \atop k}\right]$ over all k, we must get the total number of permutations:

$$\sum_{k=0}^{n} \left[{n \atop k}\right] = n!\,, \qquad \text{integer } n \geqslant 0. \tag{6.9}$$

For example, $6+11+6+1=24=4!$.

Stirling numbers are useful because the recurrence relations (6.3) and (6.8) arise in a variety of problems. For example, if we want to represent ordinary powers x^n by falling powers $x^{\underline{n}}$, we find that the first few cases are

$$\begin{aligned} x^0 &= x^{\underline{0}}; \\ x^1 &= x^{\underline{1}}; \\ x^2 &= x^{\underline{2}} + x^{\underline{1}}; \\ x^3 &= x^{\underline{3}} + 3x^{\underline{2}} + x^{\underline{1}}; \\ x^4 &= x^{\underline{4}} + 6x^{\underline{3}} + 7x^{\underline{2}} + x^{\underline{1}}. \end{aligned}$$

These coefficients look suspiciously like the numbers in Table 244, reflected between left and right; therefore we can be pretty confident that the general formula is

$$x^n = \sum_k \left\{{n \atop k}\right\} x^{\underline{k}}, \qquad \text{integer } n \geqslant 0. \tag{6.10}$$

We'd better define $\left\{{n \atop k}\right\} = \left[{n \atop k}\right] = 0$ *when* $k < 0$ *and* $n \geqslant 0$.

And sure enough, a simple proof by induction clinches the argument: We have $x \cdot x^{\underline{k}} = x^{\underline{k+1}} + kx^{\underline{k}}$, because $x^{\underline{k+1}} = x^{\underline{k}}(x-k)$; hence $x \cdot x^{n-1}$ is

$$\begin{aligned} x\sum_k \left\{{n-1 \atop k}\right\} x^{\underline{k}} &= \sum_k \left\{{n-1 \atop k}\right\} x^{\underline{k+1}} + \sum_k \left\{{n-1 \atop k}\right\} kx^{\underline{k}} \\ &= \sum_k \left\{{n-1 \atop k-1}\right\} x^{\underline{k}} + \sum_k \left\{{n-1 \atop k}\right\} kx^{\underline{k}} \\ &= \sum_k \left(k\left\{{n-1 \atop k}\right\} + \left\{{n-1 \atop k-1}\right\}\right) x^{\underline{k}} = \sum_k \left\{{n \atop k}\right\} x^{\underline{k}}. \end{aligned}$$

In other words, Stirling subset numbers are the coefficients of factorial powers that yield ordinary powers.

We can go the other way too, because Stirling cycle numbers are the coefficients of ordinary powers that yield factorial powers:

$$\begin{aligned}
x^{\overline{0}} &= x^0\,; \\
x^{\overline{1}} &= x^1\,; \\
x^{\overline{2}} &= x^2 + x^1\,; \\
x^{\overline{3}} &= x^3 + 3x^2 + 2x^1\,; \\
x^{\overline{4}} &= x^4 + 6x^3 + 11x^2 + 6x^1\,.
\end{aligned}$$

We have $(x+n-1)\cdot x^{k} = x^{k+1} + (n-1)x^{k}$, so a proof like the one just given shows that

$$(x+n-1)x^{\overline{n-1}} = (x+n-1)\sum_k \left[{n-1 \atop k}\right] x^k = \sum_k \left[{n \atop k}\right] x^k\,.$$

This leads to a proof by induction of the general formula

$$x^{\overline{n}} = \sum_k \left[{n \atop k}\right] x^k\,, \qquad \text{integer } n \geqslant 0. \tag{6.11}$$

(Setting $x = 1$ gives (6.9) again.)

But wait, you say. This equation involves rising factorial powers $x^{\overline{n}}$, while (6.10) involves falling factorials $x^{\underline{n}}$. What if we want to express $x^{\underline{n}}$ in terms of ordinary powers, or if we want to express x^n in terms of rising powers? Easy; we just throw in some minus signs and get

$$x^n = \sum_k \left\{{n \atop k}\right\} (-1)^{n-k} x^{\overline{k}}\,, \qquad \text{integer } n \geqslant 0; \tag{6.12}$$

$$x^{\underline{n}} = \sum_k \left[{n \atop k}\right] (-1)^{n-k} x^k\,, \qquad \text{integer } n \geqslant 0. \tag{6.13}$$

This works because, for example, the formula

$$x^{\underline{4}} = x(x-1)(x-2)(x-3) = x^4 - 6x^3 + 11x^2 - 6x$$

is just like the formula

$$x^{\overline{4}} = x(x+1)(x+2)(x+3) = x^4 + 6x^3 + 11x^2 + 6x$$

but with alternating signs. The general identity

$$x^{\underline{n}} = (-1)^n (-x)^{\overline{n}} \tag{6.14}$$

of exercise 2.17 converts (6.10) to (6.12) and (6.11) to (6.13) if we negate x.

Table 250 Basic Stirling number identities, for integer $n \geqslant 0$.

Recurrences:

$$\left\{ n \atop k \right\} = k\left\{ n-1 \atop k \right\} + \left\{ n-1 \atop k-1 \right\}.$$

$$\left[n \atop k \right] = (n-1)\left[n-1 \atop k \right] + \left[n-1 \atop k-1 \right].$$

Special values:

$$\left\{ n \atop 0 \right\} = \left[n \atop 0 \right] = [n=0].$$

$$\left\{ n \atop 1 \right\} = [n>0]; \qquad \left[n \atop 1 \right] = (n-1)!\,[n>0].$$

$$\left\{ n \atop 2 \right\} = (2^{n-1}-1)[n>0]; \qquad \left[n \atop 2 \right] = (n-1)!\,H_{n-1}\,[n>0].$$

$$\left\{ n \atop n-1 \right\} = \left[n \atop n-1 \right] = \binom{n}{2}.$$

$$\left\{ n \atop n \right\} = \left[n \atop n \right] = \binom{n}{n} = 1.$$

$$\left\{ n \atop k \right\} = \left[n \atop k \right] = \binom{n}{k} = 0, \qquad \text{if } k > n.$$

Converting between powers:

$$x^n = \sum_k \left\{ n \atop k \right\} x^{\underline{k}} = \sum_k \left\{ n \atop k \right\} (-1)^{n-k} x^{\overline{k}}.$$

$$x^{\underline{n}} = \sum_k \left[n \atop k \right] (-1)^{n-k} x^k;$$

$$x^{\overline{n}} = \sum_k \left[n \atop k \right] x^k.$$

Inversion formulas:

$$\sum_k \left[n \atop k \right] \left\{ k \atop m \right\} (-1)^{n-k} = [m=n];$$

$$\sum_k \left\{ n \atop k \right\} \left[k \atop m \right] (-1)^{n-k} = [m=n].$$

Table 251 Additional Stirling number identities, for integers $l, m, n \geqslant 0$.

$$\left\{ {n+1 \atop m+1} \right\} = \sum_k \binom{n}{k} \left\{ {k \atop m} \right\}. \tag{6.15}$$

$$\left[{n+1 \atop m+1} \right] = \sum_k \left[{n \atop k} \right] \binom{k}{m}. \tag{6.16}$$

$$\left\{ {n \atop m} \right\} = \sum_k \binom{n}{k} \left\{ {k+1 \atop m+1} \right\} (-1)^{n-k}. \tag{6.17}$$

$$\left[{n \atop m} \right] = \sum_k \left[{n+1 \atop k+1} \right] \binom{k}{m} (-1)^{m-k}. \tag{6.18}$$

$$m! \left\{ {n \atop m} \right\} = \sum_k \binom{m}{k} k^n (-1)^{m-k}. \tag{6.19}$$

$$\left\{ {n+1 \atop m+1} \right\} = \sum_{k=0}^{n} \left\{ {k \atop m} \right\} (m+1)^{n-k}. \tag{6.20}$$

$$\left[{n+1 \atop m+1} \right] = \sum_{k=0}^{n} \left[{k \atop m} \right] n^{\underline{n-k}} = n! \sum_{k=0}^{n} \left[{k \atop m} \right] / k!. \tag{6.21}$$

$$\left\{ {m+n+1 \atop m} \right\} = \sum_{k=0}^{m} k \left\{ {n+k \atop k} \right\}. \tag{6.22}$$

$$\left[{m+n+1 \atop m} \right] = \sum_{k=0}^{m} (n+k) \left[{n+k \atop k} \right]. \tag{6.23}$$

$$\binom{n}{m} = \sum_k \left\{ {n+1 \atop k+1} \right\} \left[{k \atop m} \right] (-1)^{m-k}. \tag{6.24}$$

$$(n-m)! \binom{n}{m} [n \geqslant m] = \sum_k \left[{n+1 \atop k+1} \right] \left\{ {k \atop m} \right\} (-1)^{m-k}. \tag{6.25}$$

$$\left\{ {n \atop n-m} \right\} = \sum_k \binom{m-n}{m+k} \binom{m+n}{n+k} \left[{m+k \atop k} \right]. \tag{6.26}$$

$$\left[{n \atop n-m} \right] = \sum_k \binom{m-n}{m+k} \binom{m+n}{n+k} \left\{ {m+k \atop k} \right\}. \tag{6.27}$$

$$\left\{ {n \atop l+m} \right\} \binom{l+m}{l} = \sum_k \left\{ {k \atop l} \right\} \left\{ {n-k \atop m} \right\} \binom{n}{k}. \tag{6.28}$$

$$\left[{n \atop l+m} \right] \binom{l+m}{l} = \sum_k \left[{k \atop l} \right] \left[{n-k \atop m} \right] \binom{n}{k}. \tag{6.29}$$

We can remember when to stick the $(-1)^{n-k}$ factor into a formula like (6.12) because there's a natural ordering of powers when x is large:

$$x^{\overline{n}} > x^n > x^{\underline{n}}, \qquad \text{for all } x > n > 1. \tag{6.30}$$

The Stirling numbers $\left[{n \atop k}\right]$ and $\left\{{n \atop k}\right\}$ are nonnegative, so we have to use minus signs when expanding a "small" power in terms of "large" ones.

We can plug (6.11) into (6.12) and get a double sum:

$$x^n = \sum_k \left\{{n \atop k}\right\}(-1)^{n-k}x^{\overline{k}} = \sum_{k,m} \left\{{n \atop k}\right\}\left[{k \atop m}\right](-1)^{n-k}x^m .$$

This holds for all x, so the coefficients of $x^0, x^1, \ldots, x^{n-1}, x^{n+1}, x^{n+2}, \ldots$ on the right must all be zero and we must have the identity

$$\sum_k \left\{{n \atop k}\right\}\left[{k \atop m}\right](-1)^{n-k} = [m=n], \qquad \text{integers } m, n \geqslant 0. \tag{6.31}$$

Stirling numbers, like binomial coefficients, satisfy many surprising identities. But these identities aren't as versatile as the ones we had in Chapter 5, so they aren't applied nearly as often. Therefore it's best for us just to list the simplest ones, for future reference when a tough Stirling nut needs to be cracked. Tables 250 and 251 contain the formulas that are most frequently useful; the principal identities we have already derived are repeated there.

When we studied binomial coefficients in Chapter 5, we found that it was advantageous to define $\binom{n}{k}$ for negative n in such a way that the identity $\binom{n}{k} = \binom{n-1}{k} + \binom{n-1}{k-1}$ is valid without any restrictions. Using that identity to extend the $\binom{n}{k}$'s beyond those with combinatorial significance, we discovered (in Table 164) that Pascal's triangle essentially reproduces itself in a rotated form when we extend it upward. Let's try the same thing with Stirling's triangles: What happens if we decide that the basic recurrences

$$\left\{{n \atop k}\right\} = k\left\{{n-1 \atop k}\right\} + \left\{{n-1 \atop k-1}\right\}$$

$$\left[{n \atop k}\right] = (n-1)\left[{n-1 \atop k}\right] + \left[{n-1 \atop k-1}\right]$$

are valid for all integers n and k? The solution becomes unique if we make the reasonable additional stipulations that

$$\left\{{0 \atop k}\right\} = \left[{0 \atop k}\right] = [k=0] \quad \text{and} \quad \left\{{n \atop 0}\right\} = \left[{n \atop 0}\right] = [n=0]. \tag{6.32}$$

Table 253 Stirling's triangles in tandem.

n	${n \brace -5}$	${n \brace -4}$	${n \brace -3}$	${n \brace -2}$	${n \brace -1}$	${n \brace 0}$	${n \brace 1}$	${n \brace 2}$	${n \brace 3}$	${n \brace 4}$	${n \brace 5}$
−5	1										
−4	10	1									
−3	35	6	1								
−2	50	11	3	1							
−1	24	6	2	1	1						
0	0	0	0	0	0	1					
1	0	0	0	0	0	0	1				
2	0	0	0	0	0	0	1	1			
3	0	0	0	0	0	0	1	3	1		
4	0	0	0	0	0	0	1	7	6	1	
5	0	0	0	0	0	0	1	15	25	10	1

In fact, a surprisingly pretty pattern emerges: Stirling's triangle for cycles appears above Stirling's triangle for subsets, and vice versa! The two kinds of Stirling numbers are related by an extremely simple law:

$$\left[{n \atop k}\right] = \left\{{-k \atop -n}\right\}, \qquad \text{integers } k, n. \tag{6.33}$$

We have "duality," something like the relations between min and max, between $\lfloor x \rfloor$ and $\lceil x \rceil$, between $x^{\underline{n}}$ and $x^{\overline{n}}$, between gcd and lcm. It's easy to check that both of the recurrences $\left[{n \atop k}\right] = (n-1)\left[{n-1 \atop k}\right] + \left[{n-1 \atop k-1}\right]$ and ${n \brace k} = k{n-1 \brace k} + {n-1 \brace k-1}$ amount to the same thing, under this correspondence.

6.2 EULERIAN NUMBERS

Another triangle of values pops up now and again, this one due to Euler [88, page 485], and we denote its elements by $\left\langle{n \atop k}\right\rangle$. The angle brackets in this case suggest "less than" and "greater than" signs; $\left\langle{n \atop k}\right\rangle$ is the number of permutations $\pi_1\pi_2\ldots\pi_n$ of $\{1, 2, \ldots, n\}$ that have k *ascents*, namely, k places where $\pi_j < \pi_{j+1}$. (Caution: This notation is even less standard than our notations $\left[{n \atop k}\right]$, ${n \brace k}$ for Stirling numbers. But we'll see that it makes good sense.)

(Knuth [175, first edition] used $\left\langle{n \atop k+1}\right\rangle$ for $\left\langle{n \atop k}\right\rangle$.)

For example, eleven permutations of $\{1, 2, 3, 4\}$ have two ascents:

$$\begin{array}{l} 1324,\quad 1423,\quad 2314,\quad 2413,\quad 3412; \\ 1243,\quad 1342,\quad 2341; \qquad 2134,\quad 3124,\quad 4123. \end{array}$$

(The first row lists the permutations with $\pi_1 < \pi_2 > \pi_3 < \pi_4$; the second row lists those with $\pi_1 < \pi_2 < \pi_3 > \pi_4$ and $\pi_1 > \pi_2 < \pi_3 < \pi_4$.) Hence $\left\langle{4 \atop 2}\right\rangle = 11$.

Table 254 Euler's triangle.

n	$\left\langle{n \atop 0}\right\rangle$	$\left\langle{n \atop 1}\right\rangle$	$\left\langle{n \atop 2}\right\rangle$	$\left\langle{n \atop 3}\right\rangle$	$\left\langle{n \atop 4}\right\rangle$	$\left\langle{n \atop 5}\right\rangle$	$\left\langle{n \atop 6}\right\rangle$	$\left\langle{n \atop 7}\right\rangle$	$\left\langle{n \atop 8}\right\rangle$	$\left\langle{n \atop 9}\right\rangle$
0	1									
1	1	0								
2	1	1	0							
3	1	4	1	0						
4	1	11	11	1	0					
5	1	26	66	26	1	0				
6	1	57	302	302	57	1	0			
7	1	120	1191	2416	1191	120	1	0		
8	1	247	4293	15619	15619	4293	247	1	0	
9	1	502	14608	88234	156190	88234	14608	502	1	0

Table 254 lists the smallest Eulerian numbers; notice that the trademark sequence is 1, 11, 11, 1 this time. There can be at most $n-1$ ascents, when $n>0$, so we have $\left\langle{n \atop n}\right\rangle = [n=0]$ on the diagonal of the triangle.

Euler's triangle, like Pascal's, is symmetric between left and right. But in this case the symmetry law is slightly different:

$$\left\langle{n \atop k}\right\rangle = \left\langle{n \atop n-1-k}\right\rangle, \qquad \text{integer } n>0; \tag{6.34}$$

The permutation $\pi_1\pi_2\ldots\pi_n$ has $n-1-k$ ascents if and only if its "reflection" $\pi_n\ldots\pi_2\pi_1$ has k ascents.

Let's try to find a recurrence for $\left\langle{n \atop k}\right\rangle$. Each permutation $\rho=\rho_1\ldots\rho_{n-1}$ of $\{1,\ldots,n-1\}$ leads to n permutations of $\{1,2,\ldots,n\}$ if we insert the new element n in all possible ways. Suppose we put n in position j, obtaining the permutation $\pi=\rho_1\ldots\rho_{j-1}\,n\,\rho_j\ldots\rho_{n-1}$. The number of ascents in π is the same as the number in ρ, if $j=1$ or if $\rho_{j-1}<\rho_j$; it's one greater than the number in ρ, if $\rho_{j-1}>\rho_j$ or if $j=n$. Therefore π has k ascents in a total of $(k+1)\left\langle{n-1 \atop k}\right\rangle$ ways from permutations ρ that have k ascents, plus a total of $\bigl((n-2)-(k-1)+1\bigr)\left\langle{n-1 \atop k-1}\right\rangle$ ways from permutations ρ that have $k-1$ ascents. The desired recurrence is

$$\left\langle{n \atop k}\right\rangle = (k+1)\left\langle{n-1 \atop k}\right\rangle + (n-k)\left\langle{n-1 \atop k-1}\right\rangle, \qquad \text{integer } n>0. \tag{6.35}$$

Once again we start the recurrence off by setting

$$\left\langle{0 \atop k}\right\rangle = [k=0], \qquad \text{integer } k, \tag{6.36}$$

and we will assume that $\left\langle{n \atop k}\right\rangle = 0$ when $k<0$.

Eulerian numbers are useful primarily because they provide an unusual connection between ordinary powers and consecutive binomial coefficients:

$$x^n = \sum_k \left\langle {n \atop k} \right\rangle \binom{x+k}{n}, \qquad \text{integer } n \geq 0. \tag{6.37}$$

(This is "Worpitzky's identity" [308].) For example, we have

$$x^2 = \binom{x}{2} + \binom{x+1}{2},$$

$$x^3 = \binom{x}{3} + 4\binom{x+1}{3} + \binom{x+2}{3},$$

$$x^4 = \binom{x}{4} + 11\binom{x+1}{4} + 11\binom{x+2}{4} + \binom{x+3}{4},$$

and so on. It's easy to prove (6.37) by induction (exercise 14).

Incidentally, (6.37) gives us yet another way to obtain the sum of the first n squares: We have $k^2 = \left\langle {2 \atop 0} \right\rangle \binom{k}{2} + \left\langle {2 \atop 1} \right\rangle \binom{k+1}{2} = \binom{k}{2} + \binom{k+1}{2}$, hence

$$\begin{aligned} 1^2 + 2^2 + \cdots + n^2 &= \left(\binom{1}{2} + \binom{2}{2} + \cdots + \binom{n}{2}\right) + \left(\binom{2}{2} + \binom{3}{2} + \cdots + \binom{n+1}{2}\right) \\ &= \binom{n+1}{3} + \binom{n+2}{3} = \tfrac{1}{6}(n+1)n\big((n-1) + (n+2)\big). \end{aligned}$$

The Eulerian recurrence (6.35) is a bit more complicated than the Stirling recurrences (6.3) and (6.8), so we don't expect the numbers $\left\langle {n \atop k} \right\rangle$ to satisfy as many simple identities. Still, there are a few:

$$\left\langle {n \atop m} \right\rangle = \sum_{k=0}^{m} \binom{n+1}{k} (m+1-k)^n (-1)^k; \tag{6.38}$$

$$m! \left\{ {n \atop m} \right\} = \sum_k \left\langle {n \atop k} \right\rangle \binom{k}{n-m}; \tag{6.39}$$

$$\left\langle {n \atop m} \right\rangle = \sum_k \left\{ {n \atop k} \right\} \binom{n-k}{m} (-1)^{n-k-m} k!. \tag{6.40}$$

If we multiply (6.39) by z^{n-m} and sum on m, we get $\sum_m \left\{ {n \atop m} \right\} m!\, z^{n-m} = \sum_k \left\langle {n \atop k} \right\rangle (z+1)^k$. Replacing z by $z-1$ and equating coefficients of z^k gives (6.40). Thus the last two of these identities are essentially equivalent. The first identity, (6.38), gives us special values when m is small:

$$\left\langle {n \atop 0} \right\rangle = 1; \quad \left\langle {n \atop 1} \right\rangle = 2^n - n - 1; \quad \left\langle {n \atop 2} \right\rangle = 3^n - (n+1)2^n + \binom{n+1}{2}.$$

Table 256 Second-order Eulerian triangle.

n	$\left\langle\!\!\left\langle {n \atop 0} \right\rangle\!\!\right\rangle$	$\left\langle\!\!\left\langle {n \atop 1} \right\rangle\!\!\right\rangle$	$\left\langle\!\!\left\langle {n \atop 2} \right\rangle\!\!\right\rangle$	$\left\langle\!\!\left\langle {n \atop 3} \right\rangle\!\!\right\rangle$	$\left\langle\!\!\left\langle {n \atop 4} \right\rangle\!\!\right\rangle$	$\left\langle\!\!\left\langle {n \atop 5} \right\rangle\!\!\right\rangle$	$\left\langle\!\!\left\langle {n \atop 6} \right\rangle\!\!\right\rangle$	$\left\langle\!\!\left\langle {n \atop 7} \right\rangle\!\!\right\rangle$	$\left\langle\!\!\left\langle {n \atop 8} \right\rangle\!\!\right\rangle$
0	1								
1	1	0							
2	1	2	0						
3	1	8	6	0					
4	1	22	58	24	0				
5	1	52	328	444	120	0			
6	1	114	1452	4400	3708	720	0		
7	1	240	5610	32120	58140	33984	5040	0	
8	1	494	19950	195800	644020	785304	341136	40320	0

We needn't dwell further on Eulerian numbers here; it's usually sufficient simply to know that they exist, and to have a list of basic identities to fall back on when the need arises. However, before we leave this topic, we should take note of yet another triangular pattern of coefficients, shown in Table 256. We call these "second-order Eulerian numbers" $\left\langle\!\left\langle {n \atop k} \right\rangle\!\right\rangle$, because they satisfy a recurrence similar to (6.35) but with n replaced by $2n-1$ in one place:

$$\left\langle\!\!\left\langle {n \atop k} \right\rangle\!\!\right\rangle = (k+1)\left\langle\!\!\left\langle {n-1 \atop k} \right\rangle\!\!\right\rangle + (2n-1-k)\left\langle\!\!\left\langle {n-1 \atop k-1} \right\rangle\!\!\right\rangle. \qquad (6.41)$$

These numbers have a curious combinatorial interpretation, first noticed by Gessel and Stanley [118]: If we form permutations of the multiset $\{1,1,2,2,\ldots,n,n\}$ with the special property that all numbers between the two occurrences of m are greater than m, for $1 \leqslant m \leqslant n$, then $\left\langle\!\left\langle {n \atop k} \right\rangle\!\right\rangle$ is the number of such permutations that have k ascents. For example, there are eight suitable single-ascent permutations of $\{1,1,2,2,3,3\}$:

113322, 133221, 221331, 221133, 223311, 233211, 331122, 331221.

Thus $\left\langle\!\left\langle {3 \atop 1} \right\rangle\!\right\rangle = 8$. The multiset $\{1,1,2,2,\ldots,n,n\}$ has a total of

$$\sum_k \left\langle\!\!\left\langle {n \atop k} \right\rangle\!\!\right\rangle = (2n-1)(2n-3)\ldots(1) = \frac{(2n)^{\underline{n}}}{2^n} \qquad (6.42)$$

suitable permutations, because the two appearances of n must be adjacent and there are $2n-1$ places to insert them within a permutation for $n-1$. For example, when $n = 3$ the permutation 1221 has five insertion points, yielding 331221, 133221, 123321, 122331, and 122133. Recurrence (6.41) can be proved by extending the argument we used for ordinary Eulerian numbers.

Second-order Eulerian numbers are important chiefly because of their connection with Stirling numbers [119]: We have, by induction on n,

$$\left\{ {x \atop x-n} \right\} = \sum_k \left\langle\!\!\left\langle {n \atop k} \right\rangle\!\!\right\rangle \binom{x+n-1-k}{2n}, \qquad \text{integer } n \geqslant 0; \tag{6.43}$$

$$\left[{x \atop x-n} \right] = \sum_k \left\langle\!\!\left\langle {n \atop k} \right\rangle\!\!\right\rangle \binom{x+k}{2n}, \qquad \text{integer } n \geqslant 0. \tag{6.44}$$

For example,

$$\left\{ {x \atop x-1} \right\} = \binom{x}{2}, \qquad \left[{x \atop x-1} \right] = \binom{x}{2};$$

$$\left\{ {x \atop x-2} \right\} = \binom{x+1}{4} + 2\binom{x}{4}, \qquad \left[{x \atop x-2} \right] = \binom{x}{4} + 2\binom{x+1}{4};$$

$$\left\{ {x \atop x-3} \right\} = \binom{x+2}{6} + 8\binom{x+1}{6} + 6\binom{x}{6},$$

$$\left[{x \atop x-3} \right] = \binom{x}{6} + 8\binom{x+1}{6} + 6\binom{x+2}{6}.$$

(We already encountered the case $n = 1$ in (6.7).) These identities hold whenever x is an integer and n is a nonnegative integer. Since the right-hand sides are polynomials in x, we can use (6.43) and (6.44) to define Stirling numbers $\left\{ {x \atop x-n} \right\}$ and $\left[{x \atop x-n} \right]$ for arbitrary real (or complex) values of x.

If $n > 0$, these polynomials $\left\{ {x \atop x-n} \right\}$ and $\left[{x \atop x-n} \right]$ are zero when $x = 0$, $x = 1$, ..., and $x = n$; therefore they are divisible by $(x-0)$, $(x-1)$, ..., and $(x-n)$. It's interesting to look at what's left after these known factors are divided out. We define the *Stirling polynomials* $\sigma_n(x)$ by the rule

$$\sigma_n(x) = \left[{x \atop x-n} \right] \Big/ \bigl(x(x-1)\ldots(x-n)\bigr). \tag{6.45}$$

(The degree of $\sigma_n(x)$ is $n-1$.) The first few cases are

So $1/x$ is a polynomial?

(Sorry about that.)

$$\begin{aligned}
\sigma_0(x) &= 1/x;\\
\sigma_1(x) &= 1/2;\\
\sigma_2(x) &= (3x-1)/24;\\
\sigma_3(x) &= (x^2-x)/48;\\
\sigma_4(x) &= (15x^3-30x^2+5x+2)/5760.
\end{aligned}$$

They can be computed via the second-order Eulerian numbers; for example,

$$\sigma_3(x) = \bigl((x-4)(x-5) + 8(x+1)(x-4) + 6(x+2)(x+1)\bigr)/6!.$$

Table 258 Stirling convolution formulas.

$$rs\sum_{k=0}^{n}\sigma_k(r)\,\sigma_{n-k}(s) = (r+s)\sigma_n(r+s) \tag{6.46}$$

$$s\sum_{k=0}^{n}k\sigma_k(r)\,\sigma_{n-k}(s) = n\sigma_n(r+s) \tag{6.47}$$

$$rs\sum_{k=0}^{n}\sigma_k(r+k)\,\sigma_{n-k}(s+n-k) = (r+s)\sigma_n(r+s+n) \tag{6.48}$$

$$s\sum_{k=0}^{n}k\sigma_k(r+k)\,\sigma_{n-k}(s+n-k) = n\sigma_n(r+s+n) \tag{6.49}$$

$$\left\{ {n \atop m} \right\} = (-1)^{n-m+1}\frac{n!}{(m-1)!}\sigma_{n-m}(-m) \tag{6.50}$$

$$\left[{n \atop m} \right] = \frac{n!}{(m-1)!}\sigma_{n-m}(n) \tag{6.51}$$

It turns out that these polynomials satisfy two very pretty identities:

$$\left(\frac{ze^z}{e^z-1}\right)^x = x\sum_{n\geqslant 0}\sigma_n(x)\,z^n\,; \tag{6.52}$$

$$\left(\frac{1}{z}\ln\frac{1}{1-z}\right)^x = x\sum_{n\geqslant 0}\sigma_n(x+n)\,z^n\,; \tag{6.53}$$

Therefore we can obtain general convolution formulas for Stirling numbers, as we did for binomial coefficients in Table 202; the results appear in Table 258. When a sum of Stirling numbers doesn't fit the identities of Table 250 or 251, Table 258 may be just the ticket. (An example appears later in this chapter, following equation (6.100). Exercise 7.19 discusses the general principles of convolutions based on identities like (6.52) and (6.53).)

6.3 HARMONIC NUMBERS

It's time now to take a closer look at harmonic numbers, which we first met back in Chapter 2:

$$H_n = 1+\frac{1}{2}+\frac{1}{3}+\cdots+\frac{1}{n} = \sum_{k=1}^{n}\frac{1}{k}\,, \qquad \text{integer } n \geqslant 0. \tag{6.54}$$

These numbers appear so often in the analysis of algorithms that computer scientists need a special notation for them. We use H_n, the 'H' standing for "harmonic," since a tone of wavelength $1/n$ is called the nth harmonic of a tone whose wavelength is 1. The first few values look like this:

n	0	1	2	3	4	5	6	7	8	9	10
H_n	0	1	$\frac{3}{2}$	$\frac{11}{6}$	$\frac{25}{12}$	$\frac{137}{60}$	$\frac{49}{20}$	$\frac{363}{140}$	$\frac{761}{280}$	$\frac{7129}{2520}$	$\frac{7381}{2520}$

Exercise 21 shows that H_n is never an integer when $n > 1$.

Here's a card trick, based on an idea by R. T. Sharp [264], that illustrates how the harmonic numbers arise naturally in simple situations. Given n cards and a table, we'd like to create the largest possible overhang by stacking the cards up over the table's edge, subject to the laws of gravity:

This must be Table 259.

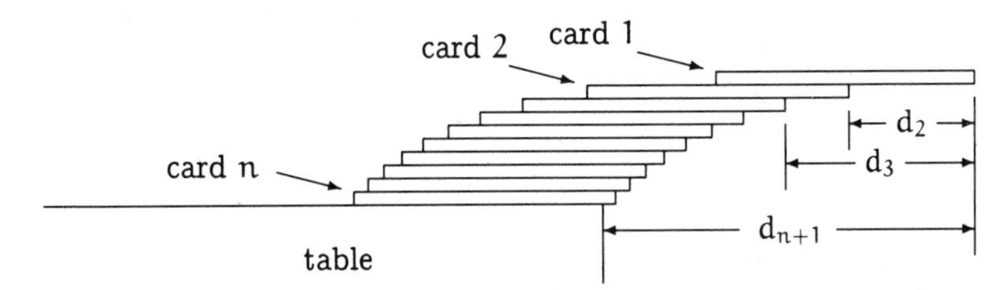

To define the problem a bit more, we require the edges of the cards to be parallel to the edge of the table; otherwise we could increase the overhang by rotating the cards so that their corners stick out a little farther. And to make the answer simpler, we assume that each card is 2 units long.

With one card, we get maximum overhang when its center of gravity is just above the edge of the table. The center of gravity is in the middle of the card, so we can create half a cardlength, or 1 unit, of overhang.

With two cards, it's not hard to convince ourselves that we get maximum overhang when the center of gravity of the top card is just above the edge of the second card, and the center of gravity of both cards combined is just above the edge of the table. The joint center of gravity of two cards will be in the middle of their common part, so we are able to achieve an additional half unit of overhang.

This pattern suggests a general method, where we place cards so that the center of gravity of the top k cards lies just above the edge of the $k+1$st card (which supports those top k). The table plays the role of the $n+1$st card. To express this condition algebraically, we can let d_k be the distance from the extreme edge of the top card to the corresponding edge of the kth card from the top. Then $d_1 = 0$, and we want to make d_{k+1} the center of gravity of the first k cards:

$$d_{k+1} = \frac{(d_1+1)+(d_2+1)+\cdots+(d_k+1)}{k}, \quad \text{for } 1 \leqslant k \leqslant n. \qquad (6.55)$$

(The center of gravity of k objects, having respective weights $w_1, \ldots, w_k$ and having respective centers of gravity at positions $p_1, \ldots p_k$, is at position $(w_1p_1 + \cdots + w_kp_k)/(w_1 + \cdots + w_k)$.) We can rewrite this recurrence in two equivalent forms

$$\begin{aligned} kd_{k+1} &= k + d_1 + \cdots + d_{k-1} + d_k\,, && k \geqslant 0;\\ (k-1)d_k &= k - 1 + d_1 + \cdots + d_{k-1}\,, && k \geqslant 1. \end{aligned}$$

Subtracting these equations tells us that

$$kd_{k+1} - (k-1)d_k = 1 + d_k\,, \qquad k \geqslant 1;$$

hence $d_{k+1} = d_k + 1/k$. The second card will be offset half a unit past the third, which is a third of a unit past the fourth, and so on. The general formula

$$d_{k+1} = H_k \tag{6.56}$$

follows by induction, and if we set $k = n$ we get $d_{n+1} = H_n$ as the total overhang when n cards are stacked as described.

Could we achieve greater overhang by holding back, not pushing each card to an extreme position but storing up "potential gravitational energy" for a later advance? No; any well-balanced card placement has

$$d_{k+1} \leqslant \frac{(1 + d_1) + (1 + d_2) + \cdots + (1 + d_k)}{k}\,, \qquad 1 \leqslant k \leqslant n.$$

Furthermore $d_1 = 0$. It follows by induction that $d_{k+1} \leqslant H_k$.

Notice that it doesn't take too many cards for the top one to be completely past the edge of the table. We need an overhang of more than one cardlength, which is 2 units. The first harmonic number to exceed 2 is $H_4 = \frac{25}{12}$, so we need only four cards.

And with 52 cards we have an H_{52}-unit overhang, which turns out to be $H_{52}/2 \approx 2.27$ cardlengths. (We will soon learn a formula that tells us how to compute an approximate value of H_n for large n without adding up a whole bunch of fractions.)

Anyone who actually tries to achieve this maximum overhang with 52 cards is probably not dealing with a full deck—or maybe he's a real joker.

An amusing problem called the "worm on the rubber band" shows harmonic numbers in another guise. A slow but persistent worm, W, starts at one end of a meter-long rubber band and crawls one centimeter per minute toward the other end. At the end of each minute, an equally persistent keeper of the band, K, whose sole purpose in life is to frustrate W, stretches it one meter. Thus after one minute of crawling, W is 1 centimeter from the start and 99 from the finish; then K stretches it one meter. During the stretching operation W maintains his relative position, 1% from the start and 99% from the finish; so W is now 2 cm from the starting point and 198 cm from the goal. After W crawls for another minute the score is 3 cm traveled and 197 to go; but K stretches, and the distances become 4.5 and 295.5. And so on. Does the worm ever reach the finish? He keeps moving, but the goal seems to move away even faster. (We're assuming an infinite longevity for K and W, an infinite elasticity of the band, and an infinitely tiny worm.)

Metric units make this problem more scientific.

Let's write down some formulas. When K stretches the rubber band, the fraction of it that W has crawled stays the same. Thus he crawls 1/100th of it the first minute, 1/200th the second, 1/300th the third, and so on. After n minutes the fraction of the band that he's crawled is

$$\frac{1}{100}\left(\frac{1}{1}+\frac{1}{2}+\frac{1}{3}+\cdots+\frac{1}{n}\right) = \frac{H_n}{100}. \tag{6.57}$$

So he reaches the finish if H_n ever surpasses 100.

We'll see how to estimate H_n for large n soon; for now, let's simply check our analysis by considering how "Superworm" would perform in the same situation. Superworm, unlike W, can crawl 50 cm per minute; so she will crawl $H_n/2$ of the band length after n minutes, according to the argument we just gave. If our reasoning is correct, Superworm should finish before n reaches 4, since $H_4 > 2$. And yes, a simple calculation shows that Superworm has only $33\frac{1}{3}$ cm left to travel after three minutes have elapsed. She finishes in 3 minutes and 40 seconds flat.

A flatworm, eh?

Harmonic numbers appear also in Stirling's triangle. Let's try to find a closed form for $\left[{n \atop 2}\right]$, the number of permutations of n objects that have exactly two cycles. Recurrence (6.8) tells us that

$$\begin{aligned}\left[{n+1 \atop 2}\right] &= n\left[{n \atop 2}\right] + \left[{n \atop 1}\right]\\ &= n\left[{n \atop 2}\right] + (n-1)!\,, \qquad \text{if } n > 0;\end{aligned}$$

and this recurrence is a natural candidate for the summation factor technique of Chapter 2:

$$\frac{1}{n!}\left[{n+1 \atop 2}\right] = \frac{1}{(n-1)!}\left[{n \atop 2}\right] + \frac{1}{n}.$$

Unfolding this recurrence tells us that $\frac{1}{n!}\left[{n+1 \atop 2}\right] = H_n$; hence

$$\left[{n+1 \atop 2}\right] = n!H_n\,. \tag{6.58}$$

We proved in Chapter 2 that the harmonic series $\sum_k 1/k$ diverges, which means that H_n gets arbitrarily large as $n \to \infty$. But our proof was indirect;

we found that a certain infinite sum (2.58) gave different answers when it was rearranged, hence $\sum_k 1/k$ could not be bounded. The fact that $H_n \to \infty$ seems counter-intuitive, because it implies among other things that a large enough stack of cards will overhang a table by a mile or more, and that the worm W will eventually reach the end of his rope. Let us therefore take a closer look at the size of H_n when n is large.

The simplest way to see that $H_n \to \infty$ is probably to group its terms according to powers of 2. We put one term into group 1, two terms into group 2, four into group 3, eight into group 4, and so on:

$$\underbrace{\frac{1}{1}}_{\text{group 1}} + \underbrace{\frac{1}{2}+\frac{1}{3}}_{\text{group 2}} + \underbrace{\frac{1}{4}+\frac{1}{5}+\frac{1}{6}+\frac{1}{7}}_{\text{group 3}} + \underbrace{\frac{1}{8}+\frac{1}{9}+\frac{1}{10}+\frac{1}{11}+\frac{1}{12}+\frac{1}{13}+\frac{1}{14}+\frac{1}{15}}_{\text{group 4}} + \cdots .$$

Both terms in group 2 are between $\frac{1}{4}$ and $\frac{1}{2}$, so the sum of that group is between $2 \cdot \frac{1}{4} = \frac{1}{2}$ and $2 \cdot \frac{1}{2} = 1$. All four terms in group 3 are between $\frac{1}{8}$ and $\frac{1}{4}$, so their sum is also between $\frac{1}{2}$ and 1. In fact, each of the 2^{k-1} terms in group k is between 2^{-k} and 2^{1-k}; hence the sum of each individual group is between $\frac{1}{2}$ and 1.

This grouping procedure tells us that if n is in group k, we must have $H_n > k/2$ and $H_n \leqslant k$ (by induction on k). Thus $H_n \to \infty$, and in fact

$$\frac{\lfloor \lg n \rfloor + 1}{2} < H_n \leqslant \lfloor \lg n \rfloor + 1 . \tag{6.59}$$

We now know H_n within a factor of 2. Although the harmonic numbers approach infinity, they approach it only logarithmically — that is, quite slowly.

We should call them the worm numbers, they're so slow.

Better bounds can be found with just a little more work and a dose of calculus. We learned in Chapter 2 that H_n is the discrete analog of the continuous function $\ln n$. The natural logarithm is defined as the area under a curve, so a geometric comparison is suggested:

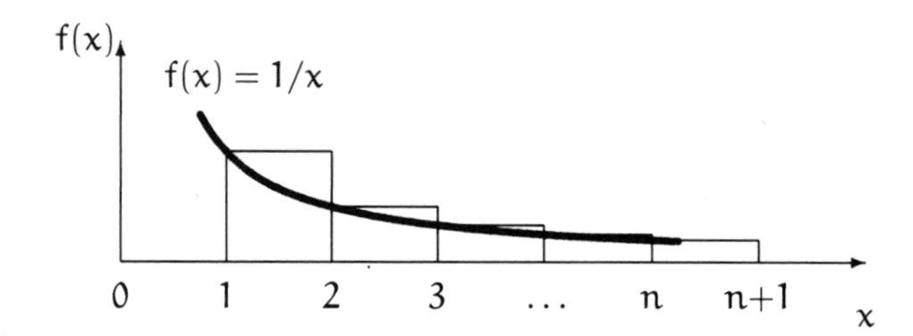

The area under the curve between 1 and n, which is $\int_1^n dx/x = \ln n$, is less than the area of the n rectangles, which is $\sum_{k=1}^{n} 1/k = H_n$. Thus $\ln n < H_n$; this is a sharper result than we had in (6.59). And by placing the rectangles

"I now see a way too how y^e aggregate of y^e termes of Musicall progressions may bee found (much after y^e same manner) by Logarithms, but y^e calculations for finding out those rules would bee still more troublesom."
—I. Newton [223]

a little differently, we get a similar upper bound:

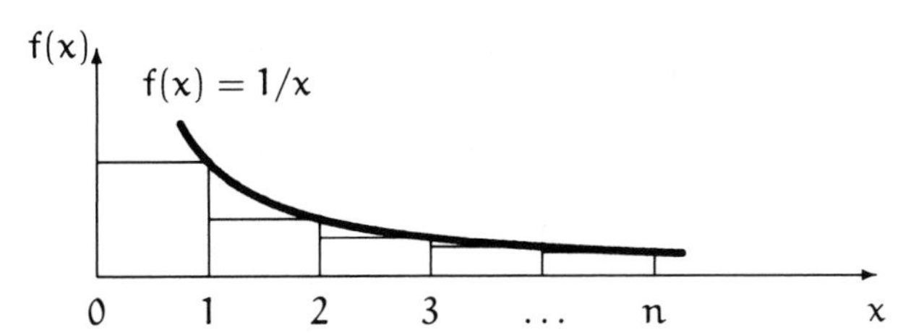

This time the area of the n rectangles, H_n, is less than the area of the first rectangle plus the area under the curve. We have proved that

$$\ln n \;<\; H_n \;<\; \ln n + 1\,, \qquad \text{for } n > 1. \tag{6.60}$$

We now know the value of H_n with an error of at most 1.

"Second order" harmonic numbers $H_n^{(2)}$ arise when we sum the squares of the reciprocals, instead of summing simply the reciprocals:

$$H_n^{(2)} \;=\; 1 + \frac{1}{4} + \frac{1}{9} + \cdots + \frac{1}{n^2} \;=\; \sum_{k=1}^{n} \frac{1}{k^2}\,.$$

Similarly, we define harmonic numbers of order r by summing $(-r)$th powers:

$$H_n^{(r)} \;=\; \sum_{k=1}^{n} \frac{1}{k^r}\,. \tag{6.61}$$

If $r > 1$, these numbers approach a limit as $n \to \infty$; we noted in Chapter 4 that this limit is conventionally called Riemann's zeta function:

$$\zeta(r) \;=\; H_\infty^{(r)} \;=\; \sum_{k\geqslant 1} \frac{1}{k^r}\,. \tag{6.62}$$

Euler discovered a neat way to use generalized harmonic numbers to approximate the ordinary ones, $H_n^{(1)}$. Let's consider the infinite series

$$\ln\left(\frac{k}{k-1}\right) \;=\; \frac{1}{k} + \frac{1}{2k^2} + \frac{1}{3k^3} + \frac{1}{4k^4} + \cdots\,, \tag{6.63}$$

which converges when $k > 1$. The left-hand side is $\ln k - \ln(k-1)$; therefore if we sum both sides for $2 \leqslant k \leqslant n$ the left-hand sum telescopes and we get

$$\begin{aligned} \ln n - \ln 1 \;&=\; \sum_{k=2}^{n}\left(\frac{1}{k} + \frac{1}{2k^2} + \frac{1}{3k^3} + \frac{1}{4k^4} + \cdots\right) \\ &=\; (H_n - 1) + \tfrac{1}{2}(H_n^{(2)} - 1) + \tfrac{1}{3}(H_n^{(3)} - 1) + \tfrac{1}{4}(H_n^{(4)} - 1) + \cdots\,. \end{aligned}$$

Rearranging, we have an expression for the difference between H_n and $\ln n$:

$$H_n - \ln n = 1 - \tfrac{1}{2}(H_n^{(2)}-1) - \tfrac{1}{3}(H_n^{(3)}-1) - \tfrac{1}{4}(H_n^{(4)}-1) - \cdots .$$

When $n \to \infty$, the right-hand side approaches the limiting value

$$1 - \tfrac{1}{2}(\zeta(2)-1) - \tfrac{1}{3}(\zeta(3)-1) - \tfrac{1}{4}(\zeta(4)-1) - \cdots ,$$

which is now known as *Euler's constant* and conventionally denoted by the Greek letter γ. In fact, $\zeta(r) - 1$ is approximately $1/2^r$, so this infinite series converges rather rapidly and we can compute the decimal value

"Huius igitur quantitatis constantis C valorem deteximus, quippe est C = 0,577218."
— L. Euler [83]

$$\gamma = 0.5772156649\ldots . \tag{6.64}$$

Euler's argument establishes the limiting relation

$$\lim_{n\to\infty}(H_n - \ln n) = \gamma ; \tag{6.65}$$

thus H_n lies about 58% of the way between the two extremes in (6.60). We are gradually homing in on its value.

Further refinements are possible, as we will see in Chapter 9. We will prove, for example, that

$$H_n = \ln n + \gamma + \frac{1}{2n} - \frac{1}{12n^2} + \frac{\epsilon_n}{120n^4}, \qquad 0 < \epsilon_n < 1. \tag{6.66}$$

This formula allows us to conclude that the millionth harmonic number is

$$H_{1000000} \approx 14.3927267228657236313811275 ,$$

without adding up a million fractions. Among other things, this implies that a stack of a million cards can overhang the edge of a table by more than seven cardlengths.

What does (6.66) tell us about the worm on the rubber band? Since H_n is unbounded, the worm will definitely reach the end, when H_n first exceeds 100. Our approximation to H_n says that this will happen when n is approximately

$$e^{100-\gamma} \approx e^{99.423} .$$

In fact, exercise 9.49 proves that the critical value of n is either $\lfloor e^{100-\gamma} \rfloor$ or $\lceil e^{100-\gamma} \rceil$. We can imagine W's triumph when he crosses the finish line at last, much to K's chagrin, some 287 decillion centuries after his long crawl began. (The rubber band will have stretched to more than 10^{27} light years long; its molecules will be pretty far apart.)

Well, they can't really go at it this long; the world will have ended much earlier, when the Tower of Brahma is fully transferred.

6.4 HARMONIC SUMMATION

Now let's look at some sums involving harmonic numbers, starting with a review of a few ideas we learned in Chapter 2. We proved in (2.36) and (2.57) that

$$\sum_{0\leqslant k<n} H_k = nH_n - n\,; \tag{6.67}$$

$$\sum_{0\leqslant k<n} kH_k = \frac{n(n-1)}{2}H_n - \frac{n(n-1)}{4}\,. \tag{6.68}$$

Let's be bold and take on a more general sum, which includes both of these as special cases: What is the value of

$$\sum_{0\leqslant k<n} \binom{k}{m} H_k\,,$$

when m is a nonnegative integer?

The approach that worked best for (6.67) and (6.68) in Chapter 2 was called *summation by parts*. We wrote the summand in the form $u(k)\Delta v(k)$, and we applied the general identity

$$\sum\nolimits_a^b u(x)\Delta v(x)\,\delta x = u(x)v(x)\Big|_a^b - \sum\nolimits_a^b v(x+1)\Delta u(x)\,\delta x\,. \tag{6.69}$$

Remember? The sum that faces us now, $\sum_{0\leqslant k<n}\binom{k}{m}H_k$, is a natural for this method because we can let

$$u(k) = H_k\,, \qquad \Delta u(k) = H_{k+1} - H_k = \frac{1}{k+1}\,;$$

$$v(k) = \binom{k}{m+1}\,, \qquad \Delta v(k) = \binom{k+1}{m+1} - \binom{k}{m+1} = \binom{k}{m}\,.$$

(In other words, harmonic numbers have a simple Δ and binomial coefficients have a simple Δ^{-1}, so we're in business.) Plugging into (6.69) yields

$$\begin{aligned}\sum_{0\leqslant k<n}\binom{k}{m}H_k = \sum\nolimits_0^n \binom{x}{m}H_x\,\delta x &= \binom{x}{m+1}H_x\Big|_0^n - \sum\nolimits_0^n \binom{x+1}{m+1}\frac{\delta x}{x+1}\\ &= \binom{n}{m+1}H_n - \sum_{0\leqslant k<n}\binom{k+1}{m+1}\frac{1}{k+1}\,.\end{aligned}$$

The remaining sum is easy, since we can absorb the $(k+1)^{-1}$ using our old standby, equation (5.5):

$$\sum_{0\leqslant k<n}\binom{k+1}{m+1}\frac{1}{k+1} = \sum_{0\leqslant k<n}\binom{k}{m}\frac{1}{m+1} = \binom{n}{m+1}\frac{1}{m+1}\,.$$

Thus we have the answer we seek:

$$\sum_{0\leqslant k<n}\binom{k}{m}H_k = \binom{n}{m+1}\left(H_n - \frac{1}{m+1}\right). \qquad (6.70)$$

(This checks nicely with (6.67) and (6.68) when $m = 0$ and $m = 1$.)

The next example sum uses division instead of multiplication: Let us try to evaluate

$$S_n = \sum_{k=1}^{n} \frac{H_k}{k}.$$

If we expand H_k by its definition, we obtain a double sum,

$$S_n = \sum_{1\leqslant j\leqslant k\leqslant n} \frac{1}{j\cdot k}.$$

Now another method from Chapter 2 comes to our aid; equation (2.33) tells us that

$$S_n = \frac{1}{2}\left(\left(\sum_{k=1}^{n}\frac{1}{k}\right)^2 + \sum_{k=1}^{n}\frac{1}{k^2}\right) = \frac{1}{2}\left(H_n^2 + H_n^{(2)}\right). \qquad (6.71)$$

It turns out that we could also have obtained this answer in another way if we had tried to sum by parts (see exercise 26).

Now let's try our hands at a more difficult problem [291], which doesn't submit to summation by parts:

$$U_n = \sum_{k\geqslant 1}\binom{n}{k}\frac{(-1)^{k-1}}{k}(n-k)^n, \qquad \text{integer } n \geqslant 1.$$

(This sum doesn't explicitly mention harmonic numbers either; but who knows when they might turn up?)

(Not to give the answer away or anything.)

We will solve this problem in two ways, one by grinding out the answer and the other by being clever and/or lucky. First, the grinder's approach. We expand $(n-k)^n$ by the binomial theorem, so that the troublesome k in the denominator will combine with the numerator:

$$\begin{aligned} U_n &= \sum_{k\geqslant 1}\binom{n}{k}\frac{(-1)^{k-1}}{k}\sum_j\binom{n}{j}(-k)^j n^{n-j} \\ &= \sum_j\binom{n}{j}(-1)^{j-1}n^{n-j}\sum_{k\geqslant 1}\binom{n}{k}(-1)^k k^{j-1}. \end{aligned}$$

This isn't quite the mess it seems, because the k^{j-1} in the inner sum is a polynomial in k, and identity (5.40) tells us that we are simply taking the

nth difference of this polynomial. Almost; first we must clean up a few things. For one, k^{j-1} isn't a polynomial if $j = 0$; so we will need to split off that term and handle it separately. For another, we're missing the term $k = 0$ from the formula for nth difference; that term is nonzero when $j = 1$, so we had better restore it (and subtract it out again). The result is

$$\begin{aligned} U_n &= \sum_{j\geq 1}\binom{n}{j}(-1)^{j-1}n^{n-j}\sum_{k\geq 0}\binom{n}{k}(-1)^k k^{j-1} \\ &\quad -\sum_{j\geq 1}\binom{n}{j}(-1)^{j-1}n^{n-j}\binom{n}{0}0^{j-1} \\ &\quad -\binom{n}{0}n^n\sum_{k\geq 1}\binom{n}{k}(-1)^k k^{-1}. \end{aligned}$$

OK, now the top line (the only remaining double sum) is zero: It's the sum of multiples of nth differences of polynomials of degree less than n, and such nth differences are zero. The second line is zero except when $j = 1$, when it equals $-n^n$. So the third line is the only residual difficulty; we have reduced the original problem to a much simpler sum:

$$U_n = n^n(T_n - 1), \qquad \text{where } T_n = \sum_{k\geq 1}\binom{n}{k}\frac{(-1)^{k-1}}{k}. \tag{6.72}$$

For example, $U_3 = \binom{3}{1}\frac{8}{1} - \binom{3}{2}\frac{1}{2} = \frac{45}{2}$; $T_3 = \binom{3}{1}\frac{1}{1} - \binom{3}{2}\frac{1}{2} + \binom{3}{3}\frac{1}{3} = \frac{11}{6}$; hence $U_3 = 27(T_3 - 1)$ as claimed.

How can we evaluate T_n? One way is to replace $\binom{n}{k}$ by $\binom{n-1}{k} + \binom{n-1}{k-1}$, obtaining a simple recurrence for T_n in terms of T_{n-1}. But there's a more instructive way: We had a similar formula in (5.41), namely

$$\sum_k \binom{n}{k}\frac{(-1)^k}{x+k} = \frac{n!}{x(x+1)\ldots(x+n)}.$$

If we subtract out the term for $k = 0$ and set $x = 0$, we get $-T_n$. So let's do it:

$$\begin{aligned} T_n &= \left(\frac{1}{x} - \frac{n!}{x(x+1)\ldots(x+n)}\right)\bigg|_{x=0} \\ &= \left(\frac{(x+1)\ldots(x+n) - n!}{x(x+1)\ldots(x+n)}\right)\bigg|_{x=0} \\ &= \left(\frac{x^n{n+1 \brack n+1} + \cdots + x{n+1 \brack 2} + {n+1 \brack 1} - n!}{x(x+1)\ldots(x+n)}\right)\bigg|_{x=0} = \frac{1}{n!}{n+1 \brack 2}. \end{aligned}$$

(We have used the expansion (6.11) of $(x+1)\ldots(x+n) = x^{\overline{n+1}}/x$; we can divide x out of the numerator because $\left[{n+1 \atop 1}\right] = n!$.) But we know from (6.58) that $\left[{n+1 \atop 2}\right] = n!\,H_n$; hence $T_n = H_n$, and we have the answer:

$$U_n = n^n(H_n - 1). \tag{6.73}$$

That's one approach. The other approach will be to try to evaluate a much more general sum,

$$U_n(x,y) = \sum_{k\geqslant 1}\binom{n}{k}\frac{(-1)^{k-1}}{k}(x+ky)^n, \qquad \text{integer } n \geqslant 0; \tag{6.74}$$

the value of the original U_n will drop out as the special case $U_n(n,-1)$. (We are encouraged to try for more generality because the previous derivation "threw away" most of the details of the given problem; somehow those details must be irrelevant, because the nth difference wiped them away.)

We could replay the previous derivation with small changes and discover the value of $U_n(x,y)$. Or we could replace $(x+ky)^n$ by $(x+ky)^{n-1}(x+ky)$ and then replace $\binom{n}{k}$ by $\binom{n-1}{k} + \binom{n-1}{k-1}$, leading to the recurrence

$$U_n(x,y) = xU_{n-1}(x,y) + x^n/n + yx^{n-1}; \tag{6.75}$$

this can readily be solved with a summation factor (exercise 5.)

But it's easiest to use another trick that worked to our advantage in Chapter 2: differentiation. The derivative of $U_n(x,y)$ with respect to y brings out a k that cancels with the k in the denominator, and the resulting sum is trivial:

$$\begin{aligned}\frac{\partial}{\partial y}U_n(x,y) &= \sum_{k\geqslant 1}\binom{n}{k}(-1)^{k-1}n(x+ky)^{n-1}\\ &= \binom{n}{0}nx^{n-1} - \sum_{k\geqslant 0}\binom{n}{k}(-1)^k n(x+ky)^{n-1} = nx^{n-1}.\end{aligned}$$

(Once again, the nth difference of a polynomial of degree $< n$ has vanished.)

We've proved that the derivative of $U_n(x,y)$ with respect to y is nx^{n-1}, independent of y. In general, if $f'(y) = c$ then $f(y) = f(0) + cy$; therefore we must have $U_n(x,y) = U_n(x,0) + nx^{n-1}y$.

The remaining task is to determine $U_n(x,0)$. But $U_n(x,0)$ is just x^n times the sum $T_n = H_n$ we've already considered in (6.72); therefore the general sum in (6.74) has the closed form

$$U_n(x,y) = x^n H_n + nx^{n-1}y. \tag{6.76}$$

In particular, the solution to the original problem is $U_n(n,-1) = n^n(H_n - 1)$.

6.5 BERNOULLI NUMBERS

The next important sequence of numbers on our agenda is named after Jakob Bernoulli (1654–1705), who discovered curious relationships while working out the formulas for sums of mth powers [22]. Let's write

$$S_m(n) = 0^m + 1^m + \cdots + (n-1)^m = \sum_{k=0}^{n-1} k^m = \sum\nolimits_0^n x^m \, \delta x. \qquad (6.77)$$

(Thus, when $m > 0$ we have $S_m(n) = H_{n-1}^{(-m)}$ in the notation of generalized harmonic numbers.) Bernoulli looked at the following sequence of formulas and spotted a pattern:

$$\begin{aligned}
S_0(n) &= n \\
S_1(n) &= \tfrac{1}{2}n^2 - \tfrac{1}{2}n \\
S_2(n) &= \tfrac{1}{3}n^3 - \tfrac{1}{2}n^2 + \tfrac{1}{6}n \\
S_3(n) &= \tfrac{1}{4}n^4 - \tfrac{1}{2}n^3 + \tfrac{1}{4}n^2 \\
S_4(n) &= \tfrac{1}{5}n^5 - \tfrac{1}{2}n^4 + \tfrac{1}{3}n^3 - \tfrac{1}{30}n \\
S_5(n) &= \tfrac{1}{6}n^6 - \tfrac{1}{2}n^5 + \tfrac{5}{12}n^4 - \tfrac{1}{12}n^2 \\
S_6(n) &= \tfrac{1}{7}n^7 - \tfrac{1}{2}n^6 + \tfrac{1}{2}n^5 - \tfrac{1}{6}n^3 + \tfrac{1}{42}n \\
S_7(n) &= \tfrac{1}{8}n^8 - \tfrac{1}{2}n^7 + \tfrac{7}{12}n^6 - \tfrac{7}{24}n^4 + \tfrac{1}{12}n^2 \\
S_8(n) &= \tfrac{1}{9}n^9 - \tfrac{1}{2}n^8 + \tfrac{2}{3}n^7 - \tfrac{7}{15}n^5 + \tfrac{2}{9}n^3 - \tfrac{1}{30}n \\
S_9(n) &= \tfrac{1}{10}n^{10} - \tfrac{1}{2}n^9 + \tfrac{3}{4}n^8 - \tfrac{7}{10}n^6 + \tfrac{1}{2}n^4 - \tfrac{3}{20}n^2 \\
S_{10}(n) &= \tfrac{1}{11}n^{11} - \tfrac{1}{2}n^{10} + \tfrac{5}{6}n^9 - n^7 + n^5 - \tfrac{1}{2}n^3 + \tfrac{5}{66}n
\end{aligned}$$

Can you see it too? The coefficient of n^{m+1} in $S_m(n)$ is always $1/(m+1)$. The coefficient of n^m is always $-1/2$. The coefficient of n^{m-1} is always ... let's see ... $m/12$. The coefficient of n^{m-2} is always zero. The coefficient of n^{m-3} is always ... let's see ... hmmm ... yes, it's $-m(m-1)(m-2)/720$. The coefficient of n^{m-4} is always zero. And it looks as if the pattern will continue, with the coefficient of n^{m-k} always being some constant times $m^{\underline{k}}$.

That was Bernoulli's discovery. In modern notation we write the coefficients in the form

$$\begin{aligned}
S_m(n) &= \frac{1}{m+1}\left(B_0 n^{m+1} + \binom{m+1}{1} B_1 n^m + \cdots + \binom{m+1}{m} B_m n\right) \\
&= \frac{1}{m+1} \sum_{k=0}^{m} \binom{m+1}{k} B_k n^{m+1-k}. \qquad (6.78)
\end{aligned}$$

Bernoulli numbers are defined by an implicit recurrence relation,

$$\sum_{j=0}^{m} \binom{m+1}{j} B_j = [m=0], \qquad \text{for all } m \geqslant 0. \tag{6.79}$$

For example, $\binom{2}{0}B_0 + \binom{2}{1}B_1 = 0$. The first few values turn out to be

n	0	1	2	3	4	5	6	7	8	9	10	11	12
B_n	1	$-\frac{1}{2}$	$\frac{1}{6}$	0	$\frac{-1}{30}$	0	$\frac{1}{42}$	0	$\frac{-1}{30}$	0	$\frac{5}{66}$	0	$\frac{-691}{2730}$

(All conjectures about a simple closed form for B_n are wiped out by the appearance of the strange fraction $-691/2730$.)

We can prove Bernoulli's formula (6.78) by induction on m, using the perturbation method (one of the ways we found $S_2(n) = \square_n$ in Chapter 2):

$$\begin{aligned} S_{m+1}(n) + n^{m+1} &= \sum_{k=0}^{n-1} (k+1)^{m+1} \\ &= \sum_{k=0}^{n-1} \sum_{j=0}^{m+1} \binom{m+1}{j} k^j = \sum_{j=0}^{m+1} \binom{m+1}{j} S_j(n). \end{aligned} \tag{6.80}$$

Let $\widehat{S}_m(n)$ be the right-hand side of (6.78); we wish to show that $S_m(n) = \widehat{S}_m(n)$, assuming that $S_j(n) = \widehat{S}_j(n)$ for $0 \leqslant j < m$. We begin as we did for $m = 2$ in Chapter 2, subtracting $S_{m+1}(n)$ from both sides of (6.80). Then we expand each $S_j(n)$ using (6.78), and regroup so that the coefficients of powers of n on the right-hand side are brought together and simplified:

$$\begin{aligned} n^{m+1} &= \sum_{j=0}^{m} \binom{m+1}{j} S_j(n) = \sum_{j=0}^{m} \binom{m+1}{j} \widehat{S}_j(n) + \binom{m+1}{m} \Delta \\ &= \sum_{j=0}^{m} \binom{m+1}{j} \frac{1}{j+1} \sum_{k=0}^{j} \binom{j+1}{k} B_k n^{j+1-k} + (m+1)\Delta \\ &= \sum_{0 \leqslant k \leqslant j \leqslant m} \binom{m+1}{j} \binom{j+1}{k} \frac{B_k}{j+1} n^{j+1-k} + (m+1)\Delta \\ &= \sum_{0 \leqslant k \leqslant j \leqslant m} \binom{m+1}{j} \binom{j+1}{j-k} \frac{B_{j-k}}{j+1} n^{k+1} + (m+1)\Delta \\ &= \sum_{0 \leqslant k \leqslant j \leqslant m} \binom{m+1}{j} \binom{j+1}{k+1} \frac{B_{j-k}}{j+1} n^{k+1} + (m+1)\Delta \\ &= \sum_{0 \leqslant k \leqslant m} \frac{n^{k+1}}{k+1} \sum_{k \leqslant j \leqslant m} \binom{m+1}{j} \binom{j}{k} B_{j-k} + (m+1)\Delta \end{aligned}$$

$$= \sum_{0\le k\le m} \frac{n^{k+1}}{k+1}\binom{m+1}{k} \sum_{k\le j\le m} \binom{m+1-k}{j-k} B_{j-k} + (m+1)\,\Delta$$

$$= \sum_{0\le k\le m} \frac{n^{k+1}}{k+1}\binom{m+1}{k} \sum_{0\le j\le m-k} \binom{m+1-k}{j} B_j + (m+1)\,\Delta$$

$$= \sum_{0\le k\le m} \frac{n^{k+1}}{k+1}\binom{m+1}{k} [m-k=0] + (m+1)\,\Delta$$

$$= \frac{n^{m+1}}{m+1}\binom{m+1}{m} + (m+1)\,\Delta$$

$$= n^{m+1} + (m+1)\,\Delta, \qquad \text{where } \Delta = S_m(n) - \widehat{S}_m(n).$$

(This derivation is a good review of the standard manipulations we learned in Chapter 5.) Thus $\Delta = 0$ and $S_m(n) = \widehat{S}_m(n)$, QED.

Here's some more neat stuff that you'll probably want to skim through the first time.
—Friendly TA

In Chapter 7 we'll use generating functions to obtain a much simpler proof of (6.78). The key idea will be to show that the Bernoulli numbers are the coefficients of the power series

$$\frac{z}{e^z-1} = \sum_{n\ge 0} B_n \frac{z^n}{n!}. \tag{6.81}$$

Start Skimming

Let's simply assume for now that equation (6.81) holds, so that we can derive some of its amazing consequences. If we add $\frac{1}{2}z$ to both sides, thereby cancelling the term $B_1 z/1! = -\frac{1}{2}z$ from the right, we get

$$\frac{z}{e^z-1} + \frac{z}{2} = \frac{z}{2}\frac{e^z+1}{e^z-1} = \frac{z}{2}\frac{e^{z/2}+e^{-z/2}}{e^{z/2}-e^{-z/2}} = \frac{z}{2}\coth\frac{z}{2}. \tag{6.82}$$

Here coth is the "hyperbolic cotangent" function, otherwise known in calculus books as $\cosh z/\sinh z$; we have

$$\sinh z = \frac{e^z - e^{-z}}{2}; \qquad \cosh z = \frac{e^z + e^{-z}}{2}. \tag{6.83}$$

Changing z to $-z$ gives $\left(\frac{-z}{2}\right)\coth\left(\frac{-z}{2}\right) = \frac{z}{2}\coth\frac{z}{2}$; hence every odd-numbered coefficient of $\frac{z}{2}\coth\frac{z}{2}$ must be zero, and we have

$$B_3 = B_5 = B_7 = B_9 = B_{11} = B_{13} = \cdots = 0. \tag{6.84}$$

Furthermore (6.82) leads to a closed form for the coefficients of coth:

$$z\coth z = \frac{2z}{e^{2z}-1} + \frac{2z}{2} = \sum_{n\ge 0} B_{2n}\frac{(2z)^{2n}}{(2n)!} = \sum_{n\ge 0} 4^n B_{2n}\frac{z^{2n}}{(2n)!}. \tag{6.85}$$

But there isn't much of a market for hyperbolic functions; people are more interested in the "real" functions of trigonometry. We can express ordinary trigonometric functions in terms of their hyperbolic cousins by using the rules

$$\sin z = -i \sinh iz, \qquad \cos z = \cosh iz; \tag{6.86}$$

the corresponding power series are

$$\sin z = \frac{z^1}{1!} - \frac{z^3}{3!} + \frac{z^5}{5!} - \cdots, \qquad \sinh z = \frac{z^1}{1!} + \frac{z^3}{3!} + \frac{z^5}{5!} + \cdots;$$
$$\cos z = \frac{z^0}{0!} - \frac{z^2}{2!} + \frac{z^4}{4!} - \cdots, \qquad \cosh z = \frac{z^0}{0!} + \frac{z^2}{2!} + \frac{z^4}{4!} + \cdots.$$

Hence $\cot z = \cos z / \sin z = i \cosh iz / \sinh iz = i \coth iz$, and we have

I see, we get "real" functions by using imaginary numbers.

$$z \cot z = \sum_{n \geqslant 0} B_{2n} \frac{(2iz)^{2n}}{(2n)!} = \sum_{n \geqslant 0} (-4)^n B_{2n} \frac{z^{2n}}{(2n)!}. \tag{6.87}$$

Another remarkable formula for $z \cot z$ was found by Euler (exercise 73):

$$z \cot z = 1 - 2 \sum_{k \geqslant 1} \frac{z^2}{k^2 \pi^2 - z^2}. \tag{6.88}$$

We can expand Euler's formula in powers of z^2, obtaining

$$\begin{aligned} z \cot z &= 1 - 2 \sum_{k \geqslant 1} \left(\frac{z^2}{k^2 \pi^2} + \frac{z^4}{k^4 \pi^4} + \frac{z^6}{k^6 \pi^6} + \cdots \right) \\ &= 1 - 2 \left(\frac{z^2}{\pi^2} H_\infty^{(2)} + \frac{z^4}{\pi^4} H_\infty^{(4)} + \frac{z^6}{\pi^6} H_\infty^{(6)} + \cdots \right). \end{aligned}$$

Equating coefficients of z^{2n} with those in our other formula, (6.87), gives us an almost miraculous closed form for infinitely many infinite sums:

$$\zeta(2n) = H_\infty^{(2n)} = (-1)^{n-1} \frac{2^{2n-1} \pi^{2n} B_{2n}}{(2n)!}, \qquad \text{integer } n > 0. \tag{6.89}$$

For example,

$$\zeta(2) = H_\infty^{(2)} = 1 + \tfrac{1}{4} + \tfrac{1}{9} + \cdots = \pi^2 B_2 = \pi^2/6; \tag{6.90}$$
$$\zeta(4) = H_\infty^{(4)} = 1 + \tfrac{1}{16} + \tfrac{1}{81} + \cdots = -\pi^4 B_4/3 = \pi^4/90. \tag{6.91}$$

Formula (6.89) is not only a closed form for $H_\infty^{(2n)}$, it also tells us the approximate size of B_{2n}, since $H_\infty^{(2n)}$ is very near 1 when n is large. And it tells us that $(-1)^{n-1} B_{2n} > 0$ for all $n > 0$; thus the nonzero Bernoulli numbers alternate in sign.

And that's not all. Bernoulli numbers also appear in the coefficients of the tangent function,

Start Skipping

$$\tan z = \frac{\sin z}{\cos z} = \sum_{n \geqslant 0} (-1)^{n-1} 4^n (4^n - 1) B_{2n} \frac{z^{2n-1}}{(2n)!}, \tag{6.92}$$

as well as other trigonometric functions (exercise 70). Formula (6.92) leads to another important fact about the Bernoulli numbers, namely that

$$T_{2n-1} = (-1)^{n-1} \frac{4^n (4^n - 1)}{2n} B_{2n} \quad \text{is a positive integer.} \tag{6.93}$$

We have, for example:

n	1	3	5	7	9	11	13
T_n	1	2	16	272	7936	353792	22368256

(The T's are called *tangent numbers.*)

One way to prove (6.93) is to consider the power series

$$\begin{aligned} \frac{\sin z + x \cos z}{\cos z - x \sin z} &= x + (1+x^2)z + (2x^3+2x)\frac{z^2}{2} + (6x^4+8x^2+2)\frac{z^3}{6} + \cdots \\ &= \sum_{n \geqslant 0} T_n(x) \frac{z^n}{n!}, \end{aligned} \tag{6.94}$$

where $T_n(x)$ is a polynomial in x; setting $x = 0$ gives $T_n(0) = T_n$, the nth tangent number. If we differentiate (6.94) with respect to x, we get

$$\frac{1}{(\cos z - x \sin z)^2} = \sum_{n \geqslant 0} T_n'(x) \frac{z^n}{n!};$$

but if we differentiate with respect to z, we get

$$\frac{1 + x^2}{(\cos z - x \sin z)^2} = \sum_{n \geqslant 1} T_n(x) \frac{z^{n-1}}{(n-1)!} = \sum_{n \geqslant 0} T_{n+1}(x) \frac{z^n}{n!}.$$

(Try it — the cancellation is very pretty.) Therefore we have

$$T_{n+1}(x) = (1 + x^2) T_n'(x), \qquad T_0(x) = x, \tag{6.95}$$

a simple recurrence from which it follows that the coefficients of $T_n(x)$ are nonnegative integers. Moreover, we can easily prove that $T_n(x)$ has degree $n + 1$, and that its coefficients are alternately zero and positive. Therefore $T_{2n+1}(0) = T_{2n+1}$ is a positive integer, as claimed in (6.93).

Recurrence (6.95) gives us a simple way to calculate Bernoulli numbers, via tangent numbers, using only simple operations on integers; by contrast, the defining recurrence (6.79) involves difficult arithmetic with fractions.

If we want to compute the sum of nth powers from a to $b-1$ instead of from 0 to $n-1$, the theory of Chapter 2 tells us that

$$\sum_{k=a}^{b-1} k^m = \sum\nolimits_a^b x^m \,\delta x = S_m(b) - S_m(a). \tag{6.96}$$

This identity has interesting consequences when we consider negative values of k: We have

$$\sum_{k=-n+1}^{-1} k^m = (-1)^m \sum_{k=0}^{n-1} k^m, \qquad \text{when } m > 0,$$

hence

$$S_m(0) - S_m(-n+1) = (-1)^m\bigl(S_m(n) - S_m(0)\bigr).$$

But $S_m(0) = 0$, so we have the identity

$$S_m(1-n) = (-1)^{m+1} S_m(n), \qquad m > 0. \tag{6.97}$$

Therefore $S_m(1) = 0$. If we write the polynomial $S_m(n)$ in factored form, it will always have the factors n and $(n-1)$, because it has the roots 0 and 1. In general, $S_m(n)$ is a polynomial of degree $m+1$ with leading term $\frac{1}{m+1}n^{m+1}$. Moreover, we can set $n = \frac{1}{2}$ in (6.97) to get $S_m(\frac{1}{2}) = (-1)^{m+1} S_m(\frac{1}{2})$; if m is even, this makes $S_m(\frac{1}{2}) = 0$, so $(n - \frac{1}{2})$ will be an additional factor. These observations explain why we found the simple factorization

$$S_2(n) = \tfrac{1}{3} n (n - \tfrac{1}{2})(n-1)$$

in Chapter 2; we could have used such reasoning to deduce the value of $S_2(n)$ without calculating it! Furthermore, (6.97) implies that the polynomial with the remaining factors, $\hat{S}_m(n) = S_m(n)/(n - \frac{1}{2})$, always satisfies

$$\hat{S}_m(1-n) = \hat{S}_m(n), \qquad m \text{ even}, \quad m > 0.$$

It follows that $S_m(n)$ can always be written in the factored form

$$S_m(n) = \begin{cases} \dfrac{1}{m+1} \displaystyle\prod_{k=1}^{\lceil m/2 \rceil} (n - \tfrac{1}{2} - \alpha_k)(n - \tfrac{1}{2} + \alpha_k), & m \text{ odd;} \\[2ex] \dfrac{(n - \frac{1}{2})}{m+1} \displaystyle\prod_{k=1}^{m/2} (n - \tfrac{1}{2} - \alpha_k)(n - \tfrac{1}{2} + \alpha_k), & m \text{ even.} \end{cases} \tag{6.98}$$

Here $\alpha_1 = \frac{1}{2}$, and $\alpha_2, \ldots, \alpha_{\lceil m/2 \rceil}$ are appropriate complex numbers whose values depend on m. For example,

$$\begin{aligned} S_3(n) &= n^2(n-1)^2/4\,; \\ S_4(n) &= n(n-\tfrac{1}{2})(n-1)(n-\tfrac{1}{2}+\sqrt{7/12}\,)(n-\tfrac{1}{2}-\sqrt{7/12}\,)/5\,; \\ S_5(n) &= n^2(n-1)^2(n-\tfrac{1}{2}+\sqrt{3/4}\,)(n-\tfrac{1}{2}-\sqrt{3/4}\,)/6\,; \\ S_6(n) &= n(n-\tfrac{1}{2})(n-1)(n-\tfrac{1}{2}+\alpha)(n-\tfrac{1}{2}-\alpha)(n-\tfrac{1}{2}+\overline{\alpha})(n-\tfrac{1}{2}-\overline{\alpha})\,, \\ &\qquad \text{where } \alpha = 2^{-5/2}3^{-1/2}31^{1/4}\bigl(\sqrt{\sqrt{31}+\sqrt{27}}+i\sqrt{\sqrt{31}-\sqrt{27}}\,\bigr). \end{aligned}$$

If m is odd and greater than 1, we have $B_m = 0$; hence $S_m(n)$ is divisible by n^2 (and by $(n-1)^2$). Otherwise the roots of $S_m(n)$ don't seem to obey a simple law.

Stop Skipping

Let's conclude our study of Bernoulli numbers by looking at how they relate to Stirling numbers. One way to compute $S_m(n)$ is to change ordinary powers to falling powers, since the falling powers have easy sums. After doing those easy sums we can convert back to ordinary powers:

$$\begin{aligned} S_m(n) &= \sum_{k=0}^{n-1} k^m = \sum_{k=0}^{n-1}\sum_{j\ge0}{m \brace j}k^{\underline{j}} = \sum_{j\ge0}{m \brace j}\sum_{k=0}^{n-1}k^{\underline{j}} \\ &= \sum_{j\ge0}{m \brace j}\frac{n^{\underline{j+1}}}{j+1} \\ &= \sum_{j\ge0}{m \brace j}\frac{1}{j+1}\sum_{k\ge0}(-1)^{j+1-k}{j+1 \brack k}n^k. \end{aligned}$$

Therefore, equating coefficients with those in (6.78), we must have the identity

$$\sum_{j\ge0}{m \brace j}{j+1 \brack k}\frac{(-1)^{j+1-k}}{j+1} = \frac{1}{m+1}\binom{m+1}{k}B_{m+1-k}\,. \tag{6.99}$$

It would be nice to prove this relation directly, thereby discovering Bernoulli numbers in a new way. But the identities in Tables 250 or 251 don't give us any obvious handle on a proof by induction that the left-hand sum in (6.99) is a constant times $m^{\underline{k-1}}$. If $k = m+1$, the left-hand sum is just ${m \brace m}{m+1 \brack m+1}/(m+1) = 1/(m+1)$, so that case is easy. And if $k = m$, the left-hand side sums to ${m \brace m-1}{m \brack m}m^{-1} - {m \brace m}{m+1 \brack m}(m+1)^{-1} = \frac{1}{2}(m-1) - \frac{1}{2}m = -\frac{1}{2}$; so that case is pretty easy too. But if $k < m$, the left-hand sum looks hairy. Bernoulli would probably not have discovered his numbers if he had taken this route.

One thing we can do is replace $\left\{ {m \atop j} \right\}$ by $\left\{ {m+1 \atop j+1} \right\} - (j+1)\left\{ {m \atop j} \right\}$. The $(j+1)$ nicely cancels with the awkward denominator, and the left-hand side becomes

$$\sum_{j \geq 0} \left\{ {m+1 \atop j+1} \right\} \left[{j+1 \atop k} \right] \frac{(-1)^{j+1-k}}{j+1} - \sum_{j \geq 0} \left\{ {m \atop j+1} \right\} \left[{j+1 \atop k} \right] (-1)^{j+1-k}.$$

The second sum is zero, when $k < m$, by (6.31). That leaves us with the first sum, which cries out for a change in notation; let's rename all variables so that the index of summation is k, and so that the other parameters are m and n. Then identity (6.99) is equivalent to

$$\sum_k \left\{ {n \atop k} \right\} \left[{k \atop m} \right] \frac{(-1)^{k-m}}{k} = \frac{1}{n} \binom{n}{m} B_{n-m} + [m = n-1]. \tag{6.100}$$

Good, we have something that looks more pleasant — although Table 251 still doesn't suggest any obvious next step.

The convolution formulas in Table 258 now come to the rescue. We can use (6.51) and (6.50) to rewrite the summand in terms of Stirling polynomials:

$$\left\{ {n \atop k} \right\} \left[{k \atop m} \right] = (-1)^{n-k+1} \frac{n!}{(k-1)!} \sigma_{n-k}(-k) \cdot \frac{k!}{(m-1)!} \sigma_{k-m}(k);$$

$$\left\{ {n \atop k} \right\} \left[{k \atop m} \right] \frac{(-1)^{k-m}}{k} = (-1)^{n+1-m} \frac{n!}{(m-1)!} \sigma_{n-k}(-k)\, \sigma_{k-m}(k).$$

Things are looking good; the convolution in (6.48) yields

$$\begin{aligned} \sum_{k=0}^{n} \sigma_{n-k}(-k)\, \sigma_{k-m}(k) &= \sum_{k=0}^{n-m} \sigma_{n-m-k}\big(-n + (n-m-k)\big)\, \sigma_k(m+k) \\ &= \frac{m-n}{(m)(-n)} \sigma_{n-m}\big(m - n + (n-m)\big). \end{aligned}$$

Formula (6.100) is now verified, and we find that Bernoulli numbers are related to the constant terms in the Stirling polynomials:

Stop Skimming

$$(-1)^{m-1} m \sigma_m(0) = \frac{B_m}{m!} + [m = 1]. \tag{6.101}$$

6.6 FIBONACCI NUMBERS

Now we come to a special sequence of numbers that is perhaps the most pleasant of all, the Fibonacci sequence $\langle F_n \rangle$:

n	0	1	2	3	4	5	6	7	8	9	10	11	12	13	14
F_n	0	1	1	2	3	5	8	13	21	34	55	89	144	233	377

Unlike the harmonic numbers and the Bernoulli numbers, the Fibonacci numbers are nice simple integers. They are defined by the recurrence

$$\begin{aligned} F_0 &= 0\,; \\ F_1 &= 1\,; \\ F_n &= F_{n-1} + F_{n-2}\,, \qquad \text{for } n > 1. \end{aligned} \tag{6.102}$$

The simplicity of this rule—the simplest possible recurrence in which each number depends on the previous two—accounts for the fact that Fibonacci numbers occur in a wide variety of situations.

The back-to-nature nature of this example is shocking. This book should be banned.

"Bee trees" provide a good example of how Fibonacci numbers can arise naturally. Let's consider the pedigree of a male bee. Each male (also known as a drone) is produced asexually from a female (also known as a queen); each female, however, has two parents, a male and a female. Here are the first few levels of the tree:

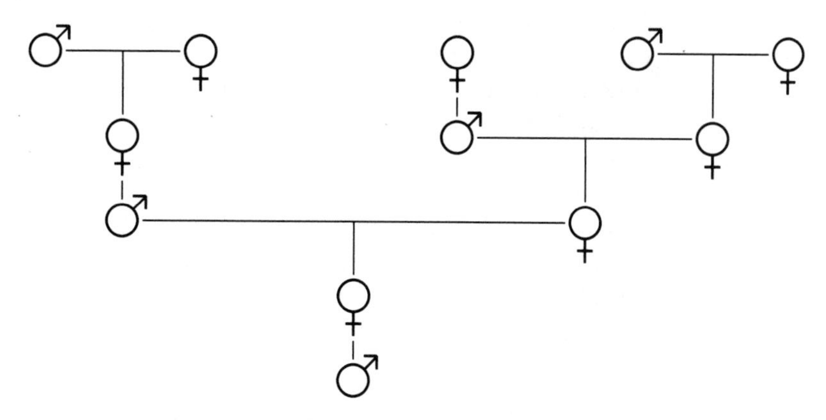

The drone has one grandfather and one grandmother; he has one great-grandfather and two great-grandmothers; he has two great-great-grandfathers and three great-great-grandmothers. In general, it is easy to see by induction that he has exactly F_{n+1} great^n-grandpas and F_{n+2} great^n-grandmas.

Fibonacci numbers are often found in nature, perhaps for reasons similar to the bee-tree law. For example, a typical sunflower has a large head that contains spirals of tightly packed florets, usually with 34 winding in one direction and 55 in another. Smaller heads will have 21 and 34, or 13 and 21; a gigantic sunflower with 89 and 144 spirals was once exhibited in England. Similar patterns are found in some species of pine cones.

Phyllotaxis, n. The love of taxis.

And here's an example of a different nature [219]: Suppose we put two panes of glass back-to-back. How many ways a_n are there for light rays to pass through or be reflected after changing direction n times? The first few cases are:

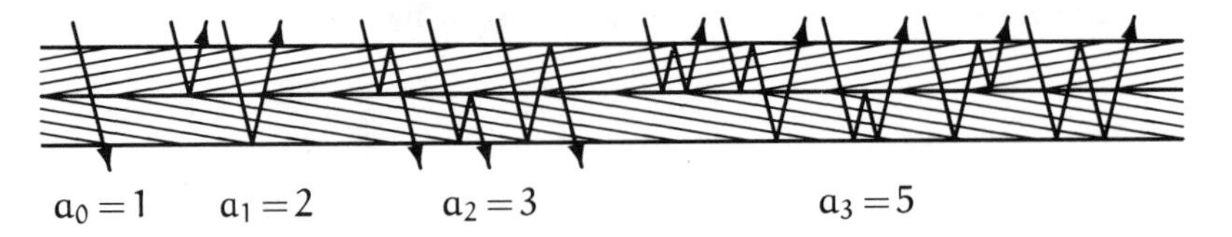

When n is even, we have an even number of bounces and the ray passes through; when n is odd, the ray is reflected and it re-emerges on the same side it entered. The a_n's seem to be Fibonacci numbers, and a little staring at the figure tells us why: For $n \geqslant 2$, the n-bounce rays either take their first bounce off the opposite surface and continue in a_{n-1} ways, or they begin by bouncing off the middle surface and then bouncing back again to finish in a_{n-2} ways. Thus we have the Fibonacci recurrence $a_n = a_{n-1} + a_{n-2}$. The initial conditions are different, but not very different, because we have $a_0 = 1 = F_2$ and $a_1 = 2 = F_3$; therefore everything is simply shifted two places, and $a_n = F_{n+2}$.

Leonardo Fibonacci introduced these numbers in 1202, and mathematicians gradually began to discover more and more interesting things about them. Édouard Lucas, the perpetrator of the Tower of Hanoi puzzle discussed in Chapter 1, worked with them extensively in the last half of the nineteenth century (in fact it was Lucas who popularized the name "Fibonacci numbers"). One of his amazing results was to use properties of Fibonacci numbers to prove that the 39-digit Mersenne number $2^{127} - 1$ is prime.

"La suite de Fibonacci possède des propriétés nombreuses fort intéressantes."
—E. Lucas [207]

One of the oldest theorems about Fibonacci numbers, due to the French astronomer Jean-Dominique Cassini in 1680 [45], is the identity

$$F_{n+1}F_{n-1} - F_n^2 = (-1)^n, \qquad \text{for } n > 0. \tag{6.103}$$

When $n = 6$, for example, Cassini's identity correctly claims that $13{\cdot}5 - 8^2 = 1$.

A polynomial formula that involves Fibonacci numbers of the form $F_{n\pm k}$ for small values of k can be transformed into a formula that involves only F_n and F_{n+1}, because we can use the rule

$$F_m = F_{m+2} - F_{m+1} \tag{6.104}$$

to express F_m in terms of higher Fibonacci numbers when $m < n$, and we can use

$$F_m = F_{m-2} + F_{m-1} \tag{6.105}$$

to replace F_m by lower Fibonacci numbers when $m > n+1$. Thus, for example, we can replace F_{n-1} by $F_{n+1} - F_n$ in (6.103) to get Cassini's identity in the form

$$F_{n+1}^2 - F_{n+1}F_n - F_n^2 = (-1)^n. \qquad (6.106)$$

Moreover, Cassini's identity reads

$$F_{n+2}F_n - F_{n+1}^2 = (-1)^{n+1}$$

when n is replaced by $n+1$; this is the same as $(F_{n+1}+F_n)F_n - F_{n+1}^2 = (-1)^{n+1}$, which is the same as (6.106). Thus Cassini(n) is true if and only if Cassini($n{+}1$) is true; equation (6.103) holds for all n by induction.

Cassini's identity is the basis of a geometrical paradox that was one of Lewis Carroll's favorite puzzles [54], [258], [298]. The idea is to take a chessboard and cut it into four pieces as shown here, then to reassemble the pieces into a rectangle:

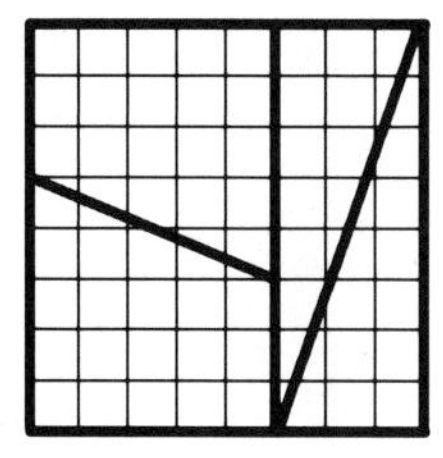

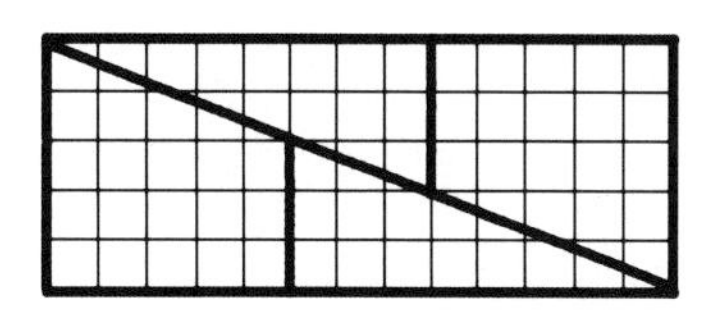

The paradox is explained because . . . well, magic tricks aren't supposed to be explained.

Presto: The original area of $8 \times 8 = 64$ squares has been rearranged to yield $5 \times 13 = 65$ squares! A similar construction dissects any $F_n \times F_n$ square into four pieces, using F_{n+1}, F_n, F_{n-1}, and F_{n-2} as dimensions wherever the illustration has 13, 8, 5, and 3 respectively. The result is an $F_{n-1} \times F_{n+1}$ rectangle; by (6.103), one square has therefore been gained or lost, depending on whether n is even or odd.

Strictly speaking, we can't apply the reduction (6.105) unless $m \geqslant 2$, because we haven't defined F_n for negative n. A lot of maneuvering becomes easier if we eliminate this boundary condition and use (6.104) and (6.105) to define Fibonacci numbers with negative indices. For example, F_{-1} turns out to be $F_1 - F_0 = 1$; then F_{-2} is $F_0 - F_{-1} = -1$. In this way we deduce the values

n	0	−1	−2	−3	−4	−5	−6	−7	−8	−9	−10	−11
F_n	0	1	−1	2	−3	5	−8	13	−21	34	−55	89

and it quickly becomes clear (by induction) that

$$F_{-n} = (-1)^{n-1}F_n, \qquad \text{integer } n. \qquad (6.107)$$

Cassini's identity (6.103) is true for *all* integers n, not just for $n > 0$, when we extend the Fibonacci sequence in this way.

The process of reducing $F_{n\pm k}$ to a combination of F_n and F_{n+1} by using (6.105) and (6.104) leads to the sequence of formulas

$$\begin{aligned} F_{n+2} &= F_{n+1} + F_n & F_{n-1} &= F_{n+1} - F_n \\ F_{n+3} &= 2F_{n+1} + F_n & F_{n-2} &= -F_{n+1} + 2F_n \\ F_{n+4} &= 3F_{n+1} + 2F_n & F_{n-3} &= 2F_{n+1} - 3F_n \\ F_{n+5} &= 5F_{n+1} + 3F_n & F_{n-4} &= -3F_{n+1} + 5F_n \end{aligned}$$

in which another pattern becomes obvious:

$$F_{n+k} = F_k F_{n+1} + F_{k-1} F_n \,. \tag{6.108}$$

This identity, easily proved by induction, holds for all integers k and n (positive, negative, or zero).

If we set $k = n$ in (6.108), we find that

$$F_{2n} = F_n F_{n+1} + F_{n-1} F_n \,; \tag{6.109}$$

hence F_{2n} is a multiple of F_n. Similarly,

$$F_{3n} = F_{2n} F_{n+1} + F_{2n-1} F_n \,,$$

and we may conclude that F_{3n} is also a multiple of F_n. By induction,

$$F_{kn} \text{ is a multiple of } F_n \,, \tag{6.110}$$

for all integers k and n. This explains, for example, why F_{15} (which equals 610) is a multiple of both F_3 and F_5 (which are equal to 2 and 5). Even more is true, in fact; exercise 27 proves that

$$\gcd(F_m, F_n) = F_{\gcd(m,n)} \,. \tag{6.111}$$

For example, $\gcd(F_{12}, F_{18}) = \gcd(144, 2584) = 8 = F_6$.

We can now prove a converse of (6.110): If $n > 2$ and if F_m is a multiple of F_n, then m is a multiple of n. For if $F_n \backslash F_m$ then $F_n \backslash \gcd(F_m, F_n) = F_{\gcd(m,n)} \leqslant F_n$. This is possible only if $F_{\gcd(m,n)} = F_n$; and our assumption that $n > 2$ makes it mandatory that $\gcd(m, n) = n$. Hence $n \backslash m$.

An extension of these divisibility ideas was used by Yuri Matijasevich in his famous proof [213] that there is no algorithm to decide if a given multivariate polynomial equation with integer coefficients has a solution in integers. Matijasevich's lemma states that, if $n > 2$, the Fibonacci number F_m is a multiple of F_n^2 if and only if m is a multiple of nF_n.

Let's prove this by looking at the sequence $\langle F_{kn} \bmod F_n^2 \rangle$ for $k = 1$, 2, 3, ..., and seeing when $F_{kn} \bmod F_n^2 = 0$. (We know that m must have the form kn if $F_m \bmod F_n = 0$.) First we have $F_n \bmod F_n^2 = F_n$; that's not zero. Next we have

$$F_{2n} = F_n F_{n+1} + F_{n-1} F_n \equiv 2F_n F_{n+1} \pmod{F_n^2},$$

by (6.108), since $F_{n+1} \equiv F_{n-1} \pmod{F_n}$. Similarly

$$F_{2n+1} = F_{n+1}^2 + F_n^2 \equiv F_{n+1}^2 \pmod{F_n^2}.$$

This congruence allows us to compute

$$\begin{aligned} F_{3n} &= F_{2n+1}F_n + F_{2n}F_{n-1} \\ &\equiv F_{n+1}^2 F_n + (2F_nF_{n+1})F_{n+1} = 3F_{n+1}^2 F_n \pmod{F_n^2}; \\ F_{3n+1} &= F_{2n+1}F_{n+1} + F_{2n}F_n \\ &\equiv F_{n+1}^3 + (2F_nF_{n+1})F_n \equiv F_{n+1}^3 \pmod{F_n^2}. \end{aligned}$$

In general, we find by induction on k that

$$F_{kn} \equiv kF_nF_{n+1}^{k-1} \quad \text{and} \quad F_{kn+1} \equiv F_{n+1}^k \pmod{F_n^2}.$$

Now F_{n+1} is relatively prime to F_n, so

$$\begin{aligned} F_{kn} \equiv 0 \pmod{F_n^2} &\iff kF_n \equiv 0 \pmod{F_n^2} \\ &\iff k \equiv 0 \pmod{F_n}. \end{aligned}$$

We have proved Matijasevich's lemma.

One of the most important properties of the Fibonacci numbers is the special way in which they can be used to represent integers. Let's write

$$j \gg k \iff j \geqslant k+2. \tag{6.112}$$

Then *every positive integer has a unique representation of the form*

$$n = F_{k_1} + F_{k_2} + \cdots + F_{k_r}, \qquad k_1 \gg k_2 \gg \cdots \gg k_r \gg 0. \tag{6.113}$$

(This is "Zeckendorf's theorem" [201], [312].) For example, the representation of one million turns out to be

$$\begin{aligned} 1000000 &= 832040 + 121393 + 46368 + 144 + 55 \\ &= F_{30} + F_{26} + F_{24} + F_{12} + F_{10}. \end{aligned}$$

We can always find such a representation by using a "greedy" approach, choosing F_{k_1} to be the largest Fibonacci number $\leqslant n$, then choosing F_{k_2} to be the largest that is $\leqslant n - F_{k_1}$, and so on. (More precisely, suppose that

$F_k \leqslant n < F_{k+1}$; then we have $0 \leqslant n - F_k < F_{k+1} - F_k = F_{k-1}$. If n is a Fibonacci number, (6.113) holds with $r = 1$ and $k_1 = k$. Otherwise $n - F_k$ has a Fibonacci representation $F_{k_2} + \cdots + F_{k_r}$, by induction on n; and (6.113) holds if we set $k_1 = k$, because the inequalities $F_{k_2} \leqslant n - F_k < F_{k-1}$ imply that $k \gg k_2$.) Conversely, any representation of the form (6.113) implies that

$$F_{k_1} \leqslant n < F_{k_1+1},$$

because the largest possible value of $F_{k_2} + \cdots + F_{k_r}$ when $k \gg k_2 \gg \cdots \gg k_r \gg 0$ is

$$F_{k-2} + F_{k-4} + \cdots + F_{k \bmod 2+2} = F_{k-1} - 1, \qquad \text{if } k \geqslant 2. \tag{6.114}$$

(This formula is easy to prove by induction on k; the left-hand side is zero when k is 2 or 3.) Therefore k_1 is the greedily chosen value described earlier, and the representation must be unique.

Any unique system of representation is a number system; therefore Zeckendorf's theorem leads to the *Fibonacci number system.* We can represent any nonnegative integer n as a sequence of 0's and 1's, writing

$$n = (b_m b_{m-1} \ldots b_2)_F \iff n = \sum_{k=2}^{m} b_k F_k. \tag{6.115}$$

This number system is something like binary (radix 2) notation, except that there never are two adjacent 1's. For example, here are the numbers from 1 to 20, expressed Fibonacci-wise:

$$\begin{array}{llll}
1 = (000001)_F & 6 = (001001)_F & 11 = (010100)_F & 16 = (100100)_F \\
2 = (000010)_F & 7 = (001010)_F & 12 = (010101)_F & 17 = (100101)_F \\
3 = (000100)_F & 8 = (010000)_F & 13 = (100000)_F & 18 = (101000)_F \\
4 = (000101)_F & 9 = (010001)_F & 14 = (100001)_F & 19 = (101001)_F \\
5 = (001000)_F & 10 = (010010)_F & 15 = (100010)_F & 20 = (101010)_F
\end{array}$$

The Fibonacci representation of a million, shown a minute ago, can be contrasted with its binary representation $2^{19} + 2^{18} + 2^{17} + 2^{16} + 2^{14} + 2^9 + 2^6$:

$$\begin{aligned}
(1000000)_{10} &= (10001010000000000010100000000)_F \\
&= (11110100001001000000)_2 .
\end{aligned}$$

The Fibonacci representation needs a few more bits because adjacent 1's are not permitted; but the two representations are analogous.

To add 1 in the Fibonacci number system, there are two cases: If the "units digit" is 0, we change it to 1; that adds $F_2 = 1$, since the units digit refers to F_2. Otherwise the two least significant digits will be 01, and we change them to 10 (thereby adding $F_3 - F_2 = 1$). Finally, we must "carry" as much as necessary by changing the digit pattern '011' to '100' until there are no two 1's in a row. (This carry rule is equivalent to replacing $F_{m+1} + F_m$ by F_{m+2}.) For example, to go from $5 = (1000)_F$ to $6 = (1001)_F$ or from $6 = (1001)_F$ to $7 = (1010)_F$ requires no carrying; but to go from $7 = (1010)_F$ to $8 = (10000)_F$ we must carry twice.

So far we've been discussing lots of properties of the Fibonacci numbers, but we haven't come up with a closed formula for them. We haven't found closed forms for Stirling numbers, Eulerian numbers, or Bernoulli numbers either; but we were able to discover the closed form $H_n = \left[{n+1 \atop 2}\right]/n!$ for harmonic numbers. Is there a relation between F_n and other quantities we know? Can we "solve" the recurrence that defines F_n?

The answer is yes. In fact, there's a simple way to solve the recurrence by using the idea of *generating function* that we looked at briefly in Chapter 5. Let's consider the infinite series

"Sit $1 + x + 2xx + 3x^3 + 5x^4 + 8x^5 + 13x^6 + 21x^7 + 34x^8$ &c Series nata ex divisione Unitatis per Trinomium $1 - x - xx$."
— A. de Moivre [64]

$$F(z) = F_0 + F_1 z + F_2 z^2 + \cdots = \sum_{n \geq 0} F_n z^n . \tag{6.116}$$

If we can find a simple formula for $F(z)$, chances are reasonably good that we can find a simple formula for its coefficients F_n.

"The quantities r, s, t, which show the relation of the terms, are the same as those in the denominator of the fraction. This property, howsoever obvious it may be, M. DeMoivre was the first that applied it to use, in the solution of problems about infinite series, which otherwise would have been very intricate."
— J. Stirling [281]

In Chapter 7 we will focus on generating functions in detail, but it will be helpful to have this example under our belts by the time we get there. The power series $F(z)$ has a nice property if we look at what happens when we multiply it by z and by z^2:

$$\begin{aligned} F(z) &= F_0 + F_1 z + F_2 z^2 + F_3 z^3 + F_4 z^4 + F_5 z^5 + \cdots , \\ zF(z) &= \phantom{F_0 + {}} F_0 z + F_1 z^2 + F_2 z^3 + F_3 z^4 + F_4 z^5 + \cdots , \\ z^2F(z) &= \phantom{F_0 + F_1 z + {}} F_0 z^2 + F_1 z^3 + F_2 z^4 + F_3 z^5 + \cdots . \end{aligned}$$

If we now subtract the last two equations from the first, the terms that involve z^2, z^3, and higher powers of z will all disappear, because of the Fibonacci recurrence. Furthermore the constant term F_0 never actually appeared in the first place, because $F_0 = 0$. Therefore all that's left after the subtraction is $(F_1 - F_0)z$, which is just z. In other words,

$$F(z) - zF(z) - z^2F(z) = z ,$$

and solving for $F(z)$ gives us the compact formula

$$F(z) = \frac{z}{1 - z - z^2} . \tag{6.117}$$

We have now boiled down all the information in the Fibonacci sequence to a simple (although unrecognizable) expression $z/(1-z-z^2)$. This, believe it or not, is progress, because we can factor the denominator and then use partial fractions to achieve a formula that we can easily expand in power series. The coefficients in this power series will be a closed form for the Fibonacci numbers.

The plan of attack just sketched can perhaps be understood better if we approach it backwards. If we have a simpler generating function, say $1/(1-\alpha z)$ where α is a constant, we know the coefficients of all powers of z, because

$$\frac{1}{1-\alpha z} = 1 + \alpha z + \alpha^2 z^2 + \alpha^3 z^3 + \cdots .$$

Similarly, if we have a generating function of the form $A/(1-\alpha z)+B/(1-\beta z)$, the coefficients are easily determined, because

$$\begin{aligned} \frac{A}{1-\alpha z} + \frac{B}{1-\beta z} &= A\sum_{n\geq 0}(\alpha z)^n + B\sum_{n\geq 0}(\beta z)^n \\ &= \sum_{n\geq 0}(A\alpha^n + B\beta^n)z^n . \end{aligned} \tag{6.118}$$

Therefore all we have to do is find constants A, B, α, and β such that

$$\frac{A}{1-\alpha z} + \frac{B}{1-\beta z} = \frac{z}{1-z-z^2},$$

and we will have found a closed form $A\alpha^n + B\beta^n$ for the coefficient F_n of z^n in $F(z)$. The left-hand side can be rewritten

$$\frac{A}{1-\alpha z} + \frac{B}{1-\beta z} = \frac{A - A\beta z + B - B\alpha z}{(1-\alpha z)(1-\beta z)},$$

so the four constants we seek are the solutions to two polynomial equations:

$$(1-\alpha z)(1-\beta z) = 1 - z - z^2; \tag{6.119}$$

$$(A+B) - (A\beta + B\alpha)z = z. \tag{6.120}$$

We want to factor the denominator of $F(z)$ into the form $(1-\alpha z)(1-\beta z)$; then we will be able to express $F(z)$ as the sum of two fractions in which the factors $(1-\alpha z)$ and $(1-\beta z)$ are conveniently separated from each other.

Notice that the denominator factors in (6.119) have been written in the form $(1-\alpha z)(1-\beta z)$, instead of the more usual form $c(z-\rho_1)(z-\rho_2)$ where ρ_1 and ρ_2 are the roots. The reason is that $(1-\alpha z)(1-\beta z)$ leads to nicer expansions in power series.

As usual, the authors can't resist a trick.

We can find α and β in several ways, one of which uses a slick trick: Let us introduce a new variable w and try to find the factorization

$$w^2 - wz - z^2 = (w - \alpha z)(w - \beta z).$$

Then we can simply set $w = 1$ and we'll have the factors of $1 - z - z^2$. The roots of $w^2 - wz - z^2 = 0$ can be found by the quadratic formula; they are

$$\frac{z \pm \sqrt{z^2 + 4z^2}}{2} = \frac{1 \pm \sqrt{5}}{2} z.$$

Therefore

$$w^2 - wz - z^2 = \left(w - \frac{1+\sqrt{5}}{2} z\right)\left(w - \frac{1-\sqrt{5}}{2} z\right)$$

and we have the constants α and β we were looking for.

The ratio of one's height to the height of one's navel is approximately 1.618, according to extensive empirical observations by European scholars [110].

The number $(1+\sqrt{5})/2 \approx 1.61803$ is important in many parts of mathematics as well as in the art world, where it has been considered since ancient times to be the most pleasing ratio for many kinds of design. Therefore it has a special name, the *golden ratio*. We denote it by the Greek letter ϕ, in honor of Phidias who is said to have used it consciously in his sculpture. The other root $(1-\sqrt{5})/2 = -1/\phi \approx -.61803$ shares many properties of ϕ, so it has the special name $\hat{\phi}$, "phi hat." These numbers are roots of the equation $w^2 - w - 1 = 0$, so we have

$$\phi^2 = \phi + 1; \qquad \hat{\phi}^2 = \hat{\phi} + 1. \tag{6.121}$$

(More about ϕ and $\hat{\phi}$ later.)

We have found the constants $\alpha = \phi$ and $\beta = \hat{\phi}$ needed in (6.119); now we merely need to find A and B in (6.120). Setting $z = 0$ in that equation tells us that $B = -A$, so (6.120) boils down to

$$-\hat{\phi}A + \phi A = 1.$$

The solution is $A = 1/(\phi - \hat{\phi}) = 1/\sqrt{5}$; the partial fraction expansion of (6.117) is therefore

$$F(z) = \frac{1}{\sqrt{5}}\left(\frac{1}{1-\phi z} - \frac{1}{1-\hat{\phi}z}\right). \tag{6.122}$$

Good, we've got $F(z)$ right where we want it. Expanding the fractions into power series as in (6.118) gives a closed form for the coefficient of z^n:

$$F_n = \frac{1}{\sqrt{5}}(\phi^n - \hat{\phi}^n). \tag{6.123}$$

(This formula was first published by Leonhard Euler [91] in 1765, but people forgot about it until it was rediscovered by Jacques Binet [25] in 1843.)

Before we stop to marvel at our derivation, we should check its accuracy. For $n = 0$ the formula correctly gives $F_0 = 0$; for $n = 1$, it gives $F_1 = (\phi - \hat{\phi})/\sqrt{5}$, which is indeed 1. For higher powers, equations (6.121) show that the numbers defined by (6.123) satisfy the Fibonacci recurrence, so they must be the Fibonacci numbers by induction. (We could also expand ϕ^n and $\hat{\phi}^n$ by the binomial theorem and chase down the various powers of $\sqrt{5}$; but that gets pretty messy. The point of a closed form is not necessarily to provide us with a fast method of calculation, but rather to tell us how F_n relates to other quantities in mathematics.)

With a little clairvoyance we could simply have guessed formula (6.123) and proved it by induction. But the method of generating functions is a powerful way to discover it; in Chapter 7 we'll see that the same method leads us to the solution of recurrences that are considerably more difficult. Incidentally, we never worried about whether the infinite sums in our derivation of (6.123) were convergent; it turns out that most operations on the coefficients of power series can be justified rigorously whether or not the sums actually converge [151]. Still, skeptical readers who suspect fallacious reasoning with infinite sums can take comfort in the fact that equation (6.123), once found by using infinite series, can be verified by a solid induction proof.

One of the interesting consequences of (6.123) is that the integer F_n is extremely close to the irrational number $\phi^n/\sqrt{5}$ when n is large. (Since $\hat{\phi}$ is less than 1 in absolute value, $\hat{\phi}^n$ becomes exponentially small and its effect is almost negligible.) For example, $F_{10} = 55$ and $F_{11} = 89$ are very near

$$\frac{\phi^{10}}{\sqrt{5}} \approx 55.00364 \quad\text{and}\quad \frac{\phi^{11}}{\sqrt{5}} \approx 88.99775\,.$$

We can use this observation to derive another closed form,

$$F_n = \left\lfloor \frac{\phi^n}{\sqrt{5}} + \frac{1}{2} \right\rfloor = \frac{\phi^n}{\sqrt{5}} \text{ rounded to the nearest integer,} \tag{6.124}$$

because $\left|\hat{\phi}^n/\sqrt{5}\right| < \frac{1}{2}$ for all $n \geqslant 0$. When n is even, F_n is a little greater than $\phi^n/\sqrt{5}$; otherwise it is a little less.

Cassini's identity (6.103) can be rewritten

$$\frac{F_{n+1}}{F_n} - \frac{F_n}{F_{n-1}} = \frac{(-1)^n}{F_{n-1}F_n}\,.$$

When n is large, $1/F_{n-1}F_n$ is very small, so F_{n+1}/F_n must be very nearly the same as F_n/F_{n-1}; and (6.124) tells us that this ratio approaches ϕ. In fact, we have

$$F_{n+1} = \phi F_n + \hat{\phi}^n\,. \tag{6.125}$$

(This identity is true by inspection when $n = 0$ or $n = 1$, and by induction when $n > 1$; we can also prove it directly by plugging in (6.123).) The ratio F_{n+1}/F_n is very close to ϕ, which it alternately overshoots and undershoots.

By coincidence, ϕ is also very nearly the number of kilometers in a mile. (The exact number is 1.609344, since 1 inch is exactly 2.54 centimeters.) This gives us a handy way to convert mentally between kilometers and miles, because a distance of F_{n+1} kilometers is (very nearly) a distance of F_n miles.

If the USA ever goes metric, our speed limit signs will go from 55 mi/hr to 89 km/hr. Or maybe the highway people will be generous and let us go 90.

Suppose we want to convert a non-Fibonacci number from kilometers to miles; what is 30 km, American style? Easy: We just use the Fibonacci number system and mentally convert 30 to its Fibonacci representation $21 + 8 + 1$ by the greedy approach explained earlier. Now we can shift each number down one notch, getting $13 + 5 + 1$. (The former '1' was F_2, since $k_r \gg 0$ in (6.113); the new '1' is F_1.) Shifting down divides by ϕ, more or less. Hence 19 miles is our estimate. (That's pretty close; the correct answer is about 18.64 miles.) Similarly, to go from miles to kilometers we can shift up a notch; 30 miles is approximately $34 + 13 + 2 = 49$ kilometers. (That's not quite as close; the correct number is about 48.28.)

It turns out that this "shift down" rule gives the correctly rounded number of miles per n kilometers for all $n \leqslant 100$, except in the cases $n = 4$, 12, 62, 75, 91, and 96, when it is off by less than 2/3 mile. And the "shift up" rule gives either the correctly rounded number of kilometers for n miles, or 1 km too many, for all $n \leqslant 126$. (The only really embarrassing case is $n = 4$, where the individual rounding errors for $n = 3 + 1$ both go the same direction instead of cancelling each other out.)

6.7 CONTINUANTS

Fibonacci numbers have important connections to the Stern–Brocot tree that we studied in Chapter 4, and they have important generalizations to a sequence of polynomials that Euler studied extensively. These polynomials are called *continuants*, because they are the key to the study of continued fractions like

$$a_0 + \cfrac{1}{a_1 + \cfrac{1}{a_2 + \cfrac{1}{a_3 + \cfrac{1}{a_4 + \cfrac{1}{a_5 + \cfrac{1}{a_6 + \cfrac{1}{a_7}}}}}}} . \tag{6.126}$$

The continuant polynomial $K_n(x_1, x_2, \ldots, x_n)$ has n parameters, and it is defined by the following recurrence:

$$\begin{aligned} K_0() &= 1\,; \\ K_1(x_1) &= x_1\,; \\ K_n(x_1, \ldots, x_n) &= K_{n-1}(x_1, \ldots, x_{n-1})x_n + K_{n-2}(x_1, \ldots, x_{n-2})\,. \end{aligned} \tag{6.127}$$

For example, the next three cases after $K_1(x_1)$ are

$$\begin{aligned} K_2(x_1, x_2) &= x_1x_2 + 1\,; \\ K_3(x_1, x_2, x_3) &= x_1x_2x_3 + x_1 + x_3\,; \\ K_4(x_1, x_2, x_3, x_4) &= x_1x_2x_3x_4 + x_1x_2 + x_1x_4 + x_3x_4 + 1\,. \end{aligned}$$

It's easy to see, inductively, that the number of terms is a Fibonacci number:

$$K_n(1, 1, \ldots, 1) = F_{n+1}\,. \tag{6.128}$$

When the number of parameters is implied by the context, we can write simply 'K' instead of 'K_n', just as we can omit the number of parameters when we use the hypergeometric functions F of Chapter 5. For example, $K(x_1, x_2) = K_2(x_1, x_2) = x_1x_2 + 1$. The subscript n is of course necessary in formulas like (6.128).

Euler observed that $K(x_1, x_2, \ldots, x_n)$ can be obtained by starting with the product $x_1x_2 \ldots x_n$ and then striking out adjacent pairs x_kx_{k+1} in all possible ways. We can represent Euler's rule graphically by constructing all "Morse code" sequences of dots and dashes having length n, where each dot contributes 1 to the length and each dash contributes 2; here are the Morse code sequences of length 4:

••••   ••–   •–•   –••   ––

These dot-dash patterns correspond to the terms of $K(x_1, x_2, x_3, x_4)$; a dot signifies a variable that's included and a dash signifies a pair of variables that's excluded. For example, •–• corresponds to x_1x_4.

A Morse code sequence of length n that has k dashes has $n-2k$ dots and $n-k$ symbols altogether. These dots and dashes can be arranged in $\binom{n-k}{k}$ ways; therefore if we replace each dot by z and each dash by 1 we get

$$K_n(z, z, \ldots, z) = \sum_{k=0}^{n} \binom{n-k}{k} z^{n-2k}\,. \tag{6.129}$$

We also know that the total number of terms in a continuant is a Fibonacci number; hence we have the identity

$$F_{n+1} = \sum_{k=0}^{n} \binom{n-k}{k}. \tag{6.130}$$

(A closed form for (6.129), generalizing the Euler–Binet formula (6.123) for Fibonacci numbers, appears in (5.74).)

The relation between continuant polynomials and Morse code sequences shows that continuants have a mirror symmetry:

$$K(x_n, \ldots, x_2, x_1) = K(x_1, x_2, \ldots, x_n). \tag{6.131}$$

Therefore they obey a recurrence that adjusts parameters at the left, in addition to the right-adjusting recurrence in definition (6.127):

$$K_n(x_1, \ldots, x_n) = x_1 K_{n-1}(x_2, \ldots, x_n) + K_{n-2}(x_3, \ldots, x_n). \tag{6.132}$$

Both of these recurrences are special cases of a more general law:

$$\begin{aligned} &K_{m+n}(x_1, \ldots, x_m, x_{m+1}, \ldots, x_{m+n}) \\ &\quad = K_m(x_1, \ldots, x_m)\, K_n(x_{m+1}, \ldots, x_{m+n}) \\ &\qquad + K_{m-1}(x_1, \ldots, x_{m-1})\, K_{n-1}(x_{m+2}, \ldots, x_{m+n}). \end{aligned} \tag{6.133}$$

This law is easily understood from the Morse code analogy: The first product $K_m K_n$ yields the terms of K_{m+n} in which there is no dash in the $[m, m+1]$ position, while the second product yields the terms in which there is a dash there. If we set all the x's equal to 1, this identity tells us that $F_{m+n+1} = F_{m+1}F_{n+1} + F_m F_n$; thus, (6.108) is a special case of (6.133).

Euler [90] discovered that continuants obey an even more remarkable law, which generalizes Cassini's identity:

$$\begin{aligned} &K_{m+n}(x_1, \ldots, x_{m+n})\, K_k(x_{m+1}, \ldots, x_{m+k}) \\ &\quad = K_{m+k}(x_1, \ldots, x_{m+k})\, K_n(x_{m+1}, \ldots, x_{m+n}) \\ &\qquad + (-1)^k K_{m-1}(x_1, \ldots, x_{m-1})\, K_{n-k-1}(x_{m+k+2}, \ldots, x_{m+n}). \end{aligned} \tag{6.134}$$

This law (proved in exercise 29) holds whenever the subscripts on the K's are all nonnegative. For example, when $k = 2$, $m = 1$, and $n = 3$, we have

$$K(x_1, x_2, x_3, x_4)\, K(x_2, x_3) = K(x_1, x_2, x_3)\, K(x_2, x_3, x_4) + 1.$$

Continuant polynomials are intimately connected with Euclid's algorithm. Suppose, for example, that the computation of $\gcd(m, n)$ finishes in four steps:

$$\begin{aligned}
\gcd(m,n) &= \gcd(n_0,n_1) & n_0 &= m, & n_1 &= n;\\
&= \gcd(n_1,n_2) & n_2 &= n_0 \bmod n_1 &&= n_0 - q_1n_1;\\
&= \gcd(n_2,n_3) & n_3 &= n_1 \bmod n_2 &&= n_1 - q_2n_2;\\
&= \gcd(n_3,n_4) & n_4 &= n_2 \bmod n_3 &&= n_2 - q_3n_3;\\
&= \gcd(n_4,0) \;=\; n_4. & 0 &= n_3 \bmod n_4 &&= n_3 - q_4n_4.
\end{aligned}$$

Then we have

$$\begin{aligned}
n_4 &= n_4 &&= K()n_4;\\
n_3 &= q_4n_4 &&= K(q_4)n_4;\\
n_2 &= q_3n_3 + n_4 &&= K(q_3,q_4)n_4;\\
n_1 &= q_2n_2 + n_3 &&= K(q_2,q_3,q_4)n_4;\\
n_0 &= q_1n_1 + n_2 &&= K(q_1,q_2,q_3,q_4)n_4.
\end{aligned}$$

In general, if Euclid's algorithm finds the greatest common divisor d in k steps, after computing the sequence of quotients $q_1, \ldots, q_k$, then the starting numbers were $K(q_1, q_2, \ldots, q_k)d$ and $K(q_2, \ldots, q_k)d$. (This fact was noticed early in the eighteenth century by Thomas Fantet de Lagny [190], who seems to have been the first person to consider continuants explicitly. Lagny pointed out that consecutive Fibonacci numbers, which occur as continuants when the q's take their minimum values, are therefore the smallest inputs that cause Euclid's algorithm to take a given number of steps.)

Continuants are also intimately connected with continued fractions, from which they get their name. We have, for example,

$$a_0 + \cfrac{1}{a_1 + \cfrac{1}{a_2 + \cfrac{1}{a_3}}} = \frac{K(a_0,a_1,a_2,a_3)}{K(a_1,a_2,a_3)}. \tag{6.135}$$

The same pattern holds for continued fractions of any depth. It is easily proved by induction; we have, for example,

$$\frac{K(a_0,a_1,a_2,a_3+1/a_4)}{K(a_1,a_2,a_3+1/a_4)} = \frac{K(a_0,a_1,a_2,a_3,a_4)}{K(a_1,a_2,a_3,a_4)},$$

because of the identity

$$\begin{aligned}
&K_n(x_1,\ldots,x_{n-1},x_n+y)\\
&\qquad = K_n(x_1,\ldots,x_{n-1},x_n) + K_{n-1}(x_1,\ldots,x_{n-1})\,y.
\end{aligned} \tag{6.136}$$

(This identity is proved and generalized in exercise 30.)

Moreover, continuants are closely connected with the Stern–Brocot tree discussed in Chapter 4. Each node in that tree can be represented as a sequence of L's and R's, say

$$R^{a_0}L^{a_1}R^{a_2}L^{a_3}\dots R^{a_{n-2}}L^{a_{n-1}}, \tag{6.137}$$

where $a_0 \geqslant 0$, $a_1 \geqslant 1$, $a_2 \geqslant 1$, $a_3 \geqslant 1$, ..., $a_{n-2} \geqslant 1$, $a_{n-1} \geqslant 0$, and n is even. Using the 2×2 matrices L and R of (4.33), it is not hard to prove by induction that the matrix equivalent of (6.137) is

$$\begin{pmatrix} K_{n-2}(a_1,\dots,a_{n-2}) & K_{n-1}(a_1,\dots,a_{n-2},a_{n-1}) \\ K_{n-1}(a_0,a_1,\dots,a_{n-2}) & K_n(a_0,a_1,\dots,a_{n-2},a_{n-1}) \end{pmatrix} \tag{6.138}$$

(The proof is part of exercise 80.) For example,

$$R^aL^bR^cL^d = \begin{pmatrix} bc+1 & bcd+b+d \\ abc+a+c & abcd+ab+ad+cd+1 \end{pmatrix}.$$

Finally, therefore, we can use (4.34) to write a closed form for the fraction in the Stern–Brocot tree whose L-and-R representation is (6.137):

$$f(R^{a_0}\dots L^{a_{n-1}}) = \frac{K_{n+1}(a_0,a_1,\dots,a_{n-1},1)}{K_n(a_1,\dots,a_{n-1},1)}. \tag{6.139}$$

(This is "Halphen's theorem" [143].) For example, to find the fraction for LRRL we have $a_0 = 0$, $a_1 = 1$, $a_2 = 2$, $a_3 = 1$, and $n = 4$; equation (6.139) gives

$$\frac{K(0,1,2,1,1)}{K(1,2,1,1)} = \frac{K(2,1,1)}{K(1,2,1,1)} = \frac{K(2,2)}{K(3,2)} = \frac{5}{7}.$$

(We have used the rule $K_n(x_1,\dots,x_{n-1},x_n+1) = K_{n+1}(x_1,\dots,x_{n-1},x_n,1)$ to absorb leading and trailing 1's in the parameter lists; this rule is obtained by setting $y = 1$ in (6.136).)

A comparison of (6.135) and (6.139) shows that the fraction corresponding to a general node (6.137) in the Stern–Brocot tree has the continued fraction representation

$$f(R^{a_0}\dots L^{a_{n-1}}) = a_0 + \cfrac{1}{a_1 + \cfrac{1}{a_2 + \cfrac{1}{\dots + \cfrac{1}{a_{n-1} + \cfrac{1}{1}}}}}. \tag{6.140}$$

Thus we can convert at sight between continued fractions and the corresponding nodes in the Stern–Brocot tree. For example,

$$f(\mathrm{LRRL}) = 0 + \cfrac{1}{1 + \cfrac{1}{2 + \cfrac{1}{1 + \cfrac{1}{1}}}} .$$

We observed in Chapter 4 that irrational numbers define infinite paths in the Stern–Brocot tree, and that they can be represented as an infinite string of L's and R's. If the infinite string for α is $\mathrm{R}^{a_0}\mathrm{L}^{a_1}\mathrm{R}^{a_2}\mathrm{L}^{a_3}\ldots$, there is a corresponding infinite continued fraction

$$\alpha = a_0 + \cfrac{1}{a_1 + \cfrac{1}{a_2 + \cfrac{1}{a_3 + \cfrac{1}{a_4 + \cfrac{1}{a_5 + \cfrac{1}{\ddots}}}}}} . \tag{6.141}$$

This infinite continued fraction can also be obtained directly: Let $\alpha_0 = \alpha$ and for $k \geqslant 0$ let

$$a_k = \lfloor \alpha_k \rfloor ; \qquad \alpha_k = a_k + \frac{1}{\alpha_{k+1}} . \tag{6.142}$$

The a's are called the "partial quotients" of α. If α is rational, say m/n, this process runs through the quotients found by Euclid's algorithm and then stops (with $\alpha_{k+1} = \infty$).

Is Euler's constant γ rational or irrational? Nobody knows. We can get partial information about this famous unsolved problem by looking for γ in the Stern–Brocot tree; if it's rational we will find it, and if it's irrational we will find all the closest rational approximations to it. The continued fraction for γ begins with the following partial quotients:

Or if they do, they're not talking.

k	0	1	2	3	4	5	6	7	8
a_k	0	1	1	2	1	2	1	4	3

Therefore its Stern–Brocot representation begins LRLLRLLRLLLLRRRL...; no pattern is evident. Calculations by Richard Brent [33] have shown that, if γ is rational, its denominator must be more than 10,000 decimal digits long.

Well, γ must be irrational, because of a little-known Einsteinian assertion: "God does not throw huge denominators at the universe."

Therefore nobody believes that γ is rational; but nobody so far has been able to prove that it isn't.

Let's conclude this chapter by proving a remarkable identity that ties a lot of these ideas together. We introduced the notion of spectrum in Chapter 3; the spectrum of α is the multiset of numbers $\lfloor n\alpha \rfloor$, where α is a given constant. The infinite series

$$\sum_{n\geqslant 1} z^{\lfloor n\phi \rfloor} = z + z^3 + z^4 + z^6 + z^8 + z^9 + \cdots$$

can therefore be said to be the generating function for the spectrum of ϕ, where $\phi = (1+\sqrt{5})/2$ is the golden ratio. The identity we will prove, discovered in 1976 by J. L. Davison [61], is an infinite continued fraction that relates this generating function to the Fibonacci sequence:

$$\cfrac{z^{F_1}}{1+\cfrac{z^{F_2}}{1+\cfrac{z^{F_3}}{1+\cfrac{z^{F_4}}{\ddots}}}} = (1-z)\sum_{n\geqslant 1} z^{\lfloor n\phi \rfloor}. \tag{6.143}$$

Both sides of (6.143) are interesting; let's look first at the numbers $\lfloor n\phi \rfloor$. If the Fibonacci representation (6.113) of n is $F_{k_1} + \cdots + F_{k_r}$, we expect $n\phi$ to be approximately $F_{k_1+1} + \cdots + F_{k_r+1}$, the number we get from shifting the Fibonacci representation left (as when converting from miles to kilometers). In fact, we know from (6.125) that

$$n\phi = F_{k_1+1} + \cdots + F_{k_r+1} - (\hat{\phi}^{k_1} + \cdots + \hat{\phi}^{k_r}).$$

Now $\hat{\phi} = -1/\phi$ and $k_1 \gg \cdots \gg k_r \gg 0$, so we have

$$\begin{aligned} |\hat{\phi}^{k_1} + \cdots + \hat{\phi}^{k_r}| &< \phi^{-k_r} + \phi^{-k_r-2} + \phi^{-k_r-4} + \cdots \\ &= \frac{\phi^{-k_r}}{1-\phi^{-2}} = \phi^{1-k_r} \leqslant \phi^{-1} < 1; \end{aligned}$$

and $\hat{\phi}^{k_1} + \cdots + \hat{\phi}^{k_r}$ has the same sign as $(-1)^{k_r}$, by a similar argument. Hence

$$\lfloor n\phi \rfloor = F_{k_1+1} + \cdots + F_{k_r+1} - [k_r(n) \text{ is even}]. \tag{6.144}$$

Let us say that a number n is *Fibonacci odd* (or F-odd for short) if its least significant Fibonacci bit is 1; this is the same as saying that $k_r(n) = 2$. Otherwise n is *Fibonacci even* (F-even). For example, the smallest F-odd numbers are 1, 4, 6, 9, 12, 14, 17, and 19. If $k_r(n)$ is even, then $n-1$ is F-even, by (6.114); similarly, if $k_r(n)$ is odd, then $n-1$ is F-odd. Therefore

$$k_r(n) \text{ is even} \quad\Longleftrightarrow\quad n-1 \text{ is F-even.}$$

Furthermore, if $k_r(n)$ is even, (6.144) implies that $k_r(\lfloor n\phi\rfloor) = 2$; if $k_r(n)$ is odd, (6.144) says that $k_r(\lfloor n\phi\rfloor) = k_r(n)+1$. Therefore $k_r(\lfloor n\phi\rfloor)$ is always even, and we have proved that

$$\lfloor n\phi\rfloor - 1 \text{ is always F-even.}$$

Conversely, if m is any F-even number, we can reverse this computation and find an n such that $m+1 = \lfloor n\phi\rfloor$. (First add 1 in F-notation as explained earlier. If no carries occur, n is $(m+2)$ shifted right; otherwise n is $(m+1)$ shifted right.) The right-hand sum of (6.143) can therefore be written

$$\sum_{n\geqslant 1} z^{\lfloor n\phi\rfloor} = z\sum_{m\geqslant 0} z^m\,[m \text{ is F-even}]\,. \tag{6.145}$$

How about the fraction on the left? Let's rewrite (6.143) so that the continued fraction looks like (6.141), with all numerators 1:

$$\cfrac{1}{z^{-F_0}+\cfrac{1}{z^{-F_1}+\cfrac{1}{z^{-F_2}+\cfrac{1}{\ddots}}}} = \frac{1-z}{z}\sum_{n\geqslant 1} z^{\lfloor n\phi\rfloor}\,. \tag{6.146}$$

(This transformation is a bit tricky! The numerator and denominator of the original fraction having z^{F_n} as numerator should be divided by $z^{F_{n-1}}$.) If we stop this new continued fraction at $1/z^{-F_n}$, its value will be a ratio of continuants,

$$\frac{K_{n+2}(0, z^{-F_0}, z^{-F_1}, \ldots, z^{-F_n})}{K_{n+1}(z^{-F_0}, z^{-F_1}, \ldots, z^{-F_n})} = \frac{K_n(z^{-F_1}, \ldots, z^{-F_n})}{K_{n+1}(z^{-F_0}, z^{-F_1}, \ldots, z^{-F_n})}\,,$$

as in (6.135). Let's look at the denominator first, in hopes that it will be tractable. Setting $Q_n = K_{n+1}(z^{-F_0}, \ldots, z^{-F_n})$, we find $Q_0 = 1$, $Q_1 = 1+z^{-1}$, $Q_2 = 1+z^{-1}+z^{-2}$, $Q_3 = 1+z^{-1}+z^{-2}+z^{-3}+z^{-4}$, and in general everything fits beautifully and gives a geometric series

$$Q_n = 1+z^{-1}+z^{-2}+\cdots+z^{-(F_{n+2}-1)}\,.$$

The corresponding numerator is $P_n = K_n(z^{-F_1}, \ldots, z^{-F_n})$; this turns out to be like Q_n but with fewer terms. For example, we have

$$P_5 = z^{-1} + z^{-2} + z^{-4} + z^{-5} + z^{-7} + z^{-9} + z^{-10} + z^{-12},$$

compared with $Q_5 = 1 + z^{-1} + \cdots + z^{-12}$. A closer look reveals the pattern governing which terms are present: We have

$$P_5 = \frac{1+z^2+z^3+z^5+z^7+z^8+z^{10}+z^{11}}{z^{12}} = z^{-12} \sum_{m=0}^{12} z^m \,[m \text{ is F-even}];$$

and in general we can prove by induction that

$$P_n = z^{1-F_{n+2}} \sum_{m=0}^{F_{n+2}-1} z^m \,[m \text{ is F-even}].$$

Therefore

$$\frac{P_n}{Q_n} = \frac{\sum_{m=0}^{F_{n+2}-1} z^m \,[m \text{ is F-even}]}{\sum_{m=0}^{F_{n+2}-1} z^m}.$$

Taking the limit as $n \to \infty$ now gives (6.146), because of (6.145).

Exercises

Warmups

1 What are the $\left[{4 \atop 2}\right] = 11$ permutations of $\{1,2,3,4\}$ that have exactly two cycles? (The cyclic forms appear in (6.4); non-cyclic forms like 2314 are desired instead.)

2 There are m^n functions from a set of n elements into a set of m elements. How many of them range over exactly k different function values?

3 Card stackers in the real world know that it's wise to allow a bit of slack so that the cards will not topple over when a breath of wind comes along. Suppose the center of gravity of the top k cards is required to be at least ϵ units from the edge of the $k+1$st card. (Thus, for example, the first card can overhang the second by at most $1-\epsilon$ units.) Can we still achieve arbitrarily large overhang, if we have enough cards?

4 Express $1/1 + 1/3 + \cdots + 1/(2n+1)$ in terms of harmonic numbers.

5 Explain how to get the recurrence (6.75) from the definition of $U_n(x,y)$ in (6.74), and solve the recurrence.

6 An explorer has left a pair of baby rabbits on an island. If baby rabbits become adults after one month, and if each pair of adult rabbits produces one pair of baby rabbits every month, how many pairs of rabbits are present after n months? (After two months there are two pairs, one of which is newborn.) Find a connection between this problem and the "bee tree" in the text.

7 Show that Cassini's identity (6.103) is a special case of (6.108), and a special case of (6.134).

8 Use the Fibonacci number system to convert 65 mi/hr into an approximate number of km/hr.

9 About how many square kilometers are in 8 square miles?

10 What is the continued fraction representation of ϕ?

Basics

11 What is $\sum_k (-1)^k \left[{n \atop k}\right]$, the row sum of Stirling's cycle-number triangle with alternating signs, when n is a nonnegative integer?

12 Prove that Stirling numbers have an inversion law analogous to (5.48):

$$g(n) = \sum_k \left\{{n \atop k}\right\}(-1)^k f(k) \iff f(n) = \sum_k \left[{n \atop k}\right](-1)^k g(k).$$

13 The differential operators $D = \frac{d}{dz}$ and $\vartheta = zD$ are mentioned in Chapters 2 and 5. We have

$$\vartheta^2 = z^2D^2 + zD,$$

because $\vartheta^2 f(z) = \vartheta z f'(z) = z\frac{d}{dz} z f'(z) = z^2 f''(z) + z f'(z)$, which is $(z^2D^2 + zD)f(z)$. Similarly it can be shown that $\vartheta^3 = z^3D^3 + 3z^2D + zD$. Prove the general formulas

$$\vartheta^n = \sum_k \left\{{n \atop k}\right\} z^k D^k,$$

$$z^n D^n = \sum_k \left[{n \atop k}\right] (-1)^{n-k} \vartheta^k,$$

for all $n \geqslant 0$. (These can be used to convert between differential expressions of the forms $\sum_k \alpha_k z^k f^{(k)}(z)$ and $\sum_k \beta_k \vartheta^k f(z)$, as in (5.109).)

14 Prove the power identity (6.37) for Eulerian numbers.

15 Prove the Eulerian identity (6.39) by taking the mth difference of (6.37).

16 What is the general solution of the double recurrence

$$A_{n,0} = a_n\,[n \geqslant 0]; \qquad A_{0,k} = 0, \qquad \text{if } k > 0;$$
$$A_{n,k} = kA_{n-1,k} + A_{n-1,k-1}, \qquad \text{integers } k, n,$$

when k and n range over the set of *all* integers?

17 Solve the following recurrences, assuming that $\left|{n \atop k}\right|$ is zero when $n < 0$ or $k < 0$:

a $$\left|{n \atop k}\right| = \left|{n-1 \atop k}\right| + n\left|{n-1 \atop k-1}\right| + [n=k=0], \qquad \text{for } n, k \geqslant 0.$$

b $$\left|{n \atop k}\right| = (n-k)\left|{n-1 \atop k}\right| + \left|{n-1 \atop k-1}\right| + [n=k=0], \qquad \text{for } n, k \geqslant 0.$$

c $$\left|{n \atop k}\right| = k\left|{n-1 \atop k}\right| + k\left|{n-1 \atop k-1}\right| + [n=k=0], \qquad \text{for } n, k \geqslant 0.$$

18 Prove that the Stirling polynomials satisfy

$$(x+1)\,\sigma_n(x+1) = (x-n)\,\sigma_n(x) + x\sigma_{n-1}(x).$$

19 Prove that the generalized Stirling numbers satisfy

$$\sum_{k=0}^{n} \left\{{x+k \atop x}\right\}\left[{x \atop x-n+k}\right](-1)^k \Big/ \binom{x+k}{n+1} = 0, \qquad \text{integer } n > 0.$$

$$\sum_{k=0}^{n} \left[{x+k \atop x}\right]\left\{{x \atop x-n+k}\right\}(-1)^k \Big/ \binom{x+k}{n+1} = 0, \qquad \text{integer } n > 0.$$

20 Find a closed form for $\sum_{k=1}^{n} H_k^{(2)}$.

21 Show that if $H_n = a_n/b_n$, where a_n and b_n are integers, the denominator b_n is a multiple of $2^{\lfloor \lg n \rfloor}$. *Hint:* Consider the number $2^{\lfloor \lg n \rfloor - 1}H_n - \frac{1}{2}$.

22 Prove that the infinite sum

$$\sum_{k \geqslant 1} \left(\frac{1}{k} - \frac{1}{k+z}\right)$$

converges for all complex numbers z, except when z is a negative integer; and show that it equals H_z when z is a nonnegative integer. (Therefore we can use this formula to define harmonic numbers H_z when z is complex.)

23 Equation (6.81) gives the coefficients of $z/(e^z - 1)$, when expanded in powers of z. What are the coefficients of $z/(e^z + 1)$? *Hint:* Consider the identity $(e^z + 1)(e^z - 1) = e^{2z} - 1$.

24 Prove that the tangent number T_{2n+1} is a multiple of 2^n. *Hint:* Prove that all coefficients of $T_{2n}(x)$ and $T_{2n+1}(x)$ are multiples of 2^n.

25 Equation (6.57) proves that the worm will eventually reach the end of the rubber band at some time N. Therefore there must come a first time n when he's closer to the end after n minutes than he was after $n-1$ minutes. Show that $n < \frac{1}{2}N$.

26 Use summation by parts to evaluate $S_n = \sum_{k=1}^{n} H_k/k$. *Hint:* Consider also the related sum $\sum_{k=1}^{n} H_{k-1}/k$.

27 Prove the gcd law (6.111) for Fibonacci numbers.

28 The *Lucas number* L_n is defined to be $F_{n+1}+F_{n-1}$. Thus, according to (6.109), we have $F_{2n} = F_n L_n$. Here is a table of the first few values:

n	0	1	2	3	4	5	6	7	8	9	10	11	12	13
L_n	2	1	3	4	7	11	18	29	47	76	123	199	322	521

a Use the repertoire method to show that the solution Q_n to the general recurrence

$$Q_0 = \alpha; \qquad Q_1 = \beta; \qquad Q_n = Q_{n-1} + Q_{n-2}, \quad n > 1$$

can be expressed in terms of F_n and L_n.

b Find a closed form for L_n in terms of ϕ and $\hat\phi$.

29 Prove Euler's identity for continuants, equation (6.134).

30 Generalize (6.136) to find an expression for the incremented continuant $K(x_1,\ldots,x_{m-1},x_m+y,x_{m+1},\ldots,x_n)$, when $1 \leqslant m \leqslant n$.

Homework exercises

31 Find a closed form for the coefficients $\left|{n \atop k}\right|$ in the representation of rising powers by falling powers:

$$x^{\overline{n}} = \sum_k \left|{n \atop k}\right| x^{\underline{k}}, \qquad \text{integer } n \geqslant 0.$$

(For example, $x^{\overline{4}} = x^{\underline{4}} + 12x^{\underline{3}} + 36x^{\underline{2}} + 24x^{\underline{1}}$, hence $\left|{4 \atop 2}\right| = 36$.).

32 In Chapter 5 we obtained the formulas

$$\sum_{k \leqslant m} \binom{n+k}{k} = \binom{n+m+1}{m} \quad \text{and} \quad \sum_{0 \leqslant k \leqslant m} \binom{k}{n} = \binom{m+1}{n+1}$$

by unfolding the recurrence $\binom{n}{k} = \binom{n-1}{k} + \binom{n-1}{k-1}$ in two ways. What identities appear when the analogous recurrence $\left\{{n \atop k}\right\} = k\left\{{n-1 \atop k}\right\} + \left\{{n-1 \atop k-1}\right\}$ is unwound?

33 Table 250 gives the values of $\left[{n \atop 2}\right]$ and $\left\{{n \atop 2}\right\}$. What are closed forms (not involving Stirling numbers) for the next cases, $\left[{n \atop 3}\right]$ and $\left\{{n \atop 3}\right\}$?

34 What are $\left\langle{-1 \atop k}\right\rangle$ and $\left\langle{-2 \atop k}\right\rangle$, if the basic recursion relation (6.35) is assumed to hold for all integers k and n, and if $\left\langle{n \atop k}\right\rangle = 0$ for all $k < 0$?

35 Prove that, for every $\epsilon > 0$, there exists an integer $n > 1$ (depending on ϵ) such that $H_n \bmod 1 < \epsilon$.

36 Is it possible to stack n bricks in such a way that the topmost brick is not above any point of the bottommost brick, yet a person who weighs the same as 100 bricks can balance on the middle of the top brick without toppling the pile?

37 Express $\sum_{k=1}^{mn} (k \bmod m)/k(k+1)$ in terms of harmonic numbers, assuming that m and n are positive integers. What is the limiting value as $n \to \infty$?

38 Find the indefinite sum $\sum \binom{r}{k}(-1)^k H_k \,\delta k$.

39 Express $\sum_{k=1}^{n} H_k^2$ in terms of n and H_n.

40 Prove that 1979 divides the numerator of $\sum_{k=1}^{1319}(-1)^{k-1}/k$, and give a similar result for 1987. *Hint:* Use Gauss's trick to obtain a sum of fractions whose numerators are 1979. See also exercise 4.

Ah! Those were prime years.

41 Evaluate the sum

$$\sum_k \binom{\lfloor (n+k)/2 \rfloor}{k}$$

in closed form, when n is an integer (possibly negative).

42 If S is a set of integers, let $S+1$ be the "shifted" set $\{x+1 \mid x \in S\}$. How many subsets of $\{1, 2, \ldots, n\}$ have the property that $S \cup (S+1) = \{1, 2, \ldots, n+1\}$?

43 Prove that the infinite sum

$$\begin{aligned} &.1 \\ +&.01 \\ +&.002 \\ +&.0003 \\ +&.00005 \\ +&.000008 \\ +&.0000013 \\ &\ \vdots \end{aligned}$$

converges to a rational number.

44 Prove the converse of Cassini's identity (6.106): If k and m are integers such that $|m^2-km-k^2| = 1$, then there is an integer n such that $k = \pm F_n$ and $m = \pm F_{n+1}$.

45 Use the repertoire method to solve the general recurrence

$$X_0 = \alpha; \qquad X_1 = \beta; \qquad X_n = X_{n-1} + X_{n-2} + \gamma n + \delta.$$

46 What are $\cos 36°$ and $\cos 72°$?

47 Show that

$$2^{n-1}F_n = \sum_k \binom{n}{2k+1} 5^k,$$

and use this identity to deduce the values of $F_p \bmod p$ and $F_{p+1} \bmod p$ when p is prime.

48 Prove that zero-valued parameters can be removed from continuant polynomials by collapsing their neighbors together:

$$\begin{aligned} &K_n(x_1, \ldots, x_{m-1}, 0, x_{m+1}, \ldots, x_n) \\ &\quad = K_{n-2}(x_1, \ldots, x_{m-2}, x_{m-1}+x_{m+1}, x_{m+2}, \ldots, x_n), \qquad 1 < m < n. \end{aligned}$$

49 Find the continued fraction representation of the number $\sum_{n \geqslant 1} 2^{-\lfloor n\phi \rfloor}$.

50 Define $f(n)$ for all positive integers n by the recurrence

$$\begin{aligned} f(1) &= 1; \\ f(2n) &= f(n); \\ f(2n+1) &= f(n) + f(n+1). \end{aligned}$$

a For which n is $f(n)$ even?

b Show that $f(n)$ can be expressed in terms of continuants.

Exam problems

51 Let p be a prime number.

a Prove that ${p \brace k} \equiv {p \brack k} \equiv 0 \pmod p$, for $1 < k < p$.

b Prove that ${p-1 \brack k} \equiv 1 \pmod p$, for $1 \leqslant k \leqslant p$.

c Prove that ${2p-2 \brace p} \equiv {2p-2 \brack p} \equiv 0 \pmod p$.

d Prove that if $p > 3$ we have ${p \brack 2} \equiv 0 \pmod{p^2}$. *Hint:* Consider $p^{\underline{p}}$.

52 Let H_n be written in lowest terms as a_n/b_n.

a Prove that $p \backslash b_n \iff p \nmid a_{\lfloor n/p \rfloor}$, if p is prime.

b Find all $n > 0$ such that a_n is divisible by 5.

53 Find a closed form for $\sum_{k=0}^{m} \binom{n}{k}^{-1}(-1)^k H_k$, when $0 \leqslant m \leqslant n$. *Hint:* Exercise 5.42 has the sum without the H_k factor.

54 Let $n > 0$. The purpose of this exercise is to show that the denominator of B_{2n} is the product of all primes p such that $(p-1)\backslash(2n)$.

a Show that $S_m(p) + [(p-1)\backslash m]$ is a multiple of p, when p is prime and $m > 0$.

b Use the result of part (a) to show that

$$B_{2n} + \sum_{p \text{ prime}} \frac{[(p-1)\backslash(2n)]}{p} = I_{2n} \quad \text{is an integer.}$$

Hint: It suffices to prove that, if p is any prime, the denominator of the fraction $B_{2n} + [(p-1)\backslash(2n)]/p$ is not divisible by p.

c Prove that the denominator of B_{2n} is always an odd multiple of 6, and it is equal to 6 for infinitely many n.

55 Prove (6.70) as a corollary of a more general identity, by summing

$$\sum_{0 \leqslant k < n} \binom{k}{m}\binom{x+k}{k}$$

and differentiating with respect to x.

56 Evaluate $\sum_{k \neq m} \binom{n}{k}(-1)^k k^{n+1}/(k-m)$ in closed form as a function of the integers m and n. (The sum is over all integers k except for the value $k = m$.)

57 The "wraparound binomial coefficients of order 5" are defined by

$$\left(\!\binom{n}{k}\!\right) = \left(\!\binom{n-1}{k}\!\right) + \left(\!\binom{n-1}{(k-1) \bmod 5}\!\right), \qquad n > 0,$$

and $\left(\!\binom{0}{k}\!\right) = [k=0]$. Let Q_n be the difference between the largest and smallest of these numbers in row n:

$$Q_n = \max_{0 \leqslant k < 5} \left(\!\binom{n}{k}\!\right) - \min_{0 \leqslant k < 5} \left(\!\binom{n}{k}\!\right).$$

Find and prove a relation between Q_n and the Fibonacci numbers.

58 Find closed forms for $\sum_{n \geqslant 0} F_n^2 z^n$ and $\sum_{n \geqslant 0} F_n^3 z^n$. What do you deduce about the quantity $F_{n+1}^3 - 4F_n^3 - F_{n-1}^3$?

59 Prove that if m and n are positive integers, there exists an integer x such that $F_x \equiv m \pmod{3^n}$.

60 Find all positive integers n such that either $F_n + 1$ or $F_n - 1$ is a prime number.

61 Prove the identity

$$\sum_{k=0}^{n} \frac{1}{F_{2^k}} = 3 - \frac{F_{2^n-1}}{F_{2^n}}, \qquad \text{integer } n \geqslant 1.$$

What is $\sum_{k=0}^{n} 1/F_{3\cdot 2^k}$?

62 Let $A_n = \phi^n + \phi^{-n}$ and $B_n = \phi^n - \phi^{-n}$.

a Find constants α and β such that $A_n = \alpha A_{n-1} + \beta A_{n-2}$ and $B_n = \alpha B_{n-1} + \beta B_{n-2}$ for all $n \geqslant 0$.

b Express A_n and B_n in terms of F_n and L_n (see exercise 28).

c Prove that $\sum_{k=1}^{n} 1/(F_{2k+1} + 1) = B_n/A_{n+1}$.

d Find a closed form for $\sum_{k=1}^{n} 1/(F_{2k+1} - 1)$.

Bonus problems

Bogus problems

63 How many permutations $\pi_1\pi_2\ldots\pi_n$ of $\{1, 2, \ldots, n\}$ have exactly k indices j such that

a $\pi_i < \pi_j$ for all $i < j$? (Such j are called "left-to-right maxima.")

b $\pi_j > j$? (Such j are called "excedances.")

64 What is the denominator of $\left[{1/2 \atop 1/2-n}\right]$, when this fraction is reduced to lowest terms?

65 Prove the identity

$$\int_0^1 \cdots \int_0^1 f(\lfloor x_1 + \cdots + x_n \rfloor)\, dx_1 \ldots dx_n = \sum_k \left\langle {n \atop k} \right\rangle \frac{f(k)}{n!}.$$

66 Show that $\left\langle\!\left\langle {n \atop 1} \right\rangle\!\right\rangle = 2\left\langle {n \atop 1} \right\rangle$, and find a closed form for $\left\langle\!\left\langle {n \atop 2} \right\rangle\!\right\rangle$.

67 Find a closed form for $\sum_{k=1}^{n} k^2 H_{n+k}$.

68 Show that the generalized harmonic numbers of exercise 22 have the power series expansion

$$H_z = \sum_{n\geqslant 2} (-1)^n H_\infty^{(n)} z^{n-1}.$$

69 Prove that the generalized factorial of equation (5.83) can be written

$$\prod_{k\geqslant 1} \left(1 + \frac{z}{k}\right) e^{-z/k} = \frac{e^{\gamma z}}{z!},$$

by considering the limit as $n \to \infty$ of the first n factors of this infinite product. Show that $\frac{d}{dz}(z!)$ is related to the general harmonic numbers of exercise 22.

70 Prove that the tangent function has the power series (6.92), and find the corresponding series for $z/\sin z$ and $\ln\bigl((\tan z)/z\bigr)$.

71 Find a relation between the numbers $T_n(1)$ and the coefficients of $1/\cos z$.

72 What is $\sum_k (-1)^k \left\langle {n \atop k} \right\rangle$, the row sum of Euler's triangle with alternating signs?

73 Prove that, for all integers $n \geqslant 1$,

$$z\cot z = \frac{z}{2^n}\cot\frac{z}{2^n} - \frac{z}{2^n}\tan\frac{z}{2^n} + \sum_{k=1}^{2^{n-1}-1}\frac{z}{2^n}\left(\cot\frac{z+k\pi}{2^n} + \cot\frac{z-k\pi}{2^n}\right),$$

and show that the limit of the kth summand is $2z^2/(z^2 - k^2\pi^2)$ for fixed k as $n \to \infty$.

74 Prove the following relation that connects Stirling numbers, Bernoulli numbers, and Catalan numbers:

$$\sum_{k=0}^{n} \left\{ {n+k \atop k} \right\} \binom{2n}{n+k} \frac{(-1)^k}{k+1} = B_n \binom{2n}{n} \frac{1}{n+1}.$$

75 Show that the four chessboard pieces of the $64 = 65$ paradox can also be reassembled to prove that $64 = 63$.

76 A sequence defined by the recurrence

$$A_1 = x, \qquad A_2 = y, \qquad A_n = A_{n-1} + A_{n-2}$$

has $A_m = 1000000$ for some m. What positive integers x and y make m as large as possible?

77 The text describes a way to change a formula involving $F_{n\pm k}$ to a formula that involves F_n and F_{n+1} only. Therefore it's natural to wonder if two such "reduced" formulas can be equal when they aren't identical in form. Let $P(x, y)$ be a polynomial in x and y with integer coefficients. Find a necessary and sufficient condition that $P(F_{n+1}, F_n) = 0$ for all $n \geqslant 0$.

78 Explain how to add positive integers, working entirely in the Fibonacci number system.

79 Is it possible that a sequence $\langle A_n \rangle$ satisfying the Fibonacci recurrence $A_n = A_{n-1} + A_{n-2}$ can contain no prime numbers, if A_0 and A_1 are relatively prime?

80 Show that continuant polynomials appear in the matrix product

$$\begin{pmatrix} 0 & 1 \\ 1 & x_1 \end{pmatrix} \begin{pmatrix} 0 & 1 \\ 1 & x_2 \end{pmatrix} \cdots \begin{pmatrix} 0 & 1 \\ 1 & x_n \end{pmatrix}$$

and in the determinant

$$\det \begin{pmatrix} x_1 & 1 & 0 & 0 & \ldots & 0 \\ -1 & x_2 & 1 & 0 & & 0 \\ 0 & -1 & x_3 & 1 & & \vdots \\ \vdots & & -1 & & & \\ & & & & \ddots & 1 \\ 0 & 0 & \ldots & & -1 & x_n \end{pmatrix}.$$

81 Generalizing (6.146), find a continued fraction related to the generating function $\sum_{n\geqslant 1} z^{\lfloor n\alpha \rfloor}$, when α is any positive irrational number.

82 Let m and n be odd, positive integers. Find closed forms for

$$S^+_{m,n} = \sum_{k\geqslant 0} \frac{1}{F_{2mk+n} + F_m}; \qquad S^-_{m,n} = \sum_{k\geqslant 0} \frac{1}{F_{2mk+n} - F_m}.$$

Hint: The sums in exercise 62 are $S^+_{1,3} - S^+_{1,2n+3}$ and $S^-_{1,3} - S^-_{1,2n+3}$.

83 Let α be an irrational number in $(0,1)$ and let a_1, a_2, a_3, ... be the partial quotients in its continued fraction representation. Show that $|D(\alpha,n)| < 2$ when $n = K(a_1,\ldots,a_m)$, where D is the discrepancy defined in Chapter 3.

84 Let Q_n be the largest denominator on level n of the Stern–Brocot tree. (Thus $\langle Q_0, Q_1, Q_2, Q_3, Q_4, \ldots\rangle = \langle 1,2,3,5,8,\ldots\rangle$ according to the diagram in Chapter 4.) Prove that $Q_n = F_{n+2}$.

85 Characterize all N such that the Fibonacci residues

$$\{F_0 \bmod N,\ F_1 \bmod N,\ F_2 \bmod N,\ \ldots\}$$

form the complete set $\{0,1,\ldots,N-1\}$. (See exercise 59.)

Research problems

86 What is the best way to extend the definition of $\left\{ {n \atop k} \right\}$ to arbitrary real values of n and k?

87 Let H_n be written in lowest terms as a_n/b_n, as in exercise 52.

a Are there infinitely many n with $11\backslash a_n$?

b Are there infinitely many n with $b_n = \text{lcm}(1,2,\ldots,n)$? (Two such values are $n = 250$ and $n = 1000$.)

88 Prove that γ and e^γ are irrational.

89 Develop a general theory of the solutions to the two-parameter recurrence

$$\left|{n \atop k}\right| = (\alpha n + \beta k + \gamma)\left|{n-1 \atop k}\right| + (\alpha' n + \beta' k + \gamma')\left|{n-1 \atop k-1}\right| + [n=k=0]\,, \qquad \text{for } n, k \geqslant 0,$$

assuming that $\left|{n \atop k}\right| = 0$ when $n < 0$ or $k < 0$. (Binomial coefficients, Stirling numbers, Eulerian numbers, and the sequences of exercises 17 and 31 are special cases.) What special values $(\alpha, \beta, \gamma, \alpha', \beta', \gamma')$ yield "fundamental solutions" in terms of which the general solution can be expressed?

7

Generating Functions

THE MOST POWERFUL WAY to deal with sequences of numbers, as far as anybody knows, is to manipulate infinite series that "generate" those sequences. We've learned a lot of sequences and we've seen a few generating functions; now we're ready to explore generating functions in depth, and to see how remarkably useful they are.

7.1 DOMINO THEORY AND CHANGE

Generating functions are important enough, and for many of us new enough, to justify a relaxed approach as we begin to look at them more closely. So let's start this chapter with some fun and games as we try to develop our intuitions about generating functions. We will study two applications of the ideas, one involving dominoes and the other involving coins.

How many ways T_n are there to completely cover a $2 \times n$ rectangle with 2×1 dominoes? We assume that the dominoes are identical (either because they're face down, or because someone has rendered them indistinguishable, say by painting them all red); thus only their orientations — vertical or horizontal — matter, and we can imagine that we're working with domino-shaped tiles. For example, there are three tilings of a 2×3 rectangle, namely ▯▯▯, ▯⊟, and ⊟▯; so $T_3 = 3$.

To find a closed form for general T_n we do our usual first thing, look at small cases. When $n = 1$ there's obviously just one tiling, ▯; and when $n = 2$ there are two, ▯▯ and ⊟.

"Let me count the ways."
— E. B. Browning

How about when $n = 0$; how many tilings of a 2×0 rectangle are there? It's not immediately clear what this question means, but we've seen similar situations before: There is one permutation of zero objects (namely the empty permutation), so $0! = 1$. There is one way to choose zero things from n things (namely to choose nothing), so $\binom{n}{0} = 1$. There is one way to partition the empty set into zero nonempty subsets, but there are no such ways to partition a nonempty set; so $\left\{ {n \atop 0} \right\} = [n=0]$. By such reasoning we can conclude that there's just one way to tile a 2×0 rectangle with dominoes, namely to use no dominoes; therefore $T_0 = 1$. (This spoils the simple pattern $T_n = n$ that holds when $n = 1$, 2, and 3; but that pattern was probably doomed anyway, since T_0 wants to be 1 according to the logic of the situation.) A proper understanding of the null case turns out to be useful whenever we want to solve an enumeration problem.

Let's look at one more small case, $n = 4$. There are two possibilities for tiling the left edge of the rectangle — we put either a vertical domino or two horizontal dominoes there. If we choose a vertical one, the partial solution is ▯▭ and the remaining 2×3 rectangle can be covered in T_3 ways. If we choose two horizontals, the partial solution ⊟▭ can be completed in T_2 ways. Thus $T_4 = T_3 + T_2 = 5$. (The five tilings are ▯▯▯▯, ▯▯⊟, ▯⊟▯, ⊟▯▯, and ⊟⊟.)

We now know the first five values of T_n:

n	0	1	2	3	4
T_n	1	1	2	3	5

These look suspiciously like the Fibonacci numbers, and it's not hard to see why: The reasoning we used to establish $T_4 = T_3 + T_2$ easily generalizes to $T_n = T_{n-1} + T_{n-2}$, for $n \geqslant 2$. Thus we have the same recurrence here as for the Fibonacci numbers, except that the initial values $T_0 = 1$ and $T_1 = 1$ are a little different. But these initial values are the consecutive Fibonacci numbers F_1 and F_2, so the T's are just Fibonacci numbers shifted up one place:

$$T_n = F_{n+1}, \qquad \text{for } n \geqslant 0.$$

(We consider this to be a closed form for T_n, because the Fibonacci numbers are important enough to be considered "known." Also, F_n itself has a closed form (6.123) in terms of algebraic operations.) Notice that this equation confirms the wisdom of setting $T_0 = 1$.

But what does all this have to do with generating functions? Well, we're about to get to that — there's another way to figure out what T_n is. This new way is based on a bold idea. Let's consider the "sum" of all possible $2 \times n$ tilings, for all $n \geqslant 0$, and call it T:

To boldly go where no tiling has gone before.

$$T = | + ▯ + ▯▯ + ⊟ + ▯▯▯ + ▯⊟ + ⊟▯ + \cdots. \qquad (7.1)$$

(The first term '|' on the right stands for the null tiling of a 2×0 rectangle.) This sum T represents lots of information. It's useful because it lets us prove things about T as a whole rather than forcing us to prove them (by induction) about its individual terms.

The terms of this sum stand for tilings, which are combinatorial objects. We won't be fussy about what's considered legal when infinitely many tilings are added together; everything can be made rigorous, but our goal right now is to expand our consciousness beyond conventional algebraic formulas.

We've added the patterns together, and we can also multiply them — by juxtaposition. For example, we can multiply the tilings ▯ and ⊟ to get the new tiling ▯⊟. But notice that multiplication is not commutative; that is, the order of multiplication counts: ▯⊟ is different from ⊟▯.

Using this notion of multiplication it's not hard to see that the null tiling plays a special role — it is the multiplicative identity. For instance, | × ⊟ = ⊟ × | = ⊟.

Now we can use domino arithmetic to manipulate the infinite sum T:

T = | + ▯ + ▯▯ + ⊟ + ▯▯▯ + ▯⊟ + ⊟▯ + ···
  = | + ▯(| + ▯ + ▯▯ + ⊟ + ···) + ⊟(| + ▯ + ▯▯ + ⊟ + ···)
  = | + ▯T + ⊟T.   (7.2)

Every valid tiling occurs exactly once in each right side, so what we've done is reasonable even though we're ignoring the cautions in Chapter 2 about "absolute convergence." The bottom line of this equation tells us that everything in T is either the null tiling, or is a vertical tile followed by something else in T, or is two horizontal tiles followed by something else in T.

I have a gut feeling that these sums must converge, as long as the dominoes are small enough.

So now let's try to solve the equation for T. Replacing the T on the left by |T and subtracting the last two terms on the right from both sides of the equation, we get

(| − ▯ − ⊟)T = | .   (7.3)

For a consistency check, here's an expanded version:

| + ▯ + ▯▯ + ⊟ + ▯▯▯ + ▯⊟ + ⊟▯ + ···
− ▯ − ▯▯ − ▯▯▯ − ▯⊟ − ▯▯▯▯ − ▯▯⊟ − ▯⊟▯ − ···
− ⊟ − ⊟▯ − ⊟▯▯ − ⊟⊟ − ⊟▯▯▯ − ⊟▯⊟ − ⊟⊟▯ − ···
―――――――――――――――――――――――――――
|

Every term in the top row, except the first, is cancelled by a term in either the second or third row, so our equation is correct.

So far it's been fairly easy to make combinatorial sense of the equations we've been working with. Now, however, to get a compact expression for T we cross a combinatorial divide. With a leap of algebraic faith we divide both sides of equation (7.3) by | − ▯ − ⊟ to get

T = | / (| − ▯ − ⊟) .   (7.4)

(Multiplication isn't commutative, so we're on the verge of cheating, by not distinguishing between left and right division. In our application it doesn't matter, because | commutes with everything. But let's not be picky, unless our wild ideas lead to paradoxes.)

The next step is to expand this fraction as a power series, using the rule

$$\frac{1}{1-z} = 1 + z + z^2 + z^3 + \cdots .$$

The null tiling |, which is the multiplicative identity for our combinatorial arithmetic, plays the part of 1, the usual multiplicative identity; and ▯ + ⊟ plays z. So we get the expansion

$$\begin{aligned}\frac{|}{|-▯-⊟} &= |+(▯+⊟)+(▯+⊟)^2+(▯+⊟)^3+\cdots\\ &= |+(▯+⊟)+(▯▯+▯⊟+⊟▯+⊟⊟)\\ &\qquad +(▯▯▯+▯▯⊟+▯⊟▯+▯⊟⊟+⊟▯▯+⊟▯⊟+⊟⊟▯+⊟⊟⊟)+\cdots .\end{aligned}$$

This is T, but the tilings are arranged in a different order than we had before. Every tiling appears exactly once in this sum; for example, ▯⊟⊟▯▯⊟▯ appears in the expansion of $(▯+⊟)^7$.

We can get useful information from this infinite sum by compressing it down, ignoring details that are not of interest. For example, we can imagine that the patterns become unglued and that the individual dominoes commute with each other; then a term like ▯⊟⊟▯▯⊟▯ becomes $▯^4 ▭^6$, because it contains four verticals and six horizontals. Collecting like terms gives us the series

$$T = | + ▯ + ▯^2 + ▭^2 + ▯^3 + 2▯▭^2 + ▯^4 + 3▯^2▭^2 + ▭^4 + \cdots .$$

The $2▯▭^2$ here represents the two terms of the old expansion, ▯⊟ and ⊟▯, that have one vertical and two horizontal dominoes; similarly $3▯^2▭^2$ represents the three terms ▯▯⊟, ▯⊟▯, and ⊟▯▯. We're essentially treating ▯ and ▭ as ordinary (commutative) variables.

We can find a closed form for the coefficients in the commutative version of T by using the binomial theorem:

$$\begin{aligned}\frac{|}{|-(▯+▭^2)} &= |+(▯+▭^2)+(▯+▭^2)^2+(▯+▭^2)^3+\cdots\\ &= \sum_{k\ge 0}(▯+▭^2)^k\\ &= \sum_{j,k\ge 0}\binom{k}{j}▯^j▭^{2k-2j}\\ &= \sum_{j,m\ge 0}\binom{j+m}{j}▯^j▭^{2m} . \qquad (7.5)\end{aligned}$$

(The last step replaces $k-j$ by m; this is legal because we have $\binom{k}{j}=0$ when $0 \leqslant k < j$.) We conclude that $\binom{j+m}{j}$ is the number of ways to tile a $2\times(j+2m)$ rectangle with j vertical dominoes and $2m$ horizontal dominoes. For example, we recently looked at the 2×10 tiling ▯⊟⊟▯▯⊟▯, which involves four verticals and six horizontals; there are $\binom{4+3}{4} = 35$ such tilings in all, so one of the terms in the commutative version of T is $35\,▯^4\,▭^6$.

We can suppress even more detail by ignoring the orientation of the dominoes. Suppose we don't care about the horizontal/vertical breakdown; we only want to know about the total number of $2 \times n$ tilings. (This, in fact, is the number T_n we started out trying to discover.) We can collect the necessary information by simply substituting a single quantity, z, for ▯ and ▭. And we might as well also replace I by 1, getting

Now I'm disoriented.

$$T = \frac{1}{1-z-z^2}\,. \tag{7.6}$$

This is the generating function (6.117) for Fibonacci numbers, except for a missing factor of z in the numerator; so we conclude that the coefficient of z^n in T is F_{n+1}.

The compact representations I/(I−▯−⊟), I/(I−▯−▭^{2}), and $1/(1-z-z^2)$ that we have deduced for T are called *generating functions*, because they generate the coefficients of interest.

Incidentally, our derivation implies that the number of $2 \times n$ domino tilings with exactly m pairs of horizontal dominoes is $\binom{n-m}{m}$. (This follows because there are $j = n - 2m$ vertical dominoes, hence there are

$$\binom{j+m}{j} = \binom{j+m}{m} = \binom{n-m}{m}$$

ways to do the tiling according to our formula.) We observed in Chapter 6 that $\binom{n-m}{m}$ is the number of Morse code sequences of length n that contain m dashes; in fact, it's easy to see that $2\times n$ domino tilings correspond directly to Morse code sequences. (The tiling ▯⊟⊟▯▯⊟▯ corresponds to '•−−••−•'.) Thus domino tilings are closely related to the continuant polynomials we studied in Chapter 6. It's a small world.

We have solved the T_n problem in two ways. The first way, guessing the answer and proving it by induction, was easier; the second way, using infinite sums of domino patterns and distilling out the coefficients of interest, was fancier. But did we use the second method only because it was amusing to play with dominoes as if they were algebraic variables? No; the real reason for introducing the second way was that the infinite-sum approach is a lot more powerful. The second method applies to many more problems, because it doesn't require us to make magic guesses.

Let's generalize up a notch, to a problem where guesswork will be beyond us. How many ways U_n are there to tile a $3 \times n$ rectangle with dominoes?

The first few cases of this problem tell us a little: The null tiling gives $U_0 = 1$. There is no valid tiling when $n = 1$, since a 2×1 domino doesn't fill a 3×1 rectangle, and since there isn't room for two. The next case, $n = 2$, can easily be done by hand; there are three tilings, [tiling], [tiling], and [tiling], so $U_2 = 3$. (Come to think of it we already knew this, because the previous problem told us that $T_3 = 3$; the number of ways to tile a 3×2 rectangle is the same as the number to tile a 2×3.) When $n = 3$, as when $n = 1$, there are no tilings. We can convince ourselves of this either by making a quick exhaustive search or by looking at the problem from a higher level: The area of a 3×3 rectangle is odd, so we can't possibly tile it with dominoes whose area is even. (The same argument obviously applies to any odd n.) Finally, when $n = 4$ there seem to be about a dozen tilings; it's difficult to be sure about the exact number without spending a lot of time to guarantee that the list is complete.

So let's try the infinite-sum approach that worked last time:

$$U = | + [tiling] + [tiling] + [tiling] + [tiling] + [tiling] + [tiling] + [tiling] + [tiling] + \cdots . \qquad (7.7)$$

Every non-null tiling begins with either [tiling] or [tiling] or [tiling]; but unfortunately the first two of these three possibilities don't simply factor out and leave us with U again. The sum of all terms in U that begin with [tiling] can, however, be written as [tiling]V, where

$$V = [tiling] + [tiling] + [tiling] + [tiling] + [tiling] + \cdots$$

is the sum of all domino tilings of a mutilated $3 \times n$ rectangle that has its lower left corner missing. Similarly, the terms of U that begin with [tiling] can be written [tiling]Λ, where

$$\Lambda = [tiling] + [tiling] + [tiling] + [tiling] + [tiling] + \cdots$$

consists of all rectangular tilings lacking their upper left corner. The series Λ is a mirror image of V. These factorizations allow us to write

$$U = | + [tiling]V + [tiling]\Lambda + [tiling]U .$$

And we can factor V and Λ as well, because such tilings can begin in only two ways:

$$V = [tiling]U + [tiling]V ,$$
$$\Lambda = [tiling]U + [tiling]\Lambda .$$

Now we have three equations in three unknowns (U, V, and Λ). We can solve them by first solving for V and Λ in terms of U, then plugging the results into the equation for U:

$$\mathrm{V} = (\mathrm{I} - \mathcal{E})^{-1}\square\mathrm{U}, \qquad \Lambda = (\mathrm{I} - \mathcal{F})^{-1}\square\mathrm{U};$$

$$\mathrm{U} = \mathrm{I} + \mathcal{L}(\mathrm{I} - \mathcal{E})^{-1}\square\mathrm{U} + \Gamma(\mathrm{I} - \mathcal{F})^{-1}\square\mathrm{U} + \equiv\mathrm{U}.$$

And the final equation can be solved for U, giving the compact formula

$$\mathrm{U} = \frac{\mathrm{I}}{\mathrm{I} - \mathcal{L}(\mathrm{I} - \mathcal{E})^{-1}\square - \Gamma(\mathrm{I} - \mathcal{F})^{-1}\square - \equiv}. \tag{7.8}$$

This expression defines the infinite sum U, just as (7.4) defines T.

The next step is to go commutative. Everything simplifies beautifully when we detach all the dominoes and use only powers of $\square$ and $\boxminus$:

$$\begin{aligned}
\mathrm{U} &= \frac{1}{1 - \square^2\boxminus(1 - \boxminus^3)^{-1} - \square^2\boxminus(1 - \boxminus^3)^{-1} - \boxminus^3} \\
&= \frac{1 - \boxminus^3}{(1 - \boxminus^3)^2 - 2\square^2\boxminus} \\
&= \frac{(1 - \boxminus^3)^{-1}}{1 - 2\square^2\boxminus(1 - \boxminus^3)^{-2}} \\
&= \frac{1}{1 - \boxminus^3} + \frac{2\square^2\boxminus}{(1 - \boxminus^3)^3} + \frac{4\square^4\boxminus^2}{(1 - \boxminus^3)^5} + \frac{8\square^6\boxminus^3}{(1 - \boxminus^3)^7} + \cdots \\
&= \sum_{k \geqslant 0} \frac{2^k\square^{2k}\boxminus^k}{(1 - \boxminus^3)^{2k+1}} \\
&= \sum_{k,m \geqslant 0} \binom{m + 2k}{m} 2^k\square^{2k}\boxminus^{k+3m}.
\end{aligned}$$

I learned in another class about "regular expressions." If I'm not mistaken, we can write

$$\mathrm{U} = (\mathcal{L}\,\mathcal{E}^*\square + \Gamma\,\mathcal{F}^*\square + \equiv)^*$$

in the language of regular expressions; so there must be some connection between regular expressions and generating functions.

(This derivation deserves careful scrutiny. The last step uses the formula $(1 - w)^{-2k-1} = \sum_m \binom{m+2k}{m} w^m$, identity (5.56).) Let's take a good look at the bottom line to see what it tells us. First, it says that every $3 \times n$ tiling uses an even number of vertical dominoes. Moreover, if there are $2k$ verticals, there must be at least k horizontals, and the total number of horizontals must be $k + 3m$ for some $m \geqslant 0$. Finally, the number of possible tilings with $2k$ verticals and $k + 3m$ horizontals is exactly $\binom{m+2k}{m}2^k$.

We now are able to analyze the 3×4 tilings that left us doubtful when we began looking at the $3 \times n$ problem. When $n = 4$ the total area is 12, so we need six dominoes altogether. There are $2k$ verticals and $k + 3m$ horizontals, for some k and m; hence $2k + k + 3m = 6$. In other words, $k + m = 2$. If we use no verticals, then $k = 0$ and $m = 2$; the number of possibilities is $\binom{2+0}{2}2^0 = 1$. (This accounts for the tiling ⊞.) If we use two verticals, then $k = 1$ and $m = 1$; there are $\binom{1+2}{1}2^1 = 6$ such tilings. And if we use four verticals, then $k = 2$ and $m = 0$; there are $\binom{0+4}{0}2^2 = 4$ such tilings, making a total of $U_4 = 11$. In general if n is even, this reasoning shows that $k + m = \frac{1}{2}n$, hence $\binom{m+2k}{m} = \binom{n/2+k}{n/2-k}$ and the total number of $3 \times n$ tilings is

$$U_n = \sum_k \binom{n/2+k}{n/2-k} 2^k = \sum_m \binom{n-m}{m} 2^{n/2-m}. \tag{7.9}$$

As before, we can also substitute z for both ▯ and ▭, getting a generating function that doesn't discriminate between dominoes of particular persuasions. The result is

$$U = \frac{1}{1 - z^3(1-z^3)^{-1} - z^3(1-z^3)^{-1} - z^3} = \frac{1-z^3}{1-4z^3+z^6}. \tag{7.10}$$

If we expand this quotient into a power series, we get

$$U = 1 + U_2\,z^3 + U_4\,z^6 + U_6\,z^9 + U_8\,z^{12} + \cdots,$$

a generating function for the numbers U_n. (There's a curious mismatch between subscripts and exponents in this formula, but it is easily explained. The coefficient of z^9, for example, is U_6, which counts the tilings of a 3×6 rectangle. This is what we want, because every such tiling contains nine dominoes.)

We could proceed to analyze (7.10) and get a closed form for the coefficients, but it's better to save that for later in the chapter after we've gotten more experience. So let's divest ourselves of dominoes for the moment and proceed to the next advertised problem, "change."

How many ways are there to pay 50 cents? We assume that the payment must be made with pennies ①, nickels ⑤, dimes ⑩, quarters ㉕, and half-dollars ㊿. George Pólya [239] popularized this problem by showing that it can be solved with generating functions in an instructive way.

Ah yes, I remember when we had half-dollars.

Let's set up infinite sums that represent all possible ways to give change, just as we tackled the domino problems by working with infinite sums that represent all possible domino patterns. It's simplest to start by working with fewer varieties of coins, so let's suppose first that we have nothing but pennies. The sum of all ways to leave some number of pennies (but just pennies) in change can be written

$$\begin{aligned} P &= \not{1} + ① + ①① + ①①① + ①①①① + \cdots \\ &= \not{1} + ① + ①^2 + ①^3 + ①^4 + \cdots. \end{aligned}$$

The first term stands for the way to leave no pennies, the second term stands for one penny, then two pennies, three pennies, and so on. Now if we're allowed to use both pennies and nickels, the sum of all possible ways is

$$\begin{aligned} \mathrm{N} &= \mathrm{P} + \text{⑤}\,\mathrm{P} + \text{⑤⑤}\,\mathrm{P} + \text{⑤⑤⑤}\,\mathrm{P} + \text{⑤⑤⑤⑤}\,\mathrm{P} + \cdots \\ &= (\not\text{⑤} + \text{⑤} + \text{⑤}^2 + \text{⑤}^3 + \text{⑤}^4 + \cdots)\,\mathrm{P}, \end{aligned}$$

since each payment has a certain number of nickels chosen from the first factor and a certain number of pennies chosen from P. (Notice that N is *not* the sum $\not\text{①} + \text{①} + \text{⑤} + (\text{①} + \text{⑤})^2 + (\text{①} + \text{⑤})^3 + \cdots$, because such a sum includes many types of payment more than once. For example, the term $(\text{①} + \text{⑤})^2 = \text{①①} + \text{①⑤} + \text{⑤①} + \text{⑤⑤}$ treats ①⑤ and ⑤① as if they were different, but we want to list each set of coins only once without respect to order.)

Similarly, if dimes are permitted as well, we get the infinite sum

$$\mathrm{D} = (\not\text{⑩} + \text{⑩} + \text{⑩}^2 + \text{⑩}^3 + \text{⑩}^4 + \cdots)\,\mathrm{N},$$

which includes terms like $\text{⑩}^3\text{⑤}^3\text{①}^5 = \text{⑩⑩⑩⑤⑤⑤①①①①①}$ when it is expanded in full. Each of these terms is a different way to make change. Adding quarters and then half-dollars to the realm of possibilities gives

Coins of the realm.

$$\begin{aligned} \mathrm{Q} &= (\not\text{㉕} + \text{㉕} + \text{㉕}^2 + \text{㉕}^3 + \text{㉕}^4 + \cdots)\,\mathrm{D}; \\ \mathrm{C} &= (\not\text{㊿} + \text{㊿} + \text{㊿}^2 + \text{㊿}^3 + \text{㊿}^4 + \cdots)\,\mathrm{Q}. \end{aligned}$$

Our problem is to find the number of terms in C worth exactly 50¢.

A simple trick solves this problem nicely: We can replace ① by z, ⑤ by z^5, ⑩ by z^{10}, ㉕ by z^{25}, and ㊿ by z^{50}. Then each term is replaced by z^n, where n is the monetary value of the original term. For example, the term ㊿⑩⑤⑤① becomes $z^{50+10+5+5+1} = z^{71}$. The four ways of paying 13 cents, namely $\text{⑩}\text{①}^3$, $\text{⑤}\text{①}^8$, $\text{⑤}^2\text{①}^3$, and ①^{13}, each reduce to z^{13}; hence the coefficient of z^{13} will be 4 after the z-substitutions are made.

Let P_n, N_n, D_n, Q_n, and C_n be the numbers of ways to pay n cents when we're allowed to use coins that are worth at most 1, 5, 10, 25, and 50 cents, respectively. Our analysis tells us that these are the coefficients of z^n in the respective power series

$$\begin{aligned} \mathrm{P} &= 1 + z + z^2 + z^3 + z^4 + \cdots, \\ \mathrm{N} &= (1 + z^5 + z^{10} + z^{15} + z^{20} + \cdots)\mathrm{P}, \\ \mathrm{D} &= (1 + z^{10} + z^{20} + z^{30} + z^{40} + \cdots)\mathrm{N}, \\ \mathrm{Q} &= (1 + z^{25} + z^{50} + z^{75} + z^{100} + \cdots)\mathrm{D}, \\ \mathrm{C} &= (1 + z^{50} + z^{100} + z^{150} + z^{200} + \cdots)\mathrm{Q}. \end{aligned}$$

How many pennies are there, really? If n is greater than, say, 10^{10}, I bet that $P_n = 0$ in the "real world."

Obviously $P_n = 1$ for all $n \geqslant 0$. And a little thought proves that we have $N_n = \lfloor n/5 \rfloor + 1$: To make n cents out of pennies and nickels, we must choose either 0 or 1 or ... or $\lfloor n/5 \rfloor$ nickels, after which there's only one way to supply the requisite number of pennies. Thus P_n and N_n are simple; but the values of D_n, Q_n, and C_n are increasingly more complicated.

One way to deal with these formulas is to realize that $1 + z^m + z^{2m} + \cdots$ is just $1/(1 - z^m)$. Thus we can write

$$\begin{aligned} P &= 1/(1-z)\,, \\ N &= P/(1-z^5)\,, \\ D &= N/(1-z^{10})\,, \\ Q &= D/(1-z^{25})\,, \\ C &= Q/(1-z^{50})\,. \end{aligned}$$

Multiplying by the denominators, we have

$$\begin{aligned} (1-z)\,P &= 1\,, \\ (1-z^5)\,N &= P\,, \\ (1-z^{10})\,D &= N\,, \\ (1-z^{25})\,Q &= D\,, \\ (1-z^{50})\,C &= Q\,. \end{aligned}$$

Now we can equate coefficients of z^n in these equations, getting recurrence relations from which the desired coefficients can quickly be computed:

$$\begin{aligned} P_n &= P_{n-1} + [n=0]\,, \\ N_n &= N_{n-5} + P_n\,, \\ D_n &= D_{n-10} + N_n\,, \\ Q_n &= Q_{n-25} + D_n\,, \\ C_n &= C_{n-50} + Q_n\,. \end{aligned}$$

For example, the coefficient of z^n in $D = (1 - z^{25})Q$ is equal to $Q_n - Q_{n-25}$; so we must have $Q_n - Q_{n-25} = D_n$, as claimed.

We could unfold these recurrences and find, for example, that $Q_n = D_n + D_{n-25} + D_{n-50} + D_{n-75} + \cdots$, stopping when the subscripts get negative. But the non-iterated form is convenient because each coefficient is computed with just one addition, as in Pascal's triangle.

Let's use the recurrences to find C_{50}. First, $C_{50} = C_0 + Q_{50}$; so we want to know Q_{50}. Then $Q_{50} = Q_{25} + D_{50}$, and $Q_{25} = Q_0 + D_{25}$; so we also want to know D_{50} and D_{25}. These D_n depend in turn on D_{40}, D_{30}, D_{20}, D_{15}, D_{10}, D_5, and on N_{50}, N_{45}, ..., N_5. A simple calculation therefore suffices to determine all the necessary coefficients:

n	0	5	10	15	20	25	30	35	40	45	50
P_n	1	1	1	1	1	1	1	1	1	1	1
N_n	1	2	3	4	5	6	7	8	9	10	11
D_n	1	2	4	6	9	12	16		25		36
Q_n	1					13					49
C_n	1										50

The final value in the table gives us our answer, C_{50}: There are exactly 50 ways to leave a 50-cent tip.

(Not counting the option of charging the tip to a credit card.)

How about a closed form for C_n? Multiplying the equations together gives us the compact expression

$$C = \frac{1}{1-z}\,\frac{1}{1-z^5}\,\frac{1}{1-z^{10}}\,\frac{1}{1-z^{25}}\,\frac{1}{1-z^{50}}\,, \qquad (7.11)$$

but it's not obvious how to get from here to the coefficient of z^n. Fortunately there is a way; we'll return to this problem later in the chapter.

More elegant formulas arise if we consider the problem of giving change when we live in a land that mints coins of every positive integer denomination (①, ②, ③, ...) instead of just the five we allowed before. The corresponding generating function is an infinite product of fractions,

$$\frac{1}{(1-z)(1-z^2)(1-z^3)\dots}\,,$$

and the coefficient of z^n when these factors are fully multiplied out is called $p(n)$, the number of *partitions* of n. A partition of n is a representation of n as a sum of positive integers, disregarding order. For example, there are seven different partitions of 5, namely

$$5 = 4+1 = 3+2 = 3+1+1 = 2+2+1 = 2+1+1+1 = 1+1+1+1+1\,;$$

hence $p(5) = 7$. (Also $p(2) = 2$, $p(3) = 3$, $p(4) = 5$, and $p(6) = 11$; it begins to look as if $p(n)$ is always a prime number. But $p(7) = 15$, spoiling the pattern.) There is no closed form for $p(n)$, but the theory of partitions is a fascinating branch of mathematics in which many remarkable discoveries have been made. For example, Ramanujan proved that $p(5n+4) \equiv 0 \pmod 5$, $p(7n+5) \equiv 0 \pmod 7$, and $p(11n+6) \equiv 0 \pmod{11}$, by making ingenious transformations of generating functions (see Andrews [11, Chapter 10]).

7.2 BASIC MANEUVERS

Now let's look more closely at some of the techniques that make power series powerful.

First a few words about terminology and notation. Our generic generating function has the form

$$G(z) \;=\; g_0 + g_1 z + g_2 z^2 + \cdots \;=\; \sum_{n\geqslant 0} g_n z^n \,, \qquad (7.12)$$

and we say that $G(z)$, or G for short, is the generating function for the sequence $\langle g_0, g_1, g_2, \ldots\rangle$, which we also call $\langle g_n\rangle$. The coefficient g_n of z^n in $G(z)$ is sometimes denoted $[z^n]\,G(z)$.

The sum in (7.12) runs over all $n \geqslant 0$, but we often find it more convenient to extend the sum over all integers n. We can do this by simply regarding $g_{-1} = g_{-2} = \cdots = 0$. In such cases we might still talk about the sequence $\langle g_0, g_1, g_2, \ldots\rangle$, as if the g_n's didn't exist for negative n.

Two kinds of "closed forms" come up when we work with generating functions. We might have a closed form for $G(z)$, expressed in terms of z; or we might have closed form for g_n, expressed in terms of n. For example, the generating function for Fibonacci numbers has the closed form $z/(1-z-z^2)$; the Fibonacci numbers themselves have the closed form $(\phi^n - \hat{\phi}^n)/\sqrt{5}$. The context will explain what kind of closed form is meant.

Now a few words about perspective. The generating function $G(z)$ appears to be two different entities, depending on how we view it. Sometimes it is a function of a complex variable z, satisfying all the standard properties proved in calculus books. And sometimes it is simply a formal power series, with z acting as a placeholder. In the previous section, for example, we used the second interpretation; we saw several examples in which z was substituted for some feature of a combinatorial object in a "sum" of such objects. The coefficient of z^n was then the number of combinatorial objects having n occurrences of that feature.

If physicists can get away with viewing light sometimes as a wave and sometimes as a particle, mathematicians should be able to view generating functions in two different ways.

When we view $G(z)$ as a function of a complex variable, its convergence becomes an issue. We said in Chapter 2 that the infinite series $\sum_{n\geqslant 0} g_n z^n$ converges (absolutely) if and only if there's a bounding constant A such that the finite sums $\sum_{0\leqslant n\leqslant N} |g_n z^n|$ never exceed A, for any N. Therefore it's easy to see that if $\sum_{n\geqslant 0} g_n z^n$ converges for some value $z = z_0$, it also converges for all z with $|z| < |z_0|$. Furthermore, we must have $\lim_{n\to\infty} |g_n z_0^n| = 0$; hence, in the notation of Chapter 9, $g_n = O\bigl(|1/z_0|^n\bigr)$ if there is convergence at z_0. And conversely if $g_n = O(M^n)$, the series $\sum_{n\geqslant 0} g_n z^n$ converges for all $|z| < 1/M$. These are the basic facts about convergence of power series.

But for our purposes convergence is usually a red herring, unless we're trying to study the asymptotic behavior of the coefficients. Nearly every operation we perform on generating functions can be justified rigorously as an operation on formal power series, and such operations are legal even when the series don't converge. (The relevant theory can be found, for example, in Bell [19], Niven [225], and Henrici [151, Chapter 1].)

Furthermore, even if we throw all caution to the winds and derive formulas without any rigorous justification, we generally can take the results of our derivation and prove them by induction. For example, the generating function for the Fibonacci numbers converges only when $|z| < 1/\phi \approx 0.618$, but we didn't need to know that when we proved the formula $F_n = (\phi^n - \hat{\phi}^n)/\sqrt{5}$. The latter formula, once discovered, can be verified directly, if we don't trust the theory of formal power series. Therefore we'll ignore questions of convergence in this chapter; it's more a hindrance than a help.

Even if we remove the tags from our mattresses.

So much for perspective. Next we look at our main tools for reshaping generating functions — adding, shifting, changing variables, differentiating, integrating, and multiplying. In what follows we assume that, unless stated otherwise, $F(z)$ and $G(z)$ are the generating functions for the sequences $\langle f_n \rangle$ and $\langle g_n \rangle$. We also assume that the f_n's and g_n's are zero for negative n, since this saves us some bickering with the limits of summation.

It's pretty obvious what happens when we add constant multiples of F and G together:

$$\begin{aligned} \alpha F(z) + \beta G(z) &= \alpha \sum_n f_n z^n + \beta \sum_n g_n z^n \\ &= \sum_n (\alpha f_n + \beta g_n) z^n . \end{aligned} \tag{7.13}$$

This gives us the generating function for the sequence $\langle \alpha f_n + \beta g_n \rangle$.

Shifting a generating function isn't much harder. To shift $G(z)$ right by m places, that is, to form the generating function for the sequence $\langle 0, \ldots, 0, g_0, g_1, \ldots \rangle = \langle g_{n-m} \rangle$ with m leading 0's, we simply multiply by z^m:

$$z^m G(z) = \sum_n g_n z^{n+m} = \sum_n g_{n-m} z^n , \qquad \text{integer } m \geqslant 0. \tag{7.14}$$

This is the operation we used (twice), along with addition, to deduce the equation $(1 - z - z^2)F(z) = z$ on our way to finding a closed form for the Fibonacci numbers in Chapter 6.

And to shift $G(z)$ left m places — that is, to form the generating function for the sequence $\langle g_m, g_{m+1}, g_{m+2}, \ldots \rangle = \langle g_{n+m} \rangle$ with the first m elements discarded — we subtract off the first m terms and then divide by z^m:

$$\frac{G(z) - g_0 - g_1 z - \cdots - g_{m-1} z^{m-1}}{z^m} = \sum_{n \geqslant m} g_n z^{n-m} = \sum_{n \geqslant 0} g_{n+m} z^n . \tag{7.15}$$

(We can't extend this last sum over all n unless $g_0 = \cdots = g_{m-1} = 0$.)

Replacing the z by a constant multiple is another of our tricks:

$$G(cz) \;=\; \sum_n g_n (cz)^n \;=\; \sum_n c^n g_n z^n \,; \tag{7.16}$$

this yields the generating function for the sequence $\langle c^n g_n \rangle$. The special case $c = -1$ is particularly useful.

I fear d generating-function dz's.

Often we want to bring down a factor of n into the coefficient. Differentiation is what lets us do that:

$$G'(z) \;=\; g_1 + 2g_2 z + 3g_3 z^2 + \cdots \;=\; \sum_n (n+1) g_{n+1}\, z^n \,. \tag{7.17}$$

Shifting this right one place gives us a form that's sometimes more useful,

$$zG'(z) \;=\; \sum_n n g_n\, z^n \,. \tag{7.18}$$

This is the generating function for the sequence $\langle n g_n \rangle$. Repeated differentiation would allow us to multiply g_n by any desired polynomial in n.

Integration, the inverse operation, lets us divide the terms by n:

$$\int_0^z G(t)\, dt \;=\; g_0 z + \frac{1}{2} g_1 z^2 + \frac{1}{3} g_2 z^3 + \cdots \;=\; \sum_{n \geqslant 1} \frac{1}{n} g_{n-1}\, z^n \,. \tag{7.19}$$

(Notice that the constant term is zero.) If we want the generating function for $\langle g_n/n \rangle$ instead of $\langle g_{n-1}/n \rangle$, we should first shift left one place, replacing $G(t)$ by $\big(G(t) - g_0\big)/t$ in the integral.

Finally, here's how we multiply generating functions together:

$$\begin{aligned} F(z)G(z) \;&=\; (f_0 + f_1 z + f_2 z^2 + \cdots)(g_0 + g_1 z + g_2 z^2 + \cdots) \\ &=\; (f_0 g_0) \;+\; (f_0 g_1 + f_1 g_0) z \;+\; (f_0 g_2 + f_1 g_1 + f_2 g_0) z^2 \;+\; \cdots \\ &=\; \sum_n \Big(\sum_k f_k g_{n-k} \Big) z^n \,. \end{aligned} \tag{7.20}$$

As we observed in Chapter 5, this gives the generating function for the sequence $\langle h_n \rangle$, the *convolution* of $\langle f_n \rangle$ and $\langle g_n \rangle$. The sum $h_n = \sum_k f_k g_{n-k}$ can also be written $h_n = \sum_{k=0}^n f_k g_{n-k}$, because $f_k = 0$ when $k < 0$ and $g_{n-k} = 0$ when $k > n$. Multiplication/convolution is a little more complicated than the other operations, but it's very useful — so useful that we will spend all of Section 7.5 below looking at examples of it.

Multiplication has several special cases that are worth considering as operations in themselves. We've already seen one of these: When $F(z) = z^m$ we get the shifting operation (7.14). In that case the sum h_n becomes the single term g_{n-m}, because all f_k's are 0 except for $f_m = 1$.

Table 320 Generating function manipulations.

$$\alpha F(z) + \beta G(z) = \sum_n (\alpha f_n + \beta g_n) z^n$$

$$z^m G(z) = \sum_n g_{n-m}\, z^n\,, \qquad \text{integer } m \geqslant 0$$

$$\frac{G(z) - g_0 - g_1 z - \cdots - g_{m-1} z^{m-1}}{z^m} = \sum_{n\geqslant 0} g_{n+m}\, z^n\,, \qquad \text{integer } m \geqslant 0$$

$$G(cz) = \sum_n c^n g_n\, z^n$$

$$G'(z) = \sum_n (n+1) g_{n+1}\, z^n$$

$$zG'(z) = \sum_n n g_n\, z^n$$

$$\int_0^z G(t)\,dt = \sum_{n\geqslant 1} \frac{1}{n} g_{n-1}\, z^n$$

$$F(z)G(z) = \sum_n \Bigl(\sum_k f_k g_{n-k}\Bigr) z^n$$

$$\frac{1}{1-z} G(z) = \sum_n \Bigl(\sum_{k\leqslant n} g_k\Bigr) z^n$$

Another useful special case arises when $F(z)$ is the familiar function $1/(1-z) = 1 + z + z^2 + \cdots$; then all f_k's (for $k \geqslant 0$) are 1 and we have the important formula

$$\frac{1}{1-z} G(z) = \sum_n \Bigl(\sum_{k\geqslant 0} g_{n-k}\Bigr) z^n = \sum_n \Bigl(\sum_{k\leqslant n} g_k\Bigr) z^n\,. \tag{7.21}$$

Multiplying a generating function by $1/(1-z)$ gives us the generating function for the cumulative sums of the original sequence.

Table 320 summarizes the operations we've discussed so far. To use all these manipulations effectively it helps to have a healthy repertoire of generating functions in stock. Table 321 lists the simplest ones; we can use those to get started and to solve quite a few problems.

Each of the generating functions in Table 321 is important enough to be memorized. Many of them are special cases of the others, and many of

Table 321 Simple sequences and their generating functions.

sequence	generating function	closed form
$\langle 1,0,0,0,0,0,\ldots\rangle$	$\sum_{n\geqslant 0}[n=0]\,z^n$	1
$\langle 0,\ldots,0,1,0,0,\ldots\rangle$	$\sum_{n\geqslant 0}[n=m]\,z^n$	z^m
$\langle 1,1,1,1,1,1,\ldots\rangle$	$\sum_{n\geqslant 0} z^n$	$\dfrac{1}{1-z}$
$\langle 1,-1,1,-1,1,-1,\ldots\rangle$	$\sum_{n\geqslant 0}(-1)^n z^n$	$\dfrac{1}{1+z}$
$\langle 1,0,1,0,1,0,\ldots\rangle$	$\sum_{n\geqslant 0}[2\backslash n]\,z^n$	$\dfrac{1}{1-z^2}$
$\langle 1,0,\ldots,0,1,0,\ldots,0,1,0,\ldots\rangle$	$\sum_{n\geqslant 0}[m\backslash n]\,z^n$	$\dfrac{1}{1-z^m}$
$\langle 1,2,3,4,5,6,\ldots\rangle$	$\sum_{n\geqslant 0}(n+1)\,z^n$	$\dfrac{1}{(1-z)^2}$
$\langle 1,2,4,8,16,32,\ldots\rangle$	$\sum_{n\geqslant 0} 2^n z^n$	$\dfrac{1}{1-2z}$
$\langle 1,4,6,4,1,0,0,\ldots\rangle$	$\sum_{n\geqslant 0}\dbinom{4}{n} z^n$	$(1+z)^4$
$\langle 1,c,\binom{c}{2},\binom{c}{3},\ldots\rangle$	$\sum_{n\geqslant 0}\dbinom{c}{n} z^n$	$(1+z)^c$
$\left\langle 1,c,\binom{c+1}{2},\binom{c+2}{3},\ldots\right\rangle$	$\sum_{n\geqslant 0}\dbinom{c+n-1}{n} z^n$	$\dfrac{1}{(1-z)^c}$
$\langle 1,c,c^2,c^3,\ldots\rangle$	$\sum_{n\geqslant 0} c^n z^n$	$\dfrac{1}{1-cz}$
$\left\langle 1,\binom{m+1}{m},\binom{m+2}{m},\binom{m+3}{m},\ldots\right\rangle$	$\sum_{n\geqslant 0}\dbinom{m+n}{m} z^n$	$\dfrac{1}{(1-z)^{m+1}}$
$\langle 0,1,\frac12,\frac13,\frac14,\ldots\rangle$	$\sum_{n\geqslant 1}\dfrac{1}{n} z^n$	$\ln\dfrac{1}{1-z}$
$\langle 0,1,-\frac12,\frac13,-\frac14,\ldots\rangle$	$\sum_{n\geqslant 1}\dfrac{(-1)^{n+1}}{n} z^n$	$\ln(1+z)$
$\langle 1,1,\frac12,\frac16,\frac1{24},\frac1{120},\ldots\rangle$	$\sum_{n\geqslant 0}\dfrac{1}{n!} z^n$	e^z

Hint: If the sequence consists of binomial coefficients, its generating function usually involves a binomial, $1\pm z$.

them can be derived quickly from the others by using the basic operations of Table 320; therefore the memory work isn't very hard.

For example, let's consider the sequence $\langle 1,2,3,4,\ldots\rangle$, whose generating function $1/(1-z)^2$ is often useful. This generating function appears near the middle of Table 321, and it's also the special case $m = 1$ of $\langle 1, \binom{m+1}{m}, \binom{m+2}{m}, \binom{m+3}{m}, \ldots \rangle$, which appears further down; it's also the special case $c = 2$ of the closely related sequence $\langle 1, c, \binom{c+1}{2}, \binom{c+2}{3}, \ldots \rangle$. We can derive it from the generating function for $\langle 1, 1, 1, 1, \ldots \rangle$ by taking cumulative sums as in (7.21); that is, by dividing $1/(1-z)$ by $(1-z)$. Or we can derive it from $\langle 1, 1, 1, 1, \ldots \rangle$ by differentiation, using (7.17).

OK, OK, I'm convinced already.

The sequence $\langle 1, 0, 1, 0, \ldots \rangle$ is another one whose generating function can be obtained in many ways. We can obviously derive the formula $\sum_n z^{2n} = 1/(1-z^2)$ by substituting z^2 for z in the identity $\sum_n z^n = 1/(1-z)$; we can also apply cumulative summation to the sequence $\langle 1, -1, 1, -1, \ldots \rangle$, whose generating function is $1/(1+z)$, getting $1/(1+z)(1-z) = 1/(1-z^2)$. And there's also a third way, which is based on a general method for extracting the even-numbered terms $\langle g_0, 0, g_2, 0, g_4, 0, \ldots \rangle$ of *any* given sequence: If we add $G(-z)$ to $G(+z)$ we get

$$G(z) + G(-z) = \sum_n g_n\bigl(1 + (-1)^n\bigr)z^n = 2\sum_n g_n[n \text{ even}]z^n ;$$

therefore

$$\frac{G(z) + G(-z)}{2} = \sum_n g_{2n}\, z^{2n} . \tag{7.22}$$

The odd-numbered terms can be extracted in a similar way,

$$\frac{G(z) - G(-z)}{2} = \sum_n g_{2n+1}\, z^{2n+1} . \tag{7.23}$$

In the special case where $g_n = 1$ and $G(z) = 1/(1-z)$, the generating function for $\langle 1, 0, 1, 0, \ldots \rangle$ is $\frac{1}{2}\bigl(G(z) + G(-z)\bigr) = \frac{1}{2}\bigl(\frac{1}{1-z} + \frac{1}{1+z}\bigr) = \frac{1}{1-z^2}$.

Let's try this extraction trick on the generating function for Fibonacci numbers. We know that $\sum_n F_n z^n = z/(1-z-z^2)$; hence

$$\begin{aligned}\sum_n F_{2n} z^{2n} &= \frac{1}{2}\left(\frac{z}{1-z-z^2} + \frac{-z}{1+z-z^2}\right)\\ &= \frac{1}{2}\left(\frac{z+z^2-z^3-z+z^2+z^3}{(1-z^2)^2-z^2}\right) = \frac{z^2}{1-3z^2+z^4} .\end{aligned}$$

This generates the sequence $\langle F_0, 0, F_2, 0, F_4, \ldots \rangle$; hence the sequence of alternate F's, $\langle F_0, F_2, F_4, F_6, \ldots \rangle = \langle 0, 1, 3, 8, \ldots \rangle$, has a simple generating function:

$$\sum_n F_{2n} z^n = \frac{z}{1-3z+z^2} . \tag{7.24}$$

7.3 SOLVING RECURRENCES

Now let's focus our attention on one of the most important uses of generating functions: the solution of recurrence relations.

Given a sequence $\langle g_n \rangle$ that satisfies a given recurrence, we seek a closed form for g_n in terms of n. A solution to this problem via generating functions proceeds in four steps that are almost mechanical enough to be programmed on a computer:

1 Write down a single equation that expresses g_n in terms of other elements of the sequence. This equation should be valid for all integers n, assuming that $g_{-1} = g_{-2} = \cdots = 0$.

2 Multiply both sides of the equation by z^n and sum over all n. This gives, on the left, the sum $\sum_n g_n z^n$, which is the generating function $G(z)$. The right-hand side should be manipulated so that it becomes some other expression involving $G(z)$.

3 Solve the resulting equation, getting a closed form for $G(z)$.

4 Expand $G(z)$ into a power series and read off the coefficient of z^n; this is a closed form for g_n.

This method works because the single function $G(z)$ represents the entire sequence $\langle g_n \rangle$ in such a way that many manipulations are possible.

Example 1: Fibonacci numbers revisited.

For example, let's rerun the derivation of Fibonacci numbers from Chapter 6. In that chapter we were feeling our way, learning a new method; now we can be more systematic. The given recurrence is

$$\begin{aligned} g_0 &= 0; \qquad g_1 = 1; \\ g_n &= g_{n-1} + g_{n-2}, \qquad \text{for } n \geqslant 2. \end{aligned}$$

We will find a closed form for g_n by using the four steps above.

Step 1 tells us to write the recurrence as a "single equation" for g_n. We could say

$$g_n = \begin{cases} 0, & \text{if } n \leqslant 0; \\ 1, & \text{if } n = 1; \\ g_{n-1} + g_{n-2}, & \text{if } n > 1; \end{cases}$$

but this is cheating. Step 1 really asks for a formula that doesn't involve a case-by-case construction. The single equation

$$g_n = g_{n-1} + g_{n-2}$$

works for $n \geqslant 2$, and it also holds when $n \leqslant 0$ (because we have $g_0 = 0$ and $g_{\text{negative}} = 0$). But when $n = 1$ we get 1 on the left and 0 on the right.

Fortunately the problem is easy to fix, since we can add $[n=1]$ to the right; this adds 1 when $n = 1$, and it makes no change when $n \neq 1$. So, we have

$$g_n = g_{n-1} + g_{n-2} + [n=1];$$

this is the equation called for in Step 1.

Step 2 now asks us to transform the equation for $\langle g_n \rangle$ into an equation for $G(z) = \sum_n g_n z^n$. The task is not difficult:

$$\begin{aligned} G(z) = \sum_n g_n z^n &= \sum_n g_{n-1} z^n + \sum_n g_{n-2} z^n + \sum_n [n=1] z^n \\ &= \sum_n g_n z^{n+1} + \sum_n g_n z^{n+2} \;+\; z \\ &= zG(z) + z^2 G(z) + z. \end{aligned}$$

Step 3 is also simple in this case; we have

$$G(z) = \frac{z}{1 - z - z^2},$$

which of course comes as no surprise.

Step 4 is the clincher. We carried it out in Chapter 6 by having a sudden flash of inspiration; let's go more slowly now, so that we can get through Step 4 safely later, when we meet problems that are more difficult. What is

$$[z^n]\, \frac{z}{1 - z - z^2},$$

the coefficient of z^n when $z/(1 - z - z^2)$ is expanded in a power series? More generally, if we are given any rational function

$$R(z) = \frac{P(z)}{Q(z)},$$

where P and Q are polynomials, what is the coefficient $[z^n]\, R(z)$?

There's one kind of rational function whose coefficients are particularly nice, namely

$$\frac{a}{(1 - \rho z)^{m+1}} = \sum_{n \geq 0} \binom{m+n}{m} a \rho^n z^n. \qquad (7.25)$$

(The case $\rho = 1$ appears in Table 321, and we can get the general formula shown here by substituting ρz for z.) A finite sum of functions like (7.25),

$$S(z) = \frac{a_1}{(1 - \rho_1 z)^{m_1+1}} + \frac{a_2}{(1 - \rho_2 z)^{m_2+1}} + \cdots + \frac{a_l}{(1 - \rho_l z)^{m_l+1}}, \qquad (7.26)$$

also has nice coefficients,

$$[z^n]\,S(z) \;=\; a_1\binom{m_1+n}{m_1}\rho_1^n + a_2\binom{m_2+n}{m_2}\rho_2^n + \cdots + a_l\binom{m_l+n}{m_l}\rho_l^n\,. \qquad (7.27)$$

We will show that every rational function $R(z)$ such that $R(0) \neq \infty$ can be expressed in the form

$$R(z) \;=\; S(z) + T(z)\,, \qquad (7.28)$$

where $S(z)$ has the form (7.26) and $T(z)$ is a polynomial. Therefore there is a closed form for the coefficients $[z^n]\,R(z)$. Finding $S(z)$ and $T(z)$ is equivalent to finding the "partial fraction expansion" of $R(z)$.

Notice that $S(z) = \infty$ when z has the values $1/\rho_1, \ldots, 1/\rho_l$. Therefore the numbers ρ_k that we need to find, if we're going to succeed in expressing $R(z)$ in the desired form $S(z) + T(z)$, must be the reciprocals of the numbers α_k where $Q(\alpha_k) = 0$. (Recall that $R(z) = P(z)/Q(z)$, where P and Q are polynomials; we have $R(z) = \infty$ only if $Q(z) = 0$.)

Suppose $Q(z)$ has the form

$$Q(z) \;=\; q_0 + q_1 z + \cdots + q_m z^m\,, \qquad \text{where } q_0 \neq 0 \text{ and } q_m \neq 0.$$

The "reflected" polynomial

$$Q^R(z) \;=\; q_0 z^m + q_1 z^{m-1} + \cdots + q_m$$

has an important relation to $Q(z)$:

$$Q^R(z) \;=\; q_0(z-\rho_1)\ldots(z-\rho_m) \iff Q(z) \;=\; q_0(1-\rho_1 z)\ldots(1-\rho_m z)\,.$$

Thus, the roots of Q^R are the reciprocals of the roots of Q, and vice versa. We can therefore find the numbers ρ_k we seek by factoring the reflected polynomial $Q^R(z)$.

For example, in the Fibonacci case we have

$$Q(z) \;=\; 1 - z - z^2\,; \qquad Q^R(z) \;=\; z^2 - z - 1\,.$$

The roots of Q^R can be found by setting $(a,b,c) = (1,-1,-1)$ in the quadratic formula $\bigl(-b \pm \sqrt{b^2-4ac}\,\bigr)/2a$; we find that they are

$$\phi \;=\; \frac{1+\sqrt5}{2} \quad\text{and}\quad \hat\phi \;=\; \frac{1-\sqrt5}{2}\,.$$

Therefore $Q^R(z) = (z-\phi)(z-\hat\phi)$ and $Q(z) = (1-\phi z)(1-\hat\phi z)$.

Once we've found the ρ's, we can proceed to find the partial fraction expansion. It's simplest if all the roots are distinct, so let's consider that special case first. We might as well state and prove the general result formally:

Rational Expansion Theorem for Distinct Roots.

If $R(z) = P(z)/Q(z)$, *where* $Q(z) = q_0(1-\rho_1 z)\ldots(1-\rho_l z)$ *and the numbers* $(\rho_1,\ldots,\rho_l)$ *are distinct, and if* $P(z)$ *is a polynomial of degree less than* l, *then*

$$[z^n]\,R(z) \;=\; a_1\rho_1^n + \cdots + a_l\rho_l^n\,, \qquad \text{where } a_k = \frac{-\rho_k P(1/\rho_k)}{Q'(1/\rho_k)}. \tag{7.29}$$

Proof: Let $a_1, \ldots, a_l$ be the stated constants. Formula (7.29) holds if $R(z) = P(z)/Q(z)$ is equal to

$$S(z) \;=\; \frac{a_1}{1-\rho_1 z} + \cdots + \frac{a_l}{1-\rho_l z}\,.$$

And we can prove that $R(z) = S(z)$ by showing that the function $T(z) = R(z) - S(z)$ is not infinite as $z \to 1/\rho_k$. For this will show that the rational function $T(z)$ is never infinite; hence $T(z)$ must be a polynomial. We also can show that $T(z) \to 0$ as $z \to \infty$; hence $T(z)$ must be zero.

Impress your parents by leaving the book open at this page.

Let $\alpha_k = 1/\rho_k$. To prove that $\lim_{z\to\alpha_k} T(z) \ne \infty$, it suffices to show that $\lim_{z\to\alpha_k}(z-\alpha_k)T(z) = 0$, because $T(z)$ is a rational function of z. Thus we want to show that

$$\lim_{z\to\alpha_k}(z-\alpha_k)R(z) \;=\; \lim_{z\to\alpha_k}(z-\alpha_k)S(z)\,.$$

The right-hand limit equals $\lim_{z\to\alpha_k} a_k(z-\alpha_k)/(1-\rho_k z) = -a_k/\rho_k$, because $(1-\rho_k z) = -\rho_k(z-\alpha_k)$ and $(z-\alpha_k)/(1-\rho_j z) \to 0$ for $j \ne k$. The left-hand limit is

$$\lim_{z\to\alpha_k}(z-\alpha_k)\frac{P(z)}{Q(z)} \;=\; P(\alpha_k)\lim_{z\to\alpha_k}\frac{z-\alpha_k}{Q(z)} \;=\; \frac{P(\alpha_k)}{Q'(\alpha_k)}\,,$$

by L'Hospital's rule. Thus the theorem is proved.

Returning to the Fibonacci example, we have $P(z) = z$ and $Q(z) = 1 - z - z^2 = (1-\phi z)(1-\hat\phi z)$; hence $Q'(z) = -1-2z$, and

$$\frac{-\rho P(1/\rho)}{Q'(1/\rho)} \;=\; \frac{-1}{-1-2/\rho} \;=\; \frac{\rho}{\rho+2}\,.$$

According to (7.29), the coefficient of ϕ^n in $[z^n]\,R(z)$ is therefore $\phi/(\phi+2) = 1/\sqrt5$; the coefficient of $\hat\phi^n$ is $\hat\phi/(\hat\phi+2) = -1/\sqrt5$. So the theorem tells us that $F_n = (\phi^n - \hat\phi^n)/\sqrt5$, as in (6.123).

When $Q(z)$ has repeated roots, the calculations become more difficult, but we can beef up the proof of the theorem and prove the following more general result:

General Expansion Theorem for Rational Generating Functions.

If $R(z) = P(z)/Q(z)$, *where* $Q(z) = q_0(1-\rho_1 z)^{d_1} \ldots (1-\rho_l z)^{d_l}$ *and the numbers* $(\rho_1, \ldots, \rho_l)$ *are distinct, and if* $P(z)$ *is a polynomial of degree less than* $d_1 + \cdots + d_l$, *then*

$$[z^n] R(z) = f_1(n)\rho_1^n + \cdots + f_l(n)\rho_l^n \qquad \text{for all } n \geqslant 0, \tag{7.30}$$

where each $f_k(n)$ *is a polynomial of degree* $d_k - 1$ *with leading coefficient*

$$\begin{aligned} a_k &= \frac{(-\rho_k)^{d_k} P(1/\rho_k) d_k}{Q^{(d_k)}(1/\rho_k)} \\ &= \frac{P(1/\rho_k)}{(d_k-1)!\, q_0 \prod_{j\neq k}(1-\rho_j/\rho_k)^{d_j}}\,. \end{aligned} \tag{7.31}$$

This can be proved by induction on $\max(d_1, \ldots, d_l)$, using the fact that

$$R(z) - \frac{a_1(d_1-1)!}{(1-\rho_1 z)^{d_1}} - \cdots - \frac{a_l(d_l-1)!}{(1-\rho_l z)^{d_l}}$$

is a rational function whose denominator polynomial is not divisible by $(1-\rho_k z)^{d_k}$ for any k.

Example 2: A more-or-less random recurrence.

Now that we've seen some general methods, we're ready to tackle new problems. Let's try to find a closed form for the recurrence

$$\begin{aligned} g_0 &= g_1 = 1; \\ g_n &= g_{n-1} + 2g_{n-2} + (-1)^n, \qquad \text{for } n \geqslant 2. \end{aligned} \tag{7.32}$$

It's always a good idea to make a table of small cases first, and the recurrence lets us do that easily:

n	0	1	2	3	4	5	6	7
$(-1)^n$	1	−1	1	−1	1	−1	1	−1
g_n	1	1	4	5	14	23	52	97

No closed form is evident, and this sequence isn't even listed in Sloane's *Handbook* [270]; so we need to go through the four-step process if we want to discover the solution.

Step 1 is easy, since we merely need to insert fudge factors to fix things when $n < 2$: The equation

$$g_n = g_{n-1} + 2g_{n-2} + (-1)^n[n \geqslant 0] + [n=1]$$

holds for all integers n. Now we can carry out Step 2:

$$\begin{aligned} G(z) = \sum_n g_n z^n &= \sum_n g_{n-1} z^n + 2\sum_n g_{n-2} z^n + \sum_{n \geqslant 0} (-1)^n z^n + \sum_{n=1} z^n \\ &= zG(z) + 2z^2 G(z) + \frac{1}{1+z} + z\,. \end{aligned}$$

N.B.: The upper index on $\sum_{n=1} z^n$ is not missing!

(Incidentally, we could also have used $\binom{-1}{n}$ instead of $(-1)^n[n \geqslant 0]$, thereby getting $\sum_n \binom{-1}{n} z^n = (1+z)^{-1}$ by the binomial theorem.) Step 3 is elementary algebra, which yields

$$G(z) = \frac{1 + z(1+z)}{(1+z)(1-z-2z^2)} = \frac{1+z+z^2}{(1-2z)(1+z)^2}\,.$$

And that leaves us with Step 4.

The squared factor in the denominator is a bit troublesome, since we know that repeated roots are more complicated than distinct roots; but there it is. We have two roots, $\rho_1 = 2$ and $\rho_2 = -1$; the general expansion theorem (7.30) tells us that

$$g_n = a_1 2^n + (a_2 n + c)(-1)^n$$

for some constant c, where

$$a_1 = \frac{1 + 1/2 + 1/4}{(1+1/2)^2} = \frac{7}{9}\,; \qquad a_2 = \frac{1 - 1 + 1}{1 - 2/(-1)} = \frac{1}{3}\,.$$

(The second formula for a_k in (7.31) is easier to use than the first one when the denominator has nice factors. We simply substitute $z = 1/\rho_k$ everywhere in $R(z)$, except in the factor where this gives zero, and divide by $(d_k - 1)!$; this gives the coefficient of $n^{d_k - 1}\rho_k^n$.) Plugging in $n = 0$ tells us that the value of the remaining constant c had better be $\frac{2}{9}$; hence our answer is

$$g_n = \tfrac{7}{9} 2^n + \left(\tfrac{1}{3} n + \tfrac{2}{9}\right)(-1)^n\,. \tag{7.33}$$

It doesn't hurt to check the cases $n = 1$ and 2, just to be sure that we didn't foul up. Maybe we should even try $n = 3$, since this formula looks weird. But it's correct, all right.

Could we have discovered (7.33) by guesswork? Perhaps after tabulating a few more values we may have observed that $g_{n+1} \approx 2g_n$ when n is large.

And with chutzpah and luck we might even have been able to smoke out the constant $\frac{7}{9}$. But it sure is simpler and more reliable to have generating functions as a tool.

Example 3: Mutually recursive sequences.

Sometimes we have two or more recurrences that depend on each other. Then we can form generating functions for both of them, and solve both by a simple extension of our four-step method.

For example, let's return to the problem of $3 \times n$ domino tilings that we explored earlier this chapter. If we want to know only the total number of ways, U_n, to cover a $3 \times n$ rectangle with dominoes, without breaking this number down into vertical dominoes versus horizontal dominoes, we needn't go into as much detail as we did before. We can merely set up the recurrences

$$\begin{aligned} &U_0 = 1\,, \qquad U_1 = 0\,; \qquad V_0 = 0\,, \qquad V_1 = 1\,; \\ &U_n = 2V_{n-1} + U_{n-2}\,, \qquad V_n = U_{n-1} + V_{n-2}\,, \qquad \text{for } n \geqslant 2. \end{aligned}$$

Here V_n is the number of ways to cover a $3 \times n$ rectangle-minus-corner, using $(3n-1)/2$ dominoes. These recurrences are easy to discover, if we consider the possible domino configurations at the rectangle's left edge, as before. Here are the values of U_n and V_n for small n:

n	0	1	2	3	4	5	6	7
U_n	1	0	3	0	11	0	41	0
V_n	0	1	0	4	0	15	0	56

(7.34)

Let's find closed forms, in four steps. First (Step 1), we have

$$U_n = 2V_{n-1} + U_{n-2} + [n=0]\,, \qquad V_n = U_{n-1} + V_{n-2}\,,$$

for all n. Hence (Step 2),

$$U(z) = 2zV(z) + z^2U(z) + 1\,, \qquad V(z) = zU(z) + z^2V(z)\,.$$

Now (Step 3) we must solve two equations in two unknowns; but these are easy, since the second equation yields $V(z) = zU(z)/(1-z^2)$; we find

$$U(z) = \frac{1-z^2}{1-4z^2+z^4}\,; \qquad V(z) = \frac{z}{1-4z^2+z^4}\,. \tag{7.35}$$

(We had this formula for $U(z)$ in (7.10), but with z^3 instead of z^2. In that derivation, n was the number of dominoes; now it's the width of the rectangle.)

The denominator $1 - 4z^2 + z^4$ is a function of z^2; this is what makes $U_{2n+1} = 0$ and $V_{2n} = 0$, as they should be. We can take advantage of this nice property of z^2 by retaining z^2 when we factor the denominator: We need not take $1-4z^2+z^4$ all the way to a product of four factors $(1-\rho_k z)$, since two factors of the form $(1-\rho_k z^2)$ will be enough to tell us the coefficients. In other words if we consider the generating function

$$W(z) = \frac{1}{1-4z+z^2} = W_0 + W_1 z + W_2 z^2 + \cdots, \qquad (7.36)$$

we will have $V(z) = zW(z^2)$ and $U(z) = (1-z^2)W(z^2)$; hence $V_{2n+1} = W_n$ and $U_{2n} = W_n - W_{n-1}$. We save time and energy by working with the simpler function $W(z)$.

The factors of $1-4z+z^2$ are $(z-2-\sqrt{3}\,)$ and $(z-2+\sqrt{3}\,)$, and they can also be written $\bigl(1-(2+\sqrt{3}\,)z\bigr)$ and $\bigl(1-(2-\sqrt{3}\,)z\bigr)$ because this polynomial is its own reflection. Thus it turns out that we have

$$\begin{aligned} V_{2n+1} = W_n &= \frac{3+2\sqrt{3}}{6}(2+\sqrt{3}\,)^n + \frac{3-2\sqrt{3}}{6}(2-\sqrt{3}\,)^n\,; \\ U_{2n} = W_n - W_{n-1} &= \frac{3+\sqrt{3}}{6}(2+\sqrt{3}\,)^n + \frac{3-\sqrt{3}}{6}(2-\sqrt{3}\,)^n \\ &= \frac{(2+\sqrt{3}\,)^n}{3-\sqrt{3}} + \frac{(2-\sqrt{3}\,)^n}{3+\sqrt{3}}\,. \end{aligned} \qquad (7.37)$$

This is the desired closed form for the number of $3 \times n$ domino tilings.

Incidentally, we can simplify the formula for U_{2n} by realizing that the second term always lies between 0 and 1. The number U_{2n} is an integer, so we have

$$U_{2n} = \left\lceil \frac{(2+\sqrt{3}\,)^n}{3-\sqrt{3}} \right\rceil, \qquad \text{for } n \geqslant 0. \qquad (7.38)$$

In fact, the other term $(2-\sqrt{3}\,)^n/(3+\sqrt{3}\,)$ is extremely small when n is large, because $2-\sqrt{3} \approx 0.268$. This needs to be taken into account if we try to use formula (7.38) in numerical calculations. For example, a fairly expensive name-brand hand calculator comes up with 413403.0005 when asked to compute $(2+\sqrt{3})^{10}/(3-\sqrt{3})$. This is correct to nine significant figures; but the true value is slightly *less* than 413403, not slightly greater. Therefore it would be a mistake to take the ceiling of 413403.0005; the correct answer, $U_{20} = 413403$, is obtained by *rounding* to the nearest integer. Ceilings can be hazardous.

I've known slippery floors too.

Example 4: A closed form for change.

When we left the problem of making change, we had just calculated the number of ways to pay 50¢. Let's try now to count the number of ways there are to change a dollar, or a million dollars—still using only pennies, nickels, dimes, quarters, and halves.

The generating function derived earlier is

$$C(z) \;=\; \frac{1}{1-z}\,\frac{1}{1-z^5}\,\frac{1}{1-z^{10}}\,\frac{1}{1-z^{25}}\,\frac{1}{1-z^{50}}\,;$$

this is a rational function of z with a denominator of degree 91. Therefore we can decompose the denominator into 91 factors and come up with a 91-term "closed form" for C_n, the number of ways to give n cents in change. But that's too horrible to contemplate. Can't we do better than the general method suggests, in this particular case?

One ray of hope suggests itself immediately, when we notice that the denominator is almost a function of z^5. The trick we just used to simplify the calculations by noting that $1-4z^2+z^4$ is a function of z^2 can be applied to $C(z)$, if we replace $1/(1-z)$ by $(1+z+z^2+z^3+z^4)/(1-z^5)$:

$$\begin{aligned} C(z) \;&=\; \frac{1+z+z^2+z^3+z^4}{1-z^5}\,\frac{1}{1-z^5}\,\frac{1}{1-z^{10}}\,\frac{1}{1-z^{25}}\,\frac{1}{1-z^{50}} \\ &=\; (1+z+z^2+z^3+z^4)\check{C}(z^5)\,, \\ \check{C}(z) \;&=\; \frac{1}{1-z}\,\frac{1}{1-z}\,\frac{1}{1-z^2}\,\frac{1}{1-z^5}\,\frac{1}{1-z^{10}}\,. \end{aligned}$$

The compressed function $\check{C}(z)$ has a denominator whose degree is only 19, so it's much more tractable than the original. This new expression for $C(z)$ shows us, incidentally, that $C_{5n} = C_{5n+1} = C_{5n+2} = C_{5n+3} = C_{5n+4}$; and indeed, this set of equations is obvious in retrospect: The number of ways to leave a 53¢ tip is the same as the number of ways to leave a 50¢ tip, because the number of pennies is predetermined modulo 5.

Now we're also getting compressed reasoning.

But $\check{C}(z)$ still doesn't have a really simple closed form based on the roots of the denominator. The easiest way to compute its coefficients of $\check{C}(z)$ is probably to recognize that each of the denominator factors is a divisor of $1-z^{10}$. Hence we can write

$$\check{C}(z) \;=\; \frac{A(z)}{(1-z^{10})^5}\,, \qquad \text{where } A(z) = A_0 + A_1 z + \cdots + A_{31}z^{31}. \qquad (7.39)$$

The actual value of $A(z)$, for the curious, is

$$\begin{aligned} &(1+z+\cdots+z^9)^2(1+z^2+\cdots+z^8)(1+z^5) \\ &\quad= 1+2z+4z^2+6z^3+9z^4+13z^5+18z^6+24z^7 \\ &\qquad +31z^8+39z^9+45z^{10}+52z^{11}+57z^{12}+63z^{13}+67z^{14}+69z^{15} \\ &\qquad +69z^{16}+67z^{17}+63z^{18}+57z^{19}+52z^{20}+45z^{21}+39z^{22}+31z^{23} \\ &\qquad +24z^{24}+18z^{25}+13z^{26}+9z^{27}+6z^{28}+4z^{29}+2z^{30}+z^{31}\,. \end{aligned}$$

Finally, since $1/(1-z^{10})^5 = \sum_{k\geqslant 0}\binom{k+4}{4}z^{10k}$, we can determine the coefficient of $\check{C}_n = [z^n]\,\check{C}(z)$ as follows, when $n = 10q + r$ and $0 \leqslant r < 10$:

$$\begin{aligned}\check{C}_{10q+r} &= \sum_{j,k} A_j\binom{k+4}{4}[10q+r=10k+j]\\ &= A_r\binom{q+4}{4} + A_{r+10}\binom{q+3}{4} + A_{r+20}\binom{q+2}{4} + A_{r+30}\binom{q+1}{4}. \qquad (7.40)\end{aligned}$$

This gives ten cases, one for each value of r; but it's a pretty good closed form, compared with alternatives that involve powers of complex numbers.

For example, we can use this expression to deduce the value of $C_{50q} = \check{C}_{10q}$. Then $r = 0$ and we have

$$C_{50q} = \binom{q+4}{4} + 45\binom{q+3}{4} + 52\binom{q+2}{4} + 2\binom{q+1}{4}.$$

The number of ways to change 50¢ is $\binom{5}{4} + 45\binom{4}{4} = 50$; the number of ways to change \$1 is $\binom{6}{4} + 45\binom{5}{4} + 52\binom{4}{4} = 292$; and the number of ways to change \$1,000,000 is

$$\begin{aligned}&\binom{2000004}{4} + 45\binom{2000003}{4} + 52\binom{2000002}{4} + 2\binom{2000001}{4}\\ &\qquad = 66666793333412666685000001.\end{aligned}$$

Example 5: A divergent series.

Now let's try to get a closed form for the numbers g_n defined by

$$\begin{aligned}g_0 &= 1;\\ g_n &= ng_{n-1}, \qquad \text{for } n > 0.\end{aligned}$$

After staring at this for a few nanoseconds we realize that g_n is just $n!$; in fact, the method of summation factors described in Chapter 2 suggests this answer immediately. But let's try to solve the recurrence with generating functions, just to see what happens. (A powerful technique should be able to handle easy recurrences like this, as well as others that have answers we can't guess so easily.)

Nowadays people are talking femtoseconds.

The equation

$$g_n = ng_{n-1} + [n=0]$$

holds for all n, and it leads to

$$G(z) = \sum_n g_n z^n = \sum_n ng_{n-1}z^n + \sum_{n=0} z^n.$$

To complete Step 2, we want to express $\sum_n ng_{n-1}z^n$ in terms of $G(z)$, and the basic maneuvers in Table 320 suggest that the derivative $G'(z) = \sum_n ng_nz^{n-1}$

is somehow involved. So we steer toward that kind of sum:

$$\begin{aligned} G(z) &= 1 + \sum_n (n+1) g_n\, z^{n+1} \\ &= 1 + \sum_n n g_n\, z^{n+1} + \sum_n g_n\, z^{n+1} \\ &= 1 + z^2 G'(z) + zG(z)\,. \end{aligned}$$

Let's check this equation, using the values of g_n for small n. Since

$$\begin{aligned} G &= 1 + z + 2z^2 + \;\; 6z^3 + 24z^4 + \cdots\,, \\ G' &= \qquad 1 + 4z \;+ 18z^2 + 96z^3 + \cdots\,, \end{aligned}$$

we have

$$\begin{aligned} z^2G' &= \qquad\quad z^2 + 4z^3 + 18z^4 + 96z^5 + \cdots\,, \\ zG &= z + z^2 + 2z^3 + \;\; 6z^4 + 24z^5 + \cdots\,, \\ 1 &= 1\,. \end{aligned}$$

These three lines add up to G, so we're fine so far. Incidentally, we often find it convenient to write 'G' instead of '$G(z)$'; the extra '(z)' just clutters up the formula when we aren't changing z.

Step 3 is next, and it's different from what we've done before because we have a differential equation to solve. But this is a differential equation that we can handle with the hypergeometric series techniques of Section 5.6; those techniques aren't too bad. (Readers who are unfamiliar with hypergeometrics needn't worry — this will be quick.)

"This will be quick." That's what the doctor said just before he stuck me with that needle. Come to think of it, "hypergeometric" sounds a lot like "hypodermic."

First we must get rid of the constant '1', so we take the derivative of both sides:

$$\begin{aligned} G' = (z^2G' + zG + 1)' &= (2zG' + z^2G'') + (G + zG') \\ &= z^2G'' + 3zG' + G\,. \end{aligned}$$

The theory in Chapter 5 tells us to rewrite this using the ϑ operator, and we know from exercise 6.13 that

$$\vartheta G = zG'\,, \qquad \vartheta^2 G = z^2G'' + zG'\,.$$

Therefore the desired form of the differential equation is

$$\vartheta G = z\vartheta^2 G + 2z\vartheta G + zG = z(\vartheta + 1)^2 G\,.$$

According to (5.109), the solution with $g_0 = 1$ is the hypergeometric series $F(1, 1; ; z)$.

Step 3 was more than we bargained for; but now that we know what the function G is, Step 4 is easy — the hypergeometric definition (5.76) gives us the power series expansion:

$$G(z) = F\left({1,1 \atop } \middle| z\right) = \sum_{n\geqslant 0} \frac{1^{\overline{n}}\,1^{\overline{n}}\,z^n}{n!} = \sum_{n\geqslant 0} n!\,z^n .$$

We've confirmed the closed form we knew all along, $g_n = n!$.

Notice that the technique gave the right answer even though $G(z)$ diverges for all nonzero z. The sequence $n!$ grows so fast, the terms $|n!\,z^n|$ approach ∞ as $n \to \infty$, unless $z = 0$. This shows that formal power series can be manipulated algebraically without worrying about convergence.

Example 6: A recurrence that goes all the way back.

Let's close this section by applying generating functions to a problem in graph theory. A *fan* of order n is a graph on the vertices $\{0, 1, \ldots, n\}$ with $2n - 1$ edges defined as follows: Vertex 0 is connected by an edge to each of the other n vertices, and vertex k is connected by an edge to vertex $k+1$, for $1 \leqslant k < n$. Here, for example, is the fan of order 4, which has five vertices and seven edges.

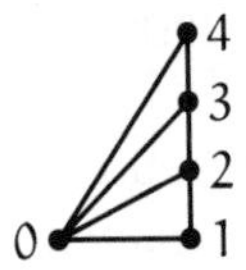

The problem of interest: How many spanning trees f_n are in such a graph? A *spanning tree* is a subgraph containing all the vertices, and containing enough edges to make the subgraph connected yet not so many that it has a cycle. It turns out that every spanning tree of a graph on $n+1$ vertices has exactly n edges. With fewer than n edges the subgraph wouldn't be connected, and with more than n it would have a cycle; graph theory books prove this.

There are $\binom{2n-1}{n}$ ways to choose n edges from among the $2n-1$ present in a fan of order n, but these choices don't always yield a spanning tree. For instance the subgraph

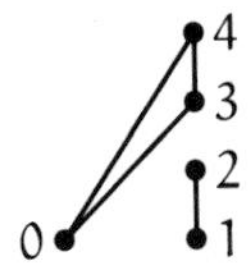

has four edges but is not a spanning tree; it has a cycle from 0 to 4 to 3 to 0, and it has no connection between $\{1, 2\}$ and the other vertices. We want to count how many of the $\binom{2n-1}{n}$ choices actually do yield spanning trees.

Let's look at some small cases. It's pretty easy to enumerate the spanning trees for $n = 1$, 2, and 3:

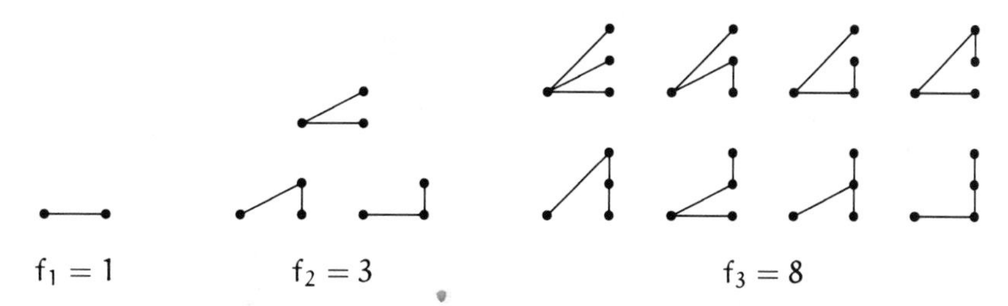

(We need not show the labels on the vertices, if we always draw vertex 0 at the left.) What about the case $n = 0$? At first it seems reasonable to set $f_0 = 1$; but we'll take $f_0 = 0$, because the existence of a fan of order 0 (which should have $2n - 1 = -1$ edges) is dubious.

Our four-step procedure tells us to find a recurrence for f_n that holds for all n. We can get a recurrence by observing how the topmost vertex (vertex n) is connected to the rest of the spanning tree. If it's not connected to vertex 0, it must be connected to vertex $n-1$, since it must be connected to the rest of the graph. In this case, any of the f_{n-1} spanning trees for the remaining fan (on the vertices 0 through $n-1$) will complete a spanning tree for the whole graph. Otherwise vertex n *is* connected to 0, and there's some number $k \leqslant n$ such that vertices $n, n-1, \ldots, k$ are connected directly but the edge between k and $k-1$ is not in the subtree. Then there can't be any edges between 0 and $\{n-1, \ldots, k\}$, or there would be a cycle. If $k = 1$, the spanning tree is therefore determined completely. And if $k > 1$, any of the f_{k-1} ways to produce a spanning tree on $\{0, 1, \ldots, k-1\}$ will yield a spanning tree on the whole graph. For example, here's what this analysis produces when $n = 4$:

$k = 4$ $\quad$ $k = 3$ $\quad$ $k = 2$ $\quad$ $k = 1$

$f_4 = f_3 + f_3 + f_2 + f_1 + 1$

The general equation, valid for $n \geqslant 1$, is

$$f_n = f_{n-1} + f_{n-1} + f_{n-2} + f_{n-3} + \cdots + f_1 + 1 .$$

(It almost seems as though the '1' on the end is f_0 and we should have chosen $f_0 = 1$; but we will doggedly stick with our choice.) A few changes suffice to make the equation valid for all integers n:

$$f_n = f_{n-1} + \sum_{k<n} f_k + [n>0] . \tag{7.41}$$

This is a recurrence that "goes all the way back" from f_{n-1} through all previous values, so it's different from the other recurrences we've seen so far in this chapter. We used a special method to get rid of a similar right-side sum in Chapter 2, when we solved the quicksort recurrence (2.12); namely, we subtracted one instance of the recurrence from another $(f_{n+1} - f_n)$. This trick would get rid of the $\sum$ now, as it did then; but we'll see that generating functions allow us to work directly with such sums. (And it's a good thing that they do, because we will be seeing much more complicated recurrences before long.)

Step 1 is finished; Step 2 is where we need to do a new thing:

$$\begin{aligned} F(z) = \sum_n f_n z^n &= \sum_n f_{n-1} z^n + \sum_{k,n} f_k z^n [k<n] + \sum_n [n>0] z^n \\ &= zF(z) + \sum_k f_k z^k \sum_n [n>k] z^{n-k} + \frac{z}{1-z} \\ &= zF(z) + F(z) \sum_{m>0} z^m + \frac{z}{1-z} \\ &= zF(z) + F(z) \frac{z}{1-z} + \frac{z}{1-z} . \end{aligned}$$

The key trick here was to change z^n to $z^k z^{n-k}$; this made it possible to express the value of the double sum in terms of $F(z)$, as required in Step 2.

Now Step 3 is simple algebra, and we find

$$F(z) = \frac{z}{1-3z+z^2} .$$

Those of us with a zest for memorization will recognize this as the generating function (7.24) for the even-numbered Fibonacci numbers. So, we needn't go through Step 4; we have found a somewhat surprising answer to the spans-of-fans problem:

$$f_n = F_{2n} , \qquad \text{for } n \geqslant 0. \tag{7.42}$$

7.4 SPECIAL GENERATING FUNCTIONS

Step 4 of the four-step procedure becomes much easier if we know the coefficients of lots of different power series. The expansions in Table 321 are quite useful, as far as they go, but many other types of closed forms are possible. Therefore we ought to supplement that table with another one, which lists power series that correspond to the "special numbers" considered in Chapter 6.

Table 337 Generating functions for special numbers.

$$\frac{1}{(1-z)^{m+1}}\ln\frac{1}{1-z} = \sum_n (H_{m+n}-H_m)\binom{m+n}{n} z^n \tag{7.43}$$

$$\frac{z}{e^z-1} = \sum_{n\geqslant 0} B_n \frac{z^n}{n!} \tag{7.44}$$

$$\frac{F_m z}{1-(F_{m-1}+F_{m+1})z+(-1)^m z^2} = \sum_n F_{mn}\, z^n \tag{7.45}$$

$$\sum_k \left\{ m \atop k \right\} \frac{k!\, z^k}{(1-z)^{k+1}} = \sum_n n^m z^n \tag{7.46}$$

$$\left(z^{-1}\right)^{\overline{-m}} = \frac{z^m}{(1-z)(1-2z)\ldots(1-mz)} = \sum_n \left\{ n \atop m \right\} z^n \tag{7.47}$$

$$z^{\overline{m}} = z(z+1)\ldots(z+m-1) = \sum_n \left[m \atop n \right] z^n \tag{7.48}$$

$$\left(e^z-1\right)^m = m! \sum_{n\geqslant 0} \left\{ n \atop m \right\} \frac{z^n}{n!} \tag{7.49}$$

$$\left(\ln\frac{1}{1-z}\right)^m = m! \sum_{n\geqslant 0} \left[n \atop m \right] \frac{z^n}{n!} \tag{7.50}$$

$$\left(\frac{z}{\ln(1+z)}\right)^m = \sum_{n\geqslant 0} \frac{z^n}{n!} \left\{ m \atop m-n \right\} \Big/ \binom{m-1}{n} \tag{7.51}$$

$$\left(\frac{z}{1-e^{-z}}\right)^m = \sum_{n\geqslant 0} \frac{z^n}{n!} \left[m \atop m-n \right] \Big/ \binom{m-1}{n} \tag{7.52}$$

$$e^{z+wz} = \sum_{m,n\geqslant 0} \binom{n}{m} w^m \frac{z^n}{n!} \tag{7.53}$$

$$e^{w(e^z-1)} = \sum_{m,n\geqslant 0} \left\{ n \atop m \right\} w^m \frac{z^n}{n!} \tag{7.54}$$

$$\frac{1}{(1-z)^w} = \sum_{m,n\geqslant 0} \left[n \atop m \right] w^m \frac{z^n}{n!} \tag{7.55}$$

$$\frac{1-w}{e^{(w-1)z}-w} = \sum_{m,n\geqslant 0} \left\langle n \atop m \right\rangle w^m \frac{z^n}{n!} \tag{7.56}$$

Table 337 is the database we need. The identities in this table are not difficult to prove, so we needn't dwell on them; this table is primarily for reference when we meet a new problem. But there's a nice proof of the first formula, (7.43), that deserves mention: We start with the identity

$$\frac{1}{(1-z)^{x+1}} = \sum_n \binom{x+n}{n} z^n$$

and differentiate it with respect to x. On the left, $(1-z)^{-x-1}$ is equal to $e^{(x+1)\ln(1/(1-z))}$, so d/dx contributes a factor of $\ln\bigl(1/(1-z)\bigr)$. On the right, the numerator of $\binom{x+n}{n}$ is $(x+n)\ldots(x+1)$, and d/dx splits this into n terms whose sum is equivalent to multiplying $\binom{x+n}{n}$ by

$$\frac{1}{x+n} + \cdots + \frac{1}{x+1} = H_{x+n} - H_x \,.$$

Replacing x by m gives (7.43). Notice that $H_{x+n} - H_x$ is meaningful even when x is not an integer.

By the way, this method of differentiating a complicated product—leaving it as a product—is usually better than expressing the derivative as a sum. For example the right side of

$$\frac{d}{dx}\bigl((x+n)^n \ldots (x+1)^1\bigr)$$
$$= (x+n)^n \ldots (x+1)^1 \left(\frac{n}{x+n} + \cdots + \frac{1}{x+1}\right)$$

would be a lot messier written out as a sum.

The general identities in Table 337 include many important special cases. For example, (7.43) simplifies to the generating function for H_n when $m = 0$:

$$\frac{1}{1-z}\ln\frac{1}{1-z} = \sum_n H_n z^n \,. \tag{7.57}$$

This equation can also be derived in other ways; for example, we can take the power series for $\ln\bigl(1/(1-z)\bigr)$ and divide it by $1-z$ to get cumulative sums.

Identities (7.51) and (7.52) involve the respective ratios $\left\{ {m \atop m-n} \right\} / \binom{m-1}{n}$ and $\left[{m \atop m-n} \right] / \binom{m-1}{n}$, which have the undefined form $0/0$ when $n \geqslant m$. However, there is a way to give them a proper meaning using the Stirling polynomials of (6.45), because we have

$$\left\{ {m \atop m-n} \right\} \Big/ \binom{m-1}{n} = (-1)^{n+1} n!\, m\sigma_n(n-m) \,; \tag{7.58}$$

$$\left[{m \atop m-n} \right] \Big/ \binom{m-1}{n} = n!\, m\sigma_n(m) \,. \tag{7.59}$$

Thus, for example, the case $m = 1$ of (7.51) should not be regarded as the power series $\sum_{n\geqslant 0}(z^n/n!)\left\{{1 \atop 1-n}\right\}/\binom{0}{n}$, but rather as

$$\frac{z}{\ln(1+z)} = -\sum_{n\geqslant 0}(-z)^n \sigma_n(n-1) = 1 + \tfrac{1}{2}z - \tfrac{1}{12}z^2 + \cdots .$$

Identities (7.53), (7.55), (7.54), and (7.56) are "double generating functions" or "super generating functions" because they have the form $G(w,z) = \sum_{m,n} g_{m,n} w^m z^n$. The coefficient of w^m is a generating function in the variable z; the coefficient of z^n is a generating function in the variable w.

7.5 CONVOLUTIONS

I always thought convolution was what happens to my brain when I try to do a proof.

The *convolution* of two given sequences $\langle f_0, f_1, \ldots\rangle = \langle f_n\rangle$ and $\langle g_0, g_1, \ldots\rangle = \langle g_n\rangle$ is the sequence $\langle f_0 g_0,\, f_0 g_1 + f_1 g_0,\, \ldots\rangle = \langle \sum_k f_k g_{n-k}\rangle$. We have observed in Sections 5.4 and 7.2 that convolution of sequences corresponds to multiplication of their generating functions. This fact makes it easy to evaluate many sums that would otherwise be difficult to handle.

Example 1: A Fibonacci convolution.

For example, let's try to evaluate $\sum_{k=0}^{n} F_k F_{n-k}$ in closed form. This is the convolution of $\langle F_n\rangle$ with itself, so the sum must be the coefficient of z^n in $F(z)^2$, where $F(z)$ is the generating function for $\langle F_n\rangle$. All we have to do is figure out the value of this coefficient.

The generating function $F(z)$ is $z/(1-z-z^2)$, a quotient of polynomials; so the general expansion theorem for rational functions tells us that the answer can be obtained from a partial fraction representation. We can use the general expansion theorem (7.30) and grind away; or we can use the fact that

$$\begin{aligned} F(z)^2 &= \left(\frac{1}{\sqrt{5}}\left(\frac{1}{1-\phi z} - \frac{1}{1-\hat\phi z}\right)\right)^2 \\ &= \frac{1}{5}\left(\frac{1}{(1-\phi z)^2} - \frac{2}{(1-\phi z)(1-\hat\phi z)} + \frac{1}{(1-\hat\phi z)^2}\right) \\ &= \frac{1}{5}\sum_{n\geqslant 0}(n+1)\phi^n z^n - \frac{2}{5}\sum_{n\geqslant 0}F_{n+1}z^n + \frac{1}{5}\sum_{n\geqslant 0}(n+1)\hat\phi^n z^n . \end{aligned}$$

Instead of expressing the answer in terms of ϕ and $\hat\phi$, let's try for a closed form in terms of Fibonacci numbers. Recalling that $\phi + \hat\phi = 1$, we have

$$\begin{aligned} \phi^n + \hat\phi^n &= [z^n]\left(\frac{1}{1-\phi z} + \frac{1}{1-\hat\phi z}\right) \\ &= [z^n]\,\frac{2-(\phi+\hat\phi)z}{(1-\phi z)(1-\hat\phi z)} = [z^n]\,\frac{2-z}{1-z-z^2} = 2F_{n+1} - F_n . \end{aligned}$$

Hence

$$F(z)^2 = \frac{1}{5}\sum_{n\geqslant 0}(n+1)(2F_{n+1}-F_n)z^n - \frac{2}{5}\sum_{n\geqslant 0}F_{n+1}\,z^n\,,$$

and we have the answer we seek:

$$\sum_{k=0}^{n} F_k F_{n-k} = \frac{2nF_{n+1}-(n+1)F_n}{5}\,. \tag{7.60}$$

For example, when $n = 3$ this formula gives $F_0F_3 + F_1F_2 + F_2F_1 + F_3F_0 = 0+1+1+0 = 2$ on the left and $(6F_4 - 4F_3)/5 = (18-8)/5 = 2$ on the right.

Example 2: Harmonic convolutions.

The efficiency of a certain computer method called "samplesort" depends on the value of the sum

$$T_{m,n} = \sum_{0\leqslant k<n}\binom{k}{m}\frac{1}{n-k}\,, \qquad \text{integers } m,n \geqslant 0.$$

Exercise 5.58 obtains the value of this sum by a somewhat intricate double induction, using summation factors. It's much easier to realize that $T_{m,n}$ is just the nth term in the convolution of $\langle\binom{0}{m},\binom{1}{m},\binom{2}{m},\ldots\rangle$ with $\langle 0,\frac{1}{1},\frac{1}{2},\ldots\rangle$. Both sequences have simple generating functions in Table 321:

$$\sum_{n\geqslant 0}\binom{n}{m}z^n = \frac{z^m}{(1-z)^{m+1}}\,; \qquad \sum_{n>0}\frac{z^n}{n} = \ln\frac{1}{1-z}\,.$$

Therefore, by (7.43),

$$\begin{aligned} T_{m,n} &= [z^n]\,\frac{z^m}{(1-z)^{m+1}}\ln\frac{1}{1-z} = [z^{n-m}]\,\frac{1}{(1-z)^{m+1}}\ln\frac{1}{1-z} \\ &= (H_n - H_m)\binom{n}{n-m}\,. \end{aligned}$$

In fact, there are many more sums that boil down to this same sort of convolution, because we have

$$\frac{1}{(1-z)^{r+1}}\ln\frac{1}{1-z}\cdot\frac{1}{(1-z)^{s+1}} = \frac{1}{(1-z)^{r+s+2}}\ln\frac{1}{1-z}$$

for all r and s. Equating coefficients of z^n gives the general identity

$$\begin{aligned} &\sum_k \binom{r+k}{k}\binom{s+n-k}{n-k}(H_{r+k}-H_r) \\ &\qquad = \binom{r+s+n+1}{n}(H_{r+s+n+1}-H_{r+s+1})\,. \end{aligned} \tag{7.61}$$

Because it's so harmonic.

This seems almost too good to be true. But it checks, at least when $n = 2$:

$$\binom{r+1}{1}\binom{s+1}{1}\frac{1}{r+1} + \binom{r+2}{2}\binom{s+0}{0}\left(\frac{1}{r+2}+\frac{1}{r+1}\right)$$
$$= \binom{r+s+3}{2}\left(\frac{1}{r+s+3}+\frac{1}{r+s+2}\right).$$

Special cases like $s = 0$ are as remarkable as the general case.

And there's more. We can use the convolution identity

$$\sum_k \binom{r+k}{k}\binom{s+n-k}{n-k} = \binom{r+s+n+1}{n}$$

to transpose H_r to the other side, since H_r is independent of k:

$$\sum_k \binom{r+k}{k}\binom{s+n-k}{n-k} H_{r+k}$$
$$= \binom{r+s+n+1}{n}(H_{r+s+n+1} - H_{r+s+1} + H_r). \qquad (7.62)$$

There's still more: If r and s are nonnegative integers l and m, we can replace $\binom{r+k}{k}$ by $\binom{l+k}{l}$ and $\binom{s+n-k}{n-k}$ by $\binom{m+n-k}{m}$; then we can change k to $k-l$ and n to $n-m-l$, getting

$$\sum_{k=0}^{n}\binom{k}{l}\binom{n-k}{m}H_k = \binom{n+1}{l+m+1}(H_{n+1} - H_{l+m+1} + H_l),$$
$$\text{integers } l, m, n \geqslant 0. \qquad (7.63)$$

Even the special case $l = m = 0$ of this identity was difficult for us to handle in Chapter 2! (See (2.36).) We've come a long way.

Example 3: Convolutions of convolutions.

If we form the convolution of $\langle f_n \rangle$ and $\langle g_n \rangle$, then convolve this with a third sequence $\langle h_n \rangle$, we get a sequence whose nth term is

$$\sum_{j+k+l=n} f_j\, g_k\, h_l\,.$$

The generating function of this three-fold convolution is, of course, the three-fold product $F(z)G(z)H(z)$. In a similar way, the m-fold convolution of a sequence $\langle g_n \rangle$ with itself has nth term equal to

$$\sum_{k_1+k_2+\cdots+k_m=n} g_{k_1}\, g_{k_2}\, \ldots\, g_{k_m}$$

and its generating function is $G(z)^m$.

We can apply these observations to the spans-of-fans problem considered earlier (Example 6 in Section 7.3). It turns out that there's another way to compute f_n, the number of spanning trees of an n-fan, based on the configurations of tree edges between the vertices $\{1, 2, \ldots, n\}$: The edge between vertex k and vertex $k + 1$ may or may not be selected for the subtree; and each of the ways to select these edges connects up certain blocks of adjacent vertices. For example, when $n = 10$ we might connect vertices $\{1, 2\}$, $\{3\}$, $\{4, 5, 6, 7\}$, and $\{8, 9, 10\}$:

Concrete blocks.

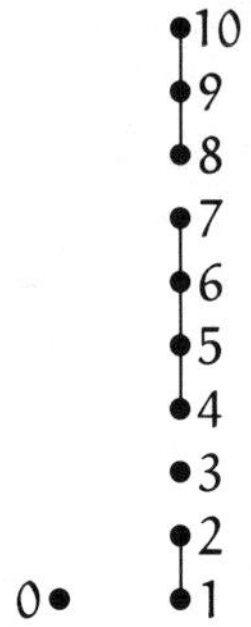

```
        •10
        |
        •9
        |
        •8

        •7
        |
        •6
        |
        •5
        |
        •4

        •3

        •2
        |
0•      •1
```

How many spanning trees can we make, by adding additional edges to vertex 0? We need to connect 0 to each of the four blocks; and there are two ways to join 0 with $\{1, 2\}$, one way to join it with $\{3\}$, four ways with $\{4, 5, 6, 7\}$, and three ways with $\{8, 9, 10\}$, or $2 \cdot 1 \cdot 4 \cdot 3 = 24$ ways altogether. Summing over all possible ways to make blocks gives us the following expression for the total number of spanning trees:

$$f_n = \sum_{m>0} \sum_{\substack{k_1+k_2+\cdots+k_m=n \\ k_1,k_2,\ldots,k_m>0}} k_1 k_2 \ldots k_m . \tag{7.64}$$

For example, $f_4 = 4 + 3 \cdot 1 + 2 \cdot 2 + 1 \cdot 3 + 2 \cdot 1 \cdot 1 + 1 \cdot 2 \cdot 1 + 1 \cdot 1 \cdot 2 + 1 \cdot 1 \cdot 1 \cdot 1 = 21$.

This is the sum of m-fold convolutions of the sequence $\langle 0, 1, 2, 3, \ldots \rangle$, for $m = 1, 2, 3, \ldots$; hence the generating function for $\langle f_n \rangle$ is

$$F(z) = G(z) + G(z)^2 + G(z)^3 + \cdots = \frac{G(z)}{1 - G(z)}$$

where $G(z)$ is the generating function for $\langle 0, 1, 2, 3, \ldots \rangle$, namely $z/(1-z)^2$. Consequently we have

$$F(z) = \frac{z}{(1-z)^2 - z} = \frac{z}{1 - 3z + z^2},$$

as before. This approach to $\langle f_n \rangle$ is more symmetrical and appealing than the complicated recurrence we had earlier.

Example 4: A convoluted recurrence.

Our next example is especially important; in fact, it's the "classic example" of why generating functions are useful in the solution of recurrences.

Suppose we have $n+1$ variables $x_0, x_1, \ldots, x_n$ whose product is to be computed by doing n multiplications. How many ways C_n are there to insert parentheses into the product $x_0 \cdot x_1 \cdot \ldots \cdot x_n$ so that the order of multiplication is completely specified? For example, when $n = 2$ there are two ways, $x_0 \cdot (x_1 \cdot x_2)$ and $(x_0 \cdot x_1) \cdot x_2$. And when $n = 3$ there are five ways,

$$x_0 \cdot (x_1 \cdot (x_2 \cdot x_3)), \quad x_0 \cdot ((x_1 \cdot x_2) \cdot x_3), \quad (x_0 \cdot x_1) \cdot (x_2 \cdot x_3),$$
$$(x_0 \cdot (x_1 \cdot x_2)) \cdot x_3, \quad ((x_0 \cdot x_1) \cdot x_2) \cdot x_3.$$

Thus $C_2 = 2$, $C_3 = 5$; we also have $C_1 = 1$ and $C_0 = 1$.

Let's use the four-step procedure of Section 7.3. What is a recurrence for the C's? The key observation is that there's exactly one '$\cdot$' operation outside all of the parentheses, when $n > 0$; this is the final multiplication that ties everything together. If this '$\cdot$' occurs between x_k and x_{k+1}, there are C_k ways to fully parenthesize $x_0 \cdot \ldots \cdot x_k$, and there are C_{n-k-1} ways to fully parenthesize $x_{k+1} \cdot \ldots \cdot x_n$; hence

$$C_n = C_0 C_{n-1} + C_1 C_{n-2} + \cdots + C_{n-1} C_0, \qquad \text{if } n > 0.$$

By now we recognize this expression as a convolution, and we know how to patch the formula so that it holds for all integers n:

$$C_n = \sum_k C_k C_{n-1-k} + [n=0]. \tag{7.65}$$

Step 1 is now complete. Step 2 tells us to multiply by z^n and sum:

$$\begin{aligned} C(z) &= \sum_n C_n z^n \\ &= \sum_{k,n} C_k C_{n-1-k} z^n + \sum_{n=0} z^n \\ &= \sum_k C_k z^k \sum_n C_{n-1-k} z^{n-k} + 1 \\ &= C(z) \cdot zC(z) + 1. \end{aligned}$$

Lo and behold, the convolution has become a product, in the generating-function world. Life is full of surprises.

The authors jest.

Step 3 is also easy. We solve for $C(z)$ by the quadratic formula:

$$C(z) = \frac{1 \pm \sqrt{1-4z}}{2z}.$$

But should we choose the + sign or the − sign? Both choices yield a function that satisfies $C(z) = zC(z)^2 + 1$, but only one of the choices is suitable for our problem. We might choose the + sign on the grounds that positive thinking is best; but we soon discover that this choice gives $C(0) = \infty$, contrary to the facts. (The correct function $C(z)$ is supposed to have $C(0) = C_0 = 1$.) Therefore we conclude that

$$C(z) = \frac{1 - \sqrt{1-4z}}{2z}.$$

Finally, Step 4. What is $[z^n]\,C(z)$? The binomial theorem tells us that

$$\sqrt{1-4z} = \sum_{k\geqslant 0} \binom{1/2}{k}(-4z)^k = 1 + \sum_{k\geqslant 1} \frac{1}{2k}\binom{-1/2}{k-1}(-4z)^k;$$

hence, using (5.37),

$$\begin{aligned} \frac{1-\sqrt{1-4z}}{2z} &= \sum_{k\geqslant 1} \frac{1}{k}\binom{-1/2}{k-1}(-4z)^{k-1} \\ &= \sum_{n\geqslant 0} \binom{-1/2}{n}\frac{(-4z)^n}{n+1} = \sum_{n\geqslant 0} \binom{2n}{n}\frac{z^n}{n+1}. \end{aligned}$$

The number of ways to parenthesize, C_n, is $\binom{2n}{n}\frac{1}{n+1}$.

We anticipated this result in Chapter 5, when we introduced the sequence of *Catalan numbers* $\langle 1, 1, 2, 5, 14, \ldots \rangle = \langle C_n \rangle$. This sequence arises in dozens of problems that seem at first to be unrelated to each other [41], because many situations have a recursive structure that corresponds to the convolution recurrence (7.65).

So the convoluted recurrence has led us to an oft-recurring convolution.

For example, let's consider the following problem: How many sequences $\langle a_1, a_2 \ldots, a_{2n} \rangle$ of +1's and −1's have the property that

$$a_1 + a_2 + \cdots + a_{2n} = 0$$

and have all their partial sums

$$a_1, \quad a_1 + a_2, \quad \ldots, \quad a_1 + a_2 + \cdots + a_{2n}$$

nonnegative? There must be n occurrences of +1 and n occurrences of −1. We can represent this problem graphically by plotting the sequence of partial sums $s_n = \sum_{k=1}^{n} a_k$ as a function of n: The five solutions for $n = 3$ are

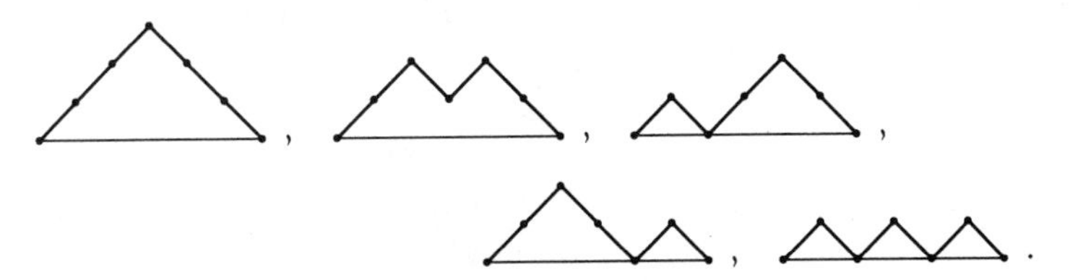

These are "mountain ranges" of width $2n$ that can be drawn with line segments of the forms ╱ and ╲. It turns out that there are exactly C_n ways to do this, and the sequences can be related to the parenthesis problem in the following way: Put an extra pair of parentheses around the entire formula, so that there are n pairs of parentheses corresponding to the n multiplications. Now replace each '$\cdot$' by $+1$ and each ')' by -1 and erase everything else. For example, the formula $x_0 \cdot \bigl((x_1 \cdot x_2) \cdot (x_3 \cdot x_4)\bigr)$ corresponds to the sequence $\langle +1, +1, -1, +1, +1, -1, -1, -1 \rangle$ by this rule. The five ways to parenthesize $x_0 \cdot x_1 \cdot x_2 \cdot x_3$ correspond to the five mountain ranges for $n = 3$ shown above.

Moreover, a slight reformulation of our sequence-counting problem leads to a surprisingly simple combinatorial solution that avoids the use of generating functions: How many sequences $\langle a_0, a_1, a_2, \ldots, a_{2n} \rangle$ of $+1$'s and -1's have the property that

$$a_0 + a_1 + a_2 + \cdots + a_{2n} = 1,$$

when all the partial sums

$$a_0, \quad a_0 + a_1, \quad a_0 + a_1 + a_2, \quad \ldots, \quad a_0 + a_1 + \cdots + a_{2n}$$

are required to be *positive*? Clearly these are just the sequences of the previous problem, with the additional element $a_0 = +1$ placed in front. But the sequences in the new problem can be enumerated by a simple counting argument, using a remarkable fact discovered by George Raney [243] in 1959: *If $\langle x_1, x_2, \ldots, x_m \rangle$ is any sequence of integers whose sum is $+1$, exactly one of the cyclic shifts*

$$\langle x_1, x_2, \ldots, x_m \rangle, \quad \langle x_2, \ldots, x_m, x_1 \rangle, \quad \ldots, \quad \langle x_m, x_1, \ldots, x_{m-1} \rangle$$

has all of its partial sums positive. For example, consider the sequence $\langle 3, -5, 2, -2, 3, 0 \rangle$. Its cyclic shifts are

$$\begin{array}{ll} \langle 3,-5,2,-2,3,0 \rangle & \langle -2,3,0,3,-5,2 \rangle \\ \langle -5,2,-2,3,0,3 \rangle & \langle 3,0,3,-5,2,-2 \rangle \ \checkmark \\ \langle 2,-2,3,0,3,-5 \rangle & \langle 0,3,-5,2,-2,3 \rangle \end{array}$$

and only the one that's checked has entirely positive partial sums.

Raney's lemma can be proved by a simple geometric argument. Let's extend the sequence periodically to get an infinite sequence

$$\langle x_1, x_2, \ldots, x_m, x_1, x_2, \ldots, x_m, x_1, x_2, \ldots \rangle ;$$

thus we let $x_{m+k} = x_k$ for all $k \geq 0$. If we now plot the partial sums $s_n = x_1 + \cdots + x_n$ as a function of n, the graph of s_n has an "average slope" of $1/m$, because $s_{m+n} = s_n + 1$. For example, the graph corresponding to our example sequence $\langle 3, -5, 2, -2, 3, 0, 3, -5, 2, \ldots \rangle$ begins as follows:

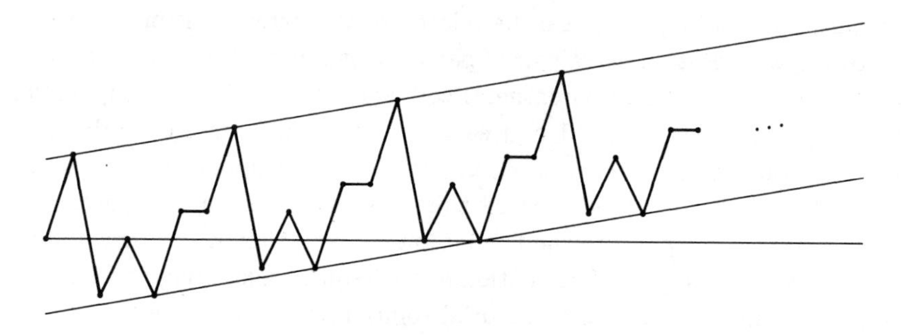

The entire graph can be contained between two lines of slope $1/m$, as shown; we have $m = 6$ in the illustration. In general these bounding lines touch the graph just once in each cycle of m points, since lines of slope $1/m$ hit points with integer coordinates only once per m units. The unique lower point of intersection is the only place in the cycle from which all partial sums will be positive, because every other point on the curve has an intersection point within m units to its right.

Ah, if stock prices would only continue to rise like this.

With Raney's lemma we can easily enumerate the sequences $\langle a_0, \ldots, a_{2n} \rangle$ of $+1$'s and -1's whose partial sums are entirely positive and whose total sum is $+1$. There are $\binom{2n+1}{n}$ sequences with n occurrences of -1 and $n+1$ occurrences of $+1$, and Raney's lemma tells us that exactly $1/(2n+1)$ of these sequences have all partial sums positive. (List all $N = \binom{2n+1}{n}$ of these sequences and all $2n+1$ of their cyclic shifts, in an $N \times (2n+1)$ array. Each row contains exactly one solution. Each solution appears exactly once in each column. So there are $N/(2n+1)$ distinct solutions in the array, each appearing $(2n+1)$ times.) The total number of sequences with positive partial sums is

$$\binom{2n+1}{n} \frac{1}{2n+1} = \binom{2n}{n} \frac{1}{n+1} = C_n .$$

(Attention, computer scientists: The partial sums in this problem represent the stack size as a function of time, when a product of $n+1$ factors is evaluated, because each "push" operation changes the size by $+1$ and each multiplication changes it by -1.)

Example 5: A recurrence with m-fold convolution.

We can generalize the problem just considered by looking at sequences $\langle a_0, \ldots, a_{mn} \rangle$ of $+1$'s and $(1-m)$'s whose partial sums are all positive and whose total sum is $+1$. Such sequences can be called m-*Raney sequences*. If there are k occurrences of $(1-m)$ and $mn+1-k$ occurrences of $+1$, we have

$$k(1-m)+(mn+1-k) = 1,$$

hence $k=n$. There are $\binom{mn+1}{n}$ sequences with n occurrences of $(1-m)$ and $mn+1-n$ occurrences of $+1$, and Raney's lemma tells us that the number of such sequences with all partial sums positive is exactly

(Attention, computer scientists: The stack interpretation now applies with respect to an m-ary operation, instead of the binary multiplication considered earlier.)

$$\binom{mn+1}{n}\frac{1}{mn+1} = \binom{mn}{n}\frac{1}{(m-1)n+1}. \tag{7.66}$$

So this is the number of m-Raney sequences. Let's call this a Fuss-Catalan number $C_n^{(m)}$, because the sequence $\langle C_n^{(m)}\rangle$ was first investigated by N. I. Fuss [109] in 1791 (many years before Catalan himself got into the act). The ordinary Catalan numbers are $C_n = C_n^{(2)}$.

Now that we know the answer, (7.66), let's play "Jeopardy" and figure out a question that leads to it. In the case $m=2$ the question was: "What numbers C_n satisfy the recurrence $C_n = \sum_k C_k C_{n-1-k} + [n=0]$?" We will try to find a similar question (a similar recurrence) in the general case.

The trivial sequence $\langle +1\rangle$ of length 1 is clearly an m-Raney sequence. If we put the number $(1-m)$ at the right of any m sequences that are m-Raney, we get an m-Raney sequence; the partial sums stay positive as they increase to $+2$, then $+3$, ..., $+m$, and $+1$. Conversely, we can show that all m-Raney sequences $\langle a_0,\ldots,a_{mn}\rangle$ arise in this way, if $n>0$: The last term a_{mn} must be $(1-m)$. The partial sums $s_j = a_0+\cdots+a_{j-1}$ are positive for $1\leqslant j\leqslant mn$, and $s_{mn}=m$ because $s_{mn}+a_{mn}=1$. Let k_1 be the largest index $\leqslant mn$ such that $s_{k_1}=1$; let k_2 be largest such that $s_{k_2}=2$; and so on. Thus $s_{k_j}=j$ and $s_k>j$, for $k_j<k\leqslant mn$ and $1\leqslant j\leqslant m$. It follows that $k_m=mn$, and we can verify without difficulty that each of the subsequences $\langle a_0,\ldots,a_{k_1-1}\rangle$, $\langle a_{k_1},\ldots,a_{k_2-1}\rangle$, ..., $\langle a_{k_{m-1}},\ldots,a_{k_m-1}\rangle$ is an m-Raney sequence. We must have $k_1=mn_1+1$, $k_2-k_1=mn_2+1$, ..., $k_m-k_{m-1}=mn_m+1$, for some positive integers n_1, n_2, ..., n_m.

Therefore $\binom{mn+1}{n}\frac{1}{mn+1}$ is the answer to the following two interesting questions: "What are the numbers $C_n^{(m)}$ defined by the recurrence

$$C_n^{(m)} = \Biggl(\sum_{n_1+n_2+\cdots+n_m=n-1} C_{n_1}^{(m)}C_{n_2}^{(m)}\ldots C_{n_m}^{(m)}\Biggr) + [n=0] \tag{7.67}$$

for all integers n?" "If $G(z)$ is a power series that satisfies

$$G(z) = z\,G(z)^m + 1, \tag{7.68}$$

what is $[z^n]\,G(z)$?"

Notice that these are not easy questions. In the ordinary Catalan case ($m = 2$), we solved (7.68) for $G(z)$ and its coefficients by using the quadratic formula and the binomial theorem; but when $m = 3$, none of the standard techniques gives any clue about how to solve the cubic equation $G = zG^3 + 1$. So it has turned out to be easier to answer this question before asking it.

Now, however, we know enough to ask even harder questions and deduce their answers. How about this one: "What is $[z^n]\,G(z)^l$, if l is a positive integer and if $G(z)$ is the power series defined by (7.68)?" The argument we just gave can be used to show that $[z^n]\,G(z)^l$ is the number of sequences of length $mn + l$ with the following three properties:

- Each element is either $+1$ or $(1 - m)$.
- The partial sums are all positive.
- The total sum is l.

For we get all such sequences in a unique way by putting together l sequences that have the m-Raney property. The number of ways to do this is

$$\sum_{n_1+n_2+\cdots+n_l=n} C^{(m)}_{n_1} C^{(m)}_{n_2} \ldots C^{(m)}_{n_l} \;=\; [z^n]\,G(z)^l\,.$$

A generalization of Raney's lemma tells us how to count these sequences: *If $\langle x_1, x_2, \ldots, x_m\rangle$ is any sequence of integers such that $x_j \leqslant 1$ for all j, and such that $x_1 + x_2 + \cdots + x_m = l > 0$, then exactly l of the cyclic shifts*

$$\langle x_1, x_2, \ldots, x_m\rangle,\quad \langle x_2, \ldots, x_m, x_1\rangle,\quad \ldots,\quad \langle x_m, x_1, \ldots, x_{m-1}\rangle$$

have all positive partial sums.

For example, we can check this statement on the sequence $\langle -2, 1, -1, 0, 1, 1, -1, 1, 1, 1\rangle$. The cyclic shifts are

$$\begin{array}{ll}
\langle -2,1,-1,0,1,1,-1,1,1,1\rangle & \langle 1,-1,1,1,1,-2,1,-1,0,1\rangle \\
\langle 1,-1,0,1,1,-1,1,1,1,-2\rangle & \langle -1,1,1,1,-2,1,-1,0,1,1\rangle \\
\langle -1,0,1,1,-1,1,1,1,-2,1\rangle & \langle 1,1,1,-2,1,-1,0,1,1,-1\rangle \ \checkmark \\
\langle 0,1,1,-1,1,1,1,-2,1,-1\rangle & \langle 1,1,-2,1,-1,0,1,1,-1,1\rangle \\
\langle 1,1,-1,1,1,1,-2,1,-1,0\rangle \ \checkmark & \langle 1,-2,1,-1,0,1,1,-1,1,1\rangle
\end{array}$$

and only the two examples marked '$\checkmark$' have all partial sums positive. This generalized lemma is proved in exercise 13.

A sequence of $+1$'s and $(1-m)$'s that has length $mn + l$ and total sum l must have exactly n occurrences of $(1 - m)$. The generalized lemma tells us that $l/(mn + l)$ of these $\binom{mn+l}{n}$ sequences have all partial sums positive;

hence our tough question has a surprisingly simple answer:

$$[z^n]\,G(z)^l \;=\; \binom{mn+l}{n}\frac{l}{mn+l}\,, \tag{7.69}$$

for all integers $l > 0$.

Readers who haven't forgotten Chapter 5 might well be experiencing *déjà vu*: "That formula looks familiar; haven't we seen it before?" Yes, indeed; equation (5.60) says that

$$[z^n]\,\mathcal{B}_t(z)^r \;=\; \binom{tn+r}{n}\frac{r}{tn+r}\,.$$

Therefore the generating function $G(z)$ in (7.68) must actually be the generalized binomial series $\mathcal{B}_m(z)$. Sure enough, equation (5.59) says

$$\mathcal{B}_m(z)^{1-m} - \mathcal{B}_m(z)^{-m} \;=\; z\,,$$

which is the same as

$$\mathcal{B}_m(z) - 1 \;=\; z\mathcal{B}_m(z)^m\,.$$

Let's switch to the notation of Chapter 5, now that we know we're dealing with generalized binomials. Chapter 5 stated a bunch of identities without proof. We have now closed part of the gap by proving that the power series $\mathcal{B}_t(z)$ defined by

$$\mathcal{B}_t(z) \;=\; \sum_n \binom{tn+1}{n}\frac{z^n}{tn+1}$$

has the remarkable property that

$$\mathcal{B}_t(z)^r \;=\; \sum_n \binom{tn+r}{n}\frac{r\,z^n}{tn+r}\,,$$

whenever t and r are positive integers.

Can we extend these results to *arbitrary* values of t and r? Yes; because the coefficients $\binom{tn+r}{n}\frac{r}{tn+r}$ are polynomials in t and r. The general rth power defined by

$$\mathcal{B}_t(z)^r \;=\; e^{r\ln\mathcal{B}_t(z)} \;=\; \sum_{n\ge0}\frac{\bigl(r\ln\mathcal{B}_t(z)\bigr)^n}{n!} \;=\; \sum_{n\ge0}\frac{r^n}{n!}\sum_{m\ge1}(-1)^{m-1}\frac{\bigl(\mathcal{B}(z)-1\bigr)^m}{m}$$

has coefficients that are polynomials in t and r; and those polynomials are equal to $\binom{tn+r}{n}\frac{r}{tn+r}$ for infinitely many values of t and r. So the two sequences of polynomials must be identically equal.

Chapter 5 also mentions the generalized exponential series

$$\mathcal{E}_t(z) = \sum_{n\geq 0} \frac{(tn+1)^{n-1}}{n!} z^n ,$$

which is said in (7.60) to have an equally remarkable property:

$$[z^n]\,\mathcal{E}_t(z)^r = \frac{r(tn+r)^{n-1}}{n!} . \tag{7.70}$$

We can prove this as a limiting case of the formulas for $\mathcal{B}_t(z)$, because it is not difficult to show that

$$\mathcal{E}_t(z)^r = \lim_{x\to\infty} \mathcal{B}_{xt}(z/x)^{xr} .$$

7.6 EXPONENTIAL GF'S

Sometimes a sequence $\langle g_n \rangle$ has a generating function whose properties are quite complicated, while the related sequence $\langle g_n/n! \rangle$ has a generating function that's quite simple. In such cases we naturally prefer to work with $\langle g_n/n! \rangle$ and then multiply by $n!$ at the end. This trick works sufficiently often that we have a special name for it: We call the power series

$$\hat{G}(z) = \sum_{n\geq 0} g_n \frac{z^n}{n!} \tag{7.71}$$

the *exponential generating function* or "egf" of the sequence $\langle g_0, g_1, g_2, \ldots \rangle$. This name arises because the exponential function e^z is the egf of $\langle 1, 1, 1, \ldots \rangle$.

Many of the generating functions in Table 337 are actually egf's. For example, equation (7.50) says that $(\ln \frac{1}{1-z})^m/m!$ is the egf for the sequence $\langle \left[{0 \atop m}\right], \left[{1 \atop m}\right], \left[{2 \atop m}\right], \ldots \rangle$. The ordinary generating function for this sequence is much more complicated (and also divergent).

Exponential generating functions have their own basic maneuvers, analogous to the operations we learned in Section 7.2. For example, if we multiply the egf of $\langle g_n \rangle$ by z, we get

$$\sum_{n\geq 0} g_n \frac{z^{n+1}}{n!} = \sum_{n\geq 1} g_{n-1} \frac{z^n}{(n-1)!} = \sum_{n\geq 0} n g_{n-1} \frac{z^n}{n!} ;$$

this is the egf of $\langle 0, g_0, 2g_1, \ldots \rangle = \langle n g_{n-1} \rangle$.

Differentiating the egf of $\langle g_0, g_1, g_2, \ldots \rangle$ with respect to z gives

Are we having fun yet?

$$\sum_{n\geq 0} n g_n \frac{z^{n-1}}{n!} = \sum_{n\geq 1} g_n \frac{z^{n-1}}{(n-1)!} = \sum_{n\geq 0} g_{n+1} \frac{z^n}{n!} ; \tag{7.72}$$

this is the egf of $\langle g_1, g_2, \ldots \rangle$. Thus differentiation on egf's corresponds to the left-shift operation $\bigl(G(z) - g_0\bigr)/z$ on ordinary gf's. (We used this left-shift property of egf's when we studied hypergeometric series, (5.106).) Integration of an egf gives

$$\int_0^z \sum_{n\geqslant 0} g_n \frac{t^n}{n!}\,dt \;=\; \sum_{n\geqslant 0} g_n \frac{z^{n+1}}{(n+1)!} \;=\; \sum_{n\geqslant 1} g_{n-1} \frac{z^n}{n!}\,; \qquad (7.73)$$

this is a right shift, the egf of $\langle 0, g_0, g_1, \ldots \rangle$.

The most interesting operation on egf's, as on ordinary gf's, is multiplication. If $\hat{F}(z)$ and $\hat{G}(z)$ are egf's for $\langle f_n \rangle$ and $\langle g_n \rangle$, then $\hat{F}(z)\hat{G}(z) = \hat{H}(z)$ is the egf for a sequence $\langle h_n \rangle$ called the *binomial convolution* of $\langle f_n \rangle$ and $\langle g_n \rangle$:

$$h_n \;=\; \sum_k \binom{n}{k} f_k\, g_{n-k}\,. \qquad (7.74)$$

Binomial coefficients appear here because $\binom{n}{k} = n!/k!\,(n-k)!$, hence

$$\frac{h_n}{n!} \;=\; \sum_{k=0}^{n} \frac{f_k}{k!} \frac{g_{n-k}}{(n-k)!}\,;$$

in other words, $\langle h_n/n! \rangle$ is the ordinary convolution of $\langle f_n/n! \rangle$ and $\langle g_n/n! \rangle$.

Binomial convolutions occur frequently in applications. For example, we defined the Bernoulli numbers in (6.79) by the implicit recurrence

$$\sum_{j=0}^{m} \binom{m+1}{j} B_j \;=\; [m=0]\,, \qquad \text{for all } m \geqslant 0;$$

this can be rewritten as a binomial convolution, if we substitute n for $m+1$ and add the term B_n to both sides:

$$\sum_k \binom{n}{k} B_k \;=\; B_n + [n=1]\,, \qquad \text{for all } n \geqslant 0. \qquad (7.75)$$

We can now relate this recurrence to power series (as promised in Chapter 6) by introducing the egf for Bernoulli numbers, $\hat{B}(z) = \sum_{n\geqslant 0} B_n z^n/n!$. The left-hand side of (7.75) is the binomial convolution of $\langle B_n \rangle$ with the constant sequence $\langle 1, 1, 1, \ldots \rangle$; hence the egf of the left-hand side is $\hat{B}(z)e^z$. The egf of the right-hand side is $\sum_{n\geqslant 0}\bigl(B_n + [n=1]\bigr)z^n/n! = \hat{B}(z) + z$. Therefore we must have $\hat{B}(z) = z/(e^z - 1)$; we have proved equation (6.81), which appears also in Table 337 as equation (7.44).

Now let's look again at a sum that has been popping up frequently in this book,

$$S_m(n) = 0^m + 1^m + 2^m + \cdots + (n-1)^m = \sum_{0 \leqslant k < n} k^m .$$

This time we will try to analyze the problem with generating functions, in hopes that it will suddenly become simpler. We will consider n to be fixed and m variable; thus our goal is to understand the coefficients of the power series

$$S(z) = S_0(n) + S_1(n)\,z + S_2(n)\,z^2 + \cdots = \sum_{m \geqslant 0} S_m(n)\,z^m .$$

We know that the generating function for $\langle 1, k, k^2, \ldots \rangle$ is

$$\frac{1}{1-kz} = \sum_{m \geqslant 0} k^m z^m ,$$

hence

$$S(z) = \sum_{m \geqslant 0} \sum_{0 \leqslant k < n} k^m z^m = \sum_{0 \leqslant k < n} \frac{1}{1-kz}$$

by interchanging the order of summation. We can put this sum in closed form,

$$\begin{aligned} S(z) &= \frac{1}{z}\left(\frac{1}{z^{-1}-0} + \frac{1}{z^{-1}-1} + \cdots + \frac{1}{z^{-1}-n+1}\right) \\ &= \frac{1}{z}\left(H_{z^{-1}} - H_{z^{-1}-n}\right); \end{aligned} \tag{7.76}$$

but we know nothing about expanding such a closed form in powers of z.

Exponential generating functions come to the rescue. The egf of our sequence $\langle S_0(n), S_1(n), S_2(n), \ldots \rangle$ is

$$\hat{S}(z,n) = S_0(n) + S_1(n)\,\frac{z}{1!} + S_2(n)\,\frac{z^2}{2!} + \cdots = \sum_{m \geqslant 0} S_m(n)\,\frac{z^m}{m!} .$$

To get these coefficients $S_m(n)$ we can use the egf for $\langle 1, k, k^2, \ldots \rangle$, namely

$$e^{kz} = \sum_{m \geqslant 0} k^m \frac{z^m}{m!} ,$$

and we have

$$\hat{S}(z,n) = \sum_{m \geqslant 0} \sum_{0 \leqslant k < n} k^m \frac{z^m}{m!} = \sum_{0 \leqslant k < n} e^{kz} .$$

And the latter sum is a geometric progression, so there's a closed form

$$\hat{S}(z,n) = \frac{e^{nz}-1}{e^z-1}. \qquad (7.77)$$

All we need to do is figure out the coefficients of this relatively simple function, and we'll know $S_m(n)$, because $S_m(n) = m!\,[z^m]\hat{S}(z,n)$.

Here's where Bernoulli numbers come into the picture. We observed a moment ago that the egf for Bernoulli numbers is

$$\hat{B}(z) = \sum_{k\geqslant 0} B_k \frac{z^k}{k!} = \frac{z}{e^z-1};$$

hence we can write

$$\begin{aligned}\hat{S}(z) &= B(z)\,\frac{e^{nz}-1}{z}\\ &= \Bigl(B_0\frac{z^0}{0!}+B_1\frac{z^1}{1!}+B_2\frac{z^2}{2!}+\cdots\Bigr)\Bigl(n\frac{z^0}{1!}+n^2\frac{z^1}{2!}+n^3\frac{z^2}{3!}+\cdots\Bigr).\end{aligned}$$

The sum $S_m(n)$ is $m!$ times the coefficient of z^m in this product. For example,

$$\begin{aligned}S_0(n) &= 0!\Bigl(B_0\frac{n}{1!\,0!}\Bigr) &&= n;\\ S_1(n) &= 1!\Bigl(B_0\frac{n^2}{2!\,0!}+B_1\frac{n}{1!\,1!}\Bigr) &&= \tfrac{1}{2}n^2-\tfrac{1}{2}n;\\ S_2(n) &= 2!\Bigl(B_0\frac{n^3}{3!\,0!}+B_1\frac{n^2}{2!\,1!}+B_2\frac{n}{1!\,2!}\Bigr) &&= \tfrac{1}{3}n^3-\tfrac{1}{2}n^2+\tfrac{1}{6}n.\end{aligned}$$

We have therefore derived the formula $\Box_n = S_2(n) = \frac{1}{3}n(n-\frac{1}{2})(n-1)$ for the umpteenth time, and this was the simplest derivation of all: In a few lines we have found the general behavior of $S_m(n)$ for all m.

The general formula can be written

$$S_{m-1}(n) = \frac{1}{m}\bigl(B_m(n)-B_m(0)\bigr), \qquad (7.78)$$

where $B_m(x)$ is the *Bernoulli polynomial* defined by

$$B_m(x) = \sum_k \binom{m}{k} B_k\, x^{m-k}. \qquad (7.79)$$

Here's why: The Bernoulli polynomial is the binomial convolution of the sequence $\langle B_0, B_1, B_2, \ldots\rangle$ with $\langle 1, x, x^2, \ldots\rangle$; hence the exponential generation function for $\langle B_0(x), B_1(x), B_2(x), \ldots\rangle$ is the product of their egf's,

$$\hat{B}(z,x) = \sum_{m\geq 0} B_m(x)\frac{z^m}{m!} = \frac{z}{e^z-1}\sum_{m\geq 0} x^m\frac{z^m}{m!} = \frac{ze^{xz}}{e^z-1}. \qquad (7.80)$$

Equation (7.78) follows because the egf for $\langle 0, S_0(n), 2S_1(n), \ldots\rangle$ is, by (7.77),

$$z\,\frac{e^{nz}-1}{e^z-1} = B(z,n) - B(z,0).$$

Let's turn now to another problem for which egf's are just the thing: How many spanning trees are possible in the *complete graph* on n vertices $\{1,2,\ldots,n\}$? Let's call this number t_n. The complete graph has $\frac{1}{2}n(n-1)$ edges, one edge joining each pair of distinct vertices; so we're essentially looking for the total number of ways to connect up n given things by drawing $n-1$ lines between them.

We have $t_1 = t_2 = 1$. Also $t_3 = 3$, because a complete graph on three vertices is a fan of order 2; we know that $f_2 = 3$. And there are sixteen spanning trees when $n = 4$:

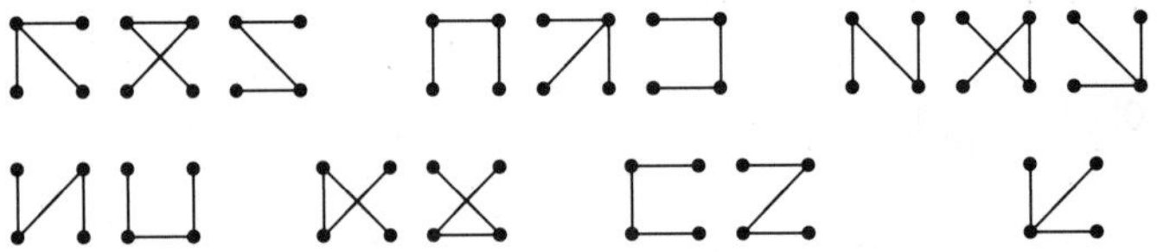

(7.81)

Hence $t_4 = 16$.

Our experience with the analogous problem for fans suggests that the best way to tackle this problem is to single out one vertex, and to look at the blocks or components that the spanning tree joins together when we ignore all edges that touch the special vertex. If the non-special vertices form m components of sizes k_1, k_2, …, k_m, then we can connect them to the special vertex in $k_1k_2\ldots k_m$ ways. For example, in the case $n = 4$, we can consider the lower left vertex to be special. The top row of (7.81) shows $3t_3$ cases where the other three vertices are joined among themselves in t_3 ways and then connected to the lower left in 3 ways. The bottom row shows $2\cdot 1\times t_2t_1\times\binom{3}{2}$ solutions where the other three vertices are divided into components of sizes 2 and 1 in $\binom{3}{2}$ ways; there's also the case 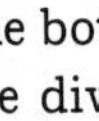 where the other three vertices are completely unconnected among themselves.

This line of reasoning leads to the recurrence

$$t_n = \sum_{m>0}\frac{1}{m!}\sum_{k_1+k_2+\cdots+k_m=n-1}\binom{n-1}{k_1,k_2,\ldots,k_m}k_1k_2\ldots k_m\,t_{k_1}t_{k_2}\ldots t_{k_m}$$

for all $n > 1$. Here's why: There are $\binom{n-1}{k_1,k_2,\ldots,k_m}$ ways to assign $n-1$ elements to a sequence of m components of respective sizes $k_1, k_2, \ldots, k_m$; there are $t_{k_1} t_{k_2} \ldots t_{k_m}$ ways to connect up those individual components with spanning trees; there are $k_1 k_2 \ldots k_m$ ways to connect vertex n to those components; and we divide by $m!$ because we want to disregard the order of the components. For example, when $n = 4$ the recurrence says that

$$t_4 = 3t_3 + \tfrac{1}{2}\left(\binom{3}{1,2}2t_1t_2 + \binom{3}{2,1}2t_2t_1\right) + \tfrac{1}{6}\left(\binom{3}{1,1,1}t_1^3\right) = 3t_3 + 2t_2t_1 + t_1^3 .$$

The recurrence for t_n looks formidable at first, possibly even frightening; but it really isn't bad, only convoluted. We can define

$$u_n = n\, t_n$$

and then everything simplifies considerably:

$$\frac{u_n}{n!} = \sum_{m>0} \frac{1}{m!} \sum_{k_1+k_2+\cdots+k_m=n-1} \frac{u_{k_1}}{k_1!}\frac{u_{k_2}}{k_2!}\cdots\frac{u_{k_m}}{k_m!}, \qquad \text{if } n > 1. \tag{7.82}$$

The inner sum is the coefficient of z^{n-1} in the egf $\hat{U}(z)$, raised to the mth power; and we obtain the correct formula also when $n = 1$, if we add in the term $\hat{U}(z)^0$ that corresponds to the case $m = 0$. So

$$\frac{u_n}{n!} = [z^{n-1}] \sum_{m\geqslant 0} \frac{1}{m!} \hat{U}(z)^m = [z^{n-1}]\, e^{\hat{U}(z)} = [z^n]\, z e^{\hat{U}(z)}$$

for all $n > 0$, and we have the equation

$$\hat{U}(z) = z e^{\hat{U}(z)} . \tag{7.83}$$

Progress! Equation (7.83) is almost like

$$\mathcal{E}(z) = e^{z\,\mathcal{E}(z)} ,$$

which defines the generalized exponential series $\mathcal{E}(z) = \mathcal{E}_1(z)$ in (5.59) and (7.70); indeed, we have

$$\hat{U}(z) = z\mathcal{E}(z) .$$

So we can read off the answer to our problem:

$$t_n = \frac{u_n}{n} = \frac{n!}{n} [z^n]\, \hat{U}(z) = (n-1)!\, [z^{n-1}]\, \mathcal{E}(z) = n^{n-2} . \tag{7.84}$$

The complete graph on $\{1, 2, \ldots, n\}$ has exactly n^{n-2} spanning trees, for all $n > 0$.

7.7 DIRICHLET GENERATING FUNCTIONS

There are many other possible ways to generate a sequence from a series; any system of "kernel" functions $K_n(z)$ such that

$$\sum_n g_n K_n(z) = 0 \quad\Longrightarrow\quad g_n = 0 \text{ for all } n$$

can be used, at least in principle. Ordinary generating functions use $K_n(z) = z^n$, and exponential generating functions use $K_n(z) = z^n/n!$; we could also try falling factorial powers $z^{\underline{n}}$, or binomial coefficients $z^{\underline{n}}/n! = \binom{z}{n}$.

The most important alternative to gf's and egf's uses the kernel functions $1/n^z$; it is intended for sequences $\langle g_1, g_2, \ldots\rangle$ that begin with $n = 1$ instead of $n = 0$:

$$\widetilde{G}(z) = \sum_{n\geqslant 1} \frac{g_n}{n^z}. \qquad (7.85)$$

This is called a *Dirichlet generating function* (dgf), because the German mathematician Gustav Lejeune Dirichlet (1805–1859) made much of it.

For example, the dgf of the constant sequence $\langle 1, 1, 1, \ldots\rangle$ is

$$\sum_{n\geqslant 1} \frac{1}{n^z} = \zeta(z). \qquad (7.86)$$

This is Riemann's *zeta function*, which we have also called the generalized harmonic number $H_\infty^{(z)}$ when $z > 1$.

The product of Dirichlet generating functions corresponds to a special kind of convolution:

$$\widetilde{F}(z)\widetilde{G}(z) = \sum_{l,m\geqslant 1} \frac{f_l}{l^z}\frac{g_m}{m^z} = \sum_{n\geqslant 1}\frac{1}{n^z}\sum_{l,m\geqslant 1} f_l\, g_m\, [l\cdot m = n].$$

Thus $\widetilde{F}(z)\widetilde{G}(z) = \widetilde{H}(z)$ is the dgf of the sequence

$$h_n = \sum_{d\backslash n} f_d\, g_{n/d}. \qquad (7.87)$$

For example, we know from (4.55) that $\sum_{d\backslash n}\mu(d) = [n=1]$; this is the Dirichlet convolution of the Möbius sequence $\langle \mu(1), \mu(2), \mu(3), \ldots\rangle$ with $\langle 1, 1, 1, \ldots\rangle$, hence

$$\widetilde{M}(z)\zeta(z) = \sum_{n\geqslant 1}\frac{[n=1]}{n^z} = 1. \qquad (7.88)$$

In other words, the dgf of $\langle \mu(1), \mu(2), \mu(3), \ldots\rangle$ is $\zeta(z)^{-1}$.

Dirichlet generating functions are particularly valuable when the sequence $\langle g_1, g_2, \ldots \rangle$ is a *multiplicative function*, namely when

$$g_{mn} = g_m g_n \qquad \text{for } m \perp n.$$

In such cases the values of g_n for all n are determined by the values of g_n when n is a power of a prime, and we can factor the dgf into a product over primes:

$$\widetilde{G}(z) = \prod_{p \text{ prime}} \left(1 + \frac{g_p}{p^z} + \frac{g_{p^2}}{p^{2z}} + \frac{g_{p^3}}{p^{3z}} + \cdots\right). \tag{7.89}$$

If, for instance, we set $g_n = 1$ for all n, we obtain a product representation of Riemann's zeta function:

$$\zeta(z) = \prod_{p \text{ prime}} \left(\frac{1}{1 - p^{-z}}\right). \tag{7.90}$$

The Möbius function has $\mu(p) = -1$ and $\mu(p^k) = 0$ for $k > 1$, hence its dgf is

$$\widetilde{M}(z) = \prod_{p \text{ prime}} (1 - p^{-z}); \tag{7.91}$$

this agrees, of course, with (7.88) and (7.90). Euler's φ function has $\varphi(p^k) = p^k - p^{k-1}$, hence its dgf has the factored form

$$\widetilde{\Phi}(z) = \prod_{p \text{ prime}} \left(1 + \frac{p - 1}{p^z - p}\right) = \prod_{p \text{ prime}} \frac{1 - p^{-z}}{1 - p^{1-z}}. \tag{7.92}$$

We conclude that $\widetilde{\Phi}(z) = \zeta(z-1)/\zeta(z)$.

Exercises

Warmups

1 An eccentric collector of $2 \times n$ domino tilings pays \$4 for each vertical domino and \$1 for each horizontal domino. How many tilings are worth exactly \$$m$ by this criterion? For example, when $m = 6$ there are three solutions: , , and .

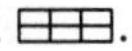

2 Give the generating function and the exponential generating function for the sequence $\langle 2, 5, 13, 35, \ldots \rangle = \langle 2^n + 3^n \rangle$ in closed form.

3 What is $\sum_{n \geq 0} H_n/10^n$?

4 The general expansion theorem for rational functions $P(z)/Q(z)$ is not completely general, because it restricts the degree of P to be less than the degree of Q. What happens if P has a larger degree than this?

5 Find a generating function $S(z)$ such that

$$[z^n]\,S(z) \;=\; \sum_k \binom{r}{k}\binom{r}{n-2k}\,.$$

Basics

6 Show that the recurrence (7.32) can be solved by the repertoire method, without using generating functions.

7 Solve the recurrence

$$\begin{aligned} g_0 &= 1\,; \\ g_n &= g_{n-1} + 2g_{n-2} + \cdots + ng_0\,, \qquad \text{for } n > 0. \end{aligned}$$

8 What is $[z^n]\,\bigl(\ln(1-z)\bigr)^2/(1-z)^{m+1}$?

9 Use the result of the previous exercise to evaluate $\sum_{k=0}^{n} H_k H_{n-k}$.

10 Set $r = s = -1/2$ in identity (7.61) and then remove all occurrences of $1/2$ by using tricks like (5.36). What amazing identity do you deduce?

I deduce that Clark Kent is really Superman.

11 This problem, whose three parts are independent, gives practice in the manipulation of generating functions. We assume that $A(z) = \sum_n a_n z^n$, $B(z) = \sum_n b_n z^n$, $C(z) = \sum_n c_n z^n$, and that the coefficients are zero for negative n.

a If $c_n = \sum_{j+2k\leqslant n} a_j b_k$, express C in terms of A and B.

b If $nb_n = \sum_{k=0}^{n} 2^k a_k/(n-k)!$, express A in terms of B.

c If r is a real number and if $a_n = \sum_{k=0}^{n} \binom{r+k}{k} b_{n-k}$, express A in terms of B; then use your formula to find coefficients $f_k(r)$ such that $b_n = \sum_{k=0}^{n} f_k(r)\,a_{n-k}$.

12 How many ways are there to put the numbers $\{1, 2, \ldots, 2n\}$ into a $2 \times n$ array so that rows and columns are in increasing order from left to right and from top to bottom? For example, one solution when $n = 5$ is

$$\begin{pmatrix} 1 & 2 & 4 & 5 & 8 \\ 3 & 6 & 7 & 9 & 10 \end{pmatrix}.$$

13 Prove Raney's generalized lemma, which is stated just before (7.69).

14 Solve the recurrence

$$\begin{aligned} g_0 &= 0\,, \qquad g_1 \;=\; 1\,, \\ g_n &= -2ng_{n-1} + \sum_k \binom{n}{k} g_k g_{n-k}\,, \qquad \text{for } n > 1, \end{aligned}$$

by using an exponential generating function.

15 The *Bell number* b_n is the number of ways to partition n things into subsets. For example, $b_3 = 5$ because we can partition $\{1,2,3\}$ in the following ways:

$$\{1,2,3\}; \quad \{1,2\}\cup\{3\}; \quad \{1,3\}\cup\{2\}; \quad \{1\}\cup\{2,3\}; \quad \{1\}\cup\{2\}\cup\{3\}.$$

Prove that $b_{n+1} = \sum_k \binom{n}{k} b_{n-k}$, and use this recurrence to find a closed form for the exponential generating function $\sum_n b_n z^n/n!$.

16 Two sequences $\langle a_n \rangle$ and $\langle b_n \rangle$ are related by the convolution formula

$$b_n = \sum_{k_1+2k_2+\cdots nk_n=n} \binom{a_1+k_1-1}{k_1}\binom{a_2+k_2-1}{k_2}\ldots\binom{a_n+k_n-1}{k_n};$$

also $a_0 = 0$ and $b_0 = 1$. Prove that the corresponding generating functions satisfy $\ln B(z) = A(z) + \frac{1}{2}A(z^2) + \frac{1}{3}A(z^3) + \cdots$.

17 Show that the exponential generating function $\hat{G}(z)$ of a sequence is related to the ordinary generating function $G(z)$ by the formula

$$\int_0^\infty \hat{G}(zt)e^{-t}\,dt = G(z),$$

if the integral exists.

18 Find the Dirichlet generating functions for the sequences

a $g_n = \sqrt{n}$;

b $g_n = \ln n$;

c $g_n = [n \text{ is squarefree}]$.

Express your answers in terms of the zeta function. (Squarefreeness is defined in exercise 4.13.)

19 Every power series $F(z) = \sum_{n\geqslant 0} f_n z^n$ with $f_0 = 1$ defines a sequence of polynomials $f_n(x)$ by the rule

$$F(z)^x = \sum_{n\geqslant 0} f_n(x) z^n,$$

where $f_n(1) = f_n$ and $f_n(0) = [n=0]$. In general, $f_n(x)$ has degree n. Show that such polynomials always satisfy the convolution formulas

$$\sum_{k=0}^{n} f_k(x) f_{n-k}(y) = f_n(x+y);$$

$$(x+y)\sum_{k=0}^{n} k f_k(x) f_{n-k}(y) = x n f_n(x+y).$$

(The identities in Tables 202 and 258 are special cases of this trick.)

20 A power series $G(z)$ is called *differentiably finite* if there exist finitely many polynomials $P_0(z), \ldots, P_m(z)$, not all zero, such that

$$P_0(z)G(z) + P_1(z)G'(z) + \cdots + P_m(z)G^{(m)}(z) = 0.$$

A sequence of numbers $\langle g_0, g_1, g_2, \ldots \rangle$ is called *polynomially recursive* if there exist finitely many polynomials $p_0(z), \ldots, p_m(z)$, not all zero, such that

$$p_0(n)g_n + p_1(n)g_{n+1} + \cdots + p_m(n)g_{n+m} = 0$$

for all integers $n \geqslant 0$. Prove that a generating function is differentiably finite if and only if its sequence of coefficients is polynomially recursive.

Homework exercises

21 A robber holds up a bank and demands \$500 in tens and twenties. He also demands to know the number of ways in which the cashier can give him the money. Find a generating function $G(z)$ for which this number is $[z^{500}]\,G(z)$, and a more compact generating function $\check{G}(z)$ for which this number is $[z^{50}]\,\check{G}(z)$. Determine the required number of ways by (a) using partial fractions; (b) using a method like (7.39).

Will he settle for $2 \times n$ domino tilings?

22 Let P be the sum of all ways to "triangulate" polygons:

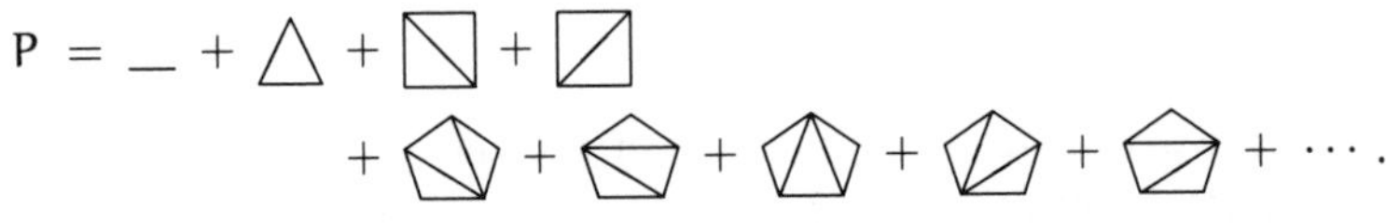

$$P = \_ + \triangle + \boxslash + \boxbslash + \cdots$$

(The first term represents a degenerate polygon with only two vertices; every other term shows a polygon that has been divided into triangles. For example, a pentagon can be triangulated in five ways.) Define a "multiplication" operation $A \triangle B$ on triangulated polygons A and B so that the equation

$$P = \_ + P \triangle P$$

is valid. Then replace each triangle by 'z'; what does this tell you about the number of ways to decompose an n-gon into triangles?

23 In how many ways can a $2 \times 2 \times n$ pillar be built out of $2 \times 1 \times 1$ bricks?

At union rates, as many as you can afford, plus a few.

24 How many spanning trees are in an n-wheel (a graph with n "outer" vertices in a cycle, each connected to an $(n+1)$st "hub" vertex), when $n \geqslant 3$?

25 Let $m \geqslant 2$ be an integer. What is a closed form for the generating function of the sequence $\langle n \bmod m \rangle$, as a function of z and m? Use this generating function to express '$n \bmod m$' in terms of the complex number $\omega = e^{2\pi i/m}$. (For example, when $m = 2$ we have $\omega = -1$ and $n \bmod 2 = \frac{1}{2} - \frac{1}{2}(-1)^n$.)

26 The second-order Fibonacci numbers $\langle \mathfrak{F}_n \rangle$ are defined by the recurrence

$$\begin{aligned} \mathfrak{F}_0 &= 0; \qquad \mathfrak{F}_1 = 1; \\ \mathfrak{F}_n &= \mathfrak{F}_{n-1} + \mathfrak{F}_{n-2} + F_n, \qquad \text{for } n > 1. \end{aligned}$$

Express $\mathfrak{F}_n$ in terms of the usual Fibonacci numbers F_n and F_{n+1}.

27 A $2 \times n$ domino tiling can also be regarded as a way to draw n disjoint lines in a $2 \times n$ array of points:

If we superimpose two such patterns, we get a set of cycles, since every point is touched by two lines. For example, if the lines above are combined with the lines

,

the result is

.

The same set of cycles is also obtained by combining

with .

But we get a unique way to reconstruct the original patterns from the superimposed ones if we assign orientations to the vertical lines by using arrows that go alternately up/down/up/down/$\cdots$ in the first pattern and alternately down/up/down/up/$\cdots$ in the second. For example,

+ = .

The number of such oriented cycle patterns must therefore be $T_n^2 = F_{n+1}^2$, and we should be able to prove this via algebra. Let Q_n be the number of oriented $2 \times n$ cycle patterns. Find a recurrence for Q_n, solve it with generating functions, and deduce algebraically that $Q_n = F_{n+1}^2$.

28 The coefficients of $A(z)$ in (7.39) satisfy $A_r + A_{r+10} + A_{r+20} + A_{r+30} = 100$ for $0 \leqslant r < 10$. Find a "simple" explanation for this.

29 What is the sum of Fibonacci products

$$\sum_{m>0} \sum_{\substack{k_1+k_2+\cdots+k_m=n \\ k_1,k_2,\ldots,k_m>0}} F_{k_1} F_{k_2} \ldots F_{k_m} \; ?$$

30 If the generating function $G(z) = 1/(1-\alpha z)(1-\beta z)$ has the partial fraction decomposition $a/(1-\alpha z)+b/(1-\beta z)$, what is the partial fraction decomposition of $G(z)^n$?

31 What function $g(n)$ of the positive integer n satisfies the recurrence

$$\sum_{d \backslash n} g(d)\, \varphi(n/d) = 1 ,$$

where φ is Euler's totient function?

32 An *arithmetic progression* is an infinite set of integers

$$\{an+b\} = \{b, a+b, 2a+b, 3a+b, \ldots\}.$$

A set of arithmetic progressions $\{a_1n+b_1\}, \ldots, \{a_mn+b_m\}$ is called an *exact cover* if every nonnegative integer occurs in one and only one of the progressions. For example, the three progressions $\{2n\}$, $\{4n+1\}$, $\{4n+3\}$ constitute an exact cover. Show that if $\{a_1n+b_1\}, \ldots, \{a_mn+b_m\}$ is an exact cover such that $2 \leqslant a_1 \leqslant \cdots \leqslant a_m$, then $a_{m-1} = a_m$. *Hint:* Use generating functions.

Exam problems

33 What is $[w^m z^n]\,(\ln(1+z))/(1-wz)$?

34 Find a closed form for the generating function $\sum_{n\geqslant 0} G_n(z)w^n$, if

$$G_n(z) = \sum_{k\leqslant n/m} \binom{n-mk}{k} z^{mk} .$$

(Here m is a fixed positive integer.)

35 Evaluate the sum $\sum_{0<k<n} 1/k(n-k)$ in two ways:

a Expand the summand in partial fractions.

b Treat the sum as a convolution and use generating functions.

36 Let $A(z)$ be the generating function for $\langle a_0, a_1, a_2, a_3, \ldots\rangle$. Express $\sum_n a_{\lfloor n/m \rfloor} z^n$ in terms of A, z, and m.

37 Let a_n be the number of ways to write the positive integer n as a sum of powers of 2, disregarding order. For example, $a_4 = 4$, since $4 = 2+2 = 2+1+1 = 1+1+1+1$. By convention we let $a_0 = 1$. Let $b_n = \sum_{k=0}^{n} a_k$ be the cumulative sum of the first a's.

a Make a table of the a's and b's up through $n = 10$. What amazing relation do you observe in your table? (Don't prove it yet.)

b Express the generating function $A(z)$ as an infinite product.

c Use the expression from part (b) to prove the result of part (a).

38 Find a closed form for the double generating function

$$M(w,z) \;=\; \sum_{m,n\geqslant 0} \min(m,n)\, w^m z^n\,.$$

Generalize your answer to obtain, for fixed $m \geqslant 2$, a closed form for

$$M(z_1,\ldots,z_m) \;=\; \sum_{n_1,\ldots,n_m\geqslant 0} \min(n_1,\ldots,n_m)\, z_1^{n_1}\ldots z_m^{n_m}\,.$$

39 Given positive integers m and n, find closed forms for

$$\sum_{1\leqslant k_1<k_2<\cdots<k_m\leqslant n} k_1k_2\ldots k_m \quad\text{and}\quad \sum_{1\leqslant k_1\leqslant k_2\leqslant\cdots\leqslant k_m\leqslant n} k_1k_2\ldots k_m\,.$$

(For example, when $m = 2$ and $n = 3$ the sums are $1\cdot2+1\cdot3+2\cdot3$ and $1\cdot1+1\cdot2+1\cdot3+2\cdot2+2\cdot3+3\cdot3$.) *Hint:* What are the coefficients of z^m in the generating functions $(1+a_1z)\ldots(1+a_nz)$ and $1/(1-a_1z)\ldots(1-a_nz)$?

40 Express $\sum_k \binom{n}{k}(kF_{k-1}-F_k)(n-k)_{\mathrm{i}}$ in closed form.

41 An *up-down permutation* of order n is an arrangement $a_1a_2\ldots a_n$ of the integers $\{1,2,\ldots,n\}$ that goes alternately up and down:

$$a_1 \;<\; a_2 \;>\; a_3 \;<\; a_4 \;>\; \cdots\,.$$

For example, 35142 is an up-down permutation of order 5. If A_n denotes the number of up-down permutations of order n, show that the exponential generating function of $\langle A_n\rangle$ is $(1+\sin z)/\cos z$.

42 A space probe has discovered that organic material on Mars has DNA composed of five symbols, denoted by (a,b,c,d,e), instead of the four components in earthling DNA. The four pairs cd, ce, ed, and ee never occur consecutively in a string of Martian DNA, but any string without forbidden pairs is possible. (Thus $bbcda$ is forbidden but $bbdca$ is OK.) How many Martian DNA strings of length n are possible? (When $n = 2$ the answer is 21, because the left and right ends of a string are distinguishable.)

43 The *Newtonian generating function* of a sequence $\langle g_n \rangle$ is defined to be

$$\dot{G}(z) = \sum_n g_n \binom{z}{n}.$$

Find a convolution formula that defines the relation between sequences $\langle f_n \rangle$, $\langle g_n \rangle$, and $\langle h_n \rangle$ whose Newtonian generating functions are related by the equation $\dot{F}(z)\dot{G}(z) = \dot{H}(z)$. Try to make your formula as simple and symmetric as possible.

44 Let q_n be the number of possible outcomes when n numbers $\{x_1, \ldots, x_n\}$ are compared with each other. For example, $q_3 = 13$ because the possibilities are

$$\begin{array}{llll} x_1 < x_2 < x_3; & x_1 < x_2 = x_3; & x_1 < x_3 < x_2; & x_1 = x_2 < x_3; \\ x_1 = x_2 = x_3; & x_1 = x_3 < x_2; & x_2 < x_1 < x_3; & \\ x_2 < x_1 = x_3; & x_2 < x_3 < x_1; & x_2 = x_3 < x_1; & \\ x_3 < x_1 < x_2; & x_3 < x_1 = x_2; & x_3 < x_2 < x_1. & \end{array}$$

Find a closed form for the egf $\hat{Q}(z) = \sum_n q_n z^n/n!$. Also find sequences $\langle a_n \rangle$, $\langle b_n \rangle$, $\langle c_n \rangle$ such that

$$q_n = \sum_{k \geq 0} k^n a_k = \sum_k {n \brace k} b_k = \sum_k \left\langle {n \atop k} \right\rangle c_k, \quad \text{for all } n > 0.$$

45 Evaluate $\sum_{m,n>0} [m \perp n]/m^2 n^2$.

46 Evaluate

$$\sum_{0 \leq k \leq n/2} \binom{n-2k}{k} \left(\frac{-4}{27}\right)^k$$

in closed form. *Hint:* $z^3 - z^2 + \frac{4}{27} = (z + \frac{1}{3})(z - \frac{2}{3})^2$.

47 Show that the numbers U_n and V_n of $3 \times n$ domino tilings, as given in (7.34), are closely related to the fractions in the Stern–Brocot tree that converge to $\sqrt{3}$.

48 A certain sequence $\langle g_n \rangle$ satisfies the recurrence

$$a g_n + b g_{n+1} + c g_{n+2} + d = 0, \qquad \text{integer } n \geq 0,$$

for some integers (a, b, c, d) with $\gcd(a, b, c, d) = 1$. It also has the closed form

$$g_n = \lfloor \alpha (1 + \sqrt{2})^n \rfloor, \qquad \text{integer } n \geq 0,$$

for some real number α between 0 and 1. Find a, b, c, d, and α.

Kissinger, take note.

49 This is a problem about powers and parity.

a Consider the sequence $\langle a_0, a_1, a_2, \ldots \rangle = \langle 2, 2, 6, \ldots \rangle$ defined by the formula

$$a_n = (1+\sqrt{2})^n + (1-\sqrt{2})^n .$$

Find a simple recurrence relation that is satisfied by this sequence.

b Prove that $\lceil (1+\sqrt{2})^n \rceil \equiv n \pmod 2$ for all integers $n > 0$.

c Find a number α of the form $(p+\sqrt{q})/2$, where p and q are positive integers, such that $\lfloor \alpha^n \rfloor \equiv n \pmod 2$ for all integers $n > 0$.

Bonus problems

50 Continuing exercise 22, consider the sum of all ways to decompose polygons into polygons:

$$Q = \_ + \triangle + \square + \boxbslash + \boxslash + ⬠ + ⬠ + ⬠ + ⬠ + ⬠ + ⬠ + ⬠ + \cdots .$$

Find a symbolic equation for Q and use it to find a generating function for the number of ways to draw nonintersecting diagonals inside a convex n-gon. (Give a closed form for the generating function as a function of z; you need not find a closed form for the coefficients.)

51 Prove that the product

$$2^{mn/2} \prod_{\substack{1\le j\le m\\ 1\le k\le n}} \left(\left(\cos^2 \frac{j\pi}{m+1}\right) \text{▯}^2 + \left(\cos^2 \frac{k\pi}{n+1}\right) \text{▭}^2 \right)$$

is the generating function for tilings of an $m \times n$ rectangle with dominoes. (There are mn factors, which we can imagine are written in the mn cells of the rectangle. If mn is odd, the middle factor is zero. The coefficient of $\text{▯}^j \text{▭}^k$ is the number of ways to do the tiling with j vertical and k horizontal dominoes.) *Hint:* This is a difficult problem, really beyond the scope of this book. You may wish to simply verify the formula in the case $m = 3$, $n = 4$.

Is this a hint or a warning?

52 Prove that the polynomials defined by the recurrence

$$p_n(y) = \left(y - \frac{1}{4}\right)^n - \sum_{k=0}^{n-1} \binom{2n}{2k} \left(\frac{-1}{4}\right)^{n-k} p_k(y), \quad \text{integer } n \ge 0,$$

have the form $p_n(y) = \sum_{m=0}^{n} \left|{n \atop m}\right| y^n$, where $\left|{n \atop m}\right|$ is a positive integer for $1 \le m \le n$. *Hint:* This exercise is very instructive but not very easy.

53 The sequence of *pentagonal numbers* $\langle 1, 5, 12, 22, \ldots \rangle$ generalizes the triangular and square numbers in an obvious way:

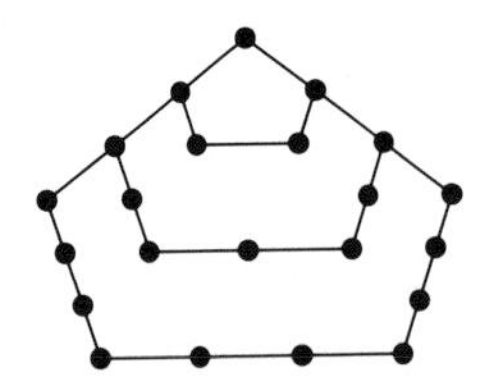

Let the nth triangular number be $T_n = n(n+1)/2$; let the nth pentagonal number be $P_n = n(3n-1)/2$; and let U_n be the $3 \times n$ domino-tiling number defined in (7.38). Prove that the triangular number $T_{(U_{4n+2}-1)/2}$ is also a pentagonal number. *Hint:* $3U_{2n}^2 = (V_{2n-1} + V_{2n+1})^2 + 2$.

54 Consider the following curious construction:

1	2	3	4	5	6	7	8	9	10	11	12	13	14	15	16	...
1	2	3	4		6	7	8	9		11	12	13	14		16	...
1	3	6	10		16	23	31	40		51	63	76	90		106	...
1	3	6			16	23	31			51	63	76			106	...
1	4	10			26	49	80			131	194	270			376	...
1	4				26	49				131	194				376	...
1	5				31	80				211	405				781	...
1					31					211					781	...
1					32					243					1024	...

(Start with a row containing all the positive integers. Then delete every mth column; here $m = 5$. Then replace the remaining entries by partial sums. Then delete every $(m-1)$st column. Then replace with partial sums again, and so on.) Use generating functions to show that the final result is the sequence of mth powers. For example, when $m = 5$ we get $\langle 1^5, 2^5, 3^5, 4^5, \ldots \rangle$ as shown.

55 Prove that if the power series $F(z)$ and $G(z)$ are differentiably finite (as defined in exercise 20), then so are $F(z) + G(z)$ and $F(z)G(z)$.

Research problems

56 Prove that there is no "simple closed form" for the coefficient of z^n in $(1 + z + z^2)^n$, as a function of n, in some large class of "simple closed forms."

57 Prove or disprove: If all the coefficients of $G(z)$ are either 0 or 1, and if all the coefficients of $G(z)^2$ are less than some constant M, then infinitely many of the coefficients of $G(z)^2$ are zero.

8

Discrete Probability

THE ELEMENT OF CHANCE enters into many of our attempts to understand the world we live in. A mathematical *theory of probability* allows us to calculate the likelihood of complex events if we assume that the events are governed by appropriate axioms. This theory has significant applications in all branches of science, and it has strong connections with the techniques we have studied in previous chapters.

Probabilities are called "discrete" if we can compute the probabilities of all events by summation instead of by integration. We are getting pretty good at sums, so it should come as no great surprise that we are ready to apply our knowledge to some interesting calculations of probabilities and averages.

8.1 DEFINITIONS

(Readers unfamiliar with probability theory will, with high probability, benefit from a perusal of Feller's classic introduction to the subject [96].)

Probability theory starts with the idea of a *probability space*, which is a set Ω of all things that can happen in a given problem together with a rule that assigns a probability $\Pr(\omega)$ to each elementary event $\omega \in \Omega$. The probability $\Pr(\omega)$ must be a nonnegative real number, and the condition

$$\sum_{\omega\in\Omega} \Pr(\omega) \;=\; 1 \qquad (8.1)$$

must hold in every discrete probability space. Thus, each value $\Pr(\omega)$ must lie in the interval $[0\,..\,1]$. We speak of Pr as a *probability distribution*, because it distributes a total probability of 1 among the events ω.

Here's an example: If we're rolling a pair of dice, the set Ω of elementary events is $\mathrm{D}^2 = \{$⚀⚀, ⚀⚁, …, ⚅⚅$\}$, where

$\mathrm{D} = \{$⚀, ⚁, ⚂, ⚃, ⚄, ⚅$\}$

Never say die.

is the set of all six ways that a given die can land. Two rolls such as ⚀⚁ and ⚁⚀ are considered to be distinct; hence this probability space has a total of $6^2 = 36$ elements.

We usually assume that dice are "fair," namely that each of the six possibilities for a particular die has probability $\frac{1}{6}$, and that each of the 36 possible rolls in Ω has probability $\frac{1}{36}$. But we can also consider "loaded" dice in which there is a different distribution of probabilities. For example, let

Careful: They might go off.

$$\Pr_1(⚀) = \Pr_1(⚅) = \tfrac{1}{4};$$
$$\Pr_1(⚁) = \Pr_1(⚂) = \Pr_1(⚃) = \Pr_1(⚄) = \tfrac{1}{8}.$$

Then $\sum_{d \in D} \Pr_1(d) = 1$, so $\Pr_1$ is a probability distribution on the set D, and we can assign probabilities to the elements of $\Omega = D^2$ by the rule

$$\Pr_{11}(d\,d') = \Pr_1(d)\,\Pr_1(d'). \tag{8.2}$$

For example, $\Pr_{11}(⚅⚂) = \frac{1}{4} \cdot \frac{1}{8} = \frac{1}{32}$. This is a valid distribution because

$$\begin{aligned}\sum_{\omega \in \Omega} \Pr_{11}(\omega) &= \sum_{dd' \in D^2} \Pr_{11}(d\,d') = \sum_{d,d' \in D} \Pr_1(d)\,\Pr_1(d') \\ &= \sum_{d \in D} \Pr_1(d) \sum_{d' \in D} \Pr_1(d') = 1 \cdot 1 = 1.\end{aligned}$$

We can also consider the case of one fair die and one loaded die,

$$\Pr_{01}(d\,d') = \Pr_0(d)\,\Pr_1(d'), \qquad \text{where } \Pr_0(d) = \tfrac{1}{6}, \tag{8.3}$$

in which case $\Pr_{01}(⚅⚂) = \frac{1}{6} \cdot \frac{1}{8} = \frac{1}{48}$. Dice in the "real world" can't really be expected to turn up equally often on each side, because there is not perfect symmetry; but $\frac{1}{6}$ is usually pretty close to the truth.

If all sides of a cube were identical, how could we tell which side is face up?

An *event* is a subset of Ω. In dice games, for example, the set

$$\{⚀⚀,\ ⚁⚁,\ ⚂⚂,\ ⚃⚃,\ ⚄⚄,\ ⚅⚅\}$$

is the event that "doubles are thrown." The individual elements ω of Ω are called *elementary events* because they cannot be decomposed into smaller subsets; we can think of ω as a one-element event $\{\omega\}$.

The probability of an event A is defined by the formula

$$\Pr(\omega \in A) = \sum_{\omega \in A} \Pr(\omega); \tag{8.4}$$

and in general if $R(\omega)$ is any statement about ω, we write '$\Pr\bigl(R(\omega)\bigr)$' for the sum of all $\Pr(\omega)$ such that $R(\omega)$ is true. Thus, for example, the probability of doubles with fair dice is $\frac{1}{36}+\frac{1}{36}+\frac{1}{36}+\frac{1}{36}+\frac{1}{36}+\frac{1}{36} = \frac{1}{6}$; but when both dice are loaded with probability distribution $\Pr_1$ it is $\frac{1}{16}+\frac{1}{64}+\frac{1}{64}+\frac{1}{64}+\frac{1}{64}+\frac{1}{16} = \frac{3}{16} > \frac{1}{6}$. Loading the dice makes the event "doubles are thrown" more probable.

(We have been using $\sum$-notation in a more general sense here than defined in Chapter 2: The sums in (8.1) and (8.4) occur over all elements ω of an arbitrary set, not over integers only. However, this new development is not really alarming; we can agree to use special notation under a $\sum$ whenever nonintegers are intended, so there will be no confusion with our ordinary conventions. The other definitions in Chapter 2 are still valid; in particular, the definition of infinite sums in that chapter gives the appropriate interpretation to our sums when the set Ω is infinite. Each probability is nonnegative, and the sum of all probabilities is bounded, so the probability of event A in (8.4) is well defined for all subsets $A \subseteq \Omega$.)

A *random variable* is a function defined on the elementary events ω of a probability space. For example, if $\Omega = D^2$ we can define $S(\omega)$ to be the sum of the spots on the dice roll ω, so that S(⚅⚂) $= 6+3 = 9$. The probability that the spots total seven is the probability of the event $S(\omega) = 7$, namely

Pr(⚀⚅) + Pr(⚁⚄) + Pr(⚂⚃)
  + Pr(⚃⚂) + Pr(⚄⚁) + Pr(⚅⚀) .

With fair dice ($\Pr = \Pr_{00}$), this happens with probability $\frac{1}{6}$; with loaded dice ($\Pr = \Pr_{11}$), it happens with probability $\frac{1}{16} + \frac{1}{64} + \frac{1}{64} + \frac{1}{64} + \frac{1}{64} + \frac{1}{16} = \frac{3}{16}$, the same as we observed for doubles.

It's customary to drop the '(ω)' when we talk about random variables, because there's usually only one probability space involved when we're working on any particular problem. Thus we say simply '$S = 7$' for the event that a 7 was rolled, and '$S = 4$' for the event { ⚀⚂, ⚁⚁, ⚂⚀ }.

A random variable can be characterized by the probability distribution of its values. Thus, for example, S takes on eleven possible values $\{2, 3, \ldots, 12\}$, and we can tabulate the probability that $S = s$ for each s in this set:

s	2	3	4	5	6	7	8	9	10	11	12
$\Pr_{00}[S=s]$	$\frac{1}{36}$	$\frac{2}{36}$	$\frac{3}{36}$	$\frac{4}{36}$	$\frac{5}{36}$	$\frac{6}{36}$	$\frac{5}{36}$	$\frac{4}{36}$	$\frac{3}{36}$	$\frac{2}{36}$	$\frac{1}{36}$
$\Pr_{11}[S=s]$	$\frac{4}{64}$	$\frac{4}{64}$	$\frac{5}{64}$	$\frac{6}{64}$	$\frac{7}{64}$	$\frac{12}{64}$	$\frac{7}{64}$	$\frac{6}{64}$	$\frac{5}{64}$	$\frac{4}{64}$	$\frac{4}{64}$

If we're working on a problem that involves only the random variable S and no other properties of dice, we can compute the answer from these probabilities alone, without regard to the details of the set $\Omega = D^2$. In fact, we could define the probability space to be the smaller set $\Omega = \{2, 3, \ldots, 12\}$, with whatever probability distribution $\Pr(s)$ is desired. Then '$S = 4$' would be an elementary event. Thus we can often ignore the underlying probability space Ω and work directly with random variables and their distributions.

If two random variables X and Y are defined over the same probability space Ω, we can characterize their behavior without knowing everything about Ω if we know the "joint distribution"

Just Say No.

$$\Pr(X{=}x \text{ and } Y{=}y)$$

for each x in the range of X and each y in the range of Y. We say that X and Y are *independent* random variables if

$$\Pr(X{=}x \text{ and } Y{=}y) \;=\; \Pr(X{=}x)\cdot\Pr(Y{=}y) \tag{8.5}$$

for all x and y. Intuitively, this means that the value of X has no effect on the value of Y.

For example, if Ω is the set of dice rolls D^2, we can let S_1 be the number of spots on the first die and S_2 the number of spots on the second. Then the random variables S_1 and S_2 are independent with respect to each of the probability distributions $\Pr_{00}$, $\Pr_{11}$, and $\Pr_{01}$ discussed earlier, because we defined the dice probability for each elementary event dd' as a product of a probability for $S_1 = d$ multiplied by a probability for $S_2 = d'$. We could have defined probabilities differently so that, say,

$$\Pr(⚀⚄)\,/\,\Pr(⚀⚅) \;\ne\; \Pr(⚁⚄)\,/\,\Pr(⚁⚅)\,;$$

A dicey inequality.

but we didn't do that, because different dice aren't supposed to influence each other. With our definitions, both of these ratios are $\Pr(S_2{=}5)/\Pr(S_2{=}6)$.

We have defined S to be the sum of the two spot values, $S_1 + S_2$. Let's consider another random variable P, the product S_1S_2. Are S and P independent? Informally, no; if we are told that $S = 2$, we know that P must be 1. Formally, no again, because the independence condition (8.5) fails spectacularly (at least in the case of fair dice): For all legal values of s and p, we have $0 < \Pr_{00}[S{=}s]\cdot\Pr_{00}[P{=}p] \leqslant \frac16\cdot\frac19$; this can't equal $\Pr_{00}[S{=}s \text{ and } P{=}p]$, which is a multiple of $\frac1{36}$.

If we want to understand the typical behavior of a given random variable, we often ask about its "average" value. But the notion of "average" is ambiguous; people generally speak about three different kinds of averages when a sequence of numbers is given:

- the *mean* (which is the sum of all values, divided by the number of values);
- the *median* (which is the middle value, numerically);
- the *mode* (which is the value that occurs most often).

For example, the mean of $(3,1,4,1,5)$ is $\frac{3+1+4+1+5}{5} = 2.8$; the median is 3; the mode is 1.

But probability theorists usually work with random variables instead of with sequences of numbers, so we want to define the notion of an "average" for random variables too. Suppose we repeat an experiment over and over again, making independent trials in such a way that each value of X occurs with a frequency approximately proportional to its probability. (For example, we might roll a pair of dice many times, observing the values of S and/or P.) We'd like to define the average value of a random variable so that such experiments will usually produce a sequence of numbers whose mean, median, or mode is approximately the same as the mean, median, or mode of X, according to our definitions.

Here's how it can be done: The *mean* of a random real-valued variable X on a probability space Ω is defined to be

$$\sum_{x\in X(\Omega)} x\cdot \Pr(X=x) \tag{8.6}$$

if this potentially infinite sum exists. (Here $X(\Omega)$ stands for the set of all values that X can assume.) The *median* of X is defined to be the set of all x such that

$$\Pr(X\leqslant x) \geqslant \frac{1}{2} \quad\text{and}\quad \Pr(X\geqslant x) \geqslant \frac{1}{2}. \tag{8.7}$$

And the *mode* of X is defined to be the set of all x such that

$$\Pr(X=x) \geqslant \Pr(X=x') \qquad \text{for all } x' \in X(\Omega). \tag{8.8}$$

In our dice-throwing example, the mean of S turns out to be $2\cdot\frac{1}{36}+3\cdot\frac{2}{36}+\cdots+12\cdot\frac{1}{36}=7$ in distribution $\Pr_{00}$, and it also turns out to be 7 in distribution $\Pr_{11}$. The median and mode both turn out to be $\{7\}$ as well, in both distributions. So S has the same average under all three definitions. On the other hand the P in distribution $\Pr_{00}$ turns out to have a mean value of $\frac{49}{4}=12.25$; its median is $\{10\}$, and its mode is $\{6,12\}$. The mean of P is unchanged if we load the dice with distribution $\Pr_{11}$, but the median drops to $\{8\}$ and the mode becomes $\{6\}$ alone.

Probability theorists have a special name and notation for the mean of a random variable: They call it the *expected value*, and write

$$EX = \sum_{\omega\in\Omega} X(\omega)\Pr(\omega). \tag{8.9}$$

In our dice-throwing example, this sum has 36 terms (one for each element of Ω), while (8.6) is a sum of only eleven terms. But both sums have the same value, because they're both equal to

$$\sum_{\substack{\omega\in\Omega\\ x\in X(\Omega)}} x\Pr(\omega)\bigl[x=X(\omega)\bigr].$$

The mean of a random variable turns out to be more meaningful in applications than the other kinds of averages, so we shall largely forget about medians and modes from now on. We will use the terms "expected value," "mean," and "average" almost interchangeably in the rest of this chapter.

I get it: On average, "average" means "mean."

If X and Y are any two random variables defined on the same probability space, then $X+Y$ is also a random variable on that space. By formula (8.9), the average of their sum is the sum of their averages:

$$E(X+Y) \;=\; \sum_{\omega\in\Omega}\bigl(X(\omega)+Y(\omega)\bigr)\Pr(\omega) \;=\; EX+EY\,. \tag{8.10}$$

Similarly, if α is any constant we have the simple rule

$$E(\alpha X) \;=\; \alpha EX\,. \tag{8.11}$$

But the corresponding rule for multiplication of random variables is more complicated in general; the expected value is defined as a sum over elementary events, and sums of products don't often have a simple form. In spite of this difficulty, there is a very nice formula for the mean of a product in the special case that the random variables are independent:

$$E(XY) \;=\; (EX)(EY), \qquad \text{if X and Y are independent.} \tag{8.12}$$

We can prove this by the distributive law for products,

$$\begin{aligned} E(XY) &= \sum_{\omega\in\Omega} X(\omega)Y(\omega)\cdot\Pr(\omega) \\ &= \sum_{\substack{x\in X(\Omega)\\ y\in Y(\Omega)}} xy\cdot\Pr(X{=}x \text{ and } Y{=}y) \\ &= \sum_{\substack{x\in X(\Omega)\\ y\in Y(\Omega)}} xy\cdot\Pr(X{=}x)\Pr(Y{=}y) \\ &= \sum_{x\in X(\Omega)} x\Pr(X{=}x)\;\cdot\sum_{y\in Y(\Omega)} y\Pr(Y{=}y) \;=\; (EX)(EY)\,. \end{aligned}$$

For example, we know that $S = S_1+S_2$ and $P = S_1S_2$, when S_1 and S_2 are the numbers of spots on the first and second of a pair of random dice. We have $ES_1 = ES_2 = \frac{7}{2}$, hence $ES = 7$; furthermore S_1 and S_2 are independent, so $EP = \frac{7}{2}\cdot\frac{7}{2} = \frac{49}{4}$, as claimed earlier. We also have $E(S+P) = ES+EP = 7+\frac{49}{4}$. But S and P are not independent, so we cannot assert that $E(SP) = 7\cdot\frac{49}{4} = \frac{343}{4}$. In fact, the expected value of SP turns out to equal $\frac{637}{6}$ in distribution $\Pr_{00}$, 112 (exactly) in distribution $\Pr_{11}$.

8.2 MEAN AND VARIANCE

The next most important property of a random variable, after we know its expected value, is its *variance*, defined as the mean square deviation from the mean:

$$VX = E\big((X - EX)^2\big). \qquad (8.13)$$

If we denote EX by μ, the variance VX is the expected value of $(X-\mu)^2$. This measures the "spread" of X's distribution.

As a simple example of variance computation, let's suppose we have just been made an offer we can't refuse: Someone has given us two gift certificates for a certain lottery. The lottery organizers sell 100 tickets for each weekly drawing. One of these tickets is selected by a uniformly random process — that is, each ticket is equally likely to be chosen — and the lucky ticket holder wins a hundred million dollars. The other 99 ticket holders win nothing.

(Slightly subtle point: There are two probability spaces, depending on what strategy we use; but EX_1 *and* EX_2 *are the same in both.)*

We can use our gift in two ways: Either we buy two tickets in the same lottery, or we buy one ticket in each of two lotteries. Which is a better strategy? Let's try to analyze this by letting X_1 and X_2 be random variables that represent the amount we win on our first and second ticket. The expected value of X_1, in millions, is

$$EX_1 = \tfrac{99}{100}\cdot 0 + \tfrac{1}{100}\cdot 100 = 1,$$

and the same holds for X_2. Expected values are additive, so our average total winnings will be

$$E(X_1 + X_2) = EX_1 + EX_2 = 2 \text{ million dollars,}$$

regardless of which strategy we adopt.

Still, the two strategies seem different. Let's look beyond expected values and study the exact probability distribution of $X_1 + X_2$:

	winnings (millions)		
	0	100	200
same drawing	.9800	.0200	
different drawings	.9801	.0198	.0001

If we buy two tickets in the same lottery we have a 98% chance of winning nothing and a 2% chance of winning \$100 million. If we buy them in different lotteries we have a 98.01% chance of winning nothing, so this is slightly more likely than before; and we have a 0.01% chance of winning \$200 million, also slightly more likely than before; and our chances of winning \$100 million are now 1.98%. So the distribution of $X_1 + X_2$ in this second situation is slightly more spread out; the middle value, \$100 million, is slightly less likely, but the extreme values are slightly more likely.

It's this notion of the spread of a random variable that the variance is intended to capture. We measure the spread in terms of the squared deviation of the random variable from its mean. In case 1, the variance is therefore

$$.98(0M - 2M)^2 + .02(100M - 2M)^2 = 196M^2;$$

in case 2 it is

$$\begin{aligned} .9801(0M - 2M)^2 + .0198(100M - 2M)^2 + .0001(200M - 2M)^2 \\ = 198M^2. \end{aligned}$$

As we expected, the latter variance is slightly larger, because the distribution of case 2 is slightly more spread out.

When we work with variances, everything is squared, so the numbers can get pretty big. (The factor M^2 is one trillion, which is somewhat imposing even for high-stakes gamblers.) To convert the numbers back to the more meaningful original scale, we often take the square root of the variance. The resulting number is called the *standard deviation*, and it is usually denoted by the Greek letter σ:

Interesting: The variance of a dollar amount is expressed in units of square dollars.

$$\sigma = \sqrt{VX}. \tag{8.14}$$

The standard deviations of the random variables $X_1 + X_2$ in our two lottery strategies are $\sqrt{196M^2} = 14.00M$ and $\sqrt{198M^2} \approx 14.071247M$. In some sense the second alternative is about \$71,247 riskier.

How does the variance help us choose a strategy? It's not clear. The strategy with higher variance is a little riskier; but do we get the most for our money by taking more risks or by playing it safe? Suppose we had the chance to buy 100 tickets instead of only two. Then we could have a guaranteed victory in a single lottery (and the variance would be zero); or we could gamble on a hundred different lotteries, with a $.99^{100} \approx .366$ chance of winning nothing but also with a nonzero probability of winning up to \$10,000,000,000. To decide between these alternatives is beyond the scope of this book; all we can do here is explain how to do the calculations.

Another way to reduce risk might be to bribe the lottery officials. I guess that's where probability becomes indiscreet.

(N.B.: Opinions expressed in these margins do not necessarily represent the opinions of the management.)

In fact, there is a simpler way to calculate the variance, instead of using the definition (8.13). (We suspect that there must be something going on in the mathematics behind the scenes, because the variances in the lottery example magically came out to be integer multiples of M^2.) We have

$$\begin{aligned} E\big((X - EX)^2\big) &= E\big(X^2 - 2X(EX) + (EX)^2\big) \\ &= E(X^2) - 2(EX)(EX) + (EX)^2, \end{aligned}$$

since (EX) is a constant; hence

$$VX = E(X^2) - (EX)^2 . \qquad (8.15)$$

"The variance is the mean of the square minus the square of the mean."

For example, the mean of $(X_1 + X_2)^2$ comes to $.98(0M)^2 + .02(100M)^2 = 200M^2$ or to $.9801(0M)^2 + .0198(100M)^2 + .0001(200M)^2 = 202M^2$ in the lottery problem. Subtracting $4M^2$ (the square of the mean) gives the results we obtained the hard way.

There's an even easier formula yet, if we want to calculate $V(X+Y)$ when X and Y are independent: We have

$$\begin{aligned} E\big((X+Y)^2\big) &= E(X^2 + 2XY + Y^2) \\ &= E(X^2) + 2(EX)(EY) + E(Y^2) , \end{aligned}$$

since we know that $E(XY) = (EX)(EY)$ in the independent case. Therefore

$$\begin{aligned} V(X+Y) &= E\big((X+Y)^2\big) - (EX+EY)^2 \\ &= E(X^2) + 2(EX)(EY) + E(Y^2) \\ &\qquad - (EX)^2 - 2(EX)(EY) - (EY)^2 \\ &= E(X^2) - (EX)^2 + E(Y^2) - (EY)^2 \\ &= VX + VY . \end{aligned} \qquad (8.16)$$

"The variance of a sum of independent random variables is the sum of their variances." For example, the variance of the amount we can win with a single lottery ticket is

$$E(X_1^2) - (EX_1)^2 = .99(0M)^2 + .01(100M)^2 - (1M)^2 = 99M^2 .$$

Therefore the variance of the total winnings of two lottery tickets in two separate (independent) lotteries is $2\times 99M^2 = 198M^2$. And the corresponding variance for n independent lottery tickets is $n \times 99M^2$.

The variance of the dice-roll sum S drops out of this same formula, since $S = S_1 + S_2$ is the sum of two independent random variables. We have

$$VS_1 = \frac{1}{6}(1^2 + 2^2 + 3^2 + 4^2 + 5^2 + 6^2) - \left(\frac{7}{2}\right)^2 = \frac{35}{12}$$

when the dice are fair; hence $VS = \frac{35}{12} + \frac{35}{12} = \frac{35}{6}$. The loaded die has

$$VS_1 = \frac{1}{8}(2\cdot 1^2 + 2^2 + 3^2 + 4^2 + 5^2 + 2\cdot 6^2) - \left(\frac{7}{2}\right)^2 = \frac{45}{12} ;$$

hence $VS = \frac{45}{6} = 7.5$ when both dice are loaded. Notice that the loaded dice give S a larger variance, although S actually assumes its average value 7 more often than it would with fair dice. If our goal is to shoot lots of lucky 7's, the variance is not our best indicator of success.

OK, we have learned how to compute variances. But we haven't really seen a good reason why the variance is a natural thing to compute. Everybody does it, but why? The main reason is *Chebyshev's inequality* ([24′] and [50′]), which states that the variance has a significant property:

If he proved it in 1867, it's a classic '67 Chebyshev.

$$\Pr\bigl((X - EX)^2 \geqslant \alpha\bigr) \;\leqslant\; VX/\alpha\,, \qquad \text{for all } \alpha > 0. \tag{8.17}$$

(This is different from the summation inequalities of Chebyshev that we encountered in Chapter 2.) Very roughly, (8.17) tells us that a random variable X will rarely be far from its mean EX if its variance VX is small. The proof is amazingly simple. We have

$$\begin{aligned} VX &= \sum_{\omega\in\Omega} \bigl(X(\omega) - EX\bigr)^2 \Pr(\omega) \\ &\geqslant \sum_{\substack{\omega\in\Omega \\ (X(\omega)-EX)^2 \geqslant \alpha}} \bigl(X(\omega) - EX\bigr)^2 \Pr(\omega) \\ &\geqslant \sum_{\substack{\omega\in\Omega \\ (X(\omega)-EX)^2 \geqslant \alpha}} \alpha \Pr(\omega) \;=\; \alpha\cdot\Pr\bigl((X - EX)^2 \geqslant \alpha\bigr)\,; \end{aligned}$$

dividing by α finishes the proof.

If we write μ for the mean and σ for the standard deviation, and if we replace α by c^2VX in (8.17), the condition $(X - EX)^2 \geqslant c^2VX$ is the same as $(X - \mu)^2 \geqslant (c\sigma)^2$; hence (8.17) says that

$$\Pr\bigl(|X - \mu| \geqslant c\sigma\bigr) \;\leqslant\; 1/c^2\,. \tag{8.18}$$

Thus, X will lie within c standard deviations of its mean value except with probability at most $1/c^2$. A random variable will lie within 2σ of μ at least 75% of the time; it will lie between $\mu - 10\sigma$ and $\mu + 10\sigma$ at least 99% of the time. These are the cases $\alpha = 4VX$ and $\alpha = 100VX$ of Chebyshev's inequality.

If we roll a pair of fair dice n times, the total value of the n rolls will almost always be near $7n$, for large n. Here's why: The variance of n independent rolls is $\frac{35}{6}n$. A variance of $\frac{35}{6}n$ means a standard deviation of only

$$\sqrt{\tfrac{35}{6}n}\,.$$

So Chebyshev's inequality tells us that the sum will be between

$$7n - 10\sqrt{\tfrac{35}{6}n} \quad \text{and} \quad 7n + 10\sqrt{\tfrac{35}{6}n}$$

at least 99% of the time. For example, the total value of a million rolls will be between 6.976 million and 7.024 million, more than 99% of the time.

In general, let X be *any* random variable over a probability space Ω, having finite mean μ and finite standard deviation σ. Then we can consider the probability space Ω^n whose elementary events are n-tuples $(\omega_1, \omega_2, \ldots, \omega_n)$ with each $\omega_k \in \Omega$, and whose probabilities are

$$\Pr(\omega_1, \omega_2, \ldots, \omega_n) = \Pr(\omega_1)\Pr(\omega_2)\ldots\Pr(\omega_n).$$

If we now define random variables X_k by the formula

$$X_k(\omega_1, \omega_2, \ldots, \omega_n) = X(\omega_k),$$

the quantity

$$X_1 + X_2 + \cdots + X_n$$

is a sum of n independent random variables, which corresponds to taking n independent "samples" of X on Ω and adding them together. The mean of $X_1+X_2+\cdots+X_n$ is $n\mu$, and the standard deviation is $\sqrt{n}\,\sigma$; hence the average of the n samples,

$$\frac{1}{n}(X_1 + X_2 + \cdots + X_n),$$

will lie between $\mu - 10\sigma/\sqrt{n}$ and $\mu + 10\sigma/\sqrt{n}$ at least 99% of the time. In other words, if we choose a large enough value of n, the average of n independent samples will almost always be very near the expected value EX. (An even stronger theorem called the Strong Law of Large Numbers is proved in textbooks of probability theory; but the simple consequence of Chebyshev's inequality that we have just derived is enough for our purposes.)

Sometimes we don't know the characteristics of a probability space, and we want to estimate the mean of a random variable X by sampling its value repeatedly. (For example, we might want to know the average temperature at noon on a January day in San Francisco; or we may wish to know the mean life expectancy of insurance agents.) If we have obtained independent empirical observations X_1, X_2, $\ldots$, X_n, we can guess that the true mean is approximately

$$\hat{E}X = \frac{X_1 + X_2 + \cdots + X_n}{n}. \tag{8.19}$$

And we can also make an estimate of the variance, using the formula

$$\hat{V}X = \frac{X_1^2 + X_2^2 + \cdots + X_n^2}{n-1} - \frac{(X_1 + X_2 + \cdots + X_n)^2}{n(n-1)}. \tag{8.20}$$

The $(n-1)$'s in this formula look like typographic errors; it seems they should be n's, as in (8.19), because the true variance VX is defined by expected values in (8.15). Yet we get a better estimate with $n-1$ instead of n here, because definition (8.20) implies that

$$E(\hat{V}X) = VX. \tag{8.21}$$

Here's why:

$$\begin{aligned}
E(\hat{V}X) &= \frac{1}{n-1} E\Bigl(\sum_{k=1}^{n} X_k^2 - \frac{1}{n}\sum_{j=1}^{n}\sum_{k=1}^{n} X_j X_k\Bigr) \\
&= \frac{1}{n-1}\Bigl(\sum_{k=1}^{n} E(X_k^2) - \frac{1}{n}\sum_{j=1}^{n}\sum_{k=1}^{n} E(X_j X_k)\Bigr) \\
&= \frac{1}{n-1}\Bigl(\sum_{k=1}^{n} E(X^2) - \frac{1}{n}\sum_{j=1}^{n}\sum_{k=1}^{n} \bigl(E(X)^2[j \neq k] + E(X^2)[j=k]\bigr)\Bigr) \\
&= \frac{1}{n-1}\Bigl(nE(X^2) - \frac{1}{n}\bigl(nE(X^2) + n(n-1)E(X)^2\bigr)\Bigr) \\
&= E(X^2) - E(X)^2 = VX.
\end{aligned}$$

(This derivation uses the independence of the observations when it replaces $E(X_j X_k)$ by $(EX)^2[j \neq k] + E(X^2)[j=k]$.)

In practice, experimental results about a random variable X are usually obtained by calculating a sample mean $\hat{\mu} = \hat{E}X$ and a sample standard deviation $\hat{\sigma} = \sqrt{\hat{V}X}$, and presenting the answer in the form '$\hat{\mu} \pm \hat{\sigma}/\sqrt{n}$'. For example, here are ten rolls of two supposedly fair dice:

The sample mean of the spot sum S is

$$\hat{\mu} = (7 + 11 + 8 + 5 + 4 + 6 + 10 + 8 + 8 + 7)/10 = 7.4;$$

the sample variance is

$$(7^2 + 11^2 + 8^2 + 5^2 + 4^2 + 6^2 + 10^2 + 8^2 + 8^2 + 7^2 - 10\hat{\mu}^2)/9 \approx 2.1^2.$$

We estimate the average spot sum of these dice to be $7.4 \pm 2.1/\sqrt{10} = 7.4 \pm 0.7$, on the basis of these experiments.

Let's work one more example of means and variances, in order to show how they can be calculated theoretically instead of empirically. One of the questions we considered in Chapter 5 was the "football victory problem," where n hats are thrown into the air and the result is a random permutation of hats. We showed in equation (5.51) that there's a probability of $n¡/n! \approx 1/e$ that nobody gets the right hat back. We also derived the formula

$$P(n,k) \;=\; \frac{1}{n!}\binom{n}{k}(n-k)¡ \;=\; \frac{1}{k!}\frac{(n-k)¡}{(n-k)!} \qquad (8.22)$$

for the probability that exactly k people end up with their own hats.

Restating these results in the formalism just learned, we can consider the probability space Π_n of all $n!$ permutations π of $\{1, 2, \ldots, n\}$, where $\Pr(\pi) = 1/n!$ for all $\pi \in \Pi_n$. The random variable

Not to be confused with a Fibonacci number.

$$F_n(\pi) \;=\; \text{number of “fixed points” of } \pi\,, \qquad \text{for } \pi \in \Pi_n,$$

measures the number of correct hat-falls in the football victory problem. Equation (8.22) gives $\Pr(F_n = k)$, but let's pretend that we don't know any such formula; we merely want to study the average value of F_n, and its standard deviation.

The average value is, in fact, extremely easy to calculate, avoiding all the complexities of Chapter 5. We simply observe that

$$\begin{aligned} F_n(\pi) \;&=\; F_{n,1}(\pi) + F_{n,2}(\pi) + \cdots + F_{n,n}(\pi)\,,\\ F_{n,k}(\pi) \;&=\; (\text{position } k \text{ of } \pi \text{ is a fixed point})\,, \qquad \text{for } \pi \in \Pi_n. \end{aligned}$$

Hence

$$EF_n \;=\; EF_{n,1} + EF_{n,2} + \cdots + EF_{n,n}\,.$$

And the expected value of $F_{n,k}$ is simply the probability that $F_{n,k} = 1$, which is $1/n$ because exactly $(n-1)!$ of the $n!$ permutations $\pi = \pi_1\pi_2\ldots\pi_n \in \Pi_n$ have $\pi_k = k$. Therefore

$$EF_n \;=\; n/n \;=\; 1\,, \qquad \text{for } n > 0. \qquad (8.23)$$

One the average.

On the average, one hat will be in its correct place. "A random permutation has one fixed point, on the average."

Now what's the standard deviation? This question is more difficult, because the $F_{n,k}$'s are not independent of each other. But we can calculate the variance by analyzing the mutual dependencies among them:

$$\begin{aligned}
E(F_n^2) &= E\Bigl(\Bigl(\sum_{k=1}^{n} F_{n,k}\Bigr)^2\Bigr) = E\Bigl(\sum_{j=1}^{n}\sum_{k=1}^{n} F_{n,j}\,F_{n,k}\Bigr)\\
&= \sum_{j=1}^{n}\sum_{k=1}^{n} E(F_{n,j}\,F_{n,k}) = \sum_{1\leqslant k\leqslant n} E(F_{n,k}^2) + 2\sum_{1\leqslant j<k\leqslant n} E(F_{n,j}\,F_{n,k})\,.
\end{aligned}$$

(We used a similar trick when we derived (2.33) in Chapter 2.) Now $F_{n,k}^2 = F_{n,k}$, since $F_{n,k}$ is either 0 or 1; hence $E(F_{n,k}^2) = EF_{n,k} = 1/n$ as before. And if $j < k$ we have $E(F_{n,j}\,F_{n,k}) = \Pr(\pi \text{ has both } j \text{ and } k \text{ as fixed points}) = (n-2)!/n! = 1/n(n-1)$. Therefore

$$E(F_n^2) = \frac{n}{n} + \binom{n}{2}\frac{2}{n(n-1)} = 2, \qquad \text{for } n \geqslant 2. \tag{8.24}$$

(As a check when $n = 3$, we have $\frac{2}{6}0^2 + \frac{3}{6}1^2 + \frac{0}{6}2^2 + \frac{1}{6}3^2 = 2$.) The variance is $E(F_n^2) - (EF_n)^2 = 1$, so the standard deviation (like the mean) is 1. "A random permutation of $n \geqslant 2$ elements has 1 ± 1 fixed points."

8.3 PROBABILITY GENERATING FUNCTIONS

If X is a random variable that takes only nonnegative integer values, we can capture its probability distribution nicely by using the techniques of Chapter 7. The *probability generating function* or pgf of X is

$$G_X(z) = \sum_{k\geqslant 0} \Pr(X=k)\,z^k\,. \tag{8.25}$$

This power series in z contains all the information about the random variable X. We can also express it in two other ways:

$$G_X(z) = \sum_{\omega\in\Omega} \Pr(\omega)\,z^{X(\omega)} = E(z^X)\,. \tag{8.26}$$

The coefficients of $G_X(z)$ are nonnegative, and they sum to 1; the latter condition can be written

$$G_X(1) = 1\,. \tag{8.27}$$

Conversely, any power series $G(z)$ with nonnegative coefficients and with $G(1) = 1$ is the pgf of some random variable.

The nicest thing about pgf's is that they usually simplify the computation of means and variances. For example, the mean is easily expressed:

$$\begin{aligned} EX &= \sum_{k\geq 0} k\cdot \Pr(X=k) \\ &= \sum_{k\geq 0} \Pr(X=k)\cdot kz^{k-1}\big|_{z=1} \\ &= G'_X(1)\,. \end{aligned} \tag{8.28}$$

We simply differentiate the pgf with respect to z and set $z = 1$.

The variance is only slightly more complicated:

$$\begin{aligned} E(X^2) &= \sum_{k\geq 0} k^2\cdot \Pr(X=k) \\ &= \sum_{k\geq 0} \Pr(X=k)\cdot \big(k(k-1)z^{k-2} + kz^{k-1}\big)\,\big|_{z=1} \;=\; G''_X(1) + G'_X(1)\,. \end{aligned}$$

Therefore

$$VX = G''_X(1) + G'_X(1) - G'_X(1)^2\,. \tag{8.29}$$

Equations (8.28) and (8.29) tell us that we can compute the mean and variance if we can compute the values of two derivatives, $G'_X(1)$ and $G''_X(1)$. We don't have to know a closed form for the probabilities; we don't even have to know a closed form for $G_X(z)$ itself.

It is convenient to write

$$\mathrm{Mean}(G) = G'(1)\,, \tag{8.30}$$

$$\mathrm{Var}(G) = G''(1) + G'(1) - G'(1)^2\,, \tag{8.31}$$

when G is any function, since we frequently want to compute these combinations of derivatives.

The second-nicest thing about pgf's is that they are comparatively simple functions of z, in many important cases. For example, let's look at the *uniform distribution* of order n, in which the random variable takes on each of the values $\{0, 1, \ldots, n-1\}$ with probability $1/n$. The pgf in this case is

$$U_n(z) = \frac{1}{n}(1 + z + \cdots + z^{n-1}) = \frac{1}{n}\frac{1-z^n}{1-z}\,, \qquad \text{for } n \geq 1. \tag{8.32}$$

We have a closed form for $U_n(z)$ because this is a geometric series.

But this closed form proves to be somewhat embarrassing: When we plug in $z = 1$ (the value of z that's most critical for the pgf), we get the undefined ratio $0/0$, even though $U_n(z)$ is a polynomial that is perfectly well defined at any value of z. The value $U_n(1) = 1$ is obvious from the non-closed form

$(1+z+\cdots+z^{n-1})/n$, yet it seems that we must resort to L'Hospital's rule to find $\lim_{z\to 1} U_n(z)$ if we want to determine $U_n(1)$ from the closed form. The determination of $U_n'(1)$ by L'Hospital's rule will be even harder, because there will be a factor of $(z-1)^2$ in the denominator; $U_n''(1)$ will be harder still.

Luckily there's a nice way out of this dilemma. If $G(z) = \sum_{n\ge 0} g_n z^n$ is any power series that converges for at least one value of z with $|z| > 1$, the power series $G'(z) = \sum_{n\ge 0} n g_n z^{n-1}$ will also have this property, and so will $G''(z)$, $G'''(z)$, etc. Therefore by Taylor's theorem we can write

$$G(1+t) \;=\; G(1) + \frac{G'(1)}{1!}t + \frac{G''(1)}{2!}t^2 + \frac{G'''(1)}{3!}t^3 + \cdots\,; \tag{8.33}$$

all derivatives of $G(z)$ at $z = 1$ will appear as coefficients, when $G(1+t)$ is expanded in powers of t.

For example, the derivatives of the uniform pgf $U_n(z)$ are easily found in this way:

$$\begin{aligned} U_n(1+t) \;&=\; \frac{1}{n}\frac{(1+t)^n - 1}{t} \\ &=\; \frac{1}{n}\binom{n}{1} + \frac{1}{n}\binom{n}{2}t + \frac{1}{n}\binom{n}{3}t^2 + \cdots + +\frac{1}{n}\binom{n}{n}t^{n-1}. \end{aligned}$$

Comparing this to (8.33) gives

$$U_n(1) \;=\; 1; \quad U_n'(1) \;=\; \frac{n-1}{2}; \quad U_n''(1) \;=\; \frac{(n-1)(n-2)}{3}; \tag{8.34}$$

and in general $U_n^{(m)}(1) = (n-1)^{\underline{m}}/(m+1)$, although we need only the cases $m = 1$ and $m = 2$ to compute the mean and the variance. The mean of the uniform distribution is

$$U_n'(1) \;=\; \frac{n-1}{2}, \tag{8.35}$$

and the variance is

$$\begin{aligned} U_n''(1) + U_n'(1) - U_n'(1)^2 \;&=\; 4\,\frac{(n-1)(n-2)}{12} + 6\,\frac{(n-1)}{12} - 3\,\frac{(n-1)^2}{12} \\ &=\; \frac{n^2-1}{12}. \end{aligned} \tag{8.36}$$

The third-nicest thing about pgf's is that the product of pgf's corresponds to the sum of independent random variables. We learned in Chapters 5 and 7 that the product of generating functions corresponds to the convolution of sequences; but it's even more important in applications to know that the convolution of probabilities corresponds to the sum of independent random variables. Indeed, if X and Y are random variables that take on nothing but integer values, the probability that $X+Y=n$ is

$$\Pr(X+Y=n) \;=\; \sum_k \Pr(X=k \text{ and } Y=n-k)\,.$$

If X and Y are independent, we now have

$$\Pr(X+Y=n) \;=\; \sum_k \Pr(X=k)\,\Pr(Y=n-k)\,,$$

a convolution. Therefore — and this is the punch line —

$$G_{X+Y}(z) \;=\; G_X(z)\,G_Y(z)\,, \qquad \text{if X and Y are independent.} \tag{8.37}$$

Earlier this chapter we observed that $V(X+Y) = VX + VY$ when X and Y are independent. Let $F(z)$ and $G(z)$ be the pgf's for X and Y, and let $H(z)$ be the pgf for $X+Y$. Then

$$H(z) \;=\; F(z)G(z)\,,$$

and our formulas (8.28) through (8.31) for mean and variance tell us that we must have

$$\text{Mean}(H) \;=\; \text{Mean}(F) + \text{Mean}(G)\,; \tag{8.38}$$

$$\text{Var}(H) \;=\; \text{Var}(F) + \text{Var}(G)\,. \tag{8.39}$$

These formulas, which are properties of the derivatives $\text{Mean}(H) = H'(1)$ and $\text{Var}(H) = H''(1) + H'(1) - H'(1)^2$, aren't valid for arbitrary function products $H(z) = F(z)G(z)$; we have

$$H'(z) \;=\; F'(z)G(z) + F(z)G'(z)\,,$$

$$H''(z) \;=\; F''(z)G(z) + 2F'(z)G'(z) + F(z)G''(z)\,.$$

But if we set $z=1$, we can see that (8.38) and (8.39) will be valid in general provided only that

$$F(1) \;=\; G(1) \;=\; 1 \tag{8.40}$$

and that the derivatives exist. The "probabilities" don't have to be in $[0\,..\,1]$ for these formulas to hold. We can normalize the functions $F(z)$ and $G(z)$ by dividing through by $F(1)$ and $G(1)$ in order to make this condition valid, whenever $F(1)$ and $G(1)$ are nonzero.

I'll graduate magna cum ulant.

Mean and variance aren't the whole story. They are merely two of an infinite series of so-called *cumulant* statistics introduced by the Danish astronomer Thorvald Nicolai Thiele [288] in 1903. The first two cumulants

κ_1 and κ_2 of a random variable are what we have called the mean and the variance; there also are higher-order cumulants that express more subtle properties of a distribution. The general formula

$$\ln G(e^t) \;=\; \frac{\kappa_1}{1!}t + \frac{\kappa_2}{2!}t^2 + \frac{\kappa_3}{3!}t^3 + \frac{\kappa_4}{4!}t^4 + \cdots \qquad (8.41)$$

defines the cumulants of all orders, when $G(z)$ is the pgf of a random variable.

Let's look at cumulants more closely. If $G(z)$ is the pgf for X, we have

$$\begin{aligned} G(e^t) \;&=\; \sum_{k\ge 0} \Pr(X=k)e^{kt} \;=\; \sum_{k,m\ge 0} \Pr(X=k)\frac{k^m t^m}{m!} \\ &=\; 1 + \frac{\mu_1}{1!}t + \frac{\mu_2}{2!}t^2 + \frac{\mu_3}{3!}t^3 + \cdots, \end{aligned} \qquad (8.42)$$

where

$$\mu_m \;=\; \sum_{k\ge 0} k^m \Pr(X=k) \;=\; E(X^m). \qquad (8.43)$$

This quantity μ_m is called the "mth moment" of X. We can take exponentials on both sides of (8.41), obtaining another formula for $G(e^t)$:

$$\begin{aligned} G(e^t) \;&=\; 1 + \frac{(\kappa_1 t + \frac{1}{2}\kappa_2 t^2 + \cdots)}{1!} + \frac{(\kappa_1 t + \frac{1}{2}\kappa_2 t^2 + \cdots)^2}{2!} + \cdots \\ &=\; 1 + \kappa_1 t + \tfrac{1}{2}(\kappa_2 + \kappa_1^2)t^2 + \cdots. \end{aligned}$$

Equating coefficients of powers of t leads to a series of formulas

$$\kappa_1 \;=\; \mu_1\,, \qquad (8.44)$$

$$\kappa_2 \;=\; \mu_2 - \mu_1^2\,, \qquad (8.45)$$

$$\kappa_3 \;=\; \mu_3 - 3\mu_1\mu_2 + 2\mu_1^3\,, \qquad (8.46)$$

$$\kappa_4 \;=\; \mu_4 - 4\mu_1\mu_3 + 12\mu_1^2\mu_2 - 3\mu_2^2 - 6\mu_1^4\,, \qquad (8.47)$$

$$\begin{aligned} \kappa_5 \;=\; \mu_5 - 5\mu_1\mu_4 + 20\mu_1^2\mu_3 - 10\mu_2\mu_3 \\ + 30\mu_1\mu_2^2 - 60\mu_1^3\mu_2 + 24\mu_1^5\,, \end{aligned} \qquad (8.48)$$

$$\vdots$$

defining the cumulants in terms of the moments. Notice that κ_2 is indeed the variance, $E(X^2) - (EX)^2$, as claimed.

Equation (8.41) makes it clear that the cumulants defined by the product $F(z)G(z)$ of two pgf's will be the sums of the corresponding cumulants of $F(z)$ and $G(z)$, because logarithms of products are sums. Therefore all cumulants of the sum of independent random variables are additive, just as the mean and variance are. This property makes cumulants more important than moments.

"For these higher half-invariants we shall propose no special names."
— T. N. Thiele [288]

If we take a slightly different tack, writing

$$G(1+t) \;=\; 1+\frac{\alpha_1}{1!}t+\frac{\alpha_2}{2!}t^2+\frac{\alpha_3}{3!}t^3+\cdots\,,$$

equation (8.33) tells us that the α's are the "factorial moments"

$$\begin{aligned}\alpha_m &= G^{(m)}(1)\\ &= \sum_{k\geqslant 0}\Pr(X=k)k^{\underline{m}}z^{k-m}\Big|_{z=1}\\ &= \sum_{k\geqslant 0}k^{\underline{m}}\Pr(X=k)\\ &= E(X^{\underline{m}})\,. \end{aligned}\qquad(8.49)$$

It follows that

$$\begin{aligned}G(e^t) &= 1+\frac{\alpha_1}{1!}(e^t-1)+\frac{\alpha_2}{2!}(e^t-1)^2+\cdots\\ &= 1+\frac{\alpha_1}{1!}(t+\tfrac{1}{2}t^2+\cdots)+\frac{\alpha_2}{2!}(t^2+t^3+\cdots)+\cdots\\ &= 1+\alpha_1 t+\tfrac{1}{2}(\alpha_2+\alpha_1)t^2+\cdots\,,\end{aligned}$$

and we can express the cumulants in terms of the derivatives $G^{(m)}(1)$:

$$\kappa_1 = \alpha_1\,, \qquad(8.50)$$

$$\kappa_2 = \alpha_2+\alpha_1-\alpha_1^2\,, \qquad(8.51)$$

$$\kappa_3 = \alpha_3+3\alpha_2+\alpha_1-3\alpha_2\alpha_1-3\alpha_1^2+2\alpha_1^3\,, \qquad(8.52)$$

$$\vdots$$

This sequence of formulas yields "additive" identities that extend (8.38) and (8.39) to all the cumulants.

Let's get back down to earth and apply these ideas to simple examples. The simplest case of a random variable is a "random constant," where X has a certain fixed value x with probability 1. In this case $G_X(z) = z^x$, and $\ln G_X(e^t) = xt$; hence the mean is x and all other cumulants are zero. It follows that the operation of multiplying any pgf by z^x increases the mean by x but leaves the variance and all other cumulants unchanged.

How do probability generating functions apply to dice? The distribution of spots on one fair die has the pgf

$$G(z) \;=\; \frac{z+z^2+z^3+z^4+z^5+z^6}{6} \;=\; zU_6(z)\,,$$

where U_6 is the pgf for the uniform distribution of order 6. The factor 'z' adds 1 to the mean, so the mean is 3.5 instead of $\frac{n-1}{2} = 2.5$ as given in (8.35); but an extra 'z' does not affect the variance (8.36), which equals $\frac{35}{12}$.

The pgf for total spots on two independent dice is the square of the pgf for spots on one die,

$$\begin{aligned} G_S(z) &= \frac{z^2+2z^3+3z^4+4z^5+5z^6+6z^7+5z^8+4z^9+3z^{10}+2z^{11}+z^{12}}{36} \\ &= z^2 U_6(z)^2 . \end{aligned}$$

If we roll a pair of fair dice n times, the probability that we get a total of k spots overall is, similarly,

$$\begin{aligned} [z^k]\, G_S(z)^n &= [z^k]\, z^{2n} U_6(z)^{2n} \\ &= [z^{k-2n}]\, U_6(z)^{2n} . \end{aligned}$$

In the hats-off-to-football-victory problem considered earlier, otherwise known as the problem of enumerating the fixed points of a random permutation, we know from (5.49) that the pgf is

Hat distribution is a different kind of uniform distribution.

$$F_n(z) = \sum_{0\le k\le n} \frac{(n-k)¡}{(n-k)!}\frac{z^k}{k!}, \qquad \text{for } n \ge 0. \tag{8.53}$$

Therefore

$$\begin{aligned} F_n'(z) &= \sum_{1\le k\le n} \frac{(n-k)¡}{(n-k)!}\frac{z^{k-1}}{(k-1)!} \\ &= \sum_{0\le k\le n-1} \frac{(n-1-k)¡}{(n-1-k)!}\frac{z^k}{k!} \\ &= F_{n-1}(z) . \end{aligned}$$

Without knowing the details of the coefficients, we can conclude from this recurrence $F_n'(z) = F_{n-1}(z)$ that $F_n^{(m)}(z) = F_{n-m}(z)$; hence

$$F_n^{(m)}(1) = F_{n-m}(1) = [n \ge m] . \tag{8.54}$$

This formula makes it easy to calculate the mean and variance; we find as before (but more quickly) that they are both equal to 1 when $n \ge 2$.

In fact, we can now show that the mth cumulant κ_m of this random variable is equal to 1 whenever $n \ge m$. For the mth cumulant depends only on $F_n'(1)$, $F_n''(1)$, ..., $F_n^{(m)}(1)$, and these are all equal to 1; hence we obtain the same answer for the mth cumulant as we do when we replace $F_n(z)$ by the limiting pgf

$$F_\infty(z) = e^{z-1}, \qquad (8.55)$$

which has $F_\infty^{(m)}(1) = 1$ for derivatives of all orders. The cumulants of F_∞ are identically equal to 1, because

$$\ln F_\infty(e^t) = \ln e^{e^t-1} = e^t - 1 = \frac{t}{1!} + \frac{t^2}{2!} + \frac{t^3}{3!} + \cdots.$$

8.4 FLIPPING COINS

Con artists know that $p \approx 0.1$ when you spin a newly minted U.S. penny on a smooth table. (The weight distribution makes Lincoln's head fall downward.)

Now let's turn to processes that have just two outcomes. If we flip a coin, there's probability p that it comes up heads and probability q that it comes up tails, where

$$p + q = 1.$$

(We assume that the coin doesn't come to rest on its edge, or fall into a hole, etc.) Throughout this section, the numbers p and q will always sum to 1. If the coin is *fair*, we have $p = q = \frac{1}{2}$; otherwise the coin is said to be *biased*.

The probability generating function for the number of heads after one toss of a coin is

$$H(z) = q + pz. \qquad (8.56)$$

If we toss the coin n times, always assuming that different coin tosses are independent, the number of heads is generated by

$$H(z)^n = (q + pz)^n = \sum_{k\geqslant 0}\binom{n}{k}p^k q^{n-k} z^k, \qquad (8.57)$$

according to the binomial theorem. Thus, the chance that we obtain exactly k heads in n tosses is $\binom{n}{k}p^k q^{n-k}$. This sequence of probabilities is called the *binomial distribution*.

Suppose we toss a coin repeatedly until heads first turns up. What is the probability that exactly k tosses will be required? We have $k = 1$ with probability p (since this is the probability of heads on the first flip); we have $k = 2$ with probability qp (since this is the probability of tails first, then heads); and for general k the probability is $q^{k-1}p$. So the generating function is

$$pz + qpz^2 + q^2pz^3 + \cdots = \frac{pz}{1-qz}. \qquad (8.58)$$

Repeating the process until n heads are obtained gives the pgf

$$\left(\frac{pz}{1-qz}\right)^n = p^n z^n \sum_k \binom{n+k-1}{k} (qz)^k$$
$$= \sum_k \binom{k-1}{k-n} p^n q^{k-n} z^k . \qquad (8.59)$$

This, incidentally, is z^n times

$$\left(\frac{p}{1-qz}\right)^n = \sum_k \binom{n+k-1}{k} p^n q^k z^k , \qquad (8.60)$$

the generating function for the *negative binomial distribution.*

The probability space in example (8.59), where we flip a coin until n heads have appeared, is different from the probability spaces we've seen earlier in this chapter, because it contains infinitely many elements. Each element is a finite sequence of heads and/or tails, containing precisely n heads in all, and ending with heads; the probability of such a sequence is $p^n q^{k-n}$, where $k-n$ is the number of tails. Thus, for example, if $n=3$ and if we write H for heads and T for tails, the sequence THTTTHH is an element of the probability space, and its probability is $qpqqqpp = p^3q^4$.

Heads I win, tails you lose.

No? OK; tails you lose, heads I win.

No? Well, then, heads you lose, tails I win.

Let X be a random variable with the binomial distribution (8.57), and let Y be a random variable with the negative binomial distribution (8.60). These distributions depend on n and p. The mean of X is $nH'(1) = np$, since its pgf is $H(z)^n$; the variance is

$$n\big(H''(1) + H'(1) - H'(1)^2\big) = n(0 + p - p^2) = npq . \qquad (8.61)$$

Thus the standard deviation is $\sqrt{npq}$: If we toss a coin n times, we expect to get heads about $np \pm \sqrt{npq}$ times. The mean and variance of Y can be found in a similar way: If we let

$$G(z) = \frac{p}{1-qz} ,$$

we have

$$G'(z) = \frac{pq}{(1-qz)^2} ,$$
$$G''(z) = \frac{2pq^2}{(1-qz)^3} ;$$

hence $G'(1) = pq/p^2 = q/p$ and $G''(1) = 2pq^2/p^3 = 2q^2/p^2$. It follows that the mean of Y is nq/p and the variance is nq/p^2.

A simpler way to derive the mean and variance of Y is to use the reciprocal generating function

$$F(z) \;=\; \frac{1-qz}{p} \;=\; \frac{1}{p}-\frac{q}{p}z\,, \tag{8.62}$$

and to write

$$G(z)^n \;=\; F(z)^{-n}\,. \tag{8.63}$$

The probability is negative that I'm getting younger.

Oh? Then it's >1 that you're getting older, or staying the same.

This polynomial $F(z)$ is not a probability generating function, because it has a negative coefficient. But it does satisfy the crucial condition $F(1) = 1$. Thus $F(z)$ is formally a binomial that corresponds to a coin for which we get heads with "probability" equal to $-q/p$; and $G(z)$ is formally equivalent to flipping such a coin -1 times(!). The negative binomial distribution with parameters (n, p) can therefore be regarded as the ordinary binomial distribution with parameters $(n', p') = (-n, -q/p)$. Proceeding formally, the mean must be $n'p' = (-n)(-q/p) = nq/p$, and the variance must be $n'p'q' = (-n)(-q/p)(1 + q/p) = nq/p^2$. This formal derivation involving negative probabilities is valid, because our derivation for ordinary binomials was based on identities between formal power series in which the assumption $0 \leqslant p \leqslant 1$ was never used.

Let's move on to another example: How many times do we have to flip a coin until we get heads twice in a row? The probability space now consists of all sequences of H's and T's that end with HH but have no consecutive H's until the final position:

$$\Omega \;=\; \{\mathtt{HH}, \mathtt{THH}, \mathtt{TTHH}, \mathtt{HTHH}, \mathtt{TTTHH}, \mathtt{THTHH}, \mathtt{HTTHH}, \ldots\}\,.$$

The probability of any given sequence is obtained by replacing H by p and T by q; for example, the sequence THTHH will occur with probability

$$\Pr(\mathtt{THTHH}) \;=\; qpqpp \;=\; p^3q^2\,.$$

We can now play with generating functions as we did at the beginning of Chapter 7, letting S be the infinite sum

$$S \;=\; \mathtt{HH} + \mathtt{THH} + \mathtt{TTHH} + \mathtt{HTHH} + \mathtt{TTTHH} + \mathtt{THTHH} + \mathtt{HTTHH} + \cdots$$

of all the elements of Ω. If we replace each H by pz and each T by qz, we get the probability generating function for the number of flips needed until two consecutive heads turn up.

There's a curious relation between S and the sum of domino tilings

$$\mathsf{T} \;=\; | + \text{▯} + \text{▯▯} + \text{⊟} + \text{▯▯▯} + \text{▯⊟} + \text{⊟▯} + \cdots$$

in equation (7.1). Indeed, we obtain S from T if we replace each ▯ by $\mathtt{T}$ and each ⊟ by $\mathtt{HT}$, then tack on an $\mathtt{HH}$ at the end. This correspondence is easy to prove because each element of Ω has the form $(\mathtt{T}+\mathtt{HT})^n\mathtt{HH}$ for some $n \geqslant 0$, and each term of T has the form $(\text{▯}+\text{⊟})^n$. Therefore by (7.4) we have

$$S \;=\; (1-\mathtt{T}-\mathtt{HT})^{-1}\mathtt{HH}\,,$$

and the probability generating function for our problem is

$$\begin{aligned} G(z) \;&=\; \bigl(1-qz-(pz)(qz)\bigr)^{-1}(pz)^2 \\ &=\; \frac{p^2z^2}{1-qz-pqz^2}\,. \end{aligned} \tag{8.64}$$

Our experience with the negative binomial distribution gives us a clue that we can most easily calculate the mean and variance of (8.64) by writing

$$G(z) \;=\; \frac{z^2}{F(z)}\,,$$

where

$$F(z) \;=\; \frac{1-qz-pqz^2}{p^2}\,,$$

and by calculating the "mean" and "variance" of this pseudo-pgf $F(z)$. (Once again we've introduced a function with $F(1)=1$.) We have

$$\begin{aligned} F'(1) \;&=\; (-q-2pq)/p^2 \;=\; 2-p^{-1}-p^{-2}\,; \\ F''(1) \;&=\; -2pq/p^2 \;=\; 2-2p^{-1}\,. \end{aligned}$$

Therefore, since $z^2 = F(z)G(z)$, $\mathrm{Mean}(z^2)=2$, and $\mathrm{Var}(z^2)=0$, the mean and variance of distribution $G(z)$ are

$$\mathrm{Mean}(G) \;=\; 2-\mathrm{Mean}(F) \;=\; p^{-2}+p^{-1}\,; \tag{8.65}$$

$$\mathrm{Var}(G) \;=\; -\mathrm{Var}(F) \;=\; p^{-4}+2p^{-3}-2p^{-2}-p^{-1}\,. \tag{8.66}$$

When $p=\frac{1}{2}$ the mean and variance are 6 and 22, respectively. (Exercise 4 discusses the calculation of means and variances by subtraction.)

Now let's try a more intricate experiment: We will flip coins until the pattern THTTH is first obtained. The sum of winning positions is now

$$S = \text{THTTH} + \text{HTHTTH} + \text{TTHTTH} + \text{HHTHTTH} + \text{HTTHTTH} + \text{THTHTTH} + \text{TTTHTTH} + \cdots;$$

this sum is more difficult to describe than the previous one. If we go back to the method by which we solved the domino problems in Chapter 7, we can obtain a formula for S by considering it as a "finite state language" defined by the following "automaton":

"'You really are an automaton—a calculating machine,' I cried. 'There is something positively inhuman in you at times.'"
—J. H. Watson [70]

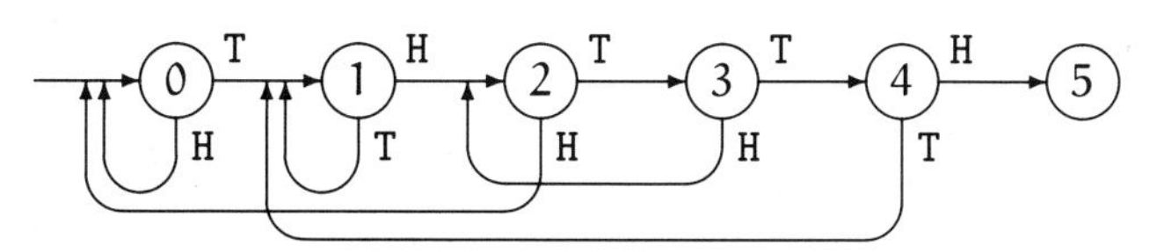

The elementary events in the probability space are the sequences of H's and T's that lead from state 0 to state 5. Suppose, for example, that we have just seen THT; then we are in state 3. Flipping tails now takes us to state 4; flipping heads in state 3 would take us to state 2 (not all the way back to state 0, since the TH we've just seen may be followed by TTH).

In this formulation, we can let S_k be the sum of all sequences of H's and T's that lead to state k; it follows that

$$\begin{aligned} S_0 &= 1 + S_0\,H + S_2\,H\,, \\ S_1 &= S_0\,T + S_1\,T + S_4\,T\,, \\ S_2 &= S_1\,H + S_3\,H\,, \\ S_3 &= S_2\,T\,, \\ S_4 &= S_3\,T\,, \\ S_5 &= S_4\,H\,. \end{aligned}$$

Now the sum S in our problem is S_5; we can obtain it by solving these six equations in the six unknowns $S_0, S_1, \ldots, S_5$. Replacing H by pz and T by qz gives generating functions where the coefficient of z^n in S_k is the probability that we are in state k after n flips.

In the same way, any diagram of transitions between states, where the transition from state j to state k occurs with given probability $p_{j,k}$, leads to a set of simultaneous linear equations whose solutions are generating functions for the state probabilities after n transitions have occurred. Systems of this kind are called *Markov processes*, and the theory of their behavior is intimately related to the theory of linear equations.

But the coin-flipping problem can be solved in a much simpler way, without the complexities of the general finite-state approach. Instead of six equations in six unknowns $S_0, S_1, \ldots, S_5$, we can characterize S with only two equations in two unknowns. The trick is to consider the auxiliary sum $N = S_0 + S_1 + S_2 + S_3 + S_4$ of all flip sequences that don't contain any occurrences of the given pattern THTTH:

$$N = 1 + \mathrm{H} + \mathrm{T} + \mathrm{HH} + \cdots + \mathrm{THTHT} + \mathrm{THTTT} + \cdots .$$

We have

$$1 + N(\mathrm{H} + \mathrm{T}) = N + S, \tag{8.67}$$

because every term on the left either ends with THTTH (and belongs to S) or doesn't (and belongs to N); conversely, every term on the right is either empty or belongs to $N\,\mathrm{H}$ or $N\,\mathrm{T}$. And we also have the important additional equation

$$N\,\mathrm{THTTH} = S + S\,\mathrm{TTH}, \tag{8.68}$$

because every term on the left completes a term of S after either the first H or the second H, and because every term on the right belongs to the left.

The solution to these two simultaneous equations is easily obtained: We have $N = (1 - S)(1 - \mathrm{H} - \mathrm{T})^{-1}$ from (8.67), hence

$$(1 - S)(1 - \mathrm{T} - \mathrm{H})^{-1}\,\mathrm{THTTH} = S(1 + \mathrm{TTH}).$$

As before, we get the probability generating function $G(z)$ for the number of flips if we replace H by pz and T by qz. A bit of simplification occurs since $p + q = 1$, and we find

$$\frac{(1 - G(z))\,p^2q^3z^5}{1 - z} = G(z)(1 + pq^2z^3);$$

hence the solution is

$$G(z) = \frac{p^2q^3z^5}{p^2q^3z^5 + (1 + pq^2z^3)(1 - z)}. \tag{8.69}$$

Notice that $G(1) = 1$, if $pq \neq 0$; we do eventually encounter the pattern THTTH, with probability 1, unless the coin is rigged so that it always comes up heads or always tails.

To get the mean and variance of the distribution (8.69), we invert $G(z)$ as we did in the previous problem, writing $G(z) = z^5/F(z)$ where F is a polynomial:

$$F(z) = \frac{p^2q^3z^5 + (1 + pq^2z^3)(1 - z)}{p^2q^3}. \tag{8.70}$$

The relevant derivatives are

$$\begin{aligned} F'(1) &= 5-(1+pq^2)/p^2q^3\,, \\ F''(1) &= 20-6pq^2/p^2q^3\,; \end{aligned}$$

and if X is the number of flips we get

$$\begin{aligned} EX &= \mathrm{Mean}(G) = 5-\mathrm{Mean}(F) = p^{-2}q^{-3}+p^{-1}q^{-1}\,; && (8.71) \\ VX &= \mathrm{Var}(G) = -\mathrm{Var}(F) \\ &= -25+p^{-2}q^{-3}+7p^{-1}q^{-1}+\mathrm{Mean}(F)^2 \\ &= (EX)^2-9p^{-2}q^{-3}-3p^{-1}q^{-1}\,. && (8.72) \end{aligned}$$

When $p=\frac{1}{2}$, the mean and variance are 36 and 996.

Let's get general: The problem we have just solved was "random" enough to show us how to analyze the case that we are waiting for the first appearance of an *arbitrary* pattern A of heads and tails. Again we let S be the sum of all winning sequences of H's and T's, and we let N be the sum of all sequences that haven't encountered the pattern A yet. Equation (8.67) will remain the same; equation (8.68) will become

$$\begin{aligned} NA = S\bigl(1 &+ A^{(1)}\,[A^{(m-1)}=A_{(m-1)}] + A^{(2)}\,[A^{(m-2)}=A_{(m-2)}] \\ &+\cdots+A^{(m-1)}\,[A^{(1)}=A_{(1)}]\bigr)\,, \qquad (8.73) \end{aligned}$$

where m is the length of A, and where $A^{(k)}$ and $A_{(k)}$ denote respectively the last k characters and the first k characters of A. For example, if A is the pattern THTTH we just studied, we have

$$\begin{aligned} A^{(1)} &= \mathtt{H}\,, & A^{(2)} &= \mathtt{TH}\,, & A^{(3)} &= \mathtt{TTH}\,, & A^{(4)} &= \mathtt{HTTH}\,; \\ A_{(1)} &= \mathtt{T}\,, & A_{(2)} &= \mathtt{TH}\,, & A_{(3)} &= \mathtt{THT}\,, & A_{(4)} &= \mathtt{THTT}\,. \end{aligned}$$

Since the only perfect match is $A^{(2)}=A_{(2)}$, equation (8.73) reduces to (8.68).

Let $\widetilde{A}$ be the result of substituting p^{-1} for H and q^{-1} for T in the pattern A. Then it is not difficult to generalize our derivation of (8.71) and (8.72) to conclude (exercise 20) that the general mean and variance are

$$EX = \sum_{k=1}^{m} \widetilde{A}_{(k)}\,[A^{(k)}=A_{(k)}]\,; \qquad (8.74)$$

$$VX = (EX)^2 - \sum_{k=1}^{m} (2k-1)\widetilde{A}_{(k)}\,[A^{(k)}=A_{(k)}]\,. \qquad (8.75)$$

In the special case $p = \frac{1}{2}$ we can interpret these formulas in a particularly simple way. Given a pattern A of m heads and tails, let

$$A{:}A \;=\; \sum_{k=1}^{m} 2^{k-1}\,[A^{(k)} = A_{(k)}]\,. \qquad (8.76)$$

We can easily find the binary representation of this number by placing a '1' under each position such that the string matches itself perfectly when it is superimposed on a copy of itself that has been shifted to start in this position:

$$A = \texttt{HTHTHHTHTH}$$
$$A{:}A = (1000010101)_2 = 512 + 16 + 4 + 1 = 533$$

```
HTHTHHTHTH          ✓
 HTHTHHTHTH
  HTHTHHTHTH
   HTHTHHTHTH
    HTHTHHTHTH
     HTHTHHTHTH     ✓
      HTHTHHTHTH
       HTHTHHTHTH   ✓
        HTHTHHTHTH
         HTHTHHTHTH ✓
```

Equation (8.74) now tells us that the expected number of flips until pattern A appears is exactly $2(A{:}A)$, if we use a fair coin, because $\widetilde{A}_{(k)} = 2^k$ when $p = q = \frac{1}{2}$. This result, first discovered by the Soviet mathematician A. D. Solov'ev in 1966 [271], seems paradoxical at first glance: Patterns with no self-overlaps occur sooner than overlapping patterns do! It takes almost twice as long to encounter HHHHH as it does to encounter HHHHT or THHHH.

"Chem bol'she periodov u nashego slova, tem pozzhe ono poi͡avli͡aetsi͡a." — A. D. Solov'ev

Now let's consider an amusing game that was invented by (of all people) Walter Penney [231] in 1969. Alice and Bill flip a coin until either HHT or HTT occurs; Alice wins if the pattern HHT comes first, Bill wins if HTT comes first. This game — now called "Penney ante" — certainly seems to be fair, if played with a fair coin, because both patterns HHT and HTT have the same characteristics if we look at them in isolation: The probability generating function for the waiting time until HHT first occurs is

$$G(z) \;=\; \frac{z^3}{z^3 - 8(z-1)}\,,$$

and the same is true for HTT. Therefore neither Alice nor Bill has an advantage, if they play solitaire.

Of course not! Who could they have an advantage over?

But there's an interesting interplay between the patterns when both are considered simultaneously. Let S_A be the sum of Alice's winning configurations, and let S_B be the sum of Bill's:

$$\begin{aligned} S_A &= \mathtt{HHT} + \mathtt{HHHT} + \mathtt{THHT} + \mathtt{HHHHT} + \mathtt{HTHHT} + \mathtt{THHHT} + \cdots\,; \\ S_B &= \mathtt{HTT} + \mathtt{THTT} + \mathtt{HTHTT} + \mathtt{TTHTT} + \mathtt{THTHTT} + \mathtt{TTTHTT} + \cdots\,. \end{aligned}$$

Also — taking our cue from the trick that worked when only one pattern was involved — let us denote by N the sum of all sequences in which neither player has won so far:

$$N = 1 + \mathtt{H} + \mathtt{T} + \mathtt{HH} + \mathtt{HT} + \mathtt{TH} + \mathtt{TT} + \mathtt{HHH} + \mathtt{HTH} + \mathtt{THH} + \cdots\,. \qquad (8.77)$$

Then we can easily verify the following set of equations:

$$\begin{aligned} 1 + N(\mathtt{H}+\mathtt{T}) &= N + S_A + S_B\,; \\ N\,\mathtt{HHT} &= S_A\,; \\ N\,\mathtt{HTT} &= S_A\,\mathtt{T} + S_B\,. \end{aligned} \qquad (8.78)$$

If we now set $\mathtt{H} = \mathtt{T} = \frac{1}{2}$, the resulting value of S_A becomes the probability that Alice wins, and S_B becomes the probability that Bill wins. The three equations reduce to

$$1 + N = N + S_A + S_B\,; \qquad \tfrac{1}{8}N = S_A\,; \qquad \tfrac{1}{8}N = \tfrac{1}{2}S_A + S_B\,;$$

and we find $S_A = \frac{2}{3}$, $S_B = \frac{1}{3}$. Alice will win about twice as often as Bill!

In a generalization of this game, Alice and Bill choose patterns A and B of heads and tails, and they flip coins until either A or B appears. The two patterns need not have the same length, but we assume that A doesn't occur within B, nor does B occur within A. (Otherwise the game would be degenerate. For example, if $A = \mathtt{HT}$ and $B = \mathtt{THTH}$, poor Bill could never win; and if $A = \mathtt{HTH}$ and $B = \mathtt{TH}$, both players might claim victory simultaneously.) Then we can write three equations analogous to (8.73) and (8.78):

$$\begin{aligned} 1 + N(\mathtt{H}+\mathtt{T}) &= N + S_A + S_B\,; \\ NA &= S_A \sum_{k=1}^{l} A^{(l-k)}\,[A^{(k)} = A_{(k)}] \;+\; S_B \sum_{k=1}^{\min(l,m)} A^{(l-k)}\,[B^{(k)} = A_{(k)}]\,; \\ NB &= S_A \sum_{k=1}^{\min(l,m)} B^{(m-k)}\,[A^{(k)} = B_{(k)}] \;+\; S_B \sum_{k=1}^{m} B^{(l-k)}\,[B^{(k)} = B_{(k)}]\,. \end{aligned}$$

(8.79)

Here l is the length of A and m is the length of B. For example, if we have $A = \text{HTTHTHTH}$ and $B = \text{THTHTTH}$, the two pattern-dependent equations are

$$\begin{aligned} N\,\text{HTTHTHTH} &= S_A\,\text{TTHTHTH} + S_A + S_B\,\text{TTHTHTH} + S_B\,\text{THTH}\,;\\ N\,\text{THTHTTH} &= S_A\,\text{THTTH} + S_A\,\text{TTH} + S_B\,\text{THTTH} + S_B\,. \end{aligned}$$

We obtain the victory probabilities by setting $H = T = \frac{1}{2}$, if we assume that a fair coin is being used; this reduces the two crucial equations to

$$\begin{aligned} N &= S_A \sum_{k=1}^{l} 2^k\,[A^{(k)} = A_{(k)}] + S_B \sum_{k=1}^{\min(l,m)} 2^k\,[B^{(k)} = A_{(k)}]\,;\\ N &= S_A \sum_{k=1}^{\min(l,m)} 2^k\,[A^{(k)} = B_{(k)}] + S_B \sum_{k=1}^{m} 2^k\,[B^{(k)} = B_{(k)}]\,. \end{aligned} \tag{8.80}$$

We can see what's going on if we generalize the A:A operation of (8.76) to a function of two independent strings A and B:

$$A{:}B \;=\; \sum_{k=1}^{\min(l,m)} 2^{k-1}\,[A^{(k)} = B_{(k)}]\,. \tag{8.81}$$

Equations (8.80) now become simply

$$S_A(A{:}A) + S_B(B{:}A) \;=\; S_A(A{:}B) + S_B(B{:}B)\,;$$

the odds in Alice's favor are

$$\frac{S_A}{S_B} \;=\; \frac{B{:}B - B{:}A}{A{:}A - A{:}B}\,. \tag{8.82}$$

(This beautiful formula was discovered by John Horton Conway [111].)

For example, if $A = \text{HTTHTHTH}$ and $B = \text{THTHTTH}$ as above, we have $A{:}A = (10000001)_2 = 129$, $A{:}B = (0001010)_2 = 10$, $B{:}A = (0001001)_2 = 9$, and $B{:}B = (1000010)_2 = 66$; so the ratio S_A/S_B is $(66-9)/(129-10) = 57/119$. Alice will win this one only 57 times out of every 176, on the average.

Strange things can happen in Penney's game. For example, the pattern HHTH wins over the pattern HTHH with 3/2 odds, and HTHH wins over THHH with 7/5 odds. So HHTH ought to be much better than THHH. Yet THHH actually wins over HHTH, with 7/5 odds! The relation between patterns is not transitive. In fact, exercise 57 proves that if Alice chooses any pattern $\tau_1\tau_2\ldots\tau_l$ of length $l \geqslant 3$, Bill can always ensure better than even chances of winning if he chooses the pattern $\overline{\tau}_2\tau_1\tau_2\ldots\tau_{l-1}$, where $\overline{\tau}_2$ is the heads/tails opposite of τ_2.

Odd, odd.

8.5 HASHING

Somehow the verb "to hash" magically became standard terminology for key transformation during the mid-1960s, yet nobody was rash enough to use such an undignified word publicly until 1967. —D. E. Knuth [175]

Let's conclude this chapter by applying probability theory to computer programming. Several important algorithms for storing and retrieving information inside a computer are based on a technique called "hashing." The general problem is to maintain a set of records that each contain a "key" value, K, and some data $D(K)$ about that key; we want to be able to find $D(K)$ quickly when K is given. For example, each key might be the name of a student, and the associated data might be that student's homework grades.

In practice, computers don't have enough capacity to set aside one memory cell for every possible key; billions of keys are possible, but comparatively few keys are actually present in any one application. One solution to the problem is to maintain two tables KEY[j] and DATA[j] for $1 \leqslant j \leqslant N$, where N is the total number of records that can be accommodated; another variable n tells how many records are actually present. Then we can search for a given key K by going through the table sequentially in an obvious way:

S1 Set $j := 1$. (We've searched through all positions $< j$.)

S2 If $j > n$, stop. (The search was unsuccessful.)

S3 If KEY[j] $= K$, stop. (The search was successful.)

S4 Increase j by 1 and return to step S2. (We'll try again.)

After a successful search, the desired data entry $D(K)$ appears in DATA[j]. After an unsuccessful search, we can insert K and $D(K)$ into the table by setting

$$n := j, \quad \texttt{KEY[n]} := K, \quad \texttt{DATA[n]} := D(K),$$

assuming that the table was not already filled to capacity.

This method works, but it can be dreadfully slow; we need to repeat step S2 a total of $n+1$ times whenever an unsuccessful search is made, and n can be quite large.

Hashing was invented to speed things up. The basic idea, in one of its popular forms, is to use m separate lists instead of one giant list. A "hash function" transforms every possible key K into a list number $h(K)$ between 1 and m. An auxiliary table FIRST[i] for $1 \leqslant i \leqslant m$ points to the first record in list i; another auxiliary table NEXT[j] for $1 \leqslant j \leqslant N$ points to the record following record j in its list. We assume that

$$\begin{aligned} \texttt{FIRST[i]} &= -1\,, && \text{if list } i \text{ is empty;} \\ \texttt{NEXT[j]} &= 0\,, && \text{if record } j \text{ is the last in its list.} \end{aligned}$$

As before, there's a variable n that tells how many records have been stored altogether.

For example, suppose the keys are names, and suppose that there are $m = 4$ lists based on the first letter of a name:

$$h(\text{name}) = \begin{cases} 1, & \text{for A–F;} \\ 2, & \text{for G–L;} \\ 3, & \text{for M–R;} \\ 4, & \text{for S–Z.} \end{cases}$$

We start with four empty lists and with $n = 0$. If, say, the first record has Nora as its key, we have $h(\text{Nora}) = 3$, so Nora becomes the key of the first item in list 3. If the next two names are Glenn and Jim, they both go into list 2. Now the tables in memory look like this:

$$\begin{aligned} &\texttt{FIRST[1]} = -1, \quad \texttt{FIRST[2]} = 2, \quad \texttt{FIRST[3]} = 1, \quad \texttt{FIRST[4]} = -1. \\ &\texttt{KEY[1]} = \text{Nora}, \qquad \texttt{NEXT[1]} = 0; \\ &\texttt{KEY[2]} = \text{Glenn}, \qquad \texttt{NEXT[2]} = 3; \\ &\texttt{KEY[3]} = \text{Jim}, \qquad \texttt{NEXT[3]} = 0; \qquad n = 3. \end{aligned}$$

(The values of DATA[1], DATA[2], and DATA[3] are confidential and will not be shown.) After 18 records have been inserted, the lists might contain the names

Let's hear it for the Concrete Math students who sat in the front rows and lent their names to this experiment.

list 1	list 2	list 3	list 4
Dianne	Glenn	Nora	Scott
Ari	Jim	Mike	Tina
Brian	Jennifer	Michael	
Fran	Joan	Ray	
Doug	Jerry	Paula	
	Jean		

and these names would appear intermixed in the KEY array with NEXT entries to keep the lists effectively separate. If we now want to search for John, we have to scan through the six names in list 2 (which happens to be the longest list); but that's not nearly as bad as looking at all 18 names.

Here's a precise specification of the algorithm that searches for key K in accordance with this scheme:

H1 Set $i := h(K)$ and $j :=$ FIRST[i].
H2 If $j \leqslant 0$, stop. (The search was unsuccessful.)
H3 If KEY[j] $= K$, stop. (The search was successful.)
H4 Set $i := j$, then set $j :=$ NEXT[i] and return to step H2. (We'll try again.)

For example, to search for Jennifer in the example given, step H1 would set $i := 2$ and $j := 2$; step H3 would find that Glenn $\neq$ Jennifer; step H4 would set $j := 3$; and step H3 would find Jim $\neq$ Jennifer.

I bet their parents are glad about that.

After a successful search, the desired data D(K) appears in `DATA[j]`, as in the previous algorithm. After an unsuccessful search, we can enter K and D(K) in the table by doing the following operations:

$$\begin{aligned}&n := n+1;\\&\textbf{if } j<0 \textbf{ then } \texttt{FIRST[i]} := n \textbf{ else } \texttt{NEXT[i]} := n;\\&\texttt{KEY[n]} := K;\quad \texttt{DATA[n]} := D(K);\quad \texttt{NEXT[n]} := 0.\end{aligned} \tag{8.83}$$

Now the table will once again be up to date.

We hope to get lists of roughly equal length, because this will make the task of searching about m times faster. The value of m is usually much greater than 4, so a factor of $1/m$ will be a significant improvement.

We don't know in advance what keys will be present, but it is generally possible to choose the hash function h so that we can consider $h(K)$ to be a random variable that is uniformly distributed between 1 and m, independent of the hash values of other keys that are present. In such cases computing the hash function is like rolling a die that has m faces. There's a chance that all the records will fall into the same list, just as there's a chance that a die will always turn up ⚅; but probability theory tells us that the lists will *almost always* be pretty evenly balanced.

Analysis of Hashing: Introduction.

"Algorithmic analysis" is a branch of computer science that derives quantitative information about the efficiency of computer methods. "Probabilistic analysis of an algorithm" is the study of an algorithm's running time, considered as a random variable that depends on assumed characteristics of the input data. Hashing is an especially good candidate for probabilistic analysis, because it is an extremely efficient method on the average, even though its worst case is too horrible to contemplate. (The worst case occurs when all keys have the same hash value.) Indeed, a computer programmer who uses hashing had better be a believer in probability theory.

Let P be the number of times step H3 is performed when the algorithm above is used to carry out a search. (Each execution of H3 is called a "probe" in the table.) If we know P, we know how often each step is performed, depending on whether the search is successful or unsuccessful:

Step	Unsuccessful search	Successful search
H1	1 time	1 time
H2	$P+1$ times	P times
H3	P times	P times
H4	P times	$P-1$ times

Thus the main quantity that governs the running time of the search procedure is the number of probes, P.

We can get a good mental picture of the algorithm by imagining that we are keeping an address book that is organized in a special way, with room for only one entry per page. On the cover of the book we note down the page number for the first entry in each of m lists; each name K determines the list $h(K)$ that it belongs to. Every page inside the book refers to the successor page in its list. The number of probes needed to find an address in such a book is the number of pages we must consult.

If n items have been inserted, their positions in the table depend only on their respective hash values, $\langle h_1, h_2, \ldots, h_n \rangle$. Each of the m^n possible sequences $\langle h_1, h_2, \ldots, h_n \rangle$ is considered to be equally likely, and P is a random variable depending on such a sequence.

Case 1: The key is not present.

Check under the doormat.

Let's consider first the behavior of P in an unsuccessful search, assuming that n records have previously been inserted into the hash table. In this case the relevant probability space consists of m^{n+1} elementary events

$$\omega = (h_1, h_2, \ldots, h_n; h_{n+1})$$

where h_j is the hash value of the jth key inserted, and where h_{n+1} is the hash value of the key for which the search is unsuccessful. We assume that the hash function h has been chosen properly so that $\Pr(\omega) = 1/m^{n+1}$ for every such ω.

For example, if $m = n = 2$, there are eight equally likely possibilities:

h_1	h_2	h_3:	P
1	1	1 :	2
1	1	2 :	0
1	2	1 :	1
1	2	2 :	1
2	1	1 :	1
2	1	2 :	1
2	2	1 :	0
2	2	2 :	2

If $h_1 = h_2 = h_3$ we make two unsuccessful probes before concluding that the new key K is not present; if $h_1 = h_2 \neq h_3$ we make none; and so on. This list of all possibilities shows that P has a probability distribution given by the pgf $(\frac{2}{8} + \frac{4}{8}z + \frac{2}{8}z^2) = (\frac{1}{2} + \frac{1}{2}z)^2$, when $m = n = 2$.

An unsuccessful search makes one probe for every item in list number h_{n+1}, so we have the general formula

$$P = [h_1 = h_{n+1}] + [h_2 = h_{n+1}] + \cdots + [h_n = h_{n+1}]. \tag{8.84}$$

The probability that $h_j = h_{n+1}$ is $1/m$, for $1 \le j \le n$; so it follows that

$$EP = E[h_1 = h_{n+1}] + E[h_2 = h_{n+1}] + \cdots + E[h_n = h_{n+1}] = \frac{n}{m}.$$

Maybe we should do that more slowly: Let X_j be the random variable

$$X_j = X_j(\omega) = [h_j = h_{n+1}].$$

Then $P = X_1 + \cdots + X_n$, and $EX_j = 1/m$ for all $j \le n$; hence

$$EP = EX_1 + \cdots + EX_n = n/m.$$

Good: As we had hoped, the average number of probes is $1/m$ times what it was without hashing. Furthermore the random variables X_j are independent, and they each have the same probability generating function

$$X_j(z) = \frac{m-1+z}{m};$$

therefore the pgf for the total number of probes in an unsuccessful search is

$$P(z) = X_1(z)\ldots X_n(z) = \left(\frac{m-1+z}{m}\right)^n. \qquad (8.85)$$

This is a binomial distribution, with $p = 1/m$ and $q = (m-1)/m$; in other words, the number of probes in an unsuccessful search behaves just like the number of heads when we toss a biased coin whose probability of heads is $1/m$ on each toss. Equation (8.61) tells us that the variance of P is therefore

$$npq = \frac{n(m-1)}{m^2}.$$

When m is large, the variance of P is approximately n/m, so the standard deviation is approximately $\sqrt{n/m}$.

Case 2: The key is present.

Now let's look at successful searches. In this case the appropriate probability space is a bit more complicated, depending on our application: We will let Ω be the set of all elementary events

$$\omega = (h_1, \ldots, h_n; k), \qquad (8.86)$$

where h_j is the hash value for the jth key as before, and where k is the index of the key being sought (the key whose hash value is h_k). Thus we have $1 \le h_j \le m$ for $1 \le j \le n$, and $1 \le k \le n$; there are $m^n \cdot n$ elementary events ω in all.

Let s_j be the probability that we are searching for the jth key that was inserted into the table. Then

$$\Pr(\omega) = s_k/m^n \tag{8.87}$$

if ω is the event (8.86). (Some applications search most often for the items that were inserted first, or for the items that were inserted last, so we will not assume that each $s_j = 1/n$.) Notice that $\sum_{\omega\in\Omega}\Pr(\omega) = \sum_{k=1}^{n} s_k = 1$, hence (8.87) defines a legal probability distribution.

The number of probes P in a successful search is p if key K was the pth key to be inserted into its list. Therefore

$$P = [h_1 = h_k] + [h_2 = h_k] + \cdots + [h_k = h_k];$$

or, if we let X_j be the random variable $[h_j = h_k]$, we have

$$P = X_1 + X_2 + \cdots + X_k. \tag{8.88}$$

Suppose, for example, that we have $m = 10$ and $n = 16$, and that the hash values have the following "random" pattern:

Where have I seen that pattern before?

$$(h_1, \ldots, h_{16}) = 3\ 1\ 4\ 1\ 5\ 9\ 2\ 6\ 5\ 3\ 5\ 8\ 9\ 7\ 9\ 3\ ;$$
$$(P_1, \ldots, P_{16}) = 1\ 1\ 1\ 2\ 1\ 1\ 1\ 1\ 2\ 2\ 3\ 1\ 2\ 1\ 3\ 3\ .$$

The number of probes P_j needed to find the jth key is shown below h_j.

Equation (8.88) represents P as a sum of random variables, but we can't simply calculate EP as $EX_1 + \cdots + EX_k$ because the quantity k itself is a random variable. What is the probability generating function for P? To answer this question we should digress a moment to talk about *conditional probability*.

Equation (8.43) was also a momentary digression.

If A and B are events in a probability space, we say that the conditional probability of A, given B, is

$$\Pr(\omega \in A \mid \omega \in B) = \frac{\Pr(\omega \in A \cap B)}{\Pr(\omega \in B)}. \tag{8.89}$$

For example, if X and Y are random variables, the conditional probability of the event $X = x$, given that $Y = y$, is

$$\Pr(X = x \mid Y = y) = \frac{\Pr(X = x \text{ and } Y = y)}{\Pr(Y = y)}. \tag{8.90}$$

For any fixed y in the range of Y, the sum of these conditional probabilities over all x in the range of X is $\Pr(Y = y)/\Pr(Y = y) = 1$; therefore (8.90) defines a probability distribution, and we can define a new random variable '$X|y$' such that $\Pr(X|y = x) = \Pr(X = x \mid Y = y)$.

If X and Y are independent, the random variable $X|y$ will be essentially the same as X, regardless of the value of y, because $\Pr(X=x \mid Y=y)$ is equal to $\Pr(X=x)$ by (8.5); that's what independence means. But if X and Y are dependent, the random variables $X|y$ and $X|y'$ need not resemble each other in any way when $y \neq y'$.

If X takes only nonnegative integer values, we can decompose its pgf into a sum of conditional pgf's with respect to any other random variable Y:

$$G_X(z) \;=\; \sum_{y \in Y(\Omega)} \Pr(Y=y)\, G_{X|y}(z)\,. \tag{8.91}$$

This holds because the coefficient of z^x on the left side is $\Pr(X=x)$, for all $x \in X(\Omega)$, and on the right it is

$$\begin{aligned}\sum_{y \in Y(\Omega)} \Pr(Y=y)\Pr(X=x \mid Y=y) \;&=\; \sum_{y \in Y(\Omega)} \Pr(X=x \text{ and } Y=y)\\ &=\; \Pr(X=x)\,.\end{aligned}$$

For example, if X is the product of the spots on two fair dice and if Y is the sum of the spots, the pgf for $X|6$ is

$$G_{X|6}(z) \;=\; \tfrac{2}{5}z^5 + \tfrac{2}{5}z^8 + \tfrac{1}{5}z^9$$

because the conditional probabilities for $Y = 6$ consist of five equally probable events {⚀⚄, ⚁⚃, ⚂⚂, ⚃⚁, ⚄⚀}. Equation (8.91) in this case reduces to

$$\begin{aligned}G_X(z) \;=\; &\tfrac{1}{36}G_{X|2}(z) + \tfrac{2}{36}G_{X|3}(z) + \tfrac{3}{36}G_{X|4}(z) + \tfrac{4}{36}G_{X|5}(z)\\ &\tfrac{5}{36}G_{X|6}(z) + \tfrac{6}{36}G_{X|7}(z) + \tfrac{5}{36}G_{X|8}(z) + \tfrac{4}{36}G_{X|9}(z)\\ &\tfrac{3}{36}G_{X|10}(z) + \tfrac{2}{36}G_{X|11}(z) + \tfrac{1}{36}G_{X|12}(z)\,,\end{aligned}$$

Oh, now I understand what mathematicians mean when they say something is "obvious," "clear," or "trivial."

a formula that is obvious once you understand it. (End of digression.)

In the case of hashing, (8.91) tells us how to write down the pgf for probes in a successful search, if we let $X = P$ and $Y = K$. For any fixed k between 1 and n, the random variable $P|k$ is defined as a sum of independent random variables $X_1 + \cdots + X_k$; this is (8.88). So it has the pgf

$$G_{P|k}(z) \;=\; \left(\frac{m-1+z}{m}\right)^{k-1} z\,.$$

Therefore the pgf for P itself is

$$\begin{aligned} G_P(z) &= \sum_{k=1}^{n} s_k G_{P|k}(z) \\ &= \sum_{k=1}^{n} s_k \Big(\frac{m-1+z}{m}\Big)^{k-1} z \\ &= z\,S\Big(\frac{m-1+z}{m}\Big), \end{aligned} \tag{8.92}$$

where

$$S(z) = s_1 + s_2 z + s_3 z^2 + \cdots + s_n z^{n-1} \tag{8.93}$$

is the pgf for the search probabilities s_k (divided by z for convenience).

Good. We have a probability generating function for P; we can now find the mean and variance by differentiation. It's somewhat easier to remove the z factor first, as we've done before, thus finding the mean and variance of $P-1$ instead:

$$\begin{aligned} F(z) &= G_P(z)/z = S\Big(\frac{m-1+z}{m}\Big); \\ F'(z) &= \frac{1}{m} S'\Big(\frac{m-1+z}{m}\Big); \\ F''(z) &= \frac{1}{m^2} S''\Big(\frac{m-1+z}{m}\Big). \end{aligned}$$

Therefore

$$EP = 1 + \mathrm{Mean}(F) = 1 + F'(1) = 1 + m^{-1}\,\mathrm{Mean}(S); \tag{8.94}$$

$$\begin{aligned} VP = \mathrm{Var}(F) &= F''(1) + F'(1) - F'(1)^2 \\ &= m^{-2}S''(1) + m^{-1}S'(1) - m^{-2}S'(1)^2 \\ &= m^{-2}\,\mathrm{Var}(S) + (m^{-1} - m^{-2})\,\mathrm{Mean}(S). \end{aligned} \tag{8.95}$$

These are general formulas expressing the mean and variance of the number of probes P in terms of the mean and variance of the assumed search distribution S.

For example, suppose we have $s_k = 1/n$ for $1 \leqslant k \leqslant n$. This means we are doing a purely "random" successful search, with all keys in the table equally likely. Then $S(z)$ is the uniform probability distribution $U_n(z)$ in

(8.32), and we have $\mathrm{Mean}(S) = (n-1)/2$, $\mathrm{Var}(S) = (n^2-1)/12$. Hence

$$EP = \frac{n-1}{2m} + 1\,; \tag{8.96}$$

$$VP = \frac{n^2-1}{12m^2} + \frac{(m-1)(n-1)}{2m^2} = \frac{(n-1)(6m+n-5)}{12m^2}\,. \tag{8.97}$$

Once again we have gained the desired speedup factor of $1/m$. If $m = n/\ln n$ and $n \to \infty$, the average number of probes per successful search in this case is about $\frac{1}{2}\ln n$, and the standard deviation is asymptotically $(\ln n)/\sqrt{12}$.

On the other hand, we might suppose that $s_k = (kH_n)^{-1}$ for $1 \leqslant k \leqslant n$; this distribution is called "Zipf's law." Then $\mathrm{Mean}(G) = n/H_n$ and $\mathrm{Var}(G) = \frac{1}{2}n(n-1)/H_n - n^2/H_n^2$. The average number of probes for $m = n/\ln n$ as $n \to \infty$ is approximately 2, with standard deviation asymptotic to $\sqrt{\ln n}$.

In both cases the analysis allows the cautious souls among us, who fear the worst case, to rest easily: Chebyshev's inequality tells us that the lists will be nice and short, except in extremely rare cases.

Case 2, continued: Variants of the variance.

We have just computed the variance of the number of probes in a successful search, by considering P to be a random variable over a probability space with $m^n \cdot n$ elements $(h_1, \ldots, h_n; k)$. But we could have adopted another point of view: Each pattern $(h_1, \ldots, h_n)$ of hash values defines a random variable $P|(h_1, \ldots, h_n)$, representing the probes we make in a successful search of a particular hash table on n given keys. The average value of $P|(h_1, \ldots, h_n)$,

OK, gang, time to put on your skim suits again.
—Friendly TA

$$A(h_1, \ldots, h_n) = \sum_{p=1}^{n} p \cdot \Pr\bigl(P|(h_1, \ldots, h_n) = p\bigr)\,, \tag{8.98}$$

can be said to represent the running time of a successful search. This quantity $A(h_1, \ldots, h_n)$ is a random variable that depends only on $(h_1, \ldots, h_n)$, not on the final component k; we can write it in the form

$$A(h_1, \ldots, h_n) = \sum_{k=1}^{n} s_k\, P(h_1, \ldots, h_n; k)\,,$$

since $P|(h_1, \ldots, h_n) = p$ with probability

$$\frac{\sum_{k=1}^{n} \Pr\bigl(P(h_1, \ldots, h_n; k) = p\bigr)}{\sum_{k=1}^{n} \Pr(h_1, \ldots, h_n; k)} = \frac{\sum_{k=1}^{n} m^{-n} s_k \bigl[P(h_1, \ldots, h_n; k) = p\bigr]}{\sum_{k=1}^{n} m^{-n} s_k}$$

$$= \sum_{k=1}^{n} s_k \bigl[P(h_1, \ldots, h_n; k) = p\bigr]\,.$$

The mean value of $A(h_1,\ldots,h_n)$, obtained by summing over all m^n possibilities $(h_1,\ldots,h_n)$ and dividing by m^n, will be the same as the mean value we obtained before in (8.94). But the *variance* of $A(h_1,\ldots,h_n)$ is something different; this is a variance of m^n averages, not a variance of $m^n\cdot n$ probe counts. For example, if $m=1$ (so that there is only one list), the "average" value $A(h_1,\ldots,h_n)=A(1,\ldots,1)$ is actually constant, so its variance VA is zero; but the number of probes in a successful search is not constant, so the variance VP is nonzero.

But the VP is nonzero only in an election year.

We can illustrate this difference between variances by carrying out the calculations for general m and n in the simplest case, when $s_k=1/n$ for $1\leqslant k\leqslant n$. In other words, we will assume temporarily that there is a uniform distribution of search keys. Any given sequence of hash values $(h_1,\ldots,h_n)$ defines m lists that contain respectively $(n_1,n_2,\ldots,n_m)$ entries for some numbers n_j, where

$$n_1+n_2+\cdots+n_m = n\,.$$

A successful search in which each of the n keys in the table is equally likely will have an average running time of

$$\begin{aligned}A(h_1,\ldots,h_n) &= \frac{(1+\cdots+n_1)+(1+\cdots+n_2)+\cdots+(1+\cdots+n_m)}{n}\\ &= \frac{n_1(n_1+1)+n_2(n_2+1)+\cdots+n_m(n_m+1)}{2n}\\ &= \frac{n_1^2+n_2^2+\cdots+n_m^2+n}{2n}\end{aligned}$$

probes. Our goal is to calculate the variance of this quantity $A(h_1,\ldots,h_n)$, over the probability space consisting of all m^n sequences $(h_1,\ldots,h_n)$.

The calculations will be simpler, it turns out, if we compute the variance of a slightly different quantity,

$$B(h_1,\ldots,h_n) = \binom{n_1}{2}+\binom{n_2}{2}+\cdots+\binom{n_m}{2}.$$

We have

$$A(h_1,\ldots,h_n) = 1+B(h_1,\ldots,h_n)/n\,,$$

hence the mean and variance of A satisfy

$$EA = 1+\frac{EB}{n}\,;\qquad VA = \frac{VB}{n^2}\,. \tag{8.99}$$

The probability that the list sizes will be $n_1, n_2, \ldots, n_m$ is the multinomial coefficient

$$\binom{n}{n_1, n_2, \ldots, n_m} = \frac{n!}{n_1!\, n_2! \ldots n_m!}$$

divided by m^n; hence the pgf for $B(h_1, \ldots, h_n)$ is

$$B_n(z) = \sum_{\substack{n_1, n_2, \ldots, n_m \geqslant 0 \\ n_1+n_2+\cdots+n_m=n}} \binom{n}{n_1, n_2, \ldots, n_m} z^{\binom{n_1}{2}+\binom{n_2}{2}+\cdots+\binom{n_m}{2}}\, m^{-n}.$$

This sum looks a bit scary to inexperienced eyes, but our experiences in Chapter 7 have taught us to recognize it as an m-fold convolution. Indeed, if we consider the exponential super-generating function

$$G(w, z) = \sum_{n \geqslant 0} B_n(z) \frac{m^n w^n}{n!},$$

we can readily verify that $G(w, z)$ is simply an mth power:

$$G(w, z) = \biggl(\sum_{k \geqslant 0} z^{\binom{k}{2}} \frac{w^k}{k!}\biggr)^m.$$

As a check, we can try setting $z = 1$; we get $G(w, 1) = (e^w)^m$, so the coefficient of $m^n w^n/n!$ is $B_n(1) = 1$.

If we knew the values of $B_n'(1)$ and $B_n''(1)$, we would be able to calculate $\mathrm{Var}(B_n)$. So we take partial derivatives of $G(w, z)$ with respect to z:

$$\begin{aligned}
\frac{\partial}{\partial z} G(w, z) &= \sum_{n \geqslant 0} B_n'(z) \frac{m^n w^n}{n!} \\
&= m \biggl(\sum_{k \geqslant 0} z^{\binom{k}{2}} \frac{w^k}{k!}\biggr)^{m-1} \sum_{k \geqslant 0} \binom{k}{2} z^{\binom{k}{2}-1} \frac{w^k}{k!}; \\
\frac{\partial^2}{\partial z^2} G(w, z) &= \sum_{n \geqslant 0} B_n''(z) \frac{m^n w^n}{n!} \\
&= m(m-1) \biggl(\sum_{k \geqslant 0} z^{\binom{k}{2}} \frac{w^k}{k!}\biggr)^{m-2} \biggl(\sum_{k \geqslant 0} \binom{k}{2} z^{\binom{k}{2}-1} \frac{w^k}{k!}\biggr)^2 \\
&\quad + m \biggl(\sum_{k \geqslant 0} z^{\binom{k}{2}} \frac{w^k}{k!}\biggr)^{m-1} \sum_{k \geqslant 0} \binom{k}{2} \biggl(\binom{k}{2} - 1\biggr) z^{\binom{k}{2}-2} \frac{w^k}{k!}.
\end{aligned}$$

Complicated, yes; but everything simplifies greatly when we set $z = 1$. For example, we have

$$\begin{aligned}\sum_{n\geq 0} B_n'(1)\frac{m^n w^n}{n!} &= me^{(m-1)w}\sum_{k\geq 2}\frac{w^k}{2(k-2)!}\\ &= me^{(m-1)w}\sum_{k\geq 0}\frac{w^{k+2}}{2k!}\\ &= \frac{mw^2e^{(m-1)w}}{2}e^w = \sum_{n\geq 0}\frac{(mw)^{n+2}}{2mn!} = \sum_{n\geq 0}\frac{n(n-1)m^n w^n}{2mn!},\end{aligned}$$

and it follows that

$$B_n'(1) = \binom{n}{2}\frac{1}{m}. \tag{8.100}$$

The expression for EA in (8.99) now gives $EA = 1+(n-1)/2m$, in agreement with (8.96).

The formula for $B_n''(1)$ involves the similar sum

$$\begin{aligned}\sum_{k\geq 0}\binom{k}{2}\left(\binom{k}{2}-1\right)\frac{w^k}{k!} &= \frac{1}{4}\sum_{k\geq 0}\frac{(k+1)k(k-1)(k-2)w^k}{k!}\\ &= \frac{1}{4}\sum_{k\geq 3}\frac{(k+1)w^k}{(k-3)!} = \frac{1}{4}\sum_{k\geq 0}\frac{(k+4)w^{k+3}}{k!} = \left(\tfrac{1}{4}w^4+w^3\right)e^w;\end{aligned}$$

hence we find that

$$\begin{aligned}\sum_{n\geq 0} B_n''(1)\frac{m^n w^n}{n!} &= m(m-1)e^{w(m-2)}\left(\tfrac{1}{2}w^2e^w\right)^2 + m\left(\tfrac{1}{4}w^4+w^3\right)e^w\\ &= me^{wm}\left(\tfrac{1}{4}mw^4+w^3\right);\\ B_n''(1) &= \binom{n}{2}\left(\binom{n}{2}-1\right)\frac{1}{m^2}.\end{aligned} \tag{8.101}$$

Now we can put all the pieces together and evaluate the desired variance VA. Massive cancellation occurs, and the result is surprisingly simple:

$$\begin{aligned}VA = \frac{VB}{n^2} &= \frac{B_n''(1)+B_n'(1)-B_n'(1)^2}{n^2}\\ &= \frac{n(n-1)}{m^2n^2}\left(\frac{(n+1)(n-2)}{4}+\frac{m}{2}-\frac{n(n-1)}{4}\right)\\ &= \frac{(m-1)(n-1)}{2m^2n}.\end{aligned} \tag{8.102}$$

When such "coincidences" occur, we suspect that there's a mathematical reason; there might be another way to attack the problem, explaining why the answer has such a simple form. And indeed, there is another approach (in exercise 60), which shows that the variance of the average successful search has the general form

$$VA = \frac{m-1}{m^2}\sum_{k=1}^{n} s_k^2(k-1) \qquad (8.103)$$

when s_k is the probability that the kth-inserted element is being sought. Equation (8.102) is the special case $s_k = 1/n$ for $1 \leqslant k \leqslant n$.

Besides the variance of the average, we might also consider the average of the variance. In other words, each sequence $(h_1, \ldots, h_n)$ that defines a hash table also defines a probability distribution for successful searching, and the variance of this probability distribution tells how spread out the number of probes will be in different successful searches. For example, let's go back to the case where we inserted $n = 16$ things into $m = 10$ lists:

Where have I seen that pattern before?

Where have I seen that graffito before?

$I\eta\nu P_\pi$.

$$(h_1, \ldots, h_{16}) = 3\ 1\ 4\ 1\ 5\ 9\ 2\ 6\ 5\ 3\ 5\ 8\ 9\ 7\ 9\ 3$$
$$(P_1, \ldots, P_{16}) = 1\ 1\ 1\ 2\ 1\ 1\ 1\ 1\ 2\ 2\ 3\ 1\ 2\ 1\ 3\ 3$$

A successful search in the resulting hash table has the pgf

$$\begin{aligned} G(3,1,4,1,\ldots,3) &= \sum_{k=1}^{n} s_k z^{P(3,1,4,1,\ldots,3;k)} \\ &= s_1z + s_2z + s_3z + s_4z^2 + \cdots + s_{16}z^3 . \end{aligned}$$

We have just considered the average number of probes in a successful search of this table, namely $A(3,1,4,1,\ldots,3) = \mathrm{Mean}\big(G(3,1,4,1,\ldots,3)\big)$. We can also consider the variance,

$$\begin{aligned} &s_1\cdot 1^2 + s_2\cdot 1^2 + s_3\cdot 1^2 + s_4\cdot 2^2 + \cdots + s_{16}\cdot 3^2 \\ &\quad - (s_1\cdot 1 + s_2\cdot 1 + s_3\cdot 1 + s_4\cdot 2 + \cdots + s_{16}\cdot 3)^2 . \end{aligned}$$

This variance is a random variable, depending on $(h_1, \ldots, h_n)$, so it is natural to consider its average value.

In other words, there are three natural kinds of variance that we may wish to know, in order to understand the behavior of a successful search: The *overall variance* of the number of probes, taken over all $(h_1, \ldots, h_n)$ and k; the *variance of the average* number of probes, where the average is taken over all k and the variance is then taken over all $(h_1, \ldots, h_n)$; and the *average of the variance* of the number of the probes, where the variance is taken over all k and the average is then taken over all $(h_1, \ldots, h_n)$. In symbols, the overall variance is

$$\begin{aligned} VP = & \sum_{1 \le h_1, \ldots, h_n \le m} \sum_{k=1}^{n} \frac{s_k}{m^n} P(h_1, \ldots, h_n; k)^2 \\ & - \Biggl(\sum_{1 \le h_1, \ldots, h_n \le m} \sum_{k=1}^{n} \frac{s_k}{m^n} P(h_1, \ldots, h_n; k) \Biggr)^2 ; \end{aligned}$$

the variance of the average is

$$\begin{aligned} VA = & \sum_{1 \le h_1, \ldots, h_n \le m} \frac{1}{m^n} \Biggl(\sum_{k=1}^{n} s_k P(h_1, \ldots, h_n; k) \Biggr)^2 \\ & - \Biggl(\sum_{1 \le h_1, \ldots, h_n \le m} \frac{1}{m^n} \sum_{k=1}^{n} s_k P(h_1, \ldots, h_n; k) \Biggr)^2 ; \end{aligned}$$

and the average of the variance is

$$\begin{aligned} AV = & \sum_{1 \le h_1, \ldots, h_n \le m} \frac{1}{m^n} \Biggl(\sum_{k=1}^{n} s_k P(h_1, \ldots, h_n; k)^2 \\ & \qquad - \Biggl(\sum_{k=1}^{n} s_k P(h_1, \ldots, h_n; k) \Biggr)^2 \Biggr) . \end{aligned}$$

It turns out that these three quantities are interrelated in a simple way:

$$VP = VA + AV . \tag{8.104}$$

In fact, conditional probability distributions always satisfy the identity

$$VX = V\bigl(E(X|Y)\bigr) + E\bigl(V(X|Y)\bigr) \tag{8.105}$$

if X and Y are random variables in any probability space and if X takes real values. (This identity is proved in exercise 22.) Equation (8.104) is the special case where X is the number of probes in a successful search and Y is the sequence of hash values $(h_1, \ldots, h_n)$.

The general equation (8.105) needs to be understood carefully, because the notation tends to conceal the different random variables and probability spaces in which expectations and variances are being calculated. For each y in the range of Y, we have defined the random variable $X|y$ in (8.90), and this random variable has an expected value $E(X|y)$ depending on y. Now $E(X|Y)$ denotes the random variable whose values are $E(X|y)$ as y ranges over all possible values of Y, and $V(E(X|Y))$ is the variance of this random variable with respect to the probability distribution of Y. Similarly, $E(V(X|Y))$ is the average of the random variables $V(X|y)$ as y varies. On the left of (8.105) is VX, the unconditional variance of X. Since variances are nonnegative, we always have

(Now is a good time to do warmup exercise 6.)

$$VX \geqslant V(E(X|Y)) \quad\text{and}\quad VX \geqslant E(V(X|Y)). \tag{8.106}$$

Case 1, again: Unsuccessful search revisited.

Let's bring our microscopic examination of hashing to a close by doing one more calculation typical of algorithmic analysis. This time we'll look more closely at the *total running time* associated with an unsuccessful search, assuming that the computer will insert the previously unknown key into its memory.

The insertion process in (8.83) has two cases, depending on whether j is negative or zero. We have $j < 0$ if and only if $P = 0$, since a negative value comes from the FIRST entry of an empty list. Thus, if the list was previously empty, we have $P = 0$ and we must set $\mathtt{FIRST}[h_{n+1}] := n+1$. (The new record will be inserted into position $n+1$.) Otherwise we have $P > 0$ and we must set a LINK entry to $n+1$. These two cases may take different amounts of time; therefore the total running time for an unsuccessful search has the form

P is still the number of probes.

$$T = \alpha + \beta P + \delta[P=0], \tag{8.107}$$

where α, β, and δ are constants that depend on the computer being used and on the way in which hashing is encoded in that machine's internal language. It would be nice to know the mean and variance of T, since such information is more relevant in practice than the mean and variance of P.

So far we have used probability generating functions only in connection with random variables that take nonnegative integer values. But it turns out that we can deal in essentially the same way with

$$G_X(z) = \sum_{\omega\in\Omega} \Pr(\omega) z^{X(\omega)}$$

when X is any real-valued random variable, because the essential characteristics of X depend only on the behavior of G_X near $z = 1$, where powers of z are well defined. For example, the running time (8.107) of an unsuccessful search is a random variable, defined on the probability space of equally likely hash values $(h_1, \ldots, h_n; h_{n+1})$ with $1 \leqslant h_j \leqslant m$; we can consider the series

$$G_T(z) = \frac{1}{m^{n+1}} \sum_{h_1=1}^{m} \cdots \sum_{h_n=1}^{m} \sum_{h_{n+1}=1}^{m} z^{\alpha+\beta P(h_1,\ldots,h_n;h_{n+1})+\delta(P(h_1,\ldots,h_n;h_{n+1})=0)}$$

to be a pgf even when α, β, and δ are not integers. (In fact, the parameters α, β, δ are physical quantities that have dimensions of time; they aren't even pure numbers! Yet we can use them in the exponent of z.) We can still calculate the mean and variance of T, by evaluating $G_T'(1)$ and $G_T''(1)$ and combining these values in the usual way.

The generating function for P instead of T is

$$P(z) \;=\; \left(\frac{m-1+z}{m}\right)^n \;=\; \sum_{p\geqslant 0} \Pr(P=p)z^p\,.$$

Therefore we have

$$\begin{aligned}
G_T(z) &= \sum_{p\geqslant 0} \Pr(P=p)z^{\alpha+\beta p+\delta[p=0]} \\
&= z^\alpha\Bigl((z^\delta-1)\Pr(P=0) + \sum_{p\geqslant 0}\Pr(P=p)z^{\beta p}\Bigr) \\
&= z^\alpha\biggl((z^\delta-1)\Bigl(\frac{m-1}{m}\Bigr)^n + \Bigl(\frac{m-1+z^\beta}{m}\Bigr)^n\biggr).
\end{aligned}$$

The determination of $\mathrm{Mean}(G_T)$ and $\mathrm{Var}(G_T)$ is now routine:

$$\mathrm{Mean}(G_T) \;=\; G_T'(1) \;=\; \alpha+\beta\frac{n}{m}+\delta\Bigl(\frac{m-1}{m}\Bigr)^n\,; \tag{8.108}$$

$$\begin{aligned}
G_T''(1) &= \alpha(\alpha-1)+2\alpha\beta\frac{n}{m}+\beta(\beta-1)\frac{n}{m}+\beta^2\frac{n(n-1)}{m^2} \\
&\qquad + 2\alpha\delta\Bigl(\frac{m-1}{m}\Bigr)^n + \delta(\delta-1)\Bigl(\frac{m-1}{m}\Bigr)^n\,; \\
\mathrm{Var}(G_T) &= G_T''(1)+G_T'(1)-G_T'(1)^2 \\
&= \beta^2\frac{n(m-1)}{m^2} - 2\beta\delta\Bigl(\frac{m-1}{m}\Bigr)^n\frac{n}{m} \\
&\qquad + \delta^2\biggl(\Bigl(\frac{m-1}{m}\Bigr)^n - \Bigl(\frac{m-1}{m}\Bigr)^{2n}\biggr).
\end{aligned} \tag{8.109}$$

In Chapter 9 we will learn how to estimate quantities like this when m and n are large. If, for example, $m=n$ and $n\to\infty$, the techniques of Chapter 9 will show that the mean and variance of T are respectively $\alpha+\beta+\delta e^{-1}+O(n^{-1})$ and $\beta^2-2\beta\delta e^{-1}+\delta^2(e^{-1}-e^{-2})+O(n^{-1})$. If $m=n/\ln n$ and $n\to\infty$ the corresponding results are

$$\begin{aligned}
\mathrm{Mean}(G_T) &= \beta\ln n+\alpha+\delta/n+O\bigl((\log n)^2/n^2\bigr)\,; \\
\mathrm{Var}(G_T) &= \beta^2\ln n - \bigl((\beta\ln n)^2+2\beta\delta\ln n\bigr)/n + O\bigl((\log n)^3/n^2\bigr)\,.
\end{aligned}$$

Exercises

Warmups

1 What's the probability of doubles in the probability distribution $\Pr_{01}$ of (8.3), when one die is fair and the other is loaded? What's the probability that $S = 7$ is rolled?

2 What's the probability that the top and bottom cards of a randomly shuffled deck are both aces? (All 52! permutations have probability 1/52!.)

3 Stanford's Concrete Math students were asked in 1979 to flip coins until they got heads twice in succession, and to report the number of flips required. The answers were

Why only ten numbers?

The other students either weren't empiricists or they were just too flipped out.

$$3,\ 2,\ 3,\ 5,\ 10,\ 2,\ 6,\ 6,\ 9,\ 2\,.$$

Princeton's Concrete Math students were asked in 1987 to do a similar thing, with the following results:

$$10,\ 2,\ 10,\ 7,\ 5,\ 2,\ 10,\ 6,\ 10,\ 2\,.$$

Estimate the mean and variance, based on (a) the Stanford sample; (b) the Princeton sample.

4 Let $H(z) = F(z)/G(z)$, where $F(1) = G(1) = 1$. Prove that

$$\begin{aligned}\mathrm{Mean}(H) &= \mathrm{Mean}(F) - \mathrm{Mean}(G)\,,\\ \mathrm{Var}(H) &= \mathrm{Var}(F) - \mathrm{Var}(G)\,,\end{aligned}$$

in analogy with (8.38) and (8.39), if the indicated derivatives exist at $z = 1$.

5 Suppose Alice and Bill play the game (8.78) with a biased coin that comes up heads with probability p. Is there a value of p for which the game becomes fair?

6 What does the conditional variance law (8.105) reduce to, when X and Y are independent random variables?

Basics

7 Show that if two dice are loaded with the same probability distribution, the probability of doubles is always at least $\frac{1}{6}$.

8 Let A and B be events such that $A \cup B = \Omega$. Prove that

$$\Pr(\omega \in A \cap B) = \Pr(\omega \in A)\Pr(\omega \in B) - \Pr(\omega \notin A)\Pr(\omega \notin B)\,.$$

9 Prove or disprove: If X and Y are independent random variables, then so are $F(X)$ and $G(Y)$, when F and G are any functions.

10 What's the maximum number of elements that can be medians of a random variable X, according to definition (8.7)?

11 Construct a random variable that has finite mean and infinite variance.

12 Let α be a constant in the range $0 < \alpha < \frac{1}{2}$. We've seen in previous chapters that there is no general closed form for the sum $\sum_{k\leqslant\alpha n}\binom{n}{k}$. Show that it *is* possible, however, to prove that

$$\sum_{k\leqslant\alpha n}\binom{n}{k} \leqslant \frac{2^{n-1}}{(1-2\alpha)^2 n}.$$

Hint: Apply Chebyshev's inequality to a sequence of n coin flips.

13 If $X_1, \ldots, X_{2n}$ are independent random variables with the same distribution, and if α is any real number whatsoever, prove that

$$\Pr\left(\left|\frac{X_1+\cdots+X_{2n}}{2n}-\alpha\right| \leqslant \left|\frac{X_1+\cdots+X_n}{n}-\alpha\right|\right) \geqslant \frac{1}{2}.$$

14 Let $F(z)$ and $G(z)$ be probability generating functions, and let

$$H(z) = p\,F(z) + q\,G(z)$$

where $p+q=1$. (This is called a *mixture* of F and G; it corresponds to flipping a coin and choosing probability distribution F or G depending on whether the coin comes up heads or tails.) Find the mean and variance of H in terms of p, q, and the mean and variance of F and G.

15 If $F(z)$ and $G(z)$ are probability generating functions, we can define another pgf $H(z)$ by "composition":

$$H(z) = F\bigl(G(z)\bigr).$$

Express $\mathrm{Mean}(H)$ and $\mathrm{Var}(H)$ in terms of $\mathrm{Mean}(F)$, $\mathrm{Var}(F)$, $\mathrm{Mean}(G)$, and $\mathrm{Var}(G)$. (Equation (8.92) is a special case.)

16 Find a closed form for the super generating function $\sum_{n\geqslant 0} F_n(z)w^n$, when $F_n(z)$ is the football-fixation generating function defined in (8.53).

17 Let $X_{n,p}$ and $Y_{n,p}$ have the binomial and negative binomial distributions, respectively, with parameters (n,p). (These distributions are defined in (8.57) and (8.60).) Prove that $\Pr(Y_{n,p}\leqslant m) = \Pr(X_{m+n,p}\geqslant n)$. What identity in binomial coefficients does this imply?

18 A random variable X is said to have the *Poisson distribution* with mean μ if $\Pr(X=k) = e^{-\mu}\mu^k/k!$ for all $k\geqslant 0$.

The distribution of fish per unit volume of water.

a What is the pgf of such a random variable?

b What are its mean, variance, and other cumulants?

19 Continuing the previous exercise, let X_1 be a random Poisson variable with mean μ_1, and let X_2 be a random Poisson variable with mean μ_2, independent of X_1.

a What is the probability that $X_1 + X_2 = n$?

b What are the mean, variance, and other cumulants of $2X_1 + 3X_2$?

20 Prove (8.74) and (8.75), the general formulas for mean and variance of the time needed to wait for a given pattern of heads and tails.

21 What does the value of N represent, if H and T are both set equal to $\frac{1}{2}$ in (8.77)?

22 Prove (8.105), the law of conditional expectations and variances.

Homework exercises

23 Let $\Pr_{00}$ be the probability distribution of two fair dice, and let $\Pr_{11}$ be the probability distribution of two loaded dice as given in (8.2). Find all events A such that $\Pr_{00}(A) = \Pr_{11}(A)$. Which of these events depend only on the random variable S? (A probability space with $\Omega = D^2$ has 2^{36} events; only 2^{11} of those events depend on S alone.)

24 Player J rolls $2n+1$ fair dice and removes those that come up ⚅. Player K then calls a number between 1 and 6, rolls the remaining dice, and removes those that show the number called. This process is repeated until no dice remain. The player who has removed the most total dice ($n+1$ or more) is the winner.

a What are the mean and variance of the total number of dice that J removes? *Hint:* The dice are independent.

b What's the probability that J wins, when $n = 2$?

25 Consider a gambling game in which you stake a given amount A and you roll a fair die. If k spots turn up, you multiply your stake by $2(k-1)/5$. (In particular, you double the stake whenever you roll ⚅, but you lose everything if you roll ⚀.) You can stop at any time and reclaim the current stake. What are the mean and variance of your stake after n rolls? (Ignore any effects of rounding to integer amounts of currency.)

26 Find the mean and variance of the number of l-cycles in a random permutation of n elements. (The football victory problem discussed in (8.23), (8.24), and (8.53) is the special case $l = 1$.)

27 Let $X_1, X_2, \ldots, X_n$ be independent samples of the random variable X. Equations (8.19) and (8.20) explain how to estimate the mean and variance of X on the basis of these observations; give an analogous formula for estimating the third cumulant κ_3. (Your formula should be an "unbiased" estimate, in the sense that its expected value should be κ_3.)

28 What is the average length of the coin-flipping game (8.78),

a given that Alice wins?

b given that Bill wins?

29 Alice, Bill, and Computer flip a fair coin until one of the respective patterns $A = \texttt{HHTH}$, $B = \texttt{HTHH}$, or $C = \texttt{THHH}$ appears for the first time. (If only two of these patterns were involved, we know from (8.82) that A would probably beat B, that B would probably beat C, and that C would probably beat A; but all three patterns are simultaneously in the game.) What are each player's chances of winning?

30 The text considers three kinds of variances associated with successful search in a hash table. Actually there are two more: We can consider the average (over k) of the variances (over $h_1, \ldots, h_n$) of $P(h_1, \ldots, h_n; k)$; and we can consider the variance (over k) of the averages (over $h_1, \ldots, h_n$). Evaluate these quantities.

31 An apple is located at vertex A of pentagon ABCDE, and a worm is located two vertices away, at C. Every day the worm crawls with equal probability to one of the two adjacent vertices. Thus after one day the worm is at vertex B with probability $\frac{1}{2}$ and at vertex D with probability $\frac{1}{2}$. After two days, the worm might be back at C again, because it has no memory of previous positions. When it reaches vertex A, it stops to dine.

Schrödinger's worm.

a What are the mean and variance of the number of days until dinner?

b What does Chebyshev's inequality say about the probability that the number of days is 100 or more?

32 Alice and Bill are in the military, stationed in one of the five states Kansas, Nebraska, Missouri, Oklahoma, or Colorado. Initially Alice is in Nebraska and Bill is in Oklahoma. Every month each person is reassigned to an adjacent state, each adjacent state being equally likely. (Here's a diagram of the adjacencies:

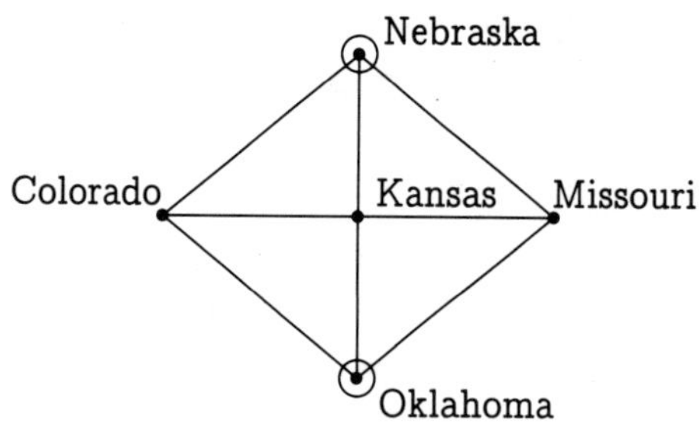

The initial states are circled.) For example, Alice is restationed after the first month to Colorado, Kansas, or Missouri, each with probability $1/3$. Find the mean and variance of the number of months it takes Alice and Bill to find each other. (You may wish to enlist a computer's help.)

Definitely a finite-state situation.

33 Are the random variables X_1 and X_2 in (8.88) independent?

34 Gina is a golfer who has probability $p = .05$ on each stroke of making a "supershot" that gains a stroke over par, probability $q = .91$ of making an ordinary shot, and probability $r = .04$ of making a "subshot" that costs her a stroke with respect to par. (Non-golfers: At each turn she advances 2, 1, or 0 steps toward her goal, with probability p, q, or r, respectively. On a par-m hole, her score is the minimum n such that she has advanced m or more steps after taking n turns. A low score is better than a high score.)

(Use a calculator for the numerical work on this problem.)

a Show that Gina wins a par-4 hole more often than she loses, when she plays against a player who shoots par. (In other words, the probability that her score is less than 4 is greater than the probability that her score is greater than 4.)

b Show that her average score on a par-4 hole is greater than 4. (Therefore she tends to lose against a "steady" player on total points, although she would tend to win in match play by holes.)

Exam problems

35 A die has been loaded with the probability distribution

$$\Pr(⚀) = p_1\,; \qquad \Pr(⚁) = p_2\,; \qquad \ldots\,; \qquad \Pr(⚅) = p_6\,.$$

Let S_n be the sum of the spots after this die has been rolled n times. Find a necessary and sufficient condition on the "loading distribution" such that the two random variables $S_n \bmod 2$ and $S_n \bmod 3$ are independent of each other, for all n.

36 The six faces of a certain die contain the spot patterns

⚀ ⚂ ⚃ ⚄ ⚅ [8 spots]

instead of the usual ⚀ through ⚅.

a Show that there is a way to assign spots to the six faces of another die so that, when these two dice are thrown, the sum of spots has the same probability distribution as the sum of spots on two ordinary dice. (Assume that all 36 face pairs are equally likely.)

b Generalizing, find all ways to assign spots to the $6n$ faces of n dice so that the distribution of spot sums will be the same as the distribution of spot sums on n ordinary dice. (Each face should receive a positive integer number of spots.)

37 Let p_n be the probability that exactly n tosses of a fair coin are needed before heads are seen twice in a row, and let $q_n = \sum_{k \geqslant n} p_k$. Find closed forms for both p_n and q_n in terms of Fibonacci numbers.

38 What is the probability generating function for the number of times you need to roll a fair die until all six faces have turned up? Generalize to m-sided fair dice: Give closed forms for the mean and variance of the number of rolls needed to see l of the m faces. What is the probability that this number will be exactly n?

39 A *Dirichlet probability generating function* has the form

$$P(z) = \sum_{n\geq 1} \frac{p_n}{n^z}.$$

Thus $P(0) = 1$. If X is a random variable with $\Pr(X=n) = p_n$, express $E(X)$, $V(X)$, and $E(\ln X)$ in terms of $P(z)$ and its derivatives.

40 The mth cumulant κ_m of the binomial distribution (8.57) has the form $nf_m(p)$, where f_m is a polynomial of degree m. (For example, $f_1(p) = p$ and $f_2(p) = p - p^2$, because the mean and variance are np and npq.)

a Find a closed form for the coefficient of p^k in $f_m(p)$.

b Prove that $f_m(\frac{1}{2}) = (2^m - 1)B_m/m + [m=1]$, where B_m is the mth Bernoulli number.

41 Let the random variable X_n be the number of flips of a fair coin until heads have turned up a total of n times. Show that $E(X_{n+1}^{-1}) = (-1)^n(\ln 2 + H_{\lfloor n/2\rfloor} - H_n)$. Use the methods of Chapter 9 to estimate this value with an absolute error of $O(n^{-3})$.

42 A certain man has a problem finding work. If he is unemployed on any given morning, there's constant probability p_h (independent of past history) that he will be hired before that evening; but if he's got a job when the day begins, there's constant probability p_f that he'll be laid off by nightfall. Find the average number of evenings on which he will have a job lined up, assuming that he is initially employed and that this process goes on for n days. (For example, if $n = 1$ the answer is $1 - p_f$.)

Does TeX choose optimal line breaks?

43 Find a closed form for the pgf $G_n(z) = \sum_{k\geq 0} p_{k,n} z^n$, where $p_{k,n}$ is the probability that a random permutation of n objects has exactly k cycles. What are the mean and standard deviation of the number of cycles?

44 The athletic department runs an intramural "knockout tournament" for 2^n tennis players as follows. In the first round, the players are paired off randomly, with each pairing equally likely, and 2^{n-1} matches are played. The winners advance to the second round, where the same process produces 2^{n-2} winners. And so on; the kth round has 2^{n-k} randomly chosen matches between the 2^{n-k+1} players who are still undefeated. The nth round produces the champion. Unbeknownst to the tournament organizers, there is actually an ordering among the players, so that x_1 is best, x_2

A peculiar set of tennis players.

is second best, ..., x_n is worst. When x_j plays x_k and $j < k$, the winner is x_j with probability p and x_k with probability $1 - p$, independent of the other matches. We assume that the same probability p applies to all j and k.

a What's the probability that x_1 wins the tournament?

b What's the probability that the nth round (the final match) is between the top two players, x_1 and x_2?

c What's the probability that the best 2^k players are the competitors in the kth-to-last round? (The previous questions were the cases $k = 0$ and $k = 1$.)

d Let $N(n)$ be the number of essentially different tournament results; two tournaments are essentially the same if the matches take place between the same players and have the same winners. Prove that $N(n) = 2^n!$.

e What's the probability that x_2 wins the tournament?

f Prove that if $\frac{1}{2} < p < 1$, the probability that x_j wins is strictly greater than the probability that x_{j+1} wins, for $1 \leqslant j < n$.

45 True sherry is made in Spain according to a multistage system called "Solera." For simplicity we'll assume that the winemaker has only three barrels, called A, B, and C. Every year a third of the wine from barrel C is bottled and replaced by wine from B; then B is topped off with a third of the wine from A; finally A is topped off with new wine. Let $A(z)$, $B(z)$, $C(z)$ be probability generating functions, where the coefficient of z^n is the fraction of n-year-old wine in the corresponding barrel just after the transfers have been made.

"A fast arithmetic computation shows that the sherry is always at least three years old. Taking computation further gives the vertigo."
— Revue du vin de France (Nov 1984)

a Assume that the operation has been going on since time immemorial, so that we have a steady state in which $A(z)$, $B(z)$, and $C(z)$ are the same at the beginning of each year. Find closed forms for these generating functions.

b Find the mean and standard deviation of the age of the wine in each barrel, under the same assumptions. What is the average age of the sherry when it is bottled? How much of it is exactly 25 years old?

c Now take the finiteness of time into account: Suppose that all three barrels contained new wine at the beginning of year 0. What is the average age of the sherry that is bottled at the beginning of year n?

46 Stefan Banach used to carry two boxes of matches, each containing n matches initially. Whenever he needed a light he chose a box at random, each with probability $\frac{1}{2}$, independent of his previous choices. After taking out a match he'd put the box back in its pocket (even if the box became empty — all famous mathematicians used to do this). When his chosen box was empty he'd throw it away and reach for the other box.

a Once he found that the other box was empty too. What's the probability that this occurs? (For $n = 1$ it happens half the time and for $n = 2$ it happens 3/8 of the time.) To answer this part, find a closed form for the generating function $P(w,z) = \sum_{m,n} p_{m,n} w^m z^n$, where $p_{m,n}$ is the probability that, starting with m matches in one box and n in the other, both boxes are empty when an empty box is first chosen. Then find a closed form for $p_{n,n}$.

b Generalizing your answer to part (a), find a closed form for the probability that exactly k matches are in the other box when an empty one is first thrown away.

c Find a closed form for the average number of matches in that other box.

And for the number in the empty box.

47 Some physicians, collaborating with some physicists, recently discovered a pair of microbes that reproduce in a peculiar way. The male microbe, called a *diphage*, has two receptors on its surface; the female microbe, called a *triphage*, has three:

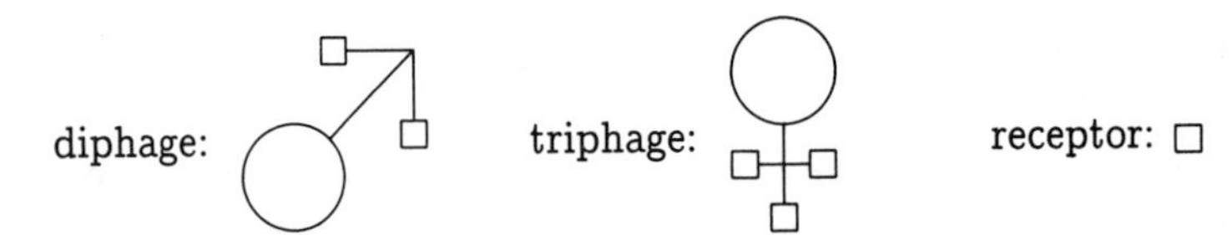

When a culture of diphages and triphages is irradiated with a psi-particle, exactly one of the receptors on one of the phages absorbs the particle; each receptor is equally likely. If it was a diphage receptor, that diphage changes to a triphage; if it was a triphage receptor, that triphage splits into two diphages. Thus if an experiment starts with one diphage, the first psi-particle changes it to a triphage, the second particle splits the triphage into two diphages, and the third particle changes one of the diphages to a triphage. The fourth particle hits either the diphage or the triphage; then there are either two triphages (probability $\frac{2}{5}$) or three diphages (probability $\frac{3}{5}$). Find a closed form for the average number of diphages present, if we begin with a single diphage and irradiate the culture n times with single psi-particles.

48 Five people stand at the vertices of a pentagon, throwing frisbees to each other.

Or, if this pentagon is in Arlington, throwing missiles at each other.

They have two frisbees, initially at adjacent vertices as shown. In each time interval, each frisbee is thrown either to the left or to the right (along an edge of the pentagon) with equal probability. This process continues until one person is the target of two frisbees simultaneously; then the game stops. (All throws are independent of past history.)

Frisbee is a trademark of Wham-O Manufacturing Company.

a Find the mean and variance of the number of pairs of throws.

b Find a closed form for the probability that the game lasts more than 100 steps, in terms of Fibonacci numbers.

49 Luke Snowwalker spends winter vacations at his mountain cabin. The front porch has m pairs of boots and the back porch has n pairs. Every time he goes for a walk he flips a (fair) coin to decide whether to leave from the front porch or the back, and he puts on a pair of boots at that porch and heads off. There's a 50/50 chance that he returns to each porch, independent of his starting point, and he leaves the boots at the porch he returns to. Thus after one walk there will be $m + [-1, 0, \text{or} +1]$ pairs on the front porch and $n - [+1, 0, \text{or} -1]$ pairs on the back porch. If all the boots pile up on one porch and if he decides to leave from the other, he goes without boots and gets frostbite, ending his vacation. Assuming that he continues his walks until the bitter end, let $P_N(m,n)$ be the probability that he completes exactly N nonfrostbitten trips, starting with m pairs on the front porch and n on the back. Thus, if both m and n are positive,

$$P_N(m,n) = \tfrac{1}{4}P_{N-1}(m-1,n+1) + \tfrac{1}{2}P_{N-1}(m,n) + \tfrac{1}{4}P_{N-1}(m+1,n-1)\,;$$

this follows because this first trip is either front/back, front/front, back/back, or back/front, each with probability $\frac{1}{4}$, and $N-1$ trips remain.

a Complete the recurrence for $P_N(m,n)$ by finding formulas that hold when $m = 0$ or $n = 0$. Use the recurrence to obtain equations that hold among the probability generating functions

$$g_{m,n}(z) = \sum_{N \geqslant 0} P_N(m,n)z^N\,.$$

b Differentiate your equations and set $z = 1$, thereby obtaining relations among the quantities $g'_{m,n}(1)$. Solve these equations, thereby determining the mean number of trips before frostbite.

c Show that $g_{m,n}$ has a closed form if we substitute $z = 1/\cos^2\theta$:

$$g_{m,n}\Big(\frac{1}{\cos^2\theta}\Big) = \frac{\sin(2m+1)\theta + \sin(2n+1)\theta}{\sin(2m+2n+2)\theta}\cos\theta\,.$$

50 Consider the function

$$H(z) = 1 + \frac{1-z}{2z}\left(z - 3 + \sqrt{(1-z)(9-z)}\right).$$

The purpose of this problem is to prove that $H(z) = \sum_{k\geq 0} h_k z^k$ is a probability generating function, and to obtain some basic facts about it.

a Let $(1-z)^{3/2}(9-z)^{1/2} = \sum_{k\geq 0} c_k z^k$. Prove that $c_0 = 3$, $c_1 = -14/3$, $c_2 = 37/27$, and $c_{3+l} = 3\sum_k \binom{l}{k}\binom{1/2}{3+k}\left(\frac{8}{9}\right)^{k+3}$ for all $l \geq 0$. *Hint:* Use the identity

$$(9-z)^{1/2} = 3(1-z)^{1/2}\left(1 + \tfrac{8}{9}z/(1-z)\right)^{1/2}$$

and expand the last factor in powers of $z/(1-z)$.

b Use part (a) and exercise 5.81 to show that the coefficients of $H(z)$ are all positive.

c Prove the amazing identity

$$\sqrt{\frac{9-H(z)}{1-H(z)}} = \sqrt{\frac{9-z}{1-z}} + 2.$$

d What are the mean and variance of H?

51 The state lottery in El Dorado uses the payoff distribution H defined in the previous problem. Each lottery ticket costs 1 doubloon, and the payoff is k doubloons with probability h_k. Your chance of winning with each ticket is completely independent of your chance with other tickets; in other words, winning or losing with one ticket does not affect your probability of winning with any other ticket you might have purchased in the same lottery.

a Suppose you start with one doubloon and play this game. If you win k doubloons, you buy k tickets in the second game; then you take the total winnings in the second game and apply all of them to the third; and so on. If none of your tickets is a winner, you're broke and you have to stop gambling. Prove that the pgf of your current holdings after n rounds of such play is

$$1 - \frac{4}{\sqrt{(9-z)/(1-z)} + 2n - 1} + \frac{4}{\sqrt{(9-z)/(1-z)} + 2n + 1}.$$

b Let g_n be the probability that you lose all your money for the first time on the nth game, and let $G(z) = g_1 z + g_2 z^2 + \cdots$. Prove that $G(1) = 1$. (This means that you're bound to lose sooner or later, with probability 1, although you might have fun playing in the meantime.) What are the mean and the variance of G?

c What is the average total number of tickets you buy, if you continue to play until going broke?

d What is the average number of games until you lose everything if you start with two doubloons instead of just one?

A doubledoubloon.

Bonus problems

52 Show that the text's definitions of median and mode for random variables correspond in some meaningful sense to the definitions of median and mode for sequences, when the probability space is finite.

53 Prove or disprove: If X, Y, and Z are random variables with the property that all three pairs (X, Y), (X, Z) and (Y, Z) are independent, then $X + Y$ is independent of Z.

54 Equation (8.20) proves that the average value of $\hat{V}X$ is VX. What is the *variance* of $\hat{V}X$?

55 A normal deck of playing cards contains 52 cards, four each with face values in the set $\{A, 2, 3, 4, 5, 6, 7, 8, 9, 10, J, Q, K\}$. Let X and Y denote the respective face values of the top and bottom cards, and consider the following algorithm for shuffling:

S1 Permute the deck randomly so that each arrangement occurs with probability $1/52!$.

S2 If $X \neq Y$, flip a biased coin that comes up heads with probability p, and go back to step S1 if heads turns up. Otherwise stop.

Each coin flip and each permutation is assumed to be independent of all the other randomizations. What value of p will make X and Y independent random variables after this procedure stops?

56 Generalize the frisbee problem of exercise 48 from a pentagon to an m-gon. What are the mean and variance of the number of collision-free throws in general, when the frisbees are initially at adjacent vertices? Show that, if m is odd, the pgf for the number of throws can be written as a product of coin-flipping distributions:

$$G_m(z) = \prod_{k=1}^{(m-1)/2} \frac{p_k z}{1 - q_k z},$$

$$\text{where } p_k = \sin^2 \frac{(2k-1)\pi}{2m}, \qquad q_k = \cos^2 \frac{(2k-1)\pi}{2m}.$$

Hint: Try the substitution $z = 1/\cos^2 \theta$.

57 Prove that the Penney-ante pattern $\tau_1 \tau_2 \ldots \tau_{l-1} \tau_l$ is always inferior to the pattern $\bar{\tau}_2 \tau_1 \tau_2 \ldots \tau_{l-1}$ when a fair coin is flipped, if $l \geqslant 3$.

58 Are there patterns A and B of heads and tails such that A is longer than B, yet A appears before B more than half the time when a fair coin is being flipped?

59 Let k and n be fixed positive integers with $k < n$.

a Find a closed form for the probability generating function

$$G(w,z) = \frac{1}{m^n}\sum_{h_1=1}^{m}\cdots\sum_{h_n=1}^{m} w^{P(h_1,\ldots,h_n;k)} z^{P(h_1,\ldots,h_n;n)}$$

for the joint distribution of the numbers of probes needed to find the kth and nth items that have been inserted into a hash table with m lists.

b Although the random variables $P(h_1,\ldots,h_n;k)$ and $P(h_1,\ldots,h_n;n)$ are dependent, show that they are somewhat independent:

$$\begin{aligned} &E\big(P(h_1,\ldots,h_n;k)P(h_1,\ldots,h_n;n)\big) \\ &\quad = \big(EP(h_1,\ldots,h_n;k)\big)\big(EP(h_1,\ldots,h_n;n)\big). \end{aligned}$$

60 Use the result of the previous exercise to prove (8.60).

61 Continuing exercise 47, find the *variance* of the number of diphages after n irradiations.

Research problems

62 The *normal distribution* is a non-discrete probability distribution characterized by having all its cumulants zero except the mean and the variance. Is there an easy way to tell if a given sequence of cumulants $\langle \kappa_1, \kappa_2, \kappa_3, \ldots\rangle$ comes from a *discrete* distribution? (All the probabilities must be "atomic" in a discrete distribution.)

63 Is there any sequence $A = \tau_1\tau_2\ldots\tau_{l-1}\tau_l$ of $l \geqslant 3$ heads and tails such that the sequences $H\tau_1\tau_2\ldots\tau_{l-1}$ and $T\tau_1\tau_2\ldots\tau_{l-1}$ both perform equally well against A in the game of Penney ante?

9

Asymptotics

EXACT ANSWERS are great when we can find them; there's something very satisfying about complete knowledge. But there's also a time when approximations are in order. If we run into a sum or a recurrence whose solution doesn't have a closed form (as far as we can tell), we still would like to know something about the answer; we don't have to insist on all or nothing. And even if we do have a closed form, our knowledge might be imperfect, since we might not know how to compare it with other closed forms.

For example, there is (apparently) no closed form for the sum

$$S_n = \sum_{k=0}^{n} \binom{3n}{k}.$$

But it is nice to know that

$$S_n \sim 2\binom{3n}{n}, \qquad \text{as } n \to \infty;$$

Uh oh ... here comes that A-word.

we say that the sum is "asymptotic to" $2\binom{3n}{n}$. It's even nicer to have more detailed information, like

$$S_n = \binom{3n}{n}\left(2 - \frac{4}{n} + O\left(\frac{1}{n^2}\right)\right), \tag{9.1}$$

which gives us a "relative error of order $1/n^2$." But even this isn't enough to tell us how big S_n is, compared with other quantities. Which is larger, S_n or the Fibonacci number F_{4n}? Answer: We have $S_2 = 22 > F_8 = 21$ when $n = 2$; but F_{4n} is eventually larger, because $F_{4n} \sim \phi^{4n}/\sqrt{5}$ and $\phi^4 \approx 6.8541$, while

$$S_n = \sqrt{\frac{3}{\pi n}}(6.75)^n\left(1 - \frac{151}{72n} + O\left(\frac{1}{n^2}\right)\right). \tag{9.2}$$

Our goal in this chapter is to learn how to understand and to derive results like this without great pain.

The word *asymptotic* stems from a Greek root meaning "not falling together." When ancient Greek mathematicians studied conic sections, they considered hyperbolas like the graph of $y = \sqrt{1+x^2}$,

Other words like 'symptom' and 'ptomaine' also come from this root.

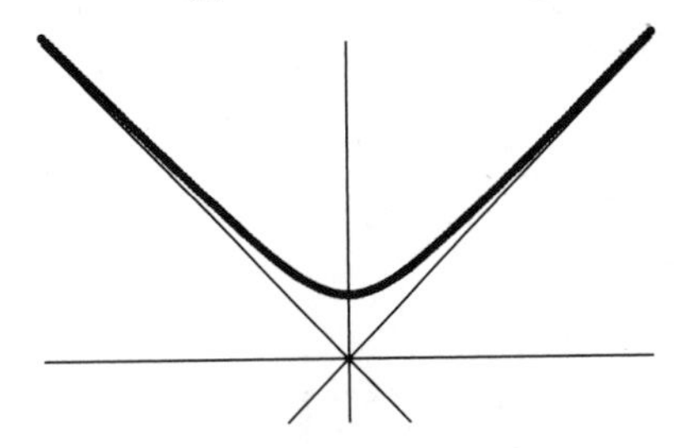

which has the lines $y = x$ and $y = -x$ as "asymptotes." The curve approaches but never quite touches these asymptotes, when $x \to \infty$. Nowadays we use "asymptotic" in a broader sense to mean any approximate value that gets closer and closer to the truth, when some parameter approaches a limiting value. For us, asymptotics means "almost falling together."

Some asymptotic formulas are very difficult to derive, well beyond the scope of this book. We will content ourselves with an introduction to the subject; we hope to acquire a suitable foundation on which further techniques can be built. We will be particularly interested in understanding the definitions of '$\sim$' and 'O' and similar symbols, and we'll study basic ways to manipulate asymptotic quantities.

9.1 A HIERARCHY

Functions of n that occur in practice usually have different "asymptotic growth ratios"; one of them will approach infinity faster than another. We formalize this by saying that

$$f(n) \prec g(n) \iff \lim_{n\to\infty} \frac{f(n)}{g(n)} = 0 . \tag{9.3}$$

This relation is transitive: If $f(n) \prec g(n)$ and $g(n) \prec h(n)$ then $f(n) \prec h(n)$. We also may write $g(n) \succ f(n)$ if $f(n) \prec g(n)$. This notation was introduced in 1871 by Paul du Bois-Reymond [29].

All functions great and small.

For example, $n \prec n^2$; informally we say that n grows more slowly than n^2. In fact,

$$n^\alpha \prec n^\beta \iff \alpha < \beta , \tag{9.4}$$

when α and β are arbitrary real numbers.

There are, of course, many functions of n besides powers of n. We can use the $\prec$ relation to rank lots of functions into an asymptotic pecking order that includes entries like this:

$$1 \prec \log\log n \prec \log n \prec n^{\epsilon} \prec n^{c} \prec n^{\log n} \prec c^{n} \prec n^{n} \prec c^{c^{n}}.$$

(Here ϵ and c are arbitrary constants with $0 < \epsilon < 1 < c$.)

All functions listed here, except 1, go to infinity as n goes to infinity. Thus when we try to place a new function in this hierarchy, we're not trying to determine *whether* it becomes infinite but rather *how fast.*

It helps to cultivate an expansive attitude when we're doing asymptotic analysis: We should THINK BIG, when imagining a variable that approaches infinity. For example, the hierarchy says that $\log n \prec n^{0.0001}$; this might seem wrong if we limit our horizons to teeny-tiny numbers like one googol, $n = 10^{100}$. For in that case, $\log n = 100$, while $n^{0.0001}$ is only $10^{0.01} \approx 1.0233$. But if we go up to a googolplex, $n = 10^{10^{100}}$, then $\log n = 10^{100}$ pales in comparison with $n^{0.0001} = 10^{10^{96}}$.

Even if ϵ is extremely small (smaller than, say, $1/10^{10^{100}}$), the value of $\log n$ will be much smaller than the value of n^{ϵ}, if n is large enough. For if we set $n = 10^{10^{2k}}$, where k is so large that $\epsilon \geqslant 10^{-k}$, we have $\log n = 10^{2k}$ but $n^{\epsilon} \geqslant 10^{10^{k}}$. The ratio $(\log n)/n^{\epsilon}$ therefore approaches zero as $n \to \infty$.

The hierarchy shown above deals with functions that go to infinity. Often, however, we're interested in functions that go to zero, so it's useful to have a similar hierarchy for those functions. We get one by taking reciprocals, because when $f(n)$ and $g(n)$ are never zero we have

A loerarchy?

$$f(n) \prec g(n) \iff \frac{1}{g(n)} \prec \frac{1}{f(n)}. \tag{9.5}$$

Thus, for example, the following functions (except 1) all go to zero:

$$\frac{1}{c^{c^{n}}} \prec \frac{1}{n^{n}} \prec \frac{1}{c^{n}} \prec \frac{1}{n^{\log n}} \prec \frac{1}{n^{c}} \prec \frac{1}{n^{\epsilon}} \prec \frac{1}{\log n} \prec \frac{1}{\log\log n} \prec 1.$$

Let's look at a few other functions to see where they fit in. The number $\pi(n)$ of primes less than or equal to n is known to be approximately $n/\ln n$. Since $1/n^{\epsilon} \prec 1/\ln n \prec 1$, multiplying by n tells us that

$$n^{1-\epsilon} \prec \pi(n) \prec n.$$

We can in fact generalize (9.4) by noticing, for example, that

$$\begin{aligned} n^{\alpha_1}(\log n)^{\alpha_2}(\log\log n)^{\alpha_3} &\prec n^{\beta_1}(\log n)^{\beta_2}(\log\log n)^{\beta_3} \\ &\iff (\alpha_1, \alpha_2, \alpha_3) < (\beta_1, \beta_2, \beta_3). \end{aligned} \tag{9.6}$$

Here '$(\alpha_1, \alpha_2, \alpha_3) < (\beta_1, \beta_2, \beta_3)$' means lexicographic order (dictionary order); in other words, either $\alpha_1 < \beta_1$, or $\alpha_1 = \beta_1$ and $\alpha_2 < \beta_2$, or $\alpha_1 = \beta_1$ and $\alpha_2 = \beta_2$ and $\alpha_3 < \beta_3$.

How about the function $e^{\sqrt{\log n}}$; where does it live in the hierarchy? We can answer questions like this by using the rule

$$e^{f(n)} \prec e^{g(n)} \iff \lim_{n\to\infty}\bigl(f(n)-g(n)\bigr) = -\infty\,, \tag{9.7}$$

which follows in two steps from definition (9.3) by taking logarithms. Consequently

$$1 \prec f(n) \prec g(n) \implies e^{|f(n)|} \prec e^{|g(n)|}\,.$$

And since $1 \prec \log\log n \prec \sqrt{\log n} \prec \epsilon \log n$, we have $\log n \prec e^{\sqrt{\log n}} \prec n^{\epsilon}$.

When two functions $f(n)$ and $g(n)$ have the *same* rate of growth, we write '$f(n) \asymp g(n)$'. The official definition is:

$$f(n) \asymp g(n) \iff |f(n)| \leqslant C|g(n)| \text{ and } |g(n)| \leqslant C|f(n)|\,,$$
$$\text{for some } C \text{ and for all sufficiently large } n. \tag{9.8}$$

This holds, for example, if $f(n)$ is constant and $g(n) = \cos n + \arctan n$. We will prove shortly that it holds whenever $f(n)$ and $g(n)$ are polynomials of the same degree. There's also a stronger relation, defined by the rule

$$f(n) \sim g(n) \iff \lim_{n\to\infty}\frac{f(n)}{g(n)} = 1\,. \tag{9.9}$$

In this case we say that "$f(n)$ is asymptotic to $g(n)$."

G. H. Hardy [148] introduced an interesting and important concept called the class of *logarithmico-exponential functions*, defined recursively as the smallest family $\mathfrak{L}$ of functions satisfying the following properties:

- The constant function $f(n) = \alpha$ is in $\mathfrak{L}$, for all real α.
- The identity function $f(n) = n$ is in $\mathfrak{L}$.
- If $f(n)$ and $g(n)$ are in $\mathfrak{L}$, so is $f(n) - g(n)$.
- If $f(n)$ is in $\mathfrak{L}$, so is $e^{f(n)}$.
- If $f(n)$ is in $\mathfrak{L}$ and is "eventually positive," then $\ln f(n)$ is in $\mathfrak{L}$.

A function $f(n)$ is called "eventually positive" if there is an integer n_0 such that $f(n) > 0$ whenever $n \geqslant n_0$.

We can use these rules to show, for example, that $f(n) + g(n)$ is in $\mathfrak{L}$ whenever $f(n)$ and $g(n)$ are, because $f(n)+g(n) = f(n) - \bigl(0 - g(n)\bigr)$. If $f(n)$ and $g(n)$ are eventually positive members of $\mathfrak{L}$, their product $f(n)g(n) = e^{\ln f(n) + \ln g(n)}$ and quotient $f(n)/g(n) = e^{\ln f(n) - \ln g(n)}$ are in $\mathfrak{L}$; so are functions like $\sqrt{f(n)} = e^{\frac{1}{2}\ln f(n)}$, etc. Hardy proved that every logarithmico-exponential function is eventually positive, eventually negative, or identically zero. Therefore the product and quotient of any two $\mathfrak{L}$-functions is in $\mathfrak{L}$, except that we cannot divide by a function that's identically zero.

Hardy's main theorem about logarithmico-exponential functions is that they form an asymptotic hierarchy: If $f(n)$ and $g(n)$ are any functions in $\mathfrak{L}$, then either $f(n) \prec g(n)$, or $f(n) \succ g(n)$, or $f(n) \asymp g(n)$. In the last case there is, in fact, a constant α such that

$$f(n) \sim \alpha\, g(n).$$

The proof of Hardy's theorem is beyond the scope of this book; but it's nice to know that the theorem exists, because almost every function we ever need to deal with is in $\mathfrak{L}$. In practice, we can generally fit a given function into a given hierarchy without great difficulty.

9.2 O NOTATION

"... wir durch das Zeichen $O(n)$ eine Größe ausdrücken, deren Ordnung in Bezug auf n die Ordnung von n nicht überschreitet; ob sie wirklich Glieder von der Ordnung n in sich enthält, bleibt bei dem bisherigen Schlußverfahren dahingestellt."
— P. Bachmann [14]

A wonderful notational convention for asymptotic analysis was introduced by Paul Bachmann in 1894 and popularized in subsequent years by Edmund Landau and others. We have seen it in formulas like

$$H_n = \ln n + \gamma + O(1/n), \tag{9.10}$$

which tells us that the nth harmonic number is equal to the natural logarithm of n plus Euler's constant, plus a quantity that is "Big Oh of 1 over n." This last quantity isn't specified exactly; but whatever it is, the notation claims that its absolute value is no more than a constant times $1/n$.

The beauty of O-notation is that it suppresses unimportant detail and lets us concentrate on salient features: The quantity $O(1/n)$ is negligibly small, if constant multiples of $1/n$ are unimportant.

Furthermore we get to use O right in the middle of a formula. If we want to express (9.10) in terms of the notations in Section 9.1, we must transpose '$\ln n + \gamma$' to the left side and specify a weaker result like

$$H_n - \ln n - \gamma \prec \frac{\log \log n}{n}$$

or a stronger result like

$$H_n - \ln n - \gamma \asymp \frac{1}{n}.$$

The Big Oh notation allows us to specify an appropriate amount of detail in place, without transposition.

The idea of imprecisely specified quantities can be made clearer if we consider some additional examples. We occasionally use the notation '± 1' to stand for something that is either $+1$ or -1; we don't know (or perhaps we don't care) which it is, yet we can manipulate it in formulas.

N. G. de Bruijn begins his book *Asymptotic Methods in Analysis* by considering a Big Ell notation that helps us understand Big Oh. If we write $L(5)$ for a number whose absolute value is less than 5 (but we don't say what the number is), then we can perform certain calculations without knowing the full truth. For example, we can deduce formulas such as $1 + L(5) = L(6)$; $L(2) + L(3) = L(5)$; $L(2)L(3) = L(6)$; $e^{L(5)} = L(e^5)$; and so on. But we cannot conclude that $L(5) - L(3) = L(2)$, since the left side might be $4 - 0$. In fact, the most we can say is $L(5) - L(3) = L(8)$.

Bachmann's O-notation is similar to L-notation but it's even less precise: $O(\alpha)$ stands for a number whose absolute value is at most a constant times $|\alpha|$. We don't say what the number is and we don't even say what the constant is. Of course the notion of a "constant" is nonsense if there is nothing variable in the picture, so we use O-notation only in contexts when there's at least one quantity (say n) whose value is varying. The formula

It's not nonsense, but it is pointless.

$$f(n) = O\bigl(g(n)\bigr) \qquad \text{for all } n \tag{9.11}$$

means in this context that there is a constant C such that

$$\bigl|f(n)\bigr| \leqslant C\bigl|g(n)\bigr| \qquad \text{for all } n; \tag{9.12}$$

and when $O\bigl(g(n)\bigr)$ stands in the middle of a formula it represents a function $f(n)$ that satisfies (9.12). The values of $f(n)$ are unknown, but we do know that they aren't too large. Similarly, de Bruijn's '$L(n)$' represents an unspecified function $f(n)$ whose values satisfy $\bigl|f(n)\bigr| < |n|$. The main difference between L and O is that O-notation involves an unspecified constant C; each appearance of O might involve a different C, but each C is independent of n.

I've got a little list — I've got a little list, Of annoying terms and details that might well be under ground, And that never would be missed — that never would be missed.

For example, we know that the sum of the first n squares is

$$\Box_n = \tfrac{1}{3}n(n+\tfrac{1}{2})(n+1) = \tfrac{1}{3}n^3 + \tfrac{1}{2}n^2 + \tfrac{1}{6}n .$$

We can write

$$\Box_n = O(n^3)$$

because $|\frac{1}{3}n^3 + \frac{1}{2}n^2 + \frac{1}{6}n| \leqslant \frac{1}{3}|n|^3 + \frac{1}{2}|n|^2 + \frac{1}{6}|n| \leqslant \frac{1}{3}|n^3| + \frac{1}{2}|n^3| + \frac{1}{6}|n^3| = |n^3|$ for all integers n. Similarly, we have the more specific formula

$$\Box_n = \tfrac{1}{3}n^3 + O(n^2);$$

we can also be sloppy and throw away information, saying that

$$\Box_n = O(n^{10}) .$$

Nothing in the definition of O requires us to give a best possible bound.

But wait a minute. What if the variable n isn't an integer? What if we have a formula like $S(x) = \frac{1}{3}x^3 + \frac{1}{2}x^2 + \frac{1}{6}x$, where x is a real number? Then we cannot say that $S(x) = O(x^3)$, because the ratio $S(x)/x^3 = \frac{1}{3} + \frac{1}{2}x^{-1} + \frac{1}{6}x^{-2}$ becomes unbounded when $x \to 0$. And we cannot say that $S(x) = O(x)$, because the ratio $S(x)/x = \frac{1}{3}x^2 + \frac{1}{2}x + \frac{1}{6}$ becomes unbounded when $x \to \infty$. So we apparently can't use O-notation with $S(x)$.

The answer to this dilemma is that variables used with O are generally subject to side conditions. For example, if we stipulate that $|x| \geqslant 1$, or that $x \geqslant \epsilon$ where ϵ is any positive constant, or that x is an integer, then we can write $S(x) = O(x^3)$. If we stipulate that $|x| \leqslant 1$, or that $|x| \leqslant c$ where c is any positive constant, then we can write $S(x) = O(x)$. The O-notation is governed by its environment, by constraints on the variables involved.

These constraints are often specified by a limiting relation. For example, we might say that

$$f(n) = O\bigl(g(n)\bigr) \qquad \text{as } n \to \infty. \tag{9.13}$$

This means that the O-condition is supposed to hold when n is "near" ∞; we don't care what happens unless n is quite large. Moreover, we don't even specify exactly what "near" means; in such cases each appearance of O implicitly asserts the existence of *two* constants C and n_0, such that

$$\bigl|f(n)\bigr| \leqslant C\bigl|g(n)\bigr| \qquad \text{whenever } n \geqslant n_0. \tag{9.14}$$

The values of C and n_0 might be different for each O, but they do not depend on n. Similarly, the notation

You are the fairest of your sex, Let me be your hero; I love you as one over x, As x approaches zero. Positively.

$$f(x) = O\bigl(g(x)\bigr) \qquad \text{as } x \to 0$$

means that there exist two constants C and ϵ such that

$$\bigl|f(x)\bigr| \leqslant C\bigl|g(x)\bigr| \qquad \text{whenever } |x| \leqslant \epsilon. \tag{9.15}$$

The limiting value does not have to be ∞ or 0; we can write

$$\ln z = z - 1 + O\bigl((z-1)^2\bigr) \qquad \text{as } z \to 1,$$

because it can be proved that $|\ln z - z + 1| \leqslant \frac{1}{4}|z-1|^2$ when $|z-1| \leqslant \frac{1}{2}$.

Our definition of O has gradually developed, over a few pages, from something that seemed pretty obvious to something that seems rather complex; we now have O representing an undefined function and either one or two unspecified constants, depending on the environment. This may seem complicated enough for any reasonable notation, but it's still not the whole story! Another subtle consideration lurks in the background. Namely, we need to realize that it's fine to write

$$\tfrac{1}{3}n^3 + \tfrac{1}{2}n^2 + \tfrac{1}{6}n \;=\; O(n^3)\,,$$

but we should *never* write this equality with the sides reversed. Otherwise we could deduce ridiculous things like $n = n^2$ from the identities $n = O(n^2)$ and $n^2 = O(n^2)$. When we work with O-notation and any other formulas that involve imprecisely specified quantities, we are dealing with *one-way equalities*. The right side of an equation does not give more information than the left side, and it may give less; the right is a "crudification" of the left.

From a strictly formal point of view, the notation $O\big(g(n)\big)$ does not stand for a single function $f(n)$, but for the *set* of all functions $f(n)$ such that $|f(n)| \le C|g(n)|$ for some constant C. An ordinary formula $g(n)$ that doesn't involve O-notation stands for the set containing a single function $f(n) = g(n)$. If S and T are sets of functions of n, the notation $S + T$ stands for the set of all functions of the form $f(n) + g(n)$, where $f(n) \in S$ and $g(n) \in T$; other notations like $S-T$, ST, S/T, $\sqrt{S}$, e^S, $\ln S$ are defined similarly. Then an "equation" between such sets of functions is, strictly speaking, a *set inclusion*; the '=' sign really means '$\subseteq$'. These formal definitions put all of our O manipulations on firm logical ground.

"And to auoide the tediouse repetition of these woordes: is equalle to: I will sette as I doe often in woorke use, a paire of parallels, or Gemove lines of one lengthe, thus: ═══, bicause noe .2. thynges, can be moare equalle."
— R. Recorde [246]

For example, the "equation"

$$\tfrac{1}{3}n^3 + O(n^2) \;=\; O(n^3)$$

means that $S_1 \subseteq S_2$, where S_1 is the set of all functions of the form $\frac{1}{3}n^3 + f_1(n)$ such that there exists a constant C_1 with $|f_1(n)| \le C_1|n^2|$, and where S_2 is the set of all functions $f_2(n)$ such that there exists a constant C_2 with $|f_2(n)| \le C_2|n^3|$. We can formally prove this "equation" by taking an arbitrary element of the left-hand side and showing that it belongs to the right-hand side: Given $\frac{1}{3}n^3 + f_1(n)$ such that $|f_1(n)| \le C_1|n^2|$, we must prove that there's a constant C_2 such that $|\frac{1}{3}n^3 + f_1(n)| \le C_2|n^3|$. The constant $C_2 = \frac{1}{3} + C_1$ does the trick, since $n^2 \le |n^3|$ for all integers n.

If '=' really means '$\subseteq$', why don't we use '$\subseteq$' instead of abusing the equals sign? There are four reasons.

First, tradition. Number theorists started using the equals sign with O-notation and the practice stuck. It's sufficiently well established by now that we cannot hope to get the mathematical community to change.

Second, tradition. Computer people are quite used to seeing equals signs abused—for years FORTRAN and BASIC programmers have been writing assignment statements like '$N = N + 1$'. One more abuse isn't much.

Third, tradition. We often read '=' as the word 'is'. For instance we verbalize the formula $H_n = O(\log n)$ by saying "H sub n is Big Oh of log n."

And in English, this 'is' is one-way. We say that a bird is an animal, but we don't say that an animal is a bird; "animal" is a crudification of "bird."

"It is obvious that the sign = is really the wrong sign for such relations, because it suggests symmetry, and there is no such symmetry. ... Once this warning has been given, there is, however, not much harm in using the sign =, and we shall maintain it, for no other reason than that it is customary."
—N. G. de Bruijn [62]

Fourth, for our purposes it's natural. If we limited our use of O-notation to situations where it occupies the whole right side of a formula—as in the harmonic number approximation $H_n = O(\log n)$, or as in the description of a sorting algorithm's running time $T(n) = O(n \log n)$—it wouldn't matter whether we used '=' or something else. But when we use O-notation in the middle of an expression, as we usually do in asymptotic calculations, our intuition is well satisfied if we think of the equals sign as an equality, and if we think of something like $O(1/n)$ as a very small quantity.

So we'll continue to use '=', and we'll continue to regard $O\big(g(n)\big)$ as an incompletely specified function, knowing that we can always fall back on the set-theoretic definition if we must.

But we ought to mention one more technicality while we're picking nits about definitions: If there are several variables in the environment, O-notation formally represents sets of functions of two or more variables, not just one. The domain of each function is every variable that is currently "free" to vary.

This concept can be a bit subtle, because a variable might be defined only in parts of an expression, when it's controlled by a $\sum$ or something similar. For example, let's look closely at the equation

$$\sum_{k=0}^{n} \big(k^2 + O(k)\big) = \tfrac{1}{3}n^3 + O(n^2), \qquad \text{integer } n \geqslant 0. \tag{9.16}$$

The expression $k^2 + O(k)$ on the left stands for the set of all two-variable functions of the form $k^2 + f(k,n)$ such that there exists a constant C with $\big|f(k,n)\big| \leqslant Ck$ for $0 \leqslant k \leqslant n$. The sum of this set of functions, for $0 \leqslant k \leqslant n$, is the set of all functions $g(n)$ of the form

$$\sum_{k=0}^{n} \big(k^2 + f(k,n)\big) = \tfrac{1}{3}n^3 + \tfrac{1}{2}n^2 + \tfrac{1}{6}n + f(0,n) + f(1,n) + \cdots + f(n,n),$$

where f has the stated property. Since we have

$$\begin{aligned} &\big|\tfrac{1}{2}n^2 + \tfrac{1}{6}n + f(0,n) + f(1,n) + \cdots + f(n,n)\big| \\ &\qquad \leqslant \tfrac{1}{2}n^2 + \tfrac{1}{6}n^2 + C\cdot 0 + C\cdot 1 + \cdots + C\cdot n \\ &\qquad < n^2 + C(n^2+n)/2 < (C+1)n^2, \end{aligned}$$

(Now is a good time to do warmup exercises 3 and 4.)

all such functions $g(n)$ belong to the right-hand side of (9.16); therefore (9.16) is true.

People sometimes abuse O-notation by assuming that it gives an exact order of growth; they use it as if it specifies a lower bound as well as an upper bound. For example, an algorithm to sort n numbers might be called inefficient "because its running time is $O(n^2)$." But a running time of $O(n^2)$ does not imply that the running time is not also $O(n)$. There's another notation, Big Omega, for lower bounds:

$$f(n) = \Omega(g(n)) \iff |f(n)| \geqslant C|g(n)| \quad \text{for some } C > 0. \tag{9.17}$$

We have $f(n) = \Omega(g(n))$ if and only if $g(n) = O(f(n))$. A sorting algorithm whose running time is $\Omega(n^2)$ is inefficient compared with one whose running time is $O(n \log n)$, if n is large enough.

Finally there's Big Theta, which specifies an exact order of growth:

Since Ω and Θ are uppercase Greek letters, the O in O-notation must be a capital Greek Omicron. After all, Greeks invented asymptotics.

$$f(n) = \Theta(g(n)) \iff \begin{array}{rl} f(n) &= O(g(n)) \\ \text{and} \quad f(n) &= \Omega(g(n)). \end{array} \tag{9.18}$$

We have $f(n) = \Theta(g(n))$ if and only if $f(n) \asymp g(n)$ in the notation we saw previously, equation (9.8).

Edmund Landau [194] invented a "little oh" notation,

$$\begin{aligned} f(n) &= o(g(n)) \\ &\iff |f(n)| \leqslant \epsilon|g(n)| \quad \begin{array}{l}\text{for all } n \geqslant n_0(\epsilon) \text{ and} \\ \text{for all constants } \epsilon > 0.\end{array} \end{aligned} \tag{9.19}$$

This is essentially the relation $f(n) \prec g(n)$ of (9.3). We also have

$$f(n) \sim g(n) \iff f(n) = g(n) + o(g(n)). \tag{9.20}$$

Many authors use 'o' in asymptotic formulas, but a more explicit 'O' expression is almost always preferable. For example, the average running time of a computer method called "bubblesort" depends on the asymptotic value of the sum $P(n) = \sum_{k=0}^{n} k^{n-k}\, k!/n!$. Elementary asymptotic methods suffice to prove that $P(n) \sim \sqrt{\pi n/2}$, which means that the ratio $P(n)/\sqrt{\pi n/2}$ approaches 1 as $n \to \infty$. However, the true behavior of $P(n)$ is best understood by considering the *difference*, $P(n) - \sqrt{\pi n/2}$, not the ratio:

n	$P(n)/\sqrt{\pi n/2}$	$P(n) - \sqrt{\pi n/2}$
1	0.798	−0.253
10	0.878	−0.484
20	0.904	−0.538
30	0.918	−0.561
40	0.927	−0.575
50	0.934	−0.585

The numerical evidence in the middle column is not very compelling; it certainly is far from a dramatic proof that $P(n)/\sqrt{\pi n/2}$ approaches 1 rapidly, if at all. But the right-hand column shows that $P(n)$ is very close indeed to $\sqrt{\pi n/2}$. Thus we can characterize the behavior of $P(n)$ much better if we can derive formulas of the form

$$P(n) = \sqrt{\pi n/2} + O(1),$$

or even sharper estimates like

$$P(n) = \sqrt{\pi n/2} - \tfrac{2}{3} + O(1/\sqrt{n}).$$

Stronger methods of asymptotic analysis are needed to prove O-results, but the additional effort required to learn these stronger methods is amply compensated by the improved understanding that comes with O-bounds.

Moreover, many sorting algorithms have running times of the form

$$T(n) = A\,n \lg n + B\,n + O(\log n)$$

for some constants A and B. Analyses that stop at $T(n) \sim A\,n\lg n$ don't tell the whole story, and it turns out to be a bad strategy to choose a sorting algorithm based just on its A value. Algorithms with a good 'A' often achieve this at the expense of a bad 'B'. Since $n \lg n$ grows only slightly faster than n, the algorithm that's faster asymptotically (the one with a slightly smaller A value) might be faster only for values of n that never actually arise in practice. Thus, asymptotic methods that allow us to go past the first term and evaluate B are necessary if we are to make the right choice of method.

Also lD, the Duraflame logarithm.

Before we go on to study O, let's talk about one more small aspect of mathematical style. Three different notations for logarithms have been used in this chapter: lg, ln, and log. We often use 'lg' in connection with computer methods, because binary logarithms are often relevant in such cases; and we often use 'ln' in purely mathematical calculations, since the formulas for natural logarithms are nice and simple. But what about 'log'? Isn't this the "common" base-10 logarithm that students learn in high school—the "common" logarithm that turns out to be very uncommon in mathematics and computer science? Yes; and many mathematicians confuse the issue by using 'log' to stand for natural logarithms or binary logarithms. There is no universal agreement here. But we can usually breathe a sigh of relief when a logarithm appears inside O-notation, because O ignores multiplicative constants. There is no difference between $O(\lg n)$, $O(\ln n)$, and $O(\log n)$, as $n \to \infty$; similarly, there is no difference between $O(\lg\lg n)$, $O(\ln\ln n)$, and $O(\log\log n)$. We get to choose whichever we please; and the one with 'log' seems friendlier because it is more pronounceable. Therefore we generally use 'log' in all contexts where it improves readability without introducing ambiguity.

Notice that $\log\log\log n$ is undefined when $n = 2$.

9.3 O MANIPULATION

Like any mathematical formalism, the O-notation has rules of manipulation that free us from the grungy details of its definition. Once we prove that the rules are correct, using the definition, we can henceforth work on a higher plane and forget about actually verifying that one set of functions is contained in another. We don't even need to calculate the constants C that are implied by each O, as long as we follow rules that guarantee the existence of such constants.

The secret of being a bore is to tell everything.
— Voltaire

For example, we can prove once and for all that

$$n^m = O(n^{m'}), \qquad \text{when } m \leqslant m'; \tag{9.21}$$

$$O\big(f(n)\big) + O\big(g(n)\big) = O\big(|f(n)| + |g(n)|\big). \tag{9.22}$$

Then we can say immediately that $\frac{1}{3}n^3 + \frac{1}{2}n^2 + \frac{1}{6}n = O(n^3) + O(n^3) + O(n^3) = O(n^3)$, without the laborious calculations in the previous section.

Here are some more rules that follow easily from the definition:

$$f(n) = O\big(f(n)\big); \tag{9.23}$$

$$c \cdot O\big(f(n)\big) = O\big(f(n)\big), \qquad \text{if } c \text{ is constant}; \tag{9.24}$$

$$O\big(O(f(n))\big) = O\big(f(n)\big); \tag{9.25}$$

$$O\big(f(n)\big)O\big(g(n)\big) = O\big(f(n)g(n)\big); \tag{9.26}$$

$$O\big(f(n)\,g(n)\big) = f(n)O\big(g(n)\big). \tag{9.27}$$

Exercise 9 proves (9.22), and the proofs of the others are similar. We can always replace something of the form on the left by what's on the right, regardless of the side conditions on the variable n.

Equations (9.27) and (9.23) allow us to derive the identity $O\big(f(n)^2\big) = O\big(f(n)\big)^2$. This sometimes helps avoid parentheses, since we can write

$$O(\log n)^2 \qquad \text{instead of} \qquad O\big((\log n)^2\big).$$

Both of these are preferable to '$O(\log^2 n)$', which is ambiguous because some authors use it to mean '$O(\log\log n)$'.

Can we also write

$$O(\log n)^{-1} \qquad \text{instead of} \qquad O\big((\log n)^{-1}\big)\,?$$

No! This is an abuse of notation, since the set of functions $1/O(\log n)$ is neither a subset nor a superset of $O(1/\log n)$. We could legitimately substitute $\Omega(\log n)^{-1}$ for $O\big((\log n)^{-1}\big)$, but this would be awkward. So we'll restrict our use of "exponents outside the O" to constant, positive integer exponents.

(Note: The formula $O(f(n))^2$ does not denote the set of all functions $g(n)^2$ where $g(n)$ is in $O(f(n))$; such functions $g(n)^2$ cannot be negative, but the set $O(f(n))^2$ includes negative functions. In general, when S is a set the notation S^2 stands for the set of all products $s_1 s_2$ with s_1 and s_2 in S, not for the set of all squares s^2 with $s \in S$.)

Power series give us some of the most useful operations of all. If the sum

$$S(z) = \sum_{n \geqslant 0} a_n z^n$$

converges absolutely for some complex number $z = z_0$, then

$$S(z) = O(1), \qquad \text{for all } |z| \leqslant |z_0|.$$

This is obvious, because

$$|S(z)| \leqslant \sum_{n \geqslant 0} |a_n||z|^n \leqslant \sum_{n \geqslant 0} |a_n||z_0|^n = C < \infty.$$

In particular, $S(z) = O(1)$ as $z \to 0$, and $S(1/n) = O(1)$ as $n \to \infty$, provided only that $S(z)$ converges for at least one nonzero value of z. We can use this principle to truncate a power series at any convenient point and estimate the remainder with O. For example, not only is $S(z) = O(1)$, but

$$\begin{aligned} S(z) &= a_0 + O(z), \\ S(z) &= a_0 + a_1 z + O(z^2), \end{aligned}$$

and so on, because

$$S(z) = \sum_{0 \leqslant k < m} a_k z^k + z^m \sum_{n \geqslant m} a_n z^{n-m}$$

and the latter sum is $O(1)$. Table 438 lists some of the most useful asymptotic formulas, half of which are simply based on truncation of power series according to this rule.

Dirichlet series, which are sums of the form $\sum_{k \geqslant 1} a_k / k^z$, can be truncated in a similar way: If a Dirichlet series converges absolutely when $z = z_0$, we can truncate it at any term and get the approximation

$$\sum_{1 \leqslant k < m} a_k / k^z + O(m^{-z}),$$

Remember that $\Re$ stands for "real part."

valid for $\Re z \geqslant \Re z_0$. The asymptotic formula for Bernoulli numbers B_n in Table 438 illustrates this principle.

On the other hand, the asymptotic formulas for H_n, $n!$, and $\pi(n)$ in Table 438 are not truncations of convergent series; if we extended them indefinitely they would diverge for all values of n. This is particularly easy to see in the case of $\pi(n)$, since we have already observed in Section 7.3, Example 5, that the power series $\sum_{k \geqslant 0} k!/(\ln n)^k$ is everywhere divergent. Yet these truncations of divergent series turn out to be useful approximations.

e 438 Asymptotic approximations, valid as $n \to \infty$ and $z \to 0$.

$$H_n = \ln n + \gamma + \frac{1}{2n} - \frac{1}{12n^2} + \frac{1}{120n^4} + O\Big(\frac{1}{n^6}\Big). \quad (9.28)$$

$$n! = \sqrt{2\pi n}\Big(\frac{n}{e}\Big)^n\Big(1 + \frac{1}{12n} + \frac{1}{288n^2} - \frac{139}{51840n^3} + O\Big(\frac{1}{n^4}\Big)\Big). \quad (9.29)$$

$$B_n = 2[n \text{ even}](-1)^{n/2-1}\frac{n!}{(2\pi)^n}\big(1 + 2^{-n} + 3^{-n} + O(4^{-n})\big). \quad (9.30)$$

$$\pi(n) = \frac{n}{\ln n} + \frac{n}{(\ln n)^2} + \frac{2!\,n}{(\ln n)^3} + \frac{3!\,n}{(\ln n)^4} + O\Big(\frac{n}{(\log n)^5}\Big). \quad (9.31)$$

$$e^z = 1 + z + \frac{z^2}{2!} + \frac{z^3}{3!} + \frac{z^4}{4!} + O(z^5). \quad (9.32)$$

$$\ln(1+z) = z - \frac{z^2}{2} + \frac{z^3}{3} - \frac{z^4}{4} + O(z^5). \quad (9.33)$$

$$\frac{1}{1-z} = 1 + z + z^2 + z^3 + z^4 + O(z^5). \quad (9.34)$$

$$(1+z)^\alpha = 1 + \alpha z + \binom{\alpha}{2}z^2 + \binom{\alpha}{3}z^3 + \binom{\alpha}{4}z^4 + O(z^5). \quad (9.35)$$

An asymptotic approximation is said to have *absolute error* $O(g(n))$ if it has the form $f(n)+O(g(n))$ where $f(n)$ doesn't involve O. The approximation has *relative error* $O(g(n))$ if it has the form $f(n)(1+O(g(n)))$ where $f(n)$ doesn't involve O. For example, the approximation for H_n in Table 438 has absolute error $O(n^{-6})$; the approximation for $n!$ has relative error $O(n^{-4})$. (The right-hand side of (9.29) doesn't actually have the required form $f(n) \times (1+O(n^{-4}))$, but we could rewrite it

$$\sqrt{2\pi n}\Big(\frac{n}{e}\Big)^n\Big(1 + \frac{1}{12n} + \frac{1}{288n^2} - \frac{139}{51840n^3}\Big)\big(1 + O(n^{-4})\big)$$

if we wanted to; a similar calculation is the subject of exercise 12.) The absolute error of this approximation is $O(n^{n-3.5}e^{-n})$. Absolute error is related to the number of correct decimal digits to the right of the decimal point if the O term is ignored; relative error corresponds to the number of correct "significant figures."

(Relative error is nice for taking reciprocals, because $1/(1+O(\epsilon)) = 1+O(\epsilon)$.)

We can use truncation of power series to prove the general laws

$$\ln(1+O(f(n))) = O(f(n)), \qquad \text{if } f(n) \prec 1; \quad (9.36)$$

$$e^{O(f(n))} = 1 + O(f(n)), \qquad \text{if } f(n) = O(1). \quad (9.37)$$

(Here we assume that $n \to \infty$; similar formulas hold for $\ln(1 + O(f(x)))$ and $e^{O(f(x))}$ as $x \to 0$.) For example, let $\ln(1 + g(n))$ be any function belonging to the left side of (9.36). Then there are constants C, n_0, and c such that

$$|g(n)| \leqslant C|f(n)| \leqslant c < 1, \qquad \text{for all } n \geqslant n_0.$$

It follows that the infinite sum

$$\ln(1 + g(n)) = g(n) \cdot \left(1 - \tfrac{1}{2}g(n) + \tfrac{1}{3}g(n)^2 - \cdots\right)$$

converges for all $n \geqslant n_0$, and the parenthesized series is bounded by the constant $1 + \frac{1}{2}c + \frac{1}{3}c^2 + \cdots$. This proves (9.36), and the proof of (9.37) is similar. Equations (9.36) and (9.37) combine to give the useful formula

$$\left(1 + O(f(n))\right)^{O(g(n))} = 1 + O\left(f(n)g(n)\right), \qquad \begin{array}{l}\text{if } f(n) \prec 1 \text{ and} \\ f(n)g(n) = O(1).\end{array} \tag{9.38}$$

Problem 1: Return to the Wheel of Fortune.

Let's try our luck now at a few asymptotic problems. In Chapter 3 we derived equation (3.13) for the number of winning positions in a certain game:

$$W = \lfloor N/K \rfloor + \tfrac{1}{2}K^2 + \tfrac{5}{2}K - 3, \qquad K = \lfloor \sqrt[3]{N} \rfloor.$$

And we promised that an asymptotic version of W would be derived in Chapter 9. Well, here we are in Chapter 9; let's try to estimate W, as $N \to \infty$.

The main idea here is to remove the floor brackets, replacing K by $N^{1/3} + O(1)$. Then we can go further and write

$$K = N^{1/3}\left(1 + O(N^{-1/3})\right);$$

this is called "pulling out the large part." (We will be using this trick a lot.) Now we have

$$\begin{aligned} K^2 &= N^{2/3}\left(1 + O(N^{-1/3})\right)^2 \\ &= N^{2/3}\left(1 + O(N^{-1/3})\right) = N^{2/3} + O(N^{1/3}) \end{aligned}$$

by (9.38) and (9.26). Similarly

$$\begin{aligned} \lfloor N/K \rfloor &= N^{1-1/3}\left(1 + O(N^{-1/3})\right)^{-1} + O(1) \\ &= N^{2/3}\left(1 + O(N^{-1/3})\right) + O(1) = N^{2/3} + O(N^{1/3}). \end{aligned}$$

It follows that the number of winning positions is

$$\begin{aligned} W &= N^{2/3} + O(N^{1/3}) + \tfrac{1}{2}\left(N^{2/3} + O(N^{1/3})\right) + O(N^{1/3}) + O(1) \\ &= \tfrac{3}{2}N^{2/3} + O(N^{1/3}). \end{aligned} \tag{9.39}$$

Notice how the O terms absorb one another until only one remains; this is typical, and it illustrates why O-notation is useful in the middle of a formula.

Problem 2: Perturbation of Stirling's formula.

Stirling's approximation for $n!$ is undoubtedly the most famous asymptotic formula of all. We will prove it later in this chapter; for now, let's just try to get better acquainted with its properties. We can write one version of the approximation in the form

$$n! = \sqrt{2\pi n}\left(\frac{n}{e}\right)^n\left(1+\frac{a}{n}+\frac{b}{n^2}+O(n^{-3})\right), \qquad \text{as } n\to\infty, \tag{9.40}$$

for certain constants a and b. Since this holds for all large n, it must also be asymptotically true when n is replaced by $n-1$:

$$\begin{aligned}(n-1)! &= \sqrt{2\pi(n-1)}\left(\frac{n-1}{e}\right)^{n-1}\\ &\quad\times\left(1+\frac{a}{n-1}+\frac{b}{(n-1)^2}+O\big((n-1)^{-3}\big)\right).\end{aligned} \tag{9.41}$$

We know, of course, that $(n-1)! = n!/n$; hence the right-hand side of this formula must simplify to the right-hand side of (9.40), divided by n.

Let us therefore try to simplify (9.41). The first factor becomes tractable if we pull out the large part:

$$\begin{aligned}\sqrt{2\pi(n-1)} &= \sqrt{2\pi n}\,(1-n^{-1})^{1/2}\\ &= \sqrt{2\pi n}\left(1-\frac{1}{2n}-\frac{1}{8n^2}+O(n^{-3})\right).\end{aligned}$$

Equation (9.35) has been used here.

Similarly we have

$$\begin{aligned}\frac{a}{n-1} &= \frac{a}{n}(1-n^{-1})^{-1} = \frac{a}{n}+\frac{a}{n^2}+O(n^{-3});\\ \frac{b}{(n-1)^2} &= \frac{b}{n^2}(1-n^{-1})^{-2} = \frac{b}{n^2}+O(n^{-3});\\ O\big((n-1)^{-3}\big) &= O\big(n^{-3}(1-n^{-1})^{-3}\big) = O(n^{-3}).\end{aligned}$$

The only thing in (9.41) that's slightly tricky to deal with is the factor $(n-1)^{n-1}$, which equals

$$n^{n-1}(1-n^{-1})^{n-1} = n^{n-1}(1-n^{-1})^n\big(1+n^{-1}+n^{-2}+O(n^{-3})\big).$$

(We are expanding everything out until we get a relative error of $O(n^{-3})$, because the relative error of a product is the sum of the relative errors of the individual factors. All of the $O(n^{-3})$ terms will coalesce.)

In order to expand $(1 - n^{-1})^n$, we first compute $\ln(1 - n^{-1})$ and then form the exponential, $e^{n\ln(1-n^{-1})}$:

$$\begin{aligned}
(1-n^{-1})^n &= \exp\bigl(n\ln(1-n^{-1})\bigr)\\
&= \exp\bigl(n(-n^{-1} - \tfrac{1}{2}n^{-2} - \tfrac{1}{3}n^{-3} + O(n^{-4}))\bigr)\\
&= \exp\bigl(-1 - \tfrac{1}{2}n^{-1} - \tfrac{1}{3}n^{-2} + O(n^{-3})\bigr)\\
&= \exp(-1)\cdot\exp(-\tfrac{1}{2}n^{-1})\cdot\exp(-\tfrac{1}{3}n^{-2})\cdot\exp\bigl(O(n^{-3})\bigr)\\
&= \exp(-1)\cdot\bigl(1 - \tfrac{1}{2}n^{-1} + \tfrac{1}{8}n^{-2} + O(n^{-3})\bigr)\\
&\qquad\cdot\bigl(1 - \tfrac{1}{3}n^{-2} + O(n^{-4})\bigr)\cdot\bigl(1 + O(n^{-3})\bigr)\\
&= e^{-1}\bigl(1 - \tfrac{1}{2}n^{-1} - \tfrac{5}{24}n^{-2} + O(n^{-3})\bigr)\,.
\end{aligned}$$

Here we use the notation $\exp z$ instead of e^z, since it allows us to work with a complicated exponent on the main line of the formula instead of in the superscript position. We must expand $\ln(1-n^{-1})$ with absolute error $O(n^{-4})$ in order to end with a relative error of $O(n^{-3})$, because the logarithm is being multiplied by n.

The right-hand side of (9.41) has now been reduced to $\sqrt{2\pi n}$ times n^{n-1}/e^n times a product of several factors:

$$\begin{aligned}
&\bigl(1 - \tfrac{1}{2}n^{-1} - \tfrac{1}{8}n^{-2} + O(n^{-3})\bigr)\\
&\quad\cdot\bigl(1 + n^{-1} + n^{-2} + O(n^{-3})\bigr)\\
&\quad\cdot\bigl(1 - \tfrac{1}{2}n^{-1} - \tfrac{5}{24}n^{-2} + O(n^{-3})\bigr)\\
&\quad\cdot\bigl(1 + an^{-1} + (a+b)n^{-2} + O(n^{-3})\bigr)\,.
\end{aligned}$$

Multiplying these out and absorbing all asymptotic terms into one $O(n^{-3})$ yields

$$1 + an^{-1} + (a + b - \tfrac{1}{12})n^{-2} + O(n^{-3})\,.$$

Hmmm; we were hoping to get $1 + an^{-1} + bn^{-2} + O(n^{-3})$, since that's what we need to match the right-hand side of (9.40). Has something gone awry? No, everything is fine; Table 438 tells us that $a = \frac{1}{12}$, hence $a + b - \frac{1}{12} = b$.

This perturbation argument doesn't prove the validity of Stirling's approximation, but it does prove something: It proves that formula (9.40) cannot be valid unless $a = \frac{1}{12}$. If we had replaced the $O(n^{-3})$ in (9.40) by $cn^{-3} + O(n^{-4})$ and carried out our calculations to a relative error of $O(n^{-4})$, we could have deduced that $b = \frac{1}{288}$. (This is not the easiest way to determine the values of a and b, but it works.)

Problem 3: The nth prime number.

Equation (9.31) is an asymptotic formula for $\pi(n)$, the number of primes that do not exceed n. If we replace n by $p = P_n$, the nth prime number, we have $\pi(p) = n$; hence

$$n = \frac{p}{\ln p} + O\Bigl(\frac{p}{(\log p)^2}\Bigr) \tag{9.42}$$

as $n \to \infty$. Let us try to "solve" this equation for p; then we will know the approximate size of the nth prime.

The first step is to simplify the O term. If we divide both sides by $p/\ln p$, we find that $n \ln p/p \to 1$; hence $p/\ln p = O(n)$ and

$$O\Bigl(\frac{p}{(\log p)^2}\Bigr) = O\Bigl(\frac{n}{\log p}\Bigr) = O\Bigl(\frac{n}{\log n}\Bigr).$$

(We have $(\log p)^{-1} \leqslant (\log n)^{-1}$ because $p \geqslant n$.)

The second step is to transpose the two sides of (9.42), except for the O term. This is legal because of the general rule

$$a_n = b_n + O\bigl(f(n)\bigr) \iff b_n = a_n + O\bigl(f(n)\bigr). \tag{9.43}$$

(Each of these equations follows from the other if we multiply both sides by -1 and then add $a_n + b_n$ to both sides.) Hence

$$\frac{p}{\ln p} = n + O\Bigl(\frac{n}{\log n}\Bigr) = n\bigl(1 + O(1/\log n)\bigr),$$

and we have

$$p = n \ln p\bigl(1 + O(1/\log n)\bigr). \tag{9.44}$$

This is an "approximate recurrence" for $p = P_n$ in terms of itself. Our goal is to change it into an "approximate closed form," and we can do this by unfolding the recurrence asymptotically. So let's try to unfold (9.44).

By taking logarithms of both sides we deduce that

$$\ln p = \ln n + \ln\ln p + O(1/\log n). \tag{9.45}$$

This value can be substituted for $\ln p$ in (9.44), but we would like to get rid of all p's on the right before making the substitution. Somewhere along the line, that last p must disappear; we can't get rid of it in the normal way for recurrences, because (9.44) doesn't specify initial conditions for small p.

One way to do the job is to start by proving the weaker result $p = O(n^2)$. This follows if we square (9.44) and divide by pn^2,

$$\frac{p}{n^2} = \frac{(\ln p)^2}{p}\bigl(1 + O(1/\log n)\bigr),$$

since the right side approaches zero as $n \to \infty$. OK, we know that $p = O(n^2)$; therefore $\log p = O(\log n)$ and $\log\log p = O(\log\log n)$. We can now conclude from (9.45) that

$$\ln p \;=\; \ln n + O(\log\log n)\,;$$

in fact, with this new estimate in hand we can conclude that $\ln\ln p = \ln\ln n + O(\log\log n/\log n)$, and (9.45) now yields

$$\ln p \;=\; \ln n + \ln\ln n + O(\log\log n/\log n)\,.$$

And we can plug this into the right-hand side of (9.44), obtaining

$$p \;=\; n\ln n + n\ln\ln n + O(n)\,.$$

This is the approximate size of the nth prime.

We can refine this estimate by using a better approximation of $\pi(n)$ in place of (9.42). The next term of (9.31) tells us that

$$n \;=\; \frac{p}{\ln p} + \frac{p}{(\ln p)^2} + O\Bigl(\frac{p}{(\log p)^3}\Bigr)\,; \tag{9.46}$$

Get out the scratch paper again, gang.

proceeding as before, we obtain the recurrence

Boo, Hiss.

$$p \;=\; n\ln p\,\bigl(1 + (\ln p)^{-1}\bigr)^{-1}\bigl(1 + O(1/\log n)^2\bigr)\,, \tag{9.47}$$

which has a relative error of $O(1/\log n)^2$ instead of $O(1/\log n)$. Taking logarithms and retaining proper accuracy (but not too much) now yields

$$\begin{aligned}
\ln p \;&=\; \ln n + \ln\ln p + O(1/\log n)\\
&=\; \ln n\Bigl(1 + \frac{\ln\ln p}{\ln n} + O(1/\log n)^2\Bigr)\,;\\
\ln\ln p \;&=\; \ln\ln n + \frac{\ln\ln n}{\ln n} + O\Bigl(\frac{\log\log n}{\log n}\Bigr)^2\,.
\end{aligned}$$

Finally we substitute these results into (9.47) and our answer finds its way out:

$$P_n \;=\; n\ln n + n\ln\ln n - n + n\frac{\ln\ln n}{\ln n} + O\Bigl(\frac{n}{\log n}\Bigr)\,. \tag{9.48}$$

For example, when $n = 10^6$ this estimate comes to $15631363.8 + O(n/\log n)$; the millionth prime is actually 15485863. Exercise 21 shows that a still more accurate approximation to P_n results if we begin with a still more accurate approximation to $\pi(n)$ in place of (9.46).

Problem 4: A sum from an old final exam.

When Concrete Mathematics was first taught at Stanford University during the 1970–1971 term, students were asked for the asymptotic value of the sum

$$S_n = \frac{1}{n^2+1} + \frac{1}{n^2+2} + \cdots + \frac{1}{n^2+n}, \qquad (9.49)$$

with an absolute error of $O(n^{-7})$. Let's imagine that we've just been given this problem on a (take-home) final; what is our first instinctive reaction?

No, we don't panic. Our first reaction is to THINK BIG. If we set $n = 10^{100}$, say, and look at the sum, we see that it consists of n terms, each of which is slightly less than $1/n^2$; hence the sum is slightly less than $1/n$. In general, we can usually get a decent start on an asymptotic problem by taking stock of the situation and getting a ballpark estimate of the answer.

Let's try to improve the rough estimate by pulling out the largest part of each term. We have

$$\frac{1}{n^2+k} = \frac{1}{n^2(1+k/n^2)} = \frac{1}{n^2}\left(1 - \frac{k}{n^2} + \frac{k^2}{n^4} - \frac{k^3}{n^6} + O\Big(\frac{k^4}{n^8}\Big)\right),$$

and so it's natural to try summing all these approximations:

$$\begin{aligned}
\frac{1}{n^2+1} &= \frac{1}{n^2} - \frac{1}{n^4} + \frac{1^2}{n^6} - \frac{1^3}{n^8} + O\Big(\frac{1^4}{n^{10}}\Big)\\
\frac{1}{n^2+2} &= \frac{1}{n^2} - \frac{2}{n^4} + \frac{2^2}{n^6} - \frac{2^3}{n^8} + O\Big(\frac{2^4}{n^{10}}\Big)\\
&\;\;\vdots\\
\frac{1}{n^2+n} &= \frac{1}{n^2} - \frac{n}{n^4} + \frac{n^2}{n^6} - \frac{n^3}{n^8} + O\Big(\frac{n^4}{n^{10}}\Big)\\
\hline
S_n &= \frac{n}{n^2} - \frac{n(n+1)}{2n^4} + \cdots.
\end{aligned}$$

It looks as if we're getting $S_n = n^{-1} - \frac{1}{2}n^{-2} + O(n^{-3})$, based on the sums of the first two columns; but the calculations are getting hairy.

If we persevere in this approach, we will ultimately reach the goal; but we won't bother to sum the other columns, for two reasons: First, the last column is going to give us terms that are $O(n^{-6})$, when $n/2 \leqslant k \leqslant n$, so we will have an error of $O(n^{-5})$; that's too big, and we will have to include yet another column in the expansion. Could the exam-giver have been so sadistic? We suspect that there must be a better way. Second, there is indeed a much better way, staring us right in the face.

Do pajamas have buttons?

Namely, we know a closed form for S_n: It's just $H_{n^2+n} - H_{n^2}$. And we know a good approximation for harmonic numbers, so we just apply it twice:

$$\begin{aligned} H_{n^2+n} &= \ln(n^2+n) + \gamma + \frac{1}{2(n^2+n)} - \frac{1}{12(n^2+n)^2} + O\Big(\frac{1}{n^8}\Big)\,; \\ H_{n^2} &= \ln n^2 + \gamma + \frac{1}{2n^2} - \frac{1}{12n^4} + O\Big(\frac{1}{n^8}\Big)\,. \end{aligned}$$

Now we can pull out large terms and simplify, as we did when looking at Stirling's approximation. We have

$$\begin{aligned} \ln(n^2+n) &= \ln n^2 + \ln\Big(1+\frac{1}{n}\Big) = \ln n^2 + \frac{1}{n} - \frac{1}{2n^2} + \frac{1}{3n^3} - \cdots\,; \\ \frac{1}{n^2+n} &= \frac{1}{n^2} - \frac{1}{n^3} + \frac{1}{n^4} - \cdots\,; \\ \frac{1}{(n^2+n)^2} &= \frac{1}{n^4} - \frac{2}{n^5} + \frac{3}{n^6} - \cdots\,. \end{aligned}$$

So there's lots of helpful cancellation, and we find

$$\begin{aligned} S_n = n^{-1} - \tfrac{1}{2}n^{-2} + \tfrac{1}{3}n^{-3} - \tfrac{1}{4}n^{-4} + \tfrac{1}{5}n^{-5} - \tfrac{1}{6}n^{-6} \\ - \tfrac{1}{2}n^{-3} + \tfrac{1}{2}n^{-4} - \tfrac{1}{2}n^{-5} + \tfrac{1}{2}n^{-6} \\ + \tfrac{1}{6}n^{-5} - \tfrac{1}{4}n^{-6} \end{aligned}$$

plus terms that are $O(n^{-7})$. A bit of arithmetic and we're home free:

$$S_n = n^{-1} - \tfrac{1}{2}n^{-2} - \tfrac{1}{6}n^{-3} + \tfrac{1}{4}n^{-4} - \tfrac{2}{15}n^{-5} + \tfrac{1}{12}n^{-6} + O(n^{-7})\,. \qquad (9.50)$$

It would be nice if we could check this answer numerically, as we did when we derived exact results in earlier chapters. Asymptotic formulas are harder to verify; an arbitrarily large constant may be hiding in a O term, so any numerical test is inconclusive. But in practice, we have no reason to believe that an adversary is trying to trap us, so we can assume that the unknown O-constants are reasonably small. With a pocket calculator we find that $S_4 = \frac{1}{17} + \frac{1}{18} + \frac{1}{19} + \frac{1}{20} = 0.2170107$; and our asymptotic estimate when $n = 4$ comes to

$$\tfrac{1}{4}\Big(1 + \tfrac{1}{4}\Big(-\tfrac{1}{2} + \tfrac{1}{4}\Big(-\tfrac{1}{6} + \tfrac{1}{4}\Big(\tfrac{1}{4} + \tfrac{1}{4}\Big(-\tfrac{2}{15} + \tfrac{1}{4}\cdot\tfrac{1}{12}\Big)\Big)\Big)\Big)\Big) = 0.2170125\,.$$

If we had made an error of, say, $\frac{1}{12}$ in the term for n^{-6}, a difference of $\frac{1}{12}\frac{1}{4096}$ would have shown up in the fifth decimal place; so our asymptotic answer is probably correct.

Problem 5: An infinite sum.

We turn now to an asymptotic question posed by Solomon Golomb [122]: What is the approximate value of

$$S_n = \sum_{k\geq 1} \frac{1}{kN_n(k)^2}, \tag{9.51}$$

where $N_n(k)$ is the number of digits required to write k in radix n notation?

First let's try again for a ballpark estimate. The number of digits, $N_n(k)$, is approximately $\log_n k = \log k/\log n$; so the terms of this sum are roughly $(\log n)^2/k(\log k)^2$. Summing on k gives $\approx (\log n)^2 \sum_{k\geq 2} 1/k(\log k)^2$, and this sum converges to a constant value because it can be compared to the integral

$$\int_2^\infty \frac{dx}{x(\ln x)^2} = -\frac{1}{\ln x}\bigg|_2^\infty = \frac{1}{\ln 2}.$$

Therefore we expect S_n to be about $C(\log n)^2$.

Hand-wavy analyses like this are useful for orientation, but we need better estimates to solve the problem. One idea is to express $N_n(k)$ exactly:

$$N_n(k) = \lfloor \log_n k \rfloor + 1. \tag{9.52}$$

Thus, for example, k has three radix n digits when $n^2 \leqslant k < n^3$, and this happens precisely when $\lfloor \log_n k \rfloor = 2$. It follows that $N_n(k) > \log_n k$, hence $S_n = \sum_{k\geq 1} 1/kN_n(k)^2 < 1 + (\log n)^2 \sum_{k\geq 2} 1/k(\log k)^2$.

Proceeding as in Problem 1, we can try to write $N_n(k) = \log_n k + O(1)$ and substitute this into the formula for S_n. The term represented here by $O(1)$ is always between 0 and 1, and it is about $\frac{1}{2}$ on the average, so it seems rather well-behaved. But still, this isn't a good enough approximation to tell us about S_n; it gives us zero significant figures (that is, high relative error) when k is small, and these are the terms that contribute the most to the sum. We need a different idea.

The key (as in Problem 4) is to use our manipulative skills to put the sum into a more tractable form, before we resort to asymptotic estimates. We can introduce a new variable of summation, $m = N_n(k)$:

$$\begin{aligned} S_n &= \sum_{k,m\geq 1} \frac{\big[m = N_n(k)\big]}{km^2} \\ &= \sum_{k,m\geq 1} \frac{[n^{m-1} \leqslant k < n^m]}{km^2} \\ &= \sum_{m\geq 1} \frac{1}{m^2}\big(H_{n^m-1} - H_{n^{m-1}-1}\big). \end{aligned}$$

This may look worse than the sum we began with, but it's actually a step forward, because we have very good approximations for the harmonic numbers.

Still, we hold back and try to simplify some more. No need to rush into asymptotics. Summation by parts allows us to group the terms for each value of H_{n^m-1} that we need to approximate:

$$S_n = \sum_{k\geqslant 1} H_{n^k-1}\Big(\frac{1}{k^2}-\frac{1}{(k+1)^2}\Big).$$

For example, H_{n^2-1} is multiplied by $1/2^2$ and then by $-1/3^2$. (We have used the fact that $H_{n^0-1} = H_0 = 0$.)

Now we're ready to expand the harmonic numbers. Our experience with estimating $(n-1)!$ has taught us that it will be easier to estimate H_{n^k} than H_{n^k-1}, since the (n^k-1)'s will be messy; therefore we write

$$\begin{aligned} H_{n^k-1} &= H_{n^k}-\frac{1}{n^k} = \ln n^k+\gamma+\frac{1}{2n^k}+O\Big(\frac{1}{n^{2k}}\Big)-\frac{1}{n^k}\\ &= k\ln n+\gamma-\frac{1}{2n^k}+O\Big(\frac{1}{n^{2k}}\Big).\end{aligned}$$

Our sum now reduces to

$$\begin{aligned} S_n &= \sum_{k\geqslant 1}\Big(k\ln n+\gamma-\frac{1}{2n^k}+O\Big(\frac{1}{n^{2k}}\Big)\Big)\Big(\frac{1}{k^2}-\frac{1}{(k+1)^2}\Big)\\ &= (\ln n)\Sigma_1+\gamma\Sigma_2-\tfrac{1}{2}\Sigma_3(n)+O\big(\Sigma_3(n^2)\big). \end{aligned} \tag{9.53}$$

There are four easy pieces left: Σ_1, Σ_2, $\Sigma_3(n)$, and $\Sigma_3(n^2)$.

Let's do the Σ_3's first, since $\Sigma_3(n^2)$ is the O term; then we'll see what sort of error we're getting. (There's no sense carrying out other calculations with perfect accuracy if they will be absorbed into a O anyway.) This sum is simply a power series,

Into a Big Oh.

$$\Sigma_3(x) = \sum_{k\geqslant 1}\Big(\frac{1}{k^2}-\frac{1}{(k+1)^2}\Big)x^{-k},$$

and the series converges when $x \geqslant 1$ so we can truncate it at any desired point. If we stop $\Sigma_3(n^2)$ at the term for $k=1$, we get $\Sigma_3(n^2) = O(n^{-2})$; hence (9.53) has an absolute error of $O(n^{-2})$. (To decrease this absolute error, we could use a better approximation to H_{n^k}; but $O(n^{-2})$ is good enough for now.) If we truncate $\Sigma_3(n)$ at the term for $k=2$, we get

$$\Sigma_3(n) = \tfrac{3}{4}n^{-1}+O(n^{-2});$$

this is all the accuracy we need.

We might as well do Σ_2 now, since it is so easy:

$$\Sigma_2 = \sum_{k\geqslant 1}\left(\frac{1}{k^2} - \frac{1}{(k+1)^2}\right).$$

This is the telescoping series $(1-\frac{1}{4}) + (\frac{1}{4}-\frac{1}{9}) + (\frac{1}{9}-\frac{1}{16}) + \cdots = 1$.

Finally, Σ_1 gives us the leading term of S_n, the coefficient of $\ln n$ in (9.53):

$$\Sigma_1 = \sum_{k\geqslant 1} k\left(\frac{1}{k^2} - \frac{1}{(k+1)^2}\right).$$

This is $(1-\frac{1}{4}) + (\frac{2}{4}-\frac{2}{9}) + (\frac{3}{9}-\frac{3}{16}) + \cdots = \frac{1}{1}+\frac{1}{4}+\frac{1}{9}+\cdots = H_\infty^{(2)} = \pi^2/6$. (If we hadn't applied summation by parts earlier, we would have seen directly that $S_n \sim \sum_{k\geqslant 1}(\ln n)/k^2$, because $H_{n^k-1} - H_{n^{k-1}-1} \sim \ln n$; so summation by parts didn't help us to evaluate the leading term, although it did make some of our other work easier.)

Now we have evaluated each of the Σ's in (9.53), so we can put everything together and get the answer to Golomb's problem:

$$S_n = \frac{\pi^2}{6}\ln n + \gamma - \frac{3}{8n} + O\left(\frac{1}{n^2}\right). \tag{9.54}$$

Notice that this grows more slowly than our original hand-wavy estimate of $C(\log n)^2$. Sometimes a discrete sum fails to obey a continuous intuition.

Problem 6: Big Phi.

Near the end of Chapter 4, we observed that the number of fractions in the Farey series $\mathcal{F}_n$ is $1+\Phi(n)$, where

$$\Phi(n) = \varphi(1) + \varphi(2) + \cdots + \varphi(n);$$

and we showed in (4.62) that

$$\Phi(n) = \frac{1}{2}\sum_{k\geqslant 1}\mu(k)\lfloor n/k\rfloor\lfloor 1+n/k\rfloor. \tag{9.55}$$

Let us now try to estimate $\Phi(n)$ when n is large. (It was sums like this that led Bachmann to invent O-notation in the first place.)

Thinking BIG tells us that $\Phi(n)$ will probably be proportional to n^2. For if the final factor were just $\lfloor n/k\rfloor$ instead of $\lfloor 1+n/k\rfloor$, we would have $|\Phi(n)| \leqslant \frac{1}{2}\sum_{k\geqslant 1}\lfloor n/k\rfloor^2 \leqslant \frac{1}{2}\sum_{k\geqslant 1}(n/k)^2 = \frac{\pi^2}{12}n^2$, because the Möbius function $\mu(k)$ is either -1, 0, or $+1$. The additional '$1+$' in that final factor adds $\sum_{k\geqslant 1}\mu(k)\lfloor n/k\rfloor$; but this is zero for $k>n$, so it cannot be more than $nH_n = O(n\log n)$ in absolute value.

This preliminary analysis indicates that we'll find it advantageous to write

$$\begin{aligned}\Phi(n) &= \frac{1}{2}\sum_{k=1}^{n}\mu(k)\Bigl(\Bigl(\frac{n}{k}\Bigr)+O(1)\Bigr)^2 = \frac{1}{2}\sum_{k=1}^{n}\mu(k)\Bigl(\Bigl(\frac{n}{k}\Bigr)^2+O\Bigl(\frac{n}{k}\Bigr)\Bigr)\\ &= \frac{1}{2}\sum_{k=1}^{n}\mu(k)\Bigl(\frac{n}{k}\Bigr)^2+\sum_{k=1}^{n}O\Bigl(\frac{n}{k}\Bigr)\\ &= \frac{1}{2}\sum_{k=1}^{n}\mu(k)\Bigl(\frac{n}{k}\Bigr)^2 + O(n\log n)\,.\end{aligned}$$

This removes the floors; the remaining problem is to evaluate the unfloored sum $\frac{1}{2}\sum_{k=1}^{n}\mu(k)n^2/k^2$ with an accuracy of $O(n\log n)$; in other words, we want to evaluate $\sum_{k=1}^{n}\mu(k)1/k^2$ with an accuracy of $O(n^{-1}\log n)$. But that's easy; we can simply run the sum all the way up to $k=\infty$, because the newly added terms are

$$\begin{aligned}\sum_{k>n}\frac{\mu(k)}{k^2} &= O\Bigl(\sum_{k>n}\frac{1}{k^2}\Bigr) = O\Bigl(\sum_{k>n}\frac{1}{k(k-1)}\Bigr)\\ &= O\Bigl(\sum_{k>n}\Bigl(\frac{1}{k-1}-\frac{1}{k}\Bigr)\Bigr) = O\Bigl(\frac{1}{n}\Bigr)\,.\end{aligned}$$

We proved in (7.88) that $\sum_{k\geqslant 1}\mu(k)/k^z = 1/\zeta(z)$. Hence $\sum_{k\geqslant 1}\mu(k)/k^2 = 1/\bigl(\sum_{k\geqslant 1}1/k^2\bigr) = 6/\pi^2$, and we have our answer:

$$\Phi(n) = \frac{3}{\pi^2}n^2 + O(n\log n)\,. \tag{9.56}$$

9.4 TWO ASYMPTOTIC TRICKS

Now that we have some facility with O manipulations, let's look at what we've done from a slightly higher perspective. Then we'll have some important weapons in our asymptotic arsenal, when we need to do battle with tougher problems.

Trick 1: Bootstrapping.

When we estimated the nth prime P_n in Problem 3 of Section 9.3, we solved an asymptotic recurrence of the form

$$P_n = n\ln P_n\bigl(1+O(1/\log n)\bigr)\,.$$

We proved that $P_n = n\ln n + O(n)$ by first using the recurrence to show the weaker result $O(n^2)$. This is a special case of a general method called *bootstrapping*, in which we solve a recurrence asymptotically by starting with a rough estimate and plugging it into the recurrence; in this way we can often derive better and better estimates, "pulling ourselves up by our bootstraps."

Here's another problem that illustrates bootstrapping nicely: What is the asymptotic value of the coefficient $g_n = [z^n]\,G(z)$ in the generating function

$$G(z) \;=\; \exp\Bigl(\sum_{k\geqslant 1}\frac{z^k}{k^2}\Bigr)\,, \tag{9.57}$$

as $n \to \infty$? If we differentiate this equation with respect to z, we find

$$G'(z) \;=\; \sum_{n=0}^{\infty} n g_n z^{n-1} \;=\; \Bigl(\sum_{k\geqslant 1}\frac{z^{k-1}}{k}\Bigr)\,G(z)\,;$$

equating coefficients of z^{n-1} on both sides gives the recurrence

$$n g_n \;=\; \sum_{0\leqslant k<n}\frac{g_k}{n-k}\,. \tag{9.58}$$

Our problem is equivalent to finding an asymptotic formula for the solution to (9.58), with the initial condition $g_0 = 1$. The first few values

n	0	1	2	3	4	5	6
g_n	1	1	$\frac{3}{4}$	$\frac{19}{36}$	$\frac{107}{288}$	$\frac{641}{2400}$	$\frac{51103}{259200}$

don't reveal much of a pattern, and the integer sequence $\langle n!^2 g_n\rangle$ doesn't appear in Sloane's *Handbook* [270]; therefore a closed form for g_n seems out of the question, and asymptotic information is probably the best we can hope to derive.

Our first handle on this problem is the observation that $0 < g_n \leqslant 1$ for all $n \geqslant 0$; this is easy to prove by induction. So we have a start:

$$g_n \;=\; O(1)\,.$$

This equation can, in fact, be used to "prime the pump" for a bootstrapping operation: Plugging it in on the right of (9.58) yields

$$n g_n \;=\; \sum_{0\leqslant k<n}\frac{O(1)}{n-k} \;=\; H_n O(1) \;=\; O(\log n)\,;$$

hence we have

$$g_n \;=\; O\Bigl(\frac{\log n}{n}\Bigr)\,, \qquad \text{for } n > 1.$$

And we can bootstrap yet again:

$$\begin{aligned} ng_n &= \frac{1}{n} + \sum_{0<k<n} \frac{O\bigl((1+\log k)/k\bigr)}{n-k} \\ &= \frac{1}{n} + \sum_{0<k<n} \frac{O(\log n)}{k(n-k)} \\ &= \frac{1}{n} + \sum_{0<k<n} \Bigl(\frac{1}{k} + \frac{1}{n-k}\Bigr)\frac{O(\log n)}{n} \\ &= \frac{1}{n} + \frac{2}{n}H_{n-1}O(\log n) \;=\; \frac{1}{n}O(\log n)^2 , \end{aligned}$$

obtaining

$$g_n = O\Bigl(\frac{\log n}{n}\Bigr)^2 . \tag{9.59}$$

Will this go on forever? Perhaps we'll have $g_n = O(n^{-1}\log n)^m$ for all m.

Actually no; we have just reached a point of diminishing returns. The next attempt at bootstrapping involves the sum

$$\begin{aligned} \sum_{0<k<n} \frac{1}{k^2(n-k)} &= \sum_{0<k<n} \Bigl(\frac{1}{nk^2} + \frac{1}{n^2k} + \frac{1}{n^2(n-k)}\Bigr) \\ &= \frac{1}{n}H_{n-1}^{(2)} + \frac{2}{n^2}H_{n-1} , \end{aligned}$$

which is $\Omega(n^{-1})$; so we cannot get an estimate for g_n that falls below $\Omega(n^{-2})$.

In fact, we now know enough about g_n to apply our old trick of pulling out the largest part:

$$\begin{aligned} ng_n &= \sum_{0\le k<n} \frac{g_k}{n} + \sum_{0\le k<n} g_k\Bigl(\frac{1}{n-k} - \frac{1}{n}\Bigr) \\ &= \frac{1}{n}\sum_{k\ge 0} g_k - \frac{1}{n}\sum_{k\ge n} g_k + \frac{1}{n}\sum_{0\le k<n} \frac{kg_k}{n-k} . \end{aligned} \tag{9.60}$$

The first sum here is $G(1) = \exp(\frac{1}{1} + \frac{1}{4} + \frac{1}{9} + \cdots) = e^{\pi^2/6}$, because $G(z)$ converges for all $|z| \le 1$. The second sum is the tail of the first; we can get an upper bound by using (9.59):

$$\sum_{k\ge n} g_k = O\Bigl(\sum_{k\ge n} \frac{(\log k)^2}{k^2}\Bigr) = O\Bigl(\frac{(\log n)^2}{n}\Bigr) .$$

This last estimate follows because, for example,

$$\sum_{k>n} \frac{(\log k)^2}{k^2} < \sum_{m\geqslant 1} \sum_{n^m<k\leqslant n^{m+1}} \frac{(\log n^{m+1})^2}{k(k-1)} < \sum_{m\geqslant 1} \frac{(m+1)^2(\log n)^2}{n^m} .$$

(Exercise 54 discusses a more general way to estimate such tails.)

The third sum in (9.60) is

$$O\biggl(\sum_{0\leqslant k<n} \frac{(\log n)^2}{k(n-k)}\biggr) = O\biggl(\frac{(\log n)^3}{n}\biggr),$$

by an argument that's already familiar. So (9.60) proves that

$$g_n = \frac{e^{\pi^2/6}}{n^2} + O\bigl(\log n/n\bigr)^3 . \qquad (9.61)$$

Finally, we can feed this formula back into the recurrence, bootstrapping once more; the result is

$$g_n = \frac{e^{\pi^2/6}}{n^2} + O(\log n/n^3) . \qquad (9.62)$$

(Exercise 23 peeks inside the remaining O term.)

Trick 2: Trading tails.

We derived (9.62) in somewhat the same way we derived the asymptotic value (9.56) of $\Phi(n)$: In both cases we started with a finite sum but got an asymptotic value by considering an infinite sum. We couldn't simply get the infinite sum by introducing O into the summand; we had to be careful to use one approach when k was small and another when k was large.

Those derivations were special cases of an important three-step asymptotic summation method we will now discuss in greater generality. Whenever we want to estimate the value of $\sum_k a_k(n)$, we can try the following approach:

(This important method was pioneered by Laplace [195'].)

1. First break the sum into two disjoint ranges, D_n and T_n. The summation over D_n should be the "dominant" part, in the sense that it includes enough terms to determine the significant digits of the sum, when n is large. The summation over the other range T_n should be just the "tail" end, which contributes little to the overall total.

2. Find an asymptotic estimate

$$a_k(n) = b_k(n) + O\bigl(c_k(n)\bigr)$$

that is valid when $k \in D_n$. The O bound need not hold when $k \in T_n$.

3 Now prove that each of the three sums

$$\Sigma_a = \sum_{k \in T_n} a_k(n), \quad \Sigma_b = \sum_{k \in T_n} b_k(n), \quad \Sigma_c = \sum_{k \in D_n} |c_k(n)| \qquad (9.63)$$

is small.

If all three steps can be completed successfully, we have a good estimate:

$$\sum_{k \in D_n \cup T_n} a_k(n) = \sum_{k \in D_n \cup T_n} b_k(n) + O(\Sigma_a(n)) + O(\Sigma_b(n)) + O(\Sigma_c(n)).$$

Here's why. We can "chop off" the tail of the given sum, getting a good estimate in the range D_n where a good estimate is necessary:

$$\sum_{k \in D_n} a_k(n) = \sum_{k \in D_n} (b_k(n) + O(c_k(n))) = \sum_{k \in D_n} b_k(n) + O(\Sigma_c(n)).$$

And we can replace the tail with another one, even though the new tail might be a terrible approximation to the old, because the tails don't really matter:

$$\begin{aligned} \sum_{k \in T_n} a_k(n) &= \sum_{k \in T_n} (b_k(n) - b_k(n) + a_k(n)) \\ &= \sum_{k \in T_n} b_k(n) + O(\Sigma_b(n)) + O(\Sigma_a(n)). \end{aligned}$$

Asymptotics is the art of knowing where to be sloppy and where to be precise.

When we evaluated the sum in (9.60), for example, we had

$$\begin{aligned} a_k(n) &= [0 \leqslant k < n] g_k/(n-k), \\ b_k(n) &= g_k/n, \\ c_k(n) &= k g_k / n(n-k); \end{aligned}$$

the ranges of summation were

$$D_n = \{0, 1, \ldots, n-1\}, \qquad T_n = \{n, n+1, \ldots\};$$

and we found that

$$\Sigma_a(n) = 0, \quad \Sigma_b(n) = O((\log n)^2/n^2), \quad \Sigma_c(n) = O((\log n)^3/n^2).$$

This led to (9.61).

Similarly, when we estimated $\Phi(n)$ in (9.55) we had

$$\begin{aligned} a_k(n) &= \mu(k)\lfloor n/k \rfloor \lfloor 1+n/k \rfloor, \quad b_k(n) = \mu(k) n^2/k^2, \quad c_k(n) = n/k; \\ D_n &= \{1, 2, \ldots, n\}, \qquad T_n = \{n+1, n+2, \ldots\}. \end{aligned}$$

We derived (9.56) by observing that $\Sigma_a(n) = 0$, $\Sigma_b(n) = O(n)$, and $\Sigma_c(n) = O(n \log n)$.

Here's another example where tail switching is effective. (Unlike our previous examples, this one illustrates the trick in its full generality, with $\Sigma_a(n) \neq 0$.) We seek the asymptotic value of

Also, horses switch their tails when feeding time approaches.

$$L_n = \sum_{k \geqslant 0} \frac{\ln(n+2^k)}{k!} .$$

The big contributions to this sum occur when k is small, because of the k! in the denominator. In this range we have

$$\ln(n+2^k) = \ln n + \frac{2^k}{n} - \frac{2^{2k}}{2n^2} + O\Big(\frac{2^{3k}}{n^3}\Big) . \qquad (9.64)$$

We can prove that this estimate holds for $0 \leqslant k < \lfloor \lg n \rfloor$, since the original terms that have been truncated with O are bounded by the convergent series

$$\sum_{m \geqslant 3} \frac{2^{km}}{mn^m} \leqslant \frac{2^{3k}}{n^3} \sum_{m \geqslant 3} \frac{2^{k(m-3)}}{n^{m-3}} \leqslant \frac{2^{3k}}{n^3}\Big(1 + \frac{1}{2} + \frac{1}{4} + \cdots\Big) = \frac{2^{3k}}{n^3} \cdot 2 .$$

(In this range, $2^k/n \leqslant 2^{\lfloor \lg n \rfloor - 1}/n \leqslant \frac{1}{2}$.)

Therefore we can apply the three-step method just described, with

$$\begin{aligned} a_k(n) &= \ln(n+2^k)/k! , \\ b_k(n) &= (\ln n + 2^k/n - 4^k/2n^2)/k! , \\ c_k(n) &= 8^k/n^3 k! ; \end{aligned}$$

$$\begin{aligned} D_n &= \{0, 1, \ldots, \lfloor \lg n \rfloor - 1\} , \\ T_n &= \{\lfloor \lg n \rfloor, \lfloor \lg n \rfloor + 1, \ldots\} . \end{aligned}$$

All we have to do is find good bounds on the three Σ's in (9.63), and we'll know that $\sum_{k \geqslant 0} a_k(n) \approx \sum_{k \geqslant 0} b_k(n)$.

The error we have committed in the dominant part of the sum, $\Sigma_c(n) = \sum_{k \in D_n} 8^k/n^3 k!$, is obviously bounded by $\sum_{k \geqslant 0} 8^k/n^3 k! = e^8/n^3$, so it can be replaced by $O(n^{-3})$. The new tail error is

$$\begin{aligned} \big|\Sigma_b(n)\big| &= \Big| \sum_{k \geqslant \lfloor \lg n \rfloor} b_k(n) \Big| \\ &< \sum_{k \geqslant \lfloor \lg n \rfloor} \frac{\ln n + 2^k + 4^k}{k!} \\ &< \frac{\ln n + 2^{\lfloor \lg n \rfloor} + 4^{\lfloor \lg n \rfloor}}{\lfloor \lg n \rfloor !} \sum_{k \geqslant 0} \frac{4^k}{k!} = O\Big(\frac{n^2}{\lfloor \lg n \rfloor !}\Big) . \end{aligned}$$

"We may not be big, but we're small."

Since $\lfloor \lg n \rfloor!$ grows faster than any power of n, this minuscule error is overwhelmed by $\Sigma_c(n) = O(n^{-3})$. The error that comes from the original tail,

$$\Sigma_a(n) \;=\; \sum_{k \geqslant \lfloor \lg n \rfloor} a_k(n) \;<\; \sum_{k \geqslant \lfloor \ln n \rfloor} \frac{k + \ln n}{k!}\,,$$

is smaller yet.

Finally, it's easy to sum $\sum_{k \geqslant 0} b_k(n)$ in closed form, and we have obtained the desired asymptotic formula:

$$\sum_{k \geqslant 0} \frac{\ln(n + 2^k)}{k!} \;=\; e \ln n + \frac{e^2}{n} - \frac{e^4}{2n^2} + O\Bigl(\frac{1}{n^3}\Bigr)\,. \qquad (9.65)$$

The method we've used makes it clear that, in fact,

$$\sum_{k \geqslant 0} \frac{\ln(n + 2^k)}{k!} \;=\; e \ln n + \sum_{k=1}^{m-1} (-1)^{k+1} \frac{e^{2^k}}{k n^k} + O\Bigl(\frac{1}{n^m}\Bigr)\,, \qquad (9.66)$$

for any fixed $m > 0$. (This is a truncation of a series that diverges for all fixed n if we let $m \to \infty$.)

There's only one flaw in our solution: We were too cautious. We derived (9.64) on the assumption that $k < \lfloor \lg n \rfloor$, but exercise 53 proves that the stated estimate is actually valid for all values of k. If we had known the stronger general result, we wouldn't have had to use the two-tail trick; we could have gone directly to the final formula! But later we'll encounter problems where exchange of tails is the only decent approach available.

9.5 EULER'S SUMMATION FORMULA

And now for our next trick — which is, in fact, the last important technique that will be discussed in this book — we turn to a general method of approximating sums that was first published by Leonhard Euler [82] in 1732. (The idea is sometimes also associated with the name of Colin Maclaurin, a professor of mathematics at Edinburgh who discovered it independently a short time later [211, page 305].)

Here's the formula:

$$\sum_{a \leqslant k < b} f(k) \;=\; \int_a^b f(x)\,dx \;+\; \sum_{k=1}^{m} \frac{B_k}{k!} f^{(k-1)}(x)\Big|_a^b \;+\; R_m\,, \qquad (9.67)$$

$$\text{where } R_m \;=\; (-1)^{m+1} \int_a^b \frac{B_m(\{x\})}{m!} f^{(m)}(x)\,dx\,, \qquad \text{integers } a \leqslant b; \;\; \text{integer } m \geqslant 1. \qquad (9.68)$$

On the left is a typical sum that we might want to evaluate. On the right is another expression for that sum, involving integrals and derivatives. If $f(x)$ is a sufficiently "smooth" function, it will have m derivatives $f'(x), \ldots, f^{(m)}(x)$, and this formula turns out to be an identity. The right-hand side is often an excellent approximation to the sum on the left, in the sense that the remainder R_m is often small. For example, we'll see that Stirling's approximation for $n!$ is a consequence of Euler's summation formula; so is our asymptotic approximation for the harmonic number H_n.

The numbers B_k in (9.67) are the Bernoulli numbers that we met in Chapter 6; the function $B_m(\{x\})$ in (9.68) is the Bernoulli polynomial that we met in Chapter 7. The notation $\{x\}$ stands for the fractional part $x - \lfloor x \rfloor$, as in Chapter 3. Euler's summation formula sort of brings everything together.

Let's recall the values of small Bernoulli numbers, since it's always handy to have them listed near Euler's general formula:

$$B_0 = 1\,, \quad B_1 = -\tfrac{1}{2}\,, \quad B_2 = \tfrac{1}{6}\,, \quad B_4 = -\tfrac{1}{30}\,, \quad B_6 = \tfrac{1}{42}\,, \quad B_8 = -\tfrac{1}{30}\,;$$
$$B_3 = B_5 = B_7 = B_9 = B_{11} = \cdots = 0\,.$$

Jakob Bernoulli discovered these numbers when studying the sums of powers of integers, and Euler's formula explains why: If we set $f(x) = x^{m-1}$, we have $f^{(m)}(x) = 0$; hence $R_m = 0$, and (9.67) reduces to

$$\sum_{a\leqslant k<b} k^{m-1} = \frac{x^m}{m}\bigg|_a^b + \sum_{k=1}^{m} \frac{B_k}{k!}(m-1)^{\underline{k-1}}x^{m-k}\bigg|_a^b$$
$$= \frac{1}{m}\sum_{k=0}^{m}\binom{m}{k}B_k\cdot(b^{m-k} - a^{m-k})\,.$$

For example, when $m = 3$ we have our favorite example of summation:

$$\sum_{0\leqslant k<n} k^2 = \frac{1}{3}\left(\binom{3}{0}B_0n^3 + \binom{3}{1}B_1n^2 + \binom{3}{2}B_2n\right) = \frac{n^3}{3} - \frac{n^2}{2} + \frac{n}{6}\,.$$

(This is the last time we shall derive this famous formula in this book.)

All good things must come to an end.

Before we prove Euler's formula, let's look at a high-level reason (due to Lagrange [192]) why such a formula ought to exist. Chapter 2 defines the difference operator Δ and explains that $\sum$ is the inverse of Δ, just as $\int$ is the inverse of the derivative operator D. We can express Δ in terms of D using Taylor's formula as follows:

$$f(x+\epsilon) = f(x) + \frac{f'(x)}{1!}\epsilon + \frac{f''(x)}{2!}\epsilon^2 + \cdots\,.$$

Setting $\epsilon = 1$ tells us that

$$\begin{aligned}\Delta f(x) &= f(x+1) - f(x)\\ &= f'(x)/1! + f''(x)/2! + f'''(x)/3! + \cdots\\ &= (D/1! + D^2/2! + D^3/3! + \cdots)\, f(x) \;=\; (e^D - 1)\, f(x). \qquad (9.69)\end{aligned}$$

Here e^D stands for the differential operation $1 + D/1! + D^2/2! + D^3/3! + \cdots$. Since $\Delta = e^D - 1$, the inverse operator $\Sigma = 1/\Delta$ should be $1/(e^D - 1)$; and we know from Table 337 that $z/(e^z - 1) = \sum_{k\geqslant 0} B_k z^k/k!$ is a power series involving Bernoulli numbers. Thus

$$\Sigma = \frac{B_0}{D} + \frac{B_1}{1!} + \frac{B_2}{2!}D + \frac{B_3}{3!}D^2 + \cdots = \textstyle\int + \displaystyle\sum_{k\geqslant 1} \frac{B_k}{k!} D^{k-1}. \qquad (9.70)$$

Applying this operator equation to $f(x)$ and attaching limits yields

$$\sum\nolimits_a^b f(x)\,\delta x = \int_a^b f(x)\,dx + \sum_{k\geqslant 1} \frac{B_k}{k!} f^{(k-1)}(x)\bigg|_a^b, \qquad (9.71)$$

which is exactly Euler's summation formula (9.67) without the remainder term. (Euler did not, in fact, consider the remainder, nor did anybody else until S. D. Poisson [236] published an important memoir about approximate summation in 1823. The remainder term is important, because the infinite sum $\sum_{k\geqslant 1}(B_k/k!) f^{(k-1)}(x)\big|_a^b$ often diverges. Our derivation of (9.71) has been purely formal, without regard to convergence.)

Now let's prove (9.67), with the remainder included. It suffices to prove the case $a = 0$ and $b = 1$, namely

$$f(0) = \int_0^1 f(x)\,dx + \sum_{k=1}^{m} \frac{B_k}{k!} f^{(k-1)}(x)\bigg|_0^1 - (-1)^m \int_0^1 \frac{B_m(x)}{m!} f^{(m)}(x)\,dx,$$

because we can then replace $f(x)$ by $f(x + l)$ for any integer l, getting

$$f(l) = \int_l^{l+1} f(x)\,dx + \sum_{k=1}^{m} \frac{B_k}{k!} f^{(k-1)}(x)\bigg|_l^{l+1} - (-1)^m \int_l^{l+1} \frac{B_m(\{x\})}{m!} f^{(m)}(x)\,dx.$$

The general formula (9.67) is just the sum of this identity over the range $a \leqslant l < b$, because intermediate terms telescope nicely.

The proof when $a = 0$ and $b = 1$ is by induction on m, starting with $m = 1$:

$$f(0) = \int_0^1 f(x)\,dx - \frac{1}{2}\big(f(1) - f(0)\big) + \int_0^1 (x - \tfrac{1}{2}) f'(x)\,dx.$$

(The Bernoulli polynomial $B_m(x)$ is defined by the equation

$$B_m(x) = \binom{m}{0}B_0x^m + \binom{m}{1}B_1x^{m-1} + \cdots + \binom{m}{m}B_mx^0 \qquad (9.72)$$

in general, hence $B_1(x) = x - \frac{1}{2}$ in particular.) In other words, we want to prove that

$$\frac{f(0)+f(1)}{2} = \int_0^1 f(x)\,dx + \int_0^1 (x-\tfrac{1}{2})f'(x)\,dx\,.$$

But this is just a special case of the formula

$$u(x)v(x)\Big|_0^1 = \int_0^1 u(x)\,dv(x) \;+\; \int_0^1 v(x)\,du(x) \qquad (9.73)$$

for integration by parts, with $u(x) = f(x)$ and $v(x) = x - \frac{1}{2}$. Hence the case $m = 1$ is easy.

To pass from $m-1$ to m and complete the induction when $m > 1$, we need to show that $R_{m-1} = (B_m/m!)f^{(m-1)}(x)\big|_0^1 + R_m$, namely that

$$(-1)^m \int_0^1 \frac{B_{m-1}(x)}{(m-1)!} f^{(m-1)}(x)\,dx$$
$$= \frac{B_m}{m!} f^{(m-1)}(x)\Big|_0^1 - (-1)^m \int_0^1 \frac{B_m(x)}{m!} f^{(m)}(x)\,dx\,.$$

This reduces to the equation

$$(-1)^m B_m f^{(m-1)}(x)\Big|_0^1 = m\int_0^1 B_{m-1}(x)f^{(m-1)}(x)\,dx + \int_0^1 B_m(x)f^{(m)}(x)\,dx\,.$$

Once again (9.73) applies to these two integrals, with $u(x) = f^{(m-1)}(x)$ and $v(x) = B_m(x)$, because the derivative of the Bernoulli polynomial (9.72) is

Will the authors never get serious?

$$\frac{d}{dx}\sum_k \binom{m}{k} B_k x^{m-k} = \sum_k \binom{m}{k}(m-k)B_k x^{m-k-1}$$
$$= m\sum_k \binom{m-1}{k} B_k x^{m-1-k} = mB_{m-1}(x)\,. \qquad (9.74)$$

(The absorption identity (5.7) was useful here.) Therefore the required formula will hold if and only if

$$(-1)^m B_m f^{(m-1)}(x)\Big|_0^1 = B_m(x)f^{(m-1)}(x)\Big|_0^1\,.$$

In other words, we need to have

$$(-1)^m B_m = B_m(1) = B_m(0), \qquad \text{for } m > 1. \tag{9.75}$$

This is a bit embarrassing, because $B_m(0)$ is obviously equal to B_m, not to $(-1)^m B_m$. But there's no problem really, because $m > 1$; we know that B_m is zero when m is odd. (Still, that was a close call.)

To complete the proof of Euler's summation formula we need to show that $B_m(1) = B_m(0)$, which is the same as saying that

$$\sum_k \binom{m}{k} B_k = B_m, \qquad \text{for } m > 1.$$

But this is just the definition of Bernoulli numbers, (6.79), so we're done.

The identity $B'_m(x) = mB_{m-1}(x)$ implies that

$$\int_0^1 B_m(x)\,dx = \frac{B_{m+1}(1) - B_{m+1}(0)}{m+1},$$

and we know now that this integral is zero when $m \geqslant 1$. Hence the remainder term in Euler's formula,

$$R_m = \frac{(-1)^{m+1}}{m!} \int_a^b B_m\bigl(\{x\}\bigr) f^{(m)}(x)\,dx,$$

multiplies $f^{(m)}(x)$ by a function $B_m\bigl(\{x\}\bigr)$ whose average value is zero. This means that R_m has a reasonable chance of being small.

Let's look more closely at $B_m(x)$ for $0 \leqslant x \leqslant 1$, since $B_m(x)$ governs the behavior of R_m. Here are the graphs for $B_m(x)$ for the first twelve values of m:

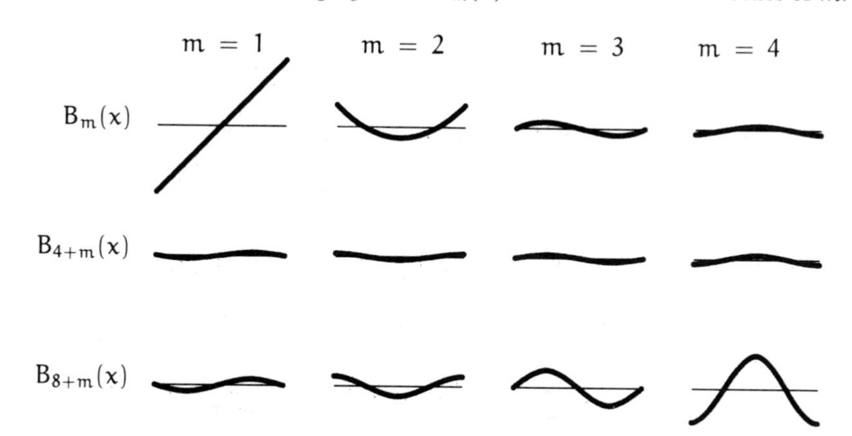

Although $B_3(x)$ through $B_9(x)$ are quite small, the Bernoulli polynomials and numbers ultimately get quite large. Fortunately R_m has a compensating factor $1/m!$, which helps to calm things down.

The graph of $B_m(x)$ begins to look very much like a sine wave when $m \geqslant 3$; exercise 58 proves that $B_m(x)$ can in fact be well approximated by a negative multiple of $\cos(2\pi x - \frac{1}{2}\pi m)$, with relative error $1/2^m$.

In general, $B_{4k+1}(x)$ is negative for $0 < x < \frac{1}{2}$ and positive for $\frac{1}{2} < x < 1$. Therefore its integral, $B_{4k+2}(x)/(4k+2)$, decreases for $0 < x < \frac{1}{2}$ and increases for $\frac{1}{2} < x < 1$. Moreover, we have

$$B_{4k+1}(1-x) = -B_{4k+1}(x), \qquad \text{for } 0 \leqslant x \leqslant 1,$$

and it follows that

$$B_{4k+2}(1-x) = B_{4k+2}(x), \qquad \text{for } 0 \leqslant x \leqslant 1.$$

The constant term B_{4k+2} causes the integral $\int_0^1 B_{4k+2}(x)\,dx$ to be zero; hence $B_{4k+2} > 0$. The integral of $B_{4k+2}(x)$ is $B_{4k+3}(x)/(4k+3)$, which must therefore be positive when $0 < x < \frac{1}{2}$ and negative when $\frac{1}{2} < x < 1$; furthermore $B_{4k+3}(1-x) = -B_{4k+3}(x)$, so $B_{4k+3}(x)$ has the properties stated for $B_{4k+1}(x)$, but negated. Therefore $B_{4k+4}(x)$ has the properties stated for $B_{4k+2}(x)$, but negated. Therefore $B_{4k+5}(x)$ has the properties stated for $B_{4k+1}(x)$; we have completed a cycle that establishes the stated properties inductively for all k.

According to this analysis, the maximum value of $B_{2m}(x)$ must occur either at $x = 0$ or at $x = \frac{1}{2}$. Exercise 17 proves that

$$B_{2m}(\tfrac{1}{2}) = (2^{1-2m} - 1)B_{2m}; \tag{9.76}$$

hence we have

$$\bigl|B_{2m}(\{x\})\bigr| \leqslant |B_{2m}|. \tag{9.77}$$

This can be used to establish a useful upper bound on the remainder in Euler's summation formula, because we know from (6.89) that

$$\frac{|B_{2m}|}{(2m)!} = \frac{2}{(2\pi)^{2m}} \sum_{k \geqslant 1} \frac{1}{k^{2m}} = O\bigl((2\pi)^{-2m}\bigr), \qquad \text{when } m > 0.$$

Therefore we can rewrite Euler's formula (9.67) as follows:

$$\sum_{a \leqslant k < b} f(k) = \int_a^b f(x)\,dx - \frac{1}{2}f(x)\Big|_a^b + \sum_{k=1}^{m} \frac{B_{2k}}{(2k)!} f^{(2k-1)}(x)\Big|_a^b + O\bigl((2\pi)^{-2m}\bigr)\int_a^b \bigl|f^{(2m)}(x)\bigr|\,dx. \tag{9.78}$$

For example, if $f(x) = e^x$, all derivatives are the same and this formula tells us that $\sum_{a \leqslant k < b} e^k = (e^b - e^a)\bigl(1 - \frac{1}{2} + B_2/2! + B_4/4! + \cdots + B_{2m}/(2m)!\bigr) +$

$O\big((2\pi)^{-2m}\big)$. Of course, we know that this sum is actually a geometric series, equal to $(e^b - e^a)/(e-1) = (e^b - e^a)\sum_{k\geqslant 0} B_k/k!$.

If $f^{(2m)}(x) \geqslant 0$ for $a \leqslant x \leqslant b$, the integral $\int_a^b |f^{(2m)}(x)|\,dx$ is just $f^{(2m-1)}(x)\big|_a^b$, so we have

$$|R_{2m}| \leqslant \left| \frac{B_{2m}}{(2m)!} f^{(2m-1)}(x)\Big|_a^b \right| ;$$

in other words, the remainder is bounded by the magnitude of the *final term* (the term just before the remainder), in this case. We can give an even better estimate if we know that

$$f^{(2m+2)}(x) \geqslant 0 \quad \text{and} \quad f^{(2m+4)}(x) \geqslant 0\,, \qquad \text{for } a \leqslant x \leqslant b. \tag{9.79}$$

For it turns out that this implies the relation

$$R_{2m} = \theta_m \frac{B_{2m+2}}{(2m+2)!} f^{(2m+1)}(x)\Big|_a^b\,, \qquad \text{for some } 0 < \theta_m < 1; \tag{9.80}$$

in other words, the remainder will then lie between 0 and the *first discarded term* in (9.78)—the term that would follow the final term if we increased m.

Here's the proof: Euler's summation formula is valid for all m, and $B_{2m+1} = 0$ when $m > 0$; hence $R_{2m} = R_{2m+1}$, and the first discarded term must be

$$R_{2m} - R_{2m+2}\,.$$

We therefore want to show that R_{2m} lies between 0 and $R_{2m} - R_{2m+2}$; and this is true if and only if R_{2m} and R_{2m+2} have opposite signs. We claim that

$$f^{(2m+2)}(x) \geqslant 0 \quad \text{for } a \leqslant x \leqslant b \quad \text{implies} \quad (-1)^m R_{2m} \geqslant 0\,. \tag{9.81}$$

This, together with (9.79), will prove that R_{2m} and R_{2m+2} have opposite signs, so the proof of (9.80) will be complete.

It's not difficult to prove (9.81) if we recall the definition of R_{2m+1} and the facts we proved about the graph of $B_{2m+1}(x)$. Namely, we have

$$R_{2m} = R_{2m+1} = \int_a^b \frac{B_{2m+1}\big(\{x\}\big)}{(2m+1)!} f^{(2m+1)}(x)\,dx\,,$$

and $f^{(2m+1)}(x)$ is increasing because its derivative $f^{(2m+2)}(x)$ is positive. (More precisely, $f^{(2m+1)}(x)$ is nondecreasing because its derivative is nonnegative.) The graph of $B_{2m+1}\big(\{x\}\big)$ looks as if $(-1)^{m+1}$ times a sine wave, so it is geometrically obvious that the second half of each sine wave is more influential than the first half when it is multiplied by an increasing function. This makes $(-1)^m R_{2m+1} \geqslant 0$, as desired. Exercise 16 proves the result formally.

9.6 FINAL SUMMATIONS

Now comes the summing up, as we prepare to conclude this book. We will apply Euler's summation formula to some interesting and important examples.

Summation 1: This one is too easy.

But first we will consider an interesting *unimportant* example, namely a sum that we already know how to do. Let's see what Euler's summation formula tells us if we apply it to the telescoping sum

$$S_n = \sum_{1 \leqslant k < n} \frac{1}{k(k+1)} = \sum_{1 \leqslant k < n} \left(\frac{1}{k} - \frac{1}{k+1} \right) = 1 - \frac{1}{n}.$$

It can't hurt to embark on our first serious application of Euler's formula with the asymptotic equivalent of training wheels.

We might as well start by writing the function $f(x) = 1/x(x+1)$ in partial fraction form,

$$f(x) = \frac{1}{x} - \frac{1}{x+1},$$

since this makes it easier to integrate and differentiate. Indeed, we have $f'(x) = -1/x^2 + 1/(x+1)^2$ and $f''(x) = 2/x^3 - 2/(x+1)^3$; in general

$$f^{(k)}(x) = (-1)^k k! \left(\frac{1}{x^{k+1}} - \frac{1}{(x+1)^{k+1}} \right), \qquad \text{for } k \geqslant 0.$$

Furthermore

$$\int_1^n f(x)\, dx = \ln x - \ln(x+1) \Big|_1^n = \ln \frac{2n}{n+1}.$$

Plugging this into the summation formula (9.67) gives

$$S_n = \ln \frac{2n}{n+1} - \sum_{k=1}^{m} (-1)^k \frac{B_k}{k} \left(\frac{1}{n^k} - \frac{1}{(n+1)^k} - 1 + \frac{1}{2^k} \right) + R_m(n),$$

$$\text{where } R_m(n) = -\int_1^n B_m(\{x\}) \left(\frac{1}{x^{m+1}} - \frac{1}{(x+1)^{m+1}} \right) dx.$$

For example, the right-hand side when $m = 4$ is

$$\ln \frac{2n}{n+1} - \frac{1}{2}\left(\frac{1}{n} - \frac{1}{n+1} - \frac{1}{2} \right) - \frac{1}{12}\left(\frac{1}{n^2} - \frac{1}{(n+1)^2} - \frac{3}{4} \right) + \frac{1}{120}\left(\frac{1}{n^4} - \frac{1}{(n+1)^4} - \frac{15}{16} \right) + R_4(n).$$

This is kind of a mess; it certainly doesn't look like the real answer $1 - n^{-1}$. But let's keep going anyway, to see what we've got. We know how to expand the right-hand terms in negative powers of n up to, say, $O(n^{-5})$:

$$\begin{aligned}
\ln\frac{n}{n+1} &= -n^{-1} + \tfrac{1}{2}n^{-2} - \tfrac{1}{3}n^{-3} + \tfrac{1}{4}n^{-4} + O(n^{-5})\,;\\
\frac{1}{n+1} &= \quad n^{-1} - \quad n^{-2} + \quad n^{-3} - \quad n^{-4} + O(n^{-5})\,;\\
\frac{1}{(n+1)^2} &= \qquad\qquad n^{-2} - 2n^{-3} + 3n^{-4} + O(n^{-5})\,;\\
\frac{1}{(n+1)^4} &= \qquad\qquad\qquad\qquad n^{-4} + O(n^{-5})\,.
\end{aligned}$$

Therefore the terms on the right of our approximation add up to

$$\begin{aligned}
&\ln 2 + \tfrac{1}{4} + \tfrac{1}{16} - \tfrac{1}{128} + \left(-1 - \tfrac{1}{2} + \tfrac{1}{2}\right)n^{-1} + \left(\tfrac{1}{2} - \tfrac{1}{2} - \tfrac{1}{12} + \tfrac{1}{12}\right)n^{-2}\\
&\qquad + \left(-\tfrac{1}{3} + \tfrac{1}{2} - \tfrac{2}{12}\right)n^{-3} + \left(\tfrac{1}{4} - \tfrac{1}{2} + \tfrac{3}{12} + \tfrac{1}{120} - \tfrac{1}{120}\right)n^{-4} + R_4(n)\\
&= \ln 2 + \tfrac{39}{128} - n^{-1} + R_4(n) + O(n^{-5})\,.
\end{aligned}$$

The coefficients of n^{-2}, n^{-3}, and n^{-4} cancel nicely, as they should.

If all were well with the world, we would be able to show that $R_4(n)$ is asymptotically small, maybe $O(n^{-5})$, and we would have an approximation to the sum. But we can't possibly show this, because we happen to know that the correct constant term is 1, not $\ln 2 + \frac{39}{128}$ (which is approximately 0.9978). So $R_4(n)$ is actually equal to $\frac{89}{128} - \ln 2 + O(n^{-4})$, but Euler's summation formula doesn't tell us this.

In other words, we lose.

One way to try fixing things is to notice that the constant terms in the approximation form a pattern, if we let m get larger and larger:

$$\ln 2 - \tfrac{1}{2}B_1 + \tfrac{1}{2}\cdot\tfrac{3}{4}B_2 - \tfrac{1}{3}\cdot\tfrac{7}{8}B_3 + \tfrac{1}{4}\cdot\tfrac{15}{16}B_4 - \tfrac{1}{5}\cdot\tfrac{31}{32}B_5 + \cdots\,.$$

Perhaps we can show that this series approaches 1 as the number of terms becomes infinite? But no; the Bernoulli numbers get very large. For example, $B_{22} = \frac{854513}{138} > 6192$; therefore $|R_{22}(n)|$ will be much larger than $|R_4(n)|$. We lose totally.

There is a way out, however, and this escape route will turn out to be important in other applications of Euler's formula. The key is to notice that $R_4(n)$ approaches a definite limit as $n \to \infty$:

$$\lim_{n\to\infty} R_4(n) \;=\; -\int_1^\infty B_4(\{x\})\left(\frac{1}{x^5} - \frac{1}{(x+1)^5}\right)dx \;=\; R_4(\infty)\,.$$

The integral $\int_1^\infty B_4(\{x\}) f^{(m)}(x)\,dx$ will exist whenever $f^{(m)}(x) = O(x^{-2})$ as $x \to \infty$, and in this case $f^{(4)}(x)$ surely qualifies. Moreover, we have

$$\begin{aligned} R_4(n) &= R_4(\infty) + \int_n^\infty B_4(\{x\})\Big(\frac{1}{x^5} - \frac{1}{(x+1)^5}\Big)\,dx \\ &= R_4(\infty) + O\Big(\int_n^\infty x^{-6}\,dx\Big) \;=\; R_4(\infty) + O(n^{-5}). \end{aligned}$$

Thus we have used Euler's summation formula to prove that

$$\begin{aligned} \sum_{1 \leqslant k < n} \frac{1}{k(k+1)} &= \ln 2 + \tfrac{39}{128} - n^{-1} + R_4(\infty) + O(n^{-5}) \\ &= C - n^{-1} + O(n^{-5}) \end{aligned}$$

for some constant C. We do not know what the constant is — some other method must be used to establish it — but Euler's summation formula is able to let us deduce that the constant exists.

Suppose we had chosen a much larger value of m. Then the same reasoning would tell us that

$$R_m(n) = R_m(\infty) + O(n^{-m-1}),$$

and we would have the formula

$$\sum_{1 \leqslant k < n} \frac{1}{k(k+1)} = C - n^{-1} + c_2 n^{-2} + c_3 n^{-3} + \cdots + c_m n^{-m} + O(n^{-m-1})$$

for certain constants c_2, c_3, ... We know that the c's happen to be zero in this case; but let's prove it, just to restore some of our confidence (in Euler's formula if not in ourselves). The term $\ln \frac{n}{n+1}$ contributes $(-1)^m/m$ to c_m; the term $(-1)^{m+1}(B_m/m)n^{-m}$ contributes $(-1)^{m+1}B_m/m$; and the term $(-1)^k(B_k/k)(n+1)^{-k}$ contributes $(-1)^m\binom{m-1}{k-1}B_k/k$. Therefore

$$\begin{aligned} (-1)^m c_m &= \frac{1}{m} - \frac{B_m}{m} + \sum_{k=1}^{m} \binom{m-1}{k-1}\frac{B_k}{k} \\ &= \frac{1}{m} - \frac{B_m}{m} + \frac{1}{m}\sum_{k=1}^{m}\binom{m}{k}B_k \;=\; \frac{1}{m}\big(1 - B_m + B_m(1) - 1\big). \end{aligned}$$

Sure enough, it's zero, when $m > 1$. We have proved that

$$\sum_{1 \leqslant k < n} \frac{1}{k(k+1)} = C - n^{-1} + O(n^{-m-1}), \qquad \text{for all } m \geqslant 1. \tag{9.82}$$

This is not enough to prove that the sum is exactly equal to $C - n^{-1}$; the actual value may be $C - n^{-1} + 2^{-n}$ or something. But Euler's summation formula does give us $O(n^{-m-1})$ for arbitrarily large m, even though we haven't evaluated any remainders explicitly.

Summation 1, again: Recapitulation and generalization.

Before we leave our training wheels, let's review what we just did from a somewhat higher perspective. We began with a sum

$$S_n = \sum_{1 \leqslant k < n} f(k)$$

and we used Euler's summation formula to write

$$S_n = F(n) - F(1) + \sum_{k=1}^{m} \bigl(T_k(n) - T_k(1)\bigr) + R_m(n), \tag{9.83}$$

where $F(x)$ was $\int f(x)\,dx$ and where $T_k(x)$ was a certain term involving B_k and $f^{(k-1)}(x)$. We also noticed that there was a constant c such that

$$f^{(m)}(x) = O(x^{c-m}) \quad \text{as } x \to \infty, \qquad \text{for all large } m.$$

(Namely, $f(k)$ was $1/k(k+1)$; $F(x)$ was $\ln\bigl(x/(x+1)\bigr)$; $T_k(x)$ was $(-1)^{k+1} \times (B_k/k)\bigl(x^{-k} - (x+1)^{-k}\bigr)$; and c was -2.) For all large enough values of m, this implied that the remainders had a small tail,

$$\begin{aligned} R'_m(n) &= R_m(\infty) - R_m(n) \\ &= (-1)^{m+1} \int_n^\infty \frac{B_m(\{x\})}{m!} f^{(m)}(x)\,dx = O(n^{c+1-m}). \end{aligned} \tag{9.84}$$

Therefore we were able to conclude that there exists a constant C such that

$$S_n = F(n) + C + \sum_{k=1}^{m} T_k(n) - R'_m(n). \tag{9.85}$$

(Notice that C nicely absorbed the $T_k(1)$ terms, which were a nuisance.)

We can save ourselves unnecessary work in future problems by simply asserting the existence of C whenever $R_m(\infty)$ exists.

Now let's suppose that $f^{(2m+2)}(x) \geqslant 0$ and $f^{(2m+4)}(x) \geqslant 0$ for $1 \leqslant x \leqslant n$. We have proved that this implies a simple bound (9.80) on the remainder,

$$R_{2m}(n) = \theta_{m,n}\bigl(T_{2m+2}(n) - T_{2m+2}(1)\bigr),$$

where $\theta_{m,n}$ lies somewhere between 0 and 1. But we don't really want bounds that involve $R_{2m}(n)$ and $T_{2m+2}(1)$; after all, we got rid of $T_k(1)$ when we introduced the constant C. What we really want is a bound like

$$-R'_{2m}(n) = \phi_{m,n} T_{2m+2}(n),$$

where $0 < \phi_{m,n} < 1$; this will allow us to conclude from (9.85) that

$$S_n = F(n) + C + T_1(n) + \sum_{k=1}^{m} T_{2k}(n) + \phi_{m,n} T_{2m+2}(n)\,, \qquad (9.86)$$

hence the remainder will truly be between zero and the first discarded term.

A slight modification of our previous argument will patch things up perfectly. Let us assume that

$$f^{(2m+2)}(x) \geqslant 0 \quad \text{and} \quad f^{(2m+4)}(x) \geqslant 0\,, \qquad \text{as } x \to \infty. \qquad (9.87)$$

The right-hand side of (9.85) is just like the negative of the right-hand side of Euler's summation formula (9.67) with $a = n$ and $b = \infty$, as far as remainder terms are concerned, and successive remainders are generated by induction on m. Therefore our previous argument can be applied.

Summation 2: Harmonic numbers harmonized.

Now that we've learned so much from a trivial (but safe) example, we can readily do a nontrivial one. Let us use Euler's summation formula to derive the approximation for H_n that we have been claiming for some time.

In this case, $f(x) = 1/x$. We already know about the integral and derivatives of f, because of Summation 1; also $f^{(m)}(x) = O(x^{-m-1})$ as $x \to \infty$. Therefore we can immediately plug into formula (9.85):

$$\sum_{1 \leqslant k < n} \frac{1}{k} = \ln n + C + B_1 n^{-1} - \sum_{k=1}^{m} \frac{B_{2k}}{2kn^{2k}} - R'_{2m}(n)\,,$$

for some constant C. The sum on the left is H_{n-1}, not H_n; but it's more convenient to work with H_{n-1} and to add $1/n$ later, than to mess around with $(n+1)$'s on the right-hand side. The $B_1 n^{-1}$ will then become $(B_1 + 1)n^{-1} = 1/(2n)$. Let us call the constant γ instead of C, since Euler's constant γ is, in fact, defined to be $\lim_{n\to\infty}(H_n - \ln n)$.

The remainder term can be estimated nicely by the theory we developed a minute ago, because $f^{(2m)}(x) = (2m)!/x^{2m+1} \geqslant 0$ for all $x > 0$. Therefore (9.86) tells us that

$$H_n = \ln n + \gamma + \frac{1}{2n} - \sum_{k=1}^{m} \frac{B_{2k}}{2kn^{2k}} + \theta_{m,n} \frac{B_{2m+2}}{(2m+2)n^{2m+2}}\,, \qquad (9.88)$$

where $\theta_{m,n}$ is some fraction between 0 and 1. This is the general formula whose first few terms are listed in Table 438. For example, when $m = 2$ we get

$$H_n = \ln n + \gamma + \frac{1}{2n} - \frac{1}{12n^2} + \frac{1}{120n^4} - \frac{\theta_{2,n}}{252n^6}\,. \qquad (9.89)$$

This equation, incidentally, gives us a good approximation to γ even when $n = 2$:

$$\gamma \;=\; H_2 - \ln 2 - \tfrac{1}{4} + \tfrac{1}{48} - \tfrac{1}{1920} + \epsilon \;=\; 0.577165\ldots + \epsilon\,,$$

where ϵ is between zero and $\frac{1}{16128}$. If we take $n = 10^4$ and $m = 250$, we get the value of γ correct to 1271 decimal places, beginning thus [171]:

$$\gamma \;=\; 0.57721\,56649\,01532\,86060\,65120\,90082\,40243\ldots\,. \tag{9.90}$$

But Euler's constant appears also in other formulas that allow it to be evaluated even more efficiently [282].

Summation 3: Stirling's approximation.

If $f(x) = \ln x$, we have $f'(x) = 1/x$, so we can evaluate the sum of logarithms using almost the same calculations as we did when summing reciprocals. Euler's summation formula yields

$$\sum_{1\leqslant k<n} \ln k \;=\; n\ln n - n + \sigma - \frac{\ln n}{2} + \sum_{k=1}^{m} \frac{B_{2k}}{2k(2k{-}1)n^{2k-1}} + \varphi_{m,n}\frac{B_{2m+2}}{(2m{+}2)(2m{+}1)n^{2m+1}}$$

where σ is a certain constant, "Stirling's constant," and $0 < \varphi_{m,n} < 1$. (In this case $f^{(2m)}(x)$ is negative, not positive; but we can still say that the remainder is governed by the first discarded term, because we could have started with $f(x) = -\ln x$ instead of $f(x) = \ln x$.) Adding $\ln n$ to both sides gives

$$\ln n! \;=\; n\ln n - n + \frac{\ln n}{2} + \sigma + \frac{1}{12n} - \frac{1}{360n^3} + \frac{\varphi_{2,n}}{1260n^5} \tag{9.91}$$

when $m = 2$. And we can get the approximation in Table 438 by taking 'exp' of both sides. (The value of e^σ turns out to be $\sqrt{2\pi}$, but we will not prove that until later. In fact, Stirling himself didn't know a closed form for this constant until several years after he had first proved its existence.)

If m is fixed and $n \to \infty$, the general formula gives a better and better approximation to $\ln n!$ in the sense of absolute error, hence it gives a better and better approximation to $n!$ in the sense of relative error. But if n is fixed and m increases, the error bound $|B_{2m+2}|/(2m+2)(2m+1)n^{2m+1}$ decreases to a certain point and then begins to increase. Therefore the approximation reaches a point beyond which a sort of uncertainty principle limits the amount by which $n!$ can be approximated.

Heisenberg may have been here.

In Chapter 5, equation (5.83), we generalized factorials to arbitrary real α by using a definition

$$\frac{1}{\alpha!} = \lim_{n\to\infty} \binom{n+\alpha}{n} n^{-\alpha}$$

suggested by Euler. Suppose α is a large number; then

$$\ln \alpha! = \lim_{n\to\infty} \Bigl(\alpha \ln n + \ln n! - \sum_{k=1}^{n} \ln(\alpha+k)\Bigr),$$

and Euler's summation formula can be used with $f(x) = \ln(x+\alpha)$ to estimate this sum:

$$\begin{aligned}
\sum_{k=1}^{n} \ln(k+\alpha) &= F_m(\alpha,n) - F_m(\alpha,0) + R_{2m}(\alpha,n), \\
F_m(\alpha,x) &= (x+\alpha)\ln(x+\alpha) - x + \frac{\ln(x+\alpha)}{2} \\
&\qquad + \sum_{k=1}^{m} \frac{B_{2k}}{2k(2k-1)(x+\alpha)^{2k-1}}, \\
R_{2m}(\alpha,n) &= \int_0^n \frac{B_{2m}(\{x\})}{2m} \frac{dx}{(x+\alpha)^{2m}}.
\end{aligned}$$

(Here we have used (9.67) with $a = 0$ and $b = n$, then added $\ln(n+\alpha) - \ln\alpha$ to both sides.) If we subtract this approximation for $\sum_{k=1}^{n} \ln(k+\alpha)$ from Stirling's approximation for $\ln n!$, then add $\alpha \ln n$ and take the limit as $n \to \infty$, we get

$$\begin{aligned}
\ln \alpha! &= \alpha \ln \alpha - \alpha + \frac{\ln \alpha}{2} + \sigma \\
&\qquad + \sum_{k=1}^{m} \frac{B_{2k}}{(2k)(2k-1)\alpha^{2k-1}} - \int_0^\infty \frac{B_{2m}(\{x\})}{2m} \frac{dx}{(x+\alpha)^{2m}},
\end{aligned}$$

because $\alpha \ln n + n \ln n - n + \frac{1}{2}\ln n - (n+\alpha)\ln(n+\alpha) + n - \frac{1}{2}\ln(n+\alpha) \to -\alpha$ and the other terms not shown here tend to zero. Thus Stirling's approximation behaves for generalized factorials (and for the Gamma function $\Gamma(\alpha+1) = \alpha!$) exactly as for ordinary factorials.

Summation 4: A bell-shaped summand.

Let's turn now to a sum that has quite a different flavor:

$$\begin{aligned}
\Theta_n &= \sum_k e^{-k^2/n} \qquad (9.92) \\
&= \cdots + e^{-9/n} + e^{-4/n} + e^{-1/n} + 1 + e^{-1/n} + e^{-4/n} + e^{-9/n} + \cdots.
\end{aligned}$$

This is a doubly infinite sum, whose terms reach their maximum value $e^0 = 1$ when $k = 0$. We call it Θ_n because it is a power series involving the quantity $e^{-1/n}$ raised to the $p(k)$th power, where $p(k)$ is a polynomial of degree 2; such power series are traditionally called "theta functions." If $n = 10^{100}$, we have

$$e^{-k^2/n} \;=\; \begin{cases} e^{-.01} \approx 0.99005, & \text{when } k = 10^{49}; \\ e^{-1} \approx 0.36788, & \text{when } k = 10^{50}; \\ e^{-100} < 10^{-43}, & \text{when } k = 10^{51}. \end{cases}$$

So the summand stays very near 1 until k gets up to about $\sqrt{n}$, when it drops off and stays very near zero. We can guess that Θ_n will be proportional to $\sqrt{n}$. Here is a graph of $e^{-k^2/n}$ when $n = 10$:

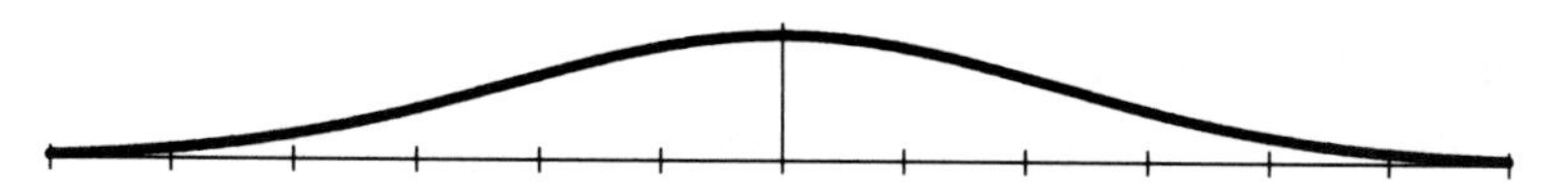

Larger values of n just stretch the graph horizontally by a factor of $\sqrt{n}$.

We can estimate Θ_n by letting $f(x) = e^{-x^2/n}$ and taking $a = -\infty$, $b = +\infty$ in Euler's summation formula. (If infinities seem too scary, let $a = -A$ and $b = +B$, then take limits as $A, B \to \infty$.) The integral of $f(x)$ is

$$\int_{-\infty}^{+\infty} e^{-x^2/n}\,dx \;=\; \sqrt{n}\int_{-\infty}^{+\infty} e^{-u^2}\,du \;=\; \sqrt{n}\,C\,,$$

if we replace x by $u\sqrt{n}$. The value of $\int_{-\infty}^{+\infty} e^{-u^2}\,du$ is well known, but we'll call it C for now and come back to it after we have finished plugging into Euler's summation formula.

The next thing we need to know is the sequence of derivatives $f'(x)$, $f''(x)$, ..., and for this purpose it's convenient to set

$$f(x) \;=\; g\bigl(x/\sqrt{n}\bigr)\,, \qquad g(x) \;=\; e^{-x^2}\,.$$

Then the chain rule of calculus says that

$$\frac{df(x)}{dx} \;=\; \frac{dg(y)}{dy}\,\frac{dy}{dx}\,, \qquad y \;=\; \frac{x}{\sqrt{n}}\,;$$

and this is the same as saying that

$$f'(x) \;=\; \frac{1}{\sqrt{n}}\,g'\bigl(x/\sqrt{n}\bigr)\,.$$

By induction we have

$$f^{(k)}(x) \;=\; n^{-k/2}g^{(k)}\bigl(x/\sqrt{n}\bigr)\,.$$

For example, we have $g'(x) = -2xe^{-x^2}$ and $g''(x) = (4x^2 - 2)e^{-x^2}$; hence

$$f'(x) = \frac{1}{\sqrt{n}}\Big(-2\frac{x}{\sqrt{n}}\Big)e^{-x^2/n}, \qquad f''(x) = \frac{1}{n}\Big(4\Big(\frac{x}{\sqrt{n}}\Big)^2 - 2\Big)e^{-x^2/n}.$$

It's easier to see what's going on if we work with the simpler function $g(x)$.

We don't have to evaluate the derivatives of $g(x)$ exactly, because we're only going to be concerned about the limiting values when $x = \pm\infty$. And for this purpose it suffices to notice that every derivative of $g(x)$ is e^{-x^2} times a polynomial in x:

$$g^{(k)}(x) = P_k(x)e^{-x^2}, \qquad \text{where } P \text{ is a polynomial of degree } k.$$

This follows by induction.

The negative exponential e^{-x^2} goes to zero much faster than $P_k(x)$ goes to infinity, when $x \to \pm\infty$, so we have

$$f^{(k)}(+\infty) = f^{(k)}(-\infty) = 0$$

for all $k \geqslant 0$. Therefore all of the terms

$$\sum_{k=1}^{\infty} \frac{B_k}{k!} f^{(k-1)}(x)\Big|_{-\infty}^{+\infty}$$

vanish, and we are left with the term from $\int f(x)\,dx$ and the remainder:

$$\begin{aligned}
\Theta_n &= C\sqrt{n} + (-1)^{m+1}\int_{-\infty}^{+\infty} \frac{B_m(\{x\})}{m!} f^{(m)}(x)\,dx \\
&= C\sqrt{n} + \frac{(-1)^{m+1}}{n^{m/2}}\int_{-\infty}^{+\infty} \frac{B_m(\{x\})}{m!} g^{(m)}\Big(\frac{x}{\sqrt{n}}\Big)\,dx \\
&= C\sqrt{n} + \frac{(-1)^{m+1}}{n^{(m-1)/2}}\int_{-\infty}^{+\infty} \frac{B_m(\{u\sqrt{n}\})}{m!} P_m(u)e^{-u^2}\,du \qquad (x = u\sqrt{n}) \\
&= C\sqrt{n} + O(n^{(1-m)/2}).
\end{aligned}$$

The O estimate here follows since $\big|B_m(\{u\sqrt{n}\})\big|$ is bounded and the integral $\int_{-\infty}^{+\infty} |P(u)|e^{-u^2}\,du$ exists whenever P is a polynomial. (The constant implied by this O depends on m.)

We have proved that $\Theta_n = C\sqrt{n} + O(n^{-M})$, for arbitrarily large M; the difference between Θ_n and $C\sqrt{n}$ is "exponentially small." Let us therefore determine the constant C that plays such a big role in the value of Θ_n.

One way to determine C is to look the integral up in a table; but we prefer to know how the value can be derived, so that we can do integrals even when they haven't been tabulated. Elementary calculus suffices to evaluate C if we are clever enough to look at the double integral

$$C^2 = \int_{-\infty}^{+\infty} e^{-x^2}\,dx \int_{-\infty}^{+\infty} e^{-y^2}\,dy = \int_{-\infty}^{+\infty}\int_{-\infty}^{+\infty} e^{-(x^2+y^2)}\,dx\,dy\,.$$

Converting to polar coordinates gives

$$\begin{aligned} C^2 &= \int_0^{2\pi}\int_0^{\infty} e^{-r^2} r\,dr\,d\theta \\ &= \frac{1}{2}\int_0^{2\pi} d\theta \int_0^{\infty} e^{-u}\,du \qquad (u = r^2) \\ &= \frac{1}{2}\int_0^{2\pi} d\theta = \pi\,. \end{aligned}$$

So $C = \sqrt{\pi}$. The fact that $x^2 + y^2 = r^2$ is the equation of a circle whose circumference is $2\pi r$ somehow explains why π gets into the act.

Another way to evaluate C is to replace x by $\sqrt{t}$ and dx by $\frac{1}{2}t^{-1/2}\,dt$:

$$C = \int_{-\infty}^{+\infty} e^{-x^2}\,dx = 2\int_0^{\infty} e^{-x^2}\,dx = \int_0^{\infty} t^{-1/2}e^{-t}\,dt\,.$$

This integral equals $\Gamma\bigl(\frac{1}{2}\bigr)$, since $\Gamma(\alpha) = \int_0^{\infty} t^{\alpha-1}e^{-t}\,dt$ according to (5.84). Therefore we have demonstrated that $\Gamma\bigl(\frac{1}{2}\bigr) = \sqrt{\pi}$.

Our final formula, then, is

$$\Theta_n = \sum_k e^{-k^2/n} = \sqrt{\pi n} + O(n^{-M})\,, \qquad \text{for all fixed } M. \tag{9.93}$$

The constant in the O depends on M; that's why we say that M is "fixed."

When $n = 2$, for example, the infinite sum Θ_2 is equal to 2.506628288; this is already an excellent approximation to $\sqrt{2\pi} = 2.506628275$, even though n is quite small. The value of Θ_{100} agrees with $10\sqrt{\pi}$ to 427 decimal places! Exercise 59 uses advanced methods to derive a rapidly convergent series for Θ_n; it turns out that

$$\Theta_n/\sqrt{\pi n} = 1 + 2e^{-n\pi^2} + O(e^{-4n\pi^2})\,. \tag{9.94}$$

Summation 5: The clincher.

Now we will do one last sum, which will turn out to tell us the value of Stirling's constant σ. This last sum also illustrates many of the other techniques of this last chapter (and of this whole book), so it will be a fitting way for us to conclude our explorations of Concrete Mathematics.

The final task seems almost absurdly easy: We will try to find the asymptotic value of

$$A_n = \sum_k \binom{2n}{k}$$

by using Euler's summation formula.

This is another case where we already know the answer (right?); but it's always interesting to try new methods on old problems, so that we can compare facts and maybe discover something new.

So we THINK BIG and realize that the main contribution to A_n comes from the middle terms, near $k = n$. It's almost always a good idea to choose notation so that the biggest contribution to a sum occurs near $k = 0$, because we can then use the tail-exchange trick to get rid of terms that have large $|k|$. Therefore we replace k by $n + k$:

$$A_n = \sum_k \binom{2n}{n+k} = \sum_k \frac{(2n)!}{(n+k)!\,(n-k)!}.$$

Things are looking reasonably good, since we know to approximate $(n \pm k)!$ when n is large and k is small.

Now we want to carry out the three-step procedure associated with the tail-exchange trick. Namely, we want to write

$$\frac{(2n)!}{(n+k)!\,(n-k)!} = a_k(n) = b_k(n) + O\bigl(c_k(n)\bigr), \qquad \text{for } k \in D_n,$$

so that we can obtain the estimate

$$A_n = \sum_k b_k(n) + O\Bigl(\sum_{k \notin D_n} a_k(n)\Bigr) + O\Bigl(\sum_{k \notin D_n} b_k(n)\Bigr) + \sum_{k \in D_n} O\bigl(c_k(n)\bigr).$$

Let us therefore try to estimate $\binom{2n}{n+k}$ in the region where $|k|$ is small. We could use Stirling's approximation as it appears in Table 438, but it's easier to work with the logarithmic equivalent in (9.91):

$$\begin{aligned} \ln a_k(n) &= \ln(2n)! - \ln(n+k)! - \ln(n-k)! \\ &= 2n\ln 2n - 2n + \tfrac{1}{2}\ln 2n + \sigma + O(n^{-1}) \\ &\qquad - (n{+}k)\ln(n{+}k) + n + k - \tfrac{1}{2}\ln(n{+}k) - \sigma + O\bigl((n{+}k)^{-1}\bigr) \\ &\qquad - (n{-}k)\ln(n{-}k) + n - k - \tfrac{1}{2}\ln(n{-}k) - \sigma + O\bigl((n{-}k)^{-1}\bigr). \end{aligned} \tag{9.95}$$

We want to convert this to a nice, simple O estimate.

The tail-exchange method allows us to work with estimates that are valid only when k is in the "dominant" set D_n. But how should we define D_n?

Actually I'm not into dominance.

We have to make D_n small enough that we can make a good estimate; for example, we had better not let k get near n, or the term $O\big((n-k)^{-1}\big)$ in (9.95) will blow up. Yet D_n must be large enough that the tail terms (the terms with $k \notin D_n$) are negligibly small compared with the overall sum. Trial and error is usually necessary to find an appropriate set D_n; in this problem the calculations we are about to make will show that it's wise to define things as follows:

$$k \in D_n \quad \Longleftrightarrow \quad |k| \leqslant n^{1/2+\epsilon}. \tag{9.96}$$

Here ϵ is a small positive constant that we can choose later, after we get to know the territory. (Our O estimates will depend on the value of ϵ.) Equation (9.95) now reduces to

$$\begin{aligned} \ln a_k(n) \;=\; & (2n+\tfrac{1}{2})\ln 2 - \sigma - \tfrac{1}{2}\ln n + O(n^{-1}) \\ & - (n+k+\tfrac{1}{2})\ln(1+k/n) - (n-k+\tfrac{1}{2})\ln(1-k/n). \end{aligned} \tag{9.97}$$

(We have pulled out the large parts of the logarithms, writing

$$\ln(n \pm k) \;=\; \ln n + \ln(1 \pm k/n),$$

and this has made a lot of $\ln n$ terms cancel out.)

Now we need to expand the terms $\ln(1 \pm k/n)$ asymptotically, until we have an error term that approaches zero as $n \to \infty$. We are multiplying $\ln(1 \pm k/n)$ by $(n \pm k + \frac{1}{2})$, so we should expand the logarithm until we reach $o(n^{-1})$, using the assumption that $|k| \leqslant n^{1/2+\epsilon}$:

$$\ln\Big(1 \pm \frac{k}{n}\Big) \;=\; \pm\frac{k}{n} - \frac{k^2}{2n^2} + O(n^{-3/2+3\epsilon}).$$

Multiplication by $n \pm k + \frac{1}{2}$ yields

$$\pm k - \frac{k^2}{2n} + \frac{k^2}{n} + O(n^{-1/2+3\epsilon}),$$

plus other terms that are absorbed in the $O(n^{-1/2+3\epsilon})$. So (9.97) becomes

$$\ln a_k(n) \;=\; (2n+\frac{1}{2})\ln 2 - \sigma - \frac{1}{2}\ln n - k^2/n + O(n^{-1/2+3\epsilon}).$$

Taking exponentials, we have

$$a_k(n) \;=\; \frac{2^{2n+1/2}}{e^{\sigma}\sqrt{n}}\, e^{-k^2/n}\big(1 + O(n^{-1/2+3\epsilon})\big). \tag{9.98}$$

This is our approximation, with

$$b_k(n) = \frac{2^{2n+1/2}}{e^\sigma \sqrt{n}}\, e^{-k^2/n}, \qquad c_k(n) = 2^{2n}\, n^{-1+3\epsilon}\, e^{-k^2/n}.$$

Notice that k enters $b_k(n)$ and $c_k(n)$ in a very simple way. We're in luck, because we will be summing over k.

The tail-exchange trick tells us that $\sum_k a_k(n)$ will be approximately $\sum_k b_k(n)$ if we have done a good job of estimation. Let us therefore evaluate

$$\sum_k b_k(n) = \frac{2^{2n+1/2}}{e^\sigma \sqrt{n}} \sum_k e^{-k^2/n} = \frac{2^{2n+1/2}}{e^\sigma \sqrt{n}} \Theta_n = \frac{2^{2n}\sqrt{2\pi}}{e^\sigma}\bigl(1 + O(n^{-M})\bigr).$$

(Another stroke of luck: We get to use the sum Θ_n from the previous example.) This is encouraging, because we know that the original sum is actually

What an amazing coincidence.

$$A_n = \sum_k \binom{2n}{k} = (1+1)^{2n} = 2^{2n}.$$

Therefore it looks as if we will have $e^\sigma = \sqrt{2\pi}$, as advertised.

But there's a catch: We still need to prove that our estimates are good enough. So let's look first at the error contributed by $c_k(n)$:

I'm tired of getting to the end of long, hard books and not even getting a word of good wishes from the author. It would be nice to read a "thanks for reading this, hope it comes in handy," instead of just running into a hard, cold, cardboard cover at the end of a long, dry proof. You know?

$$\Sigma_c(n) = \sum_{|k| \le n^{1/2+\epsilon}} 2^{2n} n^{-1+3\epsilon} e^{-k^2/n} \le 2^{2n} n^{-1+3\epsilon} \Theta_n = O(2^{2n} n^{-\frac{1}{2}+3\epsilon}).$$

Good; this is asymptotically smaller than the previous sum, if $3\epsilon < \frac{1}{2}$.

Next we must check the tails. We have

$$\sum_{k > n^{1/2+\epsilon}} e^{-k^2/n} < \exp\bigl(-\lfloor n^{1/2+\epsilon}\rfloor^2/n\bigr)\bigl(1 + e^{-1/n} + e^{-2/n} + \cdots\bigr)$$
$$= O(e^{-n^{2\epsilon}}) \cdot O(n),$$

which is $O(n^{-M})$ for all M; so $\sum_{k \notin D_n} b_k(n)$ is asymptotically negligible. (We chose the cutoff at $n^{1/2+\epsilon}$ just so that $e^{-k^2/n}$ would be exponentially small outside of D_n. Other choices like $n^{1/2}\log n$ would have been good enough too, and the resulting estimates would have been slightly sharper, but the formulas would have come out more complicated. We need not make the strongest possible estimates, since our main goal is to establish the value of the constant σ.) Similarly, the other tail

$$\sum_{k > n^{1/2+\epsilon}} \binom{2n}{n+k}$$

is bounded by $2n$ times its largest term, which occurs at the cutoff point $k \approx n^{1/2+\epsilon}$. This term is known to be approximately $b_k(n)$, which is exponentially small compared with A_n; and an exponentially small multiplier wipes out the factor of $2n$.

Thus we have successfully applied the tail-exchange trick to prove the estimate

$$2^{2n} = \sum_k \binom{2n}{k} = \frac{\sqrt{2\pi}}{e^\sigma} 2^{2n} + O(2^{2n} n^{-\frac{1}{2}+3\epsilon}), \quad \text{if } 0 < \epsilon < \tfrac{1}{6}. \qquad (9.99)$$

Thanks for reading this, hope it comes in handy.
— The authors

We may choose $\epsilon = \frac{1}{8}$ and conclude that

$$\sigma = \tfrac{1}{2} \ln 2\pi .$$

QED.

Exercises

Warmups

1 Prove or disprove: If $f_1(n) \prec g_1(n)$ and $f_2(n) \prec g_2(n)$, then we have $f_1(n) + f_2(n) \prec g_1(n) + g_2(n)$.

2 Which function grows faster:

a $n^{(\ln n)}$ or $(\ln n)^n$?

b $n^{(\ln \ln \ln n)}$ or $(\ln n)!$?

c $(n!)!$ or $((n-1)!)!\,(n-1)!^{n!}$?

d $F^2_{\lceil H_n \rceil}$ or H_{F_n}?

3 What's wrong with the following argument? "Since $n = O(n)$ and $2n = O(n)$ and so on, we have $\sum_{k=1}^{n} kn = \sum_{k=1}^{n} O(n) = O(n^2)$."

4 Give an example of a valid equation that has O-notation on the left but not on the right. (Do not use the trick of multiplying by zero; that's too easy.) *Hint:* Consider taking limits.

5 Prove or disprove: $O\big(f(n) + g(n)\big) = f(n) + O\big(g(n)\big)$, if $f(n)$ and $g(n)$ are positive for all n. (Compare with (9.27).)

6 Multiply $\big(\ln n + \gamma + O(1/n)\big)$ by $\big(n + O(\sqrt{n}\,)\big)$, and express your answer in O-notation.

7 Estimate $\sum_{k \geq 0} e^{-k/n}$ with absolute error $O(n^{-1})$.

Basics

8 Give an example of functions $f(n)$ and $g(n)$ such that none of the three relations $f(n) \prec g(n)$, $f(n) \succ g(n)$, $f(n) \asymp g(n)$ is valid, although $f(n)$ and $g(n)$ both increase monotonically to ∞.

9 Prove (9.22) rigorously by showing that the left side is a subset of the right side, according to the set-of-functions definition of O.

10 Prove or disprove: $\cos O(x) = 1 + O(x^2)$ for all real x.

11 Prove or disprove: $O(x+y)^2 = O(x^2) + O(y^2)$.

12 Prove that

$$1 + \frac{2}{n} + O(n^{-2}) = \left(1 + \frac{2}{n}\right)\left(1 + O(n^{-2})\right),$$

as $n \to \infty$.

13 Evaluate $\left(n + 2 + O(n^{-1})\right)^n$ with relative error $O(n^{-1})$.

14 Prove that $(n+\alpha)^{n+\beta} = n^{n+\beta}e^{\alpha}\left(1 + \alpha(\beta - \frac{1}{2}\alpha)n^{-1} + O(n^{-2})\right)$.

15 Give an asymptotic formula for the "middle" trinomial coefficient $\binom{3n}{n,n,n}$, correct to relative error $O(n^{-3})$.

16 Show that if $B(1-x) = -B(x) \geqslant 0$ for $0 < x < \frac{1}{2}$, we have

$$\int_a^b B(\{x\})\,f(x)\,dx \geqslant 0$$

if we assume also that $f'(x) \geqslant 0$ for $a \leqslant x \leqslant b$.

17 Use generating functions to show that $B_m(\frac{1}{2}) = (2^{1-m} - 1)B_m$, for all $m \geqslant 0$.

18 Find $\sum_k \binom{2n}{k}^{\alpha}$ with relative error $O(n^{-1/4})$, when $\alpha > 0$.

Homework exercises

19 Use a computer to compare the left and right sides of the approximations in Table 438, when $n = 10$, $z = \alpha = 0.1$, and $O\big(f(n)\big) = O\big(f(z)\big) = 0$.

20 Prove or disprove the following estimates, as $n \to \infty$:

a $$O\left(\left(\frac{n^2}{\log\log n}\right)^{1/2}\right) = O(\lfloor\sqrt{n}\rfloor^2).$$

b $$e^{(1+O(1/n))^2} = e + O(1/n).$$

c $$n! = O\Big(\big((1-1/n)^n n\big)^n\Big).$$

21 Equation (9.48) gives the nth prime with relative error $O(\log n)^{-2}$. Improve the relative error to $O(\log n)^{-3}$ by starting with another term of (9.31) in (9.46).

22 Improve (9.54) to $O(n^{-3})$.

23 Push the approximation (9.62) further, getting absolute error $O(n^{-3})$. *Hint:* Let $g_n = c/(n+1)(n+2) + h_n$; what recurrence does h_n satisfy?

24 Suppose $a_n = O(f(n))$ and $b_n = O(f(n))$. Prove or disprove that the convolution $\sum_{k=0}^{n} a_k b_{n-k}$ is also $O(f(n))$, in the following cases:

a $f(n) = n^{-\alpha}$, $\alpha > 1$.

b $f(n) = \alpha^{-n}$, $\alpha > 1$.

25 Prove (9.1) and (9.2), with which we opened this chapter.

26 Equation (9.91) shows how to evaluate $\ln 10!$ with an absolute error $< \frac{1}{126000000}$. Therefore if we take exponentials, we get $10!$ with a relative error that is less than $e^{1/126000000} - 1 < 10^{-8}$. (In fact, the approximation gives 3628799.9714.) If we now round to the nearest integer, knowing that $10!$ is an integer, we get an exact result.

Is it always possible to calculate $n!$ in a similar way, if enough terms of Stirling's approximation are computed? Estimate the value of m that gives the best approximation to $\ln n!$, when n is a fixed (large) integer. Compare the absolute error in this approximation with $n!$ itself.

27 Use Euler's summation formula to find the asymptotic value of $H_n^{(-\alpha)} = \sum_{k=1}^{n} k^\alpha$, where α is any fixed real number. (Your answer may involve a constant that you do not know in closed form.)

28 Exercise 5.13 defines the hyperfactorial function $Q_n = 1^1 2^2 \ldots n^n$. Find the asymptotic value of Q_n with relative error $O(n^{-1})$. (Your answer may involve a constant that you do not know in closed form.)

29 Estimate the function $1^{1/1} 2^{1/2} \ldots n^{1/n}$ as in the previous exercise.

30 Find the asymptotic value of $\sum_{k \geq 0} k^l e^{-k^2/n}$ with absolute error $O(n^{-3})$, when l is a fixed nonnegative integer.

31 Evaluate $\sum_{k \geq 0} 1/(c^k + c^m)$ with absolute error $O(c^{-3m})$, when $c > 1$ and m is a positive integer.

Exam problems

32 Evaluate $e^{H_n + H_n^{(2)}}$ with absolute error $O(n^{-1})$.

33 Evaluate $\sum_{k \geq 0} \binom{n}{k} / n^k$ with absolute error $O(n^{-2})$.

34 Determine values A through F such that $(1 + 1/n)^{nH_n}$ is

$$An + B(\ln n)^2 + C \ln n + D + \frac{E(\ln n)^2}{n} + \frac{F \ln n}{n} + O(n^{-1}).$$

35 Evaluate $\sum_{k=1}^{n} 1/kH_k$ with absolute error $O(1)$.

36 Evaluate $S_n = \sum_{k=1}^{n} 1/(n^2 + k^2)$ with absolute error $O(n^{-5})$.

37 Evaluate $\sum_{k=1}^{n} (n \bmod k)$ with absolute error $O(n \log n)$.

38 Evaluate $\sum_{k \geq 0} k^k \binom{n}{k}$ with relative error $O(n^{-1})$.

39 Evaluate $\sum_{0\leqslant k<n} \ln(n-k)(\ln n)^k/k!$ with absolute error $O(n^{-1})$. *Hint:* Show that the terms for $k \geqslant 10\ln n$ are negligible.

40 Let m be a (fixed) positive integer. Evaluate $\sum_{k=1}^{n}(-1)^k H_k^m$ with absolute error $O(1)$.

41 Evaluate the "Fibonacci factorial" $\prod_{k=1}^{n} F_k$ with relative error $O(n^{-1})$ or better. Your answer may involve a constant whose value you do not know in closed form.

42 Let α be a constant in the range $0 < \alpha < \frac{1}{2}$. We've seen in previous chapters that there is no general closed form for the sum $\sum_{k\leqslant \alpha n}\binom{n}{k}$. Show that there is, however, an asymptotic formula

$$\sum_{k\leqslant \alpha n}\binom{n}{k} = 2^{nH(\alpha)-\frac{1}{2}\lg n+O(1)},$$

where $H(\alpha) = \alpha\lg\frac{1}{\alpha} + (1-\alpha)\lg(\frac{1}{1-\alpha})$. *Hint:* Show that $\binom{n}{k-1} < \frac{\alpha}{1-\alpha}\binom{n}{k}$ for $0 < k \leqslant \alpha n$.

43 Show that C_n, the number of ways to change n cents (as considered in Chapter 7) is asymptotically $cn^4 + O(n^3)$ for some constant c. What is that constant?

44 Prove that

$$x^{\underline{1/2}} = x^{1/2}\left[{1/2 \atop 1/2}\right] - x^{-1/2}\left[{1/2 \atop -1/2}\right] + x^{-3/2}\left[{1/2 \atop -3/2}\right] + O(x^{-5/2})$$

as $x \to \infty$. (Recall the definition $x^{\underline{1/2}} = x!/(x-\frac{1}{2})!$ in (5.88), and the definition of generalized Stirling numbers in Table 258.)

45 Let α be an irrational number between 0 and 1. Chapter 3 discusses the quantity $D(\alpha, n)$, which measures the maximum discrepancy by which the fractional parts $\{k\alpha\}$ for $0 \leqslant k < n$ deviate from a uniform distribution. The recurrence

$$D(\alpha, n) \leqslant D(\{\alpha^{-1}\}, \lfloor \alpha n \rfloor) + \alpha^{-1} + 2$$

was proved in (3.31); we also have the obvious bounds

$$0 \leqslant D(\alpha, n) \leqslant n.$$

Prove that $\lim_{n\to\infty} D(\alpha, n)/n = 0$. *Hint:* Chapter 6 discusses continued fractions.

46 Show that the Bell number $b_n = e^{-1}\sum_{k\geqslant 0} k^n/k!$ of exercise 7.15 is asymptotically equal to

$$m(n)^n e^{m(n)-n-1/2}/\sqrt{\ln n}\,,$$

where $m(n)\ln m(n) = n - \frac{1}{2}$, and estimate the relative error in this approximation.

47 Let m be an integer $\geqslant 2$. Analyze the two sums

$$\sum_{k=1}^{n} \lfloor \log_m n \rfloor \qquad\text{and}\qquad \sum_{k=1}^{n} \lceil \log_m n \rceil\,;$$

which is asymptotically closer to $\log_m n!$?

48 Consider a table of the harmonic numbers H_k for $1 \leqslant k \leqslant n$ in decimal notation. The kth entry $\widehat{H}_k$ has been correctly rounded to d_k significant digits, where d_k is just large enough to distinguish this value from the values of H_{k-1} and H_{k+1}. For example, here is an extract from the table, showing five entries where H_k passes 10:

k	H_k	$\widehat{H}_k$	d_k
12364	9.99980041−	9.9998	5
12365	9.99988128+	9.9999	5
12366	9.99996215−	9.99996	6
12367	10.00004301−	10.0000	6
12368	10.00012386+	10.0001	6

Estimate the total number of digits in the table, $\sum_{k=1}^{n} d_k$, with an absolute error of $O(n)$.

49 In Chapter 6 we considered the tale of a worm that reaches the end of a stretching band after n seconds, where $H_{n-1} < 100 \leqslant H_n$. Prove that if n is a positive integer such that $H_{n-1} \leqslant \alpha \leqslant H_n$, then

$$\lfloor e^{\alpha-\gamma} \rfloor \;\leqslant\; n \;\leqslant\; \lceil e^{\alpha-\gamma} \rceil\,.$$

50 Venture capitalists in Silicon Valley are being offered a deal giving them a chance for an exponential payoff on their investments: For an n million dollar investment, where $n \geqslant 2$, the GKP consortium promises to pay up to N million dollars after one year, where $N = 10^n$. Of course there's some risk; the actual deal is that GKP pays k million dollars with probability $1/(k^2 H_N^{(2)})$, for each integer k in the range $1 \leqslant k \leqslant N$. (All payments are in megabucks, that is, in exact multiples of \$1,000,000; the payoff is determined by a truly random process.) Notice that an investor always gets at least a million dollars back.

a What is the asymptotic expected return after one year, if n million dollars are invested? (In other words, what is the mean value of the payment?) Your answer should be correct within an absolute error of $O(10^{-n})$ dollars.

I once earned $O(10^{-n})$ *dollars.*

b What is the asymptotic probability that you make a profit, if you invest n million? (In other words, what is the chance that you get back more than you put in?) Your answer here should be correct within an absolute error of $O(n^{-3})$.

Bonus problems

51 Prove or disprove: $\int_n^\infty O(x^{-2})\,dx = O(n^{-1})$ as $n \to \infty$.

52 Show that there exists a power series $A(z) = \sum_{k\geq 0} a_n z^n$, convergent for all complex z, such that

$$A(n) \succ n^{n^{n^{\cdot^{\cdot^{\cdot^{n}}}}}}\Big\}n\,.$$

53 Prove that if $f(x)$ is a function whose derivatives satisfy

$$f'(x) \leqslant 0\,, \quad -f''(x) \leqslant 0\,, \quad f'''(x) \leqslant 0\,, \quad \ldots, \quad (-1)^m f^{(m+1)}(x) \leqslant 0$$

for all $x \geqslant 0$, then we have

$$f(x) \;=\; f(0) + \frac{f'(0)}{1!}x + \cdots + \frac{f^{(m-1)}(0)}{(m-1)!}x^{m-1} + O(x^m)\,, \quad \text{for } x \geqslant 0.$$

In particular, the case $f(x) = -\ln(1+x)$ proves (9.64) for all $k, n > 0$.

54 Let $f(x)$ be a positive, differentiable function such that $xf'(x) \prec f(x)$ as $x \to \infty$. Prove that

$$\sum_{k\geqslant n} \frac{f(k)}{k^{1+\alpha}} \;=\; O\left(\frac{f(n)}{n^\alpha}\right), \qquad \text{if } \alpha > 0.$$

Hint: Consider the quantity $f(k-\frac{1}{2})/(k-\frac{1}{2})^\alpha - f(k+\frac{1}{2})/(k+\frac{1}{2})^\alpha$.

55 Improve (9.99) to relative error $O(n^{-3/2+5\epsilon})$.

56 The quantity $Q(n) = 1 + \frac{n-1}{n} + \frac{n-1}{n}\frac{n-2}{n} + \cdots = \sum_{k\geqslant 1} n^{\underline{k}}/n^k$ occurs in the analysis of many algorithms. Find its asymptotic value, with absolute error $o(1)$.

57 An asymptotic formula for Golomb's sum $\sum_{k\geqslant 1} 1/k\lfloor 1+\log_n k\rfloor^2$ is derived in (9.54). Find an asymptotic formula for the analogous sum without floor brackets, $\sum_{k\geqslant 1} 1/k(1+\log_n k)^2$. *Hint:* We have $\int_0^\infty ue^{-u}k^{-tu}\,du = 1/(1+t\ln k)^2$.

58 Prove that

$$B_m(\{x\}) = -2\frac{m!}{(2\pi)^m}\sum_{k\geqslant 1}\frac{\cos(2\pi kx - \frac{1}{2}\pi m)}{k^m}, \qquad \text{for } m \geqslant 2,$$

by using residue calculus, integrating

$$\frac{1}{2\pi i}\oint \frac{2\pi i\, e^{2\pi i z\theta}}{e^{2\pi i z}-1}\frac{dz}{z^m}$$

on the square contour $z = x+iy$, where $\max(|x|,|y|) = M+\frac{1}{2}$, and letting the integer M tend to ∞.

59 Let $\Theta_n(t) = \sum_k e^{-(k+t)^2/n}$, a periodic function of t. Show that the expansion of $\Theta_n(t)$ as a Fourier series is

$$\Theta_n(t) = \sqrt{\pi n}\bigl(1 + 2e^{-\pi^2 n}(\cos 2\pi t) + 2e^{-4\pi^2 n}(\cos 4\pi t) + 2e^{-9\pi^2 n}(\cos 6\pi t) + \cdots\bigr).$$

(This formula gives a rapidly convergent series for the sum $\Theta_n = \Theta_n(0)$ in equation (9.93).)

60 Explain why the coefficients in the asymptotic expansion

$$\binom{2n}{n} = \frac{4^n}{\sqrt{\pi n}}\left(1 - \frac{1}{8n} + \frac{1}{128n^2} + \frac{5}{1024n^3} - \frac{21}{32768n^4} + O(n^{-5})\right)$$

all have denominators that are powers of 2.

61 Exercise 45 proves that the discrepancy $D(\alpha, n)$ is $o(n)$ for all irrational numbers α. Exhibit an irrational α such that $D(\alpha, n)$ is *not* $O(n^{1-\epsilon})$ for any $\epsilon > 0$.

62 Given n, let $\left\{{n \atop m(n)}\right\} = \max_k \left\{{n \atop k}\right\}$ be the largest entry in row n of Stirling's subset triangle. Show that for all sufficiently large n, we have $m(n) = \lfloor \overline{m}(n) \rfloor$ or $m(n) = \lceil \overline{m}(n) \rceil$, where

$$\overline{m}(n)(\overline{m}(n)+2)\ln(\overline{m}(n)+2) = n(\overline{m}(n)+1).$$

Hint: This is difficult.

63 Prove that S. W. Golomb's self-describing sequence of exercise 2.36 satisfies $f(n) = \phi^{2-\phi}n^{\phi-1} + O(n^{\phi-1}/\log n)$.

64 Find a proof of the identity

$$\sum_{n\geqslant 1}\frac{\cos 2n\pi x}{n^2} = \pi^2\left(x^2 - x + \tfrac{1}{6}\right) \qquad \text{for } 0 \leqslant x \leqslant 1,$$

that uses only "Eulerian" (eighteenth-century) mathematics.

Research problems

65 Find a "combinatorial" proof of Stirling's approximation. (Note that n^n is the number of mappings of $\{1,2,\ldots,n\}$ into itself, and $n!$ is the number of mappings of $\{1,2,\ldots,n\}$ onto itself.)

66 Consider an $n \times n$ array of dots, $n \geqslant 3$, in which each dot has four neighbors. (At the edges we "wrap around" modulo n.) Let χ_n be the number of ways to assign the colors red, white, and blue to these dots in such a way that no neighboring dots have the same color. (Thus $\chi_3 = 12$.) Prove that

$$\chi_n \sim \left(\tfrac{4}{3}\right)^{3n^2/2} e^{-\pi/6}.$$

67 Let Q_n be the least integer m such that $H_m > n$. Find the smallest integer n such that $Q_n \neq \lfloor e^{n-\gamma} + \frac{1}{2} \rfloor$, or prove that no such n exist.

Th-th-th-that's all, folks!

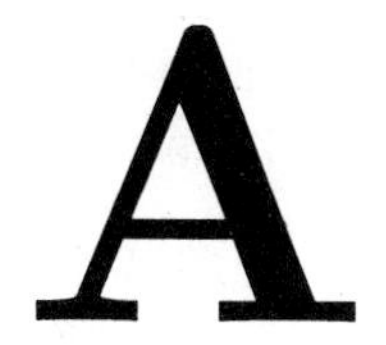

Answers to Exercises

EVERY EXERCISE is answered here (at least briefly), and some of these answers go beyond what was asked. Readers will learn best if they make a serious attempt to find their own answers BEFORE PEEKING at this appendix.

The authors will be interested to learn of any solutions (or partial solutions) to the research problems, or of any simpler (or more correct) ways to solve the non-research ones.

(The first finder of every error in this book will receive a reward of $2.56.)

Does that mean I have to find every error?

(We meant to say "any error.")

Does that mean only one person gets a reward?

(Hmmm. Try it and see.)

1.1 The proof is fine except when $n = 2$. If all sets of two horses have horses of the same color, the statement is true for any number of horses.

1.2 If X_n is the number of moves, we have $X_0 = 0$ and $X_n = X_{n-1} + 1 + X_{n-1} + 1 + X_{n-1}$ when $n > 0$. It follows (for example by adding 1 to both sides) that $X_n = 3^n - 1$. (After $\frac{1}{2}X_n$ moves, it turns out that the entire tower will be on the middle peg, halfway home!)

1.3 There are 3^n possible arrangements, since each disk can be on any of the pegs. We must hit them all, since the shortest solution takes $3^n - 1$ moves. (This construction is equivalent to a "ternary Gray code," which runs through all numbers from $(0\ldots0)_3$ to $(2\ldots2)_3$, changing only one digit at a time.)

1.4 No. If the largest disk doesn't have to move, $2^{n-1} - 1$ moves will suffice (by induction); otherwise $(2^{n-1} - 1) + 1 + (2^{n-1} - 1)$ will suffice (again by induction).

The number of intersection points turns out to give the whole story; convexity was a red herring.

1.5 No; different circles can intersect in at most two points, so the fourth circle can increase the number of regions to at most 14. However, it is possible to do the job with ovals:

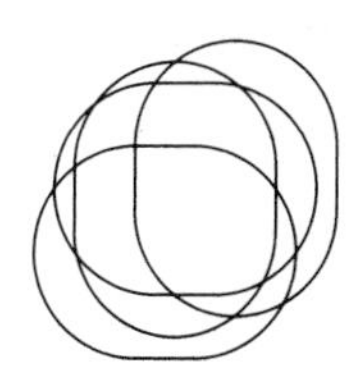

Venn [294] claimed that there is no way to do the five-set case with ellipses, but a five-set construction with ellipses was found by Grünbaum [137].

1.6 If the nth line intersects the previous lines in $k > 0$ distinct points, we get $k-1$ new bounded regions (assuming that none of the previous lines were mutually parallel) and two new infinite regions. Hence the maximum number of bounded regions is $(n-2)+(n-3)+\cdots = S_{n-2} = (n-1)(n-2)/2 = L_n - 2n$.

1.7 The basis is unproved; and in fact, $H(1) \neq 2$.

1.8 $Q_2 = (1+\beta)/\alpha$; $Q_3 = (1+\alpha+\beta)/\alpha\beta$; $Q_4 = (1+\alpha)/\beta$; $Q_5 = \alpha$; $Q_6 = \beta$. So the sequence is periodic!

1.9 (a) We get $P(n-1)$ from the inequality

$$x_1 \ldots x_{n-1}\left(\frac{x_1+\cdots+x_{n-1}}{n-1}\right) \leqslant \left(\frac{x_1+\cdots+x_{n-1}}{n-1}\right)^n .$$

(b) $x_1 \ldots x_n x_{n+1} \ldots x_{2n} \leqslant \bigl(((x_1+\cdots+x_n)/n)((x_{n+1}+\cdots+x_{2n})/n)\bigr)^n$ by $P(n)$; the product inside is $\leqslant \bigl((x_1+\cdots+x_{2n})/2n\bigr)^2$ by $P(2)$. (c) For example, $P(5)$ follows from $P(6)$ from $P(3)$ from $P(4)$ from $P(2)$.

1.10 First show that $R_n = R_{n-1} + 1 + Q_{n-1} + 1 + R_{n-1}$, when $n > 0$. Incidentally, the methods of Chapter 7 will tell us that $Q_n = \bigl((1+\sqrt{3}\,)^{n+1} - (1-\sqrt{3}\,)^{n+1}\bigr)/(2\sqrt{3}) - 1$.

1.11 (a) We cannot do better than to move a double $(n-1)$-tower, then move (and invert the order of) the two largest disks, then move the double $(n-1)$-tower again; hence $A_n = 2A_{n-1} + 2$ and $A_n = 2T_n = 2^{n+1} - 2$. This solution interchanges the two largest disks but returns the other $2n-2$ to their original order.

(b) We can produce the original order in $2A_n$ moves; but this cannot be optimum, because it takes 4 moves instead of 3 when $n = 1$. Let B_n be the minimum number of moves. Then $B_n = 2C_n + 1$, where C_n is the minimum number of moves when there are $2n-1$ disks, one of the largest size and two indistinguishable disks of every other size. And $C_n = 2A_{n-1} + 1 = 2^{n+1} - 3$. Hence $B_n = 2^{n+2} - 5$, for all $n > 0$. Curiously this is just $2A_n - 1$.

1.12 If all $n_k > 0$, then $A(n_1, \ldots, n_m) = 2A(n_1, \ldots, n_{m-1}) + n_m$. This is an equation of the "generalized Josephus" type, with solution $(n_1 \ldots n_m)_2 = 2^{m-1}n_1 + \cdots + 2n_{m-1} + n_m$. Also

$$B(n_1, \ldots, n_m) = 2A(n_1, \ldots, n_{m-1}, n_m - 1) + 1;$$

this is $A(n_1, \ldots, n_m)$ if $n_m = 1$, otherwise it's $2A(n_1, \ldots, n_m) - 1$.

1.13 Given n straight lines that define L_n regions, we can replace them by extremely narrow zig-zags with segments sufficiently long that there are nine intersections between each pair of zig-zags. This shows that $ZZ_n = ZZ_{n-1}+9n-8$, for all $n > 0$; consequently $ZZ_n = 9S_n-8n+1 = \frac{9}{2}n^2-\frac{7}{2}n+1$.

1.14 The number of new 3-dimensional regions defined by each new cut is the number of 2-dimensional regions defined in the new plane by its intersections with the previous planes. Hence $P_n = P_{n-1} + L_{n-1}$, and it turns out that $P_5 = 26$. (Six cuts in a cubical piece of cheese can make 27 cubelets, or up to $P_6 = 42$ cuts of weirder shapes.)

Incidentally, the solution to this recurrence fits into a nice pattern if we express it in terms of binomial coefficients (see Chapter 5):

$$X_n = \binom{n}{0} + \binom{n}{1};$$
$$L_n = \binom{n}{0} + \binom{n}{1} + \binom{n}{2};$$
$$P_n = \binom{n}{0} + \binom{n}{1} + \binom{n}{2} + \binom{n}{3}.$$

I bet I know what happens in four dimensions!

Here X_n is the maximum number of 1-dimensional regions definable by n points on a line.

1.15 The function I satisfies the same recurrence as J when $n > 1$, but $I(1)$ is undefined. Since $I(2) = 2$ and $I(3) = 1$, there's no value of $I(1) = \alpha$ that will allow us to use our general method; the "end game" of unfolding depends on the two leading bits in n's binary representation.

If $n = 2^m + 2^{m-1} + k$, where $0 \leqslant k < 2^{m+1} + 2^m - (2^m + 2^{m-1}) = 2^m + 2^{m-1}$, the solution is $I(n) = 2k+1$ for all $n > 2$. Another way to express this, in terms of the representation $n = 2^m + l$, is to say that

$$I(n) = \begin{cases} J(n) + 2^{m-1}, & \text{if } 0 \leqslant l < 2^{m-1}; \\ J(n) - 2^m, & \text{if } 2^{m-1} \leqslant l < 2^m. \end{cases}$$

1.16 Let $g(n) = a(n)\alpha + b(n)\beta_0 + c(n)\beta_1 + d(n)\gamma$. We know from (1.18) that $a(n)\alpha + b(n)\beta_0 + c(n)\beta_1 = (\alpha\, \beta_{b_{m-1}}\, \beta_{b_{m-2}} \ldots \beta_{b_1}\, \beta_{b_0})_3$ when $n = (1\, b_{m-1} \ldots b_1\, b_0)_2$; this defines $a(n)$, $b(n)$, and $c(n)$. Setting $g(n) = n$ in the recurrence implies that $a(n) + c(n) - d(n) = n$; hence we know everything. [Setting $g(n) = 1$ gives the additional identity $a(n)-2b(n)-2c(n) = 1$, which can be used to define $b(n)$ in terms of the simpler functions $a(n)$ and $a(n) + c(n)$.]

1.17 In general we have $W_m \leqslant 2W_{m-k} + T_k$, for $0 \leqslant k \leqslant m$. (This relation corresponds to transferring the top $n-k$, then using only three pegs to move the bottom k, then finishing with the top $n-k$.) The stated relation turns out to be based on the unique value of k that minimizes the right-hand side of this general inequality, when $m = n(n+1)/2$. (However, we cannot conclude that equality holds; many other strategies for transferring the tower are conceivable.) If we set $Y_n = (W_{n(n+1)/2} - 1)/2^n$, we find that $Y_n \leqslant Y_{n-1} + 1$; hence $W_{n(n+1)/2} \leqslant 2^n(n-1) + 1$.

1.18 It suffices to show that both of the lines from $(n^{2j}, 0)$ intersect both of the lines from $(n^{2k}, 0)$, and that all these intersection points are distinct.

A line from $(x_j, 0)$ through $(x_j - a_j, 1)$ intersects a line from $(x_k, 0)$ through $(x_k - a_k, 1)$ at the point $(x_j - ta_j, t)$ where $t = (x_k - x_j)/(a_k - a_j)$. Let $x_j = n^{2j}$ and $a_j = n^j + (0 \text{ or } n^{-n})$. Then the ratio $t = (n^{2k} - n^{2j})/(n^k - n^j + (-n^{-n} \text{ or } 0 \text{ or } n^{-n}))$ lies strictly between $n^j + n^k - 1$ and $n^j + n^k + 1$; hence the y coordinate of the intersection point uniquely identifies j and k. Also the four intersections that have the same j and k are distinct.

1.19 Not when $n > 11$. A bent line whose half-lines run at angles θ and $\theta + 30°$ from its apex can intersect four times with another whose half-lines run at angles ϕ and $\phi + 30°$ only if $|\theta - \phi| > 30°$. We can't choose more than 11 angles this far apart from each other. (Is it possible to choose 11?)

1.20 Let $h(n) = a(n)\alpha + b(n)\beta_0 + c(n)\beta_1 + d(n)\gamma_0 + e(n)\gamma_1$. We know from (1.18) that $a(n)\alpha + b(n)\beta_0 + c(n)\beta_1 = (\alpha\, \beta_{b_{m-1}}\, \beta_{b_{m-2}} \ldots \beta_{b_1}\, \beta_{b_0})_4$ when $n = (1\, b_{m-1} \ldots b_1\, b_0)_2$; this defines $a(n)$, $b(n)$, and $c(n)$. Setting $h(n) = n$ in the recurrence implies that $a(n) + c(n) - 2d(n) - 2e(n) = n$; setting $h(n) = n^2$ implies that $a(n) + c(n) + 4e(n) = n^2$. Hence $d(n) = \bigl(3a(n) + 3c(n) - n^2 - 2n\bigr)/4$; $e(n) = \bigl(n^2 - a(n) - c(n)\bigr)/4$.

1.21 We can let m be the least (or any) common multiple of $2n$, $2n-1$, ..., $n+1$. [A non-rigorous argument suggests that a "random" value of m will succeed with probability

$$\frac{n}{2n}\frac{n-1}{2n-1}\cdots\frac{1}{n+1} = 1\Big/\binom{2n}{n} \sim \frac{\sqrt{\pi n}}{4^n},$$

so we might expect to find such an m less than 4^n.]

1.22 Take a regular polygon with 2^n sides and label the sides with the elements of a "de Bruijn cycle" of length 2^n. (This is a cyclic sequence of 0's and 1's in which all n-tuples of adjacent elements are different; see [173, exercise 2.3.4.2–23] and [174, exercise 3.2.2–17].) Attach a very thin convex extension to each side that's labeled 1. The n sets are copies of the resulting polygon, rotated by the length of k sides for $k = 0, 1, \ldots, n-1$.

I once rode a de Bruijn cycle (when visiting at his home in Nuenen, Netherlands).

1.23 Yes. (We need principles of elementary number theory from Chapter 4.) Let $L(n) = \operatorname{lcm}(1, 2, \ldots, n)$. We can assume that $n > 2$; hence by

Bertrand's postulate there is a prime p between $n/2$ and n. We can also assume that $j > n/2$, since $q' = L(n) + 1 - q$ leaves $j' = n + 1 - j$ if and only if q leaves j. Choose q so that $q \equiv 1 \pmod{L(n)/p}$ and $q \equiv j + 1 - n \pmod{p}$. The people are now removed in order $1, 2, \ldots, n - p, j + 1, j + 2, \ldots, n, n - p + 1, \ldots, j - 1$.

1.24 The only known examples are: $X_n = a/X_{n-1}$, which has period 2; R. C. Lyness's recurrence of period 5 in exercise 8; H. Todd's recurrence $X_n = (1 + X_{n-1} + X_{n-2})/X_{n-3}$, which has period 8; and recurrences derived from these by substitutions of the form $Y_n = \alpha X_{mn}$. An exhaustive search by Bill Gosper turned up no nontrivial solutions of period 4 when $k = 2$. A partial theory has been developed by Lyness [210] and by Kurshan and Gopinath [189]. An interesting example of another type, with period 9 when the starting values are real, is the recurrence $X_n = |X_{n-1}| - X_{n-2}$ discovered by Morton Brown [38]. Nonlinear recurrences having any desired period $\geqslant 5$ can be based on continuants [55].

1.25 If $T^{(k)}(n)$ denotes the minimum number of moves needed to transfer n disks with k auxiliary pegs (hence $T^{(1)}(n) = T_n$ and $T^{(2)}(n) = W_n$), we have $T^{(k)}(\binom{n+1}{k}) \leqslant 2T^{(k)}(\binom{n}{k}) + T^{(k-1)}(\binom{n}{k-1})$. No examples (n, k) are known where this inequality fails to be an equality. When k is small compared with n, the formula $2^{n+1-k}\binom{n-1}{k-1}$ gives a convenient (but non-optimum) upper bound on $T^{(k)}(\binom{n}{k})$.

1.26 The execution-order permutation can be computed in $O(n \log n)$ steps for all m and n [175, exercises 5.1.1–2 and 5.1.1–5]. Bjorn Poonen [241] has proved that non-Josephus sets with exactly four "bad guys" exist whenever $n \equiv 0 \pmod 3$ and $n \geqslant 9$; in fact, the number of such sets is at least $\epsilon\binom{n}{4}$ for some $\epsilon > 0$. He also found by extensive computations that the only other $n < 24$ with non-Josephus sets is $n = 20$, which has 236 such sets with $k = 14$ and two with $k = 13$. (One of the latter is $\{1, 2, 3, 4, 5, 6, 7, 8, 11, 14, 15, 16, 17\}$; the other is its reflection with respect to 21.) There is a unique non-Josephus set with $n = 15$ and $k = 9$, namely $\{3, 4, 5, 6, 8, 10, 11, 12, 13\}$.

2.1 There's no agreement about this; three answers are defensible: (1) We can say that $\sum_{k=m}^{n} q_k$ is always equivalent to $\sum_{m \leqslant k \leqslant n} q_k$; then the stated sum is zero. (2) A person might say that the given sum is $q_4 + q_3 + q_2 + q_1 + q_0$, by summing over decreasing values of k. But this conflicts with the generally accepted convention that $\sum_{k=1}^{n} q_k = 0$ when $n = 0$. (3) We can say that $\sum_{k=m}^{n} q_k = \sum_{k \leqslant n} q_k - \sum_{k<m} q_k$; then the stated sum is $-q_1 - q_2 - q_3$. This convention may appear strange, but it obeys the useful law $\sum_{k=a}^{b} + \sum_{k=b+1}^{c} = \sum_{k=a}^{c}$ for all a, b, c.

It's best to use the notation $\sum_{k=m}^{n}$ only when $n - m \geqslant -1$; then both conventions (1) and (3) agree.

2.2 This is $|x|$. Incidentally, the quantity $\bigl([x>0]-[x<0]\bigr)$ is often called $\operatorname{sign}(x)$ or $\operatorname{signum}(x)$; it is $+1$ when $x>0$, 0 when $x=0$, and -1 when $x<0$.

2.3 The first sum is, of course, $a_0+a_1+a_2+a_3+a_4+a_5$; the second is $a_4+a_1+a_0+a_1+a_4$, because the sum is over the values $k \in \{-2,-1,0,+1,+2\}$. The commutative law doesn't hold here because the function $p(k)=k^2$ is not a permutation. Some values of n (e.g., $n=3$) have no k such that $p(k)=n$; others (e.g., $n=4$) have two such k.

2.4 (a) $\sum_{i=1}^{4}\sum_{j=i+1}^{4}\sum_{k=j+1}^{4}a_{ijk} = \sum_{i=1}^{2}\sum_{j=i+1}^{3}\sum_{k=j+1}^{4}a_{ijk} = \bigl((a_{123}+a_{124})+a_{134}\bigr)+a_{234}$.

(b) $\sum_{k=1}^{4}\sum_{j=1}^{k-1}\sum_{i=1}^{j-1}a_{ijk} = \sum_{k=3}^{4}\sum_{j=2}^{k-1}\sum_{i=1}^{j-1}a_{ijk} = a_{123}+\bigl(a_{124}+(a_{134}+a_{234})\bigr)$.

2.5 The same index 'k' is being used for two different index variables, although k is bound in the inner sum. This is a famous mistake in mathematics (and computer programming). The result turns out to be correct if $a_j=a_k$ for all j and k, $1\leqslant j,k\leqslant n$.

2.6 It's $[1\leqslant j\leqslant n](n-j+1)$. The first factor is necessary here because we should get zero when $j<1$ or $j>n$.

2.7 $mx^{\overline{m-1}}$. A version of finite calculus based on ∇ instead of Δ would therefore give special prominence to *rising* factorial powers.

2.8 0, if $m\geqslant 1$; $1/|m|!$, if $m\leqslant 0$.

2.9 $x^{\overline{m+n}} = x^{\overline{m}}\,(x+m)^{\overline{n}}$, for integers m and n. Setting $m=-n$ tells us that $x^{\overline{-n}} = 1/(x-n)^{\overline{n}} = 1/(x-1)^{\underline{n}}$.

2.10 Another possible right-hand side is $\mathrm{E}u\,\Delta v + v\,\Delta u$.

2.11 Break the left-hand side into two sums, and change k to $k+1$ in the second of these.

2.12 If $p(k)=n$ then $n+c=k+\bigl((-1)^k+1\bigr)c$ and $\bigl((-1)^k+1\bigr)$ is even; hence $(-1)^{n+c}=(-1)^k$ and $k=n-(-1)^{n+c}c$. Conversely, this value of k yields $p(k)=n$.

2.13 Let $R_0=\alpha$, and $R_n=R_{n-1}+(-1)^n(\beta+n\gamma+n^2\delta)$ for $n>0$. Then $R(n)=A(n)\alpha+B(n)\beta+C(n)\gamma+D(n)\delta$. Setting $R_n=1$ yields $A(n)=1$. Setting $R_n=(-1)^n$ yields $A(n)+2B(n)=(-1)^n$. Setting $R_n=(-1)^n n$ yields $-B(n)+2C(n)=(-1)^n n$. Setting $R_n=(-1)^n n^2$ yields $B(n)-2C(n)+2D(n)=(-1)^n n^2$. Therefore $2D(n)=(-1)^n(n^2+n)$; the stated sum is $D(n)$.

2.14 The suggested rewrite is legitimate since we have $k=\sum_{1\leqslant j\leqslant k}1$ when $1\leqslant k\leqslant n$. Sum first on k; the multiple sum reduces to

$$\sum_{1\leqslant j\leqslant n}(2^{n+1}-2^j) \;=\; n2^{n+1}-(2^{n+1}-2)\,.$$

2.15 The first step replaces $k(k+1)$ by $2\sum_{1\leqslant j\leqslant k} j$. The second step gives $\boxdot_n + \Box_n = \left(\sum_{k=1}^{n} k\right)^2 + \Box_n$.

2.16 $x^{\underline{m}}(x-m)^{\underline{n}} = x^{\underline{m+n}} = x^{\underline{n}}(x-n)^{\underline{m}}$, by (2.52).

2.17 Use induction for the first two ='s, and (2.52) for the third. The second line follows from the first.

2.18 Use the facts that $(\Re z)^+ \leqslant |z|$, $(\Re z)^- \leqslant |z|$, $(\Im z)^+ \leqslant |z|$, $(\Im z)^- \leqslant |z|$, and $|z| \leqslant (\Re z)^+ + (\Re z)^- + (\Im z)^+ + (\Im z)^-$.

2.19 Multiply both sides by $2^{n-1}/n!$ and let $S_n = 2^n T_n/n! = S_{n-1} + 3\cdot 2^{n-1} = 3(2^n - 1) + S_0$. The solution is $T_n = 3\cdot n! + n!/2^{n-1}$. (We'll see in Chapter 4 that T_n is an integer only when n is 0 or a power of 2.)

"It is a profoundly erroneous truism, repeated by all copybooks and by eminent people when they are making speeches, that we should cultivate the habit of thinking of what we are doing. The precise opposite is the case. Civilization advances by extending the number of important operations which we can perform without thinking about them. Operations of thought are like cavalry charges in a battle—they are strictly limited in number, they require fresh horses, and must only be made at decisive moments."
— *A. N. Whitehead [302]*

2.20 The perturbation method gives

$$S_n + (n+1)H_{n+1} = S_n + \Bigl(\sum_{0\leqslant k\leqslant n} H_k\Bigr) + n + 1 .$$

2.21 Extracting the final term of S_{n+1} gives $S_{n+1} = 1 - S_n$; extracting the first term gives

$$\begin{aligned} S_{n+1} &= (-1)^{n+1} + \sum_{1\leqslant k\leqslant n+1} (-1)^{n+1-k} = (-1)^{n+1} + \sum_{0\leqslant k\leqslant n} (-1)^{n-k} \\ &= (-1)^{n+1} + S_n . \end{aligned}$$

Hence $2S_n = 1 + (-1)^n$ and we have $S_n = [n \text{ is even}]$. Similarly, we find

$$T_{n+1} = n + 1 - T_n = \sum_{k=0}^{n} (-1)^{n-k}(k+1) = T_n + S_n ,$$

hence $2T_n = n + 1 - S_n$ and we have $T_n = \frac{1}{2}\bigl(n + [n \text{ is odd}]\bigr)$. Finally, the same approach yields

$$\begin{aligned} U_{n+1} = (n+1)^2 - U_n &= U_n + 2T_n + S_n \\ &= U_n + n + [n \text{ is odd}] + [n \text{ is even}] \\ &= U_n + n + 1 . \end{aligned}$$

Hence U_n is the triangular number $\frac{1}{2}(n+1)n$.

2.22 Twice the sum gives a "vanilla" sum over $1 \leqslant j,k \leqslant n$, which splits into three sums that can be handled easily.

2.23 (a) This approach gives four sums that evaluate to $2n + H_n - 2n + (H_n + \frac{1}{n+1} - 1)$. (It would have been easier to replace the summand by $1/k + 1/(k+1)$.) (b) Let $u(x) = 2x + 1$ and $\Delta v(x) = 1/x(x+1) = (x-1)^{\underline{-2}}$; then $\Delta u(x) = 2$ and $v(x) = -(x-1)^{\underline{-1}} = -1/x$. The answer is $2H_n - \frac{n}{n+1}$.

2.24 Summing by parts, $\sum x^{\underline{m}} H_x \,\delta x = x^{\underline{m+1}} H_x/(m+1) - x^{\underline{m+1}}/(m+1)^2 + C$; hence $\sum_{0 \le k < n} k^{\underline{m}} H_k = n^{\underline{m+1}}\bigl(H_n - 1/(m+1)\bigr)/(m+1) + 0^{\underline{m+1}}/(m+1)^2$. In our case $m = -2$, so the sum comes to $1 - (H_n+1)/(n+1)$.

2.25 Here are some of the basic analogies:

$$\sum_{k\in K} c a_k = c \sum_{k\in K} a_k \quad\longleftrightarrow\quad \prod_{k\in K} a_k^c = \Bigl(\prod_{k\in K} a_k\Bigr)^c$$

$$\sum_{k\in K}(a_k+b_k) = \sum_{k\in K} a_k + \sum_{k\in K} b_k \longleftrightarrow \prod_{k\in K} a_k b_k = \Bigl(\prod_{k\in K} a_k\Bigr)\Bigl(\prod_{k\in K} b_k\Bigr)$$

$$\sum_{k\in K} a_k = \sum_{p(k)\in K} a_{p(k)} \quad\longleftrightarrow\quad \prod_{k\in K} a_k = \prod_{p(k)\in K} a_{p(k)}$$

$$\sum_{\substack{j\in J\\ k\in K}} a_{j,k} = \sum_{j\in J}\sum_{k\in K} a_{j,k} \quad\longleftrightarrow\quad \prod_{\substack{j\in J\\ k\in K}} a_{j,k} = \prod_{j\in J}\prod_{k\in K} a_{j,k}$$

$$\sum_{k\in K} a_k = \sum_{k} a_k[k\in K] \quad\longleftrightarrow\quad \prod_{k\in K} a_k = \prod_{k} a_k^{[k\in K]}$$

$$\sum_{k\in K} 1 = \#K \quad\longleftrightarrow\quad \prod_{k\in K} c = c^{\#K}$$

2.26 $P^2 = \bigl(\prod_{1\le j,k\le n} a_j a_k\bigr)\bigl(\prod_{1\le j=k\le n} a_j a_k\bigr)$. The first factor is $\bigl(\prod_{k=1}^n a_k^n\bigr)^2$; the second factor is $\prod_{k=1}^n a_k^2$. Hence $P = \bigl(\prod_{k=1}^n a_k\bigr)^{n+1}$.

2.27 $\Delta(c^{\underline{x}}) = c^{\underline{x}}(c-x-1) = c^{\underline{x+2}}/(c-x)$. Setting $c=-2$ and decreasing x by 2 yields $\Delta(-(-2)^{\underline{x-2}}) = (-2)^{\underline{x}}/x$, hence the stated sum is $(-2)^{\underline{-1}} - (-2)^{\underline{n-1}} = (-1)^n n! - 1$.

2.28 The interchange of summation between the second and third lines is not justifiable; the terms of this sum do not converge absolutely. Everything else is perfectly correct, except that the result of $\sum_{k\ge1}[k=j-1]k/j$ should perhaps have been written $[j-1\ge1](j-1)/j$ and simplified explicitly.

As opposed to imperfectly correct.

2.29 Use partial fractions to get

$$\frac{k}{4k^2-1} = \frac14\left(\frac{1}{2k+1} + \frac{1}{2k-1}\right).$$

The $(-1)^k$ factor now makes the two halves of each term cancel with their neighbors. Hence the answer is $-1/4$.

2.30 $\sum_a^b x\,dx = \frac12(b^{\underline{2}} - a^{\underline{2}}) = \frac12(b-a)(b+a-1)$. So we have

$$(b-a)(b+a-1) = 2100 = 2^2\cdot3\cdot5^2\cdot7.$$

There is one solution for each way to write $2100 = x \cdot y$ where x is even and y is odd; we let $a = \frac{1}{2}|x-y| + \frac{1}{2}$ and $b = \frac{1}{2}(x+y) + \frac{1}{2}$. So the number of solutions is the number of divisors of $3 \cdot 5^2 \cdot 7$, namely 12. In general, there are $\prod_{p>2}(n_p + 1)$ ways to represent $\prod_p p^{n_p}$, where the products range over primes.

2.31 $\sum_{j,k\geqslant 2} j^{-k} = \sum_{j\geqslant 2} 1/j^2(1-1/j) = \sum_{j\geqslant 2} 1/j(j-1)$. The second sum is, similarly, 3/4.

2.32 If $2n \leqslant x < 2n+1$, the sums are $0+\cdots+n+(x-n-1)+\cdots+(x-2n) = n(x-n) = (x-1)+(x-3)+\cdots+(x-2n+1)$. If $2n-1 \leqslant x < 2n$ they are, similarly, both equal to $n(x-n)$. (Looking ahead to Chapter 3, the formula $\lfloor\frac{1}{2}(x+1)\rfloor\bigl(x - \lfloor\frac{1}{2}(x+1)\rfloor\bigr)$ covers both cases.)

2.33 If K is empty, $\bigwedge_{k\in K} a_k = \infty$. The basic analogies are:

$$\sum_{k\in K} ca_k = c\sum_{k\in K} a_k \quad\longleftrightarrow\quad \bigwedge_{k\in K}(c+a_k) = c + \bigwedge_{k\in K} a_k$$

$$\sum_{k\in K}(a_k+b_k) = \sum_{k\in K} a_k + \sum_{k\in K} b_k \quad\longleftrightarrow\quad \bigwedge_{k\in K}\min(a_k, b_k) = \min\Bigl(\bigwedge_{k\in K} a_k, \bigwedge_{k\in K} b_k\Bigr)$$

$$\sum_{k\in K} a_k = \sum_{p(k)\in K} a_{p(k)} \quad\longleftrightarrow\quad \bigwedge_{k\in K} a_k = \bigwedge_{p(k)\in K} a_{p(k)}$$

$$\sum_{\substack{j\in J\\ k\in K}} a_{j,k} = \sum_{j\in J}\sum_{k\in K} a_{j,k} \quad\longleftrightarrow\quad \bigwedge_{\substack{j\in J\\ k\in K}} a_{j,k} = \bigwedge_{j\in J}\bigwedge_{k\in K} a_{j,k}$$

$$\sum_{k\in K} a_k = \sum_k a_k[k\in K] \quad\longleftrightarrow\quad \bigwedge_{k\in K} a_k = \bigwedge_k a_k \cdot \infty^{(k\notin K)}$$

A permutation that consumes terms of one sign faster than those of the other can steer the sum toward any value that it likes.

2.34 Let $K^+ = \{k \mid a_k \geqslant 0\}$ and $K^- = \{k \mid a_k < 0\}$. Then if, for example, n is odd, we choose F_n to be $F_{n-1} \cup E_n$, where $E_n \subseteq K^-$ is sufficiently large that $\sum_{k\in(F_{n-1}\cap K^+)} a_k - \sum_{k\in E_n}(-a_k) < A^-$.

2.35 Goldbach's sum can be shown to equal

$$\sum_{m,n\geqslant 2} m^{-n} = \sum_{m\geqslant 2}\frac{1}{m(m-1)} = 1$$

as follows: By unsumming a geometric series, it equals $\sum_{k\in P,\, l\geqslant 1} k^{-l}$; therefore the proof will be complete if we can find a one-to-one correspondence between ordered pairs (m,n) with $m,n \geqslant 2$ and ordered pairs (k,l) with $k \in P$ and $l \geqslant 1$, where $m^n = k^l$ when the pairs correspond. If $m \notin P$ we let $(m,n) \longleftrightarrow (m^n, 1)$; but if $m = a^b \in P$, we let $(m,n) \longleftrightarrow (a^n, b)$.

2.36 (a) By definition, $g(n)-g(n-1)=f(n)$. (b) By part (a), $g\bigl(g(n)\bigr)-g\bigl(g(n-1)\bigr)=\sum_k f(k)\bigl[g(n-1)<k\leqslant g(n)\bigr]=n\bigl(g(n)-g(n-1)\bigr)=nf(n)$. (c) By part (a) again,

With this self-description, Golomb's sequence wouldn't do too well on the Dating Game.

$$\begin{aligned}
&g\bigl(g(g(n))\bigr)-g\bigl(g(g(n-1))\bigr)\\
&\quad=\sum_k f(k)\bigl[g(g(n-1))<k\leqslant g(g(n))\bigr]\\
&\quad=\sum_{j,k} j\,\bigl[j=f(k)\bigr]\bigl[g(g(n-1))<k\leqslant g(g(n))\bigr]\\
&\quad=\sum_{j,k} j\,\bigl[j=f(k)\bigr]\bigl[g(n-1)<j\leqslant g(n)\bigr]\\
&\quad=\sum_{j,k} j\,\bigl[g(j-1)<k\leqslant g(j)\bigr]\bigl[g(n-1)<j\leqslant g(n)\bigr]\\
&\quad=\sum_{j} j\,\bigl(g(j)-g(j-1)\bigr)\bigl[g(n-1)<j\leqslant g(n)\bigr]\\
&\quad=\sum_{j} jf(j)\,\bigl[g(n-1)<j\leqslant g(n)\bigr]\;=\;n\sum_{j} j\,\bigl[g(n-1)<j\leqslant g(n)\bigr]\,.
\end{aligned}$$

2.37 (RLG thinks they probably won't fit; DEK thinks they probably will; OP is not committing himself.)

3.1 $m=\lfloor\lg n\rfloor;\ \ l=n-2^m=n-2^{\lfloor\lg n\rfloor}$.

3.2 (a) $\lfloor x+.5\rfloor$. (b) $\lceil x-.5\rceil$.

3.3 This is $\lfloor mn-\{m\alpha\}n/\alpha\rfloor=mn-1$, since $0<\{m\alpha\}<1$.

3.4 Something where no proof is required, only a lucky guess (I guess).

3.5 We have $\lfloor nx\rfloor=n\lfloor x\rfloor \iff n\lfloor x\rfloor\leqslant\lfloor nx\rfloor<n\lfloor x\rfloor+1 \iff n\lfloor x\rfloor\leqslant nx<n\lfloor x\rfloor+1 \iff nx-n\{x\}\leqslant nx<nx-n\{x\}+1$, by (3.5(a)), (3.7(a)), (3.7(c)), and (3.8); and this is equivalent to $n\{x\}<1$, when n is a positive integer. (Notice that $n\lfloor x\rfloor\leqslant\lfloor nx\rfloor$ for all x in this case.)

3.6 $\lfloor f(x)\rfloor=\lfloor f(\lceil x\rceil)\rfloor$.

3.7 $\lfloor n/m\rfloor+n \bmod m$.

3.8 If all boxes contain $<\lceil n/m\rceil$ objects, then $n\leqslant(\lceil n/m\rceil-1)m$, so $n/m+1\leqslant\lceil n/m\rceil$, contradicting (3.5). The other proof is similar.

3.9 We have $m/n-1/q=(n \text{ mumble } m)/qn$. The process must terminate, because $0\leqslant n \text{ mumble } m<m$. The denominators of the representation are strictly increasing, hence distinct, because $qn/(n \text{ mumble } m)>q$.

3.10 $\lceil x+\frac12\rceil-\bigl[(2x+1)/4 \text{ is not an integer}\bigr]$ is the nearest integer to x, if $\{x\}\neq\frac12$; otherwise it's the nearest even integer. (See exercise 2.) Thus the formula gives an "unbiased" way to round.

3.11 If n is an integer, $\alpha < n < \beta \iff \lfloor\alpha\rfloor < n < \lceil\beta\rceil$. The number of integers satisfying $a < n < b$ when a and b are integers is $(b-a-1)[b > a]$. We would therefore get the wrong answer if $\alpha = \beta = \text{integer}$.

3.12 Subtract $\lfloor n/m\rfloor$ from both sides, by (3.6), getting $\lceil(n \bmod m)/m\rceil = \lfloor(n \bmod m + m - 1)/m\rfloor$. Both sides are now equal to $[n \bmod m > 0]$, since $0 \leqslant n \bmod m < m$.

A shorter but less direct proof simply observes that the first term in (3.24) must equal the last term in (3.25).

3.13 If they form a partition, the text's formula for $N(\alpha, n)$ implies that $1/\alpha + 1/\beta = 1$, because the coefficients of n in the equation $N(\alpha, n) + N(\beta, n) = n$ must agree if the equation is to hold for large n. Hence α and β are both rational or both irrational. If both are irrational, we do get a partition, as shown in the text. If both can be written with numerator m, the value $m-1$ occurs in neither spectrum. (However, Golomb [121] has observed that the sets $\{\lfloor n\alpha\rfloor \mid n \geqslant 1\}$ and $\{\lceil n\beta\rceil - 1 \mid n \geqslant 1\}$ always do form a partition, when $1/\alpha + 1/\beta = 1$.)

3.14 It's obvious if $ny = 0$, otherwise true by (3.21) and (3.6).

3.15 Plug in $\lceil mx\rceil$ for n in (3.24): $\lceil mx\rceil = \lceil x\rceil + \lceil x - \frac{1}{m}\rceil + \cdots + \lceil x - \frac{m-1}{m}\rceil$.

3.16 The formula $n \bmod 3 = 1 + \frac{1}{3}\big((\omega-1)\omega^n - (\omega+2)\omega^{2n}\big)$ can be verified by checking it when $0 \leqslant n < 3$.

A general formula for $n \bmod m$, when m is any positive integer, appears in exercise 7.25.

3.17 $\sum_{j,k}[0 \leqslant k < m][1 \leqslant j \leqslant x + k/m] = \sum_{j,k}[0 \leqslant k < m][1 \leqslant j \leqslant \lceil x\rceil] \times [k \geqslant m(j-x)] = \sum_{1 \leqslant j \leqslant \lceil x\rceil}\sum_k[0 \leqslant k < m] - \sum_{j=\lceil x\rceil}\sum_k\big[0 \leqslant k < m(j-x)\big] = m\lceil x\rceil - \big\lceil m(\lceil x\rceil - x)\big\rceil = -\lceil -mx\rceil = \lfloor mx\rfloor$.

3.18 We have

$$S = \sum_{0 \leqslant j < \lceil n\alpha\rceil}\ \sum_{k \geqslant n}\big[j\alpha^{-1} \leqslant k < (j+\nu)\alpha^{-1}\big].$$

If $j \leqslant n\alpha - 1 \leqslant n\alpha - \nu$, there is no contribution, because $(j+\nu)\alpha^{-1} \leqslant n$. Hence $j = \lfloor n\alpha\rfloor$ is the only case that matters, and the value in that case equals $\big\lceil(\lfloor n\alpha\rfloor + \nu)\alpha^{-1}\big\rceil - n \leqslant \lceil\nu\alpha^{-1}\rceil$.

3.19 If and only if b is an integer. (If b is an integer, $\log_b x$ is a continuous, increasing function that takes integer values only at integer points. If b is not an integer, the condition fails when $x = b$.)

3.20 We have $\sum_k kx[\alpha \leqslant kx \leqslant \beta] = x\sum_k k\big[\lceil\alpha/x\rceil \leqslant k \leqslant \lfloor\beta/x\rfloor\big]$, which sums to $\frac{1}{2}x\big(\lfloor\beta/x\rfloor\lfloor\beta/x + 1\rfloor - \lceil\alpha/x\rceil\lceil\alpha/x - 1\rceil\big)$.

3.21 If $10^n \leqslant 2^M < 10^{n+1}$, there are exactly $n+1$ such powers of 2, because there's exactly one n-digit power of 2 for each n. Therefore the answer is $1 + \lfloor M \log 2 \rfloor$.

[The number of powers of 2 with leading digit l is more difficult, when $l > 1$; it's $\sum_{0 \leqslant n \leqslant M} \bigl(\lfloor n \log 2 - \log l \rfloor - \lfloor n \log 2 - \log(l+1) \rfloor \bigr)$.]

3.22 All terms are the same for n and $n-1$ except the kth, where $n = 2^{k-1}q$ and q is odd; we have $S_n = S_{n-1} + 1$ and $T_n = T_{n-1} + 2^k q$. Hence $S_n = n$ and $T_n = n(n+1)$.

3.23 $X_n = m \iff \frac{1}{2}m(m-1) < n \leqslant \frac{1}{2}m(m+1) \iff m^2 - m + \frac{1}{4} < 2n < m^2 + m + \frac{1}{4} \iff m - \frac{1}{2} < \sqrt{2n} < m + \frac{1}{2}$.

3.24 Let $\beta = \alpha/(\alpha+1)$. Then the number of times the nonnegative integer m occurs in $\mathrm{Spec}(\beta)$ is exactly one more than the number of times it occurs in $\mathrm{Spec}(\alpha)$. Why? Because $N(\beta, n) = N(\alpha, n) + n + 1$.

3.25 Continuing the development in the text, if we could find a value of m such that $K_m \leqslant m$, we could violate the stated inequality at $n+1$ when $n = 2m+1$. (Also when $n = 3m+1$ and $n = 3m+2$.) But the existence of such an $m = n' + 1$ requires that $2K_{\lfloor n'/2 \rfloor} \leqslant n'$ or $3K_{\lfloor n'/3 \rfloor} \leqslant n'$, i.e., that

$$K_{\lfloor n'/2 \rfloor} \leqslant \lfloor n'/2 \rfloor \qquad \text{or} \qquad K_{\lfloor n'/3 \rfloor} \leqslant \lfloor n'/3 \rfloor .$$

Aha. This goes down further and further, and we'll never get to the bottom.

What we really want to prove is that K_n is strictly *greater* than n, for all $n > 0$. In fact, it's easy to prove this by induction, although it's a stronger result than the one we couldn't prove!

(This exercise teaches an important lesson. It's more an exercise about the nature of induction than about properties of the floor function.)

In trying to devise a proof by mathematical induction, you may fail for two opposite reasons. You may fail because you try to prove too much: Your $P(n)$ is too heavy a burden. Yet you may also fail because you try to prove too little: Your $P(n)$ is too weak a support. In general, you have to balance the statement of your theorem so that the support is just enough for the burden.
—G. Pólya [238]

3.26 Induction, using the stronger hypothesis

$$D_n^{(q)} \leqslant (q-1)\left(\left(\frac{q}{q-1}\right)^{n+1} - 1\right), \qquad \text{for } n \geqslant 0.$$

3.27 If $D_n^{(3)} = 2^m b - a$, where b is odd and a is 0 or 1, then $D_{n+b}^{(3)} = 3^m b - a$.

3.28 The key observation is that $a_n = m^2$ implies $a_{n+2k+1} = (m+k)^2 + m - k$ and $a_{n+2k+2} = (m+k)^2 + 2m$, for $0 \leqslant k \leqslant m$; hence $a_{n+2m+1} = (2m)^2$. The solution can therefore be written as follows, where $0 \leqslant k \leqslant 2^m$:

$$a_n = \begin{cases} (2^m + k)^2 + 2^m - k, & \text{if } n = 2^{m+1} + m - 2 + 2k + 1; \\ (2^m + k)^2 + 2^{m+1}, & \text{if } n = 2^{m+1} + m - 2 + 2k + 2. \end{cases}$$

3.29 $D(\alpha', \lfloor \alpha n \rfloor)$ is at most the maximum of the right-hand side of

$$s(\alpha', \lfloor n\alpha \rfloor, \nu') = -s(\alpha, n, \nu) + S - \epsilon - [0 \text{ or } 1] - \nu' + [0 \text{ or } 1] .$$

3.30 $X_n = \alpha^{2^n} + \alpha^{-2^n}$, by induction; and X_n is an integer.

This logic is seriously floored.

3.31 Here's an "elegant," "impressive" proof that gives no clue about how it was discovered:

$$\begin{aligned}\lfloor x\rfloor + \lfloor y\rfloor + \lfloor x+y\rfloor &= \lfloor x + \lfloor y\rfloor\rfloor + \lfloor x+y\rfloor \\ &\leqslant \lfloor x + \tfrac{1}{2}\lfloor 2y\rfloor\rfloor + \lfloor x + \tfrac{1}{2}\lfloor 2y\rfloor + \tfrac{1}{2}\rfloor \\ &= \lfloor 2x + \lfloor 2y\rfloor\rfloor \;=\; \lfloor 2x\rfloor + \lfloor 2y\rfloor\,.\end{aligned}$$

But there's also a simple, graphical proof based on the observation that we need to consider only the case $0 \leqslant x, y < 1$. Then the functions look like this in the plane:

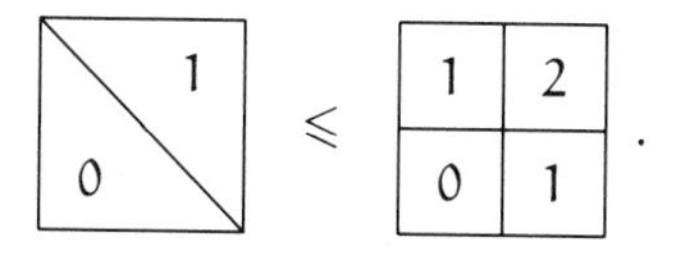

A slightly stronger result is possible, namely

$$\lceil x\rceil + \lfloor y\rfloor + \lfloor x+y\rfloor \;\leqslant\; \lceil 2x\rceil + \lfloor 2y\rfloor\,;$$

but this is stronger only when $\{x\} = \frac{1}{2}$. If we replace (x, y) by $(-x, x+y)$ in this identity and apply the reflective law (3.4), we get

$$\lfloor y\rfloor + \lfloor x+y\rfloor + \lfloor 2x\rfloor \;\leqslant\; \lfloor x\rfloor + \lfloor 2x+2y\rfloor\,.$$

3.32 Let $f(x)$ be the sum in question. Since $f(x) = f(-x)$, we may assume that $x \geqslant 0$. The terms are bounded by 2^k as $k \to -\infty$ and by $x/2^k$ as $k \to +\infty$, so the sum exists for all real x.

We have $f(2x) = 2\sum_k 2^{k-1}\|x/2^{k-1}\|^2 = 2f(x)$. Let $f(x) = l(x) + r(x)$ where $l(x)$ is the sum for $k \leqslant 0$ and $r(x)$ is the sum for $k > 0$. Then $l(x+1) = l(x)$, and $l(x) \leqslant 1/2$ for all x. When $0 \leqslant x < 1$, we have $r(x) = x^2/2 + x^2/4 + \cdots = x^2$, and $r(x+1) = (x-1)^2/2 + (x+1)^2/4 + (x+1)^2/8 + \cdots = x^2 + 1$. Hence $f(x+1) = f(x) + 1$, when $0 \leqslant x < 1$.

We can now prove by induction that $f(x+n) = f(x) + n$ for all integers $n \geqslant 0$, when $0 \leqslant x < 1$. In particular, $f(n) = n$. Therefore in general, $f(x) = 2^{-m}f(2^m x) = 2^{-m}\lfloor 2^m x\rfloor + 2^{-m}f(\{2^m x\})$. But $f(\{2^m x\}) = l(\{2^m x\}) + r(\{2^m x\}) \leqslant \frac{1}{2} + 1$; so $|f(x) - x| \leqslant |2^{-m}\lfloor 2^m x\rfloor - x| + 2^{-m}\cdot\frac{3}{2} \leqslant 2^{-m}\cdot\frac{5}{2}$ for all m.

The inescapable conclusion is that $f(x) = |x|$ for all real x.

3.33 Let $r = n - \frac{1}{2}$ be the radius of the circle. (a) There are $2n-1$ horizontal lines and $2n-1$ vertical lines between cells of the board, and the circle crosses each of these lines twice. Since r^2 is not an integer, the Pythagorean theorem tells us that the circle doesn't pass through the corner of any cell. Hence the circle passes through as many cells as there are crossing points, namely $8n - 4 = 8r$. (The same formula gives the number of cells at the edge of the board.) (b) $f(k) = 4\lfloor\sqrt{r^2 - k^2}\rfloor$.

It follows from (a) and (b) that

$$\tfrac{1}{4}\pi r^2 - 2r \leqslant \sum_{0<k<r} \lfloor\sqrt{r^2 - k^2}\rfloor \leqslant \tfrac{1}{4}\pi r^2\,, \qquad r = n - \tfrac{1}{2}.$$

The task of obtaining more precise estimates of this sum is a famous problem in number theory, investigated by Gauss and many others; see Dickson [65, volume 2, chapter 6].

3.34 (a) Let $m = \lceil \lg n \rceil$. We can add $2^m - n$ terms to simplify the calculations at the boundary:

$$\begin{aligned} f(n) + (2^m - n)m &= \sum_{k=1}^{2^m} \lceil \lg k \rceil = \sum_{j,k} j[j = \lceil \lg k \rceil][1 \leqslant k \leqslant 2^m] \\ &= \sum_{j,k} j[2^{j-1} < k \leqslant 2^j][1 \leqslant j \leqslant m] \\ &= \sum_{j=1}^{m} j\,2^{j-1} = 2^m(m-1) + 1\,. \end{aligned}$$

Consequently $f(n) = nm - 2^m + 1$.

(b) We have $\lceil n/2 \rceil = \lfloor (n+1)/2 \rfloor$, and it follows that the solution to the general recurrence $g(n) = a(n) + g(\lceil n/2 \rceil) + g(\lfloor n/2 \rfloor)$ must satisfy $\Delta g(n) = \Delta a(n) + \Delta g(\lfloor n/2 \rfloor)$. In particular, when $a(n) = n-1$, $\Delta f(n) = 1 + \Delta f(\lfloor n/2 \rfloor)$ is satisfied by the number of bits in the binary representation of n, namely $\lceil \lg(n+1) \rceil$. Now convert from Δ to Σ.

3.35 $(n+1)^2 n!\, e = A_n + (n+1)^2 + (n+1) + B_n$, where

$$A_n = \frac{(n+1)^2 n!}{0!} + \frac{(n+1)^2 n!}{1!} + \cdots + \frac{(n+1)^2 n!}{(n-1)!} \quad \text{is a multiple of } n$$

$$\begin{aligned} \text{and } B_n &= \frac{(n+1)^2 n!}{(n+2)!} + \frac{(n+1)^2 n!}{(n+3)!} + \cdots \\ &= \frac{n+1}{n+2}\left(1 + \frac{1}{n+3} + \frac{1}{(n+3)(n+4)} + \cdots\right) \\ &< \frac{n+1}{n+2}\left(1 + \frac{1}{n+3} + \frac{1}{(n+3)(n+3)} + \cdots\right) \\ &= \frac{(n+1)(n+3)}{(n+2)^2} < 1\,. \end{aligned}$$

Hence the answer is $2 \bmod n$.

3.36 The sum is

$$\sum_{k,l,m} 2^{-l}4^{-m}\big[m=\lfloor\lg l\rfloor\big]\big[l=\lfloor\lg k\rfloor\big][1<k<2^{2^n}]$$
$$= \sum_{k,l,m} 2^{-l}4^{-m}[2^m\leqslant l<2^{m+1}][2^l\leqslant k<2^{l+1}][0\leqslant m<n]$$
$$= \sum_{l,m} 4^{-m}[2^m\leqslant l<2^{m+1}][0\leqslant m<n]$$
$$= \sum_{m} 2^{-m}[0\leqslant m<n] \;=\; 2(1-2^{-n}).$$

3.37 First consider the case $m<n$, which breaks into subcases based on whether $m<\frac{1}{2}n$; then show that both sides change in the same way when m is increased by n.

This is really only a level 4 problem, in spite of the way it's stated.

3.38 At most one x_k can be noninteger. Discard all integer x_k, and suppose that n are left. When $\{x\}\neq 0$, the average of $\{mx\}$ as $m\to\infty$ lies between $\frac{1}{4}$ and $\frac{1}{2}$; hence $\{mx_1\}+\cdots+\{mx_n\}-\{mx_1+\cdots+mx_n\}$ cannot have average value zero when $n>1$.

But the argument just given relies on a difficult theorem about uniform distribution. An elementary proof is possible, sketched here for $n=2$: Let P_m be the point $\big(\{mx\},\{my\}\big)$. Divide the unit square $0\leqslant x,y<1$ into triangular regions A and B according as $x+y<1$ or $x+y\geqslant 1$. We want to show that $P_m\in B$ for some m, if $\{x\}$ and $\{y\}$ are nonzero. If $P_1\in B$, we're done. Otherwise there is a disk D of radius $\epsilon>0$ centered at P_1 such that $D\subseteq A$. By Dirichlet's box principle, the sequence $P_1,\ldots,P_N$ must contain two points with $|P_k-P_j|<\epsilon$ and $k>j$, if N is large enough.

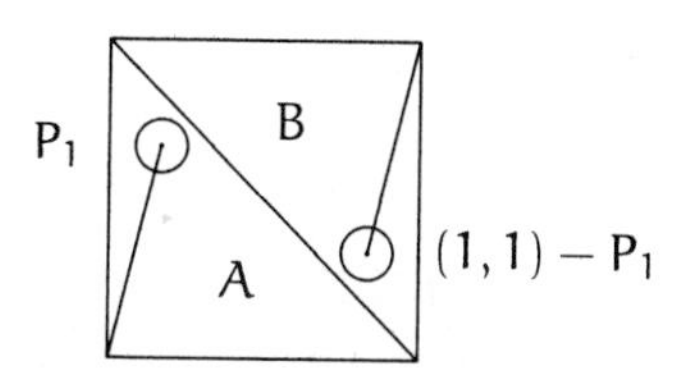

It follows that P_{k-j-1} is within ϵ of $(1,1)-P_1$; hence $P_{k-j-1}\in B$.

3.39 Replace j by $b-j$ and add the term $j=0$ to the sum, so that exercise 15 can be used for the sum on j. The result,

$$\lceil x/b^k\rceil-\lceil x/b^{k+1}\rceil+b-1\,,$$

telescopes when summed on k.

3.40 Let $\lfloor 2\sqrt{n}\rfloor = 4k + r$ where $-2 \leqslant r < 2$, and let $m = \lfloor\sqrt{n}\rfloor$. Then the following relationships can be proved by induction:

segment	r	m	x	y	if and only if
W_k	-2	$2k-1$	$m(m+1)-n-k$	k	$(2k-1)(2k-1) \leqslant n \leqslant (2k-1)(2k)$
S_k	-1	$2k-1$	$-k$	$m(m+1)-n+k$	$(2k-1)(2k) < n < (2k)(2k)$
E_k	0	$2k$	$n-m(m+1)+k$	$-k$	$(2k)(2k) \leqslant n \leqslant (2k)(2k+1)$
N_k	1	$2k$	k	$n-m(m+1)-k$	$(2k)(2k+1) < n < (2k+1)(2k+1)$

Thus, when $k \geqslant 1$, W_k is a segment of length $2k$ where the path travels west and $y(n) = k$; S_k is a segment of length $2k-2$ where the path travels south and $x(n) = -k$; etc. (a) The desired formula is therefore

$$y(n) = (-1)^m\Bigl(\bigl(n - m(m+1)\bigr)\cdot\bigl[\lfloor 2\sqrt{n}\rfloor \text{ is odd}\bigr] - \lceil\tfrac{1}{2}m\rceil\Bigr).$$

(b) On all segments, $k = \max\bigl(|x(n)|, |y(n)|\bigr)$. On segments W_k and S_k we have $x < y$ and $n + x + y = m(m+1) = (2k)^2 - 2k$; on segments E_k and N_k we have $x \geqslant y$ and $n - x - y = m(m+1) = (2k)^2 + 2k$. Hence the sign is $(-1)^{[x(n)<y(n)]}$.

3.41 Since $1/\phi + 1/\phi^2 = 1$, the stated sequences do partition the positive integers. Since the condition $g(n) = f\bigl(f(n)\bigr) + 1$ determines f and g uniquely, we need only show that $\bigl\lfloor\lfloor n\phi\rfloor\phi\bigr\rfloor + 1 = \lfloor n\phi^2\rfloor$ for all $n > 0$. This follows from exercise 3, with $\alpha = \phi$ and $n = 1$.

3.42 No; an argument like the analysis of the two-spectrum case in the text and in exercise 13 shows that a tripartition occurs if and only if $1/\alpha + 1/\beta + 1/\gamma = 1$ and

$$\left\{\frac{n+1}{\alpha}\right\} + \left\{\frac{n+1}{\beta}\right\} + \left\{\frac{n+1}{\gamma}\right\} = 1,$$

for all $n > 0$. But the average value of $\bigl\{(n+1)/\alpha\bigr\}$ is $1/2$ if α is irrational, by the theorem on uniform distribution. The parameters can't all be rational, and if $\gamma = m/n$ the average is $3/2 - 1/(2n)$. Hence γ must be an integer, but this doesn't work either. (There's also a proof of impossibility that uses only simple principles, without the theorem on uniform distribution; see [125].)

3.43 One step of unfolding the recurrence for K_n gives the minimum of the four numbers $1 + a + a\cdot b\cdot K_{\lfloor(n-1-a)/(a\cdot b)\rfloor}$, where a and b are each 2 or 3. (This simplification involves an application of (3.11) to remove floors within floors, together with the identity $x + \min(y,z) = \min(x+y, x+z)$. We must omit terms with negative subscripts; i.e., with $n - 1 - a < 0$.)

Continuing along such lines now leads to the following interpretation: K_n is the least number $> n$ in the multiset S of all numbers of the form

$$1 + a_1 + a_1a_2 + a_1a_2a_3 + \cdots + a_1a_2a_3\ldots a_m\,,$$

where $m \geqslant 0$ and each a_k is 2 or 3. Thus,

$$S = \{1, 3, 4, 7, 9, 10, 13, 15, 19, 21, 22, 27, 28, 31, 31, \ldots\};$$

the number 31 is in S "twice" because it has two representations $1 + 2 + 4 + 8 + 16 = 1 + 3 + 9 + 18$.

(Michael Fredman [108] has shown that $\lim_{n\to\infty} K_n/n = 1$; i.e., that S has no enormous gaps.)

3.44 Let $d_n^{(q)} = D_{n-1}^{(q)} \operatorname{mumble}(q-1)$, so that $D_n^{(q)} = (qD_{n-1}^{(q)} + d_n^{(q)})/(q-1)$ and $a_n^{(q)} = \lceil D_{n-1}^{(q)}/(q-1)\rceil$. Now $D_{k-1}^{(q)} \leqslant (q-1)n \iff a_k^{(q)} \leqslant n$, and the results follow. (This is the solution found by Euler [94], who determined the a's and d's sequentially without realizing that a single sequence $D_n^{(q)}$ would suffice.)

3.45 Let $\alpha > 1$ satisfy $\alpha + 1/(4\alpha) = m$. Then we find $Y_n = \alpha_n + 1/(4\alpha_n)$, where $\alpha_n = \sqrt{2}\,\alpha_{n-1}^2$. It follows that

$$Y_n = \lceil \alpha_n \rceil = \left\lceil 2^{(2^n-1)/2}\alpha^{2^n} \right\rceil.$$

3.46 The hint follows from (3.9), since $2n(n+1) = \lfloor 2(n+\frac{1}{2})^2 \rfloor$. Let $n+\theta = (\sqrt{2}^{\,l} + \sqrt{2}^{\,l-1})m$ and $n' + \theta' = (\sqrt{2}^{\,l+1} + \sqrt{2}^{\,l})m$, where $0 \leqslant \theta, \theta' < 1$. Then $\theta' = 2\theta \bmod 1 = 2\theta - d$, where d is 0 or 1. We want to prove that $n' = \lfloor \sqrt{2}(n + \frac{1}{2}) \rfloor$; this equality holds if and only if

$$0 \leqslant \theta'(2 - \sqrt{2}) + \sqrt{2}(1 - d) < 2.$$

To solve the recurrence, note that $\operatorname{Spec}(1 + 1/\sqrt{2}\,)$ and $\operatorname{Spec}(1 + \sqrt{2}\,)$ partition the positive integers; hence any positive integer a can be written uniquely in the form $a = \lfloor (\sqrt{2}^{\,l} + \sqrt{2}^{\,l-1})m \rfloor$, where l and m are integers with m odd and $l \geqslant 0$. It follows that $L_n = \lfloor (\sqrt{2}^{\,l+n} + \sqrt{2}^{\,l+n-1})m \rfloor$.

3.47 (a) $c = -\frac{1}{2}$. (b) c is an integer. (c) $c = 0$. (d) c is arbitrary. See the answer to exercise 1.2.4–40 in [173] for more general results.

3.48 $m\alpha + n\beta + p \neq 0$ for all integers m, n, and p. (In other words, α, β, and 1 must be linearly independent over the rationals.)

3.49 According to unpublished notes of William A. Veech, it is sufficient to have $\alpha\beta$, β, and 1 linearly independent over the rationals.

3.50 H. S. Wilf observes that the functional equation $f(x^2-1) = f(x)^2$ would determine $f(x)$ for all $x \geqslant \phi$ if we knew $f(x)$ on any interval $(\phi \,.\,.\, \phi + \epsilon)$.

3.51 There are infinitely many ways to partition the positive integers into three or more generalized spectra with *irrational* α_k; for example,

$$\mathrm{Spec}(2\alpha; 0) \cup \mathrm{Spec}(4\alpha; -\alpha) \cup \mathrm{Spec}(4\alpha; -3\alpha) \cup \mathrm{Spec}(\beta; 0)$$

works. But there's a precise sense in which all such partitions arise by "expanding" a basic one, $\mathrm{Spec}(\alpha) \cup \mathrm{Spec}(\beta)$; see [128]. The only known rational examples, e.g.,

$$\mathrm{Spec}(7; -3) \cup \mathrm{Spec}(\tfrac{7}{2}; -1) \cup \mathrm{Spec}(\tfrac{7}{4}; 0)\,,$$

are based on parameters like those in the stated conjecture, which is due to A. S. Fraenkel [103].

3.52 Partial results are discussed in [77, pages 30–31].

4.1 1, 2, 4, 6, 16, 12.

"Man made
the integers:
All else is
Dieudonné."
—R. K. Guy

4.2 Note that $m_p + n_p = \min(m_p, n_p) + \max(m_p, n_p)$. The recurrence $\mathrm{lcm}(m, n) = \big(n/(n \bmod m)\big)\,\mathrm{lcm}(n \bmod m, m)$ is valid but not really advisable for computing lcm's; the best way known to compute $\mathrm{lcm}(m, n)$ is to compute $\gcd(m, n)$ first and then to divide mn by the gcd.

4.3 This holds if x is an integer, but $\pi(x)$ is defined for all real x. The correct formula,

$$\pi(x) - \pi(x-1) \;=\; \big[\lfloor x \rfloor \text{ is prime}\big]\,,$$

is easy to verify.

4.4 Between $\frac{1}{0}$ and $\frac{0}{-1}$ we'd have a left-right reflected Stern–Brocot tree with all denominators negated, etc. So the result is *all* fractions m/n with $m \perp n$. The condition $m'n - mn' = 1$ still holds throughout the construction. (This is called the *Stern–Brocot wreath*, because we can conveniently regard the final $\frac{0}{1}$ as identical to the first $\frac{0}{1}$, thereby joining the trees in a cycle at the top. The Stern–Brocot wreath has interesting applications to computer graphics because it represents all rational directions in the plane.)

4.5 $L^k = \binom{1\ k}{0\ 1}$ and $R^k = \binom{1\ 0}{k\ 1}$; this holds even when $k < 0$. (We will find a general formula for any product of L's and R's in Chapter 6.)

4.6 $a = b$. (Chapter 3 defined $x \bmod 0 = x$, primarily so that this would be true.)

After all, 'mod y' sort of means "pretend y is zero." So if it already is, there's nothing to pretend.

4.7 We need $m \bmod 10 = 0$, $m \bmod 9 = k$, and $m \bmod 8 = 1$. But m can't be both even and odd.

4.8 We want $10x + 6y \equiv 10x + y \pmod{15}$; hence $5y \equiv 0 \pmod{15}$; hence $y \equiv 0 \pmod 3$. We must have $y = 3$, and x can be arbitrary.

4.9 $3^{2k+1} \bmod 4 = 3$, so $(3^{2k+1}-1)/2$ is odd. The stated number is divisible by $(3^7-1)/2$ and $(3^{11}-1)/2$ (and by other numbers).

4.10 $999(1-\frac{1}{3})(1-\frac{1}{37}) = 648$.

4.11 $\sigma(0) = 1$; $\sigma(1) = -1$; $\sigma(n) = 0$ for $n > 1$. (Generalized Möbius functions defined on arbitrary partially ordered structures have interesting and important properties, first explored by Weisner [299] and developed by many other people, notably Gian-Carlo Rota [254].)

4.12 $\sum_{d\backslash m}\sum_{k\backslash d}\mu(d/k)\,g(k) = \sum_{k\backslash m}\sum_{d\backslash(m/k)}\mu(d)\,g(k) = \sum_{k\backslash m} g(k) \times [m/k=1] = g(m)$, by (4.7) and (4.9).

4.13 (a) $n_p \leqslant 1$ for all p; (b) $\mu(n) \ne 0$.

4.14 True when $k > 0$. Use (4.12), (4.14), and (4.15).

4.15 No. For example, $e_n \bmod 5 = [2 \text{ or } 3]$; $e_n \bmod 11 = [2, 3, 7, \text{or } 10]$.

4.16 $1/e_1 + 1/e_2 + \cdots + 1/e_n = 1 - 1/\bigl(e_n(e_n-1)\bigr) = 1 - 1/(e_{n+1}-1)$.

4.17 We have $f_n \bmod f_m = 2$; hence $\gcd(f_n, f_m) = \gcd(2, f_m) = 1$. (Incidentally, the relation $f_n = f_0 f_1 \ldots f_{n-1} + 2$ is very similar to the recurrence that defines the Euclid numbers e_n.)

4.18 If $n = qm$ and q is odd, $2^n + 1 = (2^m+1)(2^{n-m} - 2^{n-2m} + \cdots - 2^m + 1)$.

4.19 Let $p_1 = 2$ and let p_n be the smallest prime greater than $2^{p_{n-1}}$. Then $2^{p_{n-1}} < p_n < 2^{p_{n-1}+1}$, and it follows that we can take $b = \lim_{n\to\infty} \lg^{(n)} p_n$ where $\lg^{(n)}$ is the function lg iterated n times. The stated numerical value comes from $p_2 = 5$, $p_3 = 37$. It turns out that $p_4 = 2^{37} + 9$, and this gives the more precise value

$$b \approx 1.2516475977905$$

(but no clue about p_5).

4.20 By Bertrand's postulate, $P_n < 10^n$. Let

$$K = \sum_{k\geqslant 1} 10^{-k^2} P_k = .200300005\ldots\,.$$

Then $10^{n^2} K \equiv P_n + \text{fraction} \pmod{10^{2n-1}}$.

4.21 The first sum is $\pi(n)$, since the summand is $[k+1 \text{ is prime}]$. The inner sum in the second is $\sum_{1\leqslant k<m} [k\backslash m]$, so it is greater than 1 if and only if m is composite; again we get $\pi(n)$. Finally $\lceil\{m/n\}\rceil = [n\nmid m]$, so the third sum is an application of Wilson's theorem. To evaluate $\pi(n)$ by any of these formulas is, of course, sheer lunacy.

4.22 $(b^{mn}-1)/(b-1) = \bigl((b^m-1)/(b-1)\bigr)(b^{mn-m}+\cdots+1)$. [The only prime numbers of the form $(10^p-1)/9$ for $p<2000$ occur when $p = 2$, 19, 23, 317, 1031.]

4.23 $\rho(2k+1)=0$; $\rho(2k)=\rho(k)+1$, for $k \geqslant 1$. By induction we can show that $\rho(n)=\rho(n-2^m)$, if $n>2^m$ and $m>\rho(n)$. The kth Hanoi move is disk $\rho(k)$, if we number the disks $0, 1, \ldots, n-1$. This is clear if k is a power of 2. And if $2^m<k<2^{m+1}$, we have $\rho(k)<m$; moves k and $k-2^m$ correspond in the sequence that transfers $k+1$ disks in T_k+1+T_k steps.

4.24 The digit that contributes dp^m to n contributes $dp^{m-1}+\cdots+d = d(p^m-1)/(p-1)$ to $\epsilon_p(n!)$, hence $\epsilon_p(n!) = \bigl(n-\nu_p(n)\bigr)/(p-1)$.

4.25 $m\backslash\backslash n \iff m_p=0$ or $m_p=n_p$, for all p. It follows that (a) is true. But (b) fails, in our favorite example $m=12$, $n=18$. (This is a common fallacy.)

4.26 Yes, since $\mathcal{G}_N$ defines a subtree of the Stern–Brocot tree.

4.27 Extend the shorter string with M's (since M lies alphabetically between L and R) until both strings are the same length, then use dictionary order. For example, the topmost levels of the tree are $LL < LM < LR < MM < RL < RM < RR$. (Another solution is to append the infinite string RL^∞ to both inputs, and to keep comparing until finding $L<R$.)

4.28 We need to use only the first part of the representation:

$$\begin{array}{cccccccccccccccc} R & R & R & L & L & L & L & L & L & L & R & R & R & R & R & R \\ \frac{1}{1}, & \frac{2}{1}, & \frac{3}{1}, & \frac{4}{1}, & \frac{7}{2}, & \frac{10}{3}, & \frac{13}{4}, & \frac{16}{5}, & \frac{19}{6}, & \frac{22}{7}, & \frac{25}{8}, & \frac{47}{15}, & \frac{69}{22}, & \frac{91}{29}, & \frac{113}{36}, & \frac{135}{43}, \ \ldots\,. \end{array}$$

The fraction $\frac{4}{1}$ appears because it's a better upper bound than $\frac{1}{0}$, not because it's closer than $\frac{3}{1}$. Similarly, $\frac{25}{8}$ is a better lower bound than $\frac{3}{1}$. The simplest upper bounds and the simplest lower bounds all appear, but the next really good approximation doesn't occur until just before the string of R's switches back to L.

4.29 $1/\alpha$. To get $1-x$ from x in binary notation, we interchange 0 and 1; to get $1/\alpha$ from α in Stern–Brocot notation, we interchange L and R. (The finite cases must also be considered, but they must work since the correspondence is order preserving.)

4.30 The m integers $x \in [A, A+m)$ are different mod m; hence their residues $(x \bmod m_1, \ldots, x \bmod m_r)$ run through all $m_1 \ldots m_r = m$ possible values, one of which must be $(a_1 \bmod m_1, \ldots, a_r \bmod m_r)$ by the pigeonhole principle.

4.31 A number in radix b notation is divisible by d if and only if the sum of its digits is divisible by d, whenever $b \equiv 1 \pmod d$. This follows because $(a_m \ldots a_0)_b = a_m b^m + \cdots + a_0 b^0 \equiv a_m + \cdots + a_0$.

4.32 The $\varphi(m)$ numbers $\{\,kn \bmod m \mid k \perp m \text{ and } 0 \leqslant k < m\,\}$ are the numbers $\{\,k \mid k \perp m \text{ and } 0 \leqslant k < m\,\}$ in some order. Multiply them together and divide by $\prod_{0 \leqslant k < m,\, k \perp m} k$.

4.33 Obviously $h(1) = 1$. If $m \perp n$ then $h(mn) = \sum_{d \backslash mn} f(d)\, g(mn/d) = \sum_{c \backslash m, d \backslash n} f(cd)\, g\bigl((m/c)(n/d)\bigr) = \sum_{c \backslash m} \sum_{d \backslash n} f(c)\, g(m/c)\, f(d)\, g(n/d)$; this is $h(m)\, h(n)$, since $c \perp d$ for every term in the sum.

4.34 $g(m) = \sum_{d \backslash m} f(d) = \sum_{d \backslash m} f(m/d) = \sum_{d \geqslant 1} f(m/d)$ if $f(x)$ is zero when x is not an integer.

4.35 The base cases are

$$I(0,n) = 0; \qquad I(m,0) = 1.$$

When $m, n > 0$, there are two rules, where the first is trivial if $m > n$ and the second is trivial if $m < n$:

$$\begin{aligned} I(m,n) &= I(m, n \bmod m) - \lfloor n/m \rfloor I(n \bmod m, m); \\ I(m,n) &= I(m \bmod n, n). \end{aligned}$$

4.36 A factorization of any of the given quantities into nonunits must have $m^2 - 10n^2 = \pm 2$ or ± 3, but this is impossible mod 10.

4.37 Let $a_n = 2^{-n} \ln(e_n - \frac{1}{2})$ and $b_n = 2^{-n} \ln(e_n + \frac{1}{2})$. Then

$$e_n = \lfloor E^{2^n} + \tfrac{1}{2} \rfloor \iff a_n \leqslant \ln E < b_n.$$

And $a_{n-1} < a_n < b_n < b_{n-1}$, so we can take $E = \lim_{n \to \infty} e^{a_n}$. In fact, it turns out that

$$E^2 = \frac{3}{2} \prod_{n \geqslant 1} \left(1 + \frac{1}{(2e_n - 1)^2}\right)^{1/2^n},$$

a product that converges rapidly to $(1.26408473530530111)^2$. But these observations don't tell us what e_n is, unless we can find another expression for E that doesn't depend on Euclid numbers.

4.38 $a^n - b^n = (a^m - b^m)(a^{n-m}b^0 + a^{n-2m}b^m + \cdots + a^{n \bmod m}b^{n-m-n \bmod m}) + b^{m\lfloor n/m \rfloor}(a^{n \bmod m} - b^{n \bmod m})$.

4.39 If $a_1 \ldots a_t$ and $b_1 \ldots b_u$ are perfect squares, so is

$$a_1 \ldots a_t b_1 \ldots b_u / c_1^2 \ldots c_v^2,$$

where $\{a_1, \ldots, a_t\} \cap \{b_1, \ldots, b_u\} = \{c_1, \ldots, c_v\}$. (It can be shown, in fact, that the sequence $\langle S(1), S(2), S(3), \ldots, \rangle$ contains every nonprime positive integer exactly once.)

4.40 Let $f(n) = \prod_{1 \leqslant k \leqslant n,\, p \nmid k} k = n!/p^{\lfloor n/p \rfloor} \lfloor n/p \rfloor!$ and $g(n) = n!/p^{\epsilon_p(n!)}$. Then

$$g(n) = f(n) f(\lfloor n/p \rfloor) f(\lfloor n/p^2 \rfloor) \ldots = f(n) g(\lfloor n/p \rfloor).$$

Also $f(n) \equiv a_0!(p-1)!^{\lfloor n/p \rfloor} \equiv a_0!(-1)^{\lfloor n/p \rfloor} \pmod p$, and $\epsilon_p(n!) = \lfloor n/p \rfloor + \epsilon_p(\lfloor n/p \rfloor!)$. These recurrences make it easy to prove the result by induction. (Several other solutions are possible.)

4.41 (a) If $n^2 \equiv -1 \pmod p$ then $(n^2)^{(p-1)/2} \equiv -1$; but Fermat says it's $+1$. (b) Let $n = ((p-1)/2)!$; we have $n \equiv (-1)^{(p-1)/2} \prod_{1 \leqslant k < p/2} (p-k) = (p-1)!/n$, hence $n^2 \equiv (p-1)!$.

4.42 First we observe that $k \perp l \iff k \perp l + ak$ for any integer a, since $\gcd(k, l) = \gcd(k, l + ak)$ by Euclid's algorithm. Now

$$\begin{aligned} m \perp n \text{ and } n' \perp n &\iff mn' \perp n \\ &\iff mn' + nm' \perp n. \end{aligned}$$

Similarly

$$m' \perp n' \text{ and } n \perp n' \iff mn' + nm' \perp n'.$$

Hence

$$m \perp n \text{ and } m' \perp n' \text{ and } n \perp n' \iff mn' + nm' \perp nn'.$$

4.43 We want to multiply by $L^{-1}R$, then by $R^{-1}L^{-1}RL$, then $L^{-1}R$, then $R^{-2}L^{-1}RL^2$, etc.; the nth multiplier is $R^{-\rho(n)}L^{-1}RL^{\rho(n)}$, since we must cancel $\rho(n)$ R's. And $R^{-m}L^{-1}RL^m = \binom{0\ \ -1}{1\ \ 2m+1}$.

4.44 We can find the simplest rational number that lies in

$$[.3155, .3165) = \left[\tfrac{631}{2000}, \tfrac{633}{2000}\right)$$

by looking at the Stern–Brocot representations of $\frac{631}{2000}$ and $\frac{633}{2000}$ and stopping just before the former has L where the latter has R:

$(m_1, n_1, m_2, n_2) := (631, 2000, 633, 2000)$;
while $m_1 > n_1$ **or** $m_2 < n_2$ **do**
  if $m_2 < n_2$ **then** (output(L); $(n_1, n_2) := (n_1, n_2) - (m_1, m_2)$)
        else (output(R); $(m_1, m_2) := (m_1, m_2) - (n_1, n_2)$).

The output is LLLRRRRR $= \frac{6}{19} \approx .3158$. Incidentally, an average of .334 implies at least 287 at bats.

4.45 $x^2 \equiv x \pmod{10^n} \iff x(x-1) \equiv 0 \pmod{2^n}$ and $x(x-1) \equiv 0 \pmod{5^n} \iff x \bmod 2^n = [0 \text{ or } 1]$ and $x \bmod 5^n = [0 \text{ or } 1]$. (The last step is justified because $x(x-1) \bmod 5 = 0$ implies that either x or $x-1$ is a multiple of 5, in which case the other factor is relatively prime to 5^n and can be divided from the congruence.)

So there are at most four solutions, of which two ($x = 0$ and $x = 1$) don't qualify for the title "n-digit number" unless $n = 1$. The other two solutions have the forms x and $10^n + 1 - x$, and at least one of these numbers is $\geqslant 10^{n-1}$. When $n = 4$ the other solution, $10001 - 9376 = 625$, is not a four-digit number. We expect to get two n-digit solutions for about 90% of all n, but this conjecture has not been proved.

(Such self-reproducing numbers have been called "automorphic.")

4.46 (a) If $j'j - k'k = \gcd(j,k)$, we have $n^{k'k} n^{\gcd(j,k)} = n^{j'j} \equiv 1$ and $n^{k'k} \equiv 1$. (b) Let $n = pq$, where p is the smallest prime divisor of n. If $2^n \equiv 1 \pmod n$ then $2^n \equiv 1 \pmod p$. Also $2^{p-1} \equiv 1 \pmod p$; hence $2^{\gcd(p-1,n)} \equiv 1 \pmod p$. But $\gcd(p-1, n) = 1$ by the definition of p.

4.47 If $n^{m-1} \equiv 1 \pmod m$ we must have $n \perp m$. If $n^k \equiv n^j$ for some $1 \leqslant j < k < m$, then $n^{k-j} \equiv 1$ because we can divide by n^j. Therefore if the numbers $n^1 \bmod m, \ldots, n^{m-1} \bmod m$ are not distinct, there is a $k < m-1$ with $n^k \equiv 1$. The least such k divides $m-1$, by exercise 46(a). But then $kq = (m-1)/p$ for some prime p and some positive integer q; this is impossible, since $n^{kq} \not\equiv 1$. Therefore the numbers $n^1 \bmod m, \ldots, n^{m-1} \bmod m$ are distinct and relatively prime to m. Therefore the numbers $1, \ldots, m-1$ are relatively prime to m, and m must be prime.

4.48 By pairing numbers up with their inverses, we can reduce the product $\pmod m$ to $\prod_{1 \leqslant n < m,\, n^2 \bmod m = 1} n$. Now we can use our knowledge of the solutions to $n^2 \bmod m = 1$. By residue arithmetic we find that the result is $m - 1$ if $m = 4$, p^k, or $2p^k$ $(p > 2)$; otherwise it's $+1$.

4.49 (a) Either $m < n$ ($\Phi(N-1)$ cases) or $m = n$ (one case) or $m > n$ ($\Phi(N-1)$ again). Hence $R(N) = 2\Phi(N-1) + 1$. (b) From (4.62) we get

$$2\Phi(N-1) + 1 = 1 + \sum_{d \geqslant 1} \mu(d) \lfloor N/d \rfloor \lfloor N/d - 1 \rfloor ;$$

hence the stated result holds if and only if

$$\sum_{d \geqslant 1} \mu(d) \lfloor N/d \rfloor = 1, \qquad \text{for } N \geqslant 1.$$

And this is a special case of (4.61) if we set $f(x) = [x \geqslant 1]$.

4.50 (a) If f is any function,

$$\begin{aligned}\sum_{0\le k<m} f(k) &= \sum_{d\backslash m}\sum_{0\le k<m} f(k)\,[d=\gcd(k,m)]\\ &= \sum_{d\backslash m}\sum_{0\le k<m} f(k)\,[k/d\perp m/d]\\ &= \sum_{d\backslash m}\sum_{0\le k<m/d} f(kd)\,[k\perp m/d]\\ &= \sum_{d\backslash m}\sum_{0\le k<d} f(km/d)\,[k\perp d]\,;\end{aligned}$$

we saw a special case of this in the derivation of (4.63). An analogous derivation holds for $\prod$ instead of $\sum$. Thus we have

$$z^m-1 = \prod_{0\le k<m}(z-\omega^k) = \prod_{d\backslash m}\prod_{\substack{0\le k<d\\ k\perp d}}(z-\omega^{km/d}) = \prod_{d\backslash m}\Psi_d(z)$$

because $\omega^{m/d}=e^{2\pi i/d}$.

Part (b) follows from part (a) by the analog of (4.56) for products instead of sums. Incidentally, this formula shows that $\Psi_m(z)$ has integer coefficients, since $\Psi_m(z)$ is obtained by multiplying and dividing polynomials whose leading coefficient is 1.

4.51 $(x_1+\cdots+x_n)^p=\sum_{k_1+\cdots+k_n=p} p!/(k_1!\ldots k_n!)x_1^{k_1}\ldots x_n^{k_n}$, and the coefficient is divisible by p unless some $k_j=p$. Hence $(x_1+\cdots+x_n)^p\equiv x_1^p+\cdots+x_n^p$ (mod p). Now we can set all the x's to 1, obtaining $n^p\equiv n$.

4.52 If $p>n$ there is nothing to prove. Otherwise $x\perp p$, so $x^{k(p-1)}\equiv 1$ (mod p); this means that at least $\lfloor (n-1)/(p-1)\rfloor$ of the given numbers are multiples of p. And $(n-1)/(p-1)\ge n/p$ since $n\ge p$.

4.53 First show that if $m\ge 6$ and m is not prime then $(m-2)!\equiv 0$ (mod m). (If $m=p^2$, the product for $(m-2)!$ includes p and 2p; otherwise it includes d and m/d where $d<m/d$.) Next consider cases:

Case 0, $n<5$. The condition holds for $n=1$ only.

Case 1, $n\ge 5$ and n is prime. Then $(n-1)!/(n+1)$ is an integer and it can't be a multiple of n.

Case 2, $n\ge 5$, n is composite, and $n+1$ is composite. Then n and $n+1$ divide $(n-1)!$, and $n\perp n+1$; hence $n(n+1)\backslash(n-1)!$.

Case 3, $n\ge 5$, n is composite, and $n+1$ is prime. Then $(n-1)!\equiv 1$ (mod $n+1$) by Wilson's theorem, and

$$\lfloor (n-1)!/(n+1)\rfloor = \bigl((n-1)!+n\bigr)/(n+1)\,;$$

this is divisible by n.

Therefore the answer is: Either $n = 1$ or $n \neq 4$ is composite.

4.54 $\epsilon_2(1000!) > 500$ and $\epsilon_5(1000!) = 249$, hence $1000! = a \cdot 10^{249}$ for some even integer a. Since $1000 = (1300)_5$, exercise 40 tells us that $a \cdot 2^{249} = 1000!/5^{249} \equiv -1 \pmod 5$. Also $2^{249} \equiv 2$, hence $a \equiv 2$, hence $a \bmod 10 = 2$ or 7; hence the answer is $2 \cdot 10^{249}$.

4.55 One way is to prove by induction that $P_{2n}/P_n^4(n+1)$ is an integer; this stronger result helps the induction go through. Another way is based on showing that each prime p divides the numerator at least as often as it divides the denominator. This reduces to proving the inequality

$$\sum_{k=1}^{2n} \lfloor k/m \rfloor \geqslant 4 \sum_{k=1}^{n} \lfloor k/m \rfloor ,$$

which follows from

$$\lfloor (2n-1)/m \rfloor + \lfloor 2n/m \rfloor \geqslant \lfloor n/m \rfloor .$$

The latter is true when $0 \leqslant n < m$, and both sides increase by 4 when n is increased by m.

4.56 Let $f(m) = \sum_{k=1}^{2n-1} \min(k, 2n-k)[m \backslash k]$, $g(m) = \sum_{k=1}^{n-1} (2n-2k-1) \times \bigl[m \backslash (2k+1)\bigr]$. The number of times p divides the numerator of the stated product is $f(p) + f(p^2) + f(p^3) + \cdots$, and the number of times p divides the denominator is $g(p) + g(p^2) + g(p^3) + \cdots$. But $f(m) = g(m)$ whenever m is odd, by exercise 2.32. The stated product therefore reduces to $2^{n(n-1)}$, by exercise 3.22.

4.57 The hint suggests a standard interchange of summation, since

$$\sum_{1 \leqslant m \leqslant n} [d \backslash m] = \sum_{0 < k \leqslant n/d} [m = dk] = \lfloor n/d \rfloor .$$

Calling the hinted sum $\Sigma(n)$, we have

$$\Sigma(m+n) - \Sigma(m) - \Sigma(n) = \sum_{d \in S(m,n)} \varphi(d) .$$

On the other hand, we know from (4.54) that $\Sigma(n) = \frac{1}{2}n(n+1)$. Hence $\Sigma(m+n) - \Sigma(m) - \Sigma(n) = mn$.

4.58 The function $f(m)$ is multiplicative, and when $m = p^k$ it equals $1 + p + \cdots + p^k$. This is a power of 2 if and only if p is a Mersenne prime and $k = 1$. For k must be odd, and in that case the sum is

$$(1+p)(1+p^2+p^4+\cdots+p^{k-1})$$

and $(k-1)/2$ must be odd, etc. The necessary and sufficient condition is that m be a product of distinct Mersenne primes.

4.59 Proof of the hint: If $n = 1$ we have $x_1 = \alpha = 2$, so there's no problem. If $n > 1$ we can assume that $x_1 \leqslant \cdots \leqslant x_n$. Case 1: $x_1^{-1} + \cdots + x_{n-1}^{-1} + (x_n - 1)^{-1} \geqslant 1$ and $x_n > x_{n-1}$. Then we can find $\beta \geqslant x_n - 1 \geqslant x_{n-1}$ such that $x_1^{-1} + \cdots + x_{n-1}^{-1} + \beta^{-1} = 1$; hence $x_n \leqslant \beta + 1 \leqslant e_n$ and $x_1 \ldots x_n \leqslant x_1 \ldots x_{n-1}(\beta + 1) \leqslant e_1 \ldots e_n$, by induction. There is a positive integer m such that $\alpha = x_1 \ldots x_n/m$; hence $\alpha \leqslant e_1 \ldots e_n = e_{n+1} - 1$, and we have $x_1 \ldots x_n(\alpha + 1) \leqslant e_1 \ldots e_n e_{n+1}$. Case 2: $x_1^{-1} + \cdots + x_{n-1}^{-1} + (x_n - 1)^{-1} \geqslant 1$ and $x_n = x_{n-1}$. Let $a = x_n$ and $a^{-1} + (a-1)^{-1} = (a-2)^{-1} + \zeta^{-1}$. Then we can show that $a \geqslant 4$ and $(a-2)(\zeta + 1) \geqslant a^2$. So there's a $\beta \geqslant \zeta$ such that $x_1^{-1} + \cdots + x_{n-2}^{-1} + (a-2)^{-1} + \beta^{-1} = 1$; it follows by induction that $x_1 \ldots x_n \leqslant x_1 \ldots x_{n-2}(a-2)(\zeta + 1) \leqslant x_1 \ldots x_{n-2}(a-2)(\beta + 1) \leqslant e_1 \ldots e_n$, and we can finish as before. Case 3: $x_1^{-1} + \cdots + x_{n-1}^{-1} + (x_n - 1)^{-1} < 1$. Let $a = x_n$, and let $a^{-1} + \alpha^{-1} = (a-1)^{-1} + \beta^{-1}$. It can be shown that $(a-1)(\beta + 1) > a(\alpha + 1)$, because this identity is equivalent to

$$a\alpha^2 - a^2\alpha + a\alpha - a^2 + \alpha + a > 0,$$

which is a consequence of $a\alpha(\alpha - a) + (1 + a)\alpha \geqslant (1 + a)\alpha > a^2 - a$. Hence we can replace x_n and α by $a - 1$ and β, repeating this transformation until cases 1 or 2 apply.

Another consequence of the hint is that $1/x_1 + \cdots + 1/x_n < 1$ implies $1/x_1 + \cdots + 1/x_n \leqslant 1/e_1 + \cdots + 1/e_n$; see exercise 16.

4.60 The main point is that $\theta < \frac{2}{3}$. Then we can take p_1 sufficiently large (to meet the conditions below) and p_n to be the least prime greater than p_{n-1}^3. With this definition let $a_n = 3^{-n} \ln p_n$ and $b_n = 3^{-n} \ln(p_n + 1)$. If we can show that $a_{n-1} \leqslant a_n < b_n \leqslant b_{n-1}$, we can take $P = \lim_{n\to\infty} e^{a_n}$ as in exercise 37. But this hypothesis is equivalent to $p_{n-1}^3 \leqslant p_n < (p_{n-1} + 1)^3$. If there's no prime p_n in this range, there must be a prime $p < p_{n-1}^3$ such that $p + cp^\theta > (p_{n-1} + 1)^3$. But this implies that $cp^\theta > 3p^{2/3}$, which is impossible when p is sufficiently large.

We can almost certainly take $p_1 = 2$, since all available evidence indicates that the known bounds on gaps between primes are much weaker than the truth (see exercise 69). Then $p_2 = 11$, $p_3 = 1361$, $p_4 = 2521008887$, and $1.306377883863 < P < 1.306377883869$.

4.61 Let $\hat{m}$ and $\hat{n}$ be the right-hand sides; observe that $\hat{m}n' - m'\hat{n} = 1$, hence $\hat{m} \perp \hat{n}$. Also $\hat{m}/\hat{n} > m'/n'$ and $N = \big((n + N)/n'\big)n' - n \geqslant \hat{n} > \big((n+N)/n' - 1\big)n' - n = N - n' \geqslant 0$. So we have $\hat{m}/\hat{n} \geqslant m''/n''$. If equality doesn't hold, we have $n'' = (\hat{m}n' - m'\hat{n})n'' = n'(\hat{m}n'' - m''\hat{n}) + \hat{n}(m''n' - m'n'') \geqslant n' - \hat{n} > N$, a contradiction.

Incidentally, this exercise implies that $(m+m'')/(n+n'') = m'/n'$, although the former fraction is not always reduced.

4.62 $2^{-1}+2^{-2}+2^{-3}-2^{-6}-2^{-7}+2^{-12}+2^{-13}-2^{-20}-2^{-21}+2^{-30}+2^{-31}-2^{-42}-2^{-43}+\cdots$ can be written

$$\frac{1}{2}+3\sum_{k\geqslant 0}\left(2^{-4k^2-6k-3}-2^{-4k^2-10k-7}\right).$$

Incidentally, this sum can be expressed in closed form using the "theta function" $\theta(z,\lambda)=\sum_k e^{-\pi\lambda k^2+2izk}$; we have

$$e \quad\leftrightarrow\quad \tfrac{1}{2}+\tfrac{3}{8}\theta(\tfrac{4}{\pi}\ln 2, 3i\ln 2)-\tfrac{3}{128}\theta(\tfrac{4}{\pi}\ln 2, 5i\ln 2).$$

I have discovered a wonderful proof of Fermat's Last Theorem, but there's no room for it here.

4.63 Any $n>2$ either has a prime divisor d or is divisible by $d=4$. In either case, a solution with exponent n implies a solution $(a^{n/d})^d+(b^{n/d})^d=(c^{n/d})^d$ with exponent d. Since $d=4$ has no solutions, d must be prime.

The hint follows from the binomial theorem, since $a^p+(x-a)^p-pa^{p-1}$ is a multiple of x when p is odd. Assume that $a\perp x$. If x is not divisible by p, x is relatively prime to c^p/x; hence $x=m^p$ for some m. If x is divisible by p, then c^p/x is divisible by p but not by p^2, and c^p has no other factors in common with x.

Therefore, if Fermat's Last Theorem is false, the universe will not be big enough to write down any numbers that disprove it.

(The values of a, b, c must, in fact, be even higher than this result indicates! Inkeri [160] has proved that

$$\min(a,b) \;>\; \left(\frac{2p^3+p}{\ln(3p)}\right)^p.$$

A sketch of his proof appears in [249, pages 228–229], a book that contains an extensive survey of progress on Fermat's Last Theorem.)

4.64 Equal fractions in $\mathcal{P}_N$ appear in "organ-pipe order":

$$\frac{2m}{2n},\ \frac{4m}{4n},\ \ldots,\ \frac{rm}{rn},\ \ldots,\ \frac{3m}{3n},\ \frac{m}{n}.$$

Suppose that $\mathcal{P}_N$ is correct; we want to prove that $\mathcal{P}_{N+1}$ is correct. This means that if kN is odd, we want to show that

$$\frac{k-1}{N+1} = \mathcal{P}_{N,kN};$$

if kN is even, we want to show that

$$\mathcal{P}_{N,kN-1}\ \ \mathcal{P}_{N,kN}\ \ \frac{k-1}{N+1}\ \ \mathcal{P}_{N,kN}\ \ \mathcal{P}_{N,kN+1}.$$

In both cases it will be helpful to know the number of fractions that are strictly less than $(k-1)/(N+1)$ in $\mathcal{P}_N$; this is

$$\sum_{n=1}^{N}\sum_{m}\left[0\leqslant\frac{m}{n}<\frac{k-1}{N+1}\right]=\sum_{n=1}^{N}\left\lceil\frac{(k-1)n}{N+1}\right\rceil=\sum_{n=0}^{N}\left\lfloor\frac{(k-1)n+N}{N+1}\right\rfloor$$
$$=\frac{(k-2)N}{2}+\frac{d-1}{2}+d\left\lfloor\frac{N}{d}\right\rfloor$$
$$=\frac{1}{2}(kN-d+1)\,,\quad d=\gcd(k{-}1,N{+}1),$$

by (3.32). Furthermore, the number of fractions equal to $(k-1)/(N+1)$ in $\mathcal{P}_N$ that should precede it in $\mathcal{P}_{N+1}$ is $\frac{1}{2}(d-1-[d \text{ even}])$, by the nature of organ-pipe order.

If kN is odd, then d is even and $(k-1)/(N+1)$ is preceded by $\frac{1}{2}(kN-1)$ elements of $\mathcal{P}_N$; this is just the correct number to make things work. If kN is even, than d is odd and $(k-1)/(N+1)$ is preceded by $\frac{1}{2}(kN)$ elements of $\mathcal{P}_N$. If $d=1$, none of these equals $(k-1)/(N+1)$ and $\mathcal{P}_{N,kN}$ is '<'; otherwise $(k-1)/(N+1)$ falls between two equal elements and $\mathcal{P}_{N,kN}$ is '='. (C. S. Peirce [230] independently discovered the Stern–Brocot tree at about the same time as he discovered $\mathcal{P}_N$.)

4.65 The analogous question for the (analogous) Fermat numbers f_n is a famous unsolved problem. This one might be easier or harder.

4.66 It is known that no square less than 8×10^{17} divides a Mersenne number. But there has still been no proof of Schinzel's conjecture that there exist infinitely many squarefree Mersenne numbers. It is not even known if there are infinitely many p such that $p\backslash\backslash(a\pm b)$, where all prime factors of a and b are $\leqslant 31$.

4.67 M. Szegedy has proved this conjecture for all large n; see [284′], [77, pp. 78–79], and [49].

4.68 This is a much weaker conjecture than the result in the following exercise.

4.69 Cramér [56] showed that this conjecture is plausible on probabilistic grounds, and computational experience bears this out: Brent [32] has shown that $P_{n+1}-P_n\leqslant 602$ for $P_{n+1}<2.686\times10^{12}$. But the much weaker bounds in exercise 60 are the best currently proved [221]. Exercise 68 has a "yes" answer if $P_{n+1}-P_n<2P_n^{1/2}$ for all sufficiently large n. According to Guy [139, problem A8], Paul Erdős offers \$10,000 for proof that there are infinitely many n such that

$$P_{n+1}-P_n>\frac{c\ln n\,\ln\ln n\,\ln\ln\ln\ln n}{(\ln\ln\ln n)^2}$$

for all $c > 0$.

4.70 This holds if and only if $\nu_2(n) = \nu_3(n)$, according to exercise 24. The methods of [78] may help to crack this conjecture.

4.71 When $k = 3$ the smallest solution is $n = 4700063497 = 19\cdot 47\cdot 5263229$; no other solutions are known in this case.

4.72 This is known to be true for infinitely many values of a, including -1 (of course) and 0 (not so obviously). Lehmer [199] has a famous conjecture that $\varphi(n)\backslash(n-1)$ if and only if n is prime.

4.73 This is known to be equivalent to the Riemann hypothesis (that all zeros of the complex zeta function with real part between 0 and 1 have real part equal to $1/2$).

4.74 Experimental evidence suggests that there are about $p(1-1/e)$ distinct values, just as if the factorials were randomly distributed modulo p.

What's 11^4 in radix 11?

5.1 $(11)_r^4 = (14641)_r$, in any number system of radix $r \geqslant 7$, because of the binomial theorem.

5.2 The ratio $\binom{n}{k+1}/\binom{n}{k} = (n-k)/(k+1)$ is $\leqslant 1$ when $k \geqslant \lfloor n/2 \rfloor$ and $\geqslant 1$ when $k < \lceil n/2 \rceil$, so the maximum occurs when $k = \lfloor n/2 \rfloor$ and $k = \lceil n/2 \rceil$.

5.3 Expand into factorials. Both products are equal to $f(n)/f(n-k)f(k)$, where $f(n) = (n+1)!\,n!\,(n-1)!$.

5.4 $\binom{-1}{k} = (-1)^k\binom{k+1-1}{k} = (-1)^k\binom{k}{k} = (-1)^k[k\geqslant 0]$.

5.5 If $0 < k < p$, there's a p in the numerator of $\binom{p}{k}$ with nothing to cancel it in the denominator. Since $\binom{p}{k} = \binom{p-1}{k} + \binom{p-1}{k-1}$, we must have $\binom{p-1}{k} \equiv (-1)^k \pmod p$, for $0 \leqslant k < p$.

5.6 The crucial step (after second down) should be

$$\begin{aligned}
\frac{1}{n+1}\sum_k \binom{n+k}{k}\binom{n+1}{k+1}(-1)^k \\
&= \frac{1}{n+1}\sum_{k\geqslant 0}\binom{n+k}{n}\binom{n+1}{k+1}(-1)^k \\
&= \frac{1}{n+1}\sum_{k}\binom{n+k}{n}\binom{n+1}{k+1}(-1)^k \\
&\quad - \frac{1}{n+1}\binom{n-1}{n}\binom{n+1}{0}(-1)^{-1}.
\end{aligned}$$

The original derivation forgot to include this extra term, which is $[n=0]$.

5.7 Yes, because $r^{\underline{-k}} = (-1)^k/(-r-1)^{\underline{k}}$. We also have

$$r^{\overline{k}}(r+\tfrac{1}{2})^{\overline{k}} = (2r)^{\overline{2k}}/2^{2k}.$$

5.8 $f(k) = (k/n-1)^n$ is a polynomial of degree n whose leading coefficient is n^{-n}. By (5.40), the sum is $n!/n^n$. When n is large, Stirling's approximation says that this is approximately $\sqrt{2\pi n}/e^n$. (This is quite different from $(1-1/e)$, which is what we get if we use the approximation $(1-k/n)^n \sim e^{-k}$, valid for fixed k as $n \to \infty$.)

5.9 $\mathcal{E}_t(z)^t = \sum_{k\geqslant 0} t(tk+t)^{k-1}z^k/k! = \sum_{k\geqslant 0}(k+1)^{k-1}(tz)^k/k! = \mathcal{E}_1(tz)$, by (5.60).

5.10 $\sum_{k\geqslant 0} 2z^k/(k+2) = F(2,1;3;z)$, since $t_{k+1}/t_k = (k+2)z/(k+3)$.

5.11 The first is Besselian and the second is Gaussian:

But not Imbesselian.

$$\begin{aligned} z^{-1}\sin z &= \textstyle\sum_{k\geqslant 0}(-1)^k z^{2k}/(2k+1)! = F(1;1,\tfrac{3}{2};-z^2/4)\,; \\ z^{-1}\arcsin z &= \textstyle\sum_{k\geqslant 0} z^{2k}(\tfrac{1}{2})^{\overline{k}}/(2k+1)k! = F(\tfrac{1}{2},\tfrac{1}{2};\tfrac{3}{2};z^2)\,. \end{aligned}$$

5.12 (a) Yes, the term ratio is n. (b) No, the value should be 1 when $k=0$; but $(k+1)^n$ works, if n is an integer. (c) Yes, the term ratio is $(k+1)(k+3)/(k+2)$. (d) No, the term ratio is $1+1/(k+1)H_k$; and $H_k \sim \ln k$ isn't a rational function. (e) Yes, the term ratio is

$$\frac{t(k+1)}{t(k)} \bigg/ \frac{T(n-k)}{T(n-k-1)}\,.$$

(f) Not always; e.g., not when $t(k) = 2^k$ and $T(k) = 1$. (g) Yes, the term ratio can be written

$$\frac{a\,t(k+1)/t(k) + b\,t(k+2)/t(k) + c\,t(k+3)/t(k)}{a + b\,t(k+1)/t(k) + c\,t(k+2)/t(k)}\,,$$

and $t(k+m)/t(k) = \big(t(k+m)/t(k+m-1)\big)\ldots\big(t(k+1)/t(k)\big)$ is a rational function of k.

5.13 $R_n = n!^{n+1}/P_n^2 = Q_n/P_n = Q_n^2/n!^{n+1}$.

5.14 The first factor in (5.25) is $\binom{l-k}{l-k-m}$ when $k \leqslant l$, and this is $(-1)^{l-k-m} \times \binom{-m-1}{l-k-m}$. The sum for $k \leqslant l$ is the sum over all k, since $m \geqslant 0$. (The condition $n \geqslant 0$ isn't really needed, although k must assume negative values if $n < 0$.)
To go from (5.25) to (5.26), first replace s by $-1-n-q$.

5.15 If n is odd, the sum is zero, since we can replace k by $n-k$. If $n = 2m$, the sum is $(-1)^m(3m)!/m!^3$, by (5.29) with $a = b = c = m$.

5.16 This is just $(2a)!\,(2b)!\,(2c)!/(a+b)!\,(b+c)!\,(c+a)!$ times (5.29), if we write the summands in terms of factorials.

5.17 $\binom{2n-1/2}{n} = \binom{4n}{2n}/2^{2n}$; $\binom{2n-1/2}{2n} = \binom{4n}{2n}/2^{4n}$; so $\binom{2n-1/2}{n} = 2^{2n}\binom{2n-1/2}{2n}$.

5.18 $\binom{3r}{3k}\binom{3k}{k,k,k}/3^{3k}$.

5.19 $\mathcal{B}_{1-t}(-z)^{-1} = \sum_{k\geqslant 0}\binom{k-tk-1}{k}\bigl(-1/(k-tk-1)\bigr)(-z)^k$, by (5.60), and this is $\sum_{k\geqslant 0}\binom{tk}{k}\bigl(1/(tk-k+1)\bigr)z^k = \mathcal{B}_t(z)$.

5.20 It equals $F(-a_1,\ldots,-a_m;-b_1,\ldots,-b_n;(-1)^{m+n}z)$; see exercise 2.17.

5.21 $\lim_{n\to\infty}(n+m)^{\underline{m}}/n^m = 1$.

5.22 Multiplying and dividing instances of (5.83) gives

$$\begin{aligned}\frac{(-1/2)!}{x!\,(x-1/2)!} &= \lim_{n\to\infty}\binom{n+x}{n}\binom{n+x-1/2}{n}n^{-2x}\Big/\binom{n-1/2}{n}\\ &= \lim_{n\to\infty}\binom{2n+2x}{2n}n^{-2x},\end{aligned}$$

by (5.34) and (5.36). Also

$$1/(2x)! = \lim_{n\to\infty}\binom{2n+2x}{2n}(2n)^{-2x}.$$

Hence, etc. The Gamma function equivalent, incidentally, is

$$\Gamma(x)\,\Gamma(x+\tfrac{1}{2}) = \Gamma(2x)\,\Gamma(\tfrac{1}{2})/2^{2x-1}.$$

5.23 $(-1)^n n¡$, see (5.50).

5.24 This sum is $\binom{n}{m}F\Bigl({m-n,-m \atop 1/2}\Bigm|1\Bigr) = \binom{2n}{2m}$, by (5.35) and (5.93).

5.25 This is equivalent to the easily proved identity

$$(a-b)\frac{a^{\overline{k}}}{(b+1)^{\overline{k}}} = a\frac{(a+1)^{\overline{k}}}{(b+1)^{\overline{k}}} - b\frac{a^{\overline{k}}}{b^{\overline{k}}}$$

as well as to the operator formula $a-b = (\vartheta+a)-(\vartheta+b)$.

Similarly, we have

$$\begin{aligned}&(a_1-a_2)\,F\Bigl({a_1,a_2,a_3,\ldots,a_m \atop b_1,\ldots,b_n}\Bigm|z\Bigr)\\ &\quad= a_1\,F\Bigl({a_1{+}1,a_2,a_3,\ldots,a_m \atop b_1,\ldots,b_n}\Bigm|z\Bigr) - a_2\,F\Bigl({a_1,a_2{+}1,a_3,\ldots,a_m \atop b_1,\ldots,b_n}\Bigm|z\Bigr),\end{aligned}$$

because $a_1 - a_2 = (a_1 + k) - (a_2 + k)$. If $a_1 - b_1$ is a nonnegative integer d, this second identity allows us to express $F(a_1, \ldots, a_m; b_1, \ldots, b_n; z)$ as a linear combination of $F(a_2 + j, a_3, \ldots, a_m; b_2, \ldots, b_n; z)$ for $0 \leqslant j \leqslant d$, thereby eliminating an upper parameter and a lower parameter. Thus, for example, we get closed forms for $F(a, b; a - 1; z)$, $F(a, b; a - 2; z)$, etc.

Gauss [116, §7] derived analogous relations between $F(a, b; c; z)$ and any two "contiguous" hypergeometrics in which a parameter has been changed by ± 1. Rainville [242′] generalized this to cases with more parameters.

5.26 If the term ratio in the original hypergeometric series is $t_{k+1}/t_k = r(k)$, the term ratio in the new one is $t_{k+2}/t_{k+1} = r(k + 1)$. Hence

$$F\left({a_1, \ldots, a_m \atop b_1, \ldots, b_n} \middle| z\right) = 1 + \frac{a_1 \ldots a_m z}{b_1 \ldots b_n} F\left({a_1+1, \ldots, a_m+1, 1 \atop b_1+1, \ldots, b_n+1, 2} \middle| z\right).$$

5.27 This is the sum of the even terms of $F(2a_1, \ldots, 2a_m; 2b_1, \ldots, 2b_m; z)$. We have $(2a)^{\overline{2k+2}}/(2a)^{\overline{2k}} = 4(k + a)(k + a + \frac{1}{2})$, etc.

5.28 We have $F\left({a, b \atop c}\middle|z\right) = (1-z)^{-a} F\left({a, c-b \atop c}\middle|\frac{-z}{1-z}\right) = (1-z)^{-a} F\left({c-b, a \atop c}\middle|\frac{-z}{1-z}\right) = (1 - z)^{c-a-b} F\left({c-a, c-b \atop c}\middle|z\right)$. (Euler proved the identity by showing that both sides satisfy the same differential equation. The reflection law is often attributed to Euler, but it does not seem to appear in his published papers.)

5.29 The coefficients of z^n are equal, by Vandermonde's convolution. (Kummer's original proof was different: He considered $\lim_{m\to\infty} F(m, b - a; b; z/m)$ in the reflection law (5.101).)

5.30 Differentiate again to get $z(1 - z)F''(z) + (2 - 3z)F'(z) - F(z) = 0$. Therefore $F(z) = F(1, 1; 2; z)$ by (5.108).

5.31 The condition $f(k) = cT(k + 1) - cT(k)$ implies that $f(k + 1)/f(k) = \bigl(T(k+2)/T(k+1) - 1\bigr)/\bigl(1 - T(k)/T(k+1)\bigr)$ is a rational function of k.

5.32 When summing a polynomial in k, Gosper's method reduces to the "method of undetermined coefficients." We have $q(k) = r(k) = 1$, and we try to solve $p(k) = s(k + 1) - s(k)$. The method suggests letting $s(k)$ be a polynomial whose degree is $d = \deg(p) + 1$.

5.33 The solution to $k = (k - 1)s(k + 1) - (k + 1)s(k)$ is $s(k) = -k + \frac{1}{2}$; hence the answer is $(1 - 2k)/2k(k - 1) + C$.

5.34 The limiting relation holds because all terms for $k > c$ vanish, and $\epsilon - c$ cancels with $-c$ in the limit of the other terms. Therefore the second partial sum is $\lim_{\epsilon\to 0} F(-m, -n; \epsilon - m; 1) = \lim_{\epsilon\to 0}(\epsilon + n - m)^{\overline{m}}/(\epsilon - m)^{\overline{m}} = (-1)^m\binom{n-1}{m}$.

5.35 (a) $2^{-n}3^n[n \geqslant 0]$. (b) $(1 - \frac{1}{2})^{-k-1}[k \geqslant 0] = 2^{k+1}[k \geqslant 0]$.

5.36 The sum of the digits of $m+n$ is the sum of the digits of m plus the sum of the digits of n, minus $p-1$ times the number of carries, because each carry decreases the digit sum by $p-1$.

5.37 Dividing the first identity by $n!$ yields $\binom{x+y}{n} = \sum_k \binom{x}{k}\binom{y}{n-k}$, Vandermonde's convolution. The second identity follows, for example, from the formula $x^{\overline{k}} = (-1)^k(-x)^{\underline{k}}$ if we negate both x and y.

5.38 Choose c as large as possible such that $\binom{c}{3} \leqslant n$. Then $0 \leqslant n - \binom{c}{3} < \binom{c+1}{3} - \binom{c}{3} = \binom{c}{2}$; replace n by $n - \binom{c}{3}$ and continue in the same fashion. Conversely, any such representation is obtained in this way. (We can do the same thing with

$$n = \binom{a_1}{1} + \binom{a_2}{2} + \cdots + \binom{a_m}{m}, \qquad 0 \leqslant a_1 < a_2 < \cdots < a_m$$

for any fixed m.)

5.39 $x^m y^n = \sum_{k=1}^m \binom{m+n-1-k}{n-1} a^n b^{m-k} x^k + \sum_{k=1}^n \binom{m+n-1-k}{m-1} a^{n-k} b^m y^k$ for all $mn > 0$, by induction on $m+n$.

The boxed sentence on the other side of this page is true.

5.40 $(-1)^{m+1} \sum_{k=1}^n \sum_{j=1}^m \binom{r}{j}\binom{m-rk-s-1}{m-j} = (-1)^{m+1} \sum_{k=1}^n \left(\binom{m-r(k-1)-s-1}{m} - \binom{m-rk-s-1}{m}\right) = (-1)^{m+1}\left(\binom{m-s-1}{m} - \binom{m-rn-s-1}{m}\right) = \binom{rn+s}{m} - \binom{s}{m}$.

5.41 $\sum_{k\geqslant 0} n!/(n-k)!\,(n+k+1)! = \bigl(n!/(2n+1)!\bigr) \sum_{k>n} \binom{2n+1}{k}$, which is $2^{2n} n!/(2n+1)!$.

5.42 We treat n as an indeterminate real variable. Gosper's method with $q(k) = k+1$ and $r(k) = k-1-n$ has the solution $s(k) = 1/(n+2)$; hence the desired indefinite sum is $(-1)^{x-1}\frac{n+1}{n+2}\big/\binom{n+1}{x}$. And

$$\sum_{k=0}^{n} (-1)^k \Big/ \binom{n}{k} = (-1)^{x-1}\frac{n+1}{n+2} \Big/ \binom{n+1}{x} \Bigg|_0^{n+1} = 2\,\frac{n+1}{n+2}\,[n \text{ even}]\,.$$

This exercise, incidentally, implies the formula

$$\frac{1}{n\binom{n-1}{k}} = \frac{1}{(n+1)\binom{n}{k+1}} + \frac{1}{(n+1)\binom{n}{k}},$$

a "dual" to the basic recurrence (5.8).

5.43 After the hinted first step we can apply (5.21) and sum on k. Then (5.21) applies again and Vandermonde's convolution finishes the job. (A combinatorial proof of this identity has been given by Andrews [10]. There's a quick way to go from this identity to a proof of (5.29), explained in [173, exercise 1.2.6–62].)

5.44 Cancellation of factorials shows that

$$\binom{m}{j}\binom{n}{k}\binom{m+n}{m} = \binom{m+n-j-k}{m-j}\binom{j+k}{j}\binom{m+n}{j+k},$$

so the second sum is $1/\binom{m+n}{m}$ times the first. We can show that the first sum is $\binom{a+b}{a}\binom{m+n-a-b}{m-b}$, whenever $n \geqslant b$, even if $m < a$: Let a and b be fixed and call the first sum $S(m,n)$. Identity (5.32) covers the case $n = b$, and we have $S(m,n) = S(m,n-1) + S(m-1,n) + (-1)^{m+n}\binom{m+n}{m}\binom{a}{m}\binom{b}{n}$ since $\binom{m+n-j-k}{m-j} = \binom{m+n-1-j-k}{m-j} + \binom{m-1+n-j-k}{m-1-j}$. The result follows by induction on $m+n$, since $\binom{b}{n} = 0$ when $n > b$ and the case $m = 0$ is trivial. By symmetry, the formula $\binom{a+b}{a}\binom{m+n-a-b}{n-a}$ holds whenever $m \geqslant a$, even if $n < b$.

5.45 According to (5.9), $\sum_{k\leqslant n}\binom{k-1/2}{k} = \binom{n+1/2}{n}$. If this form isn't "closed" enough, we can apply (5.35) and get $(2n+1)\binom{2n}{n}4^{-n}$.

5.46 By (5.69), this convolution is the negative of the coefficient of z^{2n} in $\mathcal{B}_{-1}(z)\mathcal{B}_{-1}(-z)$. Now $(2\mathcal{B}_{-1}(z)-1)(2\mathcal{B}_{-1}(-z)-1) = \sqrt{1-16z^2}$; hence $\mathcal{B}_{-1}(z)\mathcal{B}_{-1}(-z) = \frac{1}{4}\sqrt{1-16z^2} + \frac{1}{2}\mathcal{B}_{-1}(z) + \frac{1}{2}\mathcal{B}_{-1}(-z) - \frac{1}{4}$. By the binomial theorem,

The boxed sentence on the other side of this page is false.

$$(1-16z^2)^{1/2} = \sum_n \binom{1/2}{n}(-16)^n z^{2n} = -\sum_n \binom{2n}{n}\frac{4^n z^{2n}}{2n-1},$$

so the answer is $\binom{2n}{n}4^{n-1}/(2n-1) + \binom{4n-1}{2n}/(4n-1)$.

5.47 It's the coefficient of z^n in $\bigl(\mathcal{B}_r(z)^s/Q_r(z)\bigr)\bigl(\mathcal{B}_r(z)^{-s}/Q_r(z)\bigr) = 1/Q_r(z)^2$, where $Q_r(z) = 1 - r + r\mathcal{B}_r(z)^{-1}$, by (5.61).

5.48 $F(2n+2, 1; n+2; \frac{1}{2}) = 2^{2n+1}/\binom{2n+1}{n+1}$, a special case of (5.111).

5.49 Saalschütz's identity (5.97) yields

$$\binom{x+n}{n}\frac{y}{y+n}\,F\left({-x,\ -n,\ -n-y \atop -x-n,\ 1-n-y}\,\middle|\,1\right) = \frac{(y-x)^{\overline{n}}}{(y+1)^{\overline{n}}}.$$

5.50 The left-hand side is

$$\sum_{k\geqslant 0}\frac{a^{\overline{k}}\,b^{\overline{k}}}{c^{\overline{k}}}\frac{(-z)^k}{k!}\sum_{m\geqslant 0}\binom{k+a+m-1}{m}z^m$$

$$= \sum_{n\geqslant 0} z^n \sum_{k\geqslant 0}\frac{a^{\overline{k}}\,b^{\overline{k}}}{c^{\overline{k}}\,k!}(-1)^k\binom{n+a-1}{n-k}$$

and the coefficient of z^n is

$$\binom{n+a-1}{n}F\left({a,\ b,\ -n \atop c,\ a}\,\middle|\,1\right)\frac{a^{\overline{n}}}{n!} = \frac{(c-b)^{\overline{n}}}{c^{\overline{n}}}$$

by Vandermonde's convolution (5.92).

5.51 (a) Reflection gives $F(a,-n;2a;2) = (-1)^n F(a,-n;2a;2)$. (Incidentally, this formula implies the remarkable identity $\Delta^{2m+1} f(0) = 0$, when $f(n) = 2^n x^{\underline{n}}/(2x)^{\underline{n}}$.)

(b) The term-by-term limit is $\sum_{0\leqslant k\leqslant m} \binom{m}{k} \frac{2m+1}{2m+1-k}(-2)^k$ plus an additional term for $k = 2m-1$; the additional term is

$$\frac{(-m)\ldots(-1)\,(1)\ldots(m)\,(-2m+1)\ldots(-1)\,2^{2m+1}}{(-2m)\ldots(-1)\,(2m-1)!}$$

$$= (-1)^{m+1}\frac{m!\,m!\,2^{2m+1}}{(2m)!} = \frac{-2}{\binom{-1/2}{m}};$$

hence, by (5.104), this limit is $-1/\binom{-1/2}{m}$, the negative of what we had.

5.52 The terms of both series are zero for $k > N$. This identity corresponds to replacing k by $N-k$. Notice that

$$\begin{aligned} a^{\overline{N}} &= a^{\overline{N-k}}(a+N-k)^{\overline{k}} \\ &= a^{\overline{N-k}}(a+N-1)^{\underline{k}} = a^{\overline{N-k}}(1-a-N)^{\overline{k}}(-1)^k. \end{aligned}$$

5.53 When $b = -\frac{1}{2}$, the left side of (5.110) is $1-2z$ and the right side is $(1-4z+4z^2)^{1/2}$, independent of a. The right side is the formal power series

$$1 + \binom{1/2}{1}4z(z-1) + \binom{1/2}{2}16z^2(z-1)^2 + \cdots,$$

which can be expanded and rearranged to give $1-2z+0z^2+0z^3+\cdots$; but the rearrangement involves divergent series in its intermediate steps when $z=1$, so it is not legitimate.

5.54 If $m+n$ is odd, say $2N-1$, we want to show that

$$\lim_{\epsilon\to 0} F\left({N-m-\frac{1}{2},\ -N+\epsilon \atop -m+\epsilon}\,\middle|\,1\right) = 0.$$

Equation (5.92) applies, since $-m+\epsilon > -m-\frac{1}{2}+\epsilon$, and the denominator factor $\Gamma(c-b) = \Gamma(N-m)$ is infinite since $N \leqslant m$; the other factors are finite. Otherwise $m+n$ is even; setting $n = m-2N$ we have

$$\lim_{\epsilon\to 0} F\left({-N,\ N-m-\frac{1}{2}+\epsilon \atop -m+\epsilon}\,\middle|\,1\right) = \frac{(N-1/2)^{\underline{N}}}{m^{\underline{N}}}$$

by (5.93). The remaining job is to show that

$$\binom{m}{m-2N}\frac{(N-1/2)!}{(-1/2)!}\frac{(m-N)!}{m!} = \binom{m-N}{m-2N}2^{-2N},$$

and this is the case $x = N$ of exercise 22.

5.55 Let $Q(k) = (k+A_1)\dots(k+A_M)Z$ and $R(k) = (k+B_1)\dots(k+B_N)$. Then $t(k+1)/t(k) = P(k)Q(k-1)/P(k-1)R(k)$, where $P(k) = Q(k) - R(k)$ is a nonzero polynomial.

5.56 The solution to $-(k+1)(k+2) = s(k+1)+s(k)$ is $s(k) = -\frac{1}{2}k^2 - k - \frac{1}{4}$; hence $\sum \binom{-3}{k}\,\delta k = \frac{1}{8}(-1)^{k-1}(2k^2+4k+1) + C$. Also

$$\begin{aligned}(-1)^{k-1}\left\lfloor\frac{k+1}{2}\right\rfloor\left\lfloor\frac{k+2}{2}\right\rfloor &= \frac{(-1)^{k-1}}{4}\left(k+1-\frac{1+(-1)^k}{2}\right)\left(k+2-\frac{1-(-1)^k}{2}\right)\\ &= \frac{(-1)^{k-1}}{8}(2k^2+4k+1)+\frac{1}{8}.\end{aligned}$$

5.57 We have $t(k+1)/t(k) = (k-n)(k+1+\theta)(-z)/(k+1)(k+\theta)$. Therefore we let $p(k) = k+\theta$, $q(k) = (k-n)(-z)$, $r(k) = k$. The secret function $s(k)$ must be a constant α_0, and we have

$$k+\theta = \bigl(-z(k-n)-k\bigr)\,\alpha_0\,;$$

hence $\alpha_0 = -1/(1+z)$ and $\theta = -nz/(1+z)$. The sum is

$$\sum\binom{n}{k}z^k\left(k-\frac{nz}{1+z}\right)\delta k = -\frac{n}{1+z}\binom{n-1}{k-1}z^k + C.$$

(The special case $z = 1$ was mentioned in (5.18); the general case is equivalent to (5.131).)

5.58 If $m > 0$ we can replace $\binom{k}{m}$ by $\frac{k}{m}\binom{k-1}{m-1}$ and derive the formula $T_{m,n} = \frac{n}{m}T_{m-1,n-1} - \frac{1}{m}\binom{n-1}{m}$. The summation factor $\binom{n}{m}^{-1}$ is therefore appropriate:

$$\frac{T_{m,n}}{\binom{n}{m}} = \frac{T_{m-1,n-1}}{\binom{n-1}{m-1}} - \frac{1}{m} + \frac{1}{n}.$$

We can unfold this to get

$$\frac{T_{m,n}}{\binom{n}{m}} = T_{0,n-m} - H_m + H_n - H_{n-m}.$$

Finally $T_{0,n-m} = H_{n-m}$, so $T_{m,n} = \binom{n}{m}(H_n - H_m)$. (It's also possible to derive this result by using generating functions; see Example 2 in Section 7.5.)

5.59 $\sum_{j\geqslant 0,\,k\geqslant 1}\binom{n}{j}\bigl[j = \lfloor\log_m k\rfloor\bigr] = \sum_{j\geqslant 0,\,k\geqslant 1}\binom{n}{j}[m^j \leqslant k < m^{j+1}]$, which is $\sum_{j\geqslant 0}\binom{n}{j}(m^{j+1}-m^j) = (m-1)\sum_{j\geqslant 0}\binom{n}{j}m^j = (m-1)(m+1)^n$.

5.60 $\binom{2n}{n} \approx 4^n/\sqrt{\pi n}$ is the case $m = n$ of

$$\binom{m+n}{n} \approx \sqrt{\frac{1}{2\pi}\Big(\frac{1}{m}+\frac{1}{n}\Big)}\Big(1+\frac{m}{n}\Big)^n\Big(1+\frac{n}{m}\Big)^m .$$

5.61 Let $\lfloor n/p \rfloor = q$ and $n \bmod p = r$. The polynomial identity $(x+1)^p \equiv x^p + 1 \pmod p$ implies that

$$(x+1)^{pq+r} \equiv (x+1)^r (x^p+1)^q \pmod p .$$

The coefficient of x^m on the left is $\binom{n}{m}$. On the right it's $\sum_k \binom{r}{m-pk}\binom{q}{k}$, which is just $\binom{r}{m \bmod p}\binom{q}{\lfloor m/p \rfloor}$ because $0 \leqslant r < p$.

5.62 $\binom{np}{mp} = \sum_{k_1+\cdots+k_n=mp} \binom{p}{k_1}\ldots\binom{p}{k_n} \equiv \binom{n}{m} \pmod{p^2}$, because all terms of the sum are multiples of p^2 except the $\binom{n}{m}$ terms in which exactly m of the k's are equal to p. (Stanley [275, exercise 1.6(d)] shows that the congruence actually holds modulo p^3 when $p > 3$.)

The boxed sentence on the other side of this page is not a sentence.

5.63 This is $S_n = \sum_{k=0}^{n}(-4)^k\binom{n+k}{n-k} = \sum_{k=0}^{n}(-4)^{n-k}\binom{2n-k}{k}$. The denominator of (5.74) is zero when $z = -1/4$, so we can't simply plug into that formula. The recurrence $S_n = -2S_{n-1} - S_{n-2}$ leads to the solution $S_n = (-1)^n(2n+1)$.

5.64 $\sum_{k\geqslant 0}\big(\binom{n}{2k}+\binom{n}{2k+1}\big)/(k+1) = \sum_{k\geqslant 0}\binom{n+1}{2k+1}/(k+1)$, which is

$$\frac{2}{n+2}\sum_{k\geqslant 0}\binom{n+2}{2k+2} = \frac{2^{n+2}-2}{n+2} .$$

5.65 Multiply both sides by n^{n-1} and replace k by $n-1-k$ to get

$$\sum_k \binom{n-1}{k} n^k (n-k)! = (n-1)!\sum_{k=0}^{n-1}\big(n^{k+1}/k! - n^k/(k-1)!\big) = (n-1)!\,n^n/(n-1)! .$$

(The partial sums can, in fact, be found by Gosper's algorithm.) Alternatively, $\binom{n}{k}kn^{n-1-k}k!$ can be interpreted as the number of mappings of $\{1,\ldots,n\}$ into itself with $f(1), \ldots, f(k)$ distinct but $f(k+1) \in \{f(1),\ldots,f(k)\}$; summing on k must give n^n.

5.66 This is a "walk the garden path" problem where there's only one "obvious" way to proceed at every step. First replace $k-j$ by l, then replace $\lfloor\sqrt{l}\rfloor$ by k, getting

$$\sum_{j,k\geqslant 0}\binom{-1}{j-k}\binom{j}{m}\frac{2k+1}{2^j} .$$

The infinite series converges because the terms for fixed j are dominated by a polynomial in j divided by 2^j. Now sum over k, getting

$$\sum_{j\geqslant 0}\binom{j}{m}\frac{j+1}{2^j}.$$

Absorb the $j+1$ and apply (5.57) to get the answer, $4(m+1)$.

5.67 $3\binom{2n+2}{n+5}$ by (5.26), because

$$\binom{\binom{k}{2}}{2} = 3\binom{k+1}{4}.$$

5.68 Using the fact that

$$\sum_{k\leqslant n/2}\binom{n}{k} = 2^{n-1}+\binom{n}{n/2}[n \text{ is even}],$$

we get $n\bigl(2^{n-1}-\binom{n-1}{\lfloor n/2\rfloor}\bigr)$.

5.69 Since $\binom{k+1}{2}+\binom{l-1}{2}\leqslant\binom{k}{2}+\binom{l}{2} \iff k<l$, the minimum occurs when the k's are as equal as possible. Hence, by the equipartition formula of Chapter 3, the minimum is

The boxed sentence on the other side of this page is not boxed.

$$\begin{aligned}(n \bmod m)\binom{\lceil n/m\rceil}{2}&+\bigl(n-(n \bmod m)\bigr)\binom{\lfloor n/m\rfloor}{2}\\ &= n\binom{\lfloor n/m\rfloor}{2}+(n \bmod m)\left\lfloor\frac{n}{m}\right\rfloor.\end{aligned}$$

A similar result holds for any lower index in place of 2.

5.70 This is $F(-n,\frac12;1;2)$; but it's also $(-2)^{-n}\binom{2n}{n}F(-n,-n;\frac12-n;\frac12)$ if we replace k by $n-k$. Now $F(-n,-n;\frac12-n;\frac12)=F(-\frac n2,-\frac n2;\frac12-n;1)$ by Gauss's identity (5.111). (Alternatively, $F(-n,-n;\frac12-n;\frac12)=2^{-n}F(-n,\frac12;\frac12-n;-1)$ by the reflection law (5.101), and Kummer's formula (5.94) relates this to (5.55).) The answer is 0 when n is odd, $2^{-n}\binom{n}{n/2}$ when n is even. (See [134, §1.2] for another derivation. This sum arises in the study of a simple search algorithm [164].)

5.71 (a) $S(z)=\sum_{k\geqslant 0}a_k z^{m+k}/(1-z)^{m+2k+1}=z^m(1-z)^{-m-1}A\bigl(z/(1-z)^2\bigr)$. (b) Here $A(z)=\sum_{k\geqslant 0}\binom{2k}{k}(-z)^k/(k+1)=\bigl(\sqrt{1+4z}-1\bigr)/2z$, so we have $A\bigl(z/(1-z)^2\bigr)=1-z$. Thus $S_n=[z^n]\bigl(z/(1-z)\bigr)^m=\binom{n-1}{n-m}$.

5.72 The stated quantity is $m(m-n)\ldots\bigl(m-(k-1)n\bigr)n^{k-\nu(k)}/k!$. Any prime divisor p of n divides the numerator at least $k-\nu(k)$ times and divides the denominator at most $k-\nu(k)$ times, since this is the number of times 2 divides $k!$. A prime p that does not divide n must divide the product $m(m-n)\ldots\bigl(m-(k-1)n\bigr)$ at least as often as it divides $k!$, because $m(m-n)\ldots\bigl(m-(p^r-1)n\bigr)$ is a multiple of p^r for all $r \geqslant 1$ and all m.

5.73 Plugging in $X_n = n!$ yields $\alpha = \beta = 1$; plugging in $X_n = n¡$ yields $\alpha = 1$, $\beta = 0$. Therefore the general solution is $X_n = \alpha n¡ + \beta(n! - n¡)$.

5.74 $\binom{n+1}{k} - \binom{n-1}{k-1}$, for $1 \leqslant k \leqslant n$.

5.75 The recurrence $S_k(n+1) = S_k(n) + S_{(k-1) \bmod 3}(n)$ makes it possible to verify inductively that two of the S's are equal and that $S_{(-n) \bmod 3}(n)$ differs from them by $(-1)^n$. These three values split their sum $S_0(n) + S_1(n) + S_2(n) = 2^n$ as equally as possible, so there must be $2^n \bmod 3$ occurrences of $\lceil 2^n/3 \rceil$ and $3 - (2^n \bmod 3)$ occurrences of $\lfloor 2^n/3 \rfloor$.

5.76 $Q_{n,k} = (n+1)\binom{n}{k} + \binom{n}{k+1}$.

5.77 The terms are zero unless $k_1 \leqslant \cdots \leqslant k_m$, when the product is the multinomial coefficient

$$\binom{k_m}{k_1,\, k_2 - k_1,\, \ldots,\, k_m - k_{m-1}}.$$

Therefore the sum over $k_1, \ldots, k_{m-1}$ is m^{k_m}, and the final sum over k_m yields $(m^{n+1} - 1)/(m-1)$.

5.78 Extend the sum to $k = 2m^2 + m - 1$; the new terms are $\binom{1}{4} + \binom{2}{6} + \cdots + \binom{m-1}{2m} = 0$. Since $m \perp (2m+1)$, the pairs $\bigl(k \bmod m, k \bmod (2m+1)\bigr)$ are distinct. Furthermore, the numbers $(2j+1) \bmod (2m+1)$ as j varies from 0 to $2m$ are the numbers $0, 1, \ldots, 2m$ in some order. Hence the sum is

$$\sum_{\substack{0 \leqslant k < m \\ 0 \leqslant j < 2m+1}} \binom{k}{j} = \sum_{0 \leqslant k < m} 2^k = 2^m - 1.$$

5.79 The sum is 2^{2n-1}, so the gcd must be a power of 2. If $n = 2^k q$ where q is odd, $\binom{2n}{1}$ is divisible by 2^{k+1} and not by 2^{k+2}. Each $\binom{2n}{2j+1}$ is divisible by 2^{k+1} (see exercise 36), so this must be the gcd.

5.80 If $p^r \leqslant n+1 < p^{r+1}$, we get the most radix p carries by adding k to $n-k$ when $k = p^r - 1$. The number of carries in this case is $r - \epsilon_p(n+1)$, and $r = \epsilon_p\bigl(L(n+1)\bigr)$.

5.81 Let $f_{l,m,n}(x)$ be the left-hand side. It is sufficient to show that we have $f_{l,m,n}(1) > 0$ and that $f'_{l,m,n}(x) < 0$ for $0 \leqslant x \leqslant 1$. The value of $f_{l,m,n}(1)$ is $(-1)^{n-m-1}\binom{l+m+\theta}{l+n}$ by (5.23), and this is positive because the binomial coefficient has exactly $n-m-1$ negative factors. The inequality is true when $l = 0$, for the same reason. If $l > 0$, we have $f'_{l,m,n}(x) = -l f_{l-1,m,n+1}(x)$, which is negative by induction.

5.82 Let $\epsilon_p(a)$ be the exponent by which the prime p divides a, and let $m = n - k$. The identity to be proved reduces to

$$\begin{aligned}\min\bigl(\epsilon_p(m)-\epsilon_p(m+k), \epsilon_p(m+k+1)-\epsilon_p(k+1), \epsilon_p(k)-\epsilon_p(m+1)\bigr)\\ = \min\bigl(\epsilon_p(k)-\epsilon_p(m+k), \epsilon_p(m)-\epsilon_p(k+1), \epsilon_p(m+k+1)-\epsilon_p(m+1)\bigr)\,.\end{aligned}$$

For brevity let's write this as $\min(x_1, y_1, z_1) = \min(x_2, y_2, z_2)$. Notice that $x_1 + y_1 + z_1 = x_2 + y_2 + z_2$. The general relation

$$\epsilon_p(a) < \epsilon_p(b) \quad\Longrightarrow\quad \epsilon_p(a) = \epsilon_p\bigl(|a \pm b|\bigr)$$

allows us to conclude that $x_1 \neq x_2 \Longrightarrow \min(x_1, x_2) = 0$; the same holds also for (y_1, y_2) and (z_1, z_2). It's now a simple matter to complete the proof.

5.83 If $m < n$, the quantity $\binom{j+k}{j}\binom{m+n-j-k}{m-j}$ is a polynomial in k of degree less than n, for each fixed j; hence the sum over k is zero. If $m \geqslant n$ and if r is an integer in the range $n < r \leqslant m$, the quantity $\binom{j+k}{k}\binom{m+n-j-k}{n-k}$ is a polynomial in j of degree less than r, for each fixed k; hence the sum over j is zero. If $m \geqslant n$ and if $r = -d - 1$ is an integer, for $0 \leqslant d < n$, we have

$$\binom{r}{j} = (-1)^j\binom{j+d}{j} = (-1)^j\sum_l \binom{j}{l}\binom{d}{l};$$

hence the given sum can be written

$$\begin{aligned}&\sum_{j,k,l}(-1)^k\binom{j+k}{j}\binom{d}{l}\binom{j}{l}\binom{n}{k}\binom{m+n-j-k}{m-j}\\ &\quad= \sum_{j,k,l}(-1)^k\binom{n}{k}\binom{d}{l}\binom{l+k}{l}\binom{j+k}{j-l}\binom{m+n-j-k}{m-j}\\ &\quad= \sum_{j,k,l}(-1)^{k+m+l}\binom{n}{k}\binom{d}{l}\binom{l+k}{l}\binom{-l-k-1}{j-l}\binom{-n+k-1}{m-j}\\ &\quad= \sum_{k,l}(-1)^{k+m+l}\binom{n}{k}\binom{d}{l}\binom{l+k}{l}\binom{-l-n-2}{m-l}.\end{aligned}$$

This is zero since $\binom{l+k}{l}$ is a polynomial in k of degree $d < n$.

If $m \geqslant n$, we have verified the identity for m different values of r. We need consider only one more case to prove it in general. Let $r = 0$; then $j = 0$ and the sum is

$$\sum_k (-1)^k\binom{n}{k}\binom{m+n-k}{m} = \binom{m}{n}$$

by (5.25). (Is there a substantially shorter proof?)

5.84 Following the hint, we get

$$z\mathcal{B}_t(z)^{r-1}\mathcal{B}_t'(z) = \sum_{k\geqslant 0}\binom{tk+r}{k}\frac{kz^k}{tk+r},$$

and a similar formula for $\mathcal{E}_t(z)$. Thus the formulas $\bigl(zt\mathcal{B}_t^{-1}(z)\mathcal{B}_t'(z)+1\bigr)\mathcal{B}_t(z)^r$ and $\bigl(zt\mathcal{E}_t^{-1}(z)\mathcal{E}_t'(z)+1\bigr)\mathcal{E}_t(z)^r$ give the respective right-hand sides of (5.61). We must therefore prove that

$$\bigl(zt\mathcal{B}_t^{-1}(z)\mathcal{B}_t'(z)+1\bigr)\mathcal{B}_t(z)^r = \frac{1}{1-t+t\mathcal{B}_t(z)^{-1}},$$

$$\bigl(zt\mathcal{E}_t^{-1}(z)\mathcal{E}_t'(z)+1\bigr)\mathcal{E}_t(z)^r = \frac{1}{1-zt\mathcal{E}(z)^t},$$

and these follow from (5.59).

The boxed sentence on the other side of this page is self-referential.

5.85 If $f(x) = a_n x^n + \cdots + a_1 x + a_0$ is any polynomial of degree $\leqslant n$, we can prove inductively that

$$\sum_{0\leqslant\epsilon_1,\ldots,\epsilon_n\leqslant 1}(-1)^{\epsilon_1+\cdots+\epsilon_n}f(\epsilon_1x_1+\cdots+\epsilon_nx_n) = (-1)^n n!\,a_nx_1\ldots x_n\,.$$

The stated identity is the special case where $a_n = 1/n!$ and $x_k = k^3$.

5.86 (a) First expand with $n(n-1)$ index variables l_{ij} for all $i\neq j$. Setting $k_{ij} = l_{ij}-l_{ji}$ for $1\leqslant i<j<n$ and using the constraints $\sum_{i\neq j}(l_{ij}-l_{ji}) = 0$ for all $i<n$ allows us to carry out the sums on l_{jn} for $1\leqslant j<n$ and then on l_{ji} for $1\leqslant i<j<n$ by Vandermonde's convolution. (b) $f(z)-1$ is a polynomial of degree $<n$ that has n roots, so it must be zero. (c) Consider the constant terms in

$$\prod_{\substack{1\leqslant i,j\leqslant n\\ i\neq j}}\Bigl(1-\frac{z_i}{z_j}\Bigr)^{a_i} = \sum_{k=1}^{n}\prod_{\substack{1\leqslant i,j\leqslant n\\ i\neq j}}\Bigl(1-\frac{z_i}{z_j}\Bigr)^{a_i-[i=k]}.$$

5.87 The first term is $\sum_k\binom{n-k}{k}z^{mk}$, by (5.61). The summands in the second term are

$$\frac{1}{m}\sum_{k\geqslant 0}\binom{(n+1)/m+(1+1/m)k}{k}(\zeta z)^{k+n+1}$$

$$= \frac{1}{m}\sum_{k>n}\binom{(1+1/m)k-n-1}{k-n-1}(\zeta z)^k.$$

Since $\sum_{0\leqslant j<m}(\zeta^{2j+1})^k = m(-1)^l[k=ml]$, these terms sum to

$$\sum_{k>n/m}\binom{(1+1/m)mk-n-1}{mk-n-1}(-z^m)^k$$
$$=\sum_{k>n/m}\binom{(m+1)k-n-1}{k}(-z^m)^k = \sum_{k>n/m}\binom{n-mk}{k}z^{mk}.$$

Incidentally, the functions $\mathcal{B}_m(z^m)$ and $\zeta^{2j+1}z\,\mathcal{B}_{1+1/m}(\zeta^{2j+1}z)^{1/m}$ are the $m+1$ complex roots of the equation $w^{m+1}-w^m=z^m$.

5.88 Use the facts that $\int_0^\infty(e^{-t}-e^{-nt})\,dt/t=\ln n$ and $(1-e^{-t})/t\leqslant 1$. (We have $\binom{x}{k}=O(k^{-x-1})$ as $k\to\infty$, by (5.83); so this bound implies that Stirling's series $\sum_k s_k\binom{x}{k}$ converges when $x>-1$. Hermite [155] showed that the sum is $\ln\Gamma(1+x)$.)

5.89 Adding this to (5.19) gives $y^{-r}(x+y)^{m+r}$ on both sides, by the binomial theorem. Differentiation gives

The boxed sentence on the other side of this page is not self-referential.

$$\sum_{k>m}\binom{m+r}{k}\binom{m-k}{n}x^k y^{m-k-n}$$
$$=\sum_{k>m}\binom{-r}{k}\binom{m-k}{n}(-x)^k(x+y)^{m-k-n},$$

and we can replace k by $k+m+1$ and apply (5.15) to get

$$\sum_{k\geqslant 0}\binom{m+r}{m+1+k}\binom{-n-1}{k}(-x)^{m+1+k}y^{-1-k-n}$$
$$=\sum_{k\geqslant 0}\binom{-r}{m+1+k}\binom{-n-1}{k}x^{m+1+k}(x+y)^{-1-k-n}.$$

In hypergeometric form, this reduces to

$$F\left({1-r,\,n+1 \atop m+2}\,\middle|\,\frac{-x}{y}\right)=\left(1+\frac{x}{y}\right)^{-n-1}F\left({m+1+r,\,n+1 \atop m+2}\,\middle|\,\frac{x}{x+y}\right),$$

which is the special case $(a,b,c,z)=(n+1,m+1+r,m+2,-x/y)$ of the reflection law (5.101). (Thus (5.105) is related to reflection and to the formula in exercise 52.)

5.90 If r is a nonnegative integer, the sum is finite, and the derivation in the text is valid as long as none of the terms of the sum for $0\leqslant k\leqslant r$ has zero in the denominator. Otherwise the sum is infinite, and the kth term $\binom{k-r-1}{k}/\binom{k-s-1}{k}$ is approximately $k^{s-r}(-s-1)!/(-r-1)!$ by (5.83). So we need $r > s+1$ if the infinite series is going to converge. (If r and s are complex, the condition is $\Re r > \Re s + 1$, because $|k^z| = k^{\Re z}$.) The sum is

$$F\left({-r,\ 1 \atop -s}\,\middle|\,1\right) = \frac{\Gamma(r-s-1)\Gamma(-s)}{\Gamma(r-s)\Gamma(-s-1)} = \frac{s+1}{s+1-r}$$

by (5.92); this is the same formula we found when r and s were integers.

5.91 (It's best to use a program like MACSYMA for this.) Incidentally, when $c = (a+1)/2$, this reduces to an identity that's equivalent to (5.110), in view of the Pfaff's reflection law. For if $w = -z/(1-z)$ we have $4w(1-w) = -4z/(1-z)^2$, and

$$F\left({\frac12 a,\ \frac12 a+\frac12-b \atop 1+a-b}\,\middle|\,4w(1-w)\right) = F\left({a,\ a+1-2b \atop 1+a-b}\,\middle|\,\frac{-z}{1-z}\right)$$
$$= (1-z)^a\, F\left({a,\ b \atop 1+a-b}\,\middle|\,z\right).$$

Burma-Shave

5.92 The identities can be proved, as Clausen proved them more than 150 years ago, by showing that both sides satisfy the same differential equation. One way to write the resulting equations between coefficients of z^n is in terms of binomial coefficients:

$$\sum_k \frac{\binom{r}{k}\binom{s}{k}\binom{r}{n-k}\binom{s}{n-k}}{\binom{r+s-1/2}{k}\binom{r+s-1/2}{n-k}} = \frac{\binom{2r}{n}\binom{r+s}{n}\binom{2s}{n}}{\binom{2r+2s}{n}\binom{r+s-1/2}{n}};$$

$$\sum_k \frac{\binom{-1/4+r}{k}\binom{-1/4+s}{k}\binom{-1/4-r}{n-k}\binom{-1/4-s}{n-k}}{\binom{-1+r+s}{k}\binom{-1-r-s}{n-k}}$$
$$= \frac{\binom{-1/2}{n}\binom{-1/2+r-s}{n}\binom{-1/2-r+s}{n}}{\binom{-1+r+s}{n}\binom{-1-r-s}{n}}.$$

Another way is in terms of hypergeometrics:

$$F\left({a,\, b,\, \frac12-a-b-n,\, -n \atop \frac12+a+b,\, 1-a-n,\, 1-b-n}\,\middle|\,1\right) = \frac{(2a)^{\overline{n}}\,(a+b)^{\overline{n}}\,(2b)^{\overline{n}}}{(2a+2b)^{\overline{n}}\,a^{\overline{n}}\,b^{\overline{n}}};$$

$$F\left({\frac14+a,\, \frac14+b,\, a+b-n,\, -n \atop 1+a+b,\, \frac34+a-n,\, \frac34+b-n}\,\middle|\,1\right)$$
$$= \frac{(1/2)^{\overline{n}}\,(1/2+a-b)^{\overline{n}}\,(1/2-a+b)^{\overline{n}}}{(1+a+b)^{\overline{n}}\,(1/4-a)^{\overline{n}}\,(1/4-b)^{\overline{n}}}.$$

5.93 $\alpha^{-1}\prod_{j=1}^{k}\bigl(f(j)+\alpha\bigr)/f(j)$. (The special case when f is a polynomial of degree 2 is equivalent to identity (5.133).)

5.94 This is a consequence of Henrici's "friendly monster" identity,

$$f(a,z)\,f(a,\omega z)\,f(a,\omega^2 z) = F\left({\tfrac12 a-\tfrac14,\ \tfrac12 a+\tfrac14 \atop \tfrac13 a,\ \tfrac13 a+\tfrac13,\ \tfrac13 a+\tfrac23,\ \tfrac23 a-\tfrac13,\ \tfrac23 a,\ \tfrac23 a+\tfrac13,\ a}\middle|\left(\frac{4z}{9}\right)^3\right),$$

where $f(a,z) = F(;a;z)$. This identity can be proved by showing that both sides satisfy the same differential equation. If we replace $3n$ by $3n+1$ or $3n+2$, the given sum is zero.

5.95 See [78] for partial results. The computer experiments were done by V. A. Vyssotsky.

5.96 Paul Erdős conjectures that, in fact, $\max_p \epsilon_p\bigl(\binom{2n}{n}\bigr)$ tends to infinity as $n \to \infty$.

5.97 The congruence surely holds if $2n+1$ is prime. Steven Skiena has also found the example $n = 2953$, when $2n+1 = 3\cdot 11\cdot 179$.

6.1 2314, 2431, 3241, 1342, 3124, 4132, 4213, 1423, 2143, 3412, 4321.

6.2 ${n \brace k} m^{\underline{k}}$, because every such function partitions its domain into k nonempty subsets, and there are $m^{\underline{k}}$ ways to assign function values for each partition. (Summing over k gives a combinatorial proof of (6.10).)

6.3 Now $d_{k+1} \le (\text{center of gravity}) - \epsilon = 1 - \epsilon + (d_1 + \cdots + d_k)/k$. This recurrence is like (6.55) but with $1-\epsilon$ in place of 1; hence the optimum solution is $d_{k+1} = (1-\epsilon)H_k$. This is unbounded as long as $\epsilon < 1$.

6.4 $H_{2n+1} - \frac12 H_n$. (Similarly $\sum_{k=1}^{2n}(-1)^{k-1}/k = H_{2n} - H_n$.)

6.5 $U_n(x,y)$ is equal to

$$x\sum_{k\ge1}\binom{n}{k}(-1)^{k-1}k^{-1}(x+ky)^{n-1} + y\sum_{k\ge1}\binom{n}{k}(-1)^{k-1}(x+ky)^{n-1},$$

and the first sum is $U_{n-1}(x,y) + \sum_{k\ge1}\binom{n-1}{k-1}(-1)^{k-1}k^{-1}(x+ky)^{n-1}$. The remaining k^{-1} can be absorbed, and we have $\sum_{k\ge1}\binom{n}{k}(-1)^{k-1}(x+ky)^{n-1} = x^{n-1} + \sum_{k\ge0}\binom{n}{k}(-1)^{k-1}(x+ky)^{n-1} = x^{n-1}$. This proves (6.75). Let $R_n(x,y) = x^{-n}U_n(x,y)$; then $R_0(x,y) = 0$ and $R_n(x,y) = R_{n-1}(x,y) + 1/n + y/x$, hence $R_n(x,y) = H_n + ny/x$. (Incidentally, the original sum $U_n = U_n(n,-1)$ doesn't lead to a recurrence such as this; therefore the more general sum, which detaches x from its dependence on n, is easier to solve inductively than its special case. This is another instructive example where a strong induction hypothesis makes the difference between success and failure.)

The Fibonacci recurrence is additive, but the rabbits are multiplying.

6.6 Each pair of babies bb present at the end of a month becomes a pair of adults **aa** at the end of the next month; and each pair **aa** becomes an aa and a bb. Thus each bb behaves like a drone in the bee tree and each aa behaves like a queen, except that the bee tree goes backward in time while the rabbits are going forward. There are F_{n+1} pairs of rabbits after n months; F_n of them are adults and F_{n-1} are babies. (This is the context in which Fibonacci originally introduced his numbers.)

If the harmonic numbers are worm numbers, the Fibonacci numbers are rabbit numbers.

6.7 (a) Set $k = 1 - n$ and apply (6.107). (b) Set $m = 1$ and $k = n - 1$ and apply (6.128).

6.8 $55 + 8 + 2$ becomes $89 + 13 + 3 = 105$; the true value is 104.607361.

6.9 21. (We go from F_n to F_{n+2} when the units are squared. The true answer is about 20.72.)

6.10 The partial quotients a_0, a_1, a_2, ... are all equal to 1, because $\phi = 1 + 1/\phi$. (The Stern–Brocot representation is therefore RLRLRLRLRL... .)

6.11 $(-1)^{\overline{n}} = [n=0] - [n=1]$; see (6.11).

6.12 This is a consequence of (6.31) and its dual in Table 250.

6.13 The two formulas are equivalent, by exercise 12. We can use induction. Or we can observe that $z^n D^n$ applied to $f(z) = z^x$ gives $x^{\underline{n}} z^x$ while ϑ^n applied to the same function gives $x^n z^x$; therefore the sequence $\langle \vartheta^0, \vartheta^1, \vartheta^2, \dots \rangle$ must relate to $\langle z^0 D^0, z^1 D^1, z^2 D^2, \dots \rangle$ as $\langle x^0, x^1, x^2, \dots \rangle$ relates to $\langle x^{\underline{0}}, x^{\underline{1}}, x^{\underline{2}}, \dots \rangle$.

6.14 We have

$$x\binom{x+k}{n} = (k+1)\binom{x+k}{n+1} + (n-k)\binom{x+k+1}{n+1},$$

because $(n+1)x = (k+1)(x+k-n) + (n-k)(x+k+1)$. (It suffices to verify the latter identity when $k = 0$, $k = -1$, and $k = n$.)

6.15 Since $\Delta\bigl(\binom{x+k}{n}\bigr) = \binom{x+k}{n-1}$, we have the general formula

$$\sum_k \left\langle {n \atop k} \right\rangle \binom{x+k}{n-m} = \Delta^m(x^n) = \sum_j \binom{m}{j} (-1)^{m-j} (x+j)^n .$$

Set $x = 0$ and appeal to (6.19).

6.16 $A_{n,k} = \sum_{j \geq 0} a_j {n-j \brace k}$; this sum is always finite.

6.17 (a) $\left| {n \atop k} \right| = \left[{n+1 \atop n+1-k} \right]$. (b) $\left| {n \atop k} \right| = n^{\underline{n-k}} = n!\,[n \geq k]/k!$. (c) $\left| {n \atop k} \right| = k! {n \brace k}$.

6.18 This is equivalent to (6.3) or (6.8). (It follows in particular that $\sigma_n(1) = -n\sigma_n(0) = B_n/n!$ when $n > 1$.)

6.19 Use Table 258.

6.20 $\sum_{1 \leq j \leq k \leq n} 1/j^2 = \sum_{1 \leq j \leq n} (n+1-j)/j^2 = (n+1)H_n^{(2)} - H_n$.

6.21 The hinted number is a sum of fractions with odd denominators, so it has the form a/b with a and b odd. (Incidentally, Bertrand's postulate implies that b_n is also divisible by at least one odd prime, whenever $n > 2$.)

6.22 $|z/k(k+z)| \leqslant 2|z|/k^2$ when $k > 2|z|$, so the sum is well defined when the denominators are not zero. If $z = n$ we have $\sum_{k=1}^{m}(1/k - 1/(k+n)) = H_m - H_{m+n} + H_n$, which approaches H_n as $m \to \infty$. (The quantity $H_{z-1} - \gamma$ is often called the psi function $\psi(z)$.)

6.23 $z/(e^z+1) = z/(e^z-1) - 2z/(e^{2z}-1) = \sum_{n\geqslant 0}(1-2^n)B_n z^n/n!$.

6.24 When n is odd, $T_n(x)$ is a polynomial in x^2, hence its coefficients are multiplied by even numbers when we form the derivative and compute $T_{n+1}(x)$ by (6.95). (In fact we can prove more: The Bernoulli number B_{2n} always has 2 to the first power in its denominator, by exercise 54; hence $2^{2n-k}\backslash\backslash T_{2n+1} \iff 2^k\backslash\backslash(n+1)$. The odd positive integers $(n+1)T_{2n+1}/2^{2n}$ are called Genocchi numbers $\langle 1, 1, 3, 17, 155, 2073, \ldots\rangle$, after Genocchi [117].)

6.25 $100n - nH_n < 100(n-1) - (n-1)H_{n-1} \iff H_{n-1} > 99$. (The least such n is approximately $e^{99-\gamma}$, while he finishes at $N \approx e^{100-\gamma}$, about e times as long. So he is getting closer during the final 63% of his journey.)

6.26 Let $u(k) = H_{k-1}$ and $\Delta v(k) = 1/k$, so that $u(k) = v(k)$. Then we have $S_n - H_n^{(2)} = \sum_{k=1}^{n} H_{k-1}/k = H_{k-1}^2\big|_1^{n+1} - S_n = H_n^2 - S_n$.

6.27 Observe that when $m > n$ we have $\gcd(F_m, F_n) = \gcd(F_{m-n}, F_n)$ by (6.108). This yields a proof by induction.

6.28 (a) $Q_n = \alpha(L_n - F_n)/2 + \beta F_n$. (The solution can also be written $Q_n = \alpha F_{n-1} + \beta F_n$.) (b) $L_n = \phi^n + \hat\phi^n$.

6.29 When $k = 0$ the identity is (6.133). When $k = 1$ it is, essentially,

$$\begin{aligned} K(x_1, \ldots, x_n)x_m &= K(x_1, \ldots, x_m)\,K(x_m, \ldots, x_n) \\ &\quad - K(x_1, \ldots, x_{m-2})\,K(x_{m+2}, \ldots, x_n)\,; \end{aligned}$$

in Morse code terms, the second product on the right subtracts out the cases where the first product has intersecting dashes. When $k > 1$, an induction on k suffices, using both (6.127) and (6.132). (The identity is also true when one or more of the subscripts on K become -1, if we adopt the convention that $K_{-1} = 0$. When multiplication is not commutative, Euler's identity remains valid if we write it in the form

$$\begin{aligned} &K_{m+n}(x_1, \ldots, x_{m+n})\,K_k(x_{m+k}, \ldots, x_{m+1}) \\ &\quad = K_{m+k}(x_1, \ldots, x_{m+k})\,K_n(x_{m+n}, \ldots, x_{m+1}) \\ &\qquad + (-1)^k K_{m-1}(x_1, \ldots, x_{m-1})\,K_{n-k-1}(x_{m+n}, \ldots, x_{m+k+2})\,. \end{aligned}$$

For example, we obtain the somewhat surprising noncommutative factorizations

$$(abc + a + c)(1 + ba) \;=\; (ab + 1)(cba + a + c)$$

from the case $k = 2$, $m = 0$, $n = 3$.)

6.30 The derivative of $K(x_1, \ldots, x_n)$ with respect to x_m is

$$K(x_1, \ldots, x_{m-1})\, K(x_{m+1}, \ldots, x_n)\,,$$

and the second derivative is zero; hence the answer is

$$K(x_1, \ldots, x_n) + K(x_1, \ldots, x_{m-1})\, K(x_{m+1}, \ldots, x_n)\, y\,.$$

6.31 Since $x^{\overline{n}} = (x + n - 1)^{\underline{n}} = \sum_k \binom{n}{k} x^{\underline{k}} (n-1)^{\underline{n-k}}$, we have $\left|{n \atop k}\right| = \binom{n}{k}(n-1)^{\underline{n-k}}$. These coefficients, incidentally, satisfy the recurrence

$$\left|{n \atop k}\right| \;=\; (n-1+k)\left|{n-1 \atop k}\right| + \left|{n-1 \atop k-1}\right|, \qquad \text{integers } n, k > 0.$$

6.32 $\sum_{k \le m} k\left\{{n+k \atop k}\right\} = \left\{{m+n+1 \atop m}\right\}$ and $\sum_{0 \le k \le n} \left\{{k \atop m}\right\}(m+1)^{n-k} = \left\{{n+1 \atop m+1}\right\}$, both of which appear in Table 251.

6.33 If $n > 0$, we have $\left[{n \atop 3}\right] = \frac{1}{2}(n-1)!\,(H_{n-1}^2 - H_{n-1}^{(2)})$, by (6.71); $\left\{{n \atop 3}\right\} = \frac{1}{6}(3^n - 3 \cdot 2^n + 3)$, by (6.19).

6.34 We have $\left\langle{-1 \atop k}\right\rangle = 1/(k+1)$, $\left\langle{-2 \atop k}\right\rangle = H_{k+1}^{(2)}$, and in general $\left\langle{n \atop k}\right\rangle$ is given by (6.38) for all integers n.

6.35 Let n be the least integer $> 1/\epsilon$ such that $\lfloor H_n \rfloor > \lfloor H_{n-1} \rfloor$.

6.36 Now $d_{k+1} = \bigl(100 + (1 + d_1) + \cdots + (1 + d_k)\bigr)/(100 + k)$, and the solution is $d_{k+1} = H_{k+100} - H_{101} + 1$ for $k \ge 1$. This exceeds 2 when $k \ge 176$.

6.37 The sum (by parts) is $H_{mn} - \bigl(\frac{m}{m} + \frac{m}{2m} + \cdots + \frac{m}{mn}\bigr) = H_{mn} - H_n$. The infinite sum is therefore $\ln m$. (It follows that

$$\sum_{k \ge 1} \frac{\nu_m(k)}{k(k+1)} \;=\; \frac{m}{m-1} \ln m\,,$$

because $\nu_m(k) = (m-1)\sum_{j \ge 1} (k \bmod m^j)/m^j$.)

6.38 $(-1)^k\bigl(\binom{r-1}{k} r^{-1} - \binom{r-1}{k-1} H_k\bigr) + C$. (By parts, using (5.16).)

6.39 Write it as $\sum_{1 \le j \le n} j^{-1} \sum_{j \le k \le n} H_k$ and sum first on k via (6.67), to get

$$(n+1)H_n^2 - (2n+1)H_n + 2n\,.$$

6.40 If $6n-1$ is prime, the numerator of

$$\sum_{k=1}^{4n-1} \frac{(-1)^{k-1}}{k} = H_{4n-1} - H_{2n-1}$$

is divisible by $6n-1$, because the sum is

$$\sum_{k=2n}^{4n-1} \frac{1}{k} = \sum_{k=2n}^{3n-1} \left(\frac{1}{k} + \frac{1}{6n-1-k} \right) = \sum_{k=2n}^{3n-1} \frac{6n-1}{k(6n-1-k)} .$$

Similarly if $6n+1$ is prime, the numerator of $\sum_{k=1}^{4n} (-1)^{k-1}/k = H_{4k} - H_{2k}$ is a multiple of $6n+1$. For 1987 we sum up to $k = 1324$.

6.41 $S_{n+1} = \sum_k \binom{\lfloor (n+1+k)/2 \rfloor}{k} = \sum_k \binom{\lfloor (n+k)/2 \rfloor}{k-1}$, hence we have $S_{n+1} + S_n = \sum_k \binom{\lfloor (n+k)/2+1 \rfloor}{k} = S_{n+2}$. The answer is F_{n+2}.

6.42 F_n.

6.43 Set $z = \frac{1}{10}$ in $\sum_{n \geqslant 0} F_n z^n = z/(1-z-z^2)$ to get $\frac{10}{89}$. The sum is a repeating decimal with period length 44:

$$0.11235\,95505\,61797\,75280\,89887\,64044\,94382\,02247\,19101\,12359\,55+ .$$

6.44 Replace (m,k) by $(-m,-k)$ or $(k,-m)$ or $(-k,m)$, if necessary, so that $m \geqslant k \geqslant 0$. The result is clear if $m = k$. If $m > k$, we can replace (m,k) by $(m-k, m)$ and use induction.

6.45 $X_n = A(n)\alpha + B(n)\beta + C(n)\gamma + D(n)\delta$, where $B(n) = F_n$, $A(n) = F_{n-1}$, $A(n) + B(n) - D(n) = 1$, and $B(n) - C(n) + 3D(n) = n$.

6.46 $\phi/2$ and $\phi^{-1}/2$. Let $u = \cos 72°$ and $v = \cos 36°$; then $u = 2v^2 - 1$ and $v = 1 - 2\sin^2 18° = 1 - 2u^2$. Hence $u + v = 2(u+v)(v-u)$, and $4v^2 - 2v - 1 = 0$. We can pursue this investigation to find the five complex fifth roots of unity:

$$1, \quad \frac{\phi^{-1} \pm i\sqrt{2+\phi}}{2}, \quad \frac{-\phi \pm i\sqrt{3-\phi}}{2} .$$

6.47 $2^n \sqrt{5}\, F_n = (1+\sqrt{5})^n - (1-\sqrt{5})^n$, and the even powers of $\sqrt{5}$ cancel out. Now let p be an odd prime. Then $\binom{p}{2k+1} \equiv 0$ except when $k = (p-1)/2$, and $\binom{p+1}{2k+1} \equiv 0$ except when $k = 0$ or $k = (p-1)/2$; hence $F_p \equiv 5^{(p-1)/2}$ and $2F_{p+1} \equiv 1 + 5^{(p-1)/2} \pmod p$. It can be shown that $5^{(p-1)/2} \equiv 1$ when p has the form $10k \pm 1$, and $5^{(p-1)/2} \equiv -1$ when p has the form $10k \pm 3$.

"Let p be any old prime."
(See [140], p. 419.)

6.48 This must be true because (6.138) is a polynomial identity and we can set $a_m = 0$.

6.49 Set $z = \frac{1}{2}$ in (6.146); the partial quotients are $0, 2^{F_0}, 2^{F_1}, 2^{F_2}, \ldots$. (Knuth [172] noted that this number is transcendental.)

6.50 (a) $f(n)$ is even $\iff 3\backslash n$. (b) If the binary representation of n is $(1^{a_1}0^{a_2}\ldots 1^{a_{m-1}}0^{a_m})_2$, where m is even, we have $f(n) = K(a_1, a_2, \ldots, a_{m-1})$.

6.51 (a) Combinatorial proof: The arrangements of $\{1,2,\ldots,p\}$ into k subsets or cycles are divided into "orbits" of 1 or p arrangements each, if we add 1 to each element modulo p. For example,

$$\{1,2,4\}\cup\{3,5\} \to \{2,3,5\}\cup\{4,1\} \to \{3,4,1\}\cup\{5,2\}$$
$$\to \{4,5,2\}\cup\{1,3\} \to \{5,1,3\}\cup\{2,4\} \to \{1,2,4\}\cup\{3,5\}.$$

We get an orbit of size 1 only when this transformation takes an arrangement into itself; but then $k = 1$ or $k = p$. Alternatively, there's an algebraic proof: We have $x^{\overline{p}} \equiv x^{\underline{p}} + x^{\underline{1}}$ and $x^{\underline{p}} \equiv x^p - x \pmod p$, since Fermat's theorem tells us that $x^p - x$ is divisible by $(x-0)(x-1)\ldots\bigl(x-(p-1)\bigr)$.

(b) This result follows from (a) and Wilson's theorem; or we can use $x^{\overline{p-1}} \equiv x^{\overline{p}}/(x-1) \equiv (x^p - x)/(x-1) = x^{p-1} + x^{p-2} + \cdots + x$.

(c) We have $\left\{{p+1 \atop k}\right\} \equiv \left[{p+1 \atop k}\right] \equiv 0$ for $3 \leqslant k \leqslant p$, then $\left\{{p+2 \atop k}\right\} \equiv \left[{p+2 \atop k}\right] \equiv 0$ for $4 \leqslant k \leqslant p$, etc. (Similarly, we have $\left[{2p-1 \atop p}\right] \equiv -\left\{{2p-1 \atop p}\right\} \equiv 1$.)

(d) $p! = p^{\overline{p}} = \sum_k (-1)^{p-k} p^k \left[{p \atop k}\right] = p^p\left[{p \atop p}\right] - p^{p-1}\left[{p \atop p-1}\right] + \cdots + p^3\left[{p \atop 3}\right] - p^2\left[{p \atop 2}\right] + p\left[{p \atop 1}\right]$. But $p\left[{p \atop 1}\right] = p!$, so

$$\left[{p \atop 2}\right] = p\left[{p \atop 3}\right] - p^2\left[{p \atop 4}\right] + \cdots + p^{p-2}\left[{p \atop p}\right]$$

is a multiple of p^2. (This is called Wolstenholme's theorem.)

6.52 (a) Observe that $H_n = H_n^* + H_{\lfloor n/p\rfloor}/p$, where $H_n^* = \sum_{k=1}^{n}(k \perp p)/k$. (b) Working mod 5 we have $H_r = \langle 0,1,4,1,0\rangle$ for $0 \leqslant r \leqslant 4$. Thus the first solution is $n = 4$. By part (a) we know that $5\backslash a_n \implies 5\backslash a_{\lfloor n/5\rfloor}$; so the next possible range is $n = 20 + r$, $0 \leqslant r \leqslant 4$, when we have $H_n = H_n^* + \frac{1}{5}H_4 = H_{20}^* + \frac{1}{5}H_4 + H_r + \sum_{k=1}^{r} 20/k(20+k)$. The numerator of H_{20}^*, like the numerator of H_4, is divisible by 25. Hence the only solutions in this range are $n = 20$ and $n = 24$. The next possible range is $n = 100 + r$; now $H_n = H_n^* + \frac{1}{5}H_{20}$, which is $\frac{1}{5}H_{20} + H_r$ plus a fraction whose numerator is a multiple of 5. If $\frac{1}{5}H_{20} \equiv m \pmod 5$, where m is an integer, the harmonic number H_{100+r} will have a numerator divisible by 5 if and only if $m + H_r \equiv 0 \pmod 5$; hence m must be $\equiv 0$, 1, or 4. Working modulo 5 we find $\frac{1}{5}H_{20} = \frac{1}{5}H_{20}^* + \frac{1}{25}H_4 \equiv \frac{1}{25}H_4 = \frac{1}{12} \equiv 3$; hence there are no solutions for $100 \leqslant n \leqslant 104$. Similarly there are none for $120 \leqslant n \leqslant 124$; we have found all three solutions.

(By exercise 6.51(d), we always have $p^2\backslash a_{p-1}$, $p\backslash a_{p^2-p}$, and $p\backslash a_{p^2-1}$, if p is any prime $\geqslant 5$. The argument just given shows that these are the only solutions to $p\backslash a_n$ if and only if there are no solutions to $p^{-2}H_{p-1} + H_r \equiv 0 \pmod{p}$ for $0 \leqslant r < p$. The latter condition holds not only for $p = 5$ but also for $p = 13, 17, 23, 41$, and 67—perhaps for infinitely many primes. The numerator of H_n is divisible by 3 only when $n = 2, 7$, and 22; it is divisible by 7 only when $n = 6, 42, 48, 295, 299, 337, 341, 2096, 2390, 14675, 16731, 16735$, and 102728.)

(Attention, computer programmers: Here's an interesting condition to test, for as many primes as you can.)

6.53 Summation by parts yields

$$\frac{n+1}{(n+2)^2}\left(\frac{(-1)^m}{\binom{n+1}{m+1}}\bigl((n+2)H_{m+1}-1\bigr)-1\right).$$

6.54 (a) If $m \geqslant p$ we have $S_m(p) \equiv S_{m-(p-1)}(p) \pmod{p}$, since $k^{p-1} \equiv 1$ when $1 \leqslant k < p$. Also $S_{p-1}(p) \equiv p - 1 \equiv -1$. If $0 < m < p - 1$, we can write

$$S_m(p) \;=\; \sum_{j=0}^{m}\left[{m \atop j}\right](-1)^{m-j}\sum_{k=0}^{p-1}k^{\underline{j}} \;=\; \sum_{j=0}^{m}\left[{m \atop j}\right](-1)^{m-j}\frac{p^{\underline{j+1}}}{j+1} \;\equiv\; 0\,.$$

(b) The condition in the hint implies that the denominator of I_{2n} is not divisible by any prime p; hence I_{2n} must be an integer. To prove the hint, we may assume that $n>1$. Then

(The numerators of Bernoulli numbers have important connections to the known results about Fermat's Last Theorem; see Ribenboim [249].)

$$B_{2n} + \frac{[(p-1)\backslash(2n)]}{p} + \sum_{k=0}^{2n-2}\binom{2n+1}{k}B_k\frac{p^{2n-k}}{2n+1}$$

is an integer, by (6.78), (6.84), and part (a). So we want to verify that none of the fractions $\binom{2n+1}{k}B_kp^{2n-k}/(2n+1) = \binom{2n}{k}B_kp^{2n-k}/(2n-k+1)$ has a denominator divisible by p. The denominator of $\binom{2n}{k}B_kp$ isn't divisible by p, since B_k has no p^2 in its denominator (by induction); and the denominator of $p^{2n-k-1}/(2n-k+1)$ isn't divisible by p, since $2n-k+1 < p^{2n-k}$ when $k \leqslant 2n-2$; QED. (The numbers I_{2n} are tabulated in [185]. Hermite calculated them through I_{18} in 1875 [153]. It turns out that $I_2 = I_4 = I_6 = I_8 = I_{10} = I_{12} = 1$; hence there *is* actually a "simple" pattern to the Bernoulli numbers displayed in the text, including $\frac{-691}{2730}$(!). But the numbers I_{2n} don't seem to have any memorable features when $n > 6$. For example, $B_{24} = -86579 - \frac{1}{2} - \frac{1}{3} - \frac{1}{5} - \frac{1}{7} - \frac{1}{13}$, and 86579 is prime.)

(c) The numbers $2-1$ and $3-1$ always divide $2n$. If n is prime, the only divisors of $2n$ are $1, 2, n$, and $2n$, so the denominator of B_{2n} for prime $n > 2$ will be 6 unless $2n+1$ is also prime. In the latter case we can try $4n+3$, $8n+7$, $\ldots$, until we eventually hit a nonprime (since n divides $2^{n-1}n + 2^{n-1} - 1$). (This proof does not need the more difficult, but true, theorem that there are infinitely many primes of the form $6k+1$.) The denominator of B_{2n} can be 6 also when n has nonprime values, such as 49.

6.55 The stated sum is $\frac{m+1}{x+m+1}\binom{x+n}{n}\binom{n}{m+1}$, by Vandermonde's convolution. To get (6.70), differentiate and set $x = 0$.

6.56 First replace k^{n+1} by $\bigl((k-m)+m\bigr)n+1$ and expand in powers of $k-m$; simplifications occur as in the derivation of (6.72). If $m > n$ or $m < 0$, the answer is $(-1)^n n! - m^n/\binom{n-m}{n}$. Otherwise we need to take the limit of (5.41) minus the term for $k = m$, as $x \to -m$; the answer comes to $(-1)^n n! + (-1)^{m+1}\binom{n}{m} m^n (n+1+mH_{n-m} - mH_m)$.

6.57 First prove by induction that the nth row contains at most three distinct values $A_n \geqslant B_n \geqslant C_n$; if n is even they occur in the cyclic order $[C_n, B_n, A_n, B_n, C_n]$, while if n is odd they occur in the cyclic order $[C_n, B_n, A_n, A_n, B_n]$. Also

$$\begin{aligned} A_{2n+1} &= A_{2n} + B_{2n}\,; & A_{2n} &= 2A_{2n-1}\,;\\ B_{2n+1} &= B_{2n} + C_{2n}\,; & B_{2n} &= A_{2n-1} + B_{2n-1}\,;\\ C_{2n+1} &= 2C_{2n}\,; & C_{2n} &= B_{2n-1} + C_{2n-1}\,. \end{aligned}$$

It follows that $Q_n = A_n - C_n = F_{n+1}$. (See exercise 5.75 for wraparound binomial coefficients of order 3.)

6.58 (a) $\sum_{n\geqslant 0} F_n^2 z^n = z(1-z)/(1+z)(1-3z+z^2) = \frac{1}{5}\bigl((2-3z)/(1-3z+z^2) - 2/(1+z)\bigr)$. (b) $\sum_{n\geqslant 0} F_n^3 z^n = z(1-2z-z^2)/(1-4z-z^2)(1+z-z^2) = \frac{1}{5}\bigl(2z/(1-4z-z^2) + 3z/(1+z-z^2)\bigr)$. (These formulas are obtained by squaring or cubing Binet's formula (6.123) and summing on n, then combining terms so that ϕ and $\hat\phi$ disappear.) It follows that $F_{n+1}^3 - 4F_n^3 - F_{n-1}^3 = 3(-1)^n F_n$. (The corresponding recurrence for mth powers has been found by Jarden and Motzkin [163].)

6.59 Let m be fixed. We can prove by induction on n that it is, in fact, possible to find such an x with the additional condition $x \not\equiv 2 \pmod 4$. If x is such a solution, we can move up to a solution modulo 3^{n+1} because

$$F_{8\cdot 3^{n-1}} \equiv 3^n\,, \qquad F_{8\cdot 3^{n-1}-1} \equiv 3^n + 1 \pmod{3^{n+1}}\,;$$

either x or $x + 8\cdot 3^{n-1}$ or $x + 16\cdot 3^{n-1}$ will do the job.

6.60 F_1+1, F_2+1, F_3+1, F_4-1, and F_6-1 are the only cases. Otherwise the Lucas numbers of exercise 28 arise in the factorizations

$$\begin{aligned} F_{2m} + (-1)^m &= L_{m+1}F_{m-1}\,; & F_{2m+1} + (-1)^m &= L_m F_{m+1}\,;\\ F_{2m} - (-1)^m &= L_{m-1}F_{m+1}\,; & F_{2m+1} - (-1)^m &= L_{m+1}F_m\,. \end{aligned}$$

(We have $F_{m+n} - (-1)^n F_{m-n} = L_m F_n$ in general.)

6.61 $1/F_{2m} = F_{m-1}/F_m - F_{2m-1}/F_{2m}$ when m is even and positive. The second sum is $5/4 - F_{3\cdot 2^n - 1}/F_{3\cdot 2^n}$, for $n \geqslant 1$.

6.62 (a) $A_n = \sqrt{5}\,A_{n-1} - A_{n-2}$ and $B_n = \sqrt{5}\,B_{n-1} - B_{n-2}$. Incidentally, we also have $\sqrt{5}\,A_n + B_n = 2A_{n+1}$ and $\sqrt{5}\,B_n - A_n = 2B_{n-1}$. (b) A table of small values reveals that

$$A_n = \begin{cases} L_n, & n \text{ even;} \\ \sqrt{5}\,F_n, & n \text{ odd;} \end{cases} \qquad B_n = \begin{cases} \sqrt{5}\,F_n, & n \text{ even;} \\ L_n, & n \text{ odd.} \end{cases}$$

(c) $B_n/A_{n+1} - B_{n-1}/A_n = 1/(F_{2n+1}+1)$ because $B_nA_n - B_{n-1}A_{n+1} = \sqrt{5}$ and $A_nA_{n+1} = \sqrt{5}\,(F_{2n+1}+1)$. Notice that $B_n/A_{n+1} = (F_n/F_{n+1})[n \text{ even}] + (L_n/L_{n+1})[n \text{ odd}]$. (d) Similarly, $\sum_{k=1}^{n} 1/(F_{2k+1}-1) = (A_0/B_1 - A_1/B_2) + \cdots + (A_{n-1}/B_n - A_n/B_{n+1}) = 2 - A_n/B_{n+1}$. This quantity can also be expressed as $(5F_n/L_{n+1})[n \text{ even}] + (L_n/F_{n+1})[n \text{ odd}]$.

6.63 (a) $\left[{n \atop k}\right]$. There are $\left[{n-1 \atop k-1}\right]$ with $\pi_n = n$ and $(n-1)\left[{n-1 \atop k}\right]$ with $\pi_n < n$. (b) $\left\langle{n \atop k}\right\rangle$. Each permutation $\rho_1 \ldots \rho_{n-1}$ of $\{1, \ldots, n-1\}$ leads to n permutations $\pi_1\pi_2 \ldots \pi_n = \rho_1 \ldots \rho_{j-1}\, n\, \rho_{j+1} \ldots \rho_{n-1}\rho_j$. If $\rho_1 \ldots \rho_{n-1}$ has k excedances, there are $k+1$ values of j that yield k excedances in $\pi_1\pi_2 \ldots \pi_n$; the remaining $n-1-k$ values yield $k+1$. Hence the total number of ways to get k excedances in $\pi_1\pi_2 \ldots \pi_n$ is $(k+1)\left\langle{n-1 \atop k}\right\rangle + \bigl((n-1)-(k-1)\bigr)\left\langle{n-1 \atop k-1}\right\rangle = \left\langle{n \atop k}\right\rangle$.

6.64 The denominator of $\binom{1/2}{2n}$ is $2^{4n-\nu_2(n)}$, by the proof in exercise 5.72. The denominator of $\left[{1/2 \atop 1/2-n}\right]$ is the same, by (6.44), because $\left\langle\!\left\langle{n \atop 0}\right\rangle\!\right\rangle = 1$ and $\left\langle\!\left\langle{n \atop k}\right\rangle\!\right\rangle$ is even for $k > 0$.

6.65 This is equivalent to saying that $\left\langle{n \atop k}\right\rangle/n!$ is the probability that we have $\lfloor x_1 + \cdots + x_n \rfloor = k$, when $x_1, \ldots, x_n$ are independent random numbers uniformly distributed between 0 and 1. Let $y_j = (x_1 + \cdots + x_j) \bmod 1$. Then $y_1, \ldots, y_n$ are independently and uniformly distributed, and $\lfloor x_1 + \cdots + x_n \rfloor$ is the number of descents in the y's. The permutation of the y's is random, and the probability of k descents is the same as the probability of k ascents.

6.66 We have the general formula

$$\left\langle\!\!\left\langle{n \atop m}\right\rangle\!\!\right\rangle = \sum_{k=0}^{m} \binom{2n+1}{k} \left\{{n+m+1-k \atop m+1-k}\right\} (-1)^k, \qquad \text{for } n > m \geqslant 0,$$

analogous to (6.38). When $m = 2$ this equals

$$\begin{aligned} \left\langle\!\!\left\langle{n \atop 2}\right\rangle\!\!\right\rangle &= \left\{{n+3 \atop 3}\right\} - (2n+1)\left\{{n+2 \atop 2}\right\} + \binom{2n+1}{2}\left\{{n+1 \atop 1}\right\} \\ &= \tfrac{1}{2}3^{n+2} - (2n+3)2^{n+1} + \tfrac{1}{2}(4n^2+6n+3)\,. \end{aligned}$$

6.67 $\frac{1}{3}n(n+\frac{1}{2})(n+1)(2H_{2n} - H_n) - \frac{1}{36}n(10n^2+9n-1)$. (It would be nice to automate the derivation of formulas such as this.)

6.68 $1/k - 1/(k+z) = z/k^2 - z^2/k^3 + \cdots$, and everything converges when $|z| < 1$.

6.69 Note that $\prod_{k=1}^{n}(1+z/k)e^{-z/k} = \binom{n+z}{n} n^{-z} e^{(\ln n - H_n)z}$. If $f(z) = \frac{d}{dz}(z!)$ we find $f(z)/z! + \gamma = H_z$.

6.70 For $\tan z$, we can use $\tan z = \cot z - 2\cot 2z$ (which is equivalent to the identity of exercise 23). Also $z/\sin z = z\cot z + z\tan\frac{1}{2}z$ has the power series $\sum_{n\ge 0}(-1)^{n-1}(4^n-2)B_{2n}z^{2n}/(2n)!$; and

$$\begin{aligned} \ln\frac{\tan z}{z} &= \ln\frac{\sin z}{z} - \ln\cos z \\ &= \sum_{n\geqslant 1}(-1)^n\frac{4^n B_{2n} z^{2n}}{(2n)(2n)!} - \sum_{n\geqslant 1}(-1)^n\frac{4^n(4^n-1)B_{2n}z^{2n}}{(2n)(2n)!} \\ &= \sum_{n\geqslant 1}(-1)^n\frac{4^n(4^n-2)B_{2n}z^{2n}}{(2n)(2n)!}, \end{aligned}$$

because $\frac{d}{dz}\ln\sin z = \cot z$ and $\frac{d}{dz}\ln\cos z = -\tan z$.

6.71 Since $\tan 2z + \sec 2z = (\sin z + \cos z)/(\cos z - \sin z)$, setting $x = 1$ in (6.94) gives $T_n(1) = 2^n T_n$ when n is odd, $T_n(1) = 2^n E_n$ when n is even, where $1/\cos z = \sum_{n\geqslant 0} E_{2n}z^{2n}/(2n)!$. (The E_n are called *Euler numbers*, not to be confused with the Eulerian numbers $\left\langle {n \atop k} \right\rangle$.)

6.72 $2^{n+1}(2^{n+1}-1)B_{n+1}/(n+1)$, if $n > 0$. (See (7.56) and (6.92); the desired numbers are essentially the coefficients of $1 - \tanh z$.)

6.73 $\cot(z+\pi) = \cot z$ and $\cot(z+\frac{1}{2}\pi) = -\tan z$; hence the identity is equivalent to

$$\cot z = \frac{1}{2^n}\sum_{k=0}^{2^n-1}\cot\frac{z+k\pi}{2^n},$$

which follows by induction from the case $n = 1$. The stated limit follows since $z\cot z \to 1$ as $z \to 0$. It can be shown that term-by-term passage to the limit is justified, hence (6.88) is valid. (Incidentally, the general formula

$$\cot z = \frac{1}{n}\sum_{k=0}^{n-1}\cot\frac{z+k\pi}{n}$$

is also true. It can be proved from (6.88), or from

$$\frac{1}{e^{nz}-1} = \frac{1}{n}\sum_{k=0}^{n-1}\frac{1}{e^{z+2k\pi i/n}-1},$$

which is equivalent to the partial fraction expansion of $1/(z^n-1)$.)

6.74 If $p(x)$ is any polynomial of degree $\leqslant n$, we have

$$p(x) = \sum_k p(-k)\binom{-x}{k}\binom{x+n}{n-k},$$

because this equation holds for $x = 0, -1, \ldots, -n$. The stated identity is the special case where $p(x) = x\sigma_n(x)$ and $x = 1$. Incidentally, we obtain a simpler expression for Bernoulli numbers in terms of Stirling numbers by setting $k = 1$ in (6.99):

$$\sum_{k\geqslant 0} \left\{ {m \atop k} \right\} (-1)^k \frac{k!}{k+1} = B_m .$$

6.75 Sam Loyd [204, pages 288 and 378] gave the construction

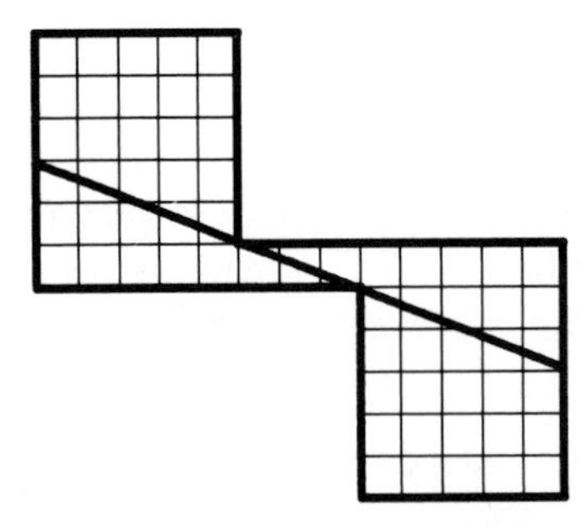

and claimed to have invented (but not published) the $64 = 65$ arrangement in 1858. (Similar paradoxes go back at least to the eighteenth century, but Loyd found better ways to present them.)

6.76 We expect $A_m/A_{m-1} \approx \phi$, so we try $A_{m-1} = 618034 + r$ and $A_{m-2} = 381966 - r$. Then $A_{m-3} = 236068 + 2r$, etc., and we find $A_{m-18} = 144 - 2584r$, $A_{m-19} = 154 + 4181r$. Hence $r = 0$, $x = 154$, $y = 144$, $m = 20$.

6.77 If $P(F_{n+1}, F_n) = 0$ for infinitely many *even* values of n, then $P(x, y)$ is divisible by $U(x,y) - 1$, where $U(x,y) = x^2 - xy - y^2$. For if t is the total degree of P, we can write

$$P(x,y) = \sum_{k=0}^{t} q_k x^k y^{t-k} + \sum_{j+k<t} r_{j,k} x^j y^k = Q(x,y) + R(x,y) .$$

Then

$$\frac{P(F_{n+1}, F_n)}{F_n^t} = \sum_{k=0}^{t} q_k \left(\frac{F_{n+1}}{F_n}\right)^k + O(1/F_n)$$

and we have $\sum_{k=0}^{t} q_k \phi^k = 0$ by taking the limit as $n \to \infty$. Hence $Q(x, y)$ is a multiple of $U(x, y)$, say $A(x, y)U(x, y)$. But $U(F_{n+1}, F_n) = (-1)^n$ and n is even, so $P_0(x, y) = P(x, y) - \big(U(x, y) - 1\big)A(x, y)$ is another polynomial such that $P_0(F_{n+1}, F_n) = 0$. The total degree of P_0 is less than t, so P_0 is a multiple of $U - 1$ by induction on t.

Similarly, $P(x, y)$ is divisible by $U(x, y) + 1$ if $P(F_{n+1}, F_n) = 0$ for infinitely many *odd* values of n. A combination of these two facts gives the desired necessary and sufficient condition: $P(x, y)$ is divisible by $U(x, y)^2 - 1$.

6.78 First add the digits without carrying, getting digits 0, 1, and 2. Then use the two carry rules

$$0\,(d{+}1)\,(e{+}1) \;\to\; 1\,d\,e\,,$$
$$0\,(d{+}2)\,0\,e \;\to\; 1\,d\,0\,(e+1)\,,$$

always applying the leftmost applicable carry. This process terminates because the binary value obtained by reading $(b_m \ldots b_2)_F$ as $(b_m \ldots b_2)_2$ increases whenever a carry is performed. But a carry might propagate to the right of the "Fibonacci point"; for example, $(1)_F + (1)_F$ becomes $(10.01)_F$. Such rightward propagation extends at most two positions; and those two digit positions can be zeroed again by using the text's "add 1" algorithm if necessary.

Incidentally, there's a corresponding "multiplication" operation on nonnegative integers: If $m = F_{j_1} + \cdots + F_{j_q}$ and $n = F_{k_1} + \cdots + F_{k_r}$ in the Fibonacci number system, let $m \circ n = \sum_{b=1}^{q} \sum_{c=1}^{r} F_{j_b + k_c}$, by analogy with multiplication of binary numbers. (This definition implies that $m \circ n \approx \sqrt{5}\,mn$ when m and n are large, although $1 \circ n \approx \phi^2 n$.) Fibonacci addition leads to a proof of the associative law $l \circ (m \circ n) = (l \circ m) \circ n$.)

6.79 Yes; for example, we can take

$$A_0 = 331635635998274737472200656430763\,;$$
$$A_1 = 1510028911088401971189590305498785\,.$$

The resulting sequence has the property that A_n is divisible by (but unequal to) p_k when $n \bmod m_k = r_k$, where the numbers (p_k, m_k, r_k) have the following 18 respective values:

$(3, 4, 1)$	$(2, 3, 2)$	$(5, 5, 1)$
$(7, 8, 3)$	$(17, 9, 4)$	$(11, 10, 2)$
$(47, 16, 7)$	$(19, 18, 10)$	$(61, 15, 3)$
$(2207, 32, 15)$	$(53, 27, 16)$	$(31, 30, 24)$
$(1087, 64, 31)$	$(109, 27, 7)$	$(41, 20, 10)$
$(4481, 64, 63)$	$(5779, 54, 52)$	$(2521, 60, 60)$

One of these triples applies to every integer n; for example, the six triples in the first column cover every odd value of n, and the middle column covers all even n that are not divisible by 6. The remainder of the proof is based on the fact that $A_{m+n} = A_m F_{n-1} + A_{m+1} F_n$, together with the congruences

$$\begin{aligned} A_0 &\equiv F_{m_k - r_k} \bmod p_k\,, \\ A_1 &\equiv F_{m_k - r_k + 1} \bmod p_k\,, \end{aligned}$$

for each of the triples (p_k, m_k, r_k). (An improved solution, in which A_0 and A_1 are numbers of "only" 17 digits each, is also possible [184].)

6.80 The matrix product is

$$\begin{pmatrix} K_{n-2}(x_2, \ldots, x_{n-1}) & K_{n-1}(x_2, \ldots, x_{n-1}, x_n) \\ K_{n-1}(x_1, x_2, \ldots, x_{n-1}) & K_n(x_1, x_2, \ldots, x_{n-1}, x_n) \end{pmatrix}.$$

This relates to products of L and R as in (6.137), because we have

$$R^a \begin{pmatrix} 0 & 1 \\ 1 & 0 \end{pmatrix} = \begin{pmatrix} 0 & 1 \\ 1 & a \end{pmatrix} = \begin{pmatrix} 0 & 1 \\ 1 & 0 \end{pmatrix} L^a.$$

The determinant is $K_n(x_1, \ldots, x_n)$; the more general tridiagonal determinant

$$\det \begin{pmatrix} x_1 & 1 & 0 & \ldots & 0 \\ y_2 & x_2 & 1 & & 0 \\ 0 & y_3 & x_3 & 1 & \vdots \\ \vdots & & & \ddots & 1 \\ 0 & 0 & \ldots & y_n & x_n \end{pmatrix}$$

satisfies the recurrence $D_n = x_n D_{n-1} - y_n D_{n-2}$.

6.81 Let $\alpha^{-1} = a_0 + 1/(a_1 + 1/(a_2 + \cdots))$ be the continued fraction representation of α^{-1}. Then we have

$$\frac{a_0}{z} + \cfrac{1}{A_0(z) + \cfrac{1}{A_1(z) + \cfrac{1}{A_2(z) + \cfrac{1}{\ddots}}}} = \frac{1-z}{z} \sum_{n \geq 1} z^{\lfloor n\alpha \rfloor},$$

where

$$A_m(z) = \frac{z^{-q_{m+1}} - z^{-q_{m-1}}}{z^{-q_m} - 1}, \qquad q_m = K_m(a_1, \ldots, a_m).$$

A proof analogous to the text's proof of (6.146) uses a generalization of Zeckendorf's theorem (Fraenkel [104, §4]). If $z = 1/b$, where b is an integer ≥ 2, this gives the continued fraction representation of the transcendental number $(b-1)\sum_{n\geqslant 1} b^{-\lfloor n\alpha\rfloor}$, as in exercise 49.

6.82 The sequences of exercise 62 satisfy $A_{-m} = A_m$, $B_{-m} = -B_m$, and

$$\begin{aligned} A_m A_n &= A_{m+n} + A_{m-n}\,; \\ A_m B_n &= B_{m+n} - B_{m-n}\,; \\ B_m B_n &= A_{m+n} - A_{m-n}\,. \end{aligned}$$

Let $f_k = B_{mk}/A_{mk+l}$ and $g_k = A_{mk}/B_{mk+l}$, where $l = \frac{1}{2}(n-m)$. Then $f_{k+1} - f_k = A_l B_m/(A_{2mk+n} + A_m)$ and $g_k - g_{k+1} = A_l B_m/(A_{2mk+n} - A_m)$; hence we have

$$\begin{aligned} S^+_{m,n} &= \frac{\sqrt{5}}{A_l B_m} \lim_{k\to\infty} (f_k - f_0) &&= \frac{\sqrt{5}}{\phi^l A_l L_m}\,; \\ S^-_{m,n} &= \frac{\sqrt{5}}{A_l B_m} \lim_{k\to\infty} (g_0 - g_k) &&= \frac{\sqrt{5}}{A_l L_m}\left(\frac{2}{B_l} - \frac{1}{\phi^l}\right) \\ & &&= \frac{2}{F_l L_l L_m} - S^+_{m,n}\,. \end{aligned}$$

6.83 Let $p = K(0, a_1, a_2, \ldots, a_m)$, so that p/n is the mth convergent to the continued fraction. Then $\alpha = p/n + (-1)^m/nq$, where $q = K(a_1, \ldots, a_m, \beta)$ and $\beta > 1$. The points $\{k\alpha\}$ for $0 \leqslant k < n$ can therefore be written

$$\frac{0}{n},\ \frac{1}{n} + \frac{(-1)^m \pi_1}{nq},\ \ldots,\ \frac{n-1}{n} + \frac{(-1)^m \pi_{n-1}}{nq},$$

where $\pi_1 \ldots \pi_{n-1}$ is a permutation of $\{1, \ldots, n-1\}$. Let $f(v)$ be the number of such points $< v$; then $f(v)$ and vn both increase by 1 when v increases from k/n to $(k+1)/n$, except when $k = 0$ or $k = n-1$, so they never differ by 2 or more.

6.84 By (6.139) and (6.136), we want to maximize $K(a_1, \ldots, a_m)$ over all sequences of positive integers whose sum is $\leqslant n+1$. The maximum occurs when all the a's are 1, for if $j \geqslant 1$ and $a \geqslant 1$ we have

$$\begin{aligned} &K_{j+k+1}(1, \ldots, 1, a+1, b_1, \ldots, b_k) \\ &\quad = K_{j+k+1}(1, \ldots, 1, a, b_1, \ldots, b_k) + K_j(1, \ldots, 1)\, K_k(b_1, \ldots, b_k) \\ &\quad \leqslant K_{j+k+1}(1, \ldots, 1, a, b_1, \ldots, b_k) + K_{j+k}(1, \ldots, 1, a, b_1, \ldots, b_k) \\ &\quad = K_{j+k+2}(1, \ldots, 1, a, b_1, \ldots, b_k)\,. \end{aligned}$$

(Motzkin and Straus [220] solve more general maximization problems on continuants.)

6.85 The property holds if and only if N has one of the seven forms 5^k, $2\cdot5^k$, $4\cdot5^k$, $3^j\cdot5^k$, $6\cdot5^k$, $7\cdot5^k$, $14\cdot5^k$.

6.86 A candidate for the case $n \bmod 1 = \frac{1}{2}$ appears in [179, section 6], although it may be best to multiply the integers discussed there by some constant involving $\sqrt{\pi}$.

6.87 (a) If there are only finitely many solutions, it is natural to conjecture that the same holds for all primes. (b) The behavior of b_n is quite strange: We have $b_n = \operatorname{lcm}(1,\ldots,n)$ for $968 \leqslant n \leqslant 1066$; on the other hand, $b_{600} = \operatorname{lcm}(1,\ldots,600)/(3^3\cdot5^2\cdot43)$. Andrew Odlyzko observes that p divides $\operatorname{lcm}(1,\ldots,n)/b_n$ if and only if $kp^m \leqslant n < (k+1)p^m$ for some $m \geqslant 1$ and some $k < p$ such that p divides the numerator of H_k. Therefore infinitely many such n exist if it can be shown, for example, that almost all primes have only one such value of k (namely $k = p-1$).

Another reason to remember 1066?

6.88 (Brent [33] found the surprisingly large partial quotient 1568705 in e^γ, but this seems to be just a coincidence. For example, Gosper has found even larger partial quotients in π: The 453,294th is 12996958 and the 11,504,931st is 878783625.)

6.89 Consider the generating function $\sum_{m,n\geqslant0}\left|{m+n \atop m}\right|w^m z^n$, which has the form $\sum_n\bigl(wF(a,b,c)+zF(a',b',c')\bigr)^n$, where $F(a,b,c)$ is the differential operator $a + b\vartheta_w + c\vartheta_z$.

7.1 Substitute z^4 for ▯ and z for ▭ in the generating function, getting $1/(1-z^4-z^2)$. This is like the generating function for T, but with z replaced by z^2. Therefore the answer is zero if m is odd, otherwise $F_{m/2+1}$.

7.2 $G(z) = 1/(1-2z) + 1/(1-3z)$; $\hat{G}(z) = e^{2z} + e^{3z}$.

7.3 Set $z = 1/10$ in the generating function, getting $\frac{10}{9}\ln\frac{10}{9}$.

7.4 Divide $P(z)$ by $Q(z)$, getting a quotient $T(z)$ and a remainder $P_0(z)$ whose degree is less than the degree of Q. The coefficients of $T(z)$ must be added to the coefficients $[z^n]\,P_0(z)/Q(z)$ for small n. (This is the polynomial $T(z)$ in (7.28).)

7.5 This is the convolution of $(1+z^2)^r$ with $(1+z)^r$, so

$$S(z) = (1+z+z^2+z^3)^r .$$

Incidentally, no simple form is known for the coefficients of this generating function; hence the stated sum probably has no simple closed form. (We can use generating functions to obtain negative results as well as positive ones.)

7.6 Let the solution to $g_0 = \alpha$, $g_1 = \beta$, $g_n = g_{n-1} + 2g_{n-2} + (-1)^n\gamma$ be $g_n = A(n)\alpha + B(n)\beta + C(n)\gamma$. The function 2^n works when $\alpha = 1$, $\beta = 2$, $\gamma = 0$; the function $(-1)^n$ works when $\alpha = 1$, $\beta = -1$, $\gamma = 0$; the function $(-1)^n n$ works when $\alpha = 0$, $\beta = -1$, $\gamma = 3$. Hence $A(n) + 2B(n) = 2^n$, $A(n) - B(n) = (-1)^n$, and $-B(n) + 3C(n) = (-1)^n n$.

7.7 $G(z) = \bigl(z/(1-z)^2\bigr)G(z) + 1$, hence

I bet that the controversial "fan of order zero" does have one spanning tree.

$$G(z) = \frac{1-2z+z^2}{1-3z+z^2} = 1 + \frac{z}{1-3z+z^2};$$

we have $g_n = F_{2n} + [n=0]$.

7.8 Differentiate $(1-z)^{-x-1}$ twice with respect to x, obtaining

$$\binom{x+n}{n}\bigl((H_{x+n} - H_x)^2 - (H^{(2)}_{x+n} - H^{(2)}_x)\bigr).$$

Now set $x = m$.

7.9 $(n+1)(H_n^2 - H_n^{(2)}) - 2n(H_n - 1)$.

7.10 The identity $H_{k-1/2} - H_{-1/2} = \frac{2}{2k-1} + \cdots + \frac{2}{1} = 2H_{2k} - H_k$ implies that $\sum_k \binom{2k}{k}\binom{2n-2k}{n-k}(2H_{2k} - H_k) = 4^n H_n$.

7.11 (a) $C(z) = A(z)B(z^2)/(1-z)$. (b) $zB'(z) = A(2z)e^z$, hence $A(z) = \frac{z}{2}e^{-z/2}B'(\frac{z}{2})$. (c) $A(z) = B(z)/(1-z)^{r+1}$, hence $B(z) = (1-z)^{r+1}A(z)$ and we have $f_k(r) = \binom{r+1}{k}(-1)^k$.

7.12 C_n. The numbers in the upper row correspond to the positions of $+1$'s in a sequence of $+1$'s and -1's that defines a "mountain range"; the numbers in the lower row correspond to the positions of -1's. For example, the given array corresponds to

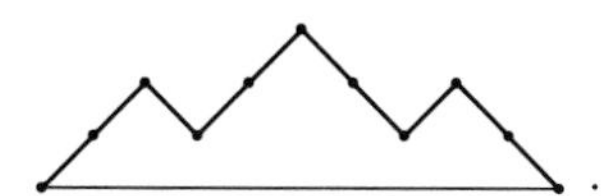

7.13 Extend the sequence periodically (let $x_{m+k} = x_k$) and define $s_n = x_1 + \cdots + x_n$. We have $s_m = l$, $s_{2m} = 2l$, etc. There must be a largest index k_j such that $s_{k_j} = j$, $s_{k_j+m} = l + j$, etc. These indices $k_1, \ldots, k_l$ (modulo m) specify the cyclic shifts in question.

For example, in the sequence $\langle -2, 1, -1, 0, 1, 1, -1, 1, 1, 1\rangle$ with $m = 10$ and $l = 2$ we have $k_1 = 17$, $k_2 = 24$.

7.14 $\hat{G}(z) = -2z\hat{G}(z) + \hat{G}(z)^2 + z$ (be careful about the final term!) leads via the quadratic formula to

$$\hat{G}(z) = \frac{1 + 2z - \sqrt{1+4z^2}}{2}.$$

Hence $g_{2n+1} = 0$ and $g_{2n} = (-1)^n (2n)!\, C_{n-1}$, for all $n > 0$.

7.15 There are $\binom{n}{k} b_{n-k}$ partitions with k other objects in the subset containing $n+1$. Hence $\hat{B}'(z) = e^z \hat{B}(z)$. The solution to this differential equation is $\hat{B}(z) = e^{e^z + c}$, and $c = -1$ since $\hat{B}(0) = 1$. (We can also get this result by summing (7.49) on m, since $b_n = \sum_m {n \brace m}$.)

7.16 One way is to take the logarithm of

$$B(z) = 1/\bigl((1-z)^{a_1}(1-z^2)^{a_2}(1-z^3)^{a_3}(1-z^4)^{a_4}\ldots\bigr),$$

then use the formula for $\ln \frac{1}{1-z}$ and interchange the order of summation.

7.17 This follows since $\int_0^\infty t^n e^{-t}\,dt = n!$. There's also a formula that goes in the other direction:

$$\hat{G}(z) = \frac{1}{2\pi}\int_{-\pi}^{+\pi} G(ze^{-i\theta})\, e^{e^{i\theta}}\, d\theta\,.$$

7.18 (a) $\zeta(z - \frac{1}{2})$; (b) $-\zeta'(z)$; (c) $\zeta(z)/\zeta(2z)$. Every positive integer is uniquely representable as $m^2 q$, where q is squarefree.

7.19 If $n > 0$, the coefficient $[z^n] \exp\bigl(x \ln F(z)\bigr)$ is a polynomial of degree n in x that's a multiple of x. The first convolution formula comes from equating coefficients of z^n in $F(z)^x F(z)^y = F(z)^{x+y}$. The second comes from equating coefficients of z^{n-1} in $F'(z)F(z)^{x-1}F(z)^y = F'(z)F(z)^{x+y-1}$, because we have

$$F'(z)F(z)^{x-1} = x^{-1}\frac{\partial}{\partial z}\bigl(F(z)^x\bigr) = x^{-1}\sum_{n\geqslant 0} n f_n(x) z^{n-1}\,.$$

(Further convolutions follow by taking $\partial/\partial x$, as in (7.43).)

7.20 Let $G(z) = \sum_{n\geqslant 0} g_n z^n$. Then

$$z^l G^{(k)}(z) = \sum_{n\geqslant 0} n^{\underline{k}} g_n z^{n-k+l} = \sum_{n\geqslant 0} (n+k-l)^{\underline{k}} g_{n+k-l} z^n$$

for all $k, l \geqslant 0$, if we regard $g_n = 0$ for $n < 0$. Hence if $P_0(z), \ldots, P_m(z)$ are polynomials, not all zero, having maximum degree d, then there are polynomials $p_0(n), \ldots, p_{m+d}(n)$ such that

$$P_0(z)G(z) + \cdots + P_m(z)G^{(m)}(z) = \sum_{n\geqslant 0}\sum_{j=0}^{m+d} p_j(n)\, g_{n+j-d} z^n\,.$$

Therefore a differentiably finite $G(z)$ implies that

$$\sum_{j=0}^{m+d} p_j(n+d)\, g_{n+j} = 0\,, \qquad \text{for all } n \geqslant 0.$$

The converse is similar. (One consequence is that $G(z)$ is differentiably finite if and only if the corresponding egf, $\hat{G}(z)$, is differentiably finite.)

7.21 This is the problem of giving change with denominations 10 and 20, so $G(z) = 1/(1-z^{10})(1-z^{20}) = \check{G}(z^{10})$, where $\check{G}(z) = 1/(1-z)(1-z^2)$. (a) The partial fraction decomposition of $\check{G}(z)$ is $\frac{1}{2}(1-z)^{-2} + \frac{1}{4}(1-z)^{-1} + \frac{1}{4}(1+z)^{-1}$, so $[z^n]\,\check{G}(z) = \frac{1}{4}\bigl(2n + 3 + (-1)^n\bigr)$. Setting $n = 50$ yields 26 ways to make the payment. (b) $\check{G}(z) = (1+z)/(1-z^2)^2 = (1+z)(1+2z^2+3z^4+\cdots)$, so $[z^n]\,\check{G}(z) = \lfloor n/2 \rfloor + 1$. (Compare this with the value $N_n = \lfloor n/5 \rfloor + 1$ in the text's coin-changing problem. The bank robber's problem is equivalent to the problem of making change with pennies and tuppences.)

This slow method of finding the answer is just the cashier's way of stalling until the police come.

The USA has two-cent pieces, but they haven't been minted since 1873.

7.22 Each polygon has a "base" (the line segment at the bottom). If A and B are triangulated polygons, let $A \triangle B$ be the result of pasting the base of A to the upper left diagonal of $\triangle$, and pasting the base of B to the upper right diagonal. Thus, for example,

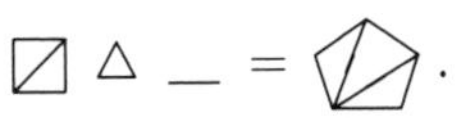

(The polygons might need to be warped a bit and/or banged into shape.) Every triangulation arises in this way, because the base line is part of a unique triangle and there are triangulated polygons A and B at its left and right.

Replacing each triangle by z gives a power series in which the coefficient of z^n is the number of triangulations with n triangles, namely the number of ways to decompose an $(n+2)$-gon into triangles. Since $P = 1 + zP^2$, this is the generating function for Catalan numbers $C_0 + C_1 z + C_2 z^2 + \cdots$; the number of ways to triangulate an n-gon is $C_{n-2} = \binom{2n-4}{n-2}/(n-1)$.

7.23 Let a_n be the stated number, and b_n the number of ways with a $2\times1\times1$ notch missing at the top. By considering the possible patterns visible on the top surface, we have

$$\begin{aligned} a_n &= 2a_{n-1} + 4b_{n-1} + a_{n-2} + [n=0]\,; \\ b_n &= a_{n-1} + b_{n-1}\,. \end{aligned}$$

Hence the generating functions satisfy $A = 2zA + 4zB + z^2A + 1$, $B = zA + zB$, and we have

$$A(z) = \frac{1-z}{(1+z)(1-4z+z^2)}\,.$$

This formula relates to the problem of $3 \times n$ domino tilings; we have $A_n = \frac{1}{3}\bigl(U_{2n} + V_{2n+1} + (-1)^n\bigr) = \frac{1}{6}(2+\sqrt{3}\,)^{n+1} + \frac{1}{6}(2-\sqrt{3}\,)^{n+1} + \frac{1}{3}(-1)^n$, which is $(2+\sqrt{3}\,)^{n+1}/6$ rounded to the nearest integer.

7.24 $n\sum_{k_1+\cdots+k_m=n} k_1 \cdot \ldots \cdot k_m/m = F_{2n+1} + F_{2n-1} - 2$. (Consider the coefficient $[z^{n-1}]\,\frac{d}{dz}\ln\bigl(1/(1-G(z))\bigr)$, where $G(z) = z/(1-z)^2$.)

7.25 The generating function is $P(z)/(1-z^m)$, where $P(z) = z + 2z^2 + \cdots + (m-1)z^{m-1} = \bigl((m-1)z^{m+1} - mz^m + z\bigr)/(1-z)^2$. The denominator is $Q(z) = 1 - z^m = (1-\omega^0 z)(1-\omega^1 z)\ldots(1-\omega^{m-1}z)$. By the rational expansion theorem for distinct roots, we obtain

$$n \bmod m \;=\; \frac{m-1}{2} + \sum_{k=1}^{m-1} \frac{\omega^{-kn}}{\omega^k - 1}\,.$$

7.26 $(1-z-z^2)\mathfrak{F}(z) = F(z)$ leads to $\mathfrak{F}_n = \bigl(2(n+1)F_n + nF_{n+1}\bigr)/5$ as in equation (7.60).

7.27 Each oriented cycle pattern begins with ↕ or ⇄ or a $2 \times k$ cycle (for some $k \geqslant 2$) oriented in one of two ways. Hence

$$Q_n \;=\; Q_{n-1} + Q_{n-2} + 2Q_{n-2} + 2Q_{n-3} + \cdots + 2Q_0$$

for $n \geqslant 2$; $Q_0 = Q_1 = 1$. The generating function is therefore

$$\begin{aligned} Q(z) \;&=\; zQ(z) + z^2Q(z) + 2z^2Q(z)/(1-z) + 1 \\ &=\; 1/\bigl(1 - z - z^2 - 2z^2/(1-z)\bigr) \\ &=\; \frac{(1-z)}{(1-2z-2z^2+z^3)} \\ &=\; \frac{\phi^2/5}{1-\phi^2 z} + \frac{\phi^{-2}/5}{1-\phi^{-2}z} + \frac{2/5}{1+z}\,, \end{aligned}$$

and $Q_n = \bigl(\phi^{2n+2} + \phi^{-2n-2} + 2(-1)^n\bigr)/5 = \bigl((\phi^{n+1} - \hat{\phi}^{n+1})/\sqrt{5}\bigr)^2 = F_{n+1}^2$.

7.28 In general if $A(z) = (1 + z + \cdots + z^{m-1})B(z)$, we have $A_r + A_{r+m} + A_{r+2m} + \cdots = B(1)$ for $0 \leqslant r < m$. In this case $m = 10$ and $B(z) = (1 + z + \cdots + z^9)(1 + z^2 + z^4 + z^6 + z^8)(1 + z^5)$.

7.29 $F(z) + F(z)^2 + F(z)^3 + \cdots = z/(1 - z - z^2 - z) = \bigl(1/(1-(1+\sqrt{2}\,)z) - (1/(1-(1-\sqrt{2}\,)z)\bigr)/\sqrt{8}$, so the answer is $\bigl((1+\sqrt{2}\,)^n - (1-\sqrt{2}\,)^n\bigr)/\sqrt{8}$.

7.30 $\sum_{k=1}^{n} \binom{2n-1-k}{n-1}\bigl(a^n b^{n-k}/(1-\alpha z)^k + a^{n-k}b^n/(1-\beta z)^k\bigr)$, by exercise 5.39.

7.31 The dgf is $\zeta(z)^2/\zeta(z-1)$; hence we find $g(n)$ is the product of $(k+1-kp)$ over all prime powers p^k that exactly divide n.

7.32 We may assume that each $b_k \geqslant 0$. A set of arithmetic progressions forms an exact cover if and only if

$$\frac{1}{1-z} \;=\; \frac{z^{b_1}}{1-z^{a_1}} + \cdots + \frac{z^{b_m}}{1-z^{a_m}}\,.$$

Subtract $z^{b_m}/(1-z^{a_m})$ from both sides and set $z = e^{2\pi i/a_m}$. The left side is infinite, and the right side will be finite unless $a_{m-1} = a_m$.

7.33 $(-1)^{n-m+1}[n>m]/(n-m)$.

7.34 We can also write $G_n(z) = \sum_{k_1+(m+1)k_{m+1}=n} \binom{k_1+k_{m+1}}{k_{m+1}}(z^m)^{k_{m+1}}$. In general, if

$$G_n \;=\; \sum_{k_1+2k_2+\cdots+rk_r=n} \binom{k_1+k_2+\cdots+k_r}{k_1,k_2,\ldots,k_r} z_1^{k_1} z_2^{k_2} \ldots z_r^{k_r}\,,$$

we have $G_n = z_1 G_{n-1} + z_2 G_{n-2} + \cdots + z_r G_{n-r} + [n=0]$, and the generating function is $1/(1 - z_1 w - z_2 w^2 - \cdots - z_r w^r)$. In the stated special case the answer is $1/(1 - w - z^m w^{m+1})$. (See (5.74) for the case $m = 1$.)

7.35 (a) $\frac{1}{n}\sum_{0<k<n}\bigl(1/k + 1/(n-k)\bigr) = \frac{2}{n}H_{n-1}$. (b) $[z^n]\left(\ln\frac{1}{1-z}\right)^2 = \frac{2!}{n!}\left[{n \atop 2}\right] = \frac{2}{n}H_{n-1}$ by (7.50) and (6.58). Another way to do part (b) is to use the rule $[z^n]\,F(z) = \frac{1}{n}[z^{n-1}]\,F'(z)$ with $F(z) = \left(\ln\frac{1}{1-z}\right)^2$.

7.36 $\frac{1-z^m}{1-z}A(z^m)$.

7.37 (a) The amazing identity $a_{2n} = a_{2n+1} = b_n$ holds in the table

n	0	1	2	3	4	5	6	7	8	9	10
a_n	1	1	2	2	4	4	6	6	10	10	14
b_n	1	2	4	6	10	14	20	26	36	46	60

(b) $A(z) = 1/\bigl((1-z)(1-z^2)(1-z^4)(1-z^8)\ldots\bigr)$. (c) $B(z) = A(z)/(1-z)$, and we want to show that $A(z) = (1+z)B(z^2)$. This follows from $A(z) = A(z^2)/(1-z)$.

7.38 $(1-wz)M(w,z) = \sum_{m,n\geqslant 1}\bigl(\min(m,n) - \min(m-1,n-1)\bigr)w^m z^n = \sum_{m,n\geqslant 1} w^m z^n = wz/(1-w)(1-z)$. In general,

$$M(z_1,\ldots,z_m) \;=\; \frac{z_1\ldots z_m}{(1-z_1)\ldots(1-z_m)(1-z_1\ldots z_m)}\,.$$

7.39 The answers to the hint are

$$\sum_{1\leqslant k_1<k_2<\cdots<k_m\leqslant n} a_{k_1}a_{k_2}\ldots a_{k_m} \quad\text{and}\quad \sum_{1\leqslant k_1\leqslant k_2\leqslant\cdots\leqslant k_m\leqslant n} a_{k_1}a_{k_2}\ldots a_{k_m}\,,$$

respectively. Therefore: (a) We want the coefficient of z^m in the product $(1+z)(1+2z)\ldots(1+nz)$. This is the reflection of $(z+1)^{\overline{n}}$, so it is $\left[{n+1 \atop n+1}\right] + \left[{n+1 \atop n}\right]z + \cdots + \left[{n+1 \atop 1}\right]z^n$ and the answer is $\left[{n+1 \atop n+1-m}\right]$. (b) The coefficient of z^m in $1/\bigl((1-z)(1-2z)\ldots(1-nz)\bigr)$ is $\left\{{m+n \atop n}\right\}$ by (7.47).

7.40 The egf for $\langle nF_{n-1} - F_n \rangle$ is $(z-1)\hat{F}(z)$ where $\hat{F}(z) = \sum_{n\geq 0} F_n z^n/n! = (e^{\phi z} - e^{\hat{\phi} z})/\sqrt{5}$. The egf for $\langle n_¡ \rangle$ is $e^{-z}/(1-z)$. The product is

$$5^{-1/2}\bigl(e^{(\hat{\phi}-1)z} - e^{(\phi-1)z}\bigr) = 5^{-1/2}(e^{-\phi z} - e^{-\hat{\phi} z}).$$

We have $\hat{F}(z)e^{-z} = -\hat{F}(-z)$. So the answer is $(-1)^n F_n$.

7.41 The number of up-down permutations with the largest element n in position $2k$ is $\binom{n-1}{2k-1} A_{2k-1} A_{n-2k}$. Similarly, the number of up-down permutations with the smallest element 1 in position $2k+1$ is $\binom{n-1}{2k} A_{2k} A_{n-2k-1}$, because down-up permutations and up-down permutations are equally numerous. Summing over all possibilities gives

$$2A_n = \sum_k \binom{n-1}{k} A_k A_{n-1-k} + 2[n=0] + [n=1].$$

The egf $\hat{A}$ therefore satisfies $2\hat{A}'(z) = \hat{A}(z)^2 + 1$ and $\hat{A}(0) = 1$; the given function solves this differential equation.

7.42 Let a_n be the number of Martian DNA strings that don't end with c or e; let b_n be the number that do. Then

$$\begin{aligned} a_n &= 3a_{n-1} + 2b_{n-1} + [n=0], & b_n &= 2a_{n-1} + b_{n-1};\\ A(z) &= 3zA(z) + 2zB(z) + 1, & B(z) &= 2zA(z) + zB(z);\\ A(z) &= \frac{1-z}{1-4z-z^2}, & B(z) &= \frac{2z}{1-4z-z^2}; \end{aligned}$$

and the total number is $[z^n](1+z)/(1-4z-z^2) = F_{3n+2}$.

7.43 By (5.45), $g_n = \Delta^n \dot{G}(0)$. The nth difference of a product can be written

$$\Delta^n A(z)B(z) = \sum_k \binom{n}{k} (\Delta^k E^{n-k} A(z))(\Delta^{n-k} B(z)),$$

and $E^{n-k} = (1+\Delta)^{n-k} = \sum_j \binom{n-k}{j} \Delta^j$. Therefore we find

$$h_n = \sum_{j,k} \binom{n}{k}\binom{n-k}{j} f_{j+k}\, g_{n-k}.$$

This is a sum over all trinomial coefficients; it can be put into the more symmetric form

$$h_n = \sum_{j+k+l=n} \binom{n}{j,k,l} f_{j+k}\, g_{k+l}.$$

The empty set is pointless.

7.44 Each partition into k nonempty subsets can be ordered in k! ways, so $b_k = k!$. Thus $\hat{Q}(z) = \sum_{n,k\geq 0} {n \brace k} k!\, z^n/n! = \sum_{k\geq 0}(e^z-1)^k = 1/(2-e^z)$. And this is the geometric series $\sum_{k\geq 0} e^{kz}/2^{k+1}$, hence $a_k = 1/2^{k+1}$. Finally, $c_k = 2^k$; consider all permutations when the x's are distinct, change each '>' between subscripts to '<' and allow each '<' between subscripts to become either '<' or '='. (For example, the permutation $x_1x_3x_2$ produces $x_1 < x_3 < x_2$ and $x_1 = x_3 < x_2$, because $1 < 3 > 2$.)

7.45 This sum is $\sum_{n\geq 1} r(n)/n^2$, where $r(n)$ is the number of ways to write n as a product of relatively prime factors. If n is divisible by t distinct primes, $r(n) = 2^t$. Hence $r(n)/n^2$ is multiplicative and the sum is

$$\prod_p \left(1 + \frac{2}{p^2} + \frac{2}{p^4}\cdots\right) = \prod_p \left(1 + \frac{2}{p^2-1}\right)$$
$$= \prod_p \left(\frac{p^2+1}{p^2-1}\right) = \zeta(2)^2/\zeta(4) = \frac{5}{2}.$$

7.46 Let $S_n = \sum_{0\leq k\leq n/2}\binom{n-2k}{k}\alpha^k$. Then $S_n = S_{n-1} + \alpha S_{n-3} + [n=0]$, and the generating function is $1/(1-z-\alpha z^3)$. When $\alpha = -\frac{4}{27}$, the hint tells us that this has a nice factorization $1/(1+\frac{1}{3}z)(1-\frac{2}{3}z)^2$. The general expansion theorem now yields $S_n = (\frac{2}{3}n + c)(\frac{2}{3})^n + \frac{1}{9}(-\frac{1}{3})^n$, and the remaining constant c turns out to be $\frac{8}{9}$.

7.47 The Stern–Brocot representation of $\sqrt{3}$ is $R(LR^2)^\infty$, because

$$\sqrt{3}+1 = 2 + \cfrac{1}{1 + \cfrac{1}{\sqrt{3}+1}}.$$

The fractions are $\frac{1}{1}, \frac{2}{1}, \frac{3}{2}, \frac{5}{3}, \frac{7}{4}, \frac{12}{7}, \frac{19}{11}, \frac{26}{15}, \ldots$; they eventually have the cyclic pattern

$$\frac{V_{2n-1}+V_{2n+1}}{U_{2n}},\quad \frac{U_{2n}+V_{2n+1}}{V_{2n+1}},\quad \frac{U_{2n+2}+V_{2n-1}}{U_{2n}+V_{2n+1}},\quad \frac{V_{2n+1}+V_{2n+3}}{U_{2n+2}},\quad \ldots.$$

7.48 We have $g_0 = 0$, and if $g_1 = m$ the generating function satisfies

$$aG(z) + bz^{-1}G(z) + cz^{-2}\bigl(G(z) - mz\bigr) + \frac{d}{1-z} = 0.$$

Hence $G(z) = P(z)/(az^2+bz+c)(1-z)$ for some polynomial $P(z)$. Let ρ_1 and ρ_2 be the roots of cz^2+bz+a, with $|\rho_1| \geq |\rho_2|$. If $b^2-4ac \leq 0$ then $|\rho_1|^2 = \rho_1\rho_2 = a/c$ is rational, contradicting the fact that $\sqrt[n]{g_n}$ approaches

$1+\sqrt{2}$. Hence $\rho_1 = (-b+\sqrt{b^2-4ca})/2c = 1+\sqrt{2}$; and this implies that $a=-c$, $b=-2c$, $\rho_2 = 1-\sqrt{2}$. The generating function now takes the form

$$\begin{aligned} G(z) &= \frac{z\bigl(m-(r+m)z\bigr)}{(1-2z-z^2)(1-z)} \\ &= \frac{-r+(m+2r)z}{2(1-2z-z^2)} + \frac{r}{2(1-z)} \;=\; mz+(2m-r)z^2+\cdots\,, \end{aligned}$$

where $r = d/c$. Since g_2 is an integer, r is an integer. We also have

$$g_n \;=\; \alpha(1+\sqrt{2}\,)^n + \hat{\alpha}(1-\sqrt{2}\,)^n + \tfrac{1}{2}r \;=\; \lfloor \alpha(1+\sqrt{2}\,)^n \rfloor\,,$$

and this can hold only if $r=-1$, because $(1-\sqrt{2}\,)^n$ alternates in sign as it approaches zero. Hence $(a,b,c,d) = \pm(1,2,-1,1)$. Now we find $\alpha = \frac{1}{4}(1+\sqrt{2}\,m)$, which is between 0 and 1 only if $0 \leqslant m \leqslant 2$. Each of these values actually gives a solution; the sequences $\langle g_n \rangle$ are $\langle 0,0,1,3,8,\ldots\rangle$, $\langle 0,1,3,8,20,\ldots\rangle$, and $\langle 0,2,5,13,32,\ldots\rangle$.

7.49 (a) The denominator of $\bigl(1/(1-(1+\sqrt{2})z) + 1/(1-(1-\sqrt{2})z)\bigr)$ is $1-2z-z^2$; hence $a_n = 2a_{n-1}+a_{n-2}$ for $n \geqslant 2$. (b) True because a_n is even and $-1 < 1-\sqrt{2} < 0$. (c) Let

$$b_n \;=\; \left(\frac{p+\sqrt{q}}{2}\right)^n + \left(\frac{p-\sqrt{q}}{2}\right)^n.$$

We would like b_n to be odd for all $n>0$, and $-1 < (p-\sqrt{q})/2 < 0$. Working as in part (a), we find $b_0 = 2$, $b_1 = p$, and $b_n = pb_{n-1} + \frac{1}{4}(q-p^2)b_{n-2}$ for $n \geqslant 2$. One satisfactory solution has $p=3$ and $q=17$.

7.50 Extending the multiplication idea of exercise 22, we have

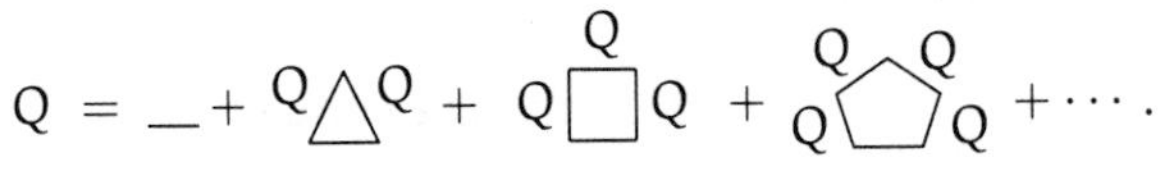

$$Q \;=\; \_\_ + Q\triangle Q + Q\,\square\,Q + Q\,\pentagon\,Q + \cdots\,.$$

Replace each n-gon by z^{n-2}. This substitution behaves properly under multiplication, because the pasting operation takes an m-gon and an n-gon into an $(m+n-2)$-gon. Thus the generating function is

$$Q \;=\; 1+zQ^2+z^2Q^3+z^3Q^4+\cdots \;=\; 1+\frac{zQ^2}{1-zQ}$$

and the quadratic formula gives $Q = \bigl(1+z-\sqrt{1-6z+z^2}\,\bigr)/2z$. The coefficient of z^{n-2} in this power series is the number of ways to put nonoverlapping diagonals into a convex n-gon. These coefficients apparently have no closed form in terms of other quantities that we have discussed in this book, but their asymptotic behavior is known [173, exercise 2.2.1–12].

Give me Legendre polynomials and I'll give you a closed form.

Incidentally, if each n-gon in Q is replaced by wz^{n-2} we get

$$Q = \frac{1+z-\sqrt{1-(4w+2)z+z^2}}{2(1+w)z},$$

a formula in which the coefficient of $w^m z^{n-2}$ is the number of ways to divide an n-gon into m polygons by nonintersecting diagonals.

7.51 The key first step is to observe that the square of the number of ways is the number of cycle patterns of a certain kind, generalizing exercise 27. These can be enumerated by evaluating the determinant of a matrix whose eigenvalues are not difficult to determine. When $m = 3$ and $n = 4$, the fact that $\cos 36° = \phi/2$ is helpful (exercise 6.46).

7.52 The first few cases are $p_0(y) = 1$, $p_1(y) = y$, $p_2(y) = y^2 + y$, $p_3(y) = y^3 + 3y^2 + 3y$. Let $p_n(y) = q_{2n}(x)$ where $y = x(1-x)$; we seek a generating function that defines $q_{2n+1}(x)$ in a convenient way. One such function is $\sum_n q_n(x)z^n/n! = 2e^{ixz}/(e^{iz}+1)$, from which it follows that $q_n(x) = i^n E_n(x)$, where $E_n(x)$ is called an Euler polynomial. We have $\sum(-1)^x x^n\,\delta x = \frac{1}{2}(-1)^{x+1}E_n(x)$, so Euler polynomials are analogous to Bernoulli polynomials, and they have factors analogous to those in (6.98). By exercise 6.23 we have $nE_{n-1}(x) = \sum_{k=0}^{n}\binom{n}{k}B_k x^{n-k}(2-2^{k+1})$; this polynomial has integer coefficients by exercise 6.54. Hence $q_{2n}(x)$, whose coefficients have denominators that are powers of 2, must have integer coefficients. Hence $p_n(y)$ has integer coefficients. Finally, the relation $(4y-1)p_n''(y) + 2p_n'(y) = 2n(2n-1)p_{n-1}(y)$ shows that

$$2m(2m-1)\left|{n \atop m}\right| = m(m+1)\left|{n \atop m+1}\right| + 2n(2n-1)\left|{n-1 \atop m-1}\right|,$$

and it follows that the $\left|{n \atop m}\right|$'s are positive. (A similar proof shows that the related quantity $(-1)^n(2n+2)E_{2n+1}(x)/(2x-1)$ has positive integer coefficients, when expressed as an nth degree polynomial in y.) It can be shown that $\left|{n \atop 1}\right|$ is the Genocchi number $(-1)^{n-1}(2^{2n+1}-2)B_{2n}$ (see exercise 6.24), and that $\left|{n \atop n-1}\right| = \binom{n}{2}$, $\left|{n \atop n-2}\right| = 2\binom{n+1}{4} + 3\binom{n}{4}$, etc.

7.53 It is $P_{(1+V_{4n+1}+V_{4n+3})/6}$. Thus, for example, $T_{20} = P_{12} = 210$; $T_{285} = P_{165} = 40755$.

7.54 Let E_k be the operation on power series that sets all coefficients to zero except those of z^n where $n \bmod m = k$. The stated construction is equivalent to the operation

$$E_0\,S\,E_0\,S\,(E_0+E_1)\,S\,\ldots\,S\,(E_0+E_1+\cdots+E_{m-1})$$

applied to $1/(1-z)$, where S means "multiply by $1/(1-z)$." There are $m!$ terms

$$E_0\, S\, E_{k_1}\, S\, E_{k_2}\, S \,\ldots\, S\, E_{k_m}$$

where $0 \leqslant k_j < j$, and every such term evaluates to $z^{rm}/(1-z^m)^{m+1}$ if r is the number of places where $k_j < k_{j+1}$. Exactly $\left\langle {m \atop r} \right\rangle$ terms have a given value of r, so the coefficient of z^{mn} is $\sum_{r=0}^{m-1} \left\langle {m \atop r} \right\rangle \binom{n+m-r}{m} = (n+1)^m$ by (6.37). (The fact that operation E_k can be expressed with complex roots of unity seems to be of no help in this problem.)

7.55 Suppose that $P_0(z)F(z) + \cdots + P_m(z)F^{(m)}(z) = Q_0(z)G(z) + \cdots + Q_n(z)G^{(n)}(z) = 0$, where $P_m(z)$ and $Q_n(z)$ are nonzero. (a) Let $H(z) = F(z) + G(z)$. Then there are rational functions $R_{k,l}(z)$ for $0 \leqslant l < m+n$ such that $H^{(k)}(z) = R_{k,0}(z)F^{(0)}(z) + \cdots + R_{k,m-1}(z)F^{(m-1)}(z) + R_{k,m}(z)G^{(0)}(z) + \cdots + R_{k,m+n-1}(z)G^{(n-1)}(z)$. The $m+n+1$ vectors $\bigl(R_{k,0}(z), \ldots, R_{k,m+n-1}(z)\bigr)$ are linearly dependent in the $(m+n)$-dimensional vector space whose components are rational functions; hence there are rational functions $S_l(z)$, not all zero, such that $S_0(z)H^{(0)}(z) + \cdots + S_{m+n}(z)H^{(m+n)}(z) = 0$. (b) Similarly, let $H(z) = F(z)G(z)$. There are rational $R_{k,l}(z)$ for $0 \leqslant l < mn$ with $H^{(k)}(z) = \sum_{i=0}^{m-1} \sum_{j=0}^{n-1} R_{k,ni+j}(z)F^{(i)}(z)G^{(j)}(z)$, hence $S_0(z)H^{(0)}(z) + \cdots + S_{mn}(z)H^{(mn)}(z) = 0$ for some rational $S_l(z)$, not all zero. (A similar proof shows that if $\langle f_n \rangle$ and $\langle g_n \rangle$ are polynomially recursive, so are $\langle f_n + g_n \rangle$ and $\langle f_n g_n \rangle$. Incidentally, there is no similar result for quotients; for example, $\cos z$ is differentiably finite, but $1/\cos z$ is not.)

7.56 Euler showed, incidentally, that this number is also $[z^n]\, 1/\sqrt{1-2z-3z^2}$, and he gave the formula $a_n = \sum_{k\geqslant 0} n^{\underline{2k}}/k!^2$. He also discovered a "memorable failure of induction" while examining these numbers: Although $3a_n - a_{n+1}$ is equal to $F_{n-1}(F_{n-1}+1)$ for $0 \leqslant n < 9$, this empirical law mysteriously breaks down when n is 9 or more!

7.57 (Paul Erdős currently offers \$500 for a solution.)

8.1 $\frac{1}{24} + \frac{1}{48} + \frac{1}{48} + \frac{1}{48} + \frac{1}{48} + \frac{1}{24} = \frac{1}{6}$. (In fact, we *always* get doubles with probability $\frac{1}{6}$ when at least one of the dice is fair.) Any two faces whose sum is 7 have the same probability in distribution $\Pr_1$, so $S = 7$ has the same probability as doubles.

8.2 There are 12 ways to specify the top and bottom cards and 50! ways to arrange the others; so the probability is $12 \cdot 50!/52! = 12/(51 \cdot 52) = \frac{1}{17 \cdot 13} = \frac{1}{221}$.

8.3 $\frac{1}{10}(3+2+\cdots+9+2) = 4.8$; $\frac{1}{9}(3^2+2^2+\cdots+9^2+2^2-10(4.8)^2) = \frac{388}{45} \approx 8.6$. The true mean and variance with a fair coin are 6 and 22, so Stanford had an unusually heads-up class. The corresponding Princeton figures are 6.4 and

$\frac{562}{45} \approx 12.5$. (This distribution has $\kappa_4 = 2974$, which is rather large. Hence the standard deviation of this variance estimate when $n = 10$ is also rather large, $\sqrt{2974/10 + 2(22)^2/9} \approx 20.1$ according to exercise 54. One cannot complain that the students cheated.)

8.4 This follows from (8.38) and (8.39), because $F(z) = G(z)H(z)$. (A similar formula holds for all the cumulants, even though $F(z)$ and $G(z)$ may have negative coefficients.)

8.5 Replace H by p and T by $q = 1 - p$. If $S_A = S_B = \frac{1}{2}$ we have $p^2qN = \frac{1}{2}$ and $pq^2N = \frac{1}{2}q + \frac{1}{2}$; the solution is $p = 1/\phi^2$, $q = 1/\phi$.

8.6 In this case $X|y$ has the same distribution as X, for all y, hence $E(X|Y) = EX$ is constant and $V\big(E(X|Y)\big) = 0$. Also $V(X|Y)$ is constant and equal to its expected value.

8.7 We have $1 = (p_1 + p_2 + \cdots + p_6)^2 \leqslant 6(p_1^2 + p_2^2 + \cdots + p_6^2)$ by Chebyshev's summation inequality of Chapter 2.

8.8 Let $p = \Pr(\omega \in A \cap B)$, $q = \Pr(\omega \notin A)$, and $r = \Pr(\omega \notin B)$. Then $p + q + r = 1$, and the identity to be proved is $p = (p + r)(p + q) - qr$.

8.9 This is true (subject to the obvious proviso that F and G are defined on the respective ranges of X and Y), because

$$\begin{aligned}\Pr\big(F(X) = f \text{ and } G(Y) = g\big) &= \sum_{\substack{x \in F^{-1}(f) \\ y \in G^{-1}(g)}} \Pr(X = x \text{ and } Y = y) \\ &= \sum_{\substack{x \in F^{-1}(f) \\ y \in G^{-1}(g)}} \Pr(X = x) \cdot \Pr(Y = y) \\ &= \Pr\big(F(X) = f\big) \cdot \Pr\big(G(y) = g\big).\end{aligned}$$

8.10 Two. Let $x_1 < x_2$ be medians; then $1 \leqslant \Pr(X \leqslant x_1) + \Pr(X \geqslant x_2) \leqslant 1$, hence equality holds. (Some discrete distributions have no median elements. For example, let Ω be the set of all fractions of the form $\pm 1/n$, with $\Pr(+1/n) = \Pr(-1/n) = \frac{\pi^2}{12} n^{-2}$.)

8.11 For example, let $K = k$ with probability $6/(k+1)(k+2)(k+3)$, for all integers $k \geqslant 0$. Then $EK = 3$, but $E(K^2) = \infty$. (Similarly we can construct random variables with finite cumulants through κ_m but with $\kappa_{m+1} = \infty$.)

8.12 Let X be the total number of heads when we toss a fair coin n times. The mean and standard deviation of X are $\mu = \frac{1}{2}n$ and $\sigma = \frac{1}{2}\sqrt{n}$. The stated sum is $\frac{1}{2}\sum_{|k - n/2| \geqslant (1/2 - \alpha)n} \binom{n}{k} = \frac{1}{2} \cdot 2^n \Pr\big(|X - \mu| \geqslant (1 - 2\alpha)\sqrt{n}\,\sigma\big)$. (A more precise estimate is obtained in exercise 9.42.)

8.13 (Solution by Boris Pittel.) Let us set $Y = (X_1 + \cdots + X_n)/n$ and $Z = (X_{n+1} + \cdots + X_{2n})/n$. Then

$$\begin{aligned}\Pr\Bigl(\Bigl|\frac{Y+Z}{2} - \alpha\Bigr| \leqslant |Y-\alpha|\Bigr) \\ &\geqslant \Pr\Bigl(\Bigl|\frac{Y-\alpha}{2}\Bigr| + \Bigl|\frac{Z-\alpha}{2}\Bigr| \leqslant |Y-\alpha|\Bigr) \\ &= \Pr\bigl(|Z-\alpha| \leqslant |Y-\alpha|\bigr) \;\geqslant\; \tfrac12 .\end{aligned}$$

The last inequality is, in fact, '$>$' in any discrete probability distribution, because $\Pr(Y=Z) > 0$.

8.14 $\mathrm{Mean}(H) = p\,\mathrm{Mean}(F) + q\,\mathrm{Mean}(G)$; $\mathrm{Var}(H) = p\,\mathrm{Var}(F) + q\,\mathrm{Var}(G) - pq\bigl(\mathrm{Mean}(F) - \mathrm{Mean}(G)\bigr)^2$. (A mixture is actually a special case of conditional probabilities: Let Y be the coin, let $X|\mathrm{H}$ be generated by $F(z)$, and let $X|\mathrm{T}$ be generated by $G(z)$. Then $VX = EV(X|Y) + VE(X|Y)$, where $EV(X|Y) = pV(X|\mathrm{H}) + qV(X|\mathrm{T})$ and $VE(X|Y)$ is the variance of $pz^{\mathrm{Mean}(F)} + qz^{\mathrm{Mean}(G)}$.)

8.15 By the chain rule, $H'(z) = G'(z)F'\bigl(G(z)\bigr)$; $H''(z) = G''(z)F'\bigl(G(z)\bigr) + G'(z)^2F''\bigl(G(z)\bigr)$. Hence

$$\begin{aligned}\mathrm{Mean}(H) &= \mathrm{Mean}(F)\,\mathrm{Mean}(G)\,;\\ \mathrm{Var}(H) &= \mathrm{Var}(F)\,\mathrm{Mean}(G)^2 + \mathrm{Mean}(F)\,\mathrm{Var}(G)\,.\end{aligned}$$

(The random variable corresponding to probability distribution H can be understood as follows: Determine a nonnegative integer n by distribution F; then add the values of n independent random variables that have distribution G. The identity for variance in this exercise is a special case of (8.105), when X has distribution H and Y has distribution F.)

8.16 $e^{w(z-1)}/(1-w)$.

8.17 $\Pr(Y_{n,p} \leqslant m) = \Pr(Y_{n,p} + n \leqslant m + n)$ = probability that we need $\leqslant m+n$ tosses to obtain n heads = probability that $m+n$ tosses yield $\geqslant n$ heads = $\Pr(X_{m+n,p} \geqslant n)$. Thus

$$\begin{aligned}\sum_{k\leqslant m}\binom{n+k-1}{k}p^nq^k &= \sum_{k\geqslant n}\binom{m+n}{k}p^kq^{m+n-k}\\ &= \sum_{k\leqslant m}\binom{m+n}{k}p^{m+n-k}q^k\,;\end{aligned}$$

and this is (5.19) with $n = r$, $x = q$, $y = p$.

8.18 (a) $G_X(z) = e^{\mu(z-1)}$. (b) The mth cumulant is μ, for all $m \geqslant 1$. (The case $\mu = 1$ is called F_∞ in (8.55).)

8.19 (a) $G_{X_1+X_2}(z) = G_{X_1}(z)G_{X_2}(z) = e^{(\mu_1+\mu_2)(z-1)}$. Hence the probability is $e^{\mu_1+\mu_2}(\mu_1+\mu_2)^n/n!$; the sum of independent Poisson variables is Poisson. (b) In general, if $K_m X$ denotes the mth cumulant of a random variable X, we have $K_m(aX_1 + bX_2) = a^m(K_m X_1) + b^m(K_m X_2)$, when $a, b \geqslant 0$. Hence the answer is $2^m\mu_1 + 3^m\mu_2$.

8.20 The general pgf will be $G(z) = z^m/F(z)$, where

$$\begin{aligned} F(z) &= z^m + (1-z)\sum_{k=1}^{m} \widetilde{A}_{(k)}[A^{(k)} = A_{(k)}]z^{m-k}\,, \\ F'(1) &= m - \sum_{k=1}^{m} \widetilde{A}_{(k)}[A^{(k)} = A_{(k)}]\,, \\ F''(1) &= m(m-1) - 2\sum_{k=1}^{m}(m-k)\widetilde{A}_{(k)}[A^{(k)} = A_{(k)}]\,. \end{aligned}$$

8.21 This is $\sum_{n\geqslant 0} q_n$, where q_n is the probability that the game between Alice and Bill is still incomplete after n flips. Let p_n be the probability that the game ends at the nth flip; then $p_n + q_n = q_{n-1}$. Hence the average time to play the game is $\sum_{n\geqslant 1} np_n = (q_0 - q_1) + 2(q_1 - q_2) + 3(q_2 - q_3) + \cdots = q_0 + q_1 + q_2 + \cdots = N$, since $\lim_{n\to\infty} nq_n = 0$.

Another way to establish this answer is to replace H and T by $\frac{1}{2}z$. Then the derivative of the first equation in (8.78) tells us that $N(1) + N'(1) = N'(1) + S_A'(1) + S_B'(1)$.

By the way, $N = \frac{16}{3}$.

8.22 By definition we have $V(X|Y) = E(X^2|Y) - \big(E(X|Y)\big)^2$ and $V\big(E(x|Y)\big) = E\big((E(X|Y))^2\big) - \big(E\big(E(X|Y)\big)\big)^2$; hence $E\big(V(X|Y)\big) + V\big(E(X|Y)\big) = E\big(E(X^2|Y)\big) - \big(E\big(E(X|Y)\big)\big)^2$. But $E\big(E(X|Y)\big) = EX$ and $E\big(E(X^2|Y)\big) = E(X^2)$, so the result is just VX.

8.23 Let $\Omega_0 = \{$ ⚀, ⚅ $\}^2$ and $\Omega_1 = \{$ ⚁, ⚂, ⚃, ⚄ $\}^2$; and let Ω_2 be the other 16 elements of Ω. Then $\Pr_{11}(\omega) - \Pr_{00}(\omega) = \frac{+20}{576}, \frac{-7}{576}, \frac{+2}{576}$ according as $\omega \in \Omega_0, \Omega_1, \Omega_2$. The events A must therefore be chosen with k_j elements from Ω_j, where (k_0, k_1, k_2) is one of the following: $(0,0,0)$, $(0,2,7)$, $(0,4,14)$, $(1,4,4)$, $(1,6,11)$, $(2,6,1)$, $(2,8,8)$, $(3,8,15)$, $(3,10,5)$, $(3,12,12)$, $(4,12,2)$, $(4,14,9)$, $(4,16,16)$. For example, there are $\binom{4}{2}\binom{16}{6}\binom{16}{1}$ events of type $(2,6,1)$. The total number of such events is $[z^0](1+z^{20})^4(1+z^{-7})^{16}(1+z^2)^{16}$, which turns out to be 1304927002. If we restrict ourselves to events that depend on S only, we get 40 solutions $S \in A$, where $A = \emptyset$, $\{\genfrac{}{}{0pt}{}{2}{12}, \genfrac{}{}{0pt}{}{4}{10}, \genfrac{}{}{0pt}{}{6}{8}\}$, $\{\genfrac{}{}{0pt}{}{2}{12}, 5, 9\}$, $\{2, 12, \genfrac{}{}{0pt}{}{4}{10}, \genfrac{}{}{0pt}{}{6}{8}, 5, 9\}$, $\{2,4,6,8,10,12\}$, $\{\genfrac{}{}{0pt}{}{3}{11}, 7, \genfrac{}{}{0pt}{}{5}{9}, 4, 10\}$, and the complements of these sets. (Here the notation '$\genfrac{}{}{0pt}{}{2}{12}$' means either 2 or 12 but not both.)

8.24 (a) Any one of the dice ends up in J's possession with probability $p = \frac{1}{6} + (\frac{5}{6})^2 p$; hence $p = \frac{6}{11}$. Let $q = \frac{5}{11}$. Then the pgf for J's total holdings is $(q + pz)^{2n+1}$, with mean $(2n+1)p$ and variance $(2n+1)pq$, by (8.61). (b) $\binom{5}{3}p^3q^2 + \binom{5}{4}p^4q + \binom{5}{5}p^5 = \frac{94176}{161051} \approx .585$.

8.25 The pgf for the current stake after n rolls is $G_n(z)$, where

$$G_0(z) = z^A;$$
$$G_n(z) = \sum_{k=1}^{6} G_{n-1}(z^{2(k-1)/5})/6, \qquad \text{for } n > 0.$$

(The noninteger exponents cause no trouble.) It follows that $\text{Mean}(G_n) = \text{Mean}(G_{n-1})$, and $\text{Var}(G_n) + \text{Mean}(G_n)^2 = \frac{22}{15}(\text{Var}(G_{n-1}) + \text{Mean}(G_{n-1})^2)$. So the mean is always A, but the variance grows to $((\frac{22}{15})^n - 1)A^2$.

This problem can perhaps be solved more easily without generating functions than with them.

8.26 The pgf $F_{l,n}(z)$ satisfies $F'_{l,n}(z) = F_{l,n-l}(z)/l$; hence $\text{Mean}(F_{l,n}) = F'_{l,n}(1) = [n \geqslant l]/l$ and $F''_{l,n}(1) = [n \geqslant 2l]/l^2$; the variance is easily computed. (In fact, we have

$$F_{l,n}(z) = \sum_{0 \leqslant k \leqslant n/l} \frac{1}{k!}\left(\frac{z-1}{l}\right)^k,$$

which approaches a Poisson distribution with mean $1/l$ as $n \to \infty$.)

8.27 $(n^2\Sigma_3 - 3n\Sigma_2\Sigma_1 + 2\Sigma_1^3)/n(n-1)(n-2)$ has the desired mean, where $\Sigma_k = X_1^k + \cdots + X_n^k$. This follows from the identities

$$E\Sigma_3 = n\mu_3;$$
$$E(\Sigma_2\Sigma_1) = n\mu_3 + n(n-1)\mu_2\mu_1;$$
$$E(\Sigma_1^3) = n\mu_3 + 3n(n-1)\mu_2\mu_1 + n(n-1)(n-2)\mu_1^3.$$

Incidentally, the third cumulant is $\kappa_3 = E((X-EX)^3)$, but the fourth cumulant does not have such a simple expression; we have $\kappa_4 = E((X-EX)^4) - 3(VX)^2$.

8.28 (The exercise implicitly calls for $p = q = \frac{1}{2}$, but the general answer is given here for completeness.) Replace H by pz and T by qz, getting $S_A(z) = p^2qz^3/(1-pz)(1-qz)(1-pqz^2)$ and $S_B(z) = pq^2z^3/(1-qz)(1-pqz^2)$. The pgf for the conditional probability that Alice wins at the nth flip, given that she wins the game, is

$$\frac{S_A(z)}{S_A(1)} = z^3 \cdot \frac{q}{1-pz} \cdot \frac{p}{1-qz} \cdot \frac{1-pq}{1-pqz^2}.$$

This is a product of pseudo-pgf's, whose mean is $3+p/q+q/p+2pq/(1-pq)$. The formulas for Bill are the same but without the factor $q/(1-pz)$, so Bill's mean is $3 + q/p + 2pq/(1-pq)$. When $p = q = \frac{1}{2}$, the answer in case (a) is

$\frac{17}{3}$; in case (b) it is $\frac{14}{3}$. Bill wins only half as often, but when he does win he tends to win sooner. The overall average number of flips is $\frac{2}{3}\cdot\frac{17}{3}+\frac{1}{3}\cdot\frac{14}{3}=\frac{16}{3}$, agreeing with exercise 21. The solitaire game for each pattern has a waiting time of 8.

8.29 Set $\mathtt{H}=\mathtt{T}=\frac{1}{2}$ in

$$\begin{aligned}1+N(\mathtt{H}+\mathtt{T}) &= N+S_A+S_B+S_C\\ N\,\mathtt{HHTH} &= S_A(1+\mathtt{HTH})+S_B(\mathtt{HTH}+\mathtt{TH}+1)+S_C(\mathtt{HTH}+\mathtt{TH})\\ N\,\mathtt{HTHH} &= S_A(\mathtt{THH}+\mathtt{H})+S_B(\mathtt{THH}+1)+S_C(\mathtt{THH})\\ N\,\mathtt{THHH} &= S_A(\mathtt{HH})+S_B(\mathtt{HH})+S_C\end{aligned}$$

to get the winning probabilities. In general we will have $S_A+S_B+S_C=1$ and

$$\begin{aligned}S_A(A{:}A)+S_B(B{:}A)+S_C(C{:}A) &= S_A(A{:}B)+S_B(B{:}B)+S_C(C{:}B)\\ &= S_A(A{:}B)+S_B(B{:}C)+S_C(C{:}C)\,.\end{aligned}$$

In particular, the equations $9S_A+3S_B+3S_C=5S_A+9S_B+S_C=2S_A+4S_B+9S_C$ imply that $S_A=\frac{16}{52}$, $S_B=\frac{17}{52}$, $S_C=\frac{19}{52}$.

8.30 The variance of $P(h_1,\ldots,h_n;k)|k$ is the variance of the shifted binomial distribution $\bigl((m-1+z)/m\bigr)^{k-1}z$, which is $(k-1)(\frac{1}{m})(1-\frac{1}{m})$ by (8.61). Hence the average of the variance is $\mathrm{Mean}(S)(m-1)/m^2$. The variance of the average is the variance of $(k-1)/m$, namely $\mathrm{Var}(S)/m^2$. According to (8.105), the sum of these two quantities should be VP, and it is. Indeed, we have just replayed the derivation of (8.95) in slight disguise. (See exercise 15.)

8.31 (a) A brute force solution would set up five equations in five unknowns: $A=\frac{1}{2}zB+\frac{1}{2}zE$; $B=\frac{1}{2}zC$; $C=1+\frac{1}{2}zB+\frac{1}{2}zD$; $D=\frac{1}{2}zC+\frac{1}{2}zE$; $E=\frac{1}{2}zD$. But positions C and D are equidistant from the goal, as are B and E, so we can lump them together. If $X=B+E$ and $Y=C+D$, there are now three equations:

$$A=\tfrac{1}{2}zX;\quad X=\tfrac{1}{2}zY;\quad Y=1+\tfrac{1}{2}zX+\tfrac{1}{2}zY\,.$$

Hence $A=z^2/(4-2z-z^2)$; we have $\mathrm{Mean}(A)=6$ and $\mathrm{Var}(A)=22$. (Rings a bell? In fact, this problem is equivalent to flipping a fair coin until getting heads twice in a row: Heads means "advance toward the apple" and tails means "go back.") (b) Chebyshev's inequality says that $\Pr(S\geqslant 100)=\Pr\bigl((S-6)^2\geqslant 94^2\bigr)\leqslant 22/94^2\approx .0025$. (The actual probability is approximately 0.0000000009, according to exercise 37. Chebyshev's inequality often gives an overestimate; that's the price we pay for its being so general and so easy to prove.)

8.32 By symmetry, we can reduce each month's situation to one of four possibilities:

> "Toto, I have a feeling we're not in Kansas anymore."
> —Dorothy

D, the states are diagonally opposite;
A, the states are adjacent and not Kansas;
K, the states are Kansas and one other;
S, the states are the same.

Considering the Markovian transitions, we get four equations

$$\begin{aligned} D &= 1 + z(\tfrac{2}{9}D + \tfrac{2}{12}K) \\ A &= z(\tfrac{4}{9}A + \tfrac{4}{12}K) \\ K &= z(\tfrac{4}{9}D + \tfrac{4}{9}A + \tfrac{4}{12}K) \\ S &= z(\tfrac{3}{9}D + \tfrac{1}{9}A + \tfrac{2}{12}K) \end{aligned}$$

whose sum is $D + K + A + S = 1 + z(D + A + K)$. The solution is

$$S = \frac{81z - 45z^2 - 4z^3}{243 - 243z + 24z^2 + 8z^3},$$

but the simplest way to find the mean and variance may be to write $z = 1 + w$ and expand in powers of w, ignoring multiples of w^2:

$$\begin{aligned} D &= \tfrac{27}{16} + \tfrac{1593}{512}w + \cdots; \\ A &= \tfrac{9}{8} + \tfrac{2115}{256}w + \cdots; \\ K &= \tfrac{15}{8} + \tfrac{2661}{256}w + \cdots. \end{aligned}$$

Now $S'(1) = \frac{27}{16} + \frac{9}{8} + \frac{15}{8} = \frac{75}{16}$, and $\frac{1}{2}S''(1) = \frac{1593}{512} + \frac{2115}{256} + \frac{2661}{256} = \frac{11145}{512}$. The mean is $\frac{75}{16}$ and the variance is $\frac{105}{4}$. (Is there a simpler way?)

8.33 First answer: Clearly yes, because the hash values $h_1, \ldots, h_n$ are independent. Second answer: Certainly no, even though the hash values $h_1, \ldots, h_n$ are independent. We have $\Pr(X_j = 0) = \sum_{k=1}^{n} s_k\bigl([j \ne k](m-1)/m\bigr) = (1 - s_j)(m-1)/m$, but $\Pr(X_1 = X_2 = 0) = \sum_{k=1}^{n} s_k[k > 2](m-1)^2/m^2 = (1 - s_1 - s_2)(m-1)^2/m^2 \ne \Pr(X_1 = 0)\Pr(X_2 = 0)$.

8.34 Let $[z^n]\,S_m(z)$ be the probability that Gina has advanced $< m$ steps after taking n turns. Then $S_m(1)$ is her average score on a par-m hole; $[z^m]\,S_m(z)$ is the probability that she loses such a hole against a steady player; and $1 - [z^{m-1}]\,S_m(z)$ is the probability that she wins it. We have the recurrence

$$\begin{aligned} S_0(z) &= 0; \\ S_m(z) &= \bigl(1 + pzS_{m-2}(z) + qzS_{m-1}(z)\bigr)/(1 - rz), \qquad \text{for } m > 0. \end{aligned}$$

To solve part (a), it suffices to compute the coefficients for $m, n \leqslant 4$; it is convenient to replace z by $100w$ so that the computations involve nothing but integers. We obtain the following tableau of coefficients:

S_0	0	0	0	0	0
S_1	1	4	16	64	256
S_2	1	95	744	4432	23552
S_3	1	100	9065	104044	819808
S_4	1	100	9975	868535	12964304

Therefore Gina wins with probability $1 - .868535 = .131465$; she loses with probability $.12964304$. (b) To find the mean number of strokes, we compute

$$S_1(1) = \tfrac{25}{24}; \quad S_2(1) = \tfrac{4675}{2304}; \quad S_3(1) = \tfrac{667825}{221184}; \quad S_4(1) = \tfrac{85134475}{21233664}.$$

(Incidentally, $S_5(1) \approx 4.9995$; she wins with respect to both holes and strokes on a par-5 hole, but loses either way when par is 3.)

8.35 The condition will be true for all n if and only if it is true for $n = 1$, by the Chinese remainder theorem. One necessary and sufficient condition is the polynomial identity

$$\begin{aligned}\bigl(p_2{+}p_4{+}p_6 + (p_1{+}p_3{+}p_5)w\bigr)\bigl(p_3{+}p_6 + (p_1{+}p_4)z + (p_2{+}p_5)z^2\bigr)\\ = \bigl(p_1wz + p_2z^2 + p_3w + p_4z + p_5wz^2 + p_6\bigr),\end{aligned}$$

but that just more-or-less restates the problem. A simpler characterization is

$$(p_2 + p_4 + p_6)(p_3 + p_6) = p_6, \qquad (p_1 + p_3 + p_5)(p_2 + p_5) = p_5,$$

which checks only two of the coefficients in the former product. The general solution has three degrees of freedom: Let $a_0 + a_1 = b_0 + b_1 + b_2 = 1$, and put $p_1 = a_1b_1$, $p_2 = a_0b_2$, $p_3 = a_1b_0$, $p_4 = a_0b_1$, $p_5 = a_1b_2$, $p_6 = a_0b_0$.

8.36 (a) ⚀ ⚁ ⚁ ⚂ ⚂ ⚃. (b) If the kth die has faces with s_1, …, s_6 spots, let $p_k(z) = z^{s_1} + \cdots + z^{s_6}$. We want to find such polynomials with $p_1(z) \ldots p_n(z) = (z + z^2 + z^3 + z^4 + z^5 + z^6)^n$. The irreducible factors of this polynomial with rational coefficients are $z^n(z+1)^n(z^2+z+1)^n(z^2-z+1)^n$; hence $p_k(z)$ must be of the form $z^{a_k}(z+1)^{b_k}(z^2+z+1)^{c_k}(z^2-z+1)^{d_k}$. We must have $a_k \geqslant 1$, since $p_k(0) = 0$; and in fact $a_k = 1$, since $a_1 + \cdots + a_n = n$. Furthermore the condition $p_k(1) = 6$ implies that $b_k = c_k = 1$. It is now easy to see that $0 \leqslant d_k \leqslant 2$, since $d_k > 2$ gives negative coefficients. When $d = 0$ and $d = 2$, we get the two dice in part (a); therefore the only solutions have k pairs of dice as in (a), plus $n - 2k$ ordinary dice, for some $k \leqslant \frac{1}{2}n$.

8.37 The number of coin-toss sequences of length n is F_{n-1}, for all $n > 0$, because of the relation between domino tilings and coin flips. Therefore the probability that exactly n tosses are needed is $F_{n-1}/2^n$, when the coin is fair. Also $q_n = F_{n+1}/2^{n-1}$, since $\sum_{k\geqslant n} F_n z^n = (F_n z^n + F_{n-1}z^{n+1})/(1-z-z^2)$. (A systematic solution via generating functions is, of course, also possible.)

8.38 When k faces have been seen, the task of rolling a new one is equivalent to flipping coins with success probability $p_k = (m-k)/m$. Hence the pgf is $\prod_{k=0}^{l-1} p_k z/(1-q_k z) = \prod_{k=0}^{l-1}(m-k)z/(m-kz)$. The mean is $\sum_{k=0}^{l-1} p_k^{-1} = m(H_m - H_{m-l})$; the variance is $m^2(H_m^{(2)} - H_{m-l}^{(2)}) - m(H_m - H_{m-l})$; and equation (7.47) provides a closed form for the requested probability, namely $m^{-n} m! {n-1 \brace l-1}/(m-1)!$. (The problem discussed in this exercise is traditionally called "coupon collecting.")

8.39 $E(X) = P(-1)$; $V(X) = P(-2) - P(-1)^2$; $E(\ln X) = -P'(0)$.

8.40 (a) We have $\kappa_m = n\bigl(0!{m \brace 1}p - 1!{m \brace 2}p^2 + 2!{m \brace 3}p^3 - \cdots\bigr)$, by (7.49). Incidentally, the third cumulant is $npq(q-p)$ and the fourth is $npq(1-6pq)$. The identity $q+pe^t = (p+qe^{-t})e^t$ shows that $f_m(p) = (-1)^m f_m(q) + [m=1]$; hence we can write $f_m(p) = g_m(pq)(q-p)^{[m \text{ odd}]}$, where g_m is a polynomial of degree $\lfloor m/2 \rfloor$, whenever $m > 1$. (b) Let $p = \frac{1}{2}$ and $F(t) = \ln(\frac{1}{2} + \frac{1}{2}e^t)$. Then $\sum_{m\geqslant 1} \kappa_m t^{m-1}/(m-1)! = F'(t) = 1 - 1/(e^t+1)$, and we can use exercise 6.23.

8.41 If $G(z)$ is the pgf for a random variable X that assumes only positive integer values, then $\int_0^1 G(z)\,dz/z = \sum_{k\geqslant 1} \Pr(X=k)/k = E(X^{-1})$. If X is the distribution of the number of flips to obtain $n+1$ heads, we have $G(z) = \bigl(pz/(1-qz)\bigr)^{n+1}$ by (8.59), and the integral is

$$\int_0^1 \left(\frac{pz}{1-qz}\right)^{n+1} \frac{dz}{z} = \int_0^1 \frac{w^n\,dw}{1+(q/p)w}$$

if we substitute $w = pz/(1-qz)$. When $p = q$ the integrand can be written $(-1)^n\bigl((1+w)^{-1} - 1 + w - w^2 + \cdots + (-1)^n w^{n-1}\bigr)$, so the integral is $(-1)^n\bigl(\ln 2 - 1 + \frac{1}{2} - \frac{1}{3} + \cdots + (-1)^n/n\bigr)$. We have $H_{2n} - H_n = \ln 2 - \frac{1}{4}n^{-1} + \frac{1}{16}n^{-2} + O(n^{-4})$ by (9.28), and it follows that $E(X_{n+1}^{-1}) = \frac{1}{2}n^{-1} - \frac{1}{4}n^{-2} + O(n^{-4})$.

8.42 Let $F_n(z)$ and $G_n(z)$ be pgf's for the number of employed evenings, if the man is initially unemployed or employed, respectively. Let $q_h = 1 - p_h$ and $q_f = 1 - p_f$. Then $F_0(z) = G_0(z) = 1$, and

$$\begin{aligned} F_n(z) &= p_h z G_{n-1}(z) + q_h F_{n-1}(z)\,; \\ G_n(z) &= p_f F_{n-1}(z) + q_f z G_{n-1}(z)\,. \end{aligned}$$

The solution is given by the super generating function

$$G(w,z) \;=\; \sum_{n\geqslant 0} G_n(z)w^n \;=\; A(w)/\bigl(1-zB(w)\bigr)\,,$$

where $B(w)=w\bigl(q_f-(q_f-p_h)w\bigr)/(1-q_h w)$ and $A(w)=\bigl(1-B(w)\bigr)/(1-w)$. Now $\sum_{n\geqslant 0} G_n'(1)w^n = \alpha w/(1-w)^2+\beta/(1-w)-\beta/\bigl(1-(q_f-p_h)w\bigr)$ where

$$\alpha \;=\; \frac{p_h}{p_h+p_f}\,, \qquad \beta \;=\; \frac{p_f(q_f-p_h)}{(p_h+p_f)^2}\,;$$

hence $G_n'(1)=\alpha n+\beta\bigl(1-(q_r-p_h)^n\bigr)$. (Similarly $G_n''(1)=\alpha^2n^2+O(n)$, so the variance is $O(n)$.)

8.43 $G_n(z) \;=\; \sum_{k\geqslant 0}\left[{n \atop k}\right]z^k/n! \;=\; z^{\overline{n}}/n!$, by (6.11). This is a product of binomial pgf's, $\prod_{k=1}^{n}\bigl((k-1+z)/k\bigr)$, where the kth has mean $1/k$ and variance $(k-1)/k^2$; hence $\mathrm{Mean}(G_n)=H_n$ and $\mathrm{Var}(G_n)=H_n-H_n^{(2)}$.

8.44 (a) The champion must be undefeated in n rounds, so the answer is p^n. (b,c) Players $x_1, \ldots, x_{2^k}$ must be "seeded" (by chance) in distinct subtournaments and they must win all $2^k(n-k)$ of their matches. The 2^n leaves of the tournament tree can be filled in $2^n!$ ways; to seed it we have $2^k!(2^{n-k})^{2^k}$ ways to place the top 2^k players, and $(2^n-2^k)!$ ways to place the others. Hence the probability is $(2p)^{2^k(n-k)}/\binom{2^n}{2^k}$. When $k=1$ this simplifies to $(2p^2)^{n-1}/(2^n-1)$. (d) Each tournament outcome corresponds to a permutation of the players: Let y_1 be the champ; let y_2 be the other finalist; let y_3 and y_4 be the players who lost to y_1 and y_2 in the semifinals; let $(y_5,\ldots,y_8)$ be those who lost respectively to $(y_1,\ldots,y_4)$ in the quarterfinals; etc. (Another proof shows that the first round has $2^n!/2^{n-1}!$ essentially different outcomes; the second round has $2^{n-1}!/2^{n-2}!$; and so on.) (e) Let S_k be the set of 2^{k-1} potential opponents of x_2 in the kth round. The conditional probability that x_2 wins, given that x_1 belongs to S_k, is

$$\begin{aligned}&\Pr(x_1 \text{ plays } x_2)\cdot p^{n-1}(1-p) \;+\; \Pr(x_1 \text{ doesn't play } x_2)\cdot p^n\\ &\quad=\; p^{k-1}p^{n-1}(1-p) \;+\; (1-p^{k-1})p^n\,.\end{aligned}$$

The chance that $x_1\in S_k$ is $2^{k-1}/(2^n-1)$; summing on k gives the answer:

$$\sum_{k=1}^{n}\frac{2^{k-1}}{2^n-1}\bigl(p^{k-1}p^{n-1}(1-p)+(1-p^{k-1})p^n\bigr) \;=\; p^n-\frac{(2p)^n-1}{2^n-1}\,p^{n-1}\,.$$

(f) Each of the $2^n!$ tournament outcomes has a certain probability of occurring, and the probability that x_j wins is the sum of these probabilities over all $(2^n-1)!$ tournament outcomes in which x_j is champion. Consider interchanging x_j with x_{j+1} in all those outcomes; this change doesn't affect the probability if x_j and x_{j+1} never meet, but it multiplies the probability by $(1-p)/p < 1$ if they do meet.

8.45 (a) $A(z) = 1/(3-2z)$; $B(z) = zA(z)^2$; $C(z) = z^2A(z)^3$. The pgf for sherry when it's bottled is $z^3A(z)^3$, which is z^3 times a negative binomial distribution with parameters $n = 3$, $p = \frac{1}{3}$. (b) $\mathrm{Mean}(A) = 2$, $\mathrm{Var}(A) = 6$; $\mathrm{Mean}(B) = 5$, $\mathrm{Var}(B) = 2\,\mathrm{Var}(A) = 12$; $\mathrm{Mean}(C) = 8$, $\mathrm{Var}(C) = 18$. The sherry is nine years old, on the average. The fraction that's 25 years old is $\binom{-3}{22}(-2)^{22}3^{-25} = \binom{24}{22}2^{22}3^{-25} = 23\cdot(\frac{2}{3})^{24} \approx .00137$. (c) Let the coefficient of w^n be the pgf for the beginning of year n. Then

$$\begin{aligned} A &= \bigl(1+\tfrac{1}{3}w/(1-w)\bigr)/(1-\tfrac{2}{3}zw)\,; \\ B &= (1+\tfrac{1}{3}zwA)/(1-\tfrac{2}{3}zw)\,; \\ C &= (1+\tfrac{1}{3}zwB)/(1-\tfrac{2}{3}zw)\,. \end{aligned}$$

Differentiate with respect to z and set $z = 1$; this makes

$$C' = \frac{8}{1-w} - \frac{1/2}{(1-\frac{2}{3}w)^3} - \frac{3/2}{(1-\frac{2}{3}w)^2} - \frac{6}{1-\frac{2}{3}w}\,.$$

The average age of bottled sherry n years after the process started is 1 greater than the coefficient of w^{n-1}, namely $9-(\frac{2}{3})^n(3n^2+21n+72)/8$. (This already exceeds 8 when $n = 11$.)

8.46 (a) $P(w,z) = 1+\frac{1}{2}\bigl(wP(w,z)+zP(w,z)\bigr) = \bigl(1-\frac{1}{2}(w+z)\bigr)^{-1}$, hence $p_{mn} = 2^{-m-n}\binom{m+n}{n}$. (b) $P_k(w,z) = \frac{1}{2}(w^k+z^k)P(w,z)$; hence

$$p_{k,m,n} = 2^{k-1-m-n}\left(\binom{m+n-k}{m} + \binom{m+n-k}{n}\right).$$

(c) $\sum_k kp_{k,n,n} = \sum_{k=0}^{n} k2^{k-2n}\binom{2n-k}{n} = \sum_{k=0}^{n}(n-k)2^{-n-k}\binom{n+k}{n}$; this can be summed using (5.20):

$$\begin{aligned} &\sum_{k=0}^{n} 2^{-n-k}\left((2n+1)\binom{n+k}{n} - (n+1)\binom{n+1+k}{n+1}\right) \\ &\qquad = (2n+1) - (n+1)2^{-n}\left(2^{n+1} - 2^{-n-1}\binom{2n+2}{n+1}\right) \\ &\qquad = \frac{2n+1}{2^{2n}}\binom{2n}{n} - 1\,. \end{aligned}$$

(The methods of Chapter 9 show that this is $2\sqrt{n/\pi} - 1 + O(n^{-1/2})$.)

8.47 After n irradiations there are $n+2$ equally likely receptors. Let the random variable X_n denote the number of diphages present; then $X_{n+1} =$

$X_n + Y_n$, where $Y_n = -1$ if the $(n+1)$st particle hits a diphage receptor (conditional probability $2X_n/(n+2)$) and $Y_n = +2$ otherwise. Hence

$$EX_{n+1} = EX_n + EY_n = EX_n - 2EX_n/(n+2) + 2\bigl(1 - 2EX_n/(n+2)\bigr).$$

The recurrence $(n+2)EX_{n+1} = (n-4)EX_n + 2n + 4$ can be solved if we multiply both sides by the summation factor $(n+1)^{\underline{5}}$; or we can guess the answer and prove it by induction: $EX_n = (2n+4)/7$ for all $n > 4$. (Incidentally, there are always two diphages and one triphage after five steps, regardless of the configuration after four.)

8.48 (a) The distance between frisbees (measured so as to make it an even number) is either 0, 2, or 4 units, initially 4. The corresponding generating functions A, B, C (where, say, $[z^n]\,C$ is the probability of distance 4 after n throws) satisfy

$$A = \tfrac{1}{4}zB\,, \quad B = \tfrac{1}{2}zB + \tfrac{1}{4}zC\,, \quad C = 1 + \tfrac{1}{4}zB + \tfrac{3}{4}zC\,.$$

It follows that $A = z^2/(16 - 20z + 5z^2) = z^2/F(z)$, and we have $\mathrm{Mean}(A) = 2 - \mathrm{Mean}(F) = 12$, $\mathrm{Var}(A) = -\mathrm{Var}(F) = 100$. (A more difficult but more amusing solution factors A as follows:

$$A = \frac{p_1 z}{1 - q_1 z}\cdot\frac{p_2 z}{1 - q_2 z} = \frac{p_2}{p_2 - p_1}\frac{p_1 z}{1 - q_1 z} + \frac{p_1}{p_1 - p_2}\frac{p_2 z}{1 - q_2 z}\,,$$

where $p_1 = \phi^2/4 = (3 + \sqrt{5}\,)/8$, $p_2 = \hat{\phi}^2/4 = (3 - \sqrt{5}\,)/8$, and $p_1 + q_1 = p_2 + q_2 = 1$. Thus, the game is equivalent to having two biased coins whose heads probabilities are p_1 and p_2; flip the coins one at a time until they have both come up heads, and the total number of flips will have the same distribution as the number of frisbee throws. The mean and variance of the waiting times for these two coins are respectively $6 \mp 2\sqrt{5}$ and $50 \mp 22\sqrt{5}$, hence the total mean and variance are 12 and 100 as before.)

(b) Expanding the generating function in partial fractions makes it possible to sum the probabilities. (Note that $\sqrt{5}/(4\phi) + \phi^2/4 = 1$, so the answer can be stated in terms of powers of ϕ.) The game will last more than n steps with probability $5^{(n-1)/2}4^{-n}(\phi^{n+2} - \phi^{-n-2})$; when n is even this is $5^{n/2}4^{-n}F_{n+2}$. So the answer is $5^{50}4^{-100}F_{102} \approx .00006$.

8.49 (a) If $n > 0$, $P_N(0,n) = \frac{1}{2}[N=0] + \frac{1}{4}P_{N-1}(0,n) + \frac{1}{4}P_{N-1}(1,n-1)$; $P_N(m,0)$ is similar; $P_N(0,0) = [N=0]$. Hence

$$\begin{aligned} g_{m,n} &= \tfrac{1}{4}zg_{m-1,n+1} + \tfrac{1}{2}zg_{m,n} + \tfrac{1}{4}zg_{m+1,n-1}\,;\\ g_{0,n} &= \tfrac{1}{2} + \tfrac{1}{4}zg_{0,n} + \tfrac{1}{4}g_{1,n-1}\,; \quad \text{etc.} \end{aligned}$$

(b) $g'_{m,n} = 1 + \frac{1}{4}g'_{m-1,n+1} + \frac{1}{2}g'_{m,n} + \frac{1}{4}g'_{m+1,n-1}$; $g'_{0,n} = \frac{1}{2} + \frac{1}{4}g'_{0,n} + \frac{1}{4}g'_{1,n-1}$; etc. By induction on m, we have $g'_{m,n} = (2m+1)g'_{0,m+n} - 2m^2$ for all $m, n \geqslant 0$.

And since $g'_{m,0} = g'_{0,m}$, we must have $g'_{m,n} = m{+}n{+}2mn$. (c) The recurrence is satisfied when $mn > 0$, because

$$\sin(2m+1)\theta \;=\; \frac{1}{\cos^2\theta}\biggl(\frac{\sin(2m-1)\theta}{4} + \frac{\sin(2m+1)\theta}{2} + \frac{\sin(2m+3)\theta}{4}\biggr);$$

this is a consequence of the identity $\sin(x-y)+\sin(x+y) = 2\sin x\cos y$. So all that remains is to check the boundary conditions.

8.50 (a) Using the hint, we get

$$3(1-z)^2\sum_k \binom{1/2}{k}\Bigl(\frac{8}{9}z\Bigr)^k(1-z)^{2-k} = 3(1-z)^2\sum_k \binom{1/2}{k}\Bigl(\frac{8}{9}\Bigr)^k\sum_j\binom{k+j-3}{j}z^{j+k};$$

now look at the coefficient of z^{3+l}. (b) $H(z) = \frac{2}{3} + \frac{5}{27}z + \frac{1}{2}\sum_{l\geqslant 0} c_{3+l}z^{2+l}$. (c) Let $r = \sqrt{(1-z)(9-z)}$. One can show that $(z-3+r)(z-3-r) = 4z$, and hence that $\bigl(r/(1-z)+2\bigr)^2 = (13-5z+4r)/(1-z) = \bigl(9-H(z)\bigr)/\bigl(1-H(z)\bigr)$. (d) Evaluating the first derivative at $z=1$ shows that $\mathrm{Mean}(H) = 1$. The second derivative diverges at $z=1$, so the variance is infinite.

8.51 (a) Let $H_n(z)$ be the pgf for your holdings after n rounds of play, with $H_0(z) = z$. The distribution for n rounds is

$$H_{n+1}(z) \;=\; H_n\bigl(H(z)\bigr)\,,$$

so the result is true by induction (using the amazing identity of the preceding problem). (b) $g_n = H_n(0) - H_{n-1}(0) = 4/n(n+1)(n+2) = 4(n-1)^{\underline{-3}}$. The mean is 2, and the variance is infinite. (c) The expected number of tickets you buy on the nth round is $\mathrm{Mean}(H_n) = 1$, by exercise 15. So the total expected number of tickets is infinite. (Thus, you almost surely lose eventually, and you expect to lose after the second game, yet you also expect to buy an infinite number of tickets.) (d) Now the pgf after n games is $H_n(z)^2$, and the method of part (b) yields a mean of $16 - \frac{4}{3}\pi^2 \approx 2.8$. (The sum $\sum_{k\geqslant 1} 1/k^2 = \pi^2/6$ shows up here.)

8.52 If ω and ω' are events with $\Pr(\omega) > \Pr(\omega')$, then a sequence of n independent experiments will encounter ω more often than ω', with high probability, because ω will occur very nearly $n\Pr(\omega)$ times. Consequently, as $n \to \infty$, the probability approaches 1 that the median or mode of the values of X in a sequence of independent trials will be a median or mode of the random variable X.

8.53 We can disprove the statement, even in the special case that each variable is 0 or 1. Let $p_0 = \Pr(X=Y=Z=0)$, $p_1 = \Pr(X = Y = \overline{Z} = 0)$, ..., $p_7 = \Pr(\overline{X} = \overline{Y} = \overline{Z} = 0)$, where $\overline{X} = 1 - X$. Then $p_0 + p_1 + \cdots + p_7 = 1$, and the variables are independent in pairs if and only if we have

$$\begin{aligned}(p_4+p_5+p_6+p_7)(p_2+p_3+p_6+p_7) &= p_6+p_7\,,\\ (p_4+p_5+p_6+p_7)(p_1+p_3+p_5+p_7) &= p_5+p_7\,,\\ (p_2+p_3+p_6+p_7)(p_1+p_3+p_5+p_7) &= p_3+p_7\,.\end{aligned}$$

But $\Pr(X+Y=Z=0) \ne \Pr(X+Y=0)\Pr(Z=0) \iff p_0 \ne (p_0+p_1)(p_0+p_2+p_4+p_6)$. One solution is

$$p_0 = p_3 = p_5 = p_6 = 1/4\,; \qquad p_1 = p_2 = p_4 = p_7 = 0\,.$$

This is equivalent to flipping two fair coins and letting X = (the first coin is heads), Y = (the second coin is heads), Z = (the coins differ). Another example, with all probabilities nonzero, is

$$\begin{aligned}p_0 &= 4/64\,, \quad p_1 = p_2 = p_4 = 5/64\,,\\ p_3 &= p_5 = p_6 = 10/64\,, \quad p_7 = 15/64\,.\end{aligned}$$

For this reason we say that n variables X_1, ..., X_n are independent if

$$\Pr(X_1=x_1 \text{ and} \cdots \text{and } X_n=x_n) = \Pr(X_1=x_1)\ldots\Pr(X_n = x_n)\,;$$

pairwise independence isn't enough to guarantee this.

8.54 (See exercise 27 for notation.) We have

$$\begin{aligned}E(\Sigma_2^2) &= n\mu_4 + n(n-1)\mu_2^2\,;\\ E(\Sigma_2\Sigma_1^2) &= n\mu_4 + 2n(n-1)\mu_3\mu_1 + n(n-1)\mu_2^2 + n(n-1)(n-2)\mu_2\mu_1^2\,;\\ E(\Sigma_1^4) &= n\mu_4 + 4n(n-1)\mu_3\mu_1 + 3n(n-1)\mu_2^2\\ &\quad + 6n(n-1)(n-2)\mu_2\mu_1^2 + n(n-1)(n-2)(n-3)\mu_1^4\,;\end{aligned}$$

it follows that $V(\hat{V}X) = \kappa_4/n + 2\kappa_2^2/(n-1)$.

8.55 There are $A = \frac{1}{17}\cdot 52!$ permutations with $X = Y$, and $B = \frac{16}{17}\cdot 52!$ permutations with $X \ne Y$. After the stated procedure, each permutation with $X = Y$ occurs with probability $\frac{1}{17}/\bigl((1-\frac{16}{17}p)A\bigr)$, because we return to step S1 with probability $\frac{16}{17}p$. Similarly, each permutation with $X \ne Y$ occurs with probability $\frac{16}{17}(1-p)/\bigl((1-\frac{16}{17}p)B\bigr)$. Choosing $p = \frac{1}{4}$ makes $\Pr(X=x \text{ and } Y=y) = \frac{1}{169}$ for all x and y. (We could therefore make two flips of a fair coin and go back to S1 if both come up heads.)

8.56 If m is even, the frisbees always stay an odd distance apart and the game lasts forever. If $m = 2l + 1$, the relevant generating functions are

$$\begin{aligned} G_m &= \tfrac{1}{4}zA_1\,; \\ A_1 &= \tfrac{1}{2}zA_1 + \tfrac{1}{4}zA_2\,, \\ A_k &= \tfrac{1}{4}zA_{k-1} + \tfrac{1}{2}zA_k + \tfrac{1}{4}zA_{k+1}\,, \qquad \text{for } 1 < k < l, \\ A_l &= \tfrac{1}{4}zA_{l-1} + \tfrac{3}{4}zA_l + 1\,. \end{aligned}$$

(The coefficient $[z^n]\,A_k$ is the probability that the distance between frisbees is $2k$ after n throws.) Taking a clue from the similar equations in exercise 49, we set $z = 1/\cos^2\theta$ and $A_1 = X\sin 2\theta$, where X is to be determined. It follows by induction (not using the equation for A_l) that $A_k = X\sin 2k\theta$. Therefore we want to choose X such that

$$\Bigl(1 - \frac{3}{4\cos^2\theta}\Bigr)\,X\,\sin 2l\theta \;=\; 1 + \frac{1}{4\cos^2\theta}\,X\,\sin(2l-2)\theta\,.$$

It turns out that $X = 2\cos^2\theta/\sin\theta\cos(2l+1)\theta$, hence

$$G_m \;=\; \frac{\cos\theta}{\cos m\theta}\,.$$

The denominator vanishes when θ is an odd multiple of $\pi/(2m)$; thus $1 - q_k z$ is a root of the denominator for $1 \leqslant k \leqslant l$, and the stated product representation must hold. To find the mean and variance we can write

Trigonometry wins again. Is there a connection with pitching pennies along the angles of the m-gon?

$$\begin{aligned} G_m &= \bigl(1 - \tfrac{1}{2}\theta^2 + \tfrac{1}{24}\theta^4 - \cdots\bigr)\big/\bigl(1 - \tfrac{1}{2}m^2\theta^2 + \tfrac{1}{24}m^4\theta^4 - \cdots\bigr) \\ &= 1 + \tfrac{1}{2}(m^2-1)\theta^2 + \tfrac{1}{24}(5m^4 - 6m^2 + 1)\theta^4 + \cdots \\ &= 1 + \tfrac{1}{2}(m^2-1)(\tan\theta)^2 + \tfrac{1}{24}(5m^4 - 14m^2 + 9)(\tan\theta)^4 + \cdots \\ &= 1 + G_m'(1)(\tan\theta)^2 + \tfrac{1}{2}G_m''(1)(\tan\theta)^4 + \cdots\,, \end{aligned}$$

because $\tan^2\theta = z - 1$ and $\tan\theta = \theta + \frac{1}{3}\theta^3 + \cdots$. So we have $\mathrm{Mean}(G_m) = \frac{1}{2}(m^2-1)$ and $\mathrm{Var}(G_m) = \frac{1}{6}m^2(m^2-1)$. (Note that this implies the identities

$$\frac{m^2-1}{2} \;=\; \sum_{k=1}^{(m-1)/2}\frac{1}{p_k} \;=\; \sum_{k=1}^{(m-1)/2}\Bigl(1\Big/\sin\frac{(2k-1)\pi}{2m}\Bigr)^2\,;$$

$$\frac{m^2(m^2-1)}{6} \;=\; \sum_{k=1}^{(m-1)/2}\Bigl(\cot\frac{(2k-1)\pi}{2m}\Big/\sin\frac{(2k-1)\pi}{2m}\Bigr)^2\,.$$

The third cumulant of this distribution is $\frac{1}{30}m^2(m^2-1)(4m^2-1)$; but the pattern of nice cumulant factorizations stops there. There's a much simpler way to derive the mean: We have $G_m + A_1 + \cdots + A_l = z(A_1 + \cdots + A_l) + 1$, hence when $z = 1$ we have $G'_m = A_1 + \cdots + A_l$. Since $G_m = 1$ when $z = 1$, an easy induction shows that $A_k = 4k$.)

8.57 We have $A{:}A \geqslant 2^{l-1}$ and $B{:}B < 2^{l-1} + 2^{l-3}$ and $B{:}A \geqslant 2^{l-2}$, hence $B{:}B - B{:}A \geqslant A{:}A - A{:}B$ is possible only if $A{:}B > 2^{l-3}$. This means that $\bar\tau_2 = \tau_3$, $\tau_1 = \tau_4$, $\tau_2 = \tau_5$, ..., $\tau_{l-3} = \tau_l$. But then $A{:}A \approx 2^{l-1} + 2^{l-4} + \cdots$, $A{:}B \approx 2^{l-3} + 2^{l-6} + \cdots$, $B{:}A \approx 2^{l-2} + 2^{l-5} + \cdots$, and $B{:}B \approx 2^{l-1} + 2^{l-4} + \cdots$; hence $B{:}B - B{:}A$ is less than $A{:}A - A{:}B$ after all. (Sharper results have been obtained by Guibas and Odlyzko [138], who show that Bill's chances are always maximized with one of the two patterns $\mathrm{H}\tau_1 \ldots \tau_{l-1}$ or $\mathrm{T}\tau_1 \ldots \tau_{l-1}$.)

8.58 According to (8.82), we want $B{:}B - B{:}A > A{:}A - A{:}B$. One solution is $A = \mathrm{TTHH}$, $B = \mathrm{HHH}$.

8.59 (a) Two cases arise depending on whether $h_k \ne h_n$ or $h_k = h_n$:

$$\begin{aligned} G(w,z) \;&=\; \frac{m-1}{m}\Bigl(\frac{m-2+w+z}{m}\Bigr)^{k-1} w \Bigl(\frac{m-1+z}{m}\Bigr)^{n-k-1} z \\ &\quad + \frac{1}{m}\Bigl(\frac{m-1+wz}{m}\Bigr)^{k-1} wz \Bigl(\frac{m-1+z}{m}\Bigr)^{n-k-1} z\,. \end{aligned}$$

(b) We can either argue algebraically, taking partial derivatives of $G(w,z)$ with respect to w and z and setting $w = z = 1$; or we can argue combinatorially: Whatever the values of h_1, ..., h_{n-1}, the expected value of $P(h_1, \ldots, h_{n-1}, h_n; n)$ is the same (averaged over h_n), because the hash sequence $(h_1, \ldots, h_{n-1})$ determines a sequence of list sizes $(n_1, n_2, \ldots, n_m)$ such that the stated expected value is $\bigl((n_1{+}1) + (n_2{+}1) + \cdots + (n_m{+}1)\bigr)/m = (n-1+m)/m$. Therefore the random variable $EP(h_1, \ldots, h_n; n)$ is independent of $(h_1, \ldots, h_{n-1})$, hence independent of $P(h_1, \ldots, h_n; k)$.

8.60 If $1 \leqslant k < l \leqslant n$, the previous exercise shows that the coefficient of $s_k s_l$ in the variance of the average is zero. Therefore we need only consider the coefficient of s_k^2, which is

$$\sum_{1\leqslant h_1,\ldots,h_n\leqslant m} \frac{P(h_1,\ldots,h_n;k)^2}{m^n} \;-\; \Biggl(\sum_{1\leqslant h_1,\ldots,h_n\leqslant m} \frac{P(h_1,\ldots,h_n;k)}{m^n}\Biggr)^2,$$

the variance of $\bigl((m-1+z)/m\bigr)^{k-1}z$; and this is $(k-1)(m-1)/m^2$ as in exercise 30.

8.61 The pgf $D_n(z)$ satisfies the recurrence

$$\begin{aligned} D_0(z) \;&=\; z\,; \\ D_n(z) \;&=\; z^2 D_{n-1}(z) \,+\, 2(1-z^3)D'_{n-1}(z)/(n+1)\,, \qquad \text{for } n > 0. \end{aligned}$$

We can now derive the recurrence

$$D''_n(1) = (n-11)D''_{n-1}(1)/(n+1) + (8n-2)/7\,,$$

which has the solution $\frac{2}{637}(n+2)(26n+15)$ for all $n \geqslant 11$ (regardless of initial conditions). Hence the variance comes to $\frac{12}{49}(n+2)(212n+123)$ for $n \geqslant 11$.

8.62 (Another question asks if a given sequence of purported cumulants comes from any distribution whatever; for example, κ_2 must be nonnegative, and $\kappa_4 + 3\kappa_2^2 = E\big((X-\mu)^4\big)$ must be at least $\big(E((X-\mu)^2)\big)^2 = \kappa_2^2$, etc. A necessary and sufficient condition for this other problem was found by Hamburger [6], [144].)

8.63 (Another question asks if there is a simple rule to tell whether H or T is preferable.) Conway conjectures that no such ties exist, and moreover that there is only one cycle in the directed graph on 2^l vertices that has an arc from each sequence to its "best beater."

9.1 True if the functions are all positive. But otherwise we might have, say, $f_1(n) = n^3 + n^2$, $f_2(n) = -n^3$, $g_1(n) = n^4 + n$, $g_2(n) = -n^4$.

9.2 (a) We have $n^{\ln n} \prec c^n \prec (\ln n)^n$, since $(\ln n)^2 \prec n \ln c \prec n \ln\ln n$. (b) $n^{\ln\ln\ln n} \prec (\ln n)! \prec n^{\ln\ln n}$. (c) Take logarithms to show that $(n!)!$ wins. (d) $F^2_{\lceil H_n \rceil} \asymp \phi^{2\ln n} = n^{2\ln\phi}$; $H_{F_n} \sim n \ln\phi$ wins because $\phi^2 = \phi + 1 < e$.

9.3 Replacing kn by $O(n)$ requires a different C for each k; but each O stands for a single C. In fact, the context of this O requires it to stand for a set of functions of two variables k and n. It would be correct to write $\sum_{k=1}^{n} kn = \sum_{k=1}^{n} O(n^2) = O(n^3)$.

9.4 For example, $\lim_{n\to\infty} O(1/n) = 0$. On the left, $O(1/n)$ is the set of all functions $f(n)$ such that there are constants C and n_0 with $|f(n)| \leqslant C/n$ for all $n \geqslant n_0$. The limit of all functions in that set is 0, so the left-hand side is the singleton set $\{0\}$. On the right, there are no variables; 0 represents $\{0\}$, the (singleton) set of all "functions of no variables, whose value is zero." (Can you see the inherent logic here? If not, come back to it next year; you probably can still manipulate O-notation even if you can't shape your intuitions into rigorous formalisms.)

9.5 Let $f(n) = n^2$ and $g(n) = 1$; then n is in the left set but not in the right, so the statement is false.

9.6 $n\ln n + \gamma n + O(\sqrt{n}\ln n)$.

9.7 $(1 - e^{-1/n})^{-1} = nB_0 - B_1 + B_2 n^{-1}/2! + \cdots = n + \frac{1}{2} + O(n^{-1})$.

9.8 For example, let $f(n) = \lfloor n/2 \rfloor!^2 + n$, $g(n) = \big(\lceil n/2 \rceil - 1\big)!\,\lceil n/2 \rceil! + n$. These functions, incidentally, satisfy $f(n) = O\big(ng(n)\big)$ and $g(n) = O\big(nf(n)\big)$; more extreme examples are clearly possible.

9.9 (For completeness, we assume that there is a side condition $n \to \infty$, so that two constants are implied by each O.) Every function on the left has the form $a(n) + b(n)$, where there exist constants m_0, B, n_0, C such that $|a(n)| \leqslant B|f(n)|$ for $n \geqslant m_0$ and $|b(n)| \leqslant C|g(n)|$ for $n \geqslant n_0$. Therefore the left-hand function is at most $\max(B,C)\bigl(|f(n)|+|g(n)|\bigr)$, for $n \geqslant \max(m_0, n_0)$, so it is a member of the right side.

9.10 If $g(x)$ belongs to the left, so that $g(x) = \cos y$ for some y, where $|y| \leqslant C|x|$ for some C, then $0 \leqslant 1 - g(x) = 2\sin^2(y/2) \leqslant \frac{1}{2}y^2 \leqslant \frac{1}{2}C^2x^2$; hence the set on the left is contained in the set on the right, and the formula is true.

9.11 The proposition is true. For if, say, $|x| \leqslant |y|$, we have $(x+y)^2 \leqslant 4y^2$. Thus $(x+y)^2 = O(x^2) + O(y^2)$. Thus $O(x+y)^2 = O\bigl((x+y)^2\bigr) = O\bigl(O(x^2) + O(y^2)\bigr) = O\bigl(O(x^2)\bigr) + O\bigl(O(y^2)\bigr) = O(x^2) + O(y^2)$.

9.12 $1 + 2/n + O(n^{-2}) = (1 + 2/n)\bigl(1 + O(n^{-2})/(1 + 2/n)\bigr)$ by (9.26), and $1/(1 + 2/n) = O(1)$; now use (9.26).

9.13 $n^n\bigl(1 + 2n^{-1} + O(n^{-2})\bigr)^n = n^n \exp\bigl(n(2n^{-1} + O(n^{-2}))\bigr) = e^2 n^n + O(n^{n-1})$.

9.14 It is $n^{n+\beta} \exp\bigl((n+\beta)(\alpha/n - \frac{1}{2}\alpha^2/n^2 + O(n^{-3}))\bigr)$.

9.15 $\ln\binom{3n}{n,n,n} = 3n\ln 3 - \ln n + \frac{1}{2}\ln 3 - \ln 2\pi + \bigl(\frac{1}{36} - \frac{1}{4}\bigr)n^{-1} + O(n^{-3})$, so the answer is

$$\frac{3^{3n+1/2}}{2\pi n}\bigl(1 - \tfrac{2}{9}n^{-1} + \tfrac{2}{81}n^{-2} + O(n^{-3})\bigr).$$

9.16 If l is any integer in the range $a \leqslant l < b$ we have

$$\begin{aligned}\int_0^1 B(x)f(l+x)\,dx &= \int_{1/2}^1 B(x)f(l+x)\,dx - \int_0^{1/2} B(1-x)f(l+x)\,dx \\ &= \int_{1/2}^1 B(x)\bigl(f(l+x) - f(l+1-x)\bigr)\,dx.\end{aligned}$$

Since $l + x \geqslant l + 1 - x$ when $x \geqslant \frac{1}{2}$, this integral is positive when $f(x)$ is nondecreasing.

9.17 $\sum_{m\geqslant 0} B_m(\frac{1}{2})z^m/m! = ze^{z/2}/(e^z - 1) = z/(e^{z/2} - 1) - z/(e^z - 1)$.

9.18 The text's derivation for the case $\alpha = 1$ generalizes to give

$$b_k(n) = \frac{2^{(2n+1/2)\alpha}}{(2\pi n)^{\alpha/2}}e^{-k^2\alpha/n}, \qquad c_k(n) = 2^{2n\alpha}n^{-(1+\alpha)/2+3\epsilon}e^{-k^2\alpha/n};$$

the answer is $2^{2n\alpha}(\pi n)^{(1-\alpha)/2}\alpha^{-1/2}\bigl(1 + O(n^{-1/2+3\epsilon})\bigr)$.

9.19 $H_{10} = 2.928968254 \approx 2.928968256$; $10! = 3628800 \approx 3628712.4$; $B_{10} = 0.075757576 \approx 0.075757494$; $\pi(10) = 4 \approx 10.0017845$; $e^{0.1} = 1.10517092 \approx 1.10517083$; $\ln 1.1 = 0.0953102 \approx 0.0953083$; $1.1111111 \approx 1.1111000$; $1.1^{0.1} = 1.00957658 \approx 1.00957643$. (The approximation to $\pi(n)$ gives more significant figures when n is larger; for example, $\pi(10^9) = 50847534 \approx 50840742$.)

9.20 (a) Yes; the left side is $o(n)$ while the right side is equivalent to $O(n)$. (b) Yes; the left side is $e \cdot e^{O(1/n)}$. (c) No; the left side is about $\sqrt{n}$ times the bound on the right.

9.21 We have $P_n = m = n\bigl(\ln m - 1 - 1/\ln m + O(1/\log n)^2\bigr)$, where

$$\ln m = \ln n + \ln\ln m - 1/\ln n + \ln\ln n/(\ln n)^2 + O(1/\log n)^2\,;$$
$$\ln\ln m = \ln\ln n + \frac{\ln\ln n}{\ln n} - \frac{(\ln\ln n)^2}{2(\ln n)^2} + \frac{\ln\ln n}{(\ln n)^2} + O(1/\log n)^2\,.$$

It follows that

$$P_n = n\biggl(\ln n + \ln\ln n - 1 + \frac{\ln\ln n - 2}{\ln n} - \frac{\frac12(\ln\ln n)^2 - 3\ln\ln n}{(\ln n)^2} + O(1/\log n)^2\biggr)\,.$$

(A slightly better approximation replaces this $O(1/\log n)^2$ by the quantity $-5/(\ln n)^2 + O(\log\log n/\log n)^3$; then we estimate $P_{1000000} \approx 15483612.4$.)

9.22 Replace $O(n^{-2k})$ by $-\frac{1}{12}n^{-2k} + O(n^{-4k})$ in the expansion of H_{n^k}; this replaces $O\bigl(\Sigma_3(n^2)\bigr)$ by $-\frac{1}{12}\Sigma_3(n^2) + O\bigl(\Sigma_3(n^4)\bigr)$ in (9.53). We have

$$\Sigma_3(n) = \tfrac34 n^{-1} + \tfrac{5}{36}n^{-2} + O(n^{-3})\,,$$

hence the term $O(n^{-2})$ in (9.54) can be replaced by $-\frac{19}{144}n^{-2} + O(n^{-3})$.

9.23 $nh_n = \sum_{0\le k<n} h_k/(n-k) + 2cH_n/(n+1)(n+2)$. Choose $c = e^{\pi^2/6} = \sum_{k\ge0} g_k$ so that $\sum_{k\ge0} h_k = 0$ and $h_n = O(\log n)/n^3$. The expansion of $\sum_{0\le k<n} h_k/(n-k)$ as in (9.60) now yields $nh_n = 2cH_n/(n+1)(n+2) + O(n^{-2})$, hence

$$g_n = e^{\pi^2/6}\left(\frac{n + 2\ln n + O(1)}{n^3}\right).$$

9.24 (a) If $\sum_{k\ge0}|f(k)| < \infty$ and if $f(n-k) = O\bigl(f(n)\bigr)$ when $0 \le k \le n/2$, we have

$$\sum_{k=0}^{n} a_k b_{n-k} = \sum_{k=0}^{n/2} O\bigl(f(k)\bigr)O\bigl(f(n)\bigr) + \sum_{k=n/2}^{n} O\bigl(f(n)\bigr)O\bigl(f(n-k)\bigr)\,,$$

which is $2O\big(f(n)\sum_{k\geqslant 0}|f(k)|\big)$, so this case is proved. (b) But in this case if $a_n = b_n = \alpha^{-n}$, the convolution $(n+1)\alpha^{-n}$ is not $O(\alpha^{-n})$.

9.25 $S_n/\binom{3n}{n} = \sum_{k=0}^{n} n^{\underline{k}}/(2n+1)^{\overline{k}}$. We may restrict the range of summation to $0 \leqslant k \leqslant (\log n)^2$, say. In this range $n^{\underline{k}} = n^k\big(1 - \binom{k}{2}/n + O(k^4/n^2)\big)$ and $(2n+1)^{\overline{k}} = (2n)^k\big(1 + \binom{k+1}{2}/2n + O(k^4/n^2)\big)$, so the summand is

$$\frac{1}{2^k}\left(1 - \frac{3k^2-k}{4n} + O\Big(\frac{k^4}{n^2}\Big)\right).$$

Hence the sum over k is $2 - 4/n + O(1/n^2)$. Stirling's approximation can now be applied to $\binom{3n}{n} = (3n)!/(2n)!\,n!$, proving (9.2).

9.26 The minimum occurs at a term $B_{2m}/(2m)(2m-1)n^{2m-1}$ where $2m \approx 2\pi n + \frac{3}{2}$, and this term is approximately equal to $1/(\pi e^{2\pi n}\sqrt{n}\,)$. The absolute error in $\ln n!$ is therefore too large to determine $n!$ exactly by rounding to an integer, when n is greater than about $e^{2\pi+1}$.

9.27 We may assume that $\alpha \neq -1$. Let $f(x) = x^\alpha$; the answer is

$$\sum_{k=1}^{n} k^\alpha \;=\; C_\alpha + \frac{n^{\alpha+1}}{\alpha+1} + \frac{n^\alpha}{2} + \sum_{k=1}^{m}\frac{B_{2k}}{2k}\binom{\alpha}{2k-1}n^{\alpha-2k+1} + O(n^{\alpha-2m-1}).$$

(The constant C_α turns out to be $\zeta(-\alpha)$, which is in fact *defined* by this formula when $\alpha > -1$.)

9.28 Take $f(x) = x\ln x$ in Euler's summation formula to get

$$A\cdot n^{n^2/2+n/2+1/12}e^{-n^2/4}\big(1 + O(n^{-2})\big),$$

where $A \approx 1.282427$ is "Glaisher's constant."

9.29 Let $f(x) = x^{-1}\ln x$. Then $f^{(2m)}(x) > 0$ for all large x, and we can write

$$\sum_{k=1}^{n}\frac{\ln k}{k} \;=\; \frac{(\ln n)^2}{2} + \ln S + \frac{\ln n}{2n} + \theta_n\frac{1-\ln n}{12n^2}, \qquad 0 < \theta_n < 1,$$

where $S \approx 0.929772$ is constant. Taking exponentials gives

$$S\sqrt{n^{\ln n}}\left(1 + \frac{\ln n}{2n} + O\Big(\frac{\log n}{n}\Big)^2\right).$$

(In general if $f(x) = x^\alpha \ln x$, Euler's summation formula applies as in exercise 27, and the resulting constant is $-\zeta'(-\alpha)$ if $\alpha \neq -1$. Thus, the theory of the zeta function gives a closed form for Glaisher's constant in the previous exercise. We have $\ln S = \gamma_1$ in the notation of answer 9.57.)

9.30 Let $g(x) = x^l e^{-x^2}$ and $f(x) = g(x/\sqrt{n})$. Then $n^{-l/2}\sum_{k\geqslant 0} k^l e^{-k^2/n}$ is

$$\int_0^\infty f(x)\,dx - \sum_{k=1}^{m}\frac{B_k}{k!}f^{(k-1)}(0) - (-1)^m\int_0^\infty \frac{B_m(\{x\})}{m!}f^{(m)}(x)\,dx$$
$$= n^{1/2}\int_0^\infty g(x)\,dx - \sum_{k=1}^{m}\frac{B_k}{k!}n^{(k-1)/2}g^{(k-1)}(0) + O(n^{-m/2}).$$

Since $g(x) = x^l - x^{2+l}/1! + x^{4+l}/2! - x^{6+l}/3! + \cdots$, the derivatives $g^{(m)}(x)$ obey a simple pattern, and the answer is

$$\frac{1}{2}n^{(l+1)/2}\Gamma\Bigl(\frac{l+1}{2}\Bigr) - \frac{B_{l+1}}{(l+1)!\,0!} + \frac{B_{l+3}n^{-1}}{(l+3)!\,1!} - \frac{B_{l+5}n^{-2}}{(l+5)!\,2!} + O(n^{-3}).$$

9.31 The somewhat surprising identity $1/(c^{m-k} + c^m) + 1/(c^{m+k} + c^m) = 1/c^m$ makes the terms for $0 \leqslant k \leqslant 2m$ sum to $(m + \frac{1}{2})/c^m$. The remaining terms are

$$\sum_{k\geqslant 1}\frac{1}{c^{2m+k} + c^m} = \sum_{k\geqslant 1}\Bigl(\frac{1}{c^{2m+k}} - \frac{1}{c^{3m+2k}} + \frac{1}{c^{4m+3k}} - \cdots\Bigr)$$
$$= \frac{1}{c^{2m+1} - c^{2m}} - \frac{1}{c^{3m+2} - c^{3m}} + \cdots,$$

and this series can be truncated at any desired point, with an error not exceeding the first omitted term.

9.32 $H_n^{(2)} = \pi^2/6 - 1/n + O(n^{-2})$ by Euler's summation formula, since we know the constant; and H_n is given by (9.89). So the answer is

$$ne^{\gamma+\pi^2/6}\bigl(1 - \tfrac{1}{2}n^{-1} + O(n^{-2})\bigr).$$

The world's top three constants, (e, π, γ), all appear in this answer.

9.33 We have $n^{\underline{k}}/n^k = 1 - \binom{k}{2}n^{-1} + O(k^4 n^{-2})$; dividing by $k!$ and summing over k yields $e - \frac{1}{2}en^{-1} + O(n^{-2})$.

9.34 $A = e^\gamma$; $B = 0$; $C = -\frac{1}{2}e^\gamma$; $D = \frac{1}{2}e^\gamma(1-\gamma)$; $E = \frac{1}{8}e^\gamma$; $F = \frac{1}{12}e^\gamma(3\gamma+1)$.

9.35 Since $1/k(\ln k + O(1)) = 1/k\ln k + O\bigl(1/k(\log k)^2\bigr)$, the given sum is $\sum_{k=2}^{n} 1/k\ln k + O(1)$. The remaining sum is $\ln\ln n + O(1)$ by Euler's summation formula.

9.36 This works out beautifully with Euler's summation formula:

$$S_n = \sum_{0\leqslant k<n}\frac{1}{n^2+k^2} + \frac{1}{n^2+x^2}\bigg|_0^n$$
$$= \int_0^n \frac{dx}{n^2+x^2} + \frac{1}{2}\frac{1}{n^2+x^2}\bigg|_0^n + \frac{B_2}{2!}\frac{-2x}{(n^2+x^2)^2}\bigg|_0^n + O(n^{-5}).$$

Hence $S_n = \frac{1}{4}\pi n^{-1} - \frac{1}{4}n^{-2} - \frac{1}{24}n^{-3} + O(n^{-5})$.

9.37 This is

$$\begin{aligned}\sum_{k,q\geqslant 1}(n-qk)\bigl[n/(q+1)<k\leqslant n/q\bigr] &= n^2 - \sum_{q\geqslant 1} q\left(\binom{\lfloor n/q\rfloor+1}{2} - \binom{\lfloor n/(q+1)\rfloor+1}{2}\right)\\ &= n^2 - \sum_{q\geqslant 1}\binom{\lfloor n/q\rfloor+1}{2}.\end{aligned}$$

The remaining sum is like (9.55) but without the factor $\mu(q)$. The same method works here as it did there, but we get $\zeta(2)$ in place of $1/\zeta(2)$, so the answer comes to $\bigl(1-\frac{\pi^2}{12}\bigr)n^2 + O(n\log n)$.

9.38 Replace k by $n-k$ and let $a_k(n) = (n-k)^{n-k}\binom{n}{k}$. Then $\ln a_k(n) = n\ln n - \ln k! - k + O(kn^{-1})$, and we can use tail-exchange with $b_k(n) = n^n e^{-k}/k!$, $c_k(n) = kb_k(n)/n$, $D_n = \{k \mid k \leqslant \ln n\}$, to get $\sum_{k=0}^{n} a_k(n) = n^n e^{1/e}\bigl(1+O(n^{-1})\bigr)$.

9.39 Tail-exchange with $b_k(n) = (\ln n - k/n - \frac{1}{2}k^2/n^2)(\ln n)^k/k!$, $c_k(n) = n^{-3}(\ln n)^{k+3}/k!$, $D_n = \{k \mid 0 \leqslant k \leqslant 10\ln n\}$. When $k \approx 10\ln n$ we have $k! \asymp \sqrt{k}\,(10/e)^k(\ln n)^k$, so the kth term is $O(n^{-10\ln(10/e)}\log n)$. The answer is $n\ln n - \ln n - \frac{1}{2}(\ln n)(1+\ln n)/n + O\bigl(n^{-2}(\log n)^3\bigr)$.

9.40 Combining terms two by two, we find that $H_{2k}^m - (H_{2k} - \frac{1}{2k})^m = \frac{m}{2k}H_{2k}^{m-1}$ plus terms whose sum over all $k \geqslant 1$ is $O(1)$. Suppose n is even. Euler's summation formula implies that

$$\sum_{k=1}^{n/2}\frac{H_{2k}^{m-1}}{k} = \sum_{k=1}^{n/2}\frac{(\ln 2e^{\gamma}k)^{m-1} + O(1/k)}{k} + O(1) = \frac{(\ln e^{\gamma}n)^m}{m} + O(1);$$

hence the sum is $\frac{1}{2}H_n^m + O(1)$. In general the answer is $(-1)^n H_n^m + O(1)$.

9.41 Let $\alpha = \hat\phi/\phi = -\phi^{-2}$. We have

$$\begin{aligned}\sum_{k=1}^{n}\ln F_k &= \sum_{k=1}^{n}\bigl(\ln\phi^k - \ln\sqrt{5} + \ln(1-\alpha^k)\bigr)\\ &= \frac{n(n+1)}{2}\ln\phi - \frac{n}{2}\ln 5 + \sum_{k\geqslant 1}\ln(1-\alpha^k) - \sum_{k>n}\ln(1-\alpha^k).\end{aligned}$$

The latter sum is $\sum_{k>n} O(\alpha^k) = O(\alpha^n)$. Hence the answer is

$$\phi^{n(n+1)/2}5^{-n/2}C + O(\phi^{n(n-3)/2}5^{-n/2}), \qquad \text{where}$$
$$C = (1-\alpha)(1-\alpha^2)(1-\alpha^3)\ldots \approx 1.226742.$$

9.42 The hint follows since $\binom{n}{k-1}/\binom{n}{k} = \frac{k}{n-k+1} \leqslant \frac{\alpha n}{n-\alpha n+1} < \frac{\alpha}{1-\alpha}$. Let $m = \lfloor \alpha n \rfloor = \alpha n - \epsilon$. Then

$$\binom{n}{m} < \sum_{k \leqslant m} \binom{n}{k}$$
$$< \binom{n}{m}\left(1 + \frac{\alpha}{1-\alpha} + \left(\frac{\alpha}{1-\alpha}\right)^2 + \cdots\right) = \binom{n}{m}\frac{1-\alpha}{1-2\alpha}.$$

So $\sum_{k \leqslant \alpha n}\binom{n}{k} = \binom{n}{m}O(1)$, and it remains to estimate $\binom{n}{m}$. By Stirling's approximation we have $\ln\binom{n}{m} = -\frac{1}{2}\ln n - (\alpha n - \epsilon)\ln(\alpha - \epsilon/n) - \big((1-\alpha)n+\epsilon\big) \times \ln(1-\alpha+\epsilon/n) + O(1) = -\frac{1}{2}\ln n - \alpha n \ln \alpha - (1-\alpha)n\ln(1-\alpha) + O(1)$.

9.43 The denominator has factors of the form $z - \omega$, where ω is a complex root of unity. Only the factor $z - 1$ occurs with multiplicity 5. Therefore by (7.31), only one of the roots has a coefficient $\Omega(n^4)$, and the coefficient is $c = 5/(5!\cdot 1\cdot 5\cdot 10\cdot 25\cdot 50) = 1/1500000$.

9.44 Stirling's approximation says that $\ln\big(x^{-\alpha}x!/(x-\alpha)!\big)$ has an asymptotic series

$$-\alpha - (x + \tfrac{1}{2} - \alpha)\ln(1-\alpha/x) - \frac{B_2}{2\cdot 1}\big(x^{-1} - (x-\alpha)^{-1}\big)$$
$$- \frac{B_4}{4\cdot 3}\big(x^{-3} - (x-\alpha)^{-3}\big) - \cdots$$

in which each coefficient of x^{-k} is a polynomial in α. Hence $x^{-\alpha}x!/(x-\alpha)! = c_0(\alpha) + c_1(\alpha)x^{-1} + \cdots + c_n(\alpha)x^{-n} + O(x^{-n-1})$ as $x \to \infty$, where $c_n(\alpha)$ is a polynomial in α. We know that $c_n(\alpha) = \left[{\alpha \atop \alpha-n}\right](-1)^n$ whenever α is an integer, and $\left[{\alpha \atop \alpha-n}\right]$ is a polynomial in α of degree $2n$; hence $c_n(\alpha) = \left[{\alpha \atop \alpha-n}\right](-1)^n$ for all real α. In other words, the asymptotic formulas

$$x^{\underline{\alpha}} = \sum_{k=0}^{n}\left[{\alpha \atop \alpha-k}\right](-1)^k x^{\alpha-k} + O(x^{\alpha-n-1}),$$
$$x^{\overline{\alpha}} = \sum_{k=0}^{n}\left[{\alpha \atop \alpha-k}\right]x^{\alpha-k} + O(x^{\alpha-n-1})$$

generalize equations (6.13) and (6.11), which hold in the all-integer case.

9.45 Let the partial quotients of α be $\langle a_1, a_2, \ldots\rangle$, and let α_m be the continued fraction $1/(a_m + \alpha_{m+1})$ for $m \geqslant 1$. Then $D(\alpha, n) = D(\alpha_1, n) < D(\alpha_2, \lfloor \alpha_1 n\rfloor) + a_1 + 3 < D(\alpha_3, \lfloor \alpha_2\lfloor \alpha_1 n\rfloor\rfloor) + a_1 + a_2 + 6 < \cdots < D(\alpha_{m+1}, \lfloor \alpha_m \lfloor \ldots \lfloor \alpha_1 n\rfloor \ldots\rfloor\rfloor) + a_1 + \cdots + a_m + 3m < \alpha_1 \ldots \alpha_m n + a_1 + \cdots + a_m + 3m$, for all m. Divide by n and let $n \to \infty$; the limit is less than $\alpha_1 \ldots \alpha_m$ for all m. Finally we have

$$\alpha_1 \ldots \alpha_m = \frac{1}{K(a_1, \ldots, a_{m-1}, a_m + \alpha_m)} < \frac{1}{F_{m+1}}.$$

9.46 For convenience we write just m instead of $m(n)$. By Stirling's approximation, the maximum value of $k^n/k!$ occurs when $k \approx m \approx n/\ln n$, so we replace k by $m + k$ and find that

$$\ln \frac{(m+k)^n}{(m+k)!} = n \ln m - m \ln m + m - \frac{\ln 2\pi m}{2} - \frac{(m+n)k^2}{2m^2} + O(k^3 m^{-2} \log n).$$

Actually we want to replace k by $\lfloor m \rfloor + k$; this adds a further $O(km^{-1} \log n)$. The tail-exchange method with $|k| \leqslant m^{1/2+\epsilon}$ now allows us to sum on k, giving a fairly sharp asymptotic estimate

A truly Bell-shaped summand.

$$b_n = \frac{e^{m-1} m^{n-m}}{\sqrt{2\pi m}} \bigl(\Theta_{2m^2/(m+n)} + O(1)\bigr) = e^{m-n-1/2} m^n \sqrt{\frac{m}{m+n}} \Bigl(1 + O\Bigl(\frac{\log n}{n^{1/2}}\Bigr)\Bigr).$$

The requested formula follows, with relative error $O(\log\log n/\log n)$.

9.47 Let $\log_m n = l + \theta$, where $0 \leqslant \theta < 1$. The floor sum is $l(n+1) + 1 - (m^{l+1} - 1)/(m-1)$; the ceiling sum is $(l+1)n - (m^{l+1} - 1)/(m-1)$; the exact sum is $(l+\theta)n - n/\ln m + O(\log n)$. Ignoring terms that are $o(n)$, the difference between ceiling and exact is $\bigl(1 - f(\theta)\bigr)n$, and the difference between exact and floor is $f(\theta)n$, where

$$f(\theta) = \frac{m^{1-\theta}}{m-1} + \theta - \frac{1}{\ln m}.$$

This function has maximum value $f(0) = f(1) = m/(m-1) - 1/\ln m$, and its minimum value is $\ln\ln m/\ln m + 1 - \bigl(\ln(m-1)\bigr)/\ln m$. The ceiling value is closer when n is nearly a power of m, but the floor value is closer when θ lies somewhere between 0 and 1.

9.48 Let $d_k = a_k + b_k$, where a_k counts digits to the left of the decimal point. Then $a_k = 1 + \lfloor \log H_k \rfloor = \log\log k + O(1)$, where 'log' denotes $\log_{10}$. To estimate b_k, let us look at the number of decimal places necessary to distinguish y from nearby numbers $y - \epsilon$ and $y + \epsilon'$: Let $\delta = 10^{-b}$ be the length of the interval of numbers that round to $\hat{y}$. We have $|y-\hat{y}| \leqslant \frac{1}{2}\delta$; also $y-\epsilon < \hat{y}-\frac{1}{2}\delta$ and $y+\epsilon' > \hat{y}+\frac{1}{2}\delta$. Therefore $\epsilon+\epsilon' > \delta$. And if $\delta < \min(\epsilon, \epsilon')$, the rounding does distinguish $\hat{y}$ from both $y-\epsilon$ and $y+\epsilon'$. Hence $10^{-b_k} < 1/(k-1)+1/k$ and $10^{1-b_k} \geqslant 1/k$; we have $b_k = \log k + O(1)$. Finally, therefore, $\sum_{k=1}^{n} d_k = \sum_{k=1}^{n}(\log k + \log\log k + O(1))$, which is $n\log n + n\log\log n + O(n)$ by Euler's summation formula.

9.49 We have $H_n > \ln n + \gamma + \frac{1}{2}n^{-1} - \frac{1}{12}n^{-2} = f(n)$, where $f(x)$ is increasing for all $x > 0$; hence if $n \geqslant e^{\alpha-\gamma}$ we have $H_n \geqslant f(e^{\alpha-\gamma}) > \alpha$. Also $H_{n-1} < \ln n + \gamma - \frac{1}{2}n^{-1} = g(n)$, where $g(x)$ is increasing for all $x > 0$; hence if $n \leqslant e^{\alpha-\gamma}$ we have $H_{n-1} \leqslant g(e^{\alpha-\gamma}) < \alpha$. Therefore $H_{n-1} \leqslant \alpha \leqslant H_n$ implies that $e^{\alpha-\gamma}+1 > n > e^{\alpha+\gamma}-1$. (Sharper results have been obtained by Boas and Wrench [27].)

9.50 (a) The expected return is $\sum_{1\leqslant k\leqslant N} k/(k^2 H_n^{(2)}) = H_N/H_N^{(2)}$, and we want the asymptotic value to $O(N^{-1})$:

$$\frac{\ln N + \gamma + O(N^{-1})}{\pi^2/6 - N^{-1} + O(N^{-2})} = \frac{6\ln 10}{\pi^2}n + \frac{6\gamma}{\pi^2} + \frac{36\ln 10}{\pi^4}\frac{n}{10^n} + O(10^{-n}).$$

The coefficient $(6\ln 10)/\pi^2 \approx 1.3998$ says that we expect about 40% profit.

(b) The probability of profit is $\sum_{n<k\leqslant N} 1/(k^2 H_N^{(2)}) = 1 - H_n^{(2)}/H_N^{(2)}$, and since $H_n^{(2)} = \frac{\pi^2}{6} - n^{-1} + \frac{1}{2}n^{-2} + O(n^{-3})$ this is

$$\frac{n^{-1} - \frac{1}{2}n^{-2} + O(n^{-3})}{\pi^2/6 + O(N^{-1})} = \frac{6}{\pi^2}n^{-1} - \frac{3}{\pi^2}n^{-2} + O(n^{-3}),$$

actually *decreasing* with n. (The expected value in (a) is high because it includes payoffs so huge that the entire world's economy would be affected if they ever had to be made.)

9.51 Strictly speaking, this is false, since the function represented by $O(x^{-2})$ might not be integrable. (It might be '$[x \in S]/x^2$', where S is not a measurable set.) But if we stipulate that $f(x)$ is an integrable function such that $f(x) = O(x^{-2})$ as $x \to \infty$, then $\left|\int_n^\infty f(x)\,dx\right| \leqslant \int_n^\infty |f(x)|\,dx \leqslant \int_n^\infty Cx^{-2}\,dx = Cn^{-1}$.

(As opposed to an execrable function.)

9.52 In fact, the stack of n's can be replaced by any function $f(n)$ that approaches infinity, however fast. Define the sequence $\langle m_0, m_1, m_2, \ldots\rangle$ by setting $m_0 = 0$ and letting m_k be the least integer $> m_{k-1}$ such that

$$\left(\frac{k+1}{k}\right)^{m_k} \geqslant f(k+1)^2.$$

Now let $A(z) = \sum_{k\geqslant 1}(z/k)^{m_k}$. This power series converges for all z, because the terms for $k > |z|$ are bounded by a geometric series. Also $A(n+1) \geqslant ((n+1)/n)^{m_n} \geqslant f(n+1)^2$, hence $\lim_{n\to\infty} f(n)/A(n) = 0$.

9.53 By induction, the O term is $(m-1)!^{-1}\int_0^x t^{m-1}f^{(m)}(t-x)\,dt$. Since $f^{(m+1)}$ has the opposite sign to $f^{(m)}$, the absolute value of this integral is bounded by $\left|f^{(m)}(0)\right|\int_0^x t^{m-1}\,dt$; so the error is bounded by the absolute value of the first discarded term.

Sounds like a nasty theorem.

9.54 Let $g(x) = f(x)/x^\alpha$. Then $g'(x) \sim -\alpha g(x)/x$ as $x \to \infty$. By the mean value theorem, $g(x-\frac12) - g(x+\frac12) = -g'(y) \sim \alpha g(y)/y$ for some y between $x-\frac12$ and $x+\frac12$. Now $g(y) = g(x)\bigl(1+O(1/x)\bigr)$, so $g(x-\frac12)-g(x+\frac12) \sim \alpha g(x)/x = \alpha f(x)/x^{1+\alpha}$. Therefore

$$\sum_{k\geqslant n}\frac{f(k)}{k^{1+\alpha}} = O\Bigl(\sum_{k\geqslant n}\bigl(g(k-\tfrac12)-g(k+\tfrac12)\bigr)\Bigr) = O\bigl(g(n-\tfrac12)\bigr).$$

9.55 The estimate of $(n+k+\frac12)\ln(1+k/n)+(n-k+\frac12)\ln(1-k/n)$ is extended to $k^2/n + k^4/6n^3 + O(n^{-3/2+5\epsilon})$, so we apparently want to have an extra factor $e^{-k^4/6n^3}$ in $b_k(n)$, and $c_k(n) = 2^{2n}n^{-2+5\epsilon}e^{-k^2/n}$. But it turns out to be better to leave $b_k(n)$ untouched and to let

$$c_k(n) = 2^{2n}n^{-2+5\epsilon}e^{-k^2/n} + 2^{2n}n^{-5+5\epsilon}k^4e^{-k^2/n},$$

thereby replacing $e^{-k^4/6n^3}$ by $1+O(k^4/n^3)$. The sum $\sum_k k^4e^{-k^2/n}$ is $O(n^{5/2})$, as shown in exercise 30.

9.56 If $k \leqslant n^{1/2+\epsilon}$ we have $\ln(n^{\underline{k}}/n^k) = -\frac12k^2/n + \frac12k/n - \frac16k^3/n^2 + O(n^{-1+4\epsilon})$ by Stirling's approximation, hence

$$n^{\underline{k}}/n^k = e^{-k^2/2n}\bigl(1+k/2n-\tfrac23k^3/(2n)^2+O(n^{-1+4\epsilon})\bigr).$$

Summing with the identity in exercise 30, and remembering to omit the term for $k=0$, gives $-1+\Theta_{2n}+\Theta_{2n}^{(1)}-\frac23\Theta_{2n}^{(3)}+O(n^{-1/2+4\epsilon}) = \sqrt{\pi n/2}-\frac13+O(n^{-1/2+4\epsilon})$.

9.57 Using the hint, the given sum becomes $\int_0^\infty ue^{-u}\zeta(1+u/\ln n)\,du$. The zeta function can be defined by the series

$$\zeta(1+z) = z^{-1}+\sum_{m\geqslant0}(-1)^m\gamma_m z^m/m!,$$

where $\gamma_0=\gamma$ and γ_m is the Stieltjes constant

$$\lim_{n\to\infty}\left(\sum_{k=1}^n\frac{(\ln k)^m}{k} - \frac{(\ln n)^{m+1}}{m+1}\right).$$

Hence the given sum is

$$\ln n+\gamma-2\gamma_1(\ln n)^{-1}+3\gamma_2(\ln n)^{-2}-\cdots.$$

9.58 Let $0 \leqslant \theta \leqslant 1$ and $f(x) = e^{2\pi i z\theta}/(e^{2\pi i z} - 1)$. We have

$$\bigl|f(z)\bigr| \;=\; \frac{e^{-2\pi y\theta}}{1+e^{-2\pi y}} \;\leqslant\; 1\,, \qquad \text{when } x \bmod 1 = \tfrac12;$$

$$\bigl|f(z)\bigr| \;\leqslant\; \frac{e^{-2\pi y\theta}}{|e^{-2\pi y}-1|} \;\leqslant\; \frac{1}{1-e^{-2\pi\epsilon}}\,, \qquad \text{when } |y| \geqslant \epsilon.$$

Therefore $\bigl|f(z)\bigr|$ is bounded on the contour, and the integral is $O(M^{1-m})$. The residue of $2\pi i f(z)/z^m$ at $z = k \neq 0$ is $e^{2\pi i k\theta}/k^m$; the residue at $z = 0$ is the coefficient of z^{-1} in

$$\frac{e^{2\pi i z\theta}}{z^{m+1}}\Bigl(B_0 + B_1\frac{2\pi i z}{1!} + \cdots\Bigr) \;=\; \frac{1}{z^{m+1}}\Bigl(B_0(\theta) + B_1(\theta)\frac{2\pi i z}{1!} + \cdots\Bigr)\,,$$

namely $(2\pi i)^m B_m(\theta)/m!$. Therefore the sum of residues inside the contour is

$$\frac{(2\pi i)^m}{m!}B_m(\theta) \;+\; 2\sum_{k=1}^{M} e^{\pi i m/2}\,\frac{\cos(2\pi k\theta - \pi m/2)}{k^m}\,.$$

This equals the contour integral $O(M^{1-m})$, so it approaches zero as $M \to \infty$.

9.59 If $F(x)$ is sufficiently well behaved, we have the general identity

$$\sum_k F(k+t) \;=\; \sum_n G(2\pi n)e^{2\pi i n t}\,,$$

where $G(y) \;=\; \int_{-\infty}^{+\infty} e^{-iyx}F(x)\,dx$. (This is "Poisson's summation formula," which can be found in standard texts such as Henrici [151, Theorem 10.6e].)

9.60 The stated formula is equivalent to

$$n^{\overline{1/2}} \;=\; n^{1/2}\Bigl(1 - \frac{1}{8n} + \frac{1}{128n^2} + \frac{5}{1024n^3} - \frac{21}{32768n^4} + O(n^{-5})\Bigr)$$

by exercise 5.22. Hence the result follows from exercises 6.64 and 9.44.

9.61 The idea is to make α "almost" rational. Let $a_k = 2^{2^{2^k}}$ be the kth partial quotient of α, and let $n = \frac12 a_{m+1}q_m$, where $q_m = K(a_1,\ldots,a_m)$ and m is even. Then $0 < \{q_m\alpha\} < 1/Q(a_1,\ldots,a_{m+1}) < 1/(2n)$, and if we take $\nu = a_{m+1}/(4n)$ we get a discrepancy $\geqslant \frac14 a_{m+1}$. If this were less than $n^{1-\epsilon}$ we would have

$$a_{m+1}^{\epsilon} \;=\; O(q_m^{1-\epsilon})\,,$$

but in fact $a_{m+1} > q_m^{2^m}$.

9.62 See Canfield [43]; see also David and Barton [60, Chapter 16] for asymptotics of Stirling numbers of both kinds.

9.63 Let $c = \phi^{2-\phi}$. The estimate $cn^{\phi-1} + o(n^{\phi-1})$ was proved by Fine [120]. Ilan Vardi observes that the sharper estimate stated can be deduced from the fact that the error term $e(n) = f(n) - cn^{\phi-1}$ satisfies the approximate recurrence $c^{\phi}n^{2-\phi}e(n) \approx -\sum_k e(k)[1 \leqslant k < cn^{\phi-1}]$. The function

$$\frac{n^{\phi-1}u(\ln\ln n/\ln\phi)}{\ln n}$$

satisfies this recurrence asymptotically, if $u(x+1) = -u(x)$. (Vardi conjectures that

$$f(n) = n^{\phi-1}\Bigl(c + u\Bigl(\frac{\ln\ln n}{\ln\phi}\Bigr)(\ln n)^{-1} + O\bigl((\log n)^{-2}\bigr)\Bigr)$$

for some such function u.) Calculations for small n show that $f(n)$ equals the nearest integer to $cn^{\phi-1}$ for $1 \leqslant n \leqslant 400$ except in one case: $f(273) = 39 > c \cdot 273^{\phi-1} \approx 38.4997$. But the small errors are eventually magnified, because of results like those in exercise 2.36. For example, $e(201636503) \approx 35.73$; $e(919986484788) \approx -1959.07$.

"The paradox is now fully established that the utmost abstractions are the true weapons with which to control our thought of concrete fact."
—A. N. Whitehead [304]

9.64 (From this identity for $B_2(x)$ we can easily derive the identity of exercise 58 by induction on m.) If $0 < x < 1$, the integral $\int_x^{1/2} \sin N\pi t\, dt/\sin\pi t$ can be expressed as a sum of N integrals that are each $O(N^{-2})$, so it is $O(N^{-1})$; the constant implied by this O may depend on x. Integrating the identity $\sum_{n=1}^{N} \cos 2n\pi t = \Re\bigl(e^{2\pi i t}(e^{2N\pi i t}-1)/(e^{2\pi i t}-1)\bigr) = -\frac{1}{2}+\frac{1}{2}\sin(2N+1)\pi t/\sin\pi t$ and letting $N \to \infty$ now gives $\sum_{n\geqslant 1}(\sin 2n\pi x)/n = \frac{\pi}{2} - \pi x$, a relation that Euler once stated without proof [85′]. Integrate again to get the desired formula. (This solution was suggested by E. M. E. Wermuth.)

9.65 The expected number of distinct elements in the sequence 1, $f(1)$, $f(f(1))$, $\ldots$, when f is a random mapping of $\{1, 2, \ldots, n\}$ into itself, is the function $Q(n)$ of exercise 56, whose value is $\frac{1}{2}\sqrt{2\pi n} + O(1)$; this might account somehow for the factor $\sqrt{2\pi n}$.

9.66 It is known that $\ln \chi_n \sim \frac{3}{2}n^2 \ln\frac{4}{3}$; the constant $e^{-\pi/6}$ has been verified empirically to eight significant digits.

9.67 This would fail if, for example, $e^{n-\gamma} = m + \frac{1}{2} + \epsilon/m$ for some integer m and some $0 < \epsilon < \frac{1}{8}$; but no counterexamples are known.

B

Bibliography

HERE ARE THE WORKS cited in this book. Numbers in the margin specify the page numbers where citations occur.

References to published problems are generally made to the places where solutions can be found, instead of to the original problem statements, unless no solution has yet appeared in print.

"This paper fills a much-needed gap in the literature."
— Math. Reviews

1 N. H. Abel, letter to B. Holmboe (1823), in his *Œuvres Complètes*, first edition, 1839, volume 2, 264–265. Reprinted in the second edition, 1881, volume 2, 254–255. *603.*

2 Milton Abramowitz and Irene A. Stegun, editors, *Handbook of Mathematical Functions.* United States Government Printing Office, 1964. Reprinted by Dover, 1965. *42.*

3 William W. Adams and J. L. Davison, "A remarkable class of continued fractions," *Proceedings of the American Mathematical Society* **65** (1977), 194–198. *604.*

4 A. V. Aho and N. J. A. Sloane, "Some doubly exponential sequences," *Fibonacci Quarterly* **11** (1973), 429–437. *602.*

5 W. Ahrens, *Mathematische Unterhaltungen und Spiele.* Teubner, Leipzig, 1901. Second edition, in two volumes, 1910 and 1918. *8, 602.*

6 Naum Il'ich Akhiezer, *Klassicheskaîa Problema Momentov i Nekotorye Voprosy Analiza, Svîazannye s Neîu.* Moscow, 1961. English translation, *The classical Moment Problem and Some Related Questions in Analysis,* Hafner, 1965. *566.*

7 R. E. Allardice and A. Y. Fraser, "La Tour d'Hanoï," *Proceedings of the Edinburgh Mathematical Society* **2** (1884), 50–53. *2.*

8 Désiré André, "Sur les permutations alternées," *Journal de Mathématiques pures et appliquées*, series 3, **7** (1881), 167–184. *604.*

215, 603. **9** George E. Andrews, "Applications of basic hypergeometric functions," *SIAM Review* **16** (1974), 441–484.

515. **10** George E. Andrews, "On sorting two ordered sets," *Discrete Mathematics* **11** (1975), 97–106.

316. **11** George E. Andrews, *The Theory of Partitions.* Addison-Wesley, 1976.

604. **12** George E. Andrews and K. Uchimura, "Identities in combinatorics IV: Differentiation and harmonic numbers," *Utilitas Mathematica* **28** (1985), 265–269.

602. **13** M. D. Atkinson, "The cyclic towers of Hanoi," *Information Processing Letters* **13** (1981), 118–119.

429. **14** Paul Bachmann, *Die analytische Zahlentheorie.* Teubner, Leipzig, 1894.

223, 603. **15** W. N. Bailey, *Generalized Hypergeometric Series.* Cambridge University Press, 1935; second edition, 1964.

602. **16** W. W. Rouse Ball and H. S. M. Coxeter, *Mathematical Recreations and Essays*, twelfth edition. University of Toronto Press, 1974. (A revision of Ball's *Mathematical Recreations and Problems*, first published by Macmillan, 1892.)

603. **17** P. Barlow, "Demonstration of a curious numerical proposition," *Journal of Natural Philosophy, Chemistry, and the Arts* **27** (1810), 193–205.

602. **18** Samuel Beatty, "Problem 3177," *American Mathematical Monthly* **34** (1927), 159–160.

318. **19** E. T. Bell, "Euler algebra," *Transactions of the American Mathematical Society* **25** (1923), 135–154.

604. **20** E. T. Bell, "Exponential numbers," *American Mathematical Monthly* **41** (1934), 411–419.

605. **21** Edward A. Bender, "Asymptotic methods in enumeration," *SIAM Review* **16** (1974), 485–515.

269. **22** Jacobi Bernoulli, *Ars Conjectandi*, opus posthumum. Basel, 1713. Reprinted in *Die Werke von Jakob Bernoulli*, volume 3, 107–286.

602. **23** J. Bertrand, "Mémoire sur le nombre de valeurs que peut prendre une fonction quand on y permute les lettres qu'elle renferme," *Journal de l'École Royale Polytechnique* **18**, cahier 30 (1845), 123–140.

42. **24** William H. Beyer, editor, *CRC Standard Mathematical Tables*, 25th edition. CRC Press, Boca Raton, Florida, 1978.

24′ J. Bienaymé, "Considérations à l'appui de la découverte de Laplace sur la loi de probabilité dans la méthode des moindres carrés," *Comptes Rendus hebdomadaires des séances de l'Académie des Sciences* (Paris) **37** (1853), 309–324. *376.*

25 J. Binet, "Mémoire sur l'intégration des équations linéaires aux différences finies, d'un ordre quelconque, à coefficients variables," *Comptes Rendus hebdomadaires des séances de l'Académie des Sciences* (Paris) **17** (1843), 559–567. *285.*

26 Gunnar Blom, "Problem E 3043: Random walk until no shoes," *American Mathematical Monthly* **94** (1987), 78–79. *605.*

27 R. P. Boas, Jr. and J. W. Wrench, Jr., "Partial sums of the harmonic series," *American Mathematical Monthly* **78** (1971), 864–870. *574, 605.*

28 P. Bohl, "Über ein in der Theorie der säkularen Störungen vorkommendes Problem," *Journal für die reine und angewandte Mathematik* **135** (1909), 189–283. *87.*

29 P. du Bois-Reymond, "Sur la grandeur relative des infinis des fonctions," *Annali di Matematica pura e applicata*, series 2, **4** (1871), 338–353. *426.*

30 Émile Borel, *Leçons sur les séries à termes positifs.* Gauthier-Villars, 1902. *605.*

31 Jonathan M. Borwein and Peter B. Borwein, *Pi and the AGM.* Wiley, 1987. *604.*

32 Richard P. Brent, "The first occurrence of large gaps between successive primes," *Mathematics of Computation* **27** (1973), 959–963. *510.*

33 Richard P. Brent, "Computation of the regular continued fraction for Euler's constant," *Mathematics of Computation* **31** (1977), 771–777. *292, 540.*

34 John Brillhart, "Some miscellaneous factorizations," *Mathematics of Computation* **17** (1963), 447–450. *602.*

35 Achille Brocot, "Calcul des rouages par approximation, nouvelle méthode," *Revue Chronométrique* **6** (1860), 186–194. (He also published a 97-page monograph with the same title in 1862.) *116.*

36 Maxey Brooke and C. R. Wall, "Problem B-14: A little surprise," *Fibonacci Quarterly* **1**, 3 (1963), 80. *604.*

37 Brother U. Alfred [Brousseau], "A mathematician's progress," *Mathematics Teacher* **59** (1966), 722–727. *602.*

38 Morton Brown, "Problem 6439: A periodic sequence," *American Mathematical Monthly* **92** (1985), 218. *487.*

39 T. Brown, "Infinite multi-variable subpolynormal Woffles which do not satisfy the lower regular Q-property (Piffles)," in *A Collection of 250 Papers on Woffle Theory Dedicated to R. S. Green on His 23rd Birthday.* Cited in A. K. Austin, "Modern research in mathematics," *The Mathematical Gazette* **51** (1967), 149–150.

602. **40** Thomas C. Brown, "Problem E 2619: Squares in a recursive sequence," *American Mathematical Monthly* **85** (1978), 52–53.

344. **41** William G. Brown, "Historical note on a recurrent combinatorial problem," *American Mathematical Monthly* **72** (1965), 973–977.

604. **42** S. A. Burr, "On moduli for which the Fibonacci sequence contains a complete system of residues," *Fibonacci Quarterly* **9** (1971), 497–504.

577, 605. **43** E. Rodney Canfield, "On the location of the maximum Stirling number(s) of the second kind," *Studies in Applied Mathematics* **59** (1978), 83–93.

31. **44** Lewis Carroll [pseudonym of C. L. Dodgson], *Through the Looking Glass and What Alice Found There.* Macmillan, 1871.

278. **45** Jean-Dominique Cassini, "Une nouvelle progression de nombres," *Histoire de l'Académie Royale des Sciences*, Paris, volume 1, 201. (Cassini's work is summarized here as one of the mathematical results presented to the academy in 1680. This volume was published in 1733.)

203. **46** E. Catalan, "Note sur une Équation aux différences finies," *Journal de Mathématiques pures et appliquées* **3** (1838), 508–516.

602. **47** Augustin-Louis Cauchy, *Cours d'analyse de l'École Royale Polytechnique.* Imprimerie Royale, Paris, 1821. Reprinted in his *Œuvres Complètes*, series 2, volume 3.

602. **48** Arnold Buffum Chace, *The Rhind Mathematical Papyrus*, volume 1. Mathematical Association of America, 1927. (Includes an excellent bibliography of Egyptian mathematics by R. C. Archibald.)

510. **49** M. Chaimovich, G. Freiman, and J. Schönheim, "On exceptions to Szegedy's theorem," *Acta Arithmetica* **49** (1987), 107–112.

602. **50** P. L. Tchebichef [Chebyshev], "Mémoire sur les nombres premiers," *Journal de Mathématiques pures et appliquées* **17** (1852), 366–390. Reprinted in his *Œuvres*, volume 1, 51–70.

376. **50′** P. L. Chebyshev, "O srednikh velichinakh," *Matematicheskiĭ Sbornik'* **2** (1867), 1–9. Reprinted in his *Polnoe Sobranie Sochineniĭ*, volume 2, 431–437. French translation, "Des valeurs moyennes," *Journal de Mathématiques pures et appliquées*, series 2, **12** (1867), 177–184; reprinted in his *Œuvres*, volume 1, 685–694.

51 Th. Clausen, "Ueber die Fälle, wenn die Reihe von der Form *603.*

$$y = 1 + \frac{\alpha}{1} \cdot \frac{\beta}{\gamma} x + \frac{\alpha . \alpha + 1}{1.2} \cdot \frac{\beta . \beta + 1}{\gamma . \gamma + 1} x^2 + \text{etc.}$$

ein Quadrat von der Form

$$z = 1 + \frac{\alpha'}{1} \cdot \frac{\beta'}{\gamma'} \cdot \frac{\delta'}{\epsilon'} x + \frac{\alpha' . \alpha' + 1}{1.2} \cdot \frac{\beta' . \beta' + 1}{\gamma' . \gamma' + 1} \cdot \frac{\delta' . \delta' + 1}{\epsilon' . \epsilon' + 1} x^2 + \text{etc.} \text{ hat,"}$$

Journal für die reine und angewandte Mathematik **3** (1828), 89–91.

52 Th. Clausen, "Beitrag zur Theorie der Reihen," *Journal für die reine und angewandte Mathematik* **3** (1828), 92–95. *603.*

53 Th. Clausen, "Theorem," *Astronomische Nachrichten* **17** (1840), columns 351–352. *604.*

54 Stuart Dodgson Collingwood, *The Lewis Carroll Picture Book.* T. Fisher Unwin, 1899. Reprinted by Dover, 1961, with the new title *Diversions and Digressions of Lewis Carroll.* *279.*

55 J. H. Conway and R. L. Graham, "Problem E 2567: A periodic recurrence," *American Mathematical Monthly* **84** (1977), 570–571. *487.*

56 Harald Cramér, "On the order of magnitude of the difference between consecutive prime numbers," *Acta Arithmetica* **2** (1937), 23–46. *510, 603.*

57 A. L. Crelle, "Démonstration élémentaire du théorème de Wilson généralisé," *Journal für die reine und angewandte Mathematik* **20** (1840), 29–56. *602.*

58 D. W. Crowe, "The n-dimensional cube and the Tower of Hanoi," *American Mathematical Monthly* **63** (1956), 29–30. *602.*

59 D. R. Curtiss, "On Kellogg's Diophantine problem," *American Mathematical Monthly* **29** (1922), 380–387. *603.*

60 F. N. David and D. E. Barton, *Combinatorial Chance.* Hafner, 1962. *577.*

61 J. L. Davison, "A series and its associated continued fraction," *Proceedings of the American Mathematical Society* **63** (1977), 29–32. *293, 604.*

62 N. G. de Bruijn, *Asymptotic Methods in Analysis.* North-Holland, 1958; third edition, 1970. Reprinted by Dover, 1981. *433, 605.*

63 N. G. de Bruijn, "Problem 9," *Nieuw Archief voor Wiskunde*, series 3, **12** (1964), 68. *604.*

64 Abraham de Moivre, *Miscellanea analytica de seriebus et quadraturis.* London, 1730. *283.*

496. **65** Leonard Eugene Dickson, *History of the Theory of Numbers.* Carnegie Institution of Washington, volume 1, 1919; volume 2, 1920; volume 3, 1923. Reprinted by Stechert, 1934, and by Chelsea, 1952, 1971.

604. **66** Edsger W. Dijkstra, *Selected Writings on Computing: A Personal Perspective.* Springer-Verlag, 1982.

602. **67** G. Lejeune Dirichlet, "Verallgemeinerung eines Satzes aus der Lehre von den Kettenbrüchen nebst einigen Anwendungen auf die Theorie der Zahlen," *Bericht über die Verhandlungen der Königlich-Preußischen Akademie der Wissenschaften zu Berlin* (1842), 93–95. Reprinted in his *Werke*, volume 1, 635–638.

603. **68** A. C. Dixon, "On the sum of the cubes of the coefficients in a certain expansion by the binomial theorem," *Messenger of Mathematics* **20** (1891), 79–80.

171. **69** John Dougall, "On Vandermonde's theorem, and some more general expansions," *Proceedings of the Edinburgh Mathematical Society* **25** (1907), 114–132.

227, 391. **70** A. Conan Doyle, "The sign of the four; or, The problem of the Sholtos," *Lippincott's Monthly Magazine* (Philadelphia) **45** (1890), 147–223.

162. **71** A. Conan Doyle, "The adventure of the final problem," *The Strand Magazine* **6** (1893), 558–570.

602. **72** Henry Ernest Dudeney, *The Canterbury Puzzles and Other Curious Problems.* E. P. Dutton, New York, 1908; 4th edition, Dover, 1958. (Dudeney had first considered the generalized Tower of Hanoi in *The Weekly Dispatch*, on 25 May 1902 and 15 March 1903.)

6. **73** G. Waldo Dunnington, *Carl Friedrich Gauss: Titan of Science.* Exposition Press, New York, 1955.

155. **74** A. W. F. Edwards, *Pascal's Arithmetical Triangle.* Oxford University Press, 1987.

202. **75** G. Eisenstein, "Entwicklung von $\alpha^{\alpha^{\alpha^{\cdot^{\cdot^{\cdot}}}}}$," *Journal für die reine und angewandte Mathematik* **28** (1844), 49–52. Reprinted in his *Mathematische Werke* **1**, 122–125.

603. **76** Erdős Pál, "Az $\frac{1}{x_1} + \frac{1}{x_2} + \cdots + \frac{1}{x_n} = \frac{a}{b}$ egyenlet egész számú megoldásairól," *Matematikai Lapok* **1** (1950), 192–209. English abstract on page 210.

500, 510, 603, 604, 605. **77** P. Erdős and R. L. Graham, *Old and New Problems and Results in Combinatorial Number Theory.* Université de Genève, L'Enseignement Mathématique, 1980.

78 P. Erdős, R. L. Graham, I. Z. Ruzsa, and E. G. Straus, "On the prime factors of $\binom{2n}{n}$," *Mathematics of Computation* **29** (1975), 83–92. *511, 526.*

79 Arulappah Eswarathasan and Eugene Levine, "p-integral harmonic sums," submitted for publication. *604.*

80 Euclid, *ΣΤΟΙΧΕΙΑ*. Ancient manuscript first printed in Basel, 1533. Scholarly edition (Greek and Latin) by J. L. Heiberg in five volumes, Teubner, Leipzig, 1883–1888. *108.*

81 Leonhard Euler, letter to Christian Goldbach (13 October 1729), in *Correspondance Mathématique et Physique de Quelques Célèbres Géomètres du XVIIIème Siècle*, edited by P. H. Fuss, St. Petersburg, 1843, volume 1, 3–7. *210, 603.*

82 Leonhard Euler, "Methodus generalis summandi progressiones," *Commentarii academiæ scientiarum Petropolitanæ* **6** (1732), 68–97. Reprinted in his *Opera Omnia*, series 1, volume 14, 42–72. *455.*

83 Leonhard Euler, "De progressionibus harmonicis observationes," *Commentarii academiæ scientiarum Petropolitanæ* **7** (1734), 150–161. Reprinted in his *Opera Omnia*, series 1, volume 14, 87–100. *264.*

84 Leonhard Euler, "De fractionibus continuis, Dissertatio," *Commentarii academiæ scientiarum Petropolitanæ* **9** (1737), 98–137. Reprinted in his *Opera Omnia*, series 1, volume 14, 187–215. *122.*

85 Leonhard Euler, "Variæ observationes circa series infinitas," *Commentarii academiæ scientiarum Petropolitanæ* **9** (1737), 160–188. Reprinted in his *Opera Omnia*, series 1, volume 14, 216–244. *602.*

85′ Leonhard Euler, letter to Christian Goldbach (4 July 1744), in *Correspondance Mathématique et Physique de Quelques Célèbres Géomètres du XVIIIème Siècle*, edited by P. H. Fuss, St. Petersburg, 1843, volume 1, 278–293. *577.*

86 Leonhard Euler, *Introductio in Analysin Infinitorum*. Tomus primus, Lausanne, 1748. Reprinted in his *Opera Omnia*, series 1, volume 8. Translated into French, 1786; German, 1788. *604.*

87 Leonhard Euler, "De partitione numerorum," *Novi commentarii academiæ scientiarum Petropolitanæ* **3** (1750), 125–169. Reprinted in his *Commentationes arithmeticæ collectæ*, volume 1, 73–101. Reprinted in his *Opera Omnia*, series 1, volume 2, 254–294. *604.*

88 Leonhard Euler, *Institutiones Calculi Differentialis cum eius usu in Analysi Finitorum ac Doctrina Serierum*. Petrograd, Academiæ Imperialis Scientiarum, 1755. Reprinted in his *Opera Omnia*, series 1, volume 10. Translated into German, 1790. *48, 253.*

133, 134. **89** Leonhard Euler, "Theoremata arithmetica nova methodo demonstrata," *Novi commentarii academiæ scientiarum Petropolitanæ* **8** (1760), 74–104. (Also presented in 1758 to the Berlin Academy.) Reprinted in his *Commentationes arithmeticæ collectæ*, volume 1, 274–286. Reprinted in his *Opera Omnia*, series 1, volume 2, 531–555.

289. **90** Leonhard Euler, "Specimen algorithmi singularis," *Novi commentarii academiæ scientiarum Petropolitanæ* **9** (1762), 53–69. (Also presented in 1757 to the Berlin Academy.) Reprinted in his *Opera Omnia*, series 1, volume 15, 31–49.

285, 605. **91** Leonhard Euler, "Observationes analyticæ," *Novi commentarii academiæ scientiarum Petropolitanæ* **11** (1765), 124–143. Reprinted in his *Opera Omnia*, series 1, volume 15, 50–69.

604. **92** Leonhard Euler, *Vollständige Anleitung zur Algebra. Erster Theil. Von den verschiedenen Rechnungs-Arten, Verhältnissen und Proportionen.* St. Petersburg, 1770. Reprinted in his *Opera Omnia*, series 1, volume 1. Translated into French, 1774; Dutch, 1778; Latin, 1790; English, 1797; Russian, 1812.

131. **93** Leonhard Euler, "Observationes circa bina biquadrata quorum summam in duo alia biquadrata resolvere liceat," *Novi commentarii academiæ scientiarum Petropolitanæ* **17** (1772), 64–69. Reprinted in his *Opera Omnia*, series 1, volume 3, 211–217.

499. **94** Leonhard Euler, "Observationes circa novum et singulare progressionum genus," *Novi commentarii academiæ scientiarum Petropolitanæ* **20** (1775), 123–139. Reprinted in his *Opera Omnia*, series 1, volume 7, 246–261.

207, 603. **95** Leonhard Euler, "Specimen transformationis singularis serierum," *Nova acta academiæ scientiarum Petropolitanæ* **12** (1794), 58–70. Submitted for publication in 1778. Reprinted in his *Opera Omnia*, series 1, volume 16(2), 41–55.

367, 605. **96** William Feller, *An Introduction to Probability Theory and Its Applications*, volume 1. Wiley, 1950; second edition, 1957; third edition, 1968.

131. **97** Pierre de Fermat, letter to Marin Mersenne (25 December 1640), in *Œuvres de Fermat*, volume 2, 212–217.

602, 603. **98** Leonardo Fibonacci [Pisano], *Liber Abaci.* First edition, 1202 (now lost); second edition 1228. Reprinted in *Scritti di Leonardo Pisano*, edited by Baldassarre Boncompagni, 1857, volume 1.

604. **99** Michael E. Fisher, "Statistical mechanics of dimers on a plane lattice," *Physical Review* **124** (1961), 1664–1672.

100 R. A. Fisher, "Moments and product moments of sampling distributions," *Proceedings of the London Mathematical Society*, series 2, **30** (1929), 199–238. *605.*

101 Pierre Forcadel, *L'arithmeticque.* Paris, 1557. *603.*

102 J. Fourier, "Refroidissement séculaire du globe terrestre," *Bulletin des Sciences par la Société philomathique de Paris*, series 3, **7** (1820), 58–70. Reprinted in *Œuvres de Fourier*, volume 2, 271–288. *22.*

103 Aviezri S. Fraenkel, "Complementing and exactly covering sequences," *Journal of Combinatorial Theory*, series A, **14** (1973), 8–20. *500, 602.*

104 Aviezri S. Fraenkel, "How to beat your Wythoff games' opponent on three fronts," *American Mathematical Monthly* **89** (1982), 353–361. *538.*

105 J. S. Frame, B. M. Stewart, and Otto Dunkel, "Partial solution to problem 3918," *American Mathematical Monthly* **48** (1941), 216–219. *602.*

106 Piero della Francesca, *Libellus de quinque corporibus regularibus.* Vatican Library, manuscript Urbinas 632. Translated into Italian by Luca Pacioli, as part 3 of Pacioli's *Diuine Proportione*, Venice, 1509. *604.*

107 W. D. Frazer and A. C. McKellar, "Samplesort: A sampling approach to minimal storage tree sorting," *Journal of the ACM* **27** (1970), 496–507. *603.*

108 Michael Lawrence Fredman, *Growth Properties of a Class of Recursively Defined Functions.* Ph.D. thesis, Stanford University, Computer Science Department, 1972. *499.*

109 Nikolaus I. Fuss, "Solutio quæstionis, quot modis polygonum n laterum in polygona m laterum, per diagonales resolvi quæat," *Nova acta academiæ scientiarum Petropolitanæ* **9** (1791), 243–251. *347.*

110 Martin Gardner, "About phi, an irrational number that has some remarkable geometrical expressions," *Scientific American* **201**, 2 (August 1959), 128–134. Reprinted with additions in his book *The 2nd Scientific American Book of Mathematical Puzzles & Diversions*, 1961, 89–103. *285.*

111 Martin Gardner, "On the paradoxical situations that arise from nontransitive relations," *Scientific American* **231**, 4 (October 1974), 120–124. Reprinted with additions in his book *Time Travel and Other Mathematical Bewilderments*, 1988, 55–69. *396.*

112 Martin Gardner, "From rubber ropes to rolling cubes, a miscellany of refreshing problems," *Scientific American* **232**, 3 (March 1975), 112–114; **232**, 4 (April 1975), 130, 133. Reprinted with additions in his book *Time Travel and Other Mathematical Bewilderments*, 1988, 111–124. *603.*

113 Martin Gardner, "On checker jumping, the amazon game, weird dice, card tricks and other playful pastimes," *Scientific American* **238**, 2 (February 1978), 19, 22, 24, 25, 30, 32. *605.*

605. **114** J. Garfunkel, "Problem E 1816: An inequality related to Stirling's formula," *American Mathematical Monthly* **74** (1967), 202.

123, 602. **115** C. F. Gauss, *Disquisitiones Arithmeticæ*. Leipzig, 1801. Reprinted in his *Werke*, volume 1.

207, 222, 514, 603. **116** Carolo Friderico Gauss, "Disquisitiones generales circa seriem infinitam

$$1+\frac{\alpha\beta}{1.\gamma}x+\frac{\alpha(\alpha+1)\beta(\beta+1)}{1\,.\,2\,.\,\gamma(\gamma+1)}xx + \frac{\alpha(\alpha+1)(\alpha+2)\beta(\beta+1)(\beta+2)}{1\,.\,2\,.\,3\,.\,\gamma(\gamma+1)(\gamma+2)}x^3+\text{etc.}$$

Pars prior," *Commentationes societatis regiæ scientiarum Gottingensis recentiores* **2** (1813). (Thesis delivered to the Royal Society in Göttingen, 20 January 1812.) Reprinted in his *Werke*, volume 3, 123–163, together with an unpublished sequel on pages 207–229.

528. **117** A. Genocchi, "Intorno all' expressioni generali di numeri Bernoulliani," *Annali di Scienze Matematiche e Fisiche* **3** (1852), 395–405.

256. **118** Ira Gessel and Richard P. Stanley, "Stirling polynomials," *Journal of Combinatorial Theory*, series A, **24** (1978), 24–33.

257. **119** Jekuthiel Ginsburg, "Note on Stirling's numbers," *American Mathematical Monthly* **35** (1928), 77–80.

577, 602. **120** Solomon W. Golomb, "Problem 5407: A nondecreasing indicator function," *American Mathematical Monthly* **74** (1967), 740–743.

493. **121** Solomon W. Golomb, "The 'Sales Tax' theorem," *Mathematics Magazine* **49** (1976), 187–189.

446. **122** Solomon W. Golomb, "Problem E 2529: An application of $\psi(x)$," *American Mathematical Monthly* **83** (1976), 487–488.

603. **123** I. J. Good, "Short proof of a conjecture by Dyson," *Journal of Mathematical Physics* **11** (1970), 1884.

224, 603. **124** R. William Gosper, Jr., "Decision procedure for indefinite hypergeometric summation," *Proceedings of the National Academy of Sciences of the United States of America* **75** (1978), 40–42.

498. **125** R. L. Graham, "On a theorem of Uspensky," *American Mathematical Monthly* **70** (1963), 407–409.

604. **126** R. L. Graham, "A Fibonacci-like sequence of composite numbers," *Mathematics Magazine* **37** (1964), 322–324.

603. **127** R. L. Graham, "Problem 5749," *American Mathematical Monthly* **77** (1970), 775.

128 Ronald L. Graham, "Covering the positive integers by disjoint sets of the form $\{ [n\alpha + \beta] : n = 1, 2, \ldots \}$," *Journal of Combinatorial Theory*, series A, **15** (1973), 354–358. *500.*

129 R. L. Graham, "Problem 1242: Bijection between integers and composites," *Mathematics Magazine* **60** (1987), 180. *602.*

130 R. L. Graham and D. E. Knuth, "Problem E 2982," *American Mathematical Monthly* **90** (1983), 54. *602.*

131 Ronald L. Graham, Donald E. Knuth, and Oren Patashnik, *Concrete Mathematics: A Foundation for Computer Science.* Addison-Wesley, 1989. (The first printing had a different Iversonian notation.) *102.*

132 R. L. Graham and H. O. Pollak, "Note on a nonlinear recurrence related to $\sqrt{2}$," *Mathematics Magazine* **43** (1970), 143–145. *602.*

133 Guido Grandi, letter to Leibniz (July 1713), in *Leibnizens mathematische Schriften*, volume 4, 215–217. *58.*

134 Daniel H. Greene and Donald E. Knuth, *Mathematics for the Analysis of Algorithms.* Birkhäuser, Boston, 1981; second edition, 1982. *520, 605.*

135 Samuel L. Greitzer, *International Mathematical Olympiads, 1959–1977.* Mathematical Association of America, 1978. *602.*

136 Oliver A. Gross, "Preferential arrangements," *American Mathematical Monthly* **69** (1962), 4–8. *604.*

137 Branko Grünbaum, "Venn diagrams and independent families of sets," *Mathematics Magazine* **48** (1975), 12–23. *484.*

138 L. J. Guibas and A. M. Odlyzko, "String overlaps, pattern matching, and nontransitive games," *Journal of Combinatorial Theory*, series A, **30** (1981), 183–208. *565, 605*

139 Richard K. Guy, *Unsolved Problems in Number Theory.* Springer-Verlag, 1981. *510.*

140 Marshall Hall, Jr., *The Theory of Groups.* Macmillan, 1959. *530.*

141 P. R. Halmos, "How to write mathematics," *L'Enseignement mathématique* **16** (1970), 123–152. Reprinted in *How to Write Mathematics*, American Mathematical Society, 1973, 19–48. *vi.*

142 Paul R. Halmos, *I Want to Be a Mathematician: An Automathography.* Springer-Verlag, 1985. Reprinted by Mathematical Association of America, 1988. *v.*

143 G. H. Halphen, "Sur des suites de fractions analogues à la suite de Farey," *Bulletin de la Société mathématique de France* **5** (1876), 170–175. Reprinted in his *Œuvres*, volume 2, 102–107. *291.*

566. **144** Hans Hamburger, "Über eine Erweiterung des Stieltjesschen Momentenproblems," *Mathematische Annalen* **81** (1920), 235–319; **82** (1921), 120–164, 168–187.

v. **145** J. M. Hammersley, "On the enfeeblement of mathematical skills by 'Modern Mathematics' and by similar soft intellectual trash in schools and universities," *Bulletin of the Institute of Mathematics and its Applications* **4**, 4 (October 1968), 66–85.

604. **146** J. M. Hammersley, "An undergraduate exercise in manipulation," preprint from Trinity College, Oxford, 1987.

42. **147** Eldon R. Hansen, *A Table of Series and Products.* Prentice-Hall, 1975.

428, 605. **148** G. H. Hardy, *Orders of Infinity: The 'Infinitärcalcül' of Paul du Bois-Reymond.* Cambridge University Press, 1910; second edition, 1924.

605. **149** G. H. Hardy, "A mathematical theorem about golf," *The Mathematical Gazette* **29** (1944), 226–227. Reprinted in his *Collected Papers*, volume 7, 488.

111, 602. **150** G. H. Hardy and E. M. Wright, *An Introduction to the Theory of Numbers.* Clarendon Press, Oxford, 1938; fifth edition, 1979.

286, 318, 576, 605. **151** Peter Henrici, *Applied and Computational Complex Analysis.* Wiley, volume 1, 1974; volume 2, 1977; volume 3, 1986.

603. **152** Peter Henrici, "De Branges' proof of the Bieberbach conjecture: A view from computational analysis," *Sitzungsberichte der Berliner Mathematischen Gesellschaft* (1987), 105–121.

532. **153** Charles Hermite, letter to C. W. Borchardt (8 September 1875), in *Journal für die reine und angewandte Mathematik* **81** (1876), 93–95. Reprinted in his *Œuvres*, volume 3, 211–214.

603. **154** Charles Hermite, *Cours de M. Hermite.* Faculté des Sciences de Paris, 1882. Third edition, 1887; fourth edition, 1891.

524, 603. **155** Charles Hermite, letter to S. Pincherle (10 May 1900), in *Annali di Matematica pura e applicata*, series 3, **5** (1901), 57–60. Reprinted in his *Œuvres*, volume 4, 529–531.

8. **156** I. N. Herstein and I. Kaplansky, *Matters Mathematical.* Harper & Row, 1974.

603. **157** A. P. Hillman and V. E. Hoggatt, Jr., "A proof of Gould's Pascal hexagon conjecture," *Fibonacci Quarterly* **10** (1972), 565–568, 598.

28. **158** C. A. R. Hoare, "Quicksort," *The Computer Journal* **5** (1962), 10–15.

603. **159** L. C. Hsu, "Note on a combinatorial algebraic identity and its application," *Fibonacci Quarterly* **11** (1973), 480–484.

160 K. Inkeri, "Abschätzungen für eventuelle Lösungen der Gleichung im Fermatschen Problem," *Annales Universitatis Turkuensis*, series A, **16**, 1 (1953), 3–9. *509.*

161 Kenneth E. Iverson, *A Programming Language*. Wiley, 1962. *24, 67, 602.*

162 C. G. J. Jacobi, *Fundamenta nova theoriæ functionum ellipticarum*. Königsberg, Bornträger, 1829. Reprinted in his *Gesammelte Werke*, volume 1, 49–239. *64.*

163 Dov Jarden and Theodor Motzkin, "The product of sequences with a common linear recursion formula of order 2," *Riveon Lematematika* **3** (1949), 25–27, 38 (Hebrew with English summary). English version reprinted in Dov Jarden, *Recurring Sequences*, Jerusalem, 1958, 42–45; second edition, Jerusalem, 1966, 30–33. *533.*

164 Arne Jonassen and Donald E. Knuth, "A trivial algorithm whose analysis isn't," *Journal of Computer and System Sciences* **16** (1978), 301–322. *520.*

165 Bush Jones, "Note on internal merging," *Software — Practice and Experience* **2** (1972), 241–243. *175.*

166 Flavius Josephus, *ΙΣΤΟΡΙΑ ΙΟΥΔΑΪΚΟΥ ΠΟΛΕΜΟΥ ΠΡΟΣ ΡΩΜΑΙΟΥΣ*. English translation, *History of the Jewish War against the Romans*, by H. St. J. Thackeray, in the Loeb Classical Library edition of Josephus's works, volumes 2 and 3, Heinemann, London, 1927–1928. (The "Josephus problem" may be based on an early manuscript now preserved only in the Slavonic version; see volume 2, page xi, and volume 3, page 654.) *8.*

167 R. Jungen, "Sur les séries de Taylor n'ayant que des singularités algébrico-logarithmiques sur leur cercle de convergence," *Commentarii Mathematici Helvetici* **3** (1931), 266–306. *604.*

168 I. Kaucký, "Problem E 2257: A harmonic identity," *American Mathematical Monthly* **78** (1971), 908. *604.*

169 Murray S. Klamkin, *International Mathematical Olympiads, 1978–1985, and Forty Supplementary Problems*. Mathematical Association of America, 1986. *602, 603.*

170 Konrad Knopp, *Theorie und Anwendung der unendlichen Reihen*. Julius Springer, Berlin, 1922; second edition, 1924. Reprinted by Dover, 1945. Fourth edition, 1947; fifth edition, 1964. English translation, *Theory and Application of Infinite Series*, 1928; second edition, 1951. *605.*

171 Donald E. Knuth, "Euler's constant to 1271 places," *Mathematics of Computation* **16** (1962), 275–281. *467.*

172 Donald Knuth, "Transcendental numbers based on the Fibonacci sequence," *Fibonacci Quarterly* **2** (1964), 43–44, 52. *531.*

168. **200** G. W. Leibniz, letter to Johann Bernoulli (May 1695), in *Leibnizens mathematische Schriften*, volume 3, 174–179.

281. **201** C. G. Lekkerkerker, "Voorstelling van natuurlijke getallen door een som van getallen van Fibonacci," *Simon Stevin* **29** (1952), 190–195.

605. **202** Elliott H. Lieb, "Residual entropy of square ice," *Physical Review* **162** (1967), 162–172.

603. **203** Calvin T. Long and Verner E. Hoggatt, Jr., "Sets of binomial coefficients with equal products," *Fibonacci Quarterly* **12** (1974), 71–79.

536. **204** Sam Loyd, *Cyclopedia of Puzzles.* Franklin Bigelow Corporation, Morningside Press, New York, 1914.

602, 603, 604. **205** E. Lucas, "Sur les rapports qui existent entre la théorie des nombres et le Calcul intégral," *Comptes Rendus hebdomadaires des séances de l'Académie des Sciences* (Paris) **82** (1876), 1303–1305.

603. **206** Édouard Lucas, "Sur les congruences des nombres eulériens et des coefficients différentiels des fonctions trigonométriques, suivant un module premier," *Bulletin de la Société mathématique de France* **6** (1878), 49–54.

278, 603. **207** Edouard Lucas, *Théorie des Nombres*, volume 1. Gauthier-Villars, Paris, 1891.

1. **208** Édouard Lucas, *Récréations mathématiques*, four volumes. Gauthier-Villars, Paris, 1891–1894. Reprinted by Albert Blanchard, Paris, 1960. (The Tower of Hanoi is discussed in volume 3, pages 55–59.)

602. **209** R. C. Lyness, "Cycles," *The Mathematical Gazette* **26** (1942), 62.

487. **210** R. C. Lyness, "Cycles," *The Mathematical Gazette* **29** (1945), 231–233.

455. **211** Colin Maclaurin, *Collected Letters*, edited by Stella Mills. Shiva Publishing, Nantwich, Cheshire, 1982.

140. **212** P. A. MacMahon, "Application of a theory of permutations in circular procession to the theory of numbers," *Proceedings of the London Mathematical Society* **23** (1892), 305–313.

280, 604. **213** I͡u. V. Matii͡asevich, "Diofantovost' perechislimykh mnozhestv," *Doklady Akademii Nauk SSSR* **191** (1970), 279–282. English translation, with amendments by the author, "Enumerable sets are diophantine," *Soviet Mathematics* **11** (1970), 354–357.

vi. **214** Z. A. Melzak, *Companion to Concrete Mathematics.* Volume 1, *Mathematical Techniques and Various Applications*, Wiley, 1973; volume 2, *Mathematical Ideas, Modeling & Applications*, Wiley, 1976.

603. **215** N. S. Mendelsohn, "Problem E 2227: Divisors of binomial coefficients," *American Mathematical Monthly* **78** (1971), 201.

216 W. H. Mills, "A prime representing function," *Bulletin of the American Mathematical Society*, series 2, **53** (1947), 604. *603.*

217 A. Moessner, "Eine Bemerkung über die Potenzen der natürlichen Zahlen," *Sitzungsberichte der Mathematisch - Naturwissenschaftliche Klasse der Bayerischen Akademie der Wissenschaften*, 1951, Heft 3, 29. *604.*

218 Peter L. Montgomery, "Problem E 2686: LCM of binomial coefficients," *American Mathematical Monthly* **86** (1979), 131. *603.*

219 Leo Moser, "Problem B-6: Some reflections," *Fibonacci Quarterly* **1**, 4 (1963), 75–76. *277.*

220 T. S. Motzkin and E. G. Straus, "Some combinatorial extremum problems," *Proceedings of the American Mathematical Society* **7** (1956), 1014–1021. *539.*

221 C. J. Mozzochi, "On the difference between consecutive primes," *Journal of Number Theory* **24** (1986), 181–187. *510.*

222 B. R. Myers, "Problem 5795: The spanning trees of an n-wheel," *American Mathematical Monthly* **79** (1972), 914–915. *604.*

223 Isaac Newton, letter to John Collins (18 February 1670), in *The Correspondence of Isaac Newton*, volume 1, 27. Excerpted in *The Mathematical Papers of Isaac Newton*, volume 3, 563. *263.*

224 Ivan Niven, *Diophantine Approximations*. Interscience, 1963. *602.*

225 Ivan Niven, "Formal power series," *American Mathematical Monthly* **76** (1969), 871–889. *318.*

226 Blaise Pascal, "De numeris multiplicibus," presented to Académie Parisienne in 1654 and published with his *Traité du triangle arithmétique* [227]. Reprinted in *Œuvres de Blaise Pascal*, volume 3, 314–339. *602.*

227 Blaise Pascal, "Traité du triangle arithmétique," in his *Traité du Triangle Arithmétique avec quelques autres petits traitez sur la mesme matiere*, Paris, 1665. Reprinted in *Œuvres de Blaise Pascal* (Hachette, 1904–1914), volume 3, 445–503; Latin editions from 1654 in volume 11, 366–390. *155, 156, 594.*

228 G. P. Patil, "On the evaluation of the negative binomial distribution with examples," *Technometrics* **2** (1960), 501–505. *605.*

229 C. S. Peirce, letter to E. S. Holden (January 1901). In *The New Elements of Mathematics*, edited by Carolyn Eisele, Mouton, The Hague, 1976, volume 1, 247–253. (See also page 211.) *603.*

230 C. S. Peirce, letter to Henry B. Fine (17 July 1903). In *The New Elements of Mathematics*, edited by Carolyn Eisele, Mouton, The Hague, 1976, volume 3, 781–784. (See also "Ordinals," an unpublished manuscript from circa 1905, in *Collected Papers of Charles Sanders Peirce*, volume 4, 268–280.) *510.*

394. **231** Walter Penney, "Problem 95: Penney-Ante," *Journal of Recreational Mathematics* **7** (1974), 321.

604. **232** J. K. Percus, *Combinatorial Methods.* Springer-Verlag, 1971.

207, 603. **233** J. F. Pfaff, "Observationes analyticæ ad *L. Euleri* institutiones calculi integralis, Vol. IV, Supplem. II & IV," *Nova acta academiæ scientiarum Petropolitanæ* **11**, Histoire section, 37–57. (This volume, printed in 1798, contains mostly proceedings from 1793, although Pfaff's memoir was actually received in 1797.)

48. **234** L. Pochhammer, "Ueber hypergeometrische Functionen n^{ter} Ordnung," *Journal für die reine und angewandte Mathematik* **71** (1870), 316–352.

605. **235** H. Poincaré, "Sur les fonctions à espaces lacunaires," *American Journal of Mathematics* **14** (1892), 201–221.

457. **236** S. D. Poisson, "Mémoire sur le calcul numérique des intégrales définies," *Mémoires de l'Académie Royale des Sciences de l'Institut de France*, series 2, **6** (1823), 571–602.

604. **237** George Pólya, "Kombinatorische Anzahlbestimmungen für Gruppen, Graphen und chemische Verbindungen," *Acta Mathematica* **68** (1937), 145–254.

vi, 16, 494, 602. **238** George Pólya, *Induction and Analogy in Mathematics.* Princeton University Press, 1954.

313, 604. **239** G. Pólya, "On picture-writing," *American Mathematical Monthly* **63** (1956), 689–697.

605. **240** George Pólya and Gabor Szegő, *Aufgaben und Lehrsätze aus der Analysis*, two volumes. Julius Springer, Berlin, 1925; fourth edition, 1970 and 1971. English translation, *Problems and Theorems in Analysis*, 1972 and 1976.

487. **241** Bjorn Poonen, "Josephus sets." Unpublished manuscript, 1987.

604. **242** R. Rado, "A note on the Bernoullian numbers," *Journal of the London Mathematical Society* **9** (1934), 88–90.

514. **242′** Earl D. Rainville, "The contiguous function relations for ${}_pF_q$ with applications to Bateman's $J_n^{u,v}$ and Rice's $H_n(\zeta, p, v)$," *Bulletin of the American Mathematical Society*, series 2, **51** (1945), 714–723.

345, 604. **243** George N. Raney, "Functional composition patterns and power series reversion," *Transactions of the American Mathematical Society* **94** (1960), 441–451.

602. **244** D. Rameswar Rao, "Problem E 2208: A divisibility problem," *American Mathematical Monthly* **78** (1971), 78–79.

245 John William Strutt, Third Baron Rayleigh, *The Theory of Sound.* First edition, 1877; second edition, 1894. (The cited material about irrational spectra is from section 92a of the second edition.) *77.*

246 Robert Recorde, *The Whetstone of Witte.* London, 1557. *432.*

247 Simeon Reich, "Problem 6056: Truncated exponential-type series," *American Mathematical Monthly* **84** (1977), 494–495. *605.*

248 Georges de Rham, "Un peu de mathématiques à propos d'une courbe plane," *Elemente der Mathematik* **2** (1947), 73–76, 89–97. Reprinted in his *Œuvres Mathématiques*, 678–689. *604.*

249 Paolo Ribenboim, *13 Lectures on Fermat's Last Theorem.* Springer-Verlag, 1979. *509, 532, 603.*

250 Bernhard Riemann, "Ueber die Darstellbarkeit einer Function durch eine trigonometrische Reihe," Habilitationsschrift, Göttingen, 1854. Published in *Abhandlungen der mathematischen Classe der Königlichen Gesellschaft der Wissenschaften zu Göttingen* **13** (1868), 87–132. Reprinted in his *Gesammelte Mathematische Werke*, 227–264. *602.*

251 Samuel Roberts, "On the figures formed by the intercepts of a system of straight lines in a plane, and on analogous relations in space of three dimensions," *Proceedings of the London Mathematical Society* **19** (1889), 405–422. *602.*

252 Øystein Rødseth, "Problem E 2273: Telescoping Vandermonde convolutions," *American Mathematical Monthly* **79** (1972), 88–89. *603.*

253 J. Barkley Rosser and Lowell Schoenfeld, "Approximate formulas for some functions of prime numbers," *Illinois Journal of Mathematics* **6** (1962), 64–94. *111.*

254 Gian-Carlo Rota, "On the foundations of combinatorial theory. I. Theory of Möbius functions," *Zeitschrift für Wahrscheinlichkeitstheorie und verwandte Gebiete* **2** (1964), 340–368. *501.*

255 Ranjan Roy, "Binomial identities and hypergeometric series," *American Mathematical Monthly* **94** (1987), 36–46. *603.*

256 Louis Saalschütz, "Eine Summationsformel," *Zeitschrift für Mathematik und Physik* **35** (1890), 186–188. *603.*

257 W. W. Sawyer, *Prelude to Mathematics.* Baltimore, Penguin, 1955. *207.*

258 O. Schlömilch, "Ein geometrisches Paradoxon," *Zeitschrift für Mathematik und Physik* **13** (1868), 162. *279.*

259 Ernst Schröder, "Vier combinatorische Probleme," *Zeitschrift für Mathematik und Physik* **15** (1870), 361–376. *604.*

604. **260** Heinrich Schröter, "Ableitung der Partialbruch- und Produkt-Entwickelungen für die trigonometrischen Funktionen," *Zeitschrift für Mathematik und Physik* **13** (1868), 254–259.

602. **261** R. S. Scorer, P. M. Grundy, and C. A. B. Smith, "Some binary games," *The Mathematical Gazette* **28** (1944), 96–103.

604. **262** J. Sedláček, "On the skeletons of a graph or digraph," in *Combinatorial Structures and their Applications*, Gordon and Breach, 1970, 387–391. (This volume contains proceedings of the Calgary International Conference of Combinatorial Structures and their Applications, 1969.)

603. **263** J. O. Shallit, "Problem 6450: Two series," *American Mathematical Monthly* **92** (1985), 513–514.

259. **264** R. T. Sharp, "Problem 52: Overhanging dominoes," *Pi Mu Epsilon Journal* **1**, 10 (1954), 411–412.

87. **265** W. Sierpiński, "Sur la valeur asymptotique d'une certaine somme," *Bulletin International Académie Polonaise des Sciences et des Lettres* (Cracovie), series A (1910), 9–11.

603. **266** W. Sierpiński, "Sur les nombres dont la somme de diviseurs est une puissance du nombre 2," *Calcutta Mathematical Society Golden Jubilee Commemorative Volume* (1958–1959), part 1, 7–9.

603. **267** Wacław Sierpiński, *A Selection of Problems in the Theory of Numbers.* Macmillan, 1964.

604. **268** David L. Silverman, "Problematical Recreations 447: Numerical links," *Aviation Week & Space Technology* **89**, 10 (1 September 1968), 71. Reprinted as Problem 147 in *Second Book of Mathematical Bafflers*, edited by Angela Fox Dunn, Dover, 1983.

223. **269** Lucy Joan Slater, *Generalized Hypergeometric Series.* Cambridge University Press, 1966.

42, 327, 450. **270** N. J. A. Sloane, *A Handbook of Integer Sequences.* Academic Press, 1973.

394. **271** A. D. Solov'ev, "Odno kombinatornoe tozhdestvo i ego primenenie k zadache o pervom nastuplenin redkogo sobytii͡a," *Teorii͡a veroi͡atnosteĭ i eë primenenii͡a* **11** (1966), 313–320. English translation, "A combinatorial identity and its application to the problem concerning the first occurrence of a rare event," *Theory of Probability and its Applications* **11** (1966), 276–282.

v. **272** William G. Spohn, Jr., "Can mathematics be saved?" *Notices of the American Mathematical Society* **16** (1969), 890–894.

605. **273** Richard P. Stanley, "Differentiably finite power series," *European Journal of Combinatorics* **1** (1980), 175–188.

274 Richard P. Stanley, "On dimer coverings of rectangles of fixed width," *Discrete Applied Mathematics* **12** (1985), 81–87. *604.*

275 Richard P. Stanley, *Enumerative Combinatorics*, volume 1. Wadsworth & Brooks/Cole, 1986. *519, 604, 605.*

276 K. G. C. von Staudt, "Beweis eines Lehrsatzes, die Bernoullischen Zahlen betreffend," *Journal für die reine und angewandte Mathematik* **21** (1840), 372–374. *604.*

277 Guy L. Steele Jr., Donald R. Woods, Raphael A. Finkel, Mark R. Crispin, Richard M. Stallman, and Geoffrey S. Goodfellow, *The Hacker's Dictionary: A Guide to the World of Computer Wizards.* Harper & Row, 1983. *124.*

278 J. Steiner, "Einige Gesetze über die Theilung der Ebene und des Raumes," *Journal für die reine und angewandte Mathematik* **1** (1826), 349–364. Reprinted in his *Gesammelte Werke*, volume 1, 77–94. *5, 602.*

279 M. A. Stern, "Ueber eine zahlentheoretische Funktion," *Journal für die reine und angewandte Mathematik* **55** (1858), 193–220. *116.*

280 L. Stickelberger, "Ueber eine Verallgemeinerung der Kreistheilung," *Mathematische Annalen* **37** (1890), 321–367. *602.*

281 James Stirling, *Methodus Differentialis.* London, 1730. English translation, *The Differential Method*, 1749. *192, 244, 2*

282 Dura W. Sweeney, "On the computation of Euler's constant," *Mathematics of Computation* **17** (1963), 170–178. *467.*

283 J. J. Sylvester, "Problem 6919," *Mathematical Questions with their Solutions from the 'Educational Times'* **37** (1882), 42–43, 80. *602.*

284 J. J. Sylvester, "On the number of fractions contained in any 'Farey series' of which the limiting number is given," *The London, Edinburgh and Dublin Philosophical Magazine and Journal of Science*, series 5, **15** (1883), 251–257. Reprinted in his *Collected Mathematical Papers*, volume 4, 101–109. *133.*

284′ M. Szegedy, "The solution of Graham's greatest common divisor problem," *Combinatorica* **6** (1986), 67–71. *510.*

285 Jonathan W. Tanner and Samuel S. Wagstaff, Jr., "New congruences for the Bernoulli numbers," *Mathematics of Computation* **48** (1987), 341–350. *131.*

286 S. Tanny, "A probabilistic interpretation of Eulerian numbers," *Duke Mathematical Journal* **40** (1973), 717–722. *604.*

287 L. Theisinger, "Bemerkung über die harmonische Reihe," *Monatshefte für Mathematik und Physik* **26** (1915), 132–134. *603.*

383, 384. **288** T. N. Thiele, *The Theory of Observations.* Charles & Edwin Layton, London, 1903. Reprinted in *The Annals of Mathematical Statistics* **2** (1931), 165–308.

605. **289** E. C. Titchmarsh, *The Theory of the Riemann Zeta-Function.* Clarendon Press, Oxford, 1951; second edition, revised by D. R. Heath-Brown, 1986.

605. **290** F. G. Tricomi and A. Erdélyi, "The asymptotic expansion of a ratio of gamma functions," *Pacific Journal of Mathematics* **1** (1951), 133–142.

266. **291** Peter Ungar, "Problem E 3052: A sum involving Stirling numbers," *American Mathematical Monthly* **94** (1987), 185–186.

602. **292** J. V. Uspensky, "On a problem arising out of the theory of a certain game," *American Mathematical Monthly* **34** (1927), 516–521.

169, 603. **293** A. Vandermonde, "Mémoire sur des irrationnelles de différens ordres avec une application au cercle," *Histoire de l'Académie Royale des Sciences* (1772), part 1, 71–72; *Mémoires de Mathématique et de Physique, Tirés des Registres de l'Académie Royale des Sciences* (1772), 489–498.

484, 602. **294** J. Venn, "On the diagrammatic and mechanical representation of propositions and reasonings," *The London, Edinburgh and Dublin Philosophical Magazine and Journal of Science*, series 5, **9** (1880), 1–18.

604. **295** John Wallis, *A Treatise of Angular Sections.* Oxford, 1684.

604. **296** Edward Waring, *Meditationes Algebraïcæ.* Cambridge, 1770; third edition, 1782.

604. **297** Frederick V. Waugh and Margaret W. Maxfield, "Side-and-diagonal numbers," *Mathematics Magazine* **40** (1967), 74–83.

279. **298** Warren Weaver, "Lewis Carroll and a geometrical paradox," *American Mathematical Monthly* **45** (1938), 234–236.

501. **299** Louis Weisner, "Abstract theory of inversion of finite series," *Transactions of the American Mathematical Society* **38** (1935), 474–484.

87. **300** Hermann Weyl, "Über die Gibbs'sche Erscheinung und verwandte Konvergenzphänomene," *Rendiconti del Circolo Matematico di Palermo* **30** (1910), 377–407.

603. **301** F. J. W. Whipple, "Some transformations of generalized hypergeometric series," *Proceedings of the London Mathematical Society*, series 2, **26** (1927), 257–272.

489. **302** Alfred North Whitehead, *An Introduction to Mathematics.* London and New York, 1911.

91. **303** Alfred North Whitehead, "Technical education and its relation to science and literature," chapter 2 in *The Organization of Thought, Educational and Scientific*, London and New York, 1917. Reprinted as chapter 4 of *The Aims of Education and Other Essays*, New York, 1929.

304 Alfred North Whitehead, *Science and the Modern World*. New York, 1925. Chapter 2 reprinted in *The World of Mathematics*, edited by James R. Newman, 1956, volume 1, 402–416. *577.*

305 H. C. Williams and H. Dubner, "The primality of R1031," *Mathematics of Computation* **47** (1986), 703–711. *602.*

306 J. Wolstenholme, "On certain properties of prime numbers," *Quarterly Journal of Pure and Applied Mathematics* **5** (1862), 35–39. *604.*

307 Derick Wood, "The Towers of Brahma and Hanoi revisited," *Journal of Recreational Mathematics* **14** (1981), 17–24. *602.*

308 J. Worpitzky, "Studien über die *Bernoulli*schen und *Euler*schen Zahlen," *Journal für die reine und angewandte Mathematik* **94** (1883), 203–232. *255.*

309 E. M. Wright, "A prime-representing function," *American Mathematical Monthly* **58** (1951), 616–618; errata in **59** (1952), 99. *602.*

310 Hermann Zapf, collected works, entitled *Hermann Zapf & His Design Philosophy*. Society of Typographic Arts, Chicago, 1987. (The AMS Euler typeface is mentioned on pages 97 and 136.) *viii.*

311 Derek A. Zave, "A series expansion involving the harmonic numbers," *Information Processing Letters* **5** (1976), 75–77. *604.*

312 E. Zeckendorf, "Représentation des nombres naturels par une somme de nombres de Fibonacci ou de nombres de Lucas," *Bulletin de la Société Royale des Sciences de Liège* **41** (1972), 179–182. *281.*

Credits for Exercises

THE EXERCISES in this book have been drawn from many sources. The authors have tried to trace the origins of all the problems that have been published before, except in cases where the exercise is so elementary that its inventor would probably not think anything was being invented.

Many of the exercises come from examinations in Stanford's Concrete Mathematics classes. The teaching assistants and instructors often devised new problems for those exams, so it is appropriate to list their names here:

The TA sessions were invaluable, I mean really great.

Keep the same instructor and the same TAs next year.

Class notes very good and useful.

I never "got" Stirling numbers.

Year	*Instructor*	*Teaching Assistant(s)*
1970	Don Knuth	Vaughan Pratt
1971	Don Knuth	Leo Guibas
1973	Don Knuth	Henson Graves, Louis Jouaillec
1974	Don Knuth	Scot Drysdale, Tom Porter
1975	Don Knuth	Mark Brown, Luis Trabb Pardo
1976	Andy Yao	Mark Brown, Lyle Ramshaw
1977	Andy Yao	Yossi Shiloach
1978	Frances Yao	Yossi Shiloach
1979	Ron Graham	Frank Liang, Chris Tong, Mark Haiman
1980	Andy Yao	Andrei Broder, Jim McGrath
1981	Ron Graham	Oren Patashnik
1982	Ernst Mayr	Joan Feigenbaum, Dave Helmbold
1983	Ernst Mayr	Anna Karlin
1984	Don Knuth	Oren Patashnik, Alex Schäffer
1985	Andrei Broder	Pang Chen, Stefan Sharkansky
1986	Don Knuth	Arif Merchant, Stefan Sharkansky

In addition, David Klarner (1971), Bob Sedgewick (1974), Leo Guibas (1975), and Lyle Ramshaw (1979) each contributed to the class by giving six or more guest lectures. Detailed lecture notes taken each year by the teaching assistants and edited by the instructors have served as the basis of this book.

1.1 Pólya [238, p. 120].
1.2 Scorer, Grundy, and Smith [261].
1.4 Ahrens [5, vol. 1, p. 61].
1.5 Venn [294].
1.6 Steiner [278]; Roberts [251].
1.8 Lyness [209].
1.9 Cauchy [47, note 2, theorem 17].
1.10 Atkinson [13].
1.11 Wood [307].
1.14 Steiner [278]; Pólya [238, chapter 3]; Brother Alfred [37].
1.17 Dudeney [72, puzzle 1].
1.21 Ball [16] credits B. A. Swinden.
1.22 Based on an idea of Peter Shor.*
1.23 Bjorn Poonen.*
1.25 Frame, Stewart, and Dunkel [105].
2.2 Iverson [161, p. 11].
2.3 [173, exercise 1.2.3–2].
2.5 [173, exercise 1.2.3–25].
2.22 Cauchy [47, note 2, theorem 16].
2.23 1982 final.
2.26 [173, exercise 1.2.3–26].
2.29 1979 midterm.
2.30 1973 midterm.
2.34 Riemann [250, section 3].
2.35 Euler [85] gave a fallacious "proof" using divergent series.
2.36 Golomb [120]; Ilan Vardi.*
2.37 Leo Moser.*
3.6 Ernst Mayr, 1982 homework.
3.8 Dirichlet [67].
3.9 Chace [48]; Fibonacci [98, pp. 77–83].
3.12 [173, exercise 1.2.4–48(a)].
3.13 Beatty [18]; Niven [224, theorem 3.7].
3.19 [173, exercise 1.2.4–34].
3.21 1975 midterm.
3.23 [173, exercise 1.2.4–41].
3.28 Brown [40].
3.30 Aho and Sloane [4].
3.31 Greitzer [135, problem 1972/3, solution 2].
3.32 [130].
3.33 1984 midterm.
3.34 1970 midterm.
3.35 1975 midterm.
3.36 1976 midterm.
3.37 1986 midterm; [181].
3.38 1974 midterm.
3.39 1971 midterm.
3.40 1980 midterm.
3.41 Klamkin [169, problem 1978/3].
3.42 Uspensky [292].
3.45 Aho and Sloane [4].
3.46 Graham and Pollak [132].
3.48 R. L. Graham and D. R. Hofstadter.*
3.51 Fraenkel [103].
3.52 S. K. Stein.*
4.4 [180, §526].
4.16 Sylvester [283].
4.19 Bertrand [23, p. 129]; Chebyshev [50]; Wright [309].
4.21 [178, pp. 148–149].
4.22 Brillhart [34]; Williams and Dubner [305].
4.23 Crowe [58].
4.24 Legendre [196, second edition, introduction].
4.26 [174, exercise 4.5.3–43].
4.31 Pascal [226].
4.36 Hardy and Wright [150, §14.5].
4.37 Aho and Sloane [4].
4.38 Lucas [205].
4.39 [129].
4.40 Stickelberger [280].
4.41 Legendre [196, §135]; Hardy and Wright [150, theorem 82].
4.42 [174, exercise 4.5.1–6].
4.44 [174, exercise 4.5.3–39].
4.45 [174, exercise 4.3.2–13].
4.47 Lehmer [197].
4.48 Gauss [115, §78]; Crelle [57].
4.52 1974 midterm.
4.53 1973 midterm, inspired by Rao [244].
4.54 1974 midterm.

4.56 Inspired by B. F. Logan.*
4.57 A special case appears in [182].
4.58 Sierpiński [266].
4.59 Curtiss [59]; Erdős [76].
4.60 Mills [216].
4.61 [173, exercise 1.3.2–19].
4.63 Barlow [17]; Abel [1].
4.64 Peirce [229].
4.66 Ribenboim [249]; Sierpiński [267, problem P^2_{10}].
4.67 [127].
4.69 Cramér [56].
4.70 P. Erdős.*
4.71 [77, p. 96].
4.72 [77, p. 103].
4.73 Landau [195, volume 2, eq. 648].
5.1 Forcadel [101].
5.3 Long and Hoggatt [203].
5.5 1983 in-class final.
5.13 1975 midterm.
5.14 [173, exercise 1.2.6–20].
5.15 Dixon [68].
5.21 Euler [81].
5.25 Gauss [116, §7].
5.28 Euler [95].
5.29 Kummer [187, eq. 26.4].
5.31 Gosper [124].
5.34 Bailey [15, §10.4].
5.36 Kummer [188, p. 116].
5.37 Vandermonde [293].
5.38 [173, exercise 1.2.6–16].
5.40 Rødseth [252].
5.43 Pfaff [233]; Saalschütz [256]; [173, exercise 1.2.6–31].
5.48 Ranjan Roy.*
5.49 Roy [255, eq. 3.13].
5.53 Gauss [116]; Richard Askey.*
5.58 Frazer and McKellar [107].
5.59 Stanford Computer Science Comprehensive Exam, Winter 1987.
5.60 [173, exercise 1.2.6–41].
5.61 Lucas [206].
5.62 1971 midterm.
5.63 1974 midterm.
5.64 1980 midterm.
5.65 1983 midterm.
5.66 1984 midterm.
5.67 1976 midterm.
5.68 1985 midterm.
5.69 Lyle Ramshaw, guest lecture in 1986.
5.70 Andrews [9, theorem 5.4].
5.71 H. S. Wilf.*
5.72 Hermite [154].
5.74 1979 midterm.
5.75 1971 midterm.
5.76 [173, exercise 1.2.6–59 (corrected)].
5.77 1986 midterm.
5.78 [176].
5.79 Mendelsohn [215].
5.80 Montgomery [218].
5.81 1986 final exam.
5.82 Hillman and Hoggatt [157].
5.85 Hsu [159].
5.86 Good [123].
5.88 Hermite [155].
5.91 Whipple [301].
5.92 Clausen [51], [52].
5.93 Gosper [124].
5.94 Henrici [152, p. 118].
5.95 [77, p. 71].
5.96 [77, p. 71].
5.97 R. William Gosper, Jr.*
6.6 Fibonacci [98, p. 283].
6.15 [175, exercise 5.1.3–2].
6.21 Theisinger [287].
6.25 Gardner [112] credits Denys Wilquin.
6.27 Lucas [205].
6.28 Lucas [207, chapter 18].
6.31 Lah [193]; R. W. Floyd.*
6.35 1977 midterm.
6.37 Shallit [263].
6.39 [173, exercise 1.2.7–15].
6.40 Klamkin [169, problem 1979/1].
6.41 1973 midterm.

6.43 Brooke and Wall [36].
6.44 Matiiasevich [213].
6.46 Francesca [106]; Wallis [295, chapter 4].
6.47 Lucas [205].
6.48 [174, exercise 4.5.3–9(c)].
6.49 Davison [61].
6.50 1985 midterm; Rham [248]; Dijkstra [66, pp. 230–232].
6.51 Waring [296]; Lagrange [191]; Wolstenholme [306].
6.52 Eswarathasan and Levine [79].
6.53 Kaucký [168] treats a special case.
6.54 Staudt [276]; Clausen [53]; Rado [242].
6.55 Andrews and Uchimura [12].
6.56 1986 midterm.
6.57 1984 midterm, suggested by R. W. Floyd.*
6.58 [173, exercise 1.2.8–30]; 1982 midterm.
6.59 Burr [42].
6.61 1976 final exam.
6.62 Borwein and Borwein [31, §3.7].
6.63 [173, section 1.2.10]; Stanley [275, proposition 1.3.12].
6.65 Tanny [286].
6.66 B. F. Logan.*
6.67 [175, exercise 6.1–13].
6.70 Euler [86, chapters 9 and 10].
6.72 [175, exercise 5.1.3–3].
6.73 Euler [86, chapters 9 and 10]; Schröter [260].
6.74 B. F. Logan.*
6.75 Comic section, *Boston Herald*, August 21, 1904.
6.76 Silverman and Dunn [268].
6.78 [183].
6.79 [126], modulo a numerical error.
6.80 [174, exercises 4.5.3–2 and 3].
6.81 Adams and Davison [3].
6.84 Lehmer [198].
6.85 Burr [42].
6.87 Part (a) is from Eswarathasan and Levine [79].
7.2 [173, exercise 1.2.9–1].
7.8 Zave [311].
7.9 [173, exercise 1.2.7–22].
7.11 1971 final exam.
7.12 [175, pp. 63–64].
7.13 Raney [243].
7.15 Bell [20].
7.16 Pólya [237, p. 149]; [173, exercise 2.3.4.4–1].
7.20 Jungen [167, p. 299] credits A. Hurwitz.
7.22 Pólya [239].
7.23 1983 homework.
7.24 Myers [222]; Sedláček [262].
7.25 [174, Carlitz's proof of lemma 3.3.3B].
7.26 [173, exercise 1.2.8–12].
7.32 [77, pp. 25–26] credits L. Mirsky and M. Newman.
7.33 1971 final exam.
7.34 Tomás Feder.*
7.36 1974 final exam.
7.37 Euler [87, paragraph 50]; 1971 final exam.
7.38 1973 final exam.
7.39 [173, exercise 1.2.9–18].
7.41 André [8]; [175, exercise 5.1.4–22].
7.42 1974 final exam.
7.44 Gross [136]; [175, exercise 5.3.1–3].
7.45 de Bruijn [63].
7.47 Waugh and Maxfield [297].
7.48 1984 final exam.
7.49 1986 final exam.
7.50 Schröder [259]; [173, exercise 2.3.4.4–31].
7.51 Fisher [99]; Percus [232, pp. 89–123]; Stanley [274].
7.52 Hammersley [146].
7.53 Euler [92, part 2, section 2, chapter 6, paragraph 91].
7.54 Moessner [217].

7.55 Stanley [273].
7.56 Euler [91].
7.57 [77, p. 48] credits P. Erdős and P. Turán.
8.13 Thomas M. Cover.*
8.15 [173, exercise 1.2.10–17].
8.17 Patil [228].
8.24 John Knuth (age 4) and DEK; 1975 final.
8.26 [173, exercise 1.3.3–18].
8.27 Fisher [100].
8.29 Guibas and Odlyzko [138].
8.32 1977 final exam.
8.34 Hardy [149] has an incorrect analysis leading to the opposite conclusion.
8.35 1981 final exam.
8.36 Gardner [113] credits George Sicherman.
8.38 [174, exercise 3.3.2–10].
8.39 [177, exercise 4.3(a)].
8.41 Feller [96, exercise IX.33].
8.43 [173, sections 1.2.10 and 1.3.3].
8.44 1984 final exam.
8.46 Feller [96] credits Hugo Steinhaus.
8.47 1974 final, suggested by "fringe analysis" of 2-3 trees.
8.48 1979 final exam.
8.49 Blom [26]; 1984 final exam.
8.50 1986 final exam.
8.51 1986 final exam.
8.53 Feller [96] credits S. N. Bernstein.
8.57 Lyle Ramshaw.*
8.63 Guibas and Odlyzko [138].
9.1 Hardy [148, 1.3(g)].
9.2 Part (c) is from Garfunkel [114].
9.3 [173, exercise 1.2.11.1–6].
9.6 [173, exercise 1.2.11.1–3].
9.8 Hardy [148, 1.2(iv)].
9.9 Landau [194, vol. 1, p. 60].
9.14 [173, exercise 1.2.11.3–6].
9.16 Knopp [170, edition $\geqslant 2$, §64C].
9.18 Bender [21, §3.1].
9.20 1971 final exam.
9.24 [134, §4.1.6].
9.27 Titchmarsh [289].
9.28 [173, exercise 1.2.11.2–7].
9.29 de Bruijn [62, section 3.7].
9.32 1976 final exam.
9.34 1973 final exam.
9.35 1975 final exam.
9.36 1980 class notes.
9.37 [174, eq. 4.5.3–21].
9.38 1977 final exam.
9.39 1975 final exam, inspired by Reich [247].
9.40 1977 final exam.
9.41 1980 final exam.
9.42 1979 final exam.
9.44 Tricomi and Erdélyi [290].
9.46 de Bruijn [62, §6.3].
9.47 1980 homework; [175, eq. 5.3.1–34].
9.48 1980 final exam.
9.49 1974 final exam.
9.50 1984 final exam.
9.51 [134, §4.2.1].
9.52 Poincaré [235]; Borel [30, p. 27].
9.53 Pólya and Szegő [240, part 1, problem 140].
9.57 Andrew M. Odlyzko.*
9.58 Henrici [151, exercise 4.9.8].
9.60 Ilan Vardi.*
9.62 Canfield [43].
9.63 Ilan Vardi.*
9.65 M. P. Schützenberger.*
9.66 Lieb [202]; Stanley [275, exercise 4.37(c)].
9.67 Boas and Wrench [27].

* Unpublished personal communication.

Index

WHEN AN INDEX ENTRY refers to a page containing a relevant exercise, the answer to that exercise (in Appendix A) might divulge further information; an answer page is not indexed here unless it refers to a topic that isn't included in the statement of the relevant exercise.

(Graffiti have been indexed too.)

List of Tables

THIS BOOK was composed at Stanford University using the TEX system for technical text developed by D. E. Knuth. The mathematics is set in a new typeface called AMS Euler, designed by Hermann Zapf for the American Mathematical Society. The text is set in a new typeface called Concrete Roman and Italic, a special version of Knuth's Computer Modern family with weights designed to blend with AMS Euler. The paper is 50-lb.-basis Finch offset, which has a neutral pH and a life expectancy of several hundred years. The offset printing and notch binding were done by Halliday Lithograph Corporation in Hanover, Massachusetts.